# The International Directory of Logicians

## Who's Who in Logic

# IDL Advisory Board

**John Corcoran**
corcoran@buffalo.edu

**Rod Downey**
rod.downey@mcs.vuw.ac.nz

**Luis Fariñas del Cerro**
farinas@irit.fr

**Melvin Fitting**
melvin.fitting@lehman.cuny.edu

**Aki Kanamori**
aki@math.bu.edu

**Ruy de Quieroz**
ruy@cin.ufpr.br

**Krister Segerberg**
krister.segerberg@filosofi.uu.se

**Jean-Paul van Bendegem**
jpvbende@vub.ac.be

**Johan van Benthem**
johan@science.uva.nl

# The International Directory of Logicians

## Who's Who in Logic

Edited by
Dov M. Gabbay
and
John Woods

© Individual author and College Publications 2009. All rights reserved.

ISBN 978-1-904987-90-1

College Publications
Scientific Director: Dov Gabbay
Managing Director: Jane Spurr
Department of Computer Science
King's College London, Strand, London WC2R 2LS, UK

http://www.collegepublications.co.uk

Original cover design by orchid creative       www.orchidcreative.co.uk
Printed by Lightning Source, Milton Keynes, UK

---

All rights reserved. No part of this publication may be reproduced, stored in a retrieval system or transmitted in any form, or by any means, electronic, mechanical, photocopying, recording or otherwise without prior permission, in writing, from the publisher.

# Preface

The International Directory of Logicians is a listing of over 260 of the world's foremost living logicians, representing the most important and influential developments of our day.

This volume provides a compact summary of each entrant's education, professional appointments, honours and awards, principal publications, as well as a description of the nature and significance of his or her contributions to logic, and a 'vision statement' concerning logic's future prospects. Logic here is understood in its broad sense, encompassing all branches of mathematical logic, philosophical and the history of logic.

Inclusion in the Directory is by invitation only, following a rigorous selection process guided by a distinguished Advisory Board. The Directory is designed to meet the needs of students and professional logicians alike. It is the most informative single-volume record of logics' present state and will serve as an invaluable historical reference for future generations of scholarship.

While the Editors and Advisory Board have taken pains to make the volume as complete as possible, doubtless there will be omissions — some occasioned by the modesty of invitees — and other grounds for correction and improvement in subsequent editions. The publisher proposes to update the Directory every three years. Suggestions from readers would be welcome, and should be directed to the Directory's administrative officer Jane Spurr at jane.spurr@kcl.ac.uk.

The editors warmly thank the Directory's entrants for their cooperation and support, members of the Advisory Board for their wise counsel on all matters pertaining to the volume's content and design, Jane Spurr in London and Carol Woods in Vancouver for invaluable technical support, and Rod Downey and Sol Feferman for their particular and most helpful efftorts on behalf of the project.

Dov M. Gabbay is Augustus De Morgan Professor of Logic, Department of Computer Science, King's College London, UK, and Special Professor at Bar-Ilan University, Israel.

John Woods is Director of the Abductive Systems Group, University of British Columbia, Canada and Charles S. Peirce Visiting Professor of Logic, Group of Logic and Computation, King's College London, UK.

# A

## ABRAMSKY, Samson

**Specialties:** Logic in computer science, game semantics, logic and categories, categorical quantum logic.

**Born:** 12 March, 1953 in London, U.K.

**Educated:** Oxford University, MA, 2001; Queen Mary College, University of London, PhD Computer Science, 1988, Cambridge (King's College), BA Philosophy, 1975, MA, 1979, Diploma in Computer Science, 1976.

**Dissertation:** *Domain Theory and the Logic of Observable Properties*; supervisor, Richard Bornat.

**Regular Academic or Research Appointments:** CHRISTOPHER STRACHEY PROFESSOR OF COMPUTING AND FELLOW OF WOLFSON COLLEGE, OXFORD UNIVERSITY, HEAD OF THEORY AND VERIFICATION RESEARCH GROUP, OXFORD UNIVERSITY COMPUTING LABORATORY, 2000–. Professor of Theoretical Computer Science, University of Edinburgh, 1995-2000; Professor of Computing Science and Head of Theory and Formal Methods Section, Department of Computing, Imperial College of Science, Technology and Medicine, London University, 1990–1995; Reader in Computing Science, Department of Computing, Imperial College of Science, Technology and Medicine, London University, 1988–1990; Lecturer, Department of Computing, Imperial College of Science and Technology, London University, 1983–1988; Lecturer, Department of Computer Science and Statistics, Queen Mary College, London University, 1980–1983; Research Student, Department of Computer Science and Statistics, Queen Mary College, London University, 1978–1980; Programmer, G.E.C. Computers Limited, working on Operating Systems development, 1976–1978.

**Visiting Academic or Research Appointments:** Visiting Professor, University of Paris VII and CNRS Laboratory PPS, 2001; Visiting Researcher, Kestrel Institute, 2000; Visiting Researcher, Kestrel Institute, 1998; Visiting Professor, Department of Mathematics, University of Pennsylvania, 1998; Nuffield Science Research Fellow, 1988–1989; Visiting Associate Professor, Department of Computer and Information Science, University of Pennsylvania, 1989; Visiting Professor, Computer Science Department, University of Nijmegen, Netherlands, 1986; Consultant, G.E.C. Hirst Research Centre 1985-1988; Visiting Lecturer, Programming Methodology Group, Chalmers University, Göteborg Sweden, 1984.

**Research Profile:** Samson Abramsky has worked on a wide range of topics in the semantics and logic of computation. He has made seminal contributions to the development of Game Semantics, and its applications to the semantics of programming languages, and of logics and type theories. His work with Jagadeesan on a fully complete, compositional game semantics for Multiplicative Linear Logic, with Jagadeesan and Malacaria on a fully abstract model for PCF, and with McCusker on applying game semantics to imperative languages, have all been fundamental to the development of Game semantics as a powerful and flexible formalism for capturing a wide range of phenomena in logic and computation.

He has also made fundamental contributions to the Geometry of Interaction, including developing an abstract categorical formulation, and applications to full completeness for various logics and type theories. Other contributions include his work on Domain theory in logical form, using Stone duality to relate programming language semantics and logics, on the lazy lamda calculus, on concurrency, and strictness analysis.

His recent work has concerned applying semantical and logical methods to quantum computation. With Coecke he has developed a novel categorical axiomatization of Quantum Mechanics. This axiomatization captures all the features which are significant for Quantum Information and Computation, and provides a basis for effective reasoning about quantum informatic processes. It can be presented in terms of a diagrammatic calculus which can be seen as a proof system for a logic, leading to a new perspective on what the right logical formulation for Quantum Mechanics should be.

**Main Publications:**

1. Observation Equivalence as a Testing Equivalence, *J. Theoretical Computer Science 53*, (1987), 225–241.
2. The Lazy $\lambda$-Calculus, in *Research Topics in Functional Programming*, D. Turner, ed., Addison Wesley 1990, 65–117.
3. Abstract Interpretation, Logical Relations and Kan Extensions, *J. Logic and Computation*, 1(1) (1990), 5–41.

4. Domain Theory in Logical Form, *Annals of Pure and Applied Logic 51*, (1991), 1–77.

5. A Domain Equation for Bisimulation, *J. Information and Computation*, 92(2) (1991), 161–218.

6. New Foundations for the Geometry of Interaction, with R. Jagadeesan, in *Symposium on Logic in Computer Science*, (Computer Society Press of the IEEE) 1992, 211–222.

7. Games and Full Completeness for Multiplicative Linear Logic, with R. Jagadeesan, in *Foundations of Software Technology and Theoretical Computer Science*, R. Shyamsundar, ed. (Springer-Verlag) 1992, 291–301.

8. Computational Interpretations of Linear logic, *J. Theoretical Computer Science*, 111 (1993), 3–57.

9. Domain Theory, with A. Jung, in *Handbook of Logic in Computer Science*, S. Abramsky, D. Gabbay and T. S. E. Maibaum, eds., Oxford University Press, 1994, 1–168.

10. New Foundations for the Geometry of Interaction, R. Jagadeesan, *Information and Computation*, 111(1), (1994), 53–119.

11. Games and Full Completeness for Multiplicative Linear Logic, with R. Jagadeesan, *Journal of Symbolic Logic*, (1994), vol. 59, no. 2, 543–574.

12. Interaction Categories and the Foundations of Typed Concurrent Programming, with S. Gay and R. Nagarajan, in *Proceedings of the 1994 Marktoberdorf Summer School on Deductive Program Design*, M. Broy, ed. (Springer-Verlag) 1996, 35–113.

13. Retracing some paths in process algebra, in *CONCUR '96: Concurrency Theory, 7th International Conference*, U. Montanari and V. Sassone, eds. (Springer-Verlag) 1996, 1–17.

14. Linearity, Sharing and State: a fully abstract game semantics for Idealized Algol, with G. McCusker, in Algol-like Languages, P. O'Hearn and R. D. Tennent, eds. (Birkhauser) 1997, 317–348.

15. A fully abstract game semantics for general references, with K. Honda and G. McCusker, in Proceedings of the Thirteenth International Symposium on Logic in Computer Science, (Computer Society Press of the IEEE) 1998, 334–344.

16. Concurrent Games and Full Completeness, with P.-A. Melliès, in *Proceedings of the Fourteenth International Symposium on Logic in Computer Science*, (Computer Society Press of the IEEE) 1999, 431–442.

17. Full Abstraction for PCF, with R. Jagadeesan and P. Malacaria, *Information and Computation*, vol. 163 (2000), 409–470.

18. Geometry of Interaction and linear combinatory algebras, with E. Haghverdi and P. Scott, in *Mathematical Structures in Computer Science* vol. 12, 2002, 625–665.

19. Applying Game Semantics to Compositional Software Modelling and verification, with D. Ghica, L. Ong and A. Murawski, in *TACAS 2004: Tools and Algorithms for the Construction and Analysis of Systems, 10 International Conference*, Springer LNCS Vol. 2988, 2004, 421–435.

20. A Categorical Semantics of Quantum Protocols, with B. Coecke, in *Proceedings of the 19th Annual IEEE Symposium on Logic in Computer Science: LICS 2004*, IEEE Computer Society, 415–425, 2004.

**Service to the Profession:** Member, *Scientific Steering Committee of the Isaac Newton Institute for the Mathematical Sciences in Cambridge*, 2003–2006; General Chair, *International Symposium on Logic in Computer Science*, 2000–2003; Member, *EACSL Ackermann Prize committee, for outstanding PhD Theses in Computer Science Logic*, 2004–2007; Editorial Board Member, *North Holland Studies in Logic and Foundations of Mathematics*, 1991–2007; Editorial Board Member, *Cambridge Tracts in Theoretical Computer Science*; Joint Editor with D. Gabbay and T. S. E. Maibaum, *Handbook of Logic in Computer Science*, Oxford University Press, 1992–2000.

**Teaching:** Twelve PhD students have completed thus far, including Luke Ong, now Professor of Computing at Oxford.

**Vision Statement:** Computer Science has opened new possibilities for logic, and greatly broadened its scope. I think now we are ready for a new wave of application of logical and semantical methods, broadly construed, to the sciences. Above all, the time is ripe to apply the power of compositional methods in the physical and biological sciences. This should go hand-in-hand with developing a comprehensive science of information – of which at present we have only fragments.

**Honours and Awards:** iCS Test-of-Time award for Domain Theory in Logical Form, 2007; EPSRC Senior Research Fellowship, 2007; Clifford Lecturer, Tulane University, March 2008; Fellow of the Royal Society, 2004–; Fellow of the Royal Society of Edinburgh, 2000–; BCS-IET Turing Lecturer, 1999; Member of Academia Europaea, 1993–.

# ADDISON, John W., Jr.

**Specialties:** Theory of definability, descriptive set theory, recursive function theory, theory of models, finite-universe logic.

**Born:** 2 April 1930 in Washington, D.C., USA.

**Educated:** University of Wisconsin, PhD Mathematics, January 1955; University of Wisconsin, MS, 1953; Princeton University, AB (Honors in Mathematics), 1951; Phillips Academy, Andover, 1946-1947. Senior Thesis: *Turing Machines*, Adviser: Alonzo Church.

**Dissertation:** *On Some Points of the Theory of Recursive Functions*, Supervisor: Stephen Kleene.

**Regular Academic or Research Appointments:**
PROFESSOR EMERITUS, MATHEMATICS, UNIVERSITY OF CALIFORNIA, BERKELEY, 1994–; Professor, Mathematics, University of California, Berkeley, 1968-1994; Associate Professor, Mathematics, University of California, Berkeley, 1962-1968; Assistant Professor, Mathematics, University of Michigan, 1957-1962; National Science Foundation Postdoctoral Fellow, 1956-1957; Instructor, Mathematics, University of Michigan, 1954-1956.

**Visiting Academic or Research Appointments:**
Visiting Professor, University of Münster, Spring 1991; Hebrew University of Jerusalem, Fall 1990; Mathematical Sciences Research Institute, Berkeley, 1989–1990; Visiting Fellow, Wolfson College, Oxford University, 1979; Visiting Professor, Mathematical Institute, Oxford University, 1972–1973; Visiting Lecturer, Mathematics, University of California, Berkeley, 1959–1960; Fellow, Instytut Matematyczny, Polska Akademia Nauk Warszawa, 1957; Member, Institute for Advanced Study, Princeton, N.J., Mentor: Kurt Gödel, Fall 1956; Technical Staff Member, Bell Telephone Laboratories, Summer 1956.

**Research Profile:** J. W. Addison has made fundamental contributions to the theory of definability, a branch of logic that overlaps set theory, recursive function theory, model theory, and finite-universe logic, and even has minor contact with proof theory. In the thirties and forties various logicians noted an analogy between the arithmetical hierarchy (of recursive function theory) and the projective hierarchy (of descriptive set theory). After the analogy was observed around 1950 to break down Addison found the 'correct' analogy, leading to a generalized theory unifying the hierarchy theories of recursive function theory and descriptive set theory. He introduced the super- and sub-scripted (lightface and boldface) $\Sigma$ and $\Pi$ notation in order to give a unified exposition of the generalized theory, which contained, for example, the Luzin-Addison separation theorem and the Kondo-Addison uniformization theorem. Around 1959 he observed that the Craig interpolation theorem was an analog of the Luzin-Addison separation theorem and this led in time to a partial unification of the hierarchy theories of pure first- and second-order logic with the above-mentioned unified theory and the discovery, for example, of numerous new interpolation theorems in pure logic. More recently he has investigated a possible still further unification with finite-universe logic, with potential applications to theoretical computer science. In 1949, Tarski had raised a series of problems concerning the definability in a language of the set of individuals definable in that language. Using the new technique of forcing, Addison showed a key case of "the undefinability of the definable": the class of arithmetical sets of natural numbers is not arithmetical. The biggest mystery left open in the early development of descriptive set theory was the status at the third and higher levels of the projective hierarchy of the so-called first separation property. Using the axiom of constructibility Addison had shown in 1956 that it held on the $\Pi$ side (and failed on the sigma side) at the third and higher levels. But he also noted that the proof that it held for the closed sets (the natural $\Pi_0^1$ projective class) "vectorized" to give a proof that it held for the $\Sigma_1^1$ (i.e. analytic) sets. Such considerations suggested to him and to Gödel that the pattern might alternate throughout the projective hierarchy. After Blackwell found a new proof using determinateness for analytic separation Addison recognized that it actually showed that separation for closed sets implied it for analytic sets and that the same argument could be used to go from the $\Pi$ side at any level to the sigma side at the next level. Moschovakis and independently Martin (who had also recognized the extendibility of Blackwell's technique) completed the alternating pattern, in part by showing that the truth on the sigma side of a more general property implied it on the following $\Pi$ side. These results led in time to increasing interest in definable determinateness as a promising new axiom, a thesis that was bolstered when it was seen that projective determinateness was implied by certain large cardinal axioms.

**Main Publications:**

1. On Some Points of the Theory of Recursive Functions, Doctoral Dissertation, University of Wisconsin, 1954, iv+92 pp. [available through University Microfilms, Inc. or on interlibrary loan]. [Summary: Summaries of Doctoral Dissertations, University of Wisconsin, vol. 16, The University of Wisconsin Press, 1956, pp. 348-349.]

2. Some unsolvable problems about cancellation semigroups. Abstract 368t, Bulletin of the American Mathematical Society, vol. 60 (1954), p. 258.

3. Analogies in the Borel, Lusin, and Kleene hierarchies. I. Abstract 139, Bulletin of the American Mathematical Society, vol. 61 (1955), p. 75.

4. Analogies in the Borel, Lusin, and Kleene hierarchies. II. Abstract 341, Bulletin of the American Mathematical Society, vol. 61 (1955), pp. 171-172.

5. A note on function quantification (with S. C. Kleene). Proceedings of the American Mathematical Society, vol. 8 (1957), pp.1003-1006.

6. Separation principles in the hierarchies of classical and effective descriptive set theory. Fundamenta Mathematicae, vol. 46 (1958), pp. 123-135.

7. Some consequences of the axiom of constructibil-

ity. Fundamenta Mathematicae, vol. 46 (1959), pp. 337-357.

8. On the Novikov-Kondo theorem. Abstract of an invited address delivered to the International Symposium on the Foundations of Mathematics: Infinitistic Methods, held at Warsaw, Poland, September 2-8, 1959 (mimeographed).

9. Hierarchies and the axiom of constructibility. Summaries of Talks Presented at the Summer Institute of Symbolic Logic, Cornell University, 1957, Second Edition, Communications Research Division, Institute for Defense Analyses, 1960, pp. 355-362.

10. The theory of hierarchies. Logic, Methodology and Philosophy of Science, Proceedings of the 1960 International Congress, edited by Ernest Nagel, Patrick Suppes, and Alfred Tarski, Stanford University Press, Stanford, 1962, pp. 26-37.

11. Some problems in hierarchy theory. Recursive Function Theory, Proceedings of Symposia in Pure Mathematics, vol. 5, American Mathematical Society, 1962, pp. 1-8.

12. The method of alternating chains. Proceedings of the 1963 International Symposium at Berkeley, edited by J. W. Addison, Leon Henkin, and Alfred Tarski, North-Holland Publishing Company, Amsterdam, 1965, pp. 1-16.

13. The undefinability of the definable (abstract). Notices of the American Mathematical Society, vol. 12 (1965), p. 347.

14. Some consequences of the axiom of definable determinateness (with Yiannis N. Moschovakis). Proceedings of the National Academy of Sciences, vol. 59 (1968), pp. 708-712.

15. Current problems in descriptive set theory. Lecture Notes Prepared in Connection with the Summer Institute on Axiomatic Set Theory Held at UCLA, Los Angeles, California, July 10–August 4, 1967, American Mathematical Society, 1967, pp. IV S1 – IV S20. [See also Proceedings of Symposia in Pure Mathematics, Volume XIII, Part II, American Mathematical Society, Providence, 1974, pp. 1-10.]

16. Tarski's theory of definability: common themes in descriptive set theory, recursive function theory, classical pure logic, and finite-universe logic. Annals of Pure and Applied Logic, vol. 126 (2004), pp. 77-92.

*Work in Progress*

17. "Infinitary Boolean Operations", a monograph covering the classification (according to their complexity or power) of infinitary analogs of the usual finitary Boolean logical (or set) operations.

18. Separation principles in difference hierarchies. (Cf. Notices of the American Mathematical Society, vol. 15 (1968), p. 238.)

19. Binary and 'double-cross' quantifiers.

20. A symmetric second-order normal form for formulas of first-order logic.

**Service to the Profession:** Chair, Department of Mathematics, University of California, Berkeley, 1968-1972, 1985-1989; Local Arrangements Committee Chair, Steering Committee Member, 1986 International Congress of Mathematicians, 1984-1986; Member, U.S. National Committee for Mathematics, National Research Council, 1982-1990; Executive Committee Member, Association for Symbolic Logic, 1977-1980; Co-editor, Proceedings of the Tarski Symposium, 1974-1975; President and Chairman, Board of Governors, Pacific Journal of Mathematics, 1971-1972; Council Member, American Association for the Advancement of Science, 1971-1974; Editor, with Leon Henkin and Alfred Tarski, Theory of Models, Proceedings of the International Symposium at Berkeley, 1963; Chair, Group in Logic and the Methodology of Science, University of California, Berkeley, 1963-1965, 1981-1985; Bibliographer of the theory of models, with Karel de Bouvere and William B. Pitt, The Theory of Models, Proceedings of the 1963 International Symposium at Berkeley, edited by J. W. Addison, Leon Henkin, and Alfred Tarski, North-Holland Publishing Co., 1965, pp. 442-492.

**Teaching:** Addison has taught undergraduate and graduate courses in logic at the University of Michigan and at the University of California, Berkeley, since 1962 (most recently in Fall 2006) and for many years has chaired the Berkeley Logic Colloquium. His doctoral supervisees include Melven Krom, Robert Barnes, Peter Hinman, Guy Benson, John LeTourneau, Stephen Garland, Dale Myers, John Steel, Robert Van Wesep, Arnold Miller, William Wadge, Samy Zafrany, and Arthur Quaife.

**Vision Statement:** Logic will play an ever-increasing role in the development of mathematics and other fields, including, for example, theoretical computer science, biology, and physics. The 'reality' of the infinite will come to be understood as simply the existence of certain common patterns in human brains. For most important questions the patterns will agree but there will be room for some interesting divergences for the consistency of which no counterexamples will ever be found.

**Honours and Awards:** Member, Golden Key International Honour Society; Member, Order of the Golden Sierra Cup; Fellow, American Association for the Advancement of Science; U.S. National Science Foundation Postdoctoral Fellow, 1956-1957; Kemper K. Knapp Resident Fellow, University of Wisconsin, 1953-1954; Member, Society of the Sigma Xi.

## ANDERSON, Curtis Anthony

**Specialties:** Intensional logic.

**Born:** 29 May 1940 in Corpus Christi, Texas, USA.

**Educated:** University of California, Los Angeles, PhD Philosophy, 1977; University of Houston, MS Mathematics, 1965; University of Houston, BS Physics and Mathematics, 1964.

**Dissertation:** *Some Models for the Logic of Sense and Denotation: Alternative (0)*; supervisor, Alonzo Church.

**Regular Academic or Research Appointments:** PROFESSOR, PHILOSOPHY, UNIVERSITY OF CALIFORNIA, SANTA BARBARA, 1993–; University of Minnesota, Minneapolis, 1978-1993; University of Texas at Austin, 1975-1978; University of Southern California, 1970-1973.

**Visiting Academic or Research Appointments:** Visiting Research Fellow, Katholieke Universiteit Leuven, Leuven, Belgium, 1996; Visiting Professor, University of Salzburg, 1992; Visiting Professor, Karl-Franzens Universität, Graz, Austria, 1990.

**Research Profile:** Anderson's work in logic has been entirely focused on intensional logics, especially those logics that hold out some promise for being able to treat the "propositional attitudes" (belief, knowledge, desire, etc.). Much of his work is concerned with the necessary philosophical clarification of fundamental issues in intensional logic.

**Main Publications:**

1. Alternative (1*): A Criterion of Identity for Intensional Entities, in *Logic, Meaning and Computation* (ed. Anderson and Zeleny) Kluwer Academic Publishers (2001) 393-427.
2. Toward a Logic of A Priori Knowledge, *Philosophical Topics* **21** (1995) 1-10.
3. Degrees of Intensionality, in *Analyomen 1, Proceedings of the 1st Conference "Perspectives in Analytic Philosophy* (ed Meggle and Wessels) de Gruyter (1994) 411-420.
4. Russellian Intensional Logic, in *Themes from Kaplan*, (ed Almog, Perry, and Wettstein) Oxford University Press (1989) 67-103.
5. Semantical Antinomies in the Logic of Sense and Denotation, *Notre Dame Journal of Formal Logic* **28** (1987) 99-114.
6. Some Difficulties Concerning Russellian Intensional Logic, *No()s* **20** (1986), 35-43.
7. The Paradox of the Knower, *Journal of Philosophy* **80** (1983) 338-55.
8. Some New Axioms for the Logic of Sense and Denotation, *No()s* **14** (1980) 217-34.

*Work in Progress*
9. General Pure Semantics: A Theory of Possible Languages as Foundation for Intensional Logic, book mss.

**Service to the Profession:** Organizing Committee, Association for Symbolic Logic, 1999-2000.

**Vision Statement:** Anderson believes that the most important logical task bearing upon philosophy is the construction of a demonstrably adequate intensional logic suitable for formalizing reasoning involving the propositional attitudes. Such a logic will not automatically solve outstanding philosophical problems but is a prerequisite for such solutions.

**Honours and Awards:** "Russellian Intensional Logic" reprinted in *Philosopher's Annual* as one of the ten best philosophical papers of 1989; Distinguished Teacher Award, College of Liberal Arts, University of Minnesota, 1984; Outstanding Graduate Student in Philosophy, UCLA, 1968-1969; NDEA Title IV Fellowship, UCLA, 1968.

## ANDRÉKA, Hajnal

**Specialties:** Mathematical logic, algebraic logic, logical foundations of spacetime theory and relativity, universal logic, philosophical logic, formal semantics, computer science and AI logics, foundation of mathematics.

**Born:** 17 November 1947 in Budapest, Hungary.

**Educated:** Hungarian Academy of Sciences, DSc with the academy Mathematics 1992; candidate's degree Mathematics 1978; Eötvös Loránd University PhD Mathematics 1975; BA Mathematics 1971.

**Dissertation:** DSc: *Complexity of equations valid in algebras of relations*; Candidate's degree: *Universal algebraic investigations in algebraic logic*, advisor: Bálint Dömölki; PhD: *Algebraic investigation of first order logic*, advisor: Rózsa Péter.

**Regular Academic or Research Appointments:** HEAD OF DEPARTMENT, RÉNYI MATHEMATICAL INSTITUTE BUDAPEST, 1978– Institute for Computer Science of the Ministry of Heavy Industries, 1971-1977.

**Visiting Academic or Research Appointments:** Visiting professorships: University of Amsterdam, 1998 spring semester; University of California at Berkeley, Mills Summer Mathematical Institute, USA, 1991 summer; Iowa State University, Iowa, USA, 1987-1988; University of Waterloo, Canada, 1983-1984.

**Research Profile:** Their (with István Németi) main research interest is logic as the scientific study or backbone of rationality, a unifying tool of the scientific method. The emphasis is on mathematical tools of logic. Connections between logic, geometry and algebra are in the center of their work. This 3-way connection (logic, geometry, algebra) appears in their work partly as algebraic logic (AL), and partly as the logical analysis of relativity theories and spacetime geometry. All parts of this 3-way connection go back to their collaboration with Alfred Tarski's group [2,8,14].

They concentrate on the dynamics of theory-formation/theory-evolution. The point is to view and study logic as being dynamic, alive and changing as opposed to viewing it as being static, eternally frozen. They use the category of cylindric algebras (CA's) for representing the category of theories and interpretations acting between these theories as morphisms. In cooperation with Johan van Benthem, Bjarni Jónsson and Vaughan Pratt, among others, they refine the Tarskian tradition in AL in order to reinforce the so-called dynamic trend in logic. Dynamic logic or the logic of actions concentrates on the dynamic (even inductive and abductive) aspects of reasoning in several ways. Besides the dynamics of theory-formation, it also provides logics for reasoning about consequences of actions relevant for artificial intelligence (AI), program verification, logics of time, spacetime, and general relativity (GR) [4,20] and work in progress [1].

They made efforts to "refresh" Tarskian algebraic logic by making the interaction between logic and algebra stronger in the algebra-to-logic direction. They view algebraization of logic as a kind of abstraction. Generally, after the step of abstraction, one studies the so obtained abstract structures (e.g., AL). They, however, emphasize that the results of this abstract study should be applied to the original "concrete" world, e.g., to logic. They initiated the study of relativized algebras, in particular relativized CA's, as an area worth studying in its own right. This study of new kinds of algebras or relations (namely relativized ones) led to new discoveries in pure logic. In particular, this catalyzed new kinds of applications in logic, and connected new trends and research interests in logic to methods of AL. They proved important properties of these relativized CA's, leading to the discovery of the very successful notion of guarded fragment of first-order logic (FOL) [15]. They used AL to obtain new purely logical results. The latter include results on the finite-variable hierarchy of FOL, results on the schema version of FOL and its extensions (decidability and definability properties, interpolation, finitizability), and results on model theory of infinitary relational structures (algebraic model theory) [2,7,8,13,16,17]. The finite-variable hierarchy is well behaved in the new, guarded fragment, while in the original FOL it is not, as proved in [15]. They apply the same kind of approach to relativity theory and its abstract counterpart spacetime geometry, by elaborating a duality theory between the observer-oriented (i.e., coordinate-dependent) and the geometrical (i.e., coordinate-independent) versions of relativity [20].

In category theoretic AL, they initiated "injectivity logic" and the injectivity approach to a very general conception of logic and applied it to the theory of partial algebras [3,5]. They contributed to the theories of cylindric, relation, polyadic, and other algebras of n-place relations establishing a new school (or at least invigorating the old one) [2,7,9–11,13–18].

They study the applications of logic in various sciences, e.g., in the semantics of natural language, algebra, computer science and AI, theory of spacetime and relativity. In relativity, one of their aims is to make (even the most "esoteric" parts of) GR accessible for the nonspecialist readers (with a background in logic). They not only apply logic to GR, but also apply GR to the foundation of logic, pursuing a research direction suggested to them by the logician László Kalmár [19]. In their application of logic to relativity, they answer the "why type" questions (e.g., which axiom of (the logicized) GR is responsible for some interesting predictions/theorems of GR), using the same method as in their earlier work in reverse mathematics, where they studied the question of which axioms of set theory are responsible for which theorems of algebra and AL [13].

**Main Publications:**

1. The generalized completeness of Horn predicate logic as a programming language. Acta Cybernetica 4,1 (Szeged 1978), 3-10. (With Németi, I.)

2. Cylindric Set Algebras. Lecture Notes in Mathematics vol. 883, Springer-Verlag, Berlin, 1981. vi +323 pp. (With Henkin, L., Monk, J. D., Tarski, A., Németi, I.)

3. A general axiomatizability theorem formulated in terms of cone-injective subcategories. In: Universal Algebra (Proc. Coll. Esztergom 1977) Colloq. Math. Soc. J. Bolyai vol. 29, North-Holland, Amsterdam, 1981. pp. 13-35. (With Németi, I.)

4. A complete logic for reasoning about programs via nonstandard model theory. Theoretical Computer Science 117 (1982). Part I in no. 2, pp. 193-212, Part II in no. 3, pp. 259-278. (With Németi, I., Sain, I.)

5. Generalization of the concept of variety and quasi-variety to partial algebras through category theory. Dissertationes Mathematicae (Rozprawy Math.) no. 204.

PWN - Polish Scientific Publishers, Warsaw, 1983. 51 pp. (With Németi, I.)

6. A system of logic for partial functions under existence-dependent Kleene equality. Journal of Symbolic Logic 53 (1988), 834-839. (With Craig, W., Németi, I.)

7. A Stone-type representation theorem for algebras of relations of higher rank. Trans. Amer. Math. Soc. 309,2 (1988), 671-682. (With Thompson, R. J.)

8. Algebraic Logic. Colloq. Math. Soc. J. Bolyai vol. 54, North-Holland, Amsterdam, 1991. vi + 746 pp. (Edited with Monk, J. D., Németi, I.)

9. Free algebras in discriminator varieties. Algebra Universalis 28 (1991), 401-447. (With Jónsson, B., Németi, I.)

10. A nonpermutational integral relation algebra. Michigan Math. J. 39 (1992), 371-384. (With Düntsch, I., Németi, I.)

11. Lambek Calculus and its relational semantics: Completeness and incompleteness. Journal of Logic, Language and Information 3 (1994), 1-37. (With Mikulás, Sz.)

12. Connections between axioms of set theory and basic theorems of universal algebra. Journal of Symbolic Logic 59,3 (1994), 912-922. (With Kurucz, Á., Németi, I.)

13. Complexity of equations valid in algebras of relations, Parts I-II. Annals of Pure and Applied Logic 89 (1997), 149-229.

14. Decision problems for equational theories of relation algebras. Memoirs of Amer. Math. Soc. vol. 126, no. 604, American Mathematical Society, Providence, Rhode Island, 1997. xiv+126pp. (With Givant, S., Németi, I.)

15. Modal languages and bounded fragments of predicate logic. Journal of Philosophical Logic 27 (1998), 217-274. (With van Benthem, J., Németi, I.)

16. Relativised quantification: some canonical varieties of sequence-set algebras. Journal of Symbolic Logic 63,1 (1998), 163-184. (With Goldblatt, R., Németi, I.)

17. Algebraic Logic. In: Handbook of Philosophical Logic, vol. 2, second edition, eds. D. M. Gabbay and F. Guenthner, Kluwer Academic Publishers, 2001. pp. 133-247. (With Németi, I., Sain, I.)

18. Operators and laws for combining preferential relations. Journal of Logic and Computation 12,1 (2002), 13-53. (With Ryan, M., Schobbens, P-Y.)

19. Can general relativistic computers break the Turing barrier? In: Logical Approaches to Computational Barriers, Second Conference on Computability in Europe, CiE 2006, Swansea, UK, July 2006, Proceedings, Arnold Beckmann, Ulrich Berger, Benedikt Löwe, and John V. Tucker (eds.), Lecture Notes in Computer Science vol. 3988, Springer-Verlag, Berlin 2006. pp.398-412. (With Németi, I.)

20. Logic of spacetime and relativity. In: Handbook of Spatial Logics. Eds: Aiello, M. Pratt-Hartman, I., and van Benthem, J. Springer Verlag, to appear. 119 pp. (With Madarász, J. X., Németi, I.)

A full list of publications is available at Andréka's homepage http://www.renyi.hu/~andreka/

*Work in Progress*

21. A twist in the geometry of rotating black holes: seeking the cause of acausality; this is an in-depth mathematical analysis of some often misinterpreted features of some exotic GR spacetimes, with Németi, I. and Wüthrich, C.

22. A book on universal logic and universal algebraic logic, with Németi, I. and Sain, I.

23. Visualization of Gödel's rotating universe, with Németi, I. and Madarász, J. X.

24. Definability theory for many-sorted FOL allowing definition of new sorts (i.e. "universes").

**Service to the Profession:** Member at Large of the Council of the Association of Symbolic Logic, 1999-2001; Member of Executive Board of European Foundation for Logic, Language and Information, 1991-1997; Member of Editorial Boards of Periodica Mathematica Hungarica 1985-1998, Studia Scientiarum Mathematicarum 1991-, Fundamenta Informatica, 1990-, Journal of Applied Logic 2004-, Studies in Universal Logic book series 2007-; Member of Advisory Board of Advanced Studies in Mathematics and Logic book series 2005-. Member of Program Committee or Organizing Committee of the following international conferences: Conference on Algebraic Logic, Budapest 1988 (organized by J. Bolyai Mathematical Society); "Algebraic Methods in Logic and Their Computer Science Applications", Warsaw 1991 (38th Semester of Stefan Banach International Banach Mathematical Center); European Summer Meeting of the Association for Symbolic Logic, Veszprém 1991 (organized jointly by J. Bolyai Mathematical Society and Symbolic Logic Department of Eötvös Loránd University); Algebraic logic and the methodology of applying it, Budapest 1994 (TEMPUS Summer School); First Southern African School and Workshop on Logic, Universal Algebra, and Theoretical Computer Science, South Africa 1999; Logic in Hungary 2005.

**Teaching:** They together with István Németi had 18 successful PhD students. In alphabetic order: Balázs Bíró (1986), Buy Huy Hien (1981), Csaba Henk (present), Eva Hoogland (1996, MA, Amsterdam), Ágnes Kurucz (1997, London), Judit X. Madarász (2002), Zsuzsanna Márkusz (1989), Maarten Marx (1995, Amsterdam), Szabolcs Mikulás (1995, Amsterdam, London), Ana Pasztor (1979, University of Miami), Ildikó Sain (1986), Gábor Sági (1999), Tarek Sayed-Ahmed (2003, Cairo), György Serény (1986), András Simon (1998), Gergely Székely (present), Jenő Szigeti (1989), Csaba Tőke (2000, MA). They positively influenced the beginning of the scientific careers of Ben Hansen (Berkeley), Miklós Ferenczi, Viktor Gyuris, Richard J. Thompson (Berkeley), and others.

**Vision Statement:** In order to keep logic healthy and evolving, we have to reach out and start applying logic to exciting new subjects in the "cutting edge of humanity". In this line, I would like to see a self-contained book introducing general relativity (GR) to the nonspecialist logician making GR fully accessible. The book should build up GR as a theory purely in (many-sorted) FOL with transparent, easily comprehensible axioms. A conceptual analysis of GR should also be included in the style of [Andréka, H., Madarász, J. X., Németi, I.: On the logical structure of relativity theories. http://www.math-inst.hu/pub/algebraic-logic/olsort.html]. Logic can serve as a main device in striving for the unity of science.

**Honours and Awards:** Alfréd Rényi Award of the Mathematical Research Institute of Hungarian Academy of Sciences, in recognition of mathematical results 1987; László Kalmár Award of John von Neumann Computer Society, in recognition of work and results on unifying theoretical mathematics and computer science 1979; Prize "Gyula Farkas" for Applying Mathematics, of the Mathematical Society János Bolyai 1978; Prize Géza Grünwald of the Mathematical Society János Bolyai 1975; scholarship for receiving a candidate's degree with the Hungarian Academy of Sciences 1974-1977.

# ANDRETTA, Alessandro

**Specialties:** Set theory

**Born:** 18 April 1960 in Torino, Italy.

**Educated:** University of California at Los Angeles, PhD Mathematics 1989; Università di Torino, Laurea in Matematica 1983.

**Dissertation:** *Iteration Trees*; supervisor, John Steel.

**Regular Academic or Research Appointments:** ASSOCIATE PROFESSOR OF MATHEMATICS, UNIVERSITÀ DI TORINO, ITALY, 2001–; Researcher in Math, Università di Torino, Italy, 1995-2001; Researcher in Math, Università di Camerino, Italy, 1992-1995; J.W.Y. Research Instructor, Dartmouth College, 1989-1991.

**Visiting Academic or Research Appointments:** Visiting Associate Professor, Mathematics, University of California at Los Angeles, USA 1998-1999; Fellow, Math Department, University of Bonn, Germany, 1998; Fellow, Italian National Council for Research (CNR) at the Math Department of UCLA, 1991-1992.

**Research Profile:** Alessandro Andretta has been working in set theory, specifically on large cardinals, the Wadge hierarchy, and applications of descriptive set theory to other areas of math.

**Main Publications:**

1. More on Wadge determinacy. Annals of Pure and Applied Logic (in press)Equivalence between Wadge and Lipschitz determinacy. Annals of Pure and Applied Logic 123 (2003), no. 1-3, 163–192.(joint work with Donald A. Martin)

2. Borel-Wadge degrees. Fundamenta Mathematicæ 177 (2003), no. 2, 175–192.(joint work with Itay Neeman and John Steel)

3. The domestic levels of K^c are iterable. Israel Journal of Mathematics. 125 (2001)(joint work with Riccardo Camerlo and Greg Hjorth)

4. Conjugacy equivalence relation on subgroups. Fundamenta Mathematicæ 167 (2001), no. 3, 189–212.(joint work with Alberto Marcone)

5. Projective sets and ordinary differential equations. Transactions of the American Mathematical Society 353 (2001), no. 1, 41–76.(joint work with Alberto Marcone)

6. Ordinary differential equations and descriptive set theory: uniqueness and globality of solutions of Cauchy problems in one dimension. Fundamenta Mathematicæ 153 (1997), no. 2, 157–190(joint work with Alberto Marcone)

7. Definability in function spaces. Real Analysis Exchange 26 (2000/01), no. 1, 285–308 (joint work with Alberto Marcone)

8. Pointwise convergence and the Wadge hierarchy. Comment. Math. Univ. Carolin. 42 (2001), no. 1, 159–172 (joint work with John Steel)

9. How to win some simple iteration games. Annals Pure Applied Logic 83 (1997), no. 2, 103–164.Building iteration trees. J. Symbolic Logic 56 (1991), no. 4, 1369–1384.

10. Large cardinals and iteration trees of height $\omega$. Annals Pure Applied Logic 54 (1991), no. 1, 1–15.

*Work in Progress*

11. (joint work with Itay Neeman and Greg Hjorth) Cardinalities in the Wadge hierarchy.

**Teaching:** Alessandro Andretta has bee teaching for many years in the Mathematics Department at the Università di Torino. He has advised, jointly with B.Velickovic (Paris VII) a PhD student, Matteo Viale, graduating in fall 2006.

**Vision Statement:** "Strong axioms of infinity are, in my opinion, one of the most intriguing topics in modern set theory and mathematical logic. As of now their influence is confined mostly within the boundary of set theory itself, but I believe that in the next decades large cardinals will turn out to be useful in concrete problems of mainstream mathematics"

## ANDREWS, Peter Bruce

**Specialties:** Higher-order logic (type theory), automated theorem proving.

**Born:** 1 November 1937 in New York, New York, U.S.A.

**Educated:** Dartmouth College A.B. (Mathematics) 1959; Princeton University Ph.D. (Mathematics) 1964.

**Dissertation:** *A Transfinite Type Theory with Type Variables*; supervisor Alonzo Church.

**Web Address:** http://gtps.math.cmu.edu/andrews.html

**Regular Academic or Research Appointments:** PROFESSOR OF MATHEMATICS, CARNEGIE MELLON UNIVERSITY, 1979–. Associate Professor of Mathematics, Carnegie Mellon University, 1968–1979; Assistant Professor of Mathematics, Carnegie Mellon University, 1963–1968.

**Research Profile:** Andrews' work has been motivated by the desire to help develop tools which will enhance the abilities of humans to reason. He has focused on automated deduction and Church's type theory, which is a rich and expressive formulation of higher-order logic in which statements from many disciplines, particularly those involving mathematics, can readily be expressed.

He drastically simplified [1] Henkin's axioms for a formulation of Church's type theory in which quantifiers and logical connectives are defined in terms of equality and the abstraction operator $\lambda$, producing a very elegant logical system which he discussed extensively in [17] and incorporated into the formulation of transfinite type theory in his thesis.

To develop sophisticated methods of searching for proofs one needs a deep understanding of the essential logical structure of the theorems to be proved. Since Herbrand's Theorem is a key tool for such understanding, Andrews undertook a close study of Herbrand's proof of the theorem, and found a serious error [2,18].

Cut-elimination is closely related to Herbrand's Theorem, and in [3] Andrews extended proofs of cut-elimination for other formulations of type theory to Church's formulation, and extended Smullyan's Unifying Principle to type theory. This Principle provides a basic tool for establishing completeness proofs for type theory, and was used in [6] to establish some basic results about the decision problem for elementary type theory.

In [4] a characterization of general models for type theory was developed and used to prove independence results. This characterization was used in [5] to construct a special nonstandard model which revealed an error in Henkin's paper on the completeness of type theory.

In a search for more sophisticated approaches to theorem-proving problems, Andrews wished to abstract away from considerations involving the rules of particular deductive calculi, and focus on the structural properties of the theorem being proved. This led to work involving *matings* [7,8] of subformulas of the theorem. This approach to automated theorem proving is closely related to the *connection method* which was developed independently by Bibel.

It is desirable to search for the basic structural ingredients of a proof in a context well suited to such a search, and then use these ingredients to construct a human-oriented proof, so Andrews developed [9] a procedure for automatically transforming acceptable general matings into proofs in natural deduction style.

Attention then turned to extending these ideas to higher-order logic. Starting from ideas presented in [10], Andrews' research assistant Dale Miller developed the concept of an *expansion proof* and proved a higher-order analogue of Herbrand's Theorem. To this Andrews added the idea of using primitive substitutions to introduce into expansion trees substitution terms for higher-order variables which could not be generated by higher-order unification, and an automatable proof procedure for higher-order logic [11] was obtained.

An automated theorem proving system called TPS [12] which implements these ideas has been developed and enhanced in collaboration with graduate students in work extending over several decades. It can operate in automatic, semi-automatic, and interactive modes. It excels [15] at proving theorems of type theory automatically, and has many features which make it very attractive for working in a combination of automatic and interactive modes [20].

A solution to the problem of deciding which instances of the definitions in a theorem to instantiate while searching for a proof of the theorem was found by Andrews, enhanced and implemented in TPS by Matthew Bishop, and is discussed in [14].

**Main Publications:**

1. "A Reduction of the Axioms for the Theory of Propositional Types". *Fundamenta Mathematicae*, **52** (1963): 345–350.

2. "False Lemmas in Herbrand", with Burton Dreben

and Stål Aanderaa. *Bulletin of the American Mathematical Society*, **69** (1963): 699–706.
3. "Resolution in Type Theory". *Journal of Symbolic Logic*, **36** (1971): 414–432.
4. "General Models, Descriptions, and Choice in Type Theory". *Journal of Symbolic Logic*, **37** (1972): 385–394.
5. "General Models and Extensionality". *Journal of Symbolic Logic*, **37** (1972): 395–397.
6. "Provability in Elementary Type Theory". *Zeitschrift fur Mathematische Logic und Grundlagen der Mathematik*, **20** (1974): 411–418.
7. "Refutations by Matings". *IEEE Transactions on Computers*, **C-25** (1976): 801–807.
8. "Theorem Proving via General Matings". *Journal of the ACM*, **28** (1981): 193–214.
9. "Transforming Matings into Natural Deduction Proofs". In W. Bibel and R. Kowalski, (eds.), *Proceedings of the 5th International Conference on Automated Deduction. Lecture Notes in Computer Science, 87*. Springer-Verlag, Les Arcs, France, 1980, 281–292.
10. "Automating Higher-Order Logic", with Dale A. Miller, Eve Longini Cohen, and Frank Pfenning. In W.W. Bledsoe and D.W. Loveland (eds.), *Automated Theorem Proving: After 25 Years. Proceedings of the Special Session on Automatic Theorem Proving, 89th Annual Meeting of the American Mathematical Society, held in Denver, Colorado, January 5–9, 1983*. American Mathematical Society, *Contemporary Mathematics Series*. Vol. 29, 1984, 169–192.
11. "On Connections and Higher-Order Logic". *Journal of Automated Reasoning*, **5** (1989): 257–291.
12. "TPS: A Theorem Proving System for Classical Type Theory", with Matthew Bishop, Sunil Issar, Dan Nesmith, Frank Pfenning, and Hongwei Xi. *Journal of Automated Reasoning*, **16** (1996): 321–353.
13. "On Sets, Types, Fixed Points, and Checkerboards", with Matthew Bishop. In Pierangelo Miglioli, Ugo Moscato, Daniele Mundici, and Mario Ornaghi, (eds.), *Theorem Proving with Analytic Tableaux and Related Methods. 5th International Workshop. (TABLEAUX '96) Lecture Notes in Artificial Intelligence, 1071*. Springer-Verlag, Terrasini, Italy, 1996, 1–15.
14. "Selectively Instantiating Definitions", with Matthew Bishop. In Claude Kirchner and Hélène Kirchner (eds.), *Proceedings of the 15th International Conference on Automated Deduction. Lecture Notes in Artificial Intelligence, 1421*. Springer-Verlag, Lindau, Germany, 1998, 365–380.
15. "System Description: TPS: A Theorem Proving System for Type Theory", with Matthew Bishop and Chad E. Brown. In David McAllester, (ed.), *Proceedings of the 17th International Conference on Automated Deduction. Lecture Notes in Artificial Intelligence, 1831*. Springer-Verlag, Pittsburgh PA, 2000, 164–169.
16. "Classical Type Theory". In Alan Robinson and Andrei Voronkov, (eds.), *Handbook of Automated Reasoning, vol. 2*. Elsevier Science, 2001, ch. 15, pp. 965–1007.
17. *An Introduction to Mathematical Logic and Type Theory: To Truth Through Proof*. Kluwer Academic Publishers, 2002. Second edition. Distributed by Springer.
18. "Herbrand Award Acceptance Speech". *Journal of Automated Reasoning*, **31** (2003): 169–187.
19. "ETPS: A System to Help Students Write Formal Proofs", (with Chad E. Brown, Frank Pfenning, Matthew Bishop, Sunil Issar, and Hongwei Xi). *Journal of Automated Reasoning*, **32** (2004): 75–92.
20. "TPS: A Hybrid Automatic-Interactive System for Developing Proofs", (with Chad E. Brown). *Journal of Applied Logic*, **4** (2006), 367–395.

**Service to the Profession:** Editorial Board, Journal of Automated Reasoning, 1985–; Editorial Board, Journal of Applied Logic, 2003–; Editorial Board, Journal of Logic and Computation, 1990–1993; Program Committee, International Conference on Automated Deduction, 1982, 1984, 1988, 1990, 1992.

**Teaching:** The textbook [17] provides a unified introduction to propositional calculus, first-order logic, and type theory. It makes nonstandard models and the incompleteness theorems easy to understand. The chapter entitled Provability and Refutability treats semantic tableaux, skolemization, cut-elimination, Herbrand's Theorem, etc.

The interactive commands for applying rules of inference in TPS are available in a program called ETPS (Educational Theorem Proving System) [19], which can be used interactively by students in logic courses to construct natural deduction proofs.

Former Ph.D. students are Edward R. Fisher, Jr., Dale A. Miller, Frank Pfenning, Sunil Issar, Matthew Bishop, and Chad E. Brown.

**Vision Statement:** Andrews looks forward to the eventual formalization of virtually all mathematical, scientific, and technical knowledge, and the development of automated reasoning tools and automated information systems which use automated reasoning to assist in storing, developing, refining, verifying, finding, and applying this knowledge. Logical research will provide intellectual foundations for these developments.

**Honours and Awards:** Herbrand Award for Distinguished Contributions to Automated Reasoning, 2003.

# APTER, Arthur William

**Specialties:** Mathematical Logic, specifically Set Theory: Large Cardinals and Forcing.

**Born:** 13 December 1954 in Brooklyn, New York USA.

**Educated:** Ph.D., Massachusetts Institute of Technology, Mathematics, 1978; B.S., Massachusetts Institute of Technology, Mathematics, 1975.

**Dissertation:** *Large Cardinals and Relative Consistency Results*; supervisor, Eugene M. Kleinberg.

**Regular Academic or Research Appointments:** PROFESSOR OF MATHEMATICS, THE CUNY GRADUATE CENTER, 2006 –; Professor of Mathematics, Baruch College of CUNY, 1991 – ; Associate Professor of Mathematics, Baruch College of CUNY, 1986 – 1990; Assistant Professor of Mathematics, Rutgers University, Newark, College of Arts and Sciences, 1981 – 1986; Assistant Professor of Mathematics, University of Miami, Florida, 1979 – 1981; Instructor of Mathematics, Massachusetts Institute of Technology, 1978 – 1979; Teaching Assistant, Mathematics, Massachusetts Institute of Technology, 1977 – 1978.

**Visiting Academic or Research Appointments:** Visitor, Mathematics Institute, Rheinische Friedrich-Wilhelms-University, January 2005; Visitor, Department of Mathematics, University of Münster, January 2005; Visitor, Institute for Logic, University of Vienna, March 2000; Visitor, Mathematics Institute, Rheinische Friedrich-Wilhelms-University, February 2000; Visitor, Department of Mathematics, University of East Anglia, February 2000; Visitor, Department of Mathematical Sciences, Carnegie Mellon University, October – November 1999; Visitor, Departments of Mathematics of Kanagawa University, Kobe University, Nagoya University, Tsukuba University, Waseda University, August – September 1999; Visitor, Department of Mathematics, Hebrew University, Fall 1992; Visitor, Department of Mathematics, Tel Aviv University, Fall 1992, August 1990, July 1985; Visiting Scholar, Department of Mathematics, Dartmouth College, Spring 1985 and Summer 1985.

**Research Profile:** Arthur W. Apter is a set theorist who works in the area of large cardinals and forcing. He has specialized primarily in the study of the structural properties of universes containing strongly compact and supercompact cardinals, and the construction of choiceless inner models satisfying a variety of large cardinal hypotheses. He maintains an interest in interactions between inner model theory and large cardinals and forcing, and equiconsistency questions concerning large cardinals.

**Main Publications:**

1. "On the Least Strongly Compact Cardinal", *Israel J. Math.* 35, 1980, 225 – 233.
2. "Some Results on Consecutive Large Cardinals", *Annals of Pure and Applied Logic* 25, 1983, 1 – 17.
3. "Successors of Singular Cardinals and Measurability", *Advances in Mathematics* 55, 1985, 228 – 241.
4. "Some Results on Consecutive Large Cardinals II: Applications of Radin Forcing", *Israel J. Math.* 52, 1985, 273 – 292.
5. "Some New Upper Bounds in Consistency Strength for Certain Choiceless Large Cardinal Patterns", *Arch. Math. Logic* 31, 1992, 201 – 205.
6. "Instances of Dependent Choice and the Measurability of $\aleph_{\omega+1}$" (with M. Magidor), *Annals of Pure and Applied Logic* 74, 1995, 203 – 219.
7. "On the Strong Equality between Supercompactness and Strong Compactness" (with S. Shelah), *Trans. Amer. Math. Soc.* 349, 1997, 103 – 128.
8. "Menas' Result is Best Possible" (with S. Shelah), *Trans. Amer. Math. Soc.* 349, 1997, 2007 – 2034.
9. "Patterns of Compact Cardinals", *Annals of Pure and Applied Logic* 89, 1997, 101 – 115.
10. "Laver Indestructibility and the Class of Compact Cardinals", *J. Symbolic Logic* 63, 1998, 149 – 157.
11. "The Least Measurable can be Strongly Compact and Indestructible" (with M. Gitik), *J. Symbolic Logic* 63, 1998, 1404 – 1412.
12. "The Calculus of Partition Sequences, Changing Cofinalities, and a Question of Woodin" (with J. Henle and S. Jackson), *Trans. Amer. Math. Soc.* 352, 2000, 969 – 1003.
13. "Identity Crises and Strong Compactness" (with J. Cummings), *J. Symbolic Logic* 65, 2000, 1895 – 1910.
14. "Identity Crises and Strong Compactness II: Strong Cardinals" (with J. Cummings), *Arch. Math. Logic* 40, 2001, 25 – 38.
15. "Aspects of Strong Compactness, Measurability, and Indestructibility", *Arch. Math. Logic* 41, 2002, 705 – 719.
16. "Indestructibility and the Level-by-Level Agreement between Strong Compactness and Supercompactness" (with J.D. Hamkins), *J. Symbolic Logic* 67, 2002, 820 – 840.
17. "Exactly Controlling the Non-Supercompact Strongly Compact Cardinals" (with J.D. Hamkins), *J. Symbolic Logic* 68, 2003, 669 – 688.
18. "Jonsson-like Partition Relations and $j : V \to V$" (with G. Sargsyan), *J. Symbolic Logic* 69, 2004, 1267 – 1281.
19. "The Consistency Strength of $\aleph_\omega$ and $\aleph_{\omega_1}$ being Rowbottom Cardinals without the Axiom of Choice" (with P. Koepke), to appear in the *Arch. Math. Logic*.
20. "Large Cardinals with Few Measures" (with J. Cummings and J.D. Hamkins), submitted for publication to *Proc. Amer. Math. Soc.*

A complete list of publications is available at http://faculty.baruch.cuny.edu/apter.
*Work in Progress*

21. Additional research in the study of different universes containing strongly compact and supercompact cardinals and large cardinals and choiceless set theory.

**Service to the Profession:** Co-editor (along with Professor Marcia Groszek of Dartmouth College) of a volume of papers to be published in *APAL* in honor of the $60^{th}$ birthday of Professor James Baumgartner of Dartmouth College; Principal Investigator of the NSF grants supporting the Mid-Atlantic Mathematical Logic Seminar (MAMLS), 1995 – ; Referee and reviewer of papers and grant proposals in set theory.

**Teaching:** Apter has taught graduate courses in set theory at MIT and Rutgers University. He has taught a graduate course in logic at the University of Miami and led a seminar in set theory for graduate students at the University of Miami. He has served as the external reader on the doctoral dissertation committees of three students at Rutgers University, along with serving on the doctoral dissertation and oral exam panels of students at CUNY. He was the faculty mentor to Grigor Sargsyan in the CUNY Baccalaureate Program, spring 2002 – spring 2003, and is slated to be Sargsyan's doctoral dissertation co-advisor at the University of California, Berkeley.

**Vision Statement:** I hope to see more interaction among the different areas of logic. In set theory, the recent advances by Woodin in the construction of an inner model for a supercompact cardinal hold great promise. I hope eventually Woodin's or other techniques will lead to the solution of the question of the equiconsistency between a strongly compact and supercompact cardinal.

**Honours and Awards:** Recipient of the Presidential Excellence Award for Distinguished Research, Baruch College, Spring 2005; Research in Pairs (RiP) Fellow (jointly with Professor James Cummings of Carnegie Mellon University) at the Mathematics Research Institute, Oberwolfach, Germany, June 8 – June 21, 1997; Recipient, Rutgers University Summer Research Fellowship, Summer 1982; Alternate in the 1979 – 1980 American Mathematical Society Postdoctoral Fellowship Competition, Spring 1979; Member, MIT Chapter of Sigma Xi. Associate Member, 1975 – 1980. Full Member, 1980 – Present.

# ÅQVIST, Lennart Ernst

**Specialties:** Philosophical logic, legal and moral philosophy, philosophy of language, history of philosophy.

**Born:** 11 March 1932 in Örebro, Sweden.

**Educated:** University of Uppsala, Swedish PhD Practical Philosophy, 1960.

**Dissertation:** The Moral Philosophy of Richard Price; supervisor, I. Hedenius.

**Regular Academic or Research Appointments:** DOCENT (ASSOCIATE PROFESSOR), PRACTICAL PHILOSOPHY, UNIVERSITY OF UPPSALA, SWEDEN, 1960–; Researcher, Swedish Council for Research in Humanities and the Social Sciences, 1991-1994, 1994-1997; Researcher, Bank of Sweden Tercentenary Foundation, 1969-1975, 1980-1986, 1986-1991; Researcher, Deutsche Forschungsgemeinschaft, with C. Rohrer, 1975-1979.

**Visiting Academic or Research Appointments:** Gastprofessor, Institut fûr Linguistik-Romanistik, Universität Stuttgart, BRD, 1974-1975; Visiting Professor, Philosophy, Brown University, USA, 1967-1969 , 1971; Substituting Professor, Philosophy, Åbo Academy, Finland, 1969, 1974; Substituting Professor, Practical Philosophy, University of Lund, Sweden, 1962.

**Research Profile:** As to Åqvist's contributions to philosophical logic and applied logic in the last century and the early $21^{st}$ one, let me mention his work in modal, epistemic, deontic, imperative, interrogative, temporal, indexical and conditional logic as well as the logic of causation and action ["agency"], especially as applied to various branches of the law, e.g. torts, property, criminal law, the law of evidence and the problem of burden of proof. As to Åqvist's work on moral philosophy [meta-ethics rather than ethics proper], he considers his main contribution to be the infinite hierarchy DHRxym of conditional dyadic preference-based deontic systems, developed in temporal logic with historical necessity and avoidability (see [47] below). As to Åqvist's contributions to the philosophy of language and linguistics, he wishes to emphasize his work on interrogatives, performatives and indexicals, especially as they appear in law and morals. Finally, as to Åqvist's contributions to the history of philosophy — seen from the standpoint of modern philosophical logic — his 1960 dissertation on Richard Price and the 2003 *Logique & Analyse* paper on Aristotle's *De Interpretatione* IX deserve particular attention.

**Main Publications:**

1. *The Moral Philosophy of Richard Price.* Uppsala: Almqvist & Wiksell, 1960. [Doctoral dissertation, *Library of Theoria* No. V].
2. *A New Approach to the Logical Theory of Interrogatives: Analysis and Formalization.* Tübingen: TBL Ver-

lag Gunter Narr, 1975. Originally published in 1965 in mimeographed form as No.3 in the Uppsala Department of Philosophy Series.

3. *Performatives and Verifiability by the Use of Language: A Study in the Applied Logic of Indexicals and Conditionals.* Uppsala: Department of Philosophy Series No. 14, 1972.

4. *Introduction to Deontic Logic and the Theory of Normative Systems.* Napoli: Bibliopolis, 1987. [No. IV in the *Indices* series of Monographs in Philosophical Logic and Formal Linguistics, ed. by Franz Guenthner and Uwe Mönnich.]

5. (Co-authored with Philip Mullock): *Causing Harm: A Logico-Legal Study.* Berlin / New York: W. de Gruyter, 1989. [In the series *Grundlagen der Kommunikation,* ed. by Roland Posner and Georg Meggle.]

Articles

6. Results Concerning Some Modal Systems That Contain S2. *Journal of Symbolic Logic* **29** (1964), 79 – 87.

7. Choice-Offering and Alternative-Presenting Disjunctive Commands. *Analysis* **25** (1964/65), 182 – 184.

8. "Next" and "Ought". Alternative Foundations for von Wright's Tense-Logic, with an Application to Deontic Logic. *Logique et Analyse* **9** (1966), 231-251.

9. Good Samaritans, Contrary-to-Duty Imperatives, and Epistemic Obligations. *Noûs* **1** (1967), 361 – 379.

10. Improved Formulations of Act-Utilitarianism. *Noûs* **3** (1969), 299–323.

11. Modal Quantification Theory with Transworld Identity. Abstract, *Journal of Symbolic Logic* **36** (1971), 698 – 700.

12. The Emotive Theory of Ethics in the Light of Recent Developments in Formal Semantics and Pragmatics. *Modality, Morality, and Other Problems of Sense and Nonsense: Essays Dedicated to Sören Halldén.* Lund: Gleerup, 1973, pp.130 – 141.

13. Modal Logic with Subjunctive Conditionals and Dispositional Predicates. *Journal of Philosophical Logic* **2** (1973), 1 – 76.

14. Music from a Set-Theoretical Point of View. *INTERFACE-Journal of New MusicResearch* **2** (1973), 1 – 22.

15. A New Approach to the Logical Theory of Actions and Causality. In Sören Stenlund (ed.), *Logical Theory and Semantic Analysis. Essays Dedicated to Stig Kanger on His Fiftieth Birthday.* Dordrecht & Boston: Reidel, 1974, pp.73 – 91.

16. (Co-authored with Franz Guenthner): Representability in QÅ of Hintikka intensional quantifiers and Keenan term quantifiers. *Theoretical Linguistics* **2** (1975), 21 – 44.

17. King Oedipus in Logical Form. *Communication & Cognition* **9** (1976), 59 – 68. With remarks on Aristotle's *Poetics.* [Contribution to an international colloquium on the Function of Tragedy, Ghent, April 1975.]

18. Formal Semantics for Verb Tenses as Analyzed by Reichenbach. In Teun A. van Dijk (ed.), *Pragmatics of Language and Literature.* Amsterdam: North-Holland, 1976, pp. 229 – 236.

19. (Co-authored with Franz Guenthner): Fundamentals of a Theory of Verb Aspect and Events within the Setting of an Improved Tense Logic. In F. Guenthner and C. Rohrer (eds.), *Studies in Formal Semantics.* Amsterdam: North-Holland, 1978, pp. 167 – 199.

20. A Conjectured Axiomatization of Two-Dimensional Reichenbachian Tense Logic. *Journal of Philosophical Logic* **8** (1979), 1 – 45.

21. (Co-authored with Jaap Hoepelman and Christian Rohrer): Adverbs of Frequency. In C. Rohrer (ed.), *Time, Tense, and Quantifiers: Proceedings of the Stuttgart Conference on the Logic of Tense and Quantification.* Tübingen: Niemayer, 1980, pp. 1 – 17.

22. On the Pure Theory of Third Party Conflicts in Dynamic Property Law. Early version: *ThD 60: Philosophical Essays Dedicated to Thorild Dahlquist on His Sixtieth Birthday.* Uppsala: Department of Philosophy Series No. 32, 1980, pp. 153 – 180. Main version: *Revue Internationale de Philosophie* **35** (1981), No. 135, *La Philosophie Scandinave,* pp. 3 – 27.

23. Predicate Calculi with Adjectives and Nouns. *Journal of Philosophical Logic* **10** (1981), 1 – 26.

24. (Co-authored with Jaap Hoepelman): Some Theorems About a "Tree" System of Deontic Tense Logic. In Risto Hilpinen (ed.), *New Studies in Deontic Logic: Norms, Actions, and the Foundations of Ethics.* Dordrecht/Boston/London: Reidel, 1981, pp. 187 – 221.

25. Deontic Logic. In Dov M. Gabbay and Franz Guenthner (eds.), *Handbook of Philosophical Logic, Vol.II* ($1^{st}$ Edition), Dordrecht/Boston/Lancaster: Reidel, 1984, pp. 605 – 714; Volume 8 ($2^{nd}$ Edition), Kluwer, 2002, pp. 147 – 264.

26. On the Logical Syntax or Linguistic Deep Structure of Certain Crime Descriptions: Prolegomena to the Doctrine of Criminal Intent. *Synthese* **65** (1985), 291 – 306.

27. The Knowledge Aspect of the Notion of Criminal Intent: An Analysis in Terms of Probability Rankings on Game-Trees (in Swedish). In N. Jareborg and P.O.Träskman (eds.), *Skuld och ansvar. Straffrättsliga studier tillägnade Alvar Nelson.* Uppsala: Iustus Förlag, 1985, pp. 76 – 98.

28. On the Logic of Causally Necessary and Sufficient Conditions: Towards a Theory of Motive-Explanations of Human Actions. *Erkenntnis* **31** (1989), 43 – 75.

29. Towards a Logical Theory of Legal Evidence: Semantic Analysis of the Bolding-Ekelöf Degrees of Evidential Strength. Early version in Sten Lindström and Wlodek Rabinowicz (eds.), *In so many Words. Philosophical Essays dedicated to Sven Danielsson on the Occasion of His Fiftieth Birthday.* Uppsala: Department of Philosophy Series No. 42, 1989, pp. 187 – 213. Main version in A.A. Martino (ed.), *Atti del III Convegno internazionale LOGICA, INFORMATICA, DIRITTO.* Firenze: Instituto per la documentazione giuridica, 1989.

30. Logical Analysis of Epistemic Modality: An Ex-

plication of the Bolding-Ekelöf Degrees of Evidential Strength. In Hannu Tapani Klami (ed.), *Rätt och sanning. Ett bevisteoretiskt symposium i Uppsala 26-27 maj 1989*. Uppsala: Iustus Förlag, 1990, pp. 43 – 54.

31. Deontic Tense Logic: Restricted Equivalence of Certain Forms of Conditional Obligation and a Solution to Chisholm's Paradox. In G. Schurz and G. Dorn (eds.), *Advances in Scientific Philosophy. Essays in Honour of PAUL WEINGARTNER on the Occasion of the $60^{th}$ Anniversary of his Birthday* (= Poznan Studies in the Philosophy of the Sciences and the Humanities, Vol. 24). Amsterdam-Atlanta,GA: Rodopi, 1991, pp. 127 – 141.

32. The Protagoras Case: An Exercise in Elementary Logic for Lawyers. With a Postscript on Leibniz. In Jes Bjarup and Mogens Blegvad (eds.), *Time, Law, and Society* [=*Archiv für Rechts- und Sozialphilosophie – Beiheft 64*]. Stuttgart: Steiner, 1995, pp. 73 – 84.

33. Discrete Tense Logic with Infinitary Inference Rules and Systematic Frame Constants: A Hilbert-Style Axiomatization. *Journal of Philosophical Logic* **25** (1996), 45 – 100.

34. Systematic Frame Constants in Defeasible Deontic Logic: A New Form of Andersonian Reduction. In Donald Nute (ed.), *Defeasible Deontic Logic: Essays in Nonmonotonic Normative Reasoning*. Dordrecht/Boston/London: Kluwer, 1997, pp. 59 – 77.

35. On Certain Extensions of von Kutschera's Preference-Based Dyadic Deontic Logic. In Wolfgang Lenzen (Hrsg.), *Das weite Spektrum der Analytischen Philosophie– Festschrift für Franz von Kutschera*. Berlin / New York: W. de Gruyter, 1997, pp. 8 – 23.

36. Prima Facie Obligations in Deontic Logic: A Chisholmian Analysis Based on Normative Preference Structures. In Christoph Fehige and Ulla Wessels (eds.), *Preferences*. Berlin / New York: W. de Gruyter, 1998, pp. 135 – 155. [From the Saarbrücken Colloquium on preferences in Saarlouis, June 24-27, 1992.]

37. On the Bearer/Counterparty Problem in Stig Kanger's Theory of Rights: A Solution in Terms of Sources-of-Law. In Lars Lindahl, Jan Odelstad and Rysiek Sliwinski (eds.), *Not Without Cause: Philosophical Essays Dedicated to Paul Needham on the Occasion of His Fiftieth Birthday*. Uppsala: Department of Philosophy Series No. 48, 1998, pp. 269 – 277.

38. Supererogation and Offence in Deontic Logic: An Analysis within Systems of Alethic Modal Logic with Levels of Perfection. In Rysiek Sliwinski (ed.), *Philosophical crumbs: Essays dedicated to Ann-Mari Henschen-Dahlquist on the occasion of her seventy-fifth birthday*. Uppsala: Department of Philosophy Series No. 49, 1999, pp. 261 – 276.

39. The Logic of Historical Necessity as Founded on Two-Dimensional Modal Tense Logic. *Journal of Philosophical Logic* **28** (1999), 329 – 369. Missing list of References appeared in *Journal of Philosophical Logic* **29** (2000), 541 – 542.

40. Three Characterizability Problems in Deontic Logic. *Nordic Journal of Philosophical Logic* **5** (2000), 65 – 82. [From the $5^{th}$ International Workshop ΔEON '00 on Deontic Logic in Computer Science, Toulouse, 20-22 January, 2000.]

41. On the Interpretation of Plato's *EUTHYPHRO*. In Erik Carlson and Rysiek Sliwinski (eds.), *Omniumgatherum. Philosophical essays dedicated to Jan Österberg on the occasion of his sixtieth birthday*. Uppsala: Department of Philosophy Series No. 50, 2001, pp. 15 – 28.

42. Old Foundations for the Logic of Agency and Action. *Studia Logica* **72** (2002), 313 – 338.

43. Conditionality and Branching Time in Deontic Logic. In John Horty and Andrew J.I. Jones (eds.), ΔEON'02, pp. 323 – 343. [From the $6^{th}$ International Workshop on Deontic Logic in Computer Science, Imperial College, London, May 2002.]

44. Future Contingents and Determinism in Aristotle's *De Interpretatione* IX: Some Logical Aspects of the so-called Second Oldest Interpretation. *Logique & Analyse* **181** (2003), 13 – 48.

45. Some Remarks on Performatives in the Law. *Artificial Intelligence and Law* **11** (2003), 105 – 124.

46. Logical Aspects of Some Burden of Proof Problems in Cases of Alleged Violations of the Right to Unionize According to Swedish Labour Law. In Peter Wahlgren (ed.), *Perspectives on Jurisprudence*: Essays in Honor of Jes Bjarup (=Scandinavian Studies in Law, Volume 48). Stockholm Institute for Scandinavian Law, Stockholm 2005, pp. 607 – 618.

47. Combinations of Tense and Deontic Modality: on the Rt Approach to Temporal Logic with Historical Necessity and Conditional Obligation. *Journal of Applied Logic* **3** (2005), 421 – 460.

48. On the Logic of Indexical Languages. Forthcoming in a *Festschrift* to Krister Segerberg in 2006. Apart from a Postscript 2005 on the logic of performatives, this paper is a translation into English of my contribution – written in Swedish in 1968 – to a collection of essays dedicated to my late friend and teacher Ingemar Hedenius.

**Service to the Profession:** Council Member, Association for Symbolic Logic, 1970-1973; Editorial Board Member, Journal of Philosophical Logic, 1972-; Editorial Board Member, Linguistics and Philosophy, 1977-1984; Consulting Editor, Theoria, 1967-.

**Teaching:** Starting with the 1960s, Åqvist's students include the philosophers Krister Segerberg, Sven Danielsson as well as the legal theorists Åke Frändberg and, from 1987 on, Torben Spaak. In 1981 I had the honour of being the co-referent on the occasion of Job van Eck's defence of his doctoral dissertation at the Rijksuniversiteit te Groningen.

**Vision Statement:** As a reminder to people worrying about the place of logic in philosophy and divers special sciences, I'd like to quote my late friend and colleague Stig Kanger: "Logic is the backbone of philosophy". And, to lawyers worrying about logic and legal theory not being 'practical' enough, I say, quoting the chemist Justus von Liebig: "Es gibt nichts Praktischeres als eine gute Theorie", i.e. "There is nothing more practical than a good theory".

**Honours and Awards:** Juris Doctor Honoris Causa, Uppsala University, 1992; Honoured with two *Festschriften*, viz (i) Tom Pauli (ed.), Philosophical essays dedicated to Lennart Åqvist on his fiftieth birthday, Uppsala University: Department of Philosophy, 1982, and (ii) K.Segerberg and R.Sliwinski (eds.), *Logic, law, morality. Thirteen essays in practical philosophy in honour of Lennart Åqvist*. Uppsala University: Department of Philosophy, 2003; Member, Royal Academy of Arts and Sciences in Uppsala, 1969-; King Oscar II's Prize, University of Uppsala, for the essay *A New Approach to the Logical Theory of Interrogatives*, 1968; Member, Law Club of Uppsala, 1960s-; Member, Philosophical Society of Uppsala, 1950s-.

# ARTEMOV, Sergei Nikolaevich

**Specialties:** Proof theory, modal and provability logic, logic of proofs, typed theories, epistemic logic, logic in computer science.

**Born:** 25 December, 1951 in Uralsk (USSR-Kazakhstan)

**Educated:** Moscow University Master Degree in Mathematics with Honors 1975; Moscow University and Steklov Mathematical Institute, Ph.D. in Mathematics 1980; Steklov Mathematical Institute, Doctor of Sciences Degree in Mathematics in 1988.

**Dissertation:** *Extensions of Arithmetic and Modal Logics*; supervisor Albert Dragalin, mentor Andrei A. Kolmogorov.

**Regular Academic or Research Appointments:** DISTINGUISHED PROFESSOR IN COMPUTER SCIENCE, MATHEMATICS, AND PHILOSOPHY, GRADUATE CENTER OF THE CITY UNIVERSITY OF NEW YORK, 2001–. Professor, Cornell University 1996–2001; Professor, Moscow University 1994–; Assistant Professor, Associate Professor, Moscow University 1984–1994; Researcher, Senior Researcher, Leading Researcher, Steklov Mathematical Institute of the USSR/Russian Academy of Sciences 1980–1997; Researcher and the head of a research group, Optimal Control Institute of the USSR Academy of Sciences 1978–1980; Lecturer, Kolmogorov College at Moscow University 1974–1984.

**Visiting Academic or Research Appointments:** Graduate School in Mathematical Logic, University of Siena, Italy 1991, 1993, 2005; Institute for Computer Science, Berne University, Switzerland 1992, 1993, 1994, 1995, 2003, 2006; Stanford University, Department of Computer Science, Center for the Study of Language and Information 1989, 1990, 1997; Department of Mathematics and Computer Science, University of Amsterdam, the Netherlands 1994; CNRS, Laboratory for Discrete Mathematics, Marseille, France 1993; CNRS, Laboratory for Informatics, Robotics and Microelectronics in Montpellier, France 1992; University Montpellier III, France 1991.

**Research Profile:** In 1978–1979, Artemov found a uniform arithmetical completeness theorem for the modal logic of provability, cf. [19] (independently discovered also by other authors). This result considerably strengthened the Solovay completeness theorem and opened up a new line of research in provability logics which eventually led to a complete classification of all propositional modal logics of formal provability.

In 1985, Artemov found a negative solution of the axiomatizability problem for the first-order provability logic, which was then the main problem in this area ([18]). He also showed that the first-order provability logic essentially depends on the choice of a specific numbering of axioms for a given theory ([16]), which demonstrated that this logic was in fact ill-defined.

In 1987m Artemov solved a problem by Kuznetsov and completed the classification of the class of modal counterparts of the intuitionistic logic modulo well-known Gödel's translation. The logic $A^*$ discovered by Artemov is now included in the list of the major modal logics.

In 1994–95, Artemov found a complete axiomatization of the Logic of Proofs LP ([8, 11]) anticipated in the works by Kolmogorov and Gödel in the early 1930s. This provided a solution to a well-known problem of connecting Gödel's provability calculus (a.k.a. the modal logic S4) to mathematical provability. Artemov's Logic of Proofs also answered a fundamental question of formalizing Brouwer-Heyting-Kolmogorov semantics for intuitionistic propositional logic. Artemov showed that LP was capable of realizing the classical modal logic by recovering explicit proof terms in any modal derivation. This provided a new semantics for modal logic: a realizability semantics of explicit proofs.

In 1997, Artemov designed (together with Nerode and Davoren) a bi-modal logic for topological dynamic systems and established its completeness with respect to the intended semantics in topological spaces with designated total function on it ([10]). A similar approach has been independently developed by Kremer and Mints. This logic provided a framework for logical studies of topological dynamic systems and hybrid control systems. This is now an active research area.

In 1999, Artemov suggested an explicit provability model for verification ([9]). The traditional implicit provability model leaves a certain loophole in the foundations of formal verification: an extension of a verification system by a verified rule, generally speaking, is not equivalent to the original system. The explicit provability model fixes this problem and provides a formal verification schema that does not extend the set of correctness assumptions about verification beyond an original trusted core.

In 2003 Artemov introduced Reflexive Combinatory Logic ([5]). This system extends the Combinatory Logic by self-referential capabilities and hence introduces self-referentiality into a prototype functional programming language.

In 2004 Artemov (in a joint work [3] with Nogina) introduced the notion of justification into formal epistemology. After the celebrated tripartite definition of knowledge as *justified true belief*, the justification condition has received the greatest attention in epistemology but lacked formal representation. Artemov-Nogina epistemic logic with justification contains assertions of the form "*F is known*," along with those of the form "*t is a justification for F.*" The "justification" part comes from the Logic of Proofs LP. This work also answered a question by van Benthem, which had remained open since 1991.

In 2004–2005 Artemov developed a theory of evidence-based common knowledge, which offered new foundations to the formal representation of knowledge ([2]). This also provides a new approach to the famous problem of logical omniscience: evidence-based systems normally have an omniscience control mechanism ([1]).

**Main Publications:**

1. S. Artemov and R. Kuznets, Logical omniscience via proof complexity. In *Computer Science Logic 2006*, pp. 135–149, Volume 420 of Lecture Notes in Computer Science, Springer, 2006.

2. S. Artemov, "Justified common knowledge," *Theoretical Computer Science* 357:1–3, 4–22, 2006.

3. S. Artemov and E. Nogina, "On epistemic logic with justification." In R. van der Meyden, editor, *Theoretical Aspects of Rationality and Knowledge. Proceedings of the Tenth Conference (TARK 2005), June 10-12, 2005, Singapore.*, pages 279–294. National University of Singapore, 2005.

4. S. Artemov and L. Beklemishev, "Provability Logic," in D. Gabbay and F. Guenthner ed. *Handbook of Philosophical Logic*, 2nd ed., v. 13, pp. 189–360, Springer, Dordrecht, 2005.

5. S. N. Artemov, "Kolmogorov's and Gödel's approach to intuitionistic logic: current developments," *Uspekhi Mat. Nauk*, 59:2 (2004), 9–36 (in Russian). English translation: *Russian Mathematical Surveys*, 59:2 (2004) 203–229.

6. S. Artemov, "Unified semantics for modality and lambda-terms via proof polynomials," in Kees Vermeulen and Ann Copestake eds. *Algebras, Diagrams and Decisions in Language, Logic and Computation*, pp. 89-119, CSLI Publications, Stanford University, 2002.

7. S. Artemov, "Operations on proofs that can be specified by means of modal logic," *Advances in Modal Logic*, v. 2, pp. 59-72, CSLI Publications, Stanford University, 2001.

8. S. Artemov, "Explicit provability and constructive semantics," the *Bulletin for Symbolic Logic*, 7:1 (2001), 1–36.

9. S. Artemov, "On explicit reflection in theorem proving and formal verification," In Springer Lecture Notes in Artificial Intelligence, v. 1632 *Automated Deduction - CADE-16. Proceedings of the 16th International Conference on Automated Deduction, Trento, Italy, July 1999* (1999), 267-281.

10. S. Artemov, J. Davoren and A. Nerode. *Modal logics and topological semantics for Hybrid Systems*, Technical Report MSI 97-05, Cornell University, 1997.

11. S. Artemov. *Operational modal logic*. Technical Report MSI 95-29, Cornell University, 1995.

12. S. Artemov. "Logic of Proofs," *Annals of Pure and Applied Logic*, 67 (1994), 29-59.

13. S. Artemov and F. Montagna. "On first order theories with provability operator," *Journal of Symbolic Logic*, 59:4 (1994), 1139–1153.

14. S. Artemov and T. Strassen. "The logic of the Gödel proof predicate." In G. Gottlob, A. Leitsch, and D. Mundici, editors, *Computational Logic and Proof Theory. Third Kurt Gödel Colloquium, KGC'93. Brno, Czech Republic, August 1993. Proceedings*, volume 713 of *Lecture Notes in Computer Science*, pages 71–82. Springer, 1993.

15. S. Artemov and T. Strassen. "The basic logic of proofs." In E. Börger, G. Jäger, H. Kleine Büning, S. Martini, and M.M. Richter, editors, *Computer Science Logic. 6th Workshop, CSL'92. San Miniato, Italy, September/October 1992. Selected Papers*, volume 702 of *Lecture Notes in Computer Science*, pages 14–28. Springer, 1992.

16. S. Artemov. "Numerically correct provability logics," *Doklady AN SSSR*, 290:6 (1986), 1289–1292 (Russian); English transl.: *Soviet Math. Doklady*, 34:2 (1987), 384–387.

17. S. Artemov. "Modal logics axiomatizing provability," *Izvestiya AN SSSR, ser. matem.* 49:6 (1985), 1123–1154 (Russian); English transl.: *Math USSR Izvestiya*, 27:2 (1986), 401–429.

18. S. Artemov. "Non-arithmeticity of truth predicate

logics of provability," *Doklady AN SSSR*, 284:2 (1985), 270-271 (Russian); English transl.: *Soviet Math. Doklady*, 32:2 (1985), 403–405.

19. S. Artemov. "Arithmetically complete modal theories," *Semiotika i Informatika, Moscow*, 14 (1980), 115-133 (Russian); English transl.: *Amer. Math. Soc. Transl.(2)*, 135 (1987), 39–54.

*Work in Progress*

20. S. Artemov and R. Iemhoff, "The basic intuitionistic logic of proofs", *Journal of Symbolic Logic*, 72:2, 439–451, 2007.

21. S. Artemov and E. Nogina. "Introducing justification into epistemic logic." *Journal of Logic and Computation*, 15:6, 1059–1073, 2005.

**Service to the Profession:** Managing Editor: *Annals of Pure and Applied Logic*, 1991–; Editor: *Studies in Logic and Foundations of Mathematics*, 1996–; Editorial Board Member: *Moscow Mathematical Journal*, 2000–; Nominating Committee Chair: Association for Symbolic Logic, 2005; Program Committee Chair: 2004 annual conference of the Association for Symbolic Logic, Carnegie-Mellon University, 2004; Elected Council member: Association for Symbolic Logic, 1995–96, 2000–2002; Principal Organizer: conferences *New Developments in Logics of Knowledge and Belief*, New York, 2004 and *Constructivism in Mathematics, Logic and Computer Science*, New York, 2003; Principal Organizer: *New York Logic Colloquium*, 2001– and *CUNY Computer Science Colloquium*, 2002–; Member of the CUNY Distinguished Professor Selection Panel, 2005–; Program Committee Member: conference *Computer Science in Russia*, 2006.

**Teaching:** Over thirty years of teaching experience in Moscow University, University of Amsterdam, Cornell University, Graduate Center of the City University of New York, over a dozen of tutorials around the world. Among Ph.D. students and postdocs advised: Lev Beklemishev, Yegor Bryukhov, Jennifer Davoren, Sergei Goryachev, Srikanth Gottiati, Georgi Japaridze, Alexei Kopylov, Victor Krivtsov, Nikolai Krupski, Roman Kuznetz, Pavel Naumov, Nikolay Pankrat'ev, Mati Pentus, Bryan Renne, Vladimir Shavrukov, Sergey Slavnov, Tyko Strassen, Alexander Yashin, Tanya Yavorskaya-Sidon, Rostik Yavorsky, Evgenii Zolin.

**Vision Statement:** Mathematical Logic mainly achieved its fundamental goal of providing Mathematics with sound foundations. The future is in using rigorous logical methods in fundamental studies of other fields. The prime areas of development right now are computer science, artificial intelligence, epistemology.

**Honours and Awards:** Keynote lecture for the Kurt Gödel Society, Computer Science Logic Symposium, European Summer School on Logic, Language and Information, Vienna, 2003; Distinguished Lecture at the New York Academy of Sciences, 2002; Clifford Lectures, 2002; Spinoza Lecture from the European Association for Logic, Language and Information, 1999; The paper "Operational Modal Logic" commended for its excellence by the Committee for the IGPL/FoLLI Prize "The Best Idea of the Year" in 1996 (split with Joe Halpern and Andrei Voronkov); President of Russia Award: "To Outstanding Scientist," 1994; Awards for the best paper of the year from the Mathematical Institute of the USSR/Russian Academy of Sciences, Moscow, 1986, 1994; Award for the best project of the year, Optimal Control Institute of the USSR Academy of Sciences, Moscow, 1979.

# AVRON, Arnon

**Specialties:** Deduction Systems, Non-classical Logics, Foundations of Mathematics.

**Born:** 10 November 1952 in Tel-Aviv, Israel.

**Educated:** Tel Aviv University, BSc Mathematics, 1973; Hebrew University of Jerusalem, MSc Mathematics, 1975; Tel Aviv University, PhD Mathematics, 1985.

**Dissertation:** *The semantics and proof theory of relevance logics and nontrivial theories containing contradictions*; supervisor Haim Gaifman.

**Regular Academic or Research Appointments:** PROFESSOR, SCHOOL OF COMPUTER SCIENCE, TEL AVIV UNIVERSITY, 1999-. Associate Professor, Computer Science Department, Tel Aviv University, 1995-1999; Senior Lecturer, Computer Science Department, Tel Aviv University, 1989-1995; Lecturer, Computer Science Department, Tel Aviv University, 1988-1989.

**Visiting Academic or Research Appointments:** Visiting Professor, Department of Mathematics, Stanford University, Stanford, California, 1992-1993; Research Associate, LFCS (Laboratory for Foundations of Computer Science), Computer Science Department, Edinburgh University, 1986-1988.

**Research Profile:** Avron's research in logic is motivated by philosophical problems on one hand, and applications in computer science and mathematics on the other. Accordingly, he has contributions in many areas of logic. These include: foundations of mathematics, mechanization of mathematics, automated reasoning, proof theory (where he introduced and successfully applied the method of hypersequents), foundations of logics, logical frameworks, substructural logics (in particular relevance logics and linear Logic), paraconsistent logics, logics for reasoning under uncertainty, many-valued logics, and database theory. Avron particularly likes to find deep connections between seemingly different subjects, to develop unified frameworks for them, and to apply in logic and mathematics ideas borrowed from computer science. Thus he has applied the idea of non-deterministic computations to introduce non-deterministic many-valued structures (a generalization of ordinary many-valued systems), and has used them (among other things) to provide modular semantics of rules, a characterization of canonical Gentzen-type systems for which the cut rule is admissible, and effective semantics for most of the logics in the Brazilian family of paraconsistent logics (also known as logics of formal inconsistency). Another example is provided by his applications of the idea of "safe queries" from database theory to set theory, computability theory, and predicative mathematics.

**Main Publications:**

1. "On Modal Systems having arithmetical interpretations", *Journal of Symbolic Logic*, 49 (1984), 935–942.
2. "Theorems on Strong Constructibility with a Compass alone", *Journal of Geometry*, 30 (1987), 28–35.
3. "The Semantics and Proof Theory of Linear Logic", *Theoretical Computer Science*, 57 (1988), 161–184.
4. "Relevance and Paraconsistency - A New Approach", *Journal of Symbolic Logic*, 55 (1990), 707-732.
5. "Simple Consequence relations" *Information and Computation*, 92 (1991), 105–139.
6. " Natural 3-valued Logics— Characterization and Proof Theory", *Journal of Symbolic Logic*, 56 (1991), 276–294.
7. "A Note on Provability, Truth and Existence", *Journal of Philosophical Logic*, 20 (1991), 403–409.
8. "Gentzen-Type Systems, Resolution and Tableaux" *Journal of Automated Reasoning*, 10 (1993), 265–281.
9. "The Method of Hypersequents in Proof Theory of Propositional Non-Classical Logics", in *Logic: Foundations to Applications* (ed. W. Hodges, M. Hyland, C. Steinhorn and J. Truss), Oxford Science Publications (1996), 1–32.
10. "Reasoning with Logical Bilattices", (with O. Arieli), *Journal of Logic, Language and Information*, 5 (1996), 25–63.
11. "The Structure of Interlaced Bilattices", *Journal of Mathematical Structures in Computer Science*, 6 (1996), 287–299.
12. "The Value of the Four Values", (with O. Arieli), *Artificial Intelligence*, 102 (1998), 97–141.
13. "Multiplicative Conjunction and an Algebraic Meaning of Contraction and Weakening", *Journal of Symbolic Logic*, 63 (1998), 831–859.
14. "On Negation, Completeness and Consistency", in *Handbook of Philosophical Logic*, (ed. D. Gabbay and F. Guenthner), Vol. 9, Kluwer Academic Publishers (2002), 287–319.
15. "Classical Gentzen-type Methods in Propositional Many-Valued Logics", in *Beyond Two: Theory and Applications of Multiple-Valued Logic*, (ed. M. Fitting and E. Orlowska), Studies in Fuzziness and Soft Computing 114, Physica Verlag (2003), 117–155.
16. "Transitive Closure and the mechanization of Mathematics", in *Thirty Five Years of Automating Mathematics*, (ed. F. Kamareddine), Kluwer Academic Publishers (2003), 149–171.
17. "Safety Signatures for First-order Languages and Their Applications", in *First-Order Logic Revisited*, (ed. V. Hendricks, F. Neuhaus, S. A. Pedersen, U. Scheffler, H. Wansing), Logos Verlag Berlin (2004), 37–58.
18. "Combining Classical Logic, Paraconsistency and Relevance", *Journal of Applied Logic*, 3 (2005), 133–160.

*Work in Progress*

19. "Non-Deterministic Multiple-valued Structures", (with I. Lev), *Journal of Logic and Computation*, 15 (2005), 241–261.
20. "Non-deterministic Matrices and Modular Semantics of Rules", in *Logica Universalis* (ed. J.-Y. Beziau), Birkhüser Verlag (2005), 149–167.

**Service to the Profession:** Chair, Department of Computer Science, Tel-Aviv University, 1996-1998; Associate editor, Studia Logica, 1997-2001.

**Teaching:** Avron attached great importance to teaching, and several of his papers are based on insights that were reached through teaching various courses in Logic and Computer Science. For one of the main courses he regularly teaches, Introduction to Discrete Mathematics, he wrote a textbook (in Hebrew) which puts a special emphasis on learning the correct language of Mathematics (in particular: it uses lambda-notation all along). Avron has also given on the Israeli Radio a series of public lectures on Gödel's theorems, and has written a corresponding popular book (in Hebrew).

**Vision Statement:** Avron believes that the most important goal of mathematical logic is still to get a deeper understanding of the foundations and nature of mathematics. A second important goal is to understand (and mechanize) reasoning in all its forms.

**Honours and Awards:** Rothschild Fellowship, 1986.

# B

## BAGARIA, Joan

**Specialties:** Set theory, mathematical logic.

**Born:** 17 August 1958 in Manlleu, Barcelona, Spain.

**Educated:** University of California at Berkeley, PhD Logic and Methodology of Science, 1991; University of Barcelona, MA Logic, 1984; University of Barcelona, BA Philosophy, 1981.

**Dissertation:** *Definable forcing and regularity properties of projective sets of reals*; supervisor, Ralph McKenzie.

**Regular Academic or Research Appointments:** ICREA RESEARCH PROFESSOR, UNIVERSITY OF BARCELONA, 2001–; *Professor Titular*, University of Barcelona, 1996-2001; Invited Professor, Pompeu Fabra University, Barcelona, 1995-1996; *Professor Titular*, Autonomous University of Barcelona, 1992-1995; Post-doctoral Researcher, University of California, Berkeley, 1991-1992.

**Visiting Academic or Research Appointments:** Invited Researcher, Instituto Venezolano de Investigaciones Científicas (IVIC), Caracas, 2006; Invited Researcher, Université Paris VII, Paris, 2006; Invited Researcher, California Institute of Technology, Pasadena, 2006Invited Researcher, Kobe University, Kobe, 2005; Invited Researcher, Instituto de Matemáticas, Morelia (Mexico), 2005; Invited Researcher, Institute for Mathematical Sciences, National University of Singapore, Singapore, 2005; Invited Researcher, Kurt Gödel Research Center for Mathematical Logic, Vienna, 2005; Invited Researcher, Center for Theoretical Study, Praha, 2003; Invited Researcher, University of California, Berkeley, 2003; Invited Researcher, Fields Institute, Toronto, 2002; Invited Researcher, Institut für Formale Logik, Vienna, 2000; Invited Researcher, Humboldt Universität, Berlin, 1998; Invited Researcher, Universidad Simón Bolívar, Caracas, 1997.

**Research Profile:** Joan Bagaria has worked mainly in Set Theory and its applications. He has made significant contributions to descriptive set theory and the set theory of the reals, especially in the analysis of regularity properties of projective sets of real numbers and in questions related to the Lebesgue measure and Baire category. One of his main interests has been the study of definable forcing and its uses in descriptive set theory and combinatorics. He obtained elegant characterizations of Martin's Axiom, and bounded forcing axioms in general, in terms of generic absoluteness, characterizations that have now become standard in the field. His work in the last years has focused on a systematic study of principles of generic absoluteness, obtaining many exact equiconsistency results, and applying them to other areas. A series of joint papers with Roger Bosch analyzed the generic absoluteness of projective formulas, as well as the theory of the constructive closure of the reals, under projective forcing extensions. As for applications, he has worked on general topology, particularly in the construction of thin-tall spaces. In Banach space theory he solved some well-known open questions about Gowers's weakly-Ramsey property for sets of block bases in infinite-dimensional separable Banach spaces, in collaboration with Jordi López-Abad. In recent work with Carles Casacuberta and Adrian R. D. Mathias, he has applied the theory of large cardinals to obtain some results of interest in category theory and homological algebra. He has also been motivated by problems in the foundations and philosophy of mathematics, especially the continuum hypothesis and the search for new set-theoretic axioms. Thus, he has worked on the analysis of different criteria for the discovery and justification of new axioms of set theory. Finally, he has written several papers on set theory for the general public and has been active in the communication and popularization of science.

**Main Publications:**

1. (with Judah, H.), Amoeba forcing, Suslin absoluteness and additivity of measure, In *Set Theory of the Continuum*, MSRI Publications. Springer-Verlag, 26 (1992), 155-173.
2. Fragments of Martin's axiom and $\Delta^1_3$ sets of reals, *Annals of Pure and Applied Logic* 69 (1994), 1-25.
3. (with Bosch, R.), Projective Forcing, *Annals of Pure and Applied Logic*, 86 (1997), 237-266.
4. A characterization of Martin's axiom in terms of absoluteness, *Journal of Symbolic Logic* 62, Number 2 (1997), 366-372.
5. (with Woodin, W.H.), $\Delta^1_n$-sets of reals, *Journal of Symbolic Logic* 62, Number 4 (1997), 1379-1428.
6. Generic absoluteness and forcing axioms, In *Models, Algebras and Proofs*. Lecture Notes in Pure and Applied Mathematics, Volume 203 (1999), 1-12.
7. Bounded forcing axioms as principles of generic absoluteness, *Archive for Mathematical Logic* 39 (2000), 393-401.

8. (with Friedman, S.), Generic absoluteness, *Annals of Pure and Applied Logic* 108 (2001), 3-13.

9. (with Asperó, D.), Bounded forcing axioms and the continuum, *Annals of Pure and Applied Logic* 109 (2001), 179-203.

10. (with López Abad, J.), Weakly-Ramsey sets in Banach spaces, *Advances in Mathematics* 160 (2001), 133-174.

11. Locally-generic Boolean algebras and cardinal sequences, *Algebra Universalis* 47 (2002), 283-302.

12. (with López Abad, J.), Determinacy and weakly-Ramsey sets in Banach spaces, *Transactions of the American Mathematical Society* 354, Number 4 (2002), 1327-1349.

13. (with Bosch, R.), Solovay models and forcing extensions, *Journal of Symbolic Logic* 69, Number 3 (2004), 742-766.

14. (with Bosch, R.), Proper forcing extensions and Solovay models, *Archive for Mathematical Logic* 43 (2004), 739-750.

15. Natural Axioms of Set Theory and the Continuum Problem. In *Proceedings of the 12th International Congress of Logic, Methodology, and Philosophy of Science*. P. Hájek, L. Valdés.Villanueva, and D. Westerståhl, Editors. King's College Publications, London (2005).

16. Axioms of Generic Absoluteness. In *Logic Colloquium, 2002*, Lecture Notes in Logic series 27 (2006), 28-47.

17. (with Castells, N.; and Larson, P.), An $\Omega$-Logic Primer. In *Set Theory. Centre de Recerca Matemàtica, 2003-2004*. Birkhäuser Verlag. Trends in Mathematics Series (2006), 1-28.

18. (with Bosch, R.), Generic absoluteness under projective forcing. *Fundamenta Mathematicae*. To appear. (2007).

19. A full list of publications is available at Bagaria's personal page http://www.icrea.es/pag.asp?id=Joan.Bagaria

*Work in Progress*

20. A joint article with C. Casacuberta and A.R.D. Mathias entitled *Epireflections and supercompact cardinals*.

21. A joint paper with C. A. Di Prisco on *Parameterized partition relations on the reals*.

22. An article on Set Theory for the *Princeton Companion to Mathematics*.

23. A book on *Generic Absoluteness*.

24. A lecture notes volume on *Mathematical Logic*.

**Service to the Profession:** Member of the Scientific Policy Committee, University of Barcelona, 2006-; ICREA Director's Scientific Advisor, 2005-; Nominating Committee Member, Association of Symbolic Logic (ASL), 2004; Program Coordinator, Set Theory and its Applications held at the Centre de Recerca Matemàtica (CRM), 2003-2004; Secretary of the Studies Counsel, Faculty of Philosophy, University of Barcelona, 2000-2001; Advisory Board Member, Scientific Board of the Centre de Recerca Matemàtica (CRM); Permanent Committee Member, Catalan Counsel for the Communication of Science (C4).

**Teaching:** Bagaria has been teaching a large number of courses and seminars in the logic program at the University of Barcelona since his appointment to the faculty in 1996. After becoming an ICREA Research Professor in 2001, a position that does not carry teaching duties, he has been teaching only graduate courses on Logic and Set theory and has given advanced seminars for graduate students. He has had 3 PhD students in the fields of Logic and Mathematics: Jordi López-Abad, David Asperó, and Roger Bosch.

**Vision Statement:** Set theory is a fascinating subject. Besides its enormous technical sophistication and its particular subject matter, namely the study of abstract infinite sets, has the appeal of living in the no-man's land between logic and mathematics. The inextricable mixture of logic, combinatorics, and metamathematics in the forcing technique or in the theory of large cardinals, lends to the subject a unique flavour that makes many of the results obtained by the use of these techniques seem almost miraculous. Moreover, many deep results in set theory have philosophical significance, and conversely, one can make fundamental questions in the philosophy of mathematics tractable by turning them into pure set-theoretical, hence mathematical problems.

**Honours and Awards:** Fulbright-La Caixa Fellow, 1985-1987.

# BARENDREGT, Hendrik (Henk) Pieter

**Specialties:** Lambda calculus, type theory, computer mathematics, phenomenology of mind.

**Born:** 18 December 1947 in Amsterdam, The Netherlands.

**Educated:** Utrecht University, PhD, 1971, MA, 1968; Montessori schools (kindergarten, elementary & grammar school), 1952-1965.

**Dissertation:** *Some extensional term models for lambda-calculi and combinatory logics*; supervisors, Georg Kreisel and Dirk van Dalen.

**Regular Academic or Research Appointments:** PROFESSOR, FOUNDATIONS OF MATHEMATICS AND COMPUTER SCIENCE, RADBOUD UNIVERSITY, NIJMEGEN, 1986–; Assistant and Associate Professor, Mathematical Logic, Utrecht University, 1972-1985.

**Visiting Academic or Research Appointments:** Adjunct Professor, Mathematics, Carnegie Mellon University, 1999-2009.

**Research Profile:** Barendregt has solved a few problems developed by others. The only exception to this is the formulation of systems of Illative Combinatory Logic (an extension of the lambda calculus or combinatory logic with operators for logical reasoning) that are conservative over ordinary first order predicate logic (joint work with M. Bunder and W. Dekkers (see items 6, 9 and 13 in Main publications below).

His career is based mainly on 1. Providing a stylish account of work by others and 2. Posing questions, some of which are solved by him, some by others. Both 1 and 2 are about topics in the lambda calculus, untyped and typed. This has inspired others to build onto these topics.

In untyped lambda calculus Barendregt introduced the notions of 'notion of reduction', solvability, $\omega$-rule, universal generators, Böhm-trees, lambda theory, lambda algebra, local and global structure of models and developed the elementary properties of these concepts.

He showed that unsolvability is the natural notion of undefinedness and that all unsolvable terms can be consistently equated. Also that in the $\lambda l$-calculus a term has a normal form if and only if it is solvable. The classical $D_\infty$ models of Dana Scott have been characterized by Hyland and Wadsworth in terms of solvability. The omega rule, stating that if two lambda terms $F$ and $G$ are equal on all closed terms $Z$, then $F, G$ are equal, was shown by him to hold for all $F, G$, except universal generators. Building on this Plotkin succeeded to give a counter-example to this rule. Barendregt used Böhm trees to display the infinite character of lambda terms, thereby presenting a new model of the lambda calculus (Böhm-like trees). The notion of lambda theory has grown into a research field using proof theory, model theory, recursion theory and universal algebra.

Barendregt made explicit notions for proof-assistants: de Bruijn criterion, Poincaré Principle, ephemeral vs petrified proof-objects. He advocated the use of reflection: coding expression and deriving properties by proving properties of the codes and then applying some soundness result.

Barendregt intends to work on a theory of mind based on experience with insight meditation. Together with colleagues in neuro-and cellular-biology he directs experiments aiming to give a foundation for this theory.

Work in progress. 1. Lambda calculi with types. 2. Research survey with W. Dekkers, R. Statman and many co-authors, on typed lambda calculi. 3. Operationalzing the notion of 'mindfulness'.

**Main Publications:**

1. Logic
2. Henk Barendregt. A characterization of terms of the $\lambda I$-calculus having a normal form. *J. Symbolic Logic*, 38:441–445, 1973.
3. H.P. Barendregt. The type-free lambda calculus.
4. H. Barendregt. Solvability in lambda calculi. In *Colloque International de Logique (Clermont-Ferrand, 1975)*, volume 249 of *Colloq. Internat. CNRS*, pages 209–219. CNRS, Paris, 1977.
5. Hendrik Pieter Barendregt. *The lambda calculus*, volume 103 of *Studies in Logic and the Foundations of Mathematics, Its syntax and semantics*. North-Holland Publishing Co., Amsterdam, 1981. Revised edition 1984, Russian edition (MIR) 1985, Chinese edition (Nanjing University Press) 1992.
6. H. P. Barendregt. Lambda calculi with types. In *Handbook of logic in computer science, Vol. 2*, Oxford Sci. Publ., pages 117–309. Oxford Univ. Press, New York, 1992.
7. Henk Barendregt, Martin Bunder, and Wil Dekkers. Systems of illative combinatory logic complete for first-order propositional and predicate calculus. *J. Symbolic Logic*, 58(3):769–788, 1993.
8. Wil Dekkers, Martin Bunder, and Henk Barendregt. Completeness of two systems of illative combinatory logic for first-order propositional and predicate calculus. *Arch. Math. Logic*, 37(5-6):327–341, 1998. Logic Colloquium '95 (Haifa).
9. Wil Dekkers, Martin Bunder, and Henk Barendregt. Completeness of the propositions-as-types interpretation of intuitionistic logic into illative combinatory logic. *J. Symbolic Logic*, 63(3):869–890, 1998.
10. Henk Barendregt. A characterization of terms of the $\lambda I$-calculus having a normal form. *J. Symbolic Logic*, 38:441–445, 1973.
11. Henk Barendregt and Erik Barendsen. Autarkic computations in formal proofs. *J. Automat. Reason.*, 28(3):321–336, 2002.
12. Henk Barendregt. The impact of the lambda calculus in logic and computer science. *Bull. Symbolic Logic*, 3(2):181–215, 1997.
13. Henk Barendregt, Martin Bunder, and Wil Dekkers. Systems of illative combinatory logic complete for first-order propositional and predicate calculus. *J. Symbolic Logic*, 58(3):769–788, 1993.
14. **Phenomenology**
15. H.P. Barendregt. Buddhist Phenomenology (Part I). In M. dalla Chiara, editor, *Proceedings of the Conference on Topics and Perspectives of Contemporary Logic and Philosophy of Science (Cesena, Italy, January 7-10, 1987)*, pages 37–55, Bologna, 1988. Clueb. URL: www.cs.ru.nl/~henk/BP/bp1.html.
16. H.P. Barendregt. Buddhist Phenomenology (Part I). In M. dalla Chiara, editor, *Proceedings of the Conference on Topics and Perspectives of Contemporary Logic*

and *Philosophy of Science (Cesena, Italy, January 7-10, 1987)*, pages 37–55, Bologna, 1988. Clueb. URL: www.cs.ru.nl/~henk/BP/bp1.html.

17. H.P. Barendregt. The Abhidhamma Model $AM_0$ of Consciousness and some of its Consequences. In M.G.T Kwee, K.J. Gergen, and F. Koshikawa, editors, *Buddhist Psychology: Practice, Research & Theory*, pages xxx–xxx. Taos Institute Publications, 2006. To appear. URL: ftp://ftp.cs.kun.nl/pub/CompMath.Found/G.pdf.

18. **Neurophysilogy**

19. M. Calle, I.E.W.M. Claassen, J.G. Veening, T. Kozicz, E.W. Roubos, and H.P. Barendregt. Opioid peptides, crf and urocortin in cerebrospinal fluid-contacting neurons in xenopus laevis. *Trends in Comparitive Edocrinology and Neurobiology*, 1040:249–252, 2005.

20. M. Calle, G.J.H. Corstens, L. Wanga, T. Kozicza, R.J. Denverc, H.P. Barendregt, and E.W. Roubos. Evidence that urocortin i acts as a neurohormone to stimulate amsh release in the toad xenopus laevis. *Brain Research*, 1040:14–28.

**Service to the Profession:** Dean, Faculty of Mathematics and Computer Science, 1990-1992; Editor, Journal of Functional Programming; Editor, Journal of Logic and Computation; Editor, Indagationes Mathematicae; Editor, Information and Computation.

**Teaching:** First order logic, metamathematics (computability, set theory, model theory, incompleteness theorems), lambda calculus, type theory. Students of note: Jan Willem Klop, Albert Visser, Marc Bezem, Bart Jacobs, Herman Geuvers.

**Vision Statement:** The concise foundations for reasoning, computing and defining structures will be molded into mathematician-friendly assistants for formalizing, in order to yield a new level of precision, force and certainty in mathematics, helping humans to learn, develop, teach, apply and referee the subject.

**Honours and Awards:** Spinoza Award, Netherlands Science Foundation, 2002; Member Royal Academy of Science (KNAW), 1997-; ‚Member Hollandsche Maatschappij voor Kunsten en Wetenschappen, 1994-; Member, Academia Europaea, 1992-.

# BENACERRAF, Paul

**Specialties:** Philosophy of Mathematics

**Born:** 26 March 1931 in Paris France.

**Educated:** Princeton University, PhD, 1960; Princeton University, AB, 1953.

**Dissertation:** *Logicism, Some Considerations*; supervisor, Hilary Putnam.

**Regular Academic or Research Appointments:** JAMES S. MC DONNELL UNIVERSITY PROFESSOR OF PHILOSOPHY, EMERITUS, 2007–; James S. McDonnell Distinguished University Professor of Philosophy, Princeton University, 1998–2007; Stuart Professor of Philosophy, Princeton University, 1979–1998; Professor, Princeton University, 1971-; Associate Professor, Princeton University, 1965–1971; Assistant Professor, Princeton University, 1961–1965; Instructor, Princeton University, 196-0-1961.

**Visiting Academic or Research Appointments:** Director's Visitor, Institute For Advanced Study, 2004–2005; Fellow, Center for Advanced Study in the Behavioral Sciences, 1979–1980.

**Research Profile:** Benacerraf's research has focused on issues in the epistemology and metaphysics of mathematics. It began with an examination of logicism — the thesis that mathematics is but a definitional extension of logic — and continued into related areas. So, "What Numbers Could Not Be" [1965] investigates what metaphysical consequences may be drawn from the Frege-Russell reduction of arithmetic. "Mathematical Truth" [1973] attempts to demonstrate a disconnect between philosophies of mathematics that privilege epistemological considerations [Formalism may be the most prominent] and those that attempt to stay true to the semantical structure of mathematical language [Logicism, in some forms] — success in one area, tends to be bought at the expense of failure in the other, or so he argues. In "What Mathematical Truth Could Not Be-I" [1996] he recaps the issues discussed in the above two papers and reviews some of the consequences of certain influential papers [e.g. Hilary Putnam's "Models and Reality"]. "What Numbers Could Not Be-II" [1999] is a discussion of an interesting and puzzling paper by George Boolos: "Must We Believe Set Theory". There are two quasi-historical papers that may also be of some interest: "Frege: the Last Logicist" [1981] and "Skolem and the Skeptic" [1985]. The first argues that Frege was not a logicist in the sense that his epistemologically-minded followers in the early 20th century would have him be, while the second argues that Skolem's 1922 paper on the foundations of set theory (and the *locus classicus* of the so-called "Skolem's Paradox" and the "Skolemite" view) is not one in which he embraces the relativistic view he was later to defend, but in which he urges considerations of relativity as *reductios* of a conventionalist view he finds in Zermelo 1908.

Throughout these papers, and others, Benacerraf is concerned with evaluating claims [which he takes to be typically exaggerated] regarding the philosophical consequences of certain logical and metamathical results: Gödel's incompleteness theorems, the Löwenheim-Skolem theorems, the definitions of number and the reductions of arithmetic to logic/set theory, the independence of Cantor's Continuum Hypothesis. In the last few years, he has been engaged on a project of reviewing and assessing Gödel's philosophical writings. But no immediate publication is in view.

**Main Publications:**

1. Tasks, SuperTasks, and the Modern Eleatics, *Journal of Philosophy*, Vol. LIX, No. 24, November 22, 1962, pp. 765784.
2. What Numbers Could Not Be, *The Philosophical Review*, Vol. LXXIV, No. 1, January 1965, pp. 4773.
3. God, the Devil, and Gödel, *The Monist*, Vol. 51, No. 1, January 1967, pp. 933.
4. Mathematical Truth, *The Journal of Philosophy*, Vol. LXX, No. 19, November 8, 1973
5. Frege: The Last Logicist in *The Foundations of Analytic Philosophy*, P. A. French, T. E. Uehling, Jr., H. K. Wettstein, eds., Midwest studies in Philosophy, Vol. VI, 1981, pp. 1735.
6. Skolem and the Skeptic, *Proceedings of the Aristotelian Society*, Supplementary Volume LIX, 1985. Oxford: Blackwell, pp. 85115.
7. Recantation, or 'Any old $\omega$-sequence would do after all', in *Philosophia Mathematica*, 4(2) May, 1996.
8. What Mathematical Truth Could Not Be – I, in *Benacerraf and His Critics*, A. Morton and S. P. Stich, eds., Blackwell's, Oxford and Cambridge [MA], 1996, pp 9-59.
9. What Mathematical Truth Could Not Be – II (Annual Logic Lecture of the British Logic Association), in *Sets and Proofs*, S. B. Cooper and J. K. Truss, eds., Cambridge University Press, Cambridge (UK)1999, pp. 27-51.
10. *Philosophy of Mathematics*, edited with Hilary Putnam, Second Edition (Revised), ivii + 600 pp., Cambridge University Press 1983, with a new Introduction, pp. 137, and a new Bibliography.

*Work in Progress*

11. A monograph on Gödel's philosophical writings.

**Service to the Profession:** Advisory Boards, Carnegie-Mellon School of Humanities and Social Science, 2003-2004; Advisory Boards, Department of Philosophy, 2002; Advisory Boards, University of Texas at Austin, 2002; Advisory Boards, University of Arizona, 2001; Advisory Boards, Harvard University, Philosophy, 1995-2000.

ASL Executive Committee, 1967-1969; American Philosophical Association, Eastern Division Secretary-Treasurer and Executive Committee 1962-1965.

**Teaching:** Benacerraf has supervised upward of 30 PhD dissertations between 1962 and the present. Although the list includes a number of very distinguished scholars, who have made important contributions to the subject, it would be inappropriate to list some but not others, and silly to list them all. Their training, and the graduate and undergraduate courses he has taught over the years, has been the most rewarding aspect of his career, and where he feels he has made the greatest contribution to the subject.

**Honours and Awards:** Fellow, American Academy of Arts and Sciences, 1998- ; Fellow American Council of Learned Societies, 1974-1975; Fellow, John Simon Guggenheim Foundation, 1967; 1970-1971.

## BENCIVENGA, Ermanno

**Specialties:** Free logic, dialectical logic, philosophy of logic.

**Born:** 2 August 1950 in Reggio Calabria, Italy.

**Educated:** University of Toronto, PhD Philosophy, 1977; University of Milan, MA Philosophy, 1976; University of Milan, Italy, BA Philosophy, 1972.

**Dissertation:** *Foundations of Free Logic*; supervisor, Bas van Fraassen.

**Regular Academic or Research Appointments:** PROFESSOR OF PHILOSOPHY, UNIVERSITY OF CALIFORNIA, IRVINE, 1987–. Associate Professor of Philosophy, University of California, Irvine, 1983–1987; Assistant Professor of Philosophy, University of California, Irvine, 1979–1983; Andrew Mellon Instructor, Rice University, 1978–1979;

**Visiting Academic or Research Appointments:** Visiting Assistant Professor, University of Pittsburgh, 1978.

**Research Profile:** Ermanno Bencivenga has proposed the only semantics for free logic that is developed at a properly quantificational level and is not committed to outer domains or nonexistent objects. He has proved the compactness of the monadic fragment of a supervaluational language; developed "weak" free logics where individual variables range over independent domains; proved the non-recursive enumerability of free logics of both indefinite and definite descriptions; developed free set theories and free arithmetics; given a finitary proof of the consistency of a free arithmetic and extended Gödel's first incompleteness theorem to it; with Peter Woodruff, developed a modal language with the lambda operator that allows for a uniform treatment of singular terms; and provided the only correct completeness proof for semantic tableaux with identity. He has discussed the relation between free logic and Kant's transcendental idealism; and developed a systematic interpretation of Hegel's dialectical logic based on the notion of meaning as narrative.

**Main Publications:**

1. Set Theory and Free Logic, *Journal of Philosophical Logic*, 5, 1976.
2. Are Arithmetical Truths Analytic? New Results in Free Set Theory, *Journal of Philosophical Logic*, 6, 1977.
3. A Semantics for a Weak Free Logic, *Notre Dame Journal of Formal Logic*, 19, 1978.
4. Free Semantics for Indefinite Descriptions. *Journal of Philosophical Logic* 7 1978.
5. Free Semantics, *Boston Studies in the Philosophy of Science* 47 1981.
6. Semantic Tableaux for a Logic with Identity, *Zeitschrift für mathematische Logik und Grundlagen der Mathematik*, 27, 1981.
7. A New Modal Language with the Lambda Operator (with Peter W. Woodruff). *Studia Logica*, 40, 1981.
8. Compactness of a Supervaluational Language, *Journal of Symbolic Logic*, 48, 1983.
9. Finitary Consistency of a Free Arithmetic, *Notre Dame Journal of Formal Logic*, 25, 1984.
10. Free Logics, In *Handbook of Philosophical Logic*, edited by D. Gabbay and F. Guenthner, vol. III. Dordrecht: Reidel, 1986.
11. Incompleteness of a Free Arithmetic. *Logique et Analyse* 31 1988.
12. Free from What?, *Erkenntnis* 33 1990.
13. What is Logic About?, *European Review of Philosophy* 4 1999.
14. *Hegel's Dialectical Logic*, New York: Oxford University Press, 2000.
15. Putting Language First: The 'Liberation' of Logic from Ontology, in *Companion to Philosophical Logic*, edited by D. Jacquette. Oxford: Blackwell, 2002.

**Vision Statement:** Logic made enormous progress early in the $20^{th}$ century; later, as the amount of research published increased tremendously, it often showed lack of vision and depth. This is a case of being a victim of one's success: later researchers felt constrained by the structure of the early groundbreaking work. Imagination and creativity must return to the forefront of logic, as well as an awareness of the continuity of the logical tradition: of how much authors like Kant and Hegel can still make important contributions to our understanding of meaning and inference.

# BÉZIAU, Jean-Yves

**Specialties:** Universal logic, paraconsistent logic, many-valued logic, modal logic, proof theory, history and philosophy of logic.

**Born:** January 15 1965 in Orléans, France.

**Educated:** University of São Paulo, Brazil, PhD Philosophy, 1996; University of Paris 7 Denis Diderot, PhD Logic and Foundations of Computer Sciences, 1995; Master Logic and Foundations of Computer Sciences, 1990; University of Paris 1 Panthéon-Sorbonne, Master Philosophy, 1988.

**Dissertation:** Philosophy: *On Logical Truth*, supervisor, Newton da Costa; Mathematics: *Researches on Universal Logic*, supervisor, Daniel Andler.

**Regular Academic or Research Appointments:** PROFESSOR OF THE SWISS NATIONAL SCIENCE FOUNDATION, INSTITUTE OF LOGIC AND SEMIOLOGY, UNIVERSITY OF NEUCHÂTEL, 2002– ; Researcher, National Laboratory for Scientific Computing, Rio de Janeiro, 1995-2002.

**Visiting Academic or Research Appointments:** Visiting Scholar, Stanford University 2000-2001; Visiting Professor, Federal Fluminense University, Brazil, Department of Mathematics, 1999; Fulbright Research Scholar, UCLA, Department of Mathematics, 1995; Research Scholar, University of São Paulo, Institute for Advanced Studies, 1994; Visiting Scholar, University of Wroclaw, Department of Logic and Philosophy of Sciences, Poland, 1992-1993.

**Research Profile:** Jean-Yves Béziau is the originator of Universal Logic, a general theory of logics considered as mathematical structures developed in order to gather in a coherent whole the numerous new logics created over the last decades. The task of universal logic is to understand how to combine, translate and identify logic structures and also to prove some general theorems and delineate their domain of validity. JYB proved a general completeness theorem connecting sequent systems and bivaluations, by application of this theorem he was able to provide a sequent system for Lukasiewicz's three-valued logic. He proposed a version of Kripke semantics without possible worlds and different semantical constructions related to it useful for the constructions of many different logics.

JYB also made important contributions in paraconsistent logic and the study of negation. He introduced the terminology "paranormal logics" to denote logics which are both paracomplete and paraconsistent. He showed that S5 and first-order classical logic are paraconsistent, and that the nameless corner of the square of opposition is a paraconsistent negation. This led him to a generalisation of the square of opposition into some three-dimensional geometrical objects, which can serve as a basis of a general theory of negation and modalities. JYB also worked in non-transitive logics and logics where the law of auto-deductibility does not hold.

He formulated and analysed several paradoxes such as the translation paradox (the fact that a logic can be translated into a weaker one), the copulation paradox (the apparition of unexpected properties when combining logics).

JYB wrote several papers in philosophy of logic pointing out the import and relevance of Polish logic for the analysis and understanding of some basic concepts of logic such as truth-functionality, extensionality, bivalence, logical form. He also investigated the notion of identity.

**Main Publications:**

1. J.-Y.Béziau (ed), *Logica Universalis*, Birkhäuser, 2005.
2. P.Suppes and J.-Y.Béziau, "Semantic computation of truth based on associations already learned", Journal of Applied Logic, 2 (2004), pp.457-467.
3. J.-Y.Béziau, "New light on the square of oppositions and its nameless corner", Logical Investigations, 10, (2003), pp.218-232.
4. J.-Y.Béziau, "Are paraconsistent negations negations ? ", in Paraconsistency: the logical way to the inconsistent, W.Carnielli et al. (eds), Marcel Dekker, New-York, 2002, pp.465-486.
5. J.-Y.Béziau, "S5 is a paraconsistent logic and so is first-order classical logic", Logical Investigations, 9, (2002), pp.301-309.
6. J.-Y.Béziau, "The philosophical import of Polish logic", in Methodology and philosophy of science at Warsaw University, M.Talasiewicz (ed.), Semper, Warsaw, 2002 pp.109-124.
7. J.-Y.Béziau, "What is classical propositional logic?", Logical Investigations, 8 (2001), pp.266-277.
8. J.-Y.Béziau, "From paraconsistent to universal logic", Sorites, 12 (2001), pp.5-32.
9. J.-Y.Béziau, "Sequents and bivaluations", Logique et Analyse, 44 (2001), pp.373-394.
10. J.-Y.Béziau, "What is paraconsistent logic?", in Frontiers of paraconsistent logic, D.Batens et al. (eds), Research Studies Press, Baldock, 2000, pp.95-111.
11. J.-Y.Béziau, "Classical negation can be expressed by one of its halves", Logic Journal of the Interest Group in Pure and Applied Logics, 7 (1999), pp.145-151.
12. J.-Y.Béziau, "The mathematical structure of logical syntax" in Advances in contemporary logic and computer science, W.A.Carnielli and I.M.L.D'Ottaviano (eds), American Mathematical Society, Providence, 1999, pp.1-17.
13. J.-Y.Béziau, "Rules, derived rules, permissible rules and the various types of systems of deduction", in Proof, types and categories, E.H.Hauesler and L.C.Pereira (eds), PUC, Rio de Janeiro, 1999, pp.159-184.
14. J.-Y.Béziau, "A sequent calculus for Lukasiewicz's three-valued logic based on Suszko's bivalent semantics", Bulletin of the Section of Logic, 28 (1999), pp.89-97.
15. J.-Y.Béziau, "Idempotent full paraconsistent negations are not algebraizable", Notre Dame Journal of Formal Logic, 39 (1998), pp.135-139.
16. J.-Y.Béziau, "What is many-valued logic ?", in Proceedings of the 27th International Symposium on Multiple-Valued Logic, IEEE Computer Society, Los Alamitos, 1997, pp.117-121.
17. J.-Y.Béziau, "Logic may be simple", Logic and Logical Philosophy, 5 (1997), pp.129-147.
18. N.C.A. da Costa & J.-Y.Béziau, "Théorie de la valuation", Logique et Analyse, 146 (1994), pp.95-117.
19. J.-Y.Béziau, "Universal logic", in Logica'94 - Proceedings of the 8th International Symposium, T.Childers & O.Majer (eds), Prague, 1994, pp.73-93.
*Work in Progress*
20. J.-Y.Béziau ,W.A.Carnielli, D.M.Gabbay (eds), *Handbook of Negation and Paraconsistency*, to be published by Elsevier, 2006.

**Service to the Profession:** Founder and Editor, *Logica Universalis* (Birkhäuser) and the book series *Studies in Universal Logic* (Birkhäuser); Organizer, First World School and Congress on Universal Logic, Montreux, Switzerland, 2005; Organizer, Third World Congress on Paraconsistency, Toulouse, France, 2003; Editor of two special issues of *Logique et Analyse* on Contemporary Brazilian Research in Logic, 1996-1997; Editor (with D.Krause) of a special issue of *Synthese* commemorating the $80^{th}$ birthday of Patrick Suppes; Translation of Newton da Costa's book in French: *Logiques classiques et non classiques*, Masson, Paris, 1997.

**Teaching:** JYB taught courses on mathematical logic, philosophy of logic, universal logic, philosophy of mathematics, semiotics, paraconsistent logic, proof theory.

**Vision Statement:** I consider that logic is a fundamental research field related to and relating many different fields (mathematics, physics, psychology, philosophy, cognitive sciences, computation, linguistics, semiotics). In my view, logic is not limited to the study of some formal systems, but concerned with the understanding of reasoning and the thought process in general.

# BIBEL, Wolfgang

**Specialties:** Artificial Intelligence, knowledge representation and reasoning, automated theorem proving, logic programming, mathematical logic.

**Born:** 28 October 1938 in Nürnberg, Germany.

**Educated:** Ludwig-Maximilians Universität München, PhD Mathematics, 1968; Universität Heidelberg, Ludwig-Maximilians Universität München, Diplom Mathematics, 1964; Universität Erlangen, Vordiplom Physics, 1961.

**Dissertation:** *Schnittelimination in einem Teilsystem der einfachen Typenlogik*; supervisor, Kurt Schütte.

**Regular Academic or Research Appointments:** PROFESSOR OF INFORMATICS, RETIRED, DARMSTADT UNIVERSITY OF TECHNOLOGY, 2004–. Professor of Informatics, Darmstadt University of Technology, 1988–2004; Professor of Computer Science, University of British Columbia, 1987–1988; Oberassistent, Technische Universität München, 1972–1987; Assistent, Technische Universität München, 1969–1972; Assistent, Ludwig-Maximilians Universität München, 1966–1969.

**Visiting Academic or Research Appointments:** Adjunct Professor, University of British Columbia, 1988– ; Visiting Professor of Informatics, Darmstadt University of Technology, 1985-1986; Visiting Associate Professor of Computer Science, Duke University, 1985; Visiting Professor, University of Roma, 1983; Visiting Professor, Universität Karlsruhe, 1978–1979; Visiting Professor, Universität des Saarlandes, 1975; Visiting Assistant Professor of Computer Science, Wayne State University, 1970–1971.

**Research Profile:** Wolfgang Bibel started his career with a proof theoretic contribution to the well-known cut-elimination problem in type theory. With this solid background he has become fascinated with the perspectives of logic in connection with the evolving computer technology and therefore has laid his research focus on applied and computational logic as a formal basis for artificial intelligence (or intellectics) throughout his career. Among the numerous possible applications in this wide field he has concentrated especially on the automation of logical deduction for theorem proving, logic programming and knowledge processing. He has developed the connection method for automated deduction in various logical calculi which allows the discovery of a derivation in terms of the structure of the given formula to be proved, in a connection-oriented way. Logically the connection method can be seen as a computationally improved version of Herbrand's theorem. In collaboration with his students numerous deductive systems have been developed on this theoretical basis including SETHEO and leanCoP. The first has become "worldmaster" in an international competition of proof systems. The second is a system consisting of a few lines of code in Prolog and thus five orders of magnitude smaller than standard provers and yet similarly efficient as those. The connection method has influenced many other researchers in their work and has proved similarly effective in non-classical logics such as intuitionistic, modal and linear logic.

Logic provides a more or less static formalism while natural human reasoning easily copes with changes occurring in the world. Bibel has tried to combine classical logic with such changes in the form of explicit transitions from one logic description to another yielding a transition logic. This work led to a logic equivalent with the conjunctive part of linear logic years before linear logic has been published. It laid the basis for solving the famous inferential frame problem for the first time. Recently an intuitively even simpler version of this transition logic has been submitted for publication. These formalisms have a particular relevance for the generation of plans for solving problems in real world scenarios. Forms of reasoning other than planning have been studied intensively like nonmonotonic reasoning. Knowledge systems based on logic formalisms have been built for various domains of applications such as for configuring legal contracts.

Deduction is closely related with computation. However, there remains a large gap between the logical specification of a computational problem and a logical version of it which is efficiently executable (like a Prolog program). Bibel has developed a strategic approach to bridge this gap and a system, called LOPS, for synthesizing programs which realized this theoretical work. In addition constructive logics have been exploited for a deductive approach to programming.

**Main Publications:**

1. Schnittelimination in einem Teilsystem der einfachen Typenlogik (Cut-elimination in a subsystem of simple type theory), *Archiv für Mathematische Logik* 12, 159-178 (1969).
2. An approach to a systematic theorem proving procedure in first-order logic, *Computing* 12, 43–55 (1974).
3. Tautology testing with a generalized matrix reduction method. *Theoretical Computer Science* 8, 31–44 (1979).
4. Syntax-directed, semantics-supported program synthesis. *Artificial Intelligence* 14, 243–261 (1980).
5. On matrices with connections. *Journal of the ACM* 28, 633–645 (1981).
6. Computationally improved versions of Herbrand's theorem. *Proceedings of the Herbrand Symposium*, Marseille, France, July 1981 (J. Stern, ed.), Studies in Logic 107, North-Holland, Amsterdam, 11–28 (1982).
7. Matings in matrices. *Communications of the ACM* 26, 844–852 (1983).

8. A deductive solution for plan generation, *New Generation Computing* 4, 115–132 (1986).

9. *Automated theorem proving*. Vieweg Verlag, Wiesbaden, 293 pp. (1982); 2nd edition 289 pp. (1987).

10. Constraint satisfaction from a deductive viewpoint. *Artificial Intelligence Journal* 35, 401–413 (1988).

11. Short proofs of the pigeonhole formulas based on the connection method, *Journal of Automated Reasoning* 6, 287–297 (1990).

12. Predicative programming, *New Generation Computing Journal* 8, 275–276 (1991).

13. SETHEO: A high-performance theorem prover, *Journal of Automated Reasoning* 8(2), 183–212 (1992), with R. Letz, J. Schumann, S. Bayerl

14. Methods and calculi for deduction, with E. Eder, in *Handbook of Logic in Artificial Intelligence and Logic Programming*, Chapter 3, Vol. 1, D.M. Gabbay, C.J. Hogger, J.A. Robinson, eds., Oxford University Press, 67–182 (1993).

15. *Wissensrepräsentation und Inferenz* (Knowledge Representation and Inference), with S. Hölldobler, T. Schaub, Vieweg, Braunschweig (1993).

16. *Deduction: Automated Logic*, Academic Press, London, 242 pp. (1993).

17. Decomposition of tautologies into regular formulas and strong completeness of connection-graph resolution, *Journal of the ACM* 44(2), 320–344 (1997). (with E. Eder)

18. A multi-level approach to program synthesis, with D. Korn, C. Kreitz, F. Kurucz, J. Otten, S. Schmitt, G. Stolpmann *Proceedings of the 7th Workshop on Logic Program Synthesis and Transformation* (LOPSTR-97), LNCS, Vol. 1463, Springer, Berlin, 1–27 (1998).

19. Let's plan it deductively!, *Artificial Intelligence* 103(1–2), 183–208 (1998).

20. leanCoP: Lean Connection-Based Theorem Proving, *Journal of Symbolic Computation* 36, 139–161 (2003). (with Jens Otten)

*Work in Progress*

21. Integration of the reduction rule FACTOR into a connection method prover which is conjectured to be as powerful as the cut, a conjecture yet to be proved correct.

22. A further elaboration and application of the transition logic.

23. Applying logic-based knowledge systems technology to applicational areas such as law.

**Service to the Profession:** Advisory Board Member, New Generation Computing, 1995–; Editor, *Computational Intelligence*, Vieweg Verlag, 1982–; Section Editor, *Artificial Intelligence Journal*, 1985–1999; Editor, *Automated Deduction – A Basis for Applications*. Vol. I–III, Kluwer, Dordrecht, 1998; Conference Chair, *Fifteenth International Conference for Automated Deduction* (CADE-98), 1998; Initiator and coordinator of the national research programme on automated deduction, *DFG*, 1992–1998; Chair ("Dekan"), Dept. of Computer Science, Darmstadt University of Technology, 1991–1992; President, IJCAII 1987–1989; Founder and Chair, *ECCAI* 1982–1986.

**Teaching:** Bibel has introduced the logic course as a requirement in the informatics program at Darmstadt University of Technology. He has supervised six Habilitations and thirty five PhD students in Computer Science, including Elmar Eder, Uwe Egly, Christoph Herrmann, Steffen Hölldobler, Holger Hoos, Christoph Kreitz, Chunping Li, Luís Paquete, Torsten Schaub, Thomas Stützle, and Michael Thielscher. A symposium honoring Bibel on the occasion of his $60^{th}$ birthday was held at Darmstadt University of Technology, Oct. 28, 1998. The symposium volume, *Intellectics and Computational Logic. Papers in Honor of Wolfgang Bibel* (S. Hölldobler, ed.), Applied Logic Series, Kluwer, Dordrecht (2000), testifies to the influence of Bibel's teaching and research over the years.

**Vision Statement:** The computational power of machines will be close to that of the human brain within a decade according to several different estimates. In order to translate this number crunching power into some form of artificial intelligence a way of integrating specific problem solving modules into extremely complex systems in a preferably conjunctive way must be found. Logic could play a vital role in this vision since it alone offers a formalism which is truly conjunctive and thus could help to overcome the complexity barrier. For this reason I forsee an even brighter future for logic in the years to come.

**Honours and Awards:** Honorary Professional Fellow, Institute of Nanotechnology, UK, 2005; Fellow, ECCAI, 1999; Donald E. Walker Distinguished Service Award, International Joint Conferences for Artificial Intelligence Inc. (IJCAII), 1999; Silver Core, International Federation for Information Processing (IFIP), 1998; Fellow, American Association for Artificial Intelligence (AAAI), 1990; Fellow, Canadian Institute for Advanced Research, 1987–1988.

# BIMBÓ, Katalin

**Specialties:** Nonclassical logics, especially, relevance, substructural and combinatory logics; Philosophy of computer science.

**Educated:** Indiana University Bloomington, PhD in Philosophy and Cognitive science, 1999; Eötvös University, Doctorate in Logic, 1994; Moscow State University, Diploma in Philosophy, 1986.

**Dissertation:** *Substructural Logics, Combinatory Logic and λ-calculus*; supervisor, J. Michael Dunn.

**Regular Academic or Research Appointments:** ASSISTANT PROFESSOR OF PHILOSOPHY, UNIVERSITY OF ALBERTA, EDMONTON, CANADA, 2008–; Research Associate, School of Informatics, Indiana University Bloomington, 2005–2008; Research Fellow, Research School of Information Sciences and Engineering, Australian National University, 2000–2003; Postdoctoral Research Fellow, School of Mathematical and Computing Sciences and Department of Philosophy, Victoria University of Wellington, 1999–2000; Assistant Professor, 1994–1999, Lecturer, 1988–1994, Junior Researcher, 1986–1988, Department of Symbolic Logic and Methodology of Sciences, Eötvös University.

**Research Profile:** Bimbó's main research results concern *relevance* and *substructural logics* as well as *combinatory logic*. She has collaborated with J. Michael Dunn on papers that defined relational semantics for various logics, from the minimal substructural logic to action logic and to certain structurally free calculi. A uniform treatment of this semantic approach is given in their forthcoming book which — at the same time — extends the range of the applicability of such semantics.

A more precise characterization of classes of structures for various nonclassical logics leads straightforwardly to topological frames. Bimbó proved *topological duality theorems* for ortho- and De Morgan lattices and for the algebra of the logic of entailment. These duality theorems provide deep logical insights, because the algebras of logics are treated abstractly as categories (in the sense of category theory), where the morphisms of the category are interpretations of the logic.

Bimbó defined new *sequent calculi* for certain nonclassical logics, including some relevance logics that lacked a similar formulation. She introduced a *purely inductive proof technique* for the cut theorem. The same proof technique can be applied to Gentzen's original sequent calculi $LK$ (for classical logic) and $LJ$ (for intuitionistic logic) to obtain direct inductive proofs of the admissibility of the single cut rule (without any considerations of mix).

Combinatory logics and λ-calculi connect straightforwardly to nonclassical logics — in more than one way. Bimbó has defined (sound and complete) *semantics* for systems of dual and symmetric combinatory logics. She proved some new results about the logic of ticket entailment relying on simple types. She also proved *Church–Rosser type theorems* for dual and symmetric combinatory logics.

She has also worked on natural language semantics, in particular, on applications of logic — primarily, of modal logics and DRT — in modeling the temporal structure of a natural language discourse.

**Main Publications:**

1. K. Bimbó and J. M. Dunn, *Generalized Galois Logics. Relational Semantics of Nonclassical Logical Calculi*, (x + 382 pp.), CSLI Lecture Notes, v. 187, CSLI Publications, CSLI, Stanford, CA, 2008. (to appear).
2. "Dual gaggle semantics for entailment," *Notre Dame Journal of Formal Logic*, **50** (2009), 20 pp. (to appear).
3. "Functorial duality for ortholattices and De Morgan lattices," *Logica Universalis*, **1** (2007), pp. 311–333.
4. "$LE^t_\to$, $LR^\circ \wedge \neg$, $LK$ and cutfree proofs," *Journal of Philosophical Logic*, **36** (2007), pp. 557–570.
5. "Relevance logics," In *Philosophy of Logic*, D. Jacquette (ed.), Handbook of the Philosophy of Science, D. Gabbay, P. Thagard and J. Woods (eds.), v. 5, Elsevier (North-Holland), Amsterdam, 2007, pp. 723–789.
6. "Admissibility of cut in $LC$ with fixed point combinator," *Studia Logica*, **81** (2005), pp. 399–423.
7. K. Bimbó and J. M. Dunn, "Relational semantics for Kleene logic and action logic," *Notre Dame Journal of Formal Logic*, **46** (2005), pp. 461–490.
8. "The Church-Rosser property in symmetric combinatory logic," *Journal of Symbolic Logic*, **70** (2005), pp. 536–556.
9. "Types of l-free hereditary right maximal terms," *Journal of Philosophical Logic*, **34** (2005), pp. 607–620.
10. "Semantics for dual and symmetric combinatory calculi," *Journal of Philosophical Logic*, **33** (2004), pp. 125–153.
11. "The Church-Rosser property in dual combinatory logic," *Journal of Symbolic Logic*, **68** (2003), pp. 132–152.
12. K. Bimbó and J. M. Dunn, "Four-valued logic," *Notre Dame Journal of Formal Logic*, **42** (2001/2003), pp. 171–192.
13. "Semantics for the structurally free logic $LC+$," *Logic Journal of IGPL*, **9** (2001), pp. 525–539.
14. K. Bimbó and J. M. Dunn, "Two extensions of the structurally free logic $LC$," *Logic Journal of IGPL*, **6** (1998), pp. 403–424.
15. "Specificity and definiteness of temporality in Hungarian," In: J. Bernard and K. Neumer (eds.), *Zeichen, Sprache, Bewußtsein. Österreichisch–Ungarische Dokumente zur Semiotik und Philosophie 2*, ÖGS–ISSS, Wien–Budapest, 1994, pp. 7–25.
16. "Are the 'and' and '.' the same sentence connectives?," In: J. Darski and Z. Vetulani (eds.), *Akten des 26. linguistischen Kolloquiums*, v. 2, Max Niemeyer, Tübingen, 1993, pp. 485–490.
17. "Translation of three-valued logics into sequent calculi," [in Russian], In: Yu. V. Ivlev (ed.), *Sovremennaya logika i metodologiya nauki*, Izdatel'stvo MGU,

Moscow, 1987, pp. 108–119.
*Work in Progress*
18. "Combinatory logic," (entry for the *Standford Encyclopedia of Philosophy*, E. Zalta (ed.)).
19. K. Bimbó and J. M. Dunn, "Symmetric gaggles," (paper).

**Service to the Profession:** Organizer and program chair of the 2002 Annual Conference of the Australasian Association for Logic, Canberra; Organizer and program chair of the 4th Symposium on Logic and Language, 1992; Reviewer for the Mathematical Reviews.

**Teaching:** First-order logic, nonclassical logics, relevance and substructural logics, computability, formal language theory, temporal and tense logics, intensional logic and Montague grammar, T<sub>E</sub>X.

**Vision Statement:** Logic influenced the development of human inquiry for thousands of years. In the mean time, logic itself has expanded and grew into a versatile discipline. In the last century, logic interacted with a wide range of fields from philosophy and mathematics to computer science, linguistics, cognitive science (and others). This process produced an abundance of new logics and new results. I hope that logic as a discipline encompassing various logics will thrive as a successful research area, and at the same time, the main results that are obtained will be increasingly disseminated in standard curricula.

**Honours and Awards:** Candidate of the Hungarian Academy of Sciences.

# BLACKBURN, Patrick Rowan

**Specialties:** Modal and hybrid logic, applications of logic in natural language, knowledge representation, and philosophy.

**Born:** 12 July 1959, Kowloon, Hong Kong, China.

**Educated:** University of Waikato BA (Philosophy), 1981, University of Sussex MSc (Logic and Scientific Method), 1986, University of Edinburgh PhD, 1990.

**Dissertation:** *Nominal Tense Logic and Other Sorted Intensional Frameworks*; supervisors, Johan van Benthem, Inge Bethke, and Barry Richards.

**Regular Academic or Research Appointments:** DIRECTEUR DE RECHERCHE, INRIA, 2000– Lecturer, Department of Computational Linguistics, University of the Saarland, 1994–2000; Research Associate, Department of Philosophy, Utrecht University, 1992–1994; Postdoctoral Fellow, Department of Mathematics and Computer Science, University of Amsterdam, 1991–1992; Postdoctoral Fellow, Centre for Cognitive Science, University of Edinburgh, 1990–1991.

**Visiting Academic or Research Appointments:** Invited Researcher, Institute for Research in Cognitive Science, University of Pennsylvania, 1998.

**Research Profile:** Patrick Blackburn is interested in logics which can be applied to problems in natural language, knowledge representation, and philosophy. His theoretical work has centered on modal logic, and in particular, hybrid logic. Hybrid logics are modal logics in which it possible to name the points of evaluation. First introduced and explored by Arthur Prior in the 1960s, hybrid logics are closely related to the description logics used in contemporary knowledge representation. Patrick Blackburn has worked on a number of topics in hybrid logic, including their model theory, proof theory, computational complexity, and their role in Arthur Prior's philosophy.

His applied work centers on logical analysis of natural language. He has worked on feature logic (logics for what computational linguists call attribute-value structures) and model-theoretic syntax (an approach to natural language syntax based on the idea that grammars are axioms that define the grammatical trees). Much of his current work centers on applications of logic in natural language semantics and pragmatics. He is particularly interested in the role inference plays in natural language and has explored the use of automated reasoning tools (theorem provers, model builders and model checkers) in handling inferential phenomena characteristic of natural language (such as presupposition). He is also interested in the semantics of tense and aspect, a topic which links up with his work on hybrid logic.

**Main Publications:**

1. *Handbook of Modal Logic*, edited with J. van Benthem and F. Wolter, Elsevier, 2007.
2. Pure Extensions, Proof Rules, and Hybrid Axiomatics, with Balder ten Cate, *Studia Logica*, 84 (2006), 277–322.
3. Arthur Prior and Hybrid Logic, *Synthese*, 50 (2006), 329–372.
4. *Representation and Inference for Natural Language: A First Course in Computational Semantics*, with J. Bos, CSLI Press, 2005.
5. Constructive Interpolation in Hybrid Logic, with M. Marx, *Journal of Symbolic Logic*, 68 (2003), 463–480.
6. Repairing the Interpolation Theorem in Quantified Modal Logic, with C. Areces and M. Marx, *Annals of Pure and Applied Logic*, 124 (2003), 287-299.
7. Hybrid Logic: Characterization, Interpolation and Complexity, with C. Areces and M. Marx, *Journal of*

*Symbolic Logic*, 66 (2001), 977–1010.
8. *Modal Logic*, with M. de Rijke and Y. Venema, Cambridge University Press, 2001.
9. Internalizing Labelled Deduction, *Journal of Logic and Computation*, 10 (2000), 137–168.
10. Hybrid Languages and Temporal Logic, with M. Tzakova, *Logic Journal of the IGPL*, 7 (1999), 27–54.
11. Hybrid Languages, with J. Seligman, *Journal of Logic, Language and Information*, 4 (1995), 251–272.
12. Tense, Temporal Reference and Tense Logic, *Journal of Semantics*, 11 (1994), 83–101.
13. Talking about Trees, with C. Gardent and W. Meyer-Viol, in *Proceedings of the Sixth Conference of the European Chapter of the Association for Computational Linguistics* (1993), 21–29.
14. A Modal Perspective on the Computational Complexity of Attribute Value Grammar, with E. Spaan, *Journal of Logic, Language and Information*, 2 (1993), 129–169.
15. Nominal Tense Logic, *Notre Dame Journal of Formal Logic*, 14 (1993), 56–83.
*Work in Progress*
16. Working with Discourse Representation Theory, with J. Bos, monograph, to appear. Draft available at www.blackburnbos.org.

**Service to the Profession:** Editor-in-Chief, *Journal of Logic, Language and Information*, 2002–2008; Editor, *Review of Symbolic Logic*, 2007– ; Editor, *Stanford Encyclopedia of Philosophy*, 2004–; Editor, *Journal of Philosophical Logic*, 2004–2007; Editor, *Journal of Logic, Language and Information*, 1997–2002; Editorial Board, *Notre Dame Journal of Formal Logic*, 2006– ; *Logique et Analyse*, 2005–; *Journal of Computational Linguistics*, 1995–1997; Co-organiser, ESSLLI 2004, 16th European Summer School in Logic, Language and Information, Nancy, France; President, *ACL Special Interest Group in Computational Semantics (SIGSEM)*, November 1999–2007; Standing Committee, *European Summer School in Logic, Language and Information*, 1996–1999.

**Teaching:** I have taught a wide variety of courses, both regular university courses and intensive summer school courses, for beginners and advanced students, on many aspects of logic and its applications. I love teaching. I like working out what students find difficult, and coming up with a way of making it clear. Good teaching should be like good writing: the goal is not to be understandable, the goal is to be impossible to misunderstand. It's not always possible to live up to this ideal, but it's the thing to aim for. Former students of mine include Johan Bos and Balder ten Cate.

**Vision Statement:** In the late 19th century, logic was closely associated with natural language, psychology, and philosophy, but it was to weak to treat these topics seriously. An important consequence of the tremendous technical developments of the 20th century is that logic is now in excellent shape to tackle some of its more traditional concerns, such as the workings of natural language and other cognitive and philosophical issues.

**Honours and Awards:** SERC (UK) Postdoctoral Fellowship in Information Technology, 1990–1992; New Zealand Junior Scholarship, 1976–1979.

## BLASS, Andreas

**Specialties:** Set theory and its applications, category-theoretic logic, logic in computer science, linear logic.

**Born:** 27 October 1947 in Nürnberg, Germany.

**Educated:** Harvard University, PhD Mathematics, 1970; University of Detroit, BS Physics, 1966.

**Dissertation:** *Orderings of Ultrafilters*; supervisor, Frank Wattenberg.

**Regular Academic or Research Appointments:** PROFESSOR, MATHEMATICS, UNIVERSITY OF MICHIGAN, 1984–; Associate Professor, Mathematics, University of Michigan, 1976-1984; Assistant Professor, Mathematics, University of Michigan, 1972-1976; Hildebrandt Research Instructor of Mathematics, University of Michigan, 1970-1972.

**Visiting Academic or Research Appointments:** Visiting Researcher, Microsoft Research, 2002-2007; Gastprofessor, Universität Heidelberg, 1989; Gastprofessor, Freie Universität Berlin, 1987; Visiting Professor, Pennsylvania State University, 1985-1986; Heidelberger Akademie der Wissenschaften, 1981, 1983; Visiting Associate Professor, University of Wisconsin, 1978-1979; Technische Universität, Berlin, 1976.

**Research Profile:** Andreas Blass's primary research area is combinatorial set theory, with emphasis on ultrafilters and on cardinal characteristics of the continuum. Much of his work involves applications of set theory to areas of mathematics outside logic; in recent years the most frequent application area has been abelian group theory. He has also done research in theoretical computer science, most of it joint with Yuri Gurevich, and much of it focusing on Gurevich's abstract state machine thesis, which says that every algorithm can be described at its natural level of abstraction by an abstract state machine. Building on set-theoretic considerations motivated by the axiom of determinacy, Blass introduced game semantics for linear logic. He has also done research in finite combinatorics and in category theory, especially topos theory.

**Main Publications:**

1. A partition theorem for perfect sets, Proc. Amer. Math. Soc., 82 (1981) 271-277.
2. Existence of bases implies the axiom of choice, in Axiomatic Set Theory, ed. J. Baumgartner, D.A. Martin, and S. Shelah, Contemp. Math. 31 (1984) 31-33.
3. Near Coherence of Filters, part I Cofinal equivalence of models of arithmetic, Notre Dame J. Formal Logic 27 (1986) 579-591; part II Applications to operator ideals, the Stone-Cech remainder of a half-line, order ideals of sequences, and slenderness of groups, Trans. Amer. Math. Soc. 300 (1987) 557-581; part III A simplified consistency proof, joint with S. Shelah, Notre Dame J. Formal Logic 30 (1989) 530-538.
4. Classifying topoi and the axiom of infinity, Algebra Universalis 26 (1989) 341-345.
5. Applications of superperfect forcing and its relatives, in Set Theory and its Applications, ed. J. Steprans and S. Watson, Springer-Verlag Lecture Notes in Mathematics 1401 (1989) 18-40.
6. A game semantics for linear logic, Ann. Pure Appl. Logic 56 (1992) 183-220.
7. Ultrafilters: where topological dynamics = algebra = combinatorics, Topology Proceedings 18 (1993) 33-56.
8. Cardinal characteristics and the product of countably many infinite cyclic groups, J. Algebra 169(1994) 512-540.
9. Seven trees in one, J. Pure Appl. Algebra 103 (1995) 1-21.
10. Questions and answers — a category arising in linear logic, complexity theory, and set theory, in Advances in Linear Logic, ed. J.-Y. Girard, Y. Lafont, and L. Regnier, London Math. Soc. Lecture Notes 222 (1995) 61-81.
11. An induction principle and pigeonhole principles for K-finite sets, J. Symbolic Logic 59 (1995) 1186-1193.
12. On the cofinality of ultrapowers, joint with H. Mildenberger, J. Symbolic Logic 64 (1999) 727-736.
13. Needed reals and recursion in generic reals, Ann. Pure Appl. Logic 109 (2001) 77-88.
14. Specker's theorem for Nöbeling's group, Proc. Amer. Math. Soc. 130 (2002) 1581-1587.
15. On polynomial time computation over unordered structures, joint with Y. Gurevich and S. Shelah, J. Symbolic Logic 67 (2002) 1093-1125.
16. Homotopy and homology of finite lattices, Electronic J. Combinatorics 10(1) (2003) R30.
17. Resource consciousness in classical logic, in Games, Logic, and Constructive Sets, ed. G. Mints and R. Muskens, CSLI Lecture Notes 161 (2003) 61-74.
18. Special families of sets and Baer-Specker groups, joint with J. Irwin, Comm. Alg. 33 (2005) 1733-1744.
19. Ordinary interactive small-step algorithms, joint with Y. Gurevich, part I, A.C.M. Trans. Comp. Logic 7 (2006) 363-419; parts II and III, to appear in A.C.M. Trans. Comp. Logic.
20. Combinatorial cardinal characteristics of the continuum, to appear in Handbook of Set Theory, ed. M. Foreman, A. Kanamori, and M. Magidor.

*Work in Progress*

21. Continuing work with Yuri Gurevich and others on the abstract state machine thesis, which says that every algorithm can be described, on its natural level of abstraction, as an abstract state machine.
22. Combinatorial set theory, especially ultrafilters, cardinal characteristics of the continuum, and their connections.
23. Book on infinite combinatorics and forcing.

**Service to the Profession:** Editorial Board Member, Journal of Pure and Applied Algebra; Editorial Board Member, Proceedings of the American Mathematical Society; Editorial Board Member, Archiv für Mathematische Logik und Grundlagenforschung (later Archive for Mathematical Logic); Editorial Board Member, Bulletin of Symbolic Logic; Editorial Board Member, Scientiae Mathematicae Japonicae; Editorial Board Member, Contemporary Mathematics; Editorial Board Member, Annals of Pure and Applied Logic; Editorial Board Member, Foundations of Mathematics e-mail list; Excecutive Committee Member, (ex officio) Council Member, Association for Symbolic Logic, Committee on prizes and awards; Committee Member, American Mathematical Society, to select the winner of the Steele prize.

**Teaching:** Supervised 20 PhD dissertations.

**Vision Statement:** I expect two sorts of developments in mathematical logic. First, there are internal developments. All the major branches of logic are healthy, making progress on difficult problems. In the part of logic that I know best, much has been learned recently about topics ranging from large cardinals (and the related inner model program) to the combinatorial structure of the reals (using novel sorts of forcing constructions) and other small uncountable sets. There is every reason to expect such progress to continue. Second, there are connections to other fields of mathematics. Although some such connections have existed for a long time (e.g., between set theory and general topology and real analysis), many new ones have arisen lately. Methods from set theory have entered Banach space theory and several parts of algebra. Model theory is applied to algebraic and analytic geometry. Recursion theory plays an important role in computer science (and methods from other parts of logic have also been useful in my work in computer science). It is reasonable to expect that more such connections will arise. For example, more areas of mathematics will encounter set-theoretic independence results as they attack more difficult and less constructive problems. It seems that, at the moment, the second direction, reaching out to other fields, is more popular, but I expect both directions to continue to develop vigorously.

# BOERGER, Egon

**Specialties:** Logic and computer science.

**Educated:** Sorbonne, Institut Superieur de Philosophie; University of Louvain/Belgium; University of Muenster, Germany, Doctoral Degree Mathematics, 1971; University of Muenster, Habiliation in Mathematical Logic, 1976.

**Regular Academic or Research Appointments:** PROFESSOR OF LOGIC AND COMPUTER SCIENCE, UNIVERSITY OF PISA, ITALY, 1985–. Professor, University of Udine, Italy, 1982–1983; University of Dortmund, Germany, 1978-1985; University of Muenster, Germany, 1976–1978; University of Salerno, Italy, 1972–1976.

**Visiting Academic or Research Appointments:** SAP Researche, Karlsruhe, 2005; Chair of Software Engineering, CS Department of ETH Zurich, 2004; Microsoft Research Redmond, 2000; IRIN Nantes, 1998; Software Technology Dept., GMD FIRST Berlin, 1996; Siemens AG Corporate Research and Development Munich, 1996, 1999; DIMACS, Rutgers University, 1995; BRICS, University of Aarhus, 1995; IIG, University of Freiburg, 1994; CIS, University of Munich, 1994; Computer Science Department, University of Paderborn, 1993, 1995; EECS Department, University of Michigan at Ann Arbor, 1991; Researcher, IBM Scientific Center Heidelberg, 1989-1990.

**Research Profile:** A pioneer of applications of logical methods in theoretical and applied computer science. Co-founder of the Computer Science Logic (CSL) conference series and of the Abstract State Machines (ASM) workshop series. Editor of over a dozen books and organizer of over two dozens international conferences/workshops/schools in logic and computer science. Co-author of five books and over 100 research papers ranging from foundational and epistemological studies to the analysis of software/hardware systems.

**Main Publications:**

1. *Computability, Complexity, Logic* (German, English, Italian editions), Vieweg Verlag, North-Holland, Boringhieri 1985-1992.
2. *The Classical Decision Problem*, with E. Graedel and Y. Gurevich, Springer-Verlag 1997, 2001.
3. *Java and the Java Virtual Machine: Definition, Verification, Validation*, with R.Staerk and J.Schmid, Springer-Verlag, 2001.
4. *Abstract State Machines. A Method for High-Level System Design and Analysis*, with R.Staerk, Springer-Verlag, 2003.
5. *Fondamenti di Informatica*, with A. Maggiolo-Schettini, ETS 1988.
6. Modeling of the .NET CLR Exception Handling Mechanism for a Mathematical Analysis, with N.G.Fruja, *J. Object Technology*, 2006.
7. A Compositional Framework for Service Interaction Patterns and Interaction Flows, with M. Barros, Proc. ICFEM, 2005.
8. An Abstract Model for Process Mediation, with M. Altenhofen and J. Lemcke, *Proc. ICFEM*, 2005.
9. A High-Level Modular Definition of the Semantics of C♯, with G. Fruja, V. Gervasi, R. Staerk), *Theoretical Computer Science*, 2005.
10. Abstract State Machines: A Unifying View of Models of Computation and of System Design Frameworks, *APAL 133*, 2005.
11. The ASM Ground Model Method as a Foundation for Requirements Engineering, Springer, LNCS 2772, 2004.
12. The ASM Refinement Method, *Formal Aspects of Computing*, 15, 2003.
13. The Origins and the Development of the ASM Method for High Level System Design and Analysis., *J. Universal Computer Sci.*, 8, 2002.
14. Discrete Systems Modeling, *Encyclopedia of Physical Science and Technology*, 7, 2001.
15. Integrating ASMs into the Software Development Life Cycle, with L. Mearelli, *J. Universal Computer Science*, 3, 1997.
16. Specification and correctness proof of a WAM extension with abstract type constraints, with C. Beierle, *Formal Aspects of Computing*, 8, 1996.
17. Correctness of Compiling Occam to Transputer Code, with I. Durdanovic, *The Computer Journal*, 39, 1996.
18. A Mathematical Definition of Full Prolog, with D. Rosenzweig, *Science of Computer Programming*, 24, 1995.
19. Logical Decision Problems and Complexity of Logic Programs, with U. Loewen, *Fundamenta Informaticae*, X, (1987).
20. Decision problems in predicate logic, *Proc. Logic Colloquium*, 1982.
21. Prefix classes of Krom formulae with identity, with S. O. Aanderaa and Y. Gurevich, *Archiv Math. Logik u. Grundlagenforschung*, 22, 1982.
22. The equivalence of Horn and network complexity of Boolean functions, with S.O.Aanderaa, *Acta Informatica*, 15, 1981.
23. Conservative reduction classes of Krom formulas, with S. O. Aanderaa and H. R. Lewis, JSL 47, 1980.
24. The r. e. complexity of decision problems for commutative semi-Thue Systems with recursive rule set,

ZMLG, 26, 1980.

25. The reachability problem for Petri nets and decision problems for Skolem arithmetic, with H. Kleine Büning, TCS, 11, 1980.

26. Two new reduction classes in Krom formulae with predicate and function symbols, JSL, 42, 1977.

27. A new general approach to the theory of the many-one equivalence of decision problems for algorithmic systems, ZMLG, 25, 1979.

28. Beitrag zur Reduktion des Entscheidungsproblems auf Klassen von Hornformeln mit kurzen Alternationen, AMLG, 16, 1974.

For further details visit http://www.di.unipi.it/~boerger

*Work in Progress*

His current upcoming work concerns logical methods and their industrial applications for the development and maintenance (specification, design, mathematical verification and experimental validation) of hardware/software systems.

# BOYER, Robert S.

**Specialties:** Automated Reasoning and Logic

**Regular Academic or Research Appointments:** PROFESSOR, COMPUTER SCIENCES, MATHEMATICS, AND PHILOSOPHY DEPARTMENTS, UNIVERSITY OF TEXAS AT AUSTIN.

**Visiting Academic or Research Appointments:** Senior Computing Research Scientist, Computational Logic, Inc., 1993-1995; Skolem Lecture, University of Oslo, 1989; Senior Member Technical Statt, Microelectronics and Computer Technology Corporation, AI Program, 1985-1987; Staff Scientist 1981, Senior Research Mathematician 1979-1981, Research Mathematician 1973-1978, SRI International; IBM Chaire Internationale d'Informatique, University de Liege, Belgium, 1980 (32 lectures); Research Fellow, University of Edinburgh 1971-1973.

**Research Profile:** Developed (with J S. Moore) the now widely used Boyer-Moore theorem prover Nqthm, which is also the predecessor of the ACL2 theorem prover (developed by J S. Moore and M. Kaufmann); devised efficient implementation strategies that ultimately led to the widely used logic programming language Prolog; presented (with his colleagues) a set of cluases for Gödel's axioms, providing a foundation for the expression of most theorrems of mathematics in a form acceptable to automated theorem provers; formulated (with J S. Moore) a new exact pattern-matching algorithm with superior logic.

**Main Publications:**

Books

1. R. Boyer and J S. Moore, A Computational Logic, Academic Press, New York, 1979
2. R. Boyer and J S. Moore, A Computational Logic Handbook. Academic Press, New York, 1988; 2nd ed. 1998
3. 

Articles

4. *R. S. Boyer and J S. Moore, Proving theorems about LISP functions, J.Association for Computing Machinery, 22(1) 129-144, 1975
5. R. S. Boyer and J S. Moore, A fast string searching algorithm, Comm. Association for Computing Machinery, 20(10) 762-772, 1977
6. R. S. Boyer and J S. Moore, Proof checking the RSA public key encryption algorithm, American Math. Monthly, 91:181-189, 1984
7. R. Boyer, E. Lusk, W. McCune, R. Overbeek, M. Stickel, and L. Wos. Set theory in first-order logic: Clauses for Gödel's axioms. J. Automated Reasoning, 2(3):287-327, 1986.
8. R. S. Boyer and J S. Moore, The addition of bounded quantification and partial functions to a computational logic and its theorem prover, J. Automated Reasoning, 4(2):117-172, 1988
9. R. Boyer and J Moore, Integrating decision procedures into heuristic theorem provers: A case study of linear arithmetic, in Machine Intelligence 11, Oxford, Oxford University Press, 1988, pp. 83-124
10. H. Ait-Kaci, R. Boyer, P. Lincoln, and R. Nasr, Efficient Implementation of Lattice Operations, ACM Trans. Programming Languages and Systems, 11(1):115-146, 1989
11. *R. Boyer and J S. Moore, MJRTY — a fast majority vote algorithm, in Automated Reasoning: Essays in Honor of Woody Bledsoe, edited by R. Boyer, Kluwer Academic, 1991, pp. 105-117
12. *R. Boyer, M. Kaufmann, and J S. Moore, The Boyer-Moore theorem prover and its interactive environment, Computers and Mathematics with Applications, 5(2):27-62, 1995
13. Y. Yu and R. S. Boyer, Automated Proofs of Object Code for a Widely Used Microprocessor, J. Association for Computing Machinery, 43(1) 166-192, 1996
14. R. S. Boyer and J S. Moore, Mechanized formal reasoning about programs and computing machines, in Automated Reasoning and Its Applications: Essays in Honor of Larry Wos, edited by R. Veroff, MIT Press, 1997, pp. 147-176
15. R. S. Boyer and J S. Moore, Single-threaded objects in ACL2, in Practical Aspects of Declarative Languages 2002, edited by S. Krishnamurthi and C. R. Ramakrishnan, Lecture Notes in Computer Science 2257, Springer-Verlag, 2002, pp. 9-27
16. R. S. Boyer, ed., Automated Reasoning: Essays in Honor of Woody Bledsoe, Kluwer Academic Press, 1991
17. R. S. Boyer and J S. Moore, edds., The Correctness Problem in Computer Science, Academic Press, London, 1981

**Service to the Profession:** Member, Editorial

Board, Journal of Automated Reasoning 1989-2000; Kluwer Series in Automated Reasoning 1990-present; Journal of Logic and Computation 1991-1993; Journal of Symbolic Computation 1984-1987; Journal of Artificial Intelligence 1977-1994.

**Honours and Awards:** Honors include 6th Herbrand Award (with J S. Moore) for exceptional contributions to automated deduction 1999; Current Prize in Automatic Theorem Proving by the American Mathematical Society (with J S. Moore) 1991; Current Prize in Automatic Theorem Proving by the American Mathematical Society (with J S. Moore) 1991; Fellow of the American Association for Artificial Intelligence 1991; Best of Austin Award in category "best argument for tenure" 1996; John McCarthy Prize for Program Verification (with J S. Moore) 1983

# BRENDLE, Jörg

**Specialties:** Set theory.

**Born:** 6 October 1964 in Nürtingen, Baden-Württemberg, Germany.

**Educated:** Eberhard-Karls-Universität Tübingen, Habilitation Mathematics, 1995; Eberhard-Karls-Universität Tübingen, Dr. rer. nat., 1991; Eberhard-Karls-Universität Tübingen, Diplom Mathematics, 1989.

**Dissertation:** *Some contributions to combinatorial set theory and its applications*; supervisor, Ulrich Felgner.

**Regular Academic or Research Appointments:** ASSOCIATE PROFESSOR, GRADUDATE SCHOOL OF ENGINEERING, KOBE UNIVERSITY, 2007–; Associate Professor, Graduate School of Science and Technology, Kobe University, 1998–2007; John Wesley Young Research Instructor, Dartmouth College, 1995-1997; Assistant, Eberhard-Karls-Universität Tübingen, 1989.

**Visiting Academic or Research Appointments:** Visiting Professor, CRM Barcelona, 2004; Visiting Professor, Fields Institute, Toronto, 2002; DFG Fellow, Eberhard-Karls-Universität Tübingen, 1993-1995; MINERVA Fellow, Bar-Ilan University, Ramat-Gan Israel, 1991-1993.

**Research Profile:** Jörg Brendle has made contributions to set theory and has applied set-theoretic methods to obtain results in other areas of pure mathematics such as group theory and general topology. His work mainly focuses on the investigation of the combinatorial structure of the real numbers from the point of view of set theory and, in particular, on measure and category on the real line (e.g. item 4 below) and on cardinal invariants of the continuum (e.g. item 14). Since independence results figure prominently in this area, forcing theory plays a major role in his research, and he has worked extensively both on refining current forcing methods and on developing new iteration techniques for forcing. Important contributions in this area include his elegant treatment of Shelah's technique of iterations along templates (item 9) which he applied to show that the almost-disjointness number can consistently have countable cofinality (item 10), as well as his development of shattered iterations which are used to prove that covering numbers related to measure and category may simultaneously have arbitrary regular values (item 3 in work in progress). Further results in combinatorial set theory deal with maximal almost disjoint families (e.g. item 12), with ultrafilters on the natural numbers (items 5 and 6), and with distributivity properties of Boolean algebras (items 13 and 15). His research is closely related to descriptive set theory, and he has also worked on measurability properties in low levels of the projective hierarchy. Finally, he has contributed to applying methods of infinitary combinatorics outside of set theory proper; in particular, he has investigated combinatorial properties of the group of all permutations of the natural numbers (items 8 and 11).

**Main Publications:**

1. Set-theoretic aspects of periodic FC-groups – extraspecial p-groups and Kurepa trees, *Journal of Algebra* **162** (1993), 259-286.
2. Combinatorial properties of classical forcing notions, *Annals of Pure and Applied Logic* **73** (1995), 143-170.
3. Strolling through Paradise, *Fundamenta Mathematicae* **148** (1995), 1-25.
4. Nicely generated and chaotic ideals, *Proceedings of the American Mathematical Society* **124** (1996), 2533-2538.
5. (with S. Shelah) Ultrafilters on $\omega$ – their ideals and their cardinal characteristics, *Transactions of the American Mathematical Society* **351** (1999), 2643-2674.
6. Between P-points and nowhere dense ultrafilters, *Israel Journal of Mathematics* **113** (1999), 205-230.
7. How small can the set of generics be? in: *Logic Colloquium '98* (S. Buss et al., eds.), Lecture Notes in Logic **13**, A K Peters (2000), 109-126.
8. (with O. Spinas and Y. Zhang) Uniformity of the meager ideal and maximal cofinitary groups, *Journal of Algebra* **232** (2000), 209-225.
9. Mad families and iteration theory, in: *Logic and Algebra* (Y. Zhang, ed.), Contemporary Mathematics **302**, American Mathematical Society (2002), 1-31.
10. The almost disjointness number may have count-

able cofinality, *Transactions of the American Mathematical Society* **355** (2003), 2633-2649.

11. (with M. Losada) The cofinality of the infinite symmetric group and groupwise density, *The Journal of Symbolic Logic* **68** (2003), 1354-1361.

12. (with S. Yatabe) Forcing indestructibility of mad families, *Annals of Pure and Applied Logic* **132** (2005), 271-312.

13. Van Douwen's diagram for dense sets of rationals, *Annals of Pure and Applied Logic* **143** (2006), 54-69.

14. Cardinal invariants of the continuum and combinatorics on uncountable cardinals, *Annals of Pure and Applied Logic* **144** (2006), 43-72.

15. Independence for distributivity numbers, in: *Algebra, Logic, Set Theory. Festschrift für Ulrich Felgner zum 65. Geburtstag* (F. Haug et al., eds.), Studies in Logic 4, College Publications (2007), 63-83.

16. (with S. Fuchino) Coloring ordinals by reals. *Fundamenta Mathematicae* 196 (2007), 151–195.

*Work in Progress*

17. (with P. Larson and S. Todorcevic) Rectangular axioms, perfect set properties, and decomposition.

18. Shattered iterations.

19. Mad families on singular cardinals.

**Teaching:** Graduate education is a main focus in the *Group of Logic, Statistics & Informatics* at Kobe University, and Brendle is regularly teaching advanced graduate courses in set theory on topics such as recent forcing techniques, descriptive set theory, Ramsey theory and Banach space geometry, set-theoretic topology etc. His past and present PhD students are Shunsuke Yatabe, Teruyuki Yorioka, and Hiroaki Minami.

**Vision Statement:** (1) Since Cohen's invention of forcing in the sixties, set theory has undergone a tremendous internal development, leading to the creation of powerful techniques such as proper forcing. Recent years have seen a shift towards building links with other areas of mathematics like real and functional analysis. I believe that this interaction with other fields needs to be intensified.

(2) While forcing models in which the continuum has size at most $\aleph_2$ are well-understood, this is not so if the continuum is larger. I think that, on the technical side, one of the main challenges of set theory is to create iteration techniques which address this issue.

# BUCHHOLZ, Wilfried

**Specialties:** Proof theory.

**Born:** 1 April 1948 in Berlin, Germany.

**Educated:** LMU (Ludwig-Maximilians-Universität München) Dr.rer.nat. 1974.

**Dissertation:** *Rekursive Bezeichnungssysteme für Ordinalzahlen auf der Grundlage der Feferman-Aczelschen Normalfunktionen Theta_alpha*; supervisor, Kurt Schütte.

**Regular Academic or Research Appointments:** PROFESSOR, LMU, 1980-; Privatdozent, LMU, 1978-1980; Wissenschaftlicher Assistent, LMU, 1974-1978.

**Visiting Academic or Research Appointments:** Visiting Professor, Carnegie-Mellon-University, 1990-1991.

**Research Profile:** Wilfried Buchholz has made fundamental contributions to proof theory, especially ordinal analysis. He has introduced several new concepts and methods which since then have proved useful and influential in many respects. Among others these achievements are: simplified collapsing functions, the (capital) Omega-rule, distinguished sets (ausgezeichnete Mengen), operator controlled derivations, notation systems for infinitary derivations, a surprisingly close intrinsic connection between ordinal assignments in Schütte-style on one side and Gentzen-Takeuti-style on the other.

**Main Publications:**

1. Normalfunktionen und konstruktive Systeme von Ordinalzahlen. Proof theory symposium, Kiel 1974. Springer Lecture Notes in Mathematics 500 (1975), 4-25.

2. (with S. Feferman, W. Pohlers, and W. Sieg), Iterated Inductive Definitions and Subsystems of Analysis: Recent Proof Theoretical Studies. Springer Lecture Notes in Mathematics 897, Berlin-Heidelberg-New York 1981.

3. A new system of proof-theoretic ordinal functions. Annals of Pure and Applied Logic 32 (1986), 195-207.

4. An independence result for ($\Pi^1_1$-CA)+BI. Annals of Pure and Applied Logic 33 (1987), 131-155.

5. (with K. Schütte), Proof Theory of Impredicative Subsystems of Analysis. Studies in Proof Theory Monographs 2, Bibliopolis, Napoli 1988.

6. Notation systems for infinitary derivations. Arch. Math. Logic 30 (1991), 277-296.

7. A simplified version of local predicativity. In: Proof Theory, Leeds 1990. Aczel, Simmons, Wainer (eds.) Cambridge University Press (1992), 115-147.

8. Explaining Gentzen's Consistency Proof within Infinitary Proof Theory. In: Computational Logic and Proof Theory. 5th Kurt G"odel Colloquium, KGC'97. G. Gottlob, A. Leitsch, D. Mundici (eds.) Springer Lecture Notes in Computer Science, Vol.1289 (1997), 4-17.

9. Explaining the Gentzen-Takeuti reduction steps: a second order system. Arch. Math. Logic 40 (2001), 255-272.

**Service to the Profession:** Advisory Member, American Mathematical Society Committee on Translations, 1999-2002; Editor, Journal of Symbolic Logic, 1992-1998;Editor, Annals of Pure and Applied Logic, 1991-.

**Teaching:** Buchholz has had two PhD students, Anton Setzer and Klaus Aehlig.

## BUKOVSKÝ, Lev

**Specialties:** Mathematical Logic, set theory, topology, real analysis.

**Born:** 9 September 1939 in Podkriváò, Slovakia.

**Educated:** Institute of Mathematics of Czechoslovak Academy of Sciences in Prague, PhD Mathematics, 1966; Faculty of Sciences, Komenius University, Bratislava, Master Mathematics, 1961.

**Dissertation:** *Applications of Syntactic Models of Set Theory (slovak)*; (informal) supervisor, Petr Vopìnka. *Doctor of Sciences DrSc:* Charles University, Prague, 1983.

**Dissertation:** *Extensions of Models of Set Theory (slovak)*.

**Regular Academic or Research Appointments:** PROFESSOR, MATHEMATICS, FACULTY OF SIENCES, P. J. ǍAFÁRIK UNIVERSITY, KOŠICE UNIVERSITY, KOŠICE, 1984–; Associate Professor, Mathematics, P. J. Šafárik University, Košice 1968-1984, Assistant Professor, Mathematics, P. J. Šafárik University, Košice, 1965-1968.

**Visiting Academic or Research Appointments:** Visiting Professor, Mathematics, Faculté des Sciences, Paris VII, 1971; Visiting Lecturer, Mathematics, Leeds University, 1969.

**Research Profile:** Lev Bukovský has made contributions to cardinal arithmetic, models of set theory and applications of set theory into topology and real analysis. He is noted for the systematic work on the exponentiations of cardinals. Independently of K. Namba, he constructed important type of forcing notion (Boolean algebra), that became the main counterexample in R. Jensen's work on fine structure of the set theory. He contributed to general characterizations of generic extensions (extensions by forcing) of models of set theory. Bukovský also worked on properties of thin sets of harmonic analysis, solving some fundamental problems using methods of combinatorial set theory. Finally, he has contributed to the study of peculiar topological spaces using methods of set theory.

**Main Publications:**

1. The continuum problem and powers of alephs, *Comment. Math. Univ. Carolinae* **6** (1965), 181 – 197

2. Borel subsets of metric separable spaces, in: *General Topology and its Relations to Modern Analysis and Algebra II*, Editor J. Novák, Academia Praha 1967, 83 – 86

3. Consistency theorems connected with some combinatorial problems, *Comment. Math. Univ. Carolinae* **7** (1966), 495 – 499

4. Nabla-model and distributivity in Boolean algebras, *Comment. Math. Univ. Carolinae* **9** (968), 595 – 612

5. Ensembles générique d'entiers, *C. R. Acad. Sci. Paris Sér. I Math.* **273** (1971), 753 – 755

6. Characterization of generic extensions of models of set theory, *Fund. Math.* **83** (1973), 35 – 46

7. Changing cofinality of a measurable cardinal (An alternative proof), *Comment. Math. Univ. Carolinae* **14** (1973), 689 – 698

8. Changing cofinality of aleph_2, in: *Set Theory and Hierarchy Theory, A Memorial Tribute to Andrzej Mostowski*, Editors W. Marek, M. Srebrny and A. Zarach, Lecture Notes in Math. 537, Springer Berlin 1976, 37 – 49

9. Iterated ultrapower and Prikry's forcing, *Comment .Math. Univ. Carolinae* **18** (1977), 77 – 85

10. Random forcing, in: *Set Theory and Hierarchy Theory V*, Lecture Notes in Math. 619 Editors A. Lachlan, M. Srebrny and A. Zarach, Springer Berlin 1977, 101 – 117

11. Any Partition into Lebesgue Measure Zero Sets Produces a Non–measurable Set, *Bull. Acad. Polon. Sci. Sér. Sci. Math.* **27** (1979), 431 – 435.

12. The Structure of the Real Line (slovak), 228 pp.,Veda, Publishing House of Slovak Academy of Sciences, Bratislava 1979.

13. (with E. Coplákova—Hartová), Minimal Collapsing Extensions of Models of ZFC, *Annals Pure Appl. Logic* **46** (1990), 265 – 298

14. (with N. N. Kholshchevnikova and M. Repický), Thin sets of harmonic analysis an infinite combinatorics, *Real Anal. Exchange* **20** (1994-95), 454-509.

15. (with I. Reclaw I. and M. Repický), Spaces not distinguishing pointwise and quasinormal convergence of real functions, *Topology Appl.* **41** (1991), 25 – 40

16. (with Z. Bukovská), Adding small sets to an N-set, *Proc. Amer. Math. Soc.* **123** (1995), 3867 – 3873

17. (with I. Reclaw I. and M. Repický), Spaces not distinguishing convergences of real– valued functions, *Topology Appl.* **112** (2001), 13 – 40

18. Cardinality of Bases and Towers of Trigonometric Thin Sets, *Real Anal. Exchange* **29** (2003/04), 147-153.

19. (with J. Haleš), On Hurewicz Properties, *Topology Appl.* **132** (2003), 71-79.

20. (with J. Haleš), QN-spaces, wQN-spaces and covering properties, *Topology Appl.* **154** (2007), 848-858.

A full list of publication is available at Bukovský's home page http://ais.upjs.sk/~bukovsky
*Work in Progress*

21. A book *The Structure of the Real Line*, essentially a reworked English edition of 12.

**Service to the Profession:** Chair, Accreditation Committee of the Government of Slovak Republic, 1999-2002; President, Slovak Mathematical Society, 1996-2000; Rector, P. J. Šafárik University, Košice, 1991-1996; Chair,Department of Computer Science, P. J. Šafárik University, Košice, 1985-1991.

**Teaching:** L. Bukovský was intensively involved in teaching of both basic and facultative courses in mathematics and theoretical computer sciences at P. J. Šafárik University. He has had eight PhD students in Mathematics, including E. Butkovièová, E. Coplákova – Hartová, P. Eliaš, J. Haleš and M. Repický. He gave great effort to the education of secondary school teachers of mathematics. He was also involved in preparing contents of the competition of secondary school students Czechoslovak Mathematical Olympiad, including the international part, 1977-1983.

**Vision Statement:** In spite of the belief that the word around is essentially finite, mathematics, physics and some other natural sciences cannot exist without the concept of infinity. Consequences of a strict logical deduction are the only criteria for acceptance of the possible basic properties of introducing such an infinity. Anyway, the logic of a finity may differ from the logic of an infinity.

**Honours and Awards:** Fellow, Learned Society of the Slovak Academy of Sciences, 2006-; Fellow, Academia Scientiarum et Artium Europaea, Salzburg, 1996-; Jur Hronec Golden Medal of Slovak Academy of Science, 1994.

# BURGESS, John Patton

**Specialties:** Philosophical Logic, Foundations and Philosophy of Mathematics

**Born:** 5 June, 1948, Berea, Ohio, U.S.A.

**Educated:** Princeton University, A.B. in Mathematics 1969; Ohio State University, M.S. in Mathematics 1970; University of California at Berkeley, Ph.D. in Logic 1974.

**Dissertation:** *Infinitary Languages and Descriptive Set Theory*; supervisor, Jack Silver.

**Regular Academic or Research Appointments:** DIRECTOR OF UNDERGRADUATE STUDIES, DEPARTMENT OF PHILOSOPHY, PRINCETON UNIVERSITY, 1990– AND PROFESSOR, DEPARTMENT OF PHILOSOPHY, PRINCETON UNIVERSITY, 1986–; Associate Professor, Department of Philosophy, Princeton University, 1981–1986; Assistant Professor, Department of Philosophy, Princeton University, 1975–1986; Post-Doctoral Fellow, Department of Mathematics, University of Wisconsin at Madison, 1974–1975.

**Research Profile:** Burgess's work has gradually moved from predominantly mathematical towards predominantly philosophical logic, though philosophical interests were there from the beginning and there are still mathematical results in his latest work. His earliest work was on the connections of descriptive and combinatorial set with model theory for infinitary and generalized-quantifier logics, and on problems in descriptive set theory both pure (related to the continuum problem) and applied (measurable selection theorems). Perhaps his most cited work is in tense logic, where he was a major advocate of the step-by-step method of model construction, and worked on combinations of tense and modality (and the problem of determinism vs indeterminism), the tense logic of the real line, and the axiomatization of tense logics with since and until operators. He has also worked on other intensional logics (intuitionistic, conditional, relevance/relevant) and their philosophical motivations, topics in the theory of truth both technical (complexity computations) and philosophical (deflationism), technical and philosophical aspects of nominalism in philosophy of mathematics, and more recently on plural logic and its relation to set theory. His recently completed work in Frege studies again combines philosophical and technical features. He is currently at work on a book reexamining the formalism of modal sentential and predicate logic in the light of the distinction between logical and metaphysical modality.

**Main Publications:**

1. "Equivalence Relations Generated by Families of Borel Sets". *American Mathematical Society Proceedings*, 69 (1978): 323–326.
2. "A Reflection Phenomenon in Descriptive Set Theory". *Fundamenta Mathematicae*, 104 (1979): 127–139.
3. "A Selection Theorem for Group Actions". *Pacific Journal of Mathematics*, 80 (1979): 333-336.
4. "Decidability & Branching Time". In K. Segerberg, (ed.), *Trends in Modal Logic, Studia Logica*, 39 (1980): 203–218.
5. "Quick Completeness Proofs for Some Logics of Conditionals". *Notre Dame Journal of Formal Logic*, 22 (1981): 76–84.
6. "Relevance: A Fallacy?". *Notre Dame Journal of Formal Logic*, 22 (1974): 97–104.
7. "Axioms for Tense Logic, I. Since & Until", *Notre Dame Journal of Formal Logic* 23 (1982): 367–374.
8. "Classical Hierarchies from a Modern Standpoint, II. R-Sets", *Fundamenta Mathematicae* 115 (1983): 97–105.
9. "Dummett's Case for Intuitionism", *History & Philosophy of Logic* 5 (1973): 177–194.

10. with Y. Gurevich, "The Decision Problem for Linear Temporal Logic", *Notre Dame Journal of Formal Logic* 26 (1985): 115–128. Hebrew University Logic Year issue.
11. "The Truth Is Never Simple". *Journal of Symbolic Logic*, 51 (1986): 663–681.
12. *A Subject with No Object: Strategies for Nominalistic Reconstrual of Mathematics*, with G. Rosen. Oxford: Oxford University Press, 1999.
13. "*Quinus ab Omni Naevo Vindicatus*". In A.A. Kazmi (ed.), *Meaning and Reference: Canadian Journal of Philosophy Supplement*, 23 (1997) 25–65.
14. "Which Modal Logic Is the Right One?", *Notre Dame Journal of Formal Logic: Special Issue on George S. Boolos* 40 (1999): 81–93.
15. "E Pluribus Unum: Plural Logic and Set Theory". *Philosophia Mathematica*, 12 (2004): 193–221.
16. *Fixing Frege*. In *Princeton Monographs in Philosophy*. Princeton University Press, 2005.

**Service to the Profession:** Editorial Board, *Philosophia Mathematica*, 1992–; Association for Symbolic Logic, Committee on Awards and Prizes, 2000–04, Council, 1991–93, Executive Committee, 1998–90; Editorial Board, *Bulletin of Symbolic Logic*, 1998–2003; Editor for Logic, Stanford Encyclopedia of Philosophy, 1997–2003; Editorial Board, *Notre Dame Journal of Formal Logic*, 1989–95; *Journal of Symbolic Logic*, Editor for Surveys, 1988–93, Consulting Editor, 1982–88.

**Teaching:** Burgess's books and several of his papers have emerged directly out of his undergraduate and especially graduate teaching. His first doctoral student was Penelope Maddy, a founder of the Department of Logic and Philosophy of Science, University of California at Irvine. Other doctoral students have included John Barker (semantic paradoxes), Nick Smith (vagueness), Reina Hayaki (modal fictionalism), and Feng Ye (finitism).

**Vision Statement:** Burgess would like to see the development of university departments of logic, so that our discipline will have a home of its own, rather than remain a not always heartily welcome guest in mathematics, philosophy, or computer science departments. To achieve this goal, greater communication and cooperation among workers in different branches of logic will be crucial.

## BURRIS, Stanley Neal

**Specialties:** Universal algebra, decidability, complexity, logical limit laws, abstract number systems, combinatorial asymptotics.

**Born:** 20 February 1943 in Chelsea, Oklahoma, USA.

**Educated:** University of Oklahoma, PhD, 1968, MA, 1966, BS, 1963.

**Dissertation:** *Theory of PreClosures*; supervisor, Allen S. Davis.

**Regular Academic or Research Appointments:** PROFESSOR EMERITUS, UNIVERSITY OF WATERLOO, 1996–; Professor of Mathematics, University of Waterloo, 1978; Associate Professor of Mathematics, University of Waterloo, 1974; Assistant Professor of Mathematics, University of Waterloo, 1968.

**Visiting Academic or Research Appointments:** Visiting Professor of Mathematics, Department of Mathematics and Computer Science, Lanzhou University, Lanzhou, China, 1997; Visiting Professor of Mathematics for the Special Year in Logic and Universal Algebra, University of Illinois at Chicago, 1991-1992; DAAD and Alexander von Humboldt Fellow, Technische Hochschule Darmstadt, 1975-1976; Research Associate to Alfred Tarski, Dept. of Logic and Methodology of Science, University of California at Berkeley, 1971.

**Research Profile:** Stanley Burris was one of the early members of a group of young mathematicians who greatly expanded the subject of Universal Algebra under the leadership of Alfred Tarski and Garrett Birkhoff. The primary goal of Burris was always to give a subject a solid, simple and elegant foundation. With these goals he worked on structure and decidability for equational classes (varieties) of algebras, complexity for algebraic structures, generalized number systems and logical limit laws, combinatorial asymptotics and the history of mathematics (especially logic in the $18^{th}$ century). Two of his favorite achievements were: (a) working with Heinrich Werner to create the Boolean product construction (crystallized out of sheaf theory), a basic component of universal algebra and a powerful tool for analyzing discriminator varieties; and (b) the subsequent work with Ralph McKenzie that made a substantial first step into the classification of congruence modular varieties with decidable first-order theories.

**Main Publications:**

with R. McKenzie *Decidability and Boolean Representations*. Memoirs A.M.S. No. 246, July 1981. [MR 83j : 03024]

with H.P. Sankappanavar *A Course in Universal Algebra*. Graduate Texts in Mathematics No. 78, Springer-Verlag, 1981. [MR 83k : 08001]
* (The Hungarian translation appeared in 1988)
* (The Chinese edition appeared in 1988)
* The On-Line Millenium Edition (free at www.thoralf.uwaterloo.ca)

1. textit. *Logic for Mathematics and Computer Science.* Prentice-Hall, 1998. ISBN 0-13-285974-2
2. . *Number Theoretic Density and Logical Limit Laws.* Amer. Math. Soc. 2001 ISBN 0-8218-26666-2
  with Jason P. Bell *Asymptotics for Logical Limit Laws. When the Growth of the Components is in an RT Class.* Trans. Amer. Math. Soc. **355** (2003), 3777–3794.
  with Jason P. Bell *Partition Identities I. Sandwich Theorems and 0-1 Laws* Electron. J. Combin. **11** (2004), no. 1, Research Paper 49, 25 pp. (electronic). (2004)
*Work in Progress*
  with Jason P. Bell and Karen A. Yeats *Counting Rooted Trees: The Universal Law $t(n) \sim C \rho^{-n} n^{-3/2}$*.
Except for highly specialized families of trees there are few results on the asymptotics for finitely axiomatizable monadic second order classes of structures. A primary goal at present is to work on general results in this area.

**Service to the Profession:** Editor for Algebra Universalis, 1980–1996.

**Teaching:** Four PhD students including H.P. Sankappanavar and Ross Willard. Undergraduate research supervisor for several students including Karen Yeats who placed second in the 2004 Moore competition for outstanding undergraduate researcher in North America.

**Vision Statement:** Logic has for several decades been moving towards becoming more connected with mainline mathematical questions, such as the possibility of resolving unsolved problems with current techniques, the complexity of various algorithms, and the quest for better or alternate proofs, most recently in geometry. This is a direction that I expect to continue.

# BUSS, Samuel R.

**Specialties:** Proof theory, complexity theory, algorithms and circuits, numerical methods.

**Born:** 6 August 1957 in New Haven, Connecticut, USA.

**Educated:** Emory University, BA Mathematics and BA Physics, 1979; Princeton University, MA and Ph.D, Mathematics, 1975.

**Dissertation:** *Bounded Arithmetic*; supervisor, Simon Kochen.

**Regular Academic or Research Appointments:** PROFESSOR, MATHEMATICS AND COMPUTER SCIENCE ENGINEERING, UNIVERSITY OF CALIFORNIA, SAN DIEGO, 1988–; Instructor, Mathematics, Berkeley, 1986-1988; Member, Mathematical Sciences Research Institute (MSRI), 1985-1986.

**Visiting Academic or Research Appointments:** See previous entry.

**Research Profile:** Buss's work in logic has been on bounded arithmetic, proof complexity, algorithms and circuits.

**Main Publications:**

1. *Bounded Arithmetic,* Ph.D. thesis, Department of Mathematics, Princeton University, 1985. 188+v pages
2. *A conservation result concerning bounded theories and the collection axiom.* Proceedings of the American Mathematic Society 100 (1987) 709-716.
3. (with Gyorgy Turan). *Resolution proofs of generalized pigeonhole principles.* Theoretical Computer Science 62 (1988) 311-317.
4. *Axiomatizations and conservation results for fragments of bounded arithmetic.* In *Logic and Computation, Proceedings of a Workshop held at Carnegie Mellon University, June 30-July 2, 1987.* AMS Contemporary Mathematics 106 (1990) 57-84.
5. *The Boolean formula value problem is in ALOGTIME.* In *Proceedings of the 19th Annual ACM Symposium on Theory of Computing (STOC'87),* 1987, pp. 123-131.
6. *Polynomial size proofs of the propositional pigeonhole principle.* Journal of Symbolic Logic 52 (1987) 916-927.
7. *Propositional consistency proofs.* Annals of Pure and Applied Logic 52 (1991) 3-29.
8. *The undecidability of k-provability.* Annals of Pure and Applied Logic 53 (1991) 75-102
9. (with Louise Hay). *On truth-table reducibility to SAT.* Information and Computation 91 (1991) 86-102.
10. *Intuitionistic validity in T-normal Kripke structures.* Annals of Pure and Applied Logic 59 (1993) 159-173.
11. *The witness function method and provably recursive functions of Peano arithmetic.* In *Proceedings of Ninth International Congress on Logic, Methodology and Philosophy of Science,* D. Prawitz, B. Skyrms and D. Westerstahl (eds), Elsevier Science North-Holland, 1994, pp. 29-68.
12. *Algorithms for Boolean formula evaluation and for tree-contraction.* In *Proof Theory, Complexity, and Arithmetic,* P. Clote and J. Krajicek (eds), Oxford University Press, 1993, pp. 95-115.
13. (with Jan Krajicek) *An application of Boolean complexity to separation problems in bounded arithmetic.* Proceedings of the London Mathematical Society 69 (1994) 1-21.
14. (with Jan Krajicek and Gaisi Takeuti). *Provably total functions in the bounded arithmetic theories $R_3^i$, $U_2^i$, and $V_2^i$.* In *Proof Theory, Arithmetic, and Complexity,* P. Clote and J. Krajicek (eds), Oxford University Press, 1993, pp. 116-161.
15. (with Aleksandar Ignjatovic) *Unprovability of Consistency Statements in Fragments of Bounded Arithmetic,* Annals of Pure and Applied Logic 74 (1995) 221-244.

16. *Alogtime algorithms for tree isomorphism, comparison and canonization*, In Computational Logic and Proof Theory, Proceedings 5th Gödel Colloquium '97. Lecture Notes in Computer Science #1289, Springer-Verlag, Berlin, 1997, pp. 18-33.

17. *Handbook of Proof Theory*, (S. Buss editor. Also, author of two articles.) Studies in Logic and the Foundations of Mathematics #137, Elsevier, Amsterdam, 1998, x+811 pages.

18. (with Pavel Pudlák). *On the computational content of intuitionistic propositional proofs*, Annals of Pure and Applied Logic 109 (2001) 9-14.

19. *Accurate and efficient simulations of rigid body rotations*. Journal of Computational Physics 164 (2000) 377-406.

20. *Nelson's Work on Logic and Foundations and Other Reflections on Foundations of Mathematics* To appear in *Diffusion, Quantum Theory and Radically Elementary Mathematics*, edited by W. Faris, Princeton University Press, to appear.

See `http://www.math.ucsd.edu/~sbuss/ResearchWeb/index.html` for a complete list of Buss's publications.

**Service to the Profession:** Vice-chair, Publisher, ASL; Editor, ACM Transactions on Logic; Editor, Journal of Applied Logic; Associate Editor, Annals of Pure and Applied Logic; Editor, Journal of Mathematical Logic; Advisory Editor, Logical Methods of Computer Science; Editor, Archive for Mathematical Logic; Managing Editor, ASL Lecture Notes in Logic; Undergraduate Vice-Chair, Math; Former Graduate, Vice-Chair, Math.

**Teaching:** PhD students: Nate Segerlind (winner Ackermann Award and Sacks Prize, co-advisor with Russell Impagliazzo), Marie Luisa Bonet, Steve Bloch, David Robinson.

**Vision Statement:** My interests have always been motivated by computational concerns. Logic is interesting in large part because of the way it combines syntactic expressiveness, semantic content, and computational methods.

**Honours and Awards:** NSF Postdoctoral Fellowship.

# BUSZKOWSKI, Wojciech

**Specialties:** Mathematical logic, computational logic, logical methods in linguistics.

**Born:** 17 October 1950 in Poznan, Poland.

**Educated:** State Title of Professor, 1997; Adam Mickiewicz University in Poznan, Habilitation Mathematics 1988; Adam Mickiewicz University in Poznan, PhD Mathematics, 1982; Adam Mickiewicz University in Poznan, MSc Mathematics, 1973.

**Dissertation:** *Studies on the Completeness and Logical Complexity of Strong Calculi of Syntactic Types*; supervisor, Tadeusz Batog.

**Regular Academic or Research Appointments:** FULL PROFESSOR OF MATHEMATICS, FACULTY OF MATHEMATICS AND COMPUTER SCIENCE, ADAM MICKIEWICZ UNIVERSITY IN POZNAN, 2000–; FULL PROFESSOR OF MATHEAMTICS, FACULTY OF MATHEAMTICS AND COMPUTER SCIENCE, UNIVERSITY OF WARMIA AND MAZURY IN OLSZTYN, 2000–; HEAD OF THE DEPARTMENT OF COMPUTATION THEORY, ADAM MICKIEWICZ UNIVERISTY, 1993–; CHAIR OF LOGIC AND COMPUTATION, UNIVERSITY OF WARMIA AND MAZURY, 2000–; Professor of Mathematics, Adam Mickiewicz University, 1991-1999; Associate Professor of Mathematics, Adam Mickiewicz University, 1989-1991; Adiunkt, Adam Mickiewicz University, 1982-1988; Assistant, Adam Mickiewicz University, 1973-1981; Adiunkt, The Mathematical Institute of Polish Academy of Sciences, 1977-1980.

**Visiting Academic or Research Appointments:** Visiting Researcher, Rovira i Virgili University, Tarragona, 2007; ECO-COST Fellowship, University of Amsterdam, 1993; Visiting Professor, University of Munich, 1991; Alexander von Humboldt Fellowship, University of Saarbruecken, 1989-1990; Visiting Professor, University of Chicago, Department of Computer Science, 1989.

**Research Profile:** Wojciech Buszkowski is a logician working on the borderline of logic, formal linguistics and computation theory. He is mainly known as a specialist in type logics (the Lambek calculus and its variants) and categorial grammars (formal grammars based on type logics), but he also works in knowledge representation theory, substructural logics, modal and tense logics, the logic of questions, natural logic, automata theory and foundations of mathematics.

In the late 1970s he started a systematic research in proof-theoretic and model-theoretic aspects of the Lambek calculus and generative capacity of grammars based on different variants of this calculus and weaker systems (classical categorial grammars, based on the reduction procedure of Ajdukiewicz and Bar-Hillel). He studied different axiomatizations of these systems, their algebraic models (residuated semigroups and monoids and special classes, e.g. powerset frames over semigroups), finite models, complexity and cut-elimination, connections with lambda calculus, compatibility of grammars and models, strong and weak generative capacity of categorial grammars as compared with Chomsky grammars and tree automata. Some concepts and methods used in this research anticipate certain methods of substructural logics and linear logics. They are closely related to studies on logical semantics of natural language (J. van Benthem, M. Moortgat) and labeled deductive systems (D. Gabbay).

In the late 1980s he proposed a unification-based learning algorithm for classical categorial grammars (the last version elaborated jointly with Gerald Penn), which was further studied from the perspective of Gold-style learning theory by M. Kanazawa and other authors. Since 1990-ties, he investigated fine aspects of type logics and their connections with linear and dynamic logics. In particular, he studied logics corresponding to Kleene and action algebras and fragments of multiplicative-additive linear logic.

Among other topics, he elaborated a relational calculus of data dependencies (with Ewa Orlowska), worked on rough sets and communication logic (following some ideas of Zdzislaw Pawlak), modal logics and algebras, and Ackermann-style systems of set theory.

**Main Publications:**

1. Logical complexity of some classes of tree languages generated by multiple tree automata, *Zeitschrift f. mathematische Logik und Grundlagen d. Mathematik* 26 (1980), 41-49.
2. Some decision problems in the theory of syntactic categories, *Zeitschrift f. mathematische Logik und Grundlagen d. Mathematik* 28 (1982), 539-548.
3. Concerning the axioms of Ackermann's set theory, *Zeitschrift f. mathematische Logik und Grundlagen d. Mathematik* 31 (1985), 63-70.
4. The equivalence of unidirectional Lambek categorial grammars and context-free grammars, *Zeitschrift f. mathematische Logik und Grundlagen d. Mathematik* 31 (1985), 369-384.
5. Generative capacity of nonassociative Lambek calculus, *Bull. Polish Academy of Sciences Math.* 34 (1986), 507-516.
6. Completeness results for Lambek Syntactic Calculus, *Zeitschrift f. mathematische Logik und Grundlagen d. Mathematik* 32 (1986), 13-28.
7. Generative power of categorial grammars, in: *Categorial Grammars and Natural Language Structures*, (R.T. Oehrle, E. Bach and D. Wheeler, eds.), D. Reidel, Dordrecht, 1988, 69-94.
8. Presuppositional completeness, *Studia Logica* 48 (1989), 24-34.
9. (with G. Penn), Categorial grammars determined from linguistic data by unification, *Studia Logica* 49 (1990), 431-454.
10. The finite model property for BCI and related systems, *Studia Logica* 57 (1996), 303-323.
11. Mathematical Linguistics and Proof Theory, in: *Handbook of Logic and Language*, (J. van Benthem and A. ter Meulen, eds.), Elsevier, Amsterdam, The MIT Press, Cambridge Mass., 1997, 683-736.
12. (with E. Orlowska), Indiscernibility based formalization of dependencies in information systems, in: *Incomplete Information: Rough Set Analysis*, (E. Orlowska, ed.), Springer, 1998, 293-315.
13. Algebraic structures in categorial grammars, *Theoretical Computer Science* 199 (1998), 5-24.
14. Lambek grammars based on pregroups, in: *Logical Aspects of Computational Linguistics*, (P. de Groote, G. Morrill and C. Retore, eds.), LNAI 2099, Springer, 2001, 95-109.
15. Finite models of some substructural logics, *Mathematical Logic Quarterly* 48 (2002), 63-72.
16. Sequent systems for compact bilinear logic, *Mathematical Logic Quarterly* 49 (2003), 467-474.
17. Relational models of Lambek logics, in: *Theory and Application of Relational Structures as Knowledge Instruments*, (H.C.M. de Swart, E. Orlowska, G. Schmidt and M. Roubens, eds.), LNCS 2929, Springer, 2003, 196-213.
18. A representation theorem for co-diagonalizable algebras, *Reports on Mathematical Logic* 38 (2004), 13-22.
19. Lambek calculus with nonlogical axioms, in: *Language and Grammar: Studies in Mathematical Linguistics and Natural Language*, (C. Casadio, P. Scott and R. Seely, eds.), CSLI Lecture Notes 168, Stanford, 2005, 77-93.
20. On action logic: equational theories of action algebras, *Journal of Logic and Computation* 17.1 (2007), 199–217.

*Work in Progress*

21. Systems of natural logic based on type hierarchies of Boolean algebras.
22. Complexity of fragments of the Lambek calculus and linear logics.
23. Research on action logic.
24. Learning systems based on logical tools.

**Service to the Profession:** Managing Editor, Studia Logica, 2001-; Editoral Board Member, Journal of Applied Logic, 2003-; Editorial Board Member, Journal of Applied Non-Classical Logics, 1990–2000; Member of Editorial Board , Grammars, 1998-; Editorial Board Member, Papers in Formal Linguistics and Logic, 1998-; Editor, *Categorial Grammar* (with J. van Benthem and W. Marciszewski), 1998; Editor of special issues of Studia Logica, *The Lambek calculus in logic and linguistics* (with M. Moortgat), 2002, Categorial grammars and pregroups (with A. Preller) 2007; President, Polish Association of Logic and Philosophy of Science, 1993-1996; Past-President of this Association, 1996-1999.

**Teaching:** Courses and classes on Mathematical Logic, Computation Theory, Foundations of Mathematics, Discrete Mathematics, Logic of Language; Promotor of PhD Theses: Jacek Marciniec, Miroslawa Kolowska-Gawiejnowicz, Marek Szczerba, Barbara Dziemidowicz, Aleksandra Kislak-Malinowska, Ewa Palka, Maciej Farulewski. Member of Promotion Committee of: E. Aarts, K. Versmissen (University of Utrecht), T. van der Wouden (University of Groningen); Co-Promotor of C. Costa Florencio (University of Utrecht); Referee of Habilitation of: Roman Murawski (Poznan), Maciej Wygralak (Poznan), Marcin Mostowski (Warsaw), Marek Zaionc (Cracow); Referee of Professorship of Maria Semeniuk-Polkowska (Warsaw), Witold Lukaszewicz (Warsaw, Olsztyn).

**Vision Statement:** Mathematical logic should not be restricted to foundations of mathematics but it could be viewed as a mathematical discipline, studying formal languages, deductive systems and their models with broad applications in mathematics, computer science, linguistics, natural sciences and humanities.

**Honours and Awards:** Medal of National Education from The State President of Poland, 2005; Gold Cross of Achievement from The State President of Poland, 2000; Price of The Ministry of Education in Poland, 1986, 1990; Fellow, Alexander von Humboldt Fellowship, 1989-1990.

# C

## CARLSON, Timothy John

**Specialties:** Mathematical logic, set theory, proof theory.

**Born:** 31 October 1951 in Lafayette, Indiana, USA.

**Educated:** University of Minnesota, PhD Mathematics, 1978; University of Minnesota, BS Mathematics, 1973.

**Dissertation:** *Some Results on Finitely Additive Measures and Infinitary Combinatorics*; supervisor: Karel Prikry.

**Regular Academic or Research Appointments:** PROFESSOR, MATHEMATICS, OHIO STATE UNIVERSITY, 2005–; Associate Professor, Mathematics, Ohio State University, 1986-2005; Assistant Professor, Mathematics, Ohio State University, 1983-1986.

**Visiting Academic or Research Appointments:** Visiting Assistant Professor, Mathematics, Ohio State University, 1982-1983; Lecturer, University of California, Berkeley, 1980-1982; Assistant Professor, University of Colorado, Boulder, 1978-1980.

**Research Profile:** Carlson has made significant contributions to several areas of mathematical logic and related areas including set theory, proof theory and combinatorics. Carlson's early work was primarily in set theory, focusing on infinitary combinatorics and the structure of the reals. In particular, he is known for establishing the consistency of the Pin-up Conjecture and for his study of special sets of reals. Carlson's work in Ramsey theory, both individual and with Stephen Simpson, was highly influential in the growth of the area. He discovered combinatorial principles which unified many earlier results, both in the finite and countably infinite domains. Carlson's work in proof theory falls into two areas: modal logics and ordinal-analysis. He established one of the fundamental results in the study of bimodal provability logics and invented a new technique for establishing consistency results in first-order modal logics which lead to his proof of the consistency of Reinhardt's Strong Mechanistic Thesis. His work in ordinal-analysis is centered on a method for constructing ordinal notations based on patterns of embeddings in conjunction with logics for the study of well-founded structures.

**Main Publications:**
1. "The pin-up conjecture", in *Axiomatic Set Theory* (eds. J. Baumgartner, D.Martin, S.Shelah), Contemporary Mathematics 31 (1984), pp. 41-62.
2. "A dual form of Ramsey's theorem", (with S. Simpson) Advances in Mathematics 53 (1984), pp. 265-290.
3. "Modal logics with several operators and provability interpretations", Israel Journal of Mathematics 54 (1986), pp. 14-24.
4. "An infinitary version of the Graham-Leeb-Rothschild theorem", Journal of Combinatorial Theory (Series A) 44 (1987), pp. 22-33.
5. "Some unifying principles in Ramsey theory", Discrete Mathematics 68 (1988), pp. 117-169.
6. "Strong measure zero and strongly meager sets of reals", Proceedings of the AMS 118 (1993), pp. 577-586.
7. "Ordinal arithmetic and $\Sigma_{1}$-elementarity", Archive for Mathematical Logic 38 (1999), pp. 449-460.
8. "Knowledge, machines, and the consistency of Reinhardt's Strong Mechanistic Thesis", Annals of Pure and Applied Logic 105 (2000), 51-82.
9. "Elementary patterns of resemblance", Annals of Pure and Applied Logic 108 (2001), pp. 19-77.
10. "The Graham-Rothschild theorem and the algebra of $\beta W$", (with N. Hindman and D. Strauss), Topology Proceedings 28 (2004), pp. 361-399.

*Work in Progress*
11. Logics for the theory of well-founded structures.
12. Ordinal-analysis based on patterns of embeddings.
13. Combinatorial independence.
14. Modal theories of computation, uncertainty and knowledge.

**Service to the Profession:** Program Committee Member, 2007 Winter Meeting of the Association for Symbolic Logic (in conjunction with the Joint Mathematical Meetings), New Orleans, 2007; Organizing Committee Member, panel discussion *Contemporary Perspectives on Hilbert's Second Problem and the Gödel Incompleteness Theorems*, Joint Mathematical Meetings, New Orleans, 2007; Program Committee Member, Special Session in Proof Theory, Logic Colloquium 2004, Turin, Italy, 2004; Program Committee Member, 2003 Annual Meeting of the Association for Symbolic Logic, University of Illinois, Chicago, 2003; Organizer, AMS Special Session "Proof Theory and the Foundations of Mathematics", Meeting of the American Mathematical Society, Ohio State University, 2001; Program Committee Member, 1998 Annual Meeting of the Association for Symbolic Logic, University of Toronto, 1998.

**Teaching:** Carlson had one official PhD student: Gregory Bishop, PhD 1992 from Ohio State University. He had a substantial advising role in the PhD work of Gunnar Wilken, PhD, 2004, University of Muenster, under Wolfram Pohlers.

**Vision Statement:** Logic and the techniques it has fostered will continue to find applications in a large number of diverse areas. The outstanding problem in logic per se is the extent to which Goedel's Incompleteness Theorems have practical importance. A greater understanding of this issue will be gained as the areas of proof theory and set theory merge.

**Honours and Awards:** Alfred P. Sloan Research Fellowship, 1984-1986.

# CARNIELLI, Walter

**Specialties:** Logic and foundations of mathematics, philosophical logic, non-classical logics.

**Born:** 11 January 1952 in Campinas, SP, Brazil.

**Educated:** State University of Campinas, PhD Mathematics, 1982, MSc Mathematics, 1978.

**Dissertation:** *Systematization of the finite many-valued logics through the method of tableaux*; supervisor, Newton C. A. da Costa.

**Regular Academic or Research Appointments:** FULL PROFESSOR OF LOGIC, DEPARTMENT OF PHILOSOPHY, STATE UNIVERSITY OF CAMPINAS, BRAZIL, 1994–; Director of the Centre for Logic, Epistemology and the History of Science, 1998-2004; Full Member of the Security and Quantum Information Group, Institute of Telecommunications, Lisbon, Portugal, 2006–.

**Visiting Academic or Research Appointments:** Department of Computer Science and Communications, University of Luxembourg, 2008; Instituto Superior Técnico, Lisbon, Portugal, 2004; Universitat de Barcelona, Catalonia, Spain, 1999; Universitat de Lleida, Catalonia, Spain, 1999; Dipartimento di Filosofia e Scienze Sociali, Universitá degli Studi di Siena, Italy, 1996; Universidad Nacional de Colombia, Bogotá, Colombia, 1994; Institut de Recherche en Informatique de Toulouse, Univ. Paul Sabatier, Toulouse, France, 1992; Univ. de Paris -Sud, Orsay, France, 1992; Istituto di Informatica, Università degli Studi di Torino, Italy, 1988; Inst. für mathematische Logik und Grundlagenforschung, Universität Münster,Germany, 1988-1989; Instituto Venezolano de Investigaciones Científicas, Dept. of Mathematics, Caracas, Venezuela, 1987; Institute of Mathematics, University of California, Berkeley, USA, 1984.

**Research Profile:** Carnielli made substantial contributions to the proof theory and semantics for contemporary heterodox (non-classical) logics. Of special significance are his contributions to many-valued logics, paraconsistent logics and combination of logics. Concerning many-valued logics, he provided a generic tableau proof system for multiple-value first-order logic with arbitrary connectives and generalized quantifiers. With his students and collaborators Carnielli introduced the *possible-translations semantics*, which led to a revival in the semantic interpretation of paraconsistent logics, and the concept *of logics of formal inconsistency* which systematize a great number of extant paraconsistent logics, opening the way to applications of paraconsistency to computer science and to philosophical investigations around the topic. Carnielli has also worked on finite and infinite combinatorics, and shaped, with collaborators, the concept of *modulated logics*, a wide class of logics dedicated to formalize quantified uncertain reasoning, with philosophical implications. Current work by Carnielli also includes the role of non-classical logics in the theory of classical computation, with consequences for quantum computation.

**Main Publications:**

1. *On coloring and covering problems for rook domains*, Discrete Mathematics 57 (1985), pp. 9-16.
2. *Systematization of the finite many-valued logics through the method of tableaux*. The Journal of Symbolic Logic 52 (2), 1987, pp. 73-493.
3. *Paraconsistent deontic logics*. (with Newton C. A. da Costa) Philosophia – The Philos. Quarterly of Israel vol.16 numbers 3 and 4 (1988), pp. 293-305.
4. *Methods and applications of mathematical logic-Proceedings of the VII Latin-American Symp. on Math. Logic*. (Editor, with Luiz. P. de Alcantara). Contemporary Mathematics 69, American Mathematical Society, 1988.
5. *Hyper-rook domain inequalities*. Studies in Applied Mathematics (Massachusetts Institute of Technology) 82, n.1 (1990), pp. 59-69.
6. *Some results on polarized partition relations of higher dimension*. (with Carlos A. Di Prisco). Mathematical Logic Quarterly 39 (1993) pp. 461-474.
7. *Ultrafilter logic and generic reasoning*. (With Pualo A.S. Veloso). In Computational Logic and Proof Theory (Vienna, 1997), pp. 34-53, Lecture Notes in Computer. Science 1289, Springer, Berlin, 1997.
8. *Advances in Contemporary Logic and Computer Science*. (Editor, with Marcelo E. Coniglio and Itala M.L. D'Ottaviano). American Mathematical Society, Series Contemporary Mathematics, Volume 235, 1999.

9. *Possible-translations semantics for paraconsistent logics*. In: Frontiers in paraconsistent logic: Proceedings of the I World Congress on Paraconsistency, Ghent, 1998, pp. 159-72 , edited by D. Batens et al., Kings College Publications, 2000.

10. *K2,2-K1,n and K2,n-K2,n bipartite Ramsey numbers*. (With Emerson L. Monte Carmelo). Discrete Mathematics, Vol. 223 (1-3), 2000, pp. 83-92.

11. *Modulated fibring and the collapsing problem*. (With Cristina . Sernadas and João Rasga). The Journal of Symbolic Logic . 67(4) 2002 pp. 1541-1569.

12. *Computability: computable functions, logic and the foundations of mathematics, with the timeline Computability and Undecidability*. (With Ricahrd L. Epstin). Second edition. Wadsworth/Thomson Learning,Belmont, CA, 2000.

13. *Modalità e multimodalità*. (With Claudio Pizzi. Franco Angeli, Milan, 2001 (In Italian).

14. *Paraconsistency: The Logical Way to the Inconsistent*. Proceedings of the II World Congress on Paraconsistency (Editor, with Marcelo E. Coniglio and Itala M.L. D'Ottaviano ) Marcel Dekker Inc., New York, 2002.

15. *A taxonomy of C- systems* (With João Marcos). In: Paraconsistency- the Logical Way to the Inconsistent, Lecture Notes in Pure and Applied Mathematics, Vol. 228, pp. 01-94 2002.

16. *Two's company: The humbug of many logical values*. (With Carlos Caleiro Marcelo E. Coniglio and João Marcos). In: Logica Universalis (Editor Jean-Yves Béziau) Basel: Birkhäuser, 2005, v., p. 169-189.

17. *Splitting Logics*. (With Marcelo E. Coniglio). In: We Will Show Them: Essays in Honour of Dov Gabbay. (Editors S. Artemov, H. Barringer, A. S. Avila Garcez, L. C. Lamb and J. Woods). Londres: King's College Publications, 2005, v. 1, p. 389-414.

18. *Logics of Formal Inconsistency*. In Handbook of Philosophical Logic, vol. 14, pp. 15–107. Eds.: D. gabbay and F. Guenthner. Springer, 2007 (with Marcelo E. Coniglio and João Marcos).

19. *Analysis and Synthesis of Logics: How to Cut and Paste Reasoning Systems*. Volume 35 in the Applied Logic Series, Springer, 2008. ISBN 978-1-4020-6781-5 (with Marcelo E. Coniglio, Dov Gabbay, Paula Gouveia and Cristina Sernadas).

*Work in Progress*

20. *Modalities and Multimodalities* (with Claudio Pizzi) to appear by Springer-Verlag.

For further details visit http://www.cle.unicamp.br/prof/carnielli

**Service to the Profession:** Editor or member of editorial boards of major journals: Reviews Editor, Editorial Board, *Logic and Logical Philosophy*; Editorial Board, *Journal of Applied Logic*; Editor, *CLE e-Prints*; Editorial Board, *Reports on Mathematical Logic*; Editorial Board, *Studia Logica*; Associate Editor, *Journal of Applied Non-Classical Logic*; Reviewer, *Mathematical Reviews*, 1985–.

**Teaching:** Carnielli's teaching focuses primarily on computability, non-classical lgoics, logic and reasoning, and combinatorial aspects of logic. A recognised lecturer, his (with R. L. Epstein) book *Computability: Computable Functions, Logic and the Foundations of Mathematics with the timeline Computability and Undecidability* Second eidtion, Wadsworth/Thomson Learning, Belmont CA, 2000) is currently adopted by more than 40 universities in 15 countries, and is a part of the syllabus of several courses in Computer Science, Philosophy, Linguistics and Mathematics. The Brazilian version of this book was, in 2007, awarded the "Jabuti Prize", the most prestigious literary prize in Brazil.

**Vision Statement:** I would like to see accomplished a return to logic as whole, independent of specializations driven by mere method or subject matter, as "epistemic logic", "linear logic", "many-valued logics", "non-monotonic logics" or even "non-classical logics". Logicians should be able to identify in Logic (not necessarily by informal methods) parameters as assumptions, postulates, reasoning, argumentation, subject, context, etc, and build a bigger theory of which specializations would be just particular cases.

**Honours and Awards:** Alexander von Humboldt Fellow, Germany; Honorary Member, Deutsche Vereinigung für Mathematische Logik und für Grundlagen der Exakten Wissenschaften, Germany; Honorary Member, Polskie Towarzystwo Logiki i Filozofii Nauki, Poland. Jabuti Award winner, for *Computabilidade: Funções Computáveis, Lógica e os Fundamentos da Matemática* (Brazilian version of *Computability: Computable Functions, Logic and the Foundatiosn of Matheamtics, with the Timeline Computability and Undecidability*. Second edition. Wadsworth/Thomson Learning, Belmaont, CA, 2000.)

# CHAGROV, Alexander

**Specialties:** Mathematical logic (non-classical propositional logics), algebraic logic (equational theories), decidability and complexity.

**Born:** 02 May 1957 in Kalinin (now Tver), Kalinin region, USSR.

**Educated:** Institute for Information Transmission Problems, Russian Academy of Sciences, Russia, Doctor of Physical-Mathematical Science PHD-II, 1998; Institute of Mathematics, Academy of Sciences of Moldavian SSR, USSR, Candidate of Science PhD-I, 1987; Kalinin State University, Mathematician and Teacher of Mathematics, 1978.

**Dissertation:** PhD 1: *Complexity of approximability of modal and superintuitionistic logics*, supervisor, Max Kanovich; PhD 2: *Modeling of computational processes by means of propositional logics*.

**Regular Academic or Research Appointments:**
PROFESSOR OF MATHEMATICS, CHAIRMAN OF DEPARTMENT OF AGLEBRA AND MATHEMATICAL LOGIC, TVER STATE UNIVERSITY, 20004– ; Professor, Mathematics, Tver State University, 1999-2004; Associate Professor (Docent), Mathematics, Tver State University, 1991-1999; Assistent Professor (Lecturer) of Mathematics, Tver State University, 1989-1999; Instructor, Mathematics, Tver State University, 1978-1989.

**Research Profile:** Alexander Chagrov has made fundamental contributions to the research of propositional logics. In 1980 he has discovered that the intuitionistic logic is embedded by Godel-Tarski-McKinsey translation not only in normal logic S4, as it was traditional by that time, but and in logics, not being normal – in the logic S3 (the first modal system from ones formulated by K. I. Lewis in the beginning of the XX$^{th}$ century, it is a proper sublogic of S4), in proper extensions of the Grzegorczyk logic and many others. Complexity problems, in particular – interrelations the standard classes of complexity NP, PSPACE, etc., with the complexity of description of logics by models have been the second direction of Chagrov's research (the first dissertation has been devoted to this direction). The problems of decidability of logics and logic properties are the third basic direction of Chagrov's research and form the research subject in his second dissertation. To this research they proved (highly nontrivial, as a rule) only positive results mostly in this direction. Chagrov has proved that decidable properties of logics from many natural classes occur very seldom – almost all the reasonable properties are undecidable; for this he has proposed some schemes of the proofs of undecidability of properties. Many results Chagrov has obtained in co-operation with other researchers, for examples: undecidability of disjunction property (with M. Zakharyaschev); undecidability of many formula properties connected with their behaviour on finite structures (with L. Chagrova); undecidability of the family of normal modal logics with finite complete sets of nonequivalent modalities (with A. Chagrova); the existence of modal and superintuitionistic logics, not having independent axiomatizations (with M. Zakharyaschev), and many others. The significant part of the monograph "Modal Logic" (with M. Zakharyaschev) is connected with scientific interests of Chagrov.

**Main Publications:**

1. On non-normal companions of Int, *Automata, Algorithms, Languages,* Kalinin, Kalinin State University, 1982, p. 138–148. (In Russian)
2. On complexity of propositional logics, *Complexity problems of mathematical logic*, Kalinin, Kalinin State University, 1985, p. 80–90. (In Russian)
3. Varieties of logical matrices, *Algebra and Logic* (Novosibirsk, Russia), 1985, v. 24, N 4, p. 426–489. (In Russian. English translation in: Algebra and Logic, 24: 278–325.)
4. A lower bound of cardinality of approximating Kripke frames, *Logical Methods of Constructing of Effective Algorithms*, Kalinin, Kalinin State University, 1986, p. 96–125. (In Russian)
5. On bound of the set of modal companions of intuitionistic logic, *Non-classical logics and its application*, Moscow, Institute of Philosophy, Academy of Sciences, USSR, 1989, p. 74—81. (In Russian)
6. Undecidable properties of extensions of provability logics, *Algebra and Logic* (Novosibirsk, Russia), 1990, v. 29, N 3, p. 350–367. (In Russian. English translation in: Algebra and Logic, 24: 231–243.)
7. Undecidable properties of extensions of provability logics, II, *Algebra and Logic* (Novosibirsk, Russia), 1990, v. 29, N 5, p. 613–623. (In Russian. English translation in: Algebra and Logic, 24: 406–413.)
8. (With M. Zakharyashchev) The disjunction property of intermediate propositional logics, *Studia Logica*, 1991, v. 50, No 2, p. 189–216.
9. (With M. Zakharyashchev) Modal Companions of Intermediate Propositional Logics, *Studia Logica*, 1991, v. 52, No 1, p. 49–82.
10. Continuality of the set of maximal superintuitionistic logics with the disjunction property, *Mathematical Notes*, 1992, v. 51, No 2. p. 117—123. (In Russian. English translation: Mathematical Notes, 1992, 51: 188—193.)
11. (With M. Zakharyaschev) The undecidability of the disjunction property of propositional logics and other related problems, *J. Symb. Log.*, 1993, v. 58, No p. 967–1002.
12. Undecidable properties of superintuitionistic logics, *Mathematical Problems of Cybernetics*, 1994, v. 5, p. 67—108. (In Russian.)
13. (With M. Zakharyaschev) On the independent axiomatizability of modal and intermediate logics, *J. Logic Computat.*, 1995, vol. 5, No. 3, pp. 287–302.
14. (With L. A.Chagrova) Algorithmic problems concerning first-order definability of modal formulas on the class of all finite frames, *Studia Logica*, 1995, v. 55, No 3, p. 421–448.
15. (With V. B. Shehtman) Algorithmic aspects of tense logics, Computer Science Logic, 8th Workshop, CSL '94 (L.Pacholski, J.Tiuryn (eds.)), Lecture Notes in Computer Science, 1995, v. 933, pp. 442–455.
16. (With M.Zakharyaschev) *Modal Logic*. Oxford University Press, 1997, 603 p.
17. (With M. Zakharyaschev, F.Wolter) Advanced

Modal Logic, *Handbook of Philosophical Logic*, $2^{nd}$ edition (D.M.Gabbay, F.Guenthner (eds.)), v. 3, Kluwer Academic Publishers, 2001, p. 83–266.

18. (With A. A. Chagrova) Normal modal logics with (in)finite sets of nonequivalent modalities, *Russian Mathematics – Three Hundred Years*, Tver, Tver State University, 2002, p. 127–139.

19. (With M. N. Rybakov) How Many Variables Does One Need to Prove PSPACE-hardness of Modal Logics, *Advances in Modal Logic*, London, King's College Publicions, 2003, v. 4, p. 71–82.

20. (With L. A. Chagrova) The Truth About Algorithmic Problems in Correspondence Theory, *Advances in Modal Logic*, College Publications, 2006, v. 6, p. 121–138.

**Teaching:** Since 1980 Chagrov works at the University of Tver (Russia) and has created the group of researchers, who have graduated from this University, in the field of non-classical logics; at the present they are – A. Chagrov, L. Chagrova (1989 PhD), I. Gorbunov (2006 PhD), M. Rybakov (2005 PhD). This group co-operates with many logicians of other scientific centers.

**Vision Statement:** Non-classical logics have as its sources the ideas of ancient thinkers (for an example, Aristotle's modal syllogistics), more near scientists to present time (Leibniz), and so "non-classical" is only the comfortable term for the designation of an important part of modern logic. Its results help to understand problems of foundations of mathematics, computer science, linguistics, etc., not to mention on logic-philosophical problems.

Kian's amended Vision Statement (Please check if this is correct): *Vision Statement*: Non-classical logics have as their sources the ideas of ancient thinkers (for an example, Aristotle's modal syllogisms) and modern scientists (Leibniz), so "non-classical" is the only comfortable term for the designation of an important part of modern logic. Its results help to understand problems of foundations of mathematics, computer science, linguistics, etc., not to mention logical philosophical problems.

# CHANG, C. C.

**Specialties:** Mathematical logic.

**Born:** 13 Oct 1927 in Tientsin, China. *Educated*: University of California at Berkeley, PhD, 1955; MA, 1950; Harvard, BA, 1949. *Dissertation*: ?.

**Regular Academic or Research Appointments:** UNIVERSITY OF CALIFORNIA AT LOS ANGELES, PROFESSOR, 1964–; Associate Professor, 1961-1964; Assistant Professor, Mathematics, 1958-1961; Instructor, Cornell University, 1955-1956.

**Visiting Academic or Research Appointments:** All Souls College; Member, American Mathematical Society; Association for Symbolic Logic.

**Main Publications:**

1. C. C. Chang: The Writing of the MV-algebras. Studia Logica 61(1): 3-6 (1998).
2. C. C. Chang: Omitting Types of Prenex Formulas. J. Symb. Log. 32(1): 61-74 (1967)
3. Lawrence Peter Belluce, C. C. Chang: A Weak Completeness Theorem for Infinite Valued First-Order Logic. J. Symb. Log. 28(1): 43-50 (1963)
4. C. C. Chang, H. Jerome Keisler: An Improved Prenex Normal Form. J. Symb. Log. 27(3): 317-326 (1962)
5. C. C. Chang, Anne C. Morel: On Closure Under Direct Product. J. Symb. Log. 23(2): 149-154 (1958)

**Service to the Profession:** Contributor, several professional journals; Editor, Annals of Mathematical Logic, 1969-; Consulting Editor, Journal for Symbolic Logic, 1968-;.

**Honours and Awards:** Senior Fulbright Research Scholar, Oxford University, 1966-1967; Senior Postdoctoral Fellow NSF, 1962-1963; Phi Beta Kappa; Sigma Xi; Pi Mu Epsilon.

# CHELLAS, Brian Farrell

**Specialties:** Modal logic, deontic logic, logic of action.

**Born:** 7 October 1941 in New York City, New York, USA.

**Educated:** Stanford University, PhD Philosophy, 1969; Florida State University, BA Philosophy, 1962.

**Dissertation:** *The Logical Form of Imperatives*; supervisor, Dana Scott.

**Regular Academic or Research Appointments:** PROFESSOR EMERITUS OF PHILOSOPHY, UNIVERSITY OF CALGARY, 1997–; Dean, Faculty of Humanities, University of Calgary, 1984-1989; Professor of Philosophy, University of Calgary, 1981-1997; Head, Department of Philosophy, University of Calgary, 1979-1982, 1983-1984; Associate Professor of Philosophy, University of Calgary, 1976-1981; Assistant Professor of Philosophy, University of Pennsylvania, 1968-1975; Instructor in Philosophy, Florida State University, 1964-1965.

**Visiting Academic or Research Appointments:** Visiting Associate Professor of Philosophy, University of Michigan, 1975-1976; Visiting Professor of Philosophy, University of Uppsala, 1970.

**Research Profile:** Much of Chellas's early research was in deontic logic. Among the results were the papers "The story of ○" (unpublished) and "Conditional obligation". He was interested also in other kinds of modal logic, such as conditional logics (his "Basic conditional logic" is fundamental in the subject), monotonic logics (a paper with Audrey McKinney gives the first correct possible worlds completeness proof for these), and $S1$-type modal logics (a paper with Krister Segerberg provides the first correct such proofs for systems of this sort). Toward the end of his career he returned to the topic of action and an operator introduced in his dissertation, which has become known as *cstit*. He also wrote two textbooks, *Modal Logic: An Introduction* and *Elementary Formal Logic*.

**Main Publications:**

1. *The Logical Form of Imperatives*. Stanford: Perry Lane Press, 1969, v+115 pp.
2. Imperatives. *Theoria* 37 (1971) 114-129.
3. Notions of relevance. *Journal of Philosophical Logic* 1 (1972) 287-293.
4. Conditional obligation. *Logical Theory and Semantic Analysis: Essays Dedicated to Stig Kanger on his Fiftieth Birthday*, ed. Sören Stenlund, pp. 23-33 Dordrecht: D. Reidel Publishing Company, 1974.
5. The completeness of monotonic modal logics. *Zeitschrift für mathematische Logik und Grundlagen der Mathematik* 21 (1975) 379-383 (with Audrey McKinney).
6. Basic conditional logic. *Journal of Philosophical Logic* 4 (1975) 133-153. Italian translation by Claudio Pizzi: Logica condizionale fondamentale. *Leggi di Natura, Modalità, Ipotesi: La Logica del Ragionamento Controfattuale*, ed. Claudio Pizzi, pp. 282-302. Milan: Giangiacomo Feltrinelli Editore, 1978.
7. Quantity and quantification. *Synthese* 31 (1975) 487-491.
8. Modalities in normal systems containing the $S5$ axiom. *Intention and Intentionality*, ed. Cora Diamond and Jenny Teichman, pp. 261-265. Hassocks: Harvester Press, 1979.
9. Another proof of the decidability of four modal logics. *Philosophia* 9 (1980) 251-264.
10. *Modal Logic: An Introduction*. Cambridge and New York: Cambridge University Press, 1980; reprinted 1984, 1988 (with corrections), 1990, 1993, 1995, xii+295 pp. Chinese translation (of 1980/1984 printings) by Zheng Wenhui and Zhang Yisheng, with a foreword by Mo Shaokui: *Mo Tai Luo Ji: Dao Lun*. Guangzhou: Zhongshan University Press, 1989, xiii+353 pp.
11. $KG^{k,l,m,n}$ and the efmp. *Logique et analyse*, n.s. 103-104 (1983) 255-262.
12. Time and modality in the logic of agency. *Studia Logica* 51 (1992) 485-517.
13. Modally finite extensions of $K4$. *Vicinae Deviae: Essays in Honour of Raymond Earl Jennings*, ed. Martin Hahn, pp. 164-170. Burnaby, British Columbia: Department of Philosophy, Simon Fraser University, 1993.
14. Modal logics with the MacIntosh rule. *Journal of Philosophical Logic* 23 (1994) 67-86 (with Krister Segerberg).
15. On bringing it about. *Journal of Philosophical Logic* 24 (1995) 563-571.
16. *Elementary Formal Logic*. Calgary: Perry Lane Press, 1996, x+448 pp.
17. Modal logics in the vicinity of $S1$. *Notre Dame Journal of Formal Logic* 37 (1996) 1-24 (with Krister Segerberg).

**Work in Progress** Unfinished at the time of Chellas's retirement were papers "A generalization of Bull's theorem" and "Impossibility interpretations of necessity" (both with Krister Segerberg) and "Uniformity and the finite model property".

**Service to the Profession:** Member, editorial board, *Journal of Philosophical Logic*, 1981-2003; Reviewer, *Mathematical Reviews*, 1975-2002; President, Society for Exact Philosophy, 1982-1984 (founding member 1970); Executive editor, *Canadian Journal of Philosophy*, 1977-1982.

**Teaching:** Chellas's teaching was primarily in logic, including modal logic, deontic logic, and the logic of action. Its quality was recognized by a teaching excellence award in 1992.

**Vision Statement:** I believe that puzzles in deontic logic and the logic of action are susceptible of logical analysis. But I more and more wonder how much help logic can ultimately be in advancing our understanding of ethical concepts. A fuller idea of my thinking can be found in my contribution to the volume *Formal Philosophy*, edited by Vincent F. Hendricks and John Symons (United Kingdom: Automatic Press, 2005).

**Honours and Awards:** Swedish Institute Fellowship, 1994; University of Calgary Students' Union Teaching Excellence Award, 1992; Social Sciences and Humanities Research Council of Canada Leave Fellowship, 1982-1983; National Endowment for the Humanities Summer Stipend, 1974; John W. Hill Foundation Scholarship, 1963-1964; Woodrow Wilson Fellowship, 1962-1963; Phi Beta Kappa, Phi Eta Sigma, Phi Kappa Phi, Phi Sigma Tau, all during 1962.

# CHOLAK, Peter

**Specialties:** Mathematical logic, computability theory, reverse mathematics.

**Born:** 6 October 1962 in Hamilton, Ohio, USA.

**Educated:** University of Wisconsin–Madison, PhD Mathematics, 1991; University of Wisconsin–Madison, MA Mathematics, MA in Computer Science, 1988; Union College, Schenectady, New York, BA Mathematics, 1984.

**Dissertation:** *Automorphisms of the Lattice of Recursively Enumerable Sets*; supervisor, Terry Millar/

**Regular Academic or Research Appointments:** PROFESSOR, UNIVERSITY OF NOTRE DAME, 2004–; Associate Professor, University of Notre Dame, 2000-2004; John and Margaret McAndrews Assistant Professor, University of Notre Dame, 1994-2000; University of Michigan, Assistant Professor, 1992-1993.

**Visiting Academic or Research Appointments:** Cornell University, Visiting Scholar in Mathematics, 1993-1994; National Science Foundation Postdoctoral Research Fellowship, 1992-1996; Victoria University of Wellington (New Zealand) Postdoctoral Fellowship, 1992-1993.

**Research Profile:** Peter Cholak's best work to date has been in several areas of Computability Theory. Cholak has a number of papers that focus on the computably enumerable sets. For example, in his PhD thesis, Cholak showed that the orbit of every non computable c.e. set (within the c.e. sets under inclusion) must contain a high set; with Downey and Stob that the orbit of promptly simple sets must contain a complete set; with Harrington that the double jump is invariant in the c.e. sets, and again with Harrington that there is an orbit such that membership in that orbit has the highest possible complexity. Cholak, with Coles, Downey, and Herrmann, have related work in the lattice of $\Pi^0_1$ classes. They show that the perfect thin classes are definable and form an orbit. With Jockusch and Slaman, Cholak showed that Ramsey Theorem for Pairs with 2 colors is $\Pi^1_1$ conservative over $RCA_0$ plus induction for $Simga^0_2$ formulas.

**Main Publications:**

1. Peter A. Cholak and Rod Downey. Invariance and noninvariance in the lattice of Pi^0_1 classes. *J. London Math. Soc. (2)*, 70(3):735–749, 2004.
2. Peter Cholak, Rod Downey, and Micheal Stob. Automorphisms of the lattice of recursively enumerable sets: promptly simple sets. *Trans. Amer. Math. Soc.*, 332(2):555–570, 1992.
3. Peter Cholak. The translation theorem. *Arch. Math. Logic*, 33(2):87–108, 1994.
4. Peter Cholak. Automorphisms of the lattice of recursively enumerable sets. *Mem. Amer. Math. Soc.*, 113(541):viii+151, 1995.
5. Peter Cholak, Sergey Goncharov, Bakhadyr Khoussainov, and Richard A. Shore. Computably categorical structures and expansions by constants. *J. Symbolic Logic*, 64(1):13–37, 1999.
6. Peter Cholak and Leo A. Harrington. Definable encodings in the computably enumerable sets. *Bull. Symbolic Logic*, 6(2):185–196, 2000.
7. Peter Cholak, Rod Downey, and Eberhard Herrmann. Some orbits for E. *Ann. Pure Appl. Logic*, 107(1-3): 193–226, 2001.
8. Peter Cholak, Carl G. Jockusch, and Theodore A. Slaman. On the strength of Ramsey's theorem for pairs. *J. Symbolic Logic*, 66(1):1–55, 2001.
9. Peter Cholak, Richard Coles, Rod Downey, and Eberhard Herrmann. Automorphisms of the lattice of Pi^0_1 classes: perfect thin classes and anc degrees.
10. Peter Cholak, Rod Downey, and Stephen Walk. Maximal contiguous degrees. *J. Symbolic Logic*, 67(1): 409–437, 2002Peter Cholak and Leo A. Harrington. On the definability of the double jump in the computably enumerable sets. *J. Math. Log.*, 2(2):261–296, 2002.
11. Peter Cholak and Leo A. Harrington. Isomorphisms of splits of computably enumerable sets. *J. Symbolic Logic*, 68(3):1044–1064, 2003.
12. Peter A. Cholak, Mariagnese Giusto, Jeffry L. Hirst, and Carl G. Jockusch, Jr. Free sets and reverse mathematics. In *Reverse mathematics 2001*, volume 21 of *Lect. Notes Log.*, pages 104–119. Assoc. Symbol. Logic, La Jolla, CA, 2005.
13. Peter Cholak, Noam Greenberg, and Joseph Miller. Uniform Almost Everywhere Domination. *J. Symbolic Logic*, 71(3):1057–1072, 2006.

*Work in Progress*

14. Peter Cholak and Leo A. Harrington. Extension theorems, orbits, and automorphisms of the computably enumerable sets. To appear in Trans. Amer. Math. Soc.
15. Peter Cholak, Rod Downey, and Noam Greenberg, Triviality and Jump Traceability.

**Service to the Profession:** Editor, Journal of Symbolic Logic, 2004–; Editor, Notre Dame Journal of Formal Logic, 2000-; Chair of the Meeting Committee of the Association for Symbolic Logic, 2001-2005.

**Teaching:** Peter Cholak has had 3 PhD students as of 2006. Steve Walk now at St. Cloud State, Charlie McCoy now at Holy Cross seminary at Notre Dame (joint with Julia Knight), Rebecca Weber now at Dartmouth College. One notable undergraduate: Sami Assaf who won an National Science Foundation Graduate Fellowship among other awards to go to Berekeley for graduate school.

**Vision Statement:** Interesting problems drive mathematical and mathematical logic.

# CHONG, Chi Tat

**Specialties:** Mathematical logic, recursion theory.

**Born:** 6 September 1949 in Hong Kong.

**Educated:** Yale University, PhD Mathematics, 1973; Iowa State University, BS Mathematics, 1969.

**Dissertation:** *Tame $\Sigma_2$ Functions in $\alpha$-recursion Theory*; supervisor, Manuel Lerman.

**Regular Academic or Research Appointments:** UNIVERSITY PROFESSOR OF MATHEMATICS AND PHILOSOPHY, UNIVERSITY OF SINGAPORE, 2004–; Professor of Mathematics, National University of Singapore, 1989–2004; Associate Professor of Mathematics, 1985-1988; Senior Lecturer, University of Singapore, 1980–1984; National University of Singapore; Lecturer, University of Singapore, 1974–1979.

**Visiting Academic or Research Appointments:** Indam-GNSAGA Visiting Professor, University of Sienna, Italy, 2006; Visiting Scholar, University of California, Berkeley, 2005; Concurrent professor, Nanjing University, China; Member, Mathematical Science Research Institute, Berkeley, 1990; Japan Society for the Promotion of Science, Visitor, University of Tokyo, 1984; Visiting Scholar, Massachusetts Institute of Technology, 1980–1981.

**Research Profile:** Chong has worked in higher recursion theory, especially the structure of $\alpha$-recursively enumerable degrees. In classical recursion theory, he has contributed to the study of the structure of 1-generic degrees and their relationship with minimal degrees. He has also done work on computation theory over the reals, especially with respect to the degree of unsolvability of Julia sets in complex dynamics. Chong has also worked on problems in reverse recursion theory, in particular studying the proof-theoretic strength of degree-theoretic and set-theoretic statements in fragments of Peano arithmetic. His current interest is in Ramsey Theorem type problems in the context of reverse mathematics.

**Main Publications:**

1. Generic sets and minimal $\alpha$-degrees, *Transactions of American Mathematical Society*, 254 (1979), 15-169.
2. *Techniques of Admissible Recursion Theory*, Lecture Notes in Mathematics 1106, Springer Verlag, 1984.
3. (with Rod Downey) Minimal degrees recursive in 1-generic degrees, *Annals of Pure and Applied Logic*, 48 (1990), 215–225.
4. (with Joseph Mourad) $\Sigma_n$ definable sets without $\Sigma_n$ induction, *Transactions American Mathematical Society*, 334 (1992), 349–363.
5. Positive reducibility of the interior of filled Julia sets, *Journal of Complexity*, 10 (1994), 437–444.
6. (with Yue Yang) $\Sigma_2$ induction and infinite injury priority arguments, Part II: Tame $\Sigma_2$ coding and the jump operator, *Annals of Pure and Applied Logic*, 87 (1997), 103-111.
7. (with Lei Qian, Theodore A Slaman and Yue Yang) $\Sigma_2$ induction and infinite injury priority arguments, Part III. Minimal pairs and Shoenfield's conjecture, *Israel Journal of Mathematics*m 121 (2001), 1–28
8. (with Richard Shore and Yue Yang). Interpreting arithmetic in the r.e. degrees under $\Sigma_4$ induction, in: *Reverse mathematics 2001*, 120–146, Lecture Notes in Logic. volume 21, Assoc. Symbolic. Logic, La Jolla, CA, 2005.
9. (with Liang Yu) . Maximal chains in the Turing degrees., *Journal of Symbolic Logic*, 72 (2007), no. 4, 1219–1227.
10. (with Yue Yang) . The jump of a $\Sigma_n$-cut, *Journal of London. Mathematical Society*, 2 75 (2007), no. 3, 690–704
11. (with Liang Yu) A $\Pi_1^1$ uniformization principle for reals, *Transactions of American Mathematical Society*, to appear
12. (with Liang Yu and Andre Nies) Higher randomness notions and their lowness properties, *Israel Journal of Mathematics*, to appear.

A list of recent publications is available at http://www.math.nus.edu.sg/~/chongct

**Service to the Profession:** Managing Editor, *Journal of Mathematical Logic*; Editor, *Proceedings of Asian Logic Conference,* 1980 (then called Southeast Asian Logic Conference), 1993, 1996; Secretary, Singapore Mathematical Society, 1977-1980, President, Singapore Mathematical Society, 1991-1993; Chairman, Committee on Logic in East Asia, Association for Symbolic Logic, 1988-1994; Council member, Association for Symbolic Logic, 1996-2000; member of Nomination Committee, Association for Symbolic Logic, 2000; Program Committee, European Summer Meeting of the Association for Symbolic Logic, 2004; Member of Executive Committee, Council of Association for Symbolic Logic, 2007– , Deputy Vice Chancellor and Provost, National University of Singapore, 1996-2004; Chair, Department of Information Systems and Computer Science, National University of Singapore, 1993-1996; Vice Dean of Science, National University of Singapore, 1985-1996.

), 1993, 1996; Secretary, *Singapore Mathematical Society*, 1977–1980.

**Vision Statement:** I see the development of reverse mathematics and computation theory on the reals as important applications of recursion theory to mathematics and the philosophy of mathematics. While these are not the only two, connecting logic with the rest of mathematics is crucial for the health and growth of the subject.

**Honours and Awards:** Public Administrative Service Medal (Gold), Republic of Singapore.

# CLARKE, Edmund M.

**Specialties:** Temporal logic model checking in computing, automatic theorem proving and symbolic computation.

**Educated:** Cornell University, Ithaca, NY, PhD Computer Science, 1976; Cornell University, Ithaca, NY, MS Computer Science, 1974; Duke University, Durham NC, MA Mathematics, 1968; University of Virginia, Charlottesville, VA, BA Mathematics, 1967.

**Dissertation:** *Completeness and Incompleteness Theorems for Hoare-like Axiom Systems*; supervisor, Robert Constable.

**Regular Academic or Research Appointments:** FORE SYSTEMS PROFESSORSHIP CHAIR, COMPUTER SCIENCE, CARNEGIE-MELLON, 1995– ; Professor, Computer Science, Carnegie-Mellon, 1990-1995, Associate Professor, Computer Science, Carnegie-Mellon, 1982-1990; Assistant Professor, Computer Science, Harvard University, 1978-1982; Lecturer, Computer Science, Duke University, 1976-1978.

**Research Profile:** Dr. Clarke's interests include software and hardware verification and automatic theorem proving. In his PhD thesis he proved that certain programming language control structures did not have good Hoare style proof systems. In 1981 he and his PhD student Allen Emerson first proposed the use of Model Checking as a verification technique for finite state concurrent systems. His research group pioneered the use of Model Checking for hardware verification. Symbolic Model Checking using BDDs was also developed by his group. This important technique was the subject of Kenneth McMillan's PhD thesis, which received an ACM Doctoral Dissertation Award. In addition, his resarch group developed the first parallel resolution theorem prover (Parthenon) and the first theorem prover to be based on a symbolic computation system (Analytica).

Logical errors in sequential circuit designs and communication protocols are an important problem for system designers. They can delay getting a new product on the market or cause the failure of some critical device that is already in use. My research group has developed a verification method called temporal logic model checking for this class of systems. In this approach specifications are expressed in a propositional temporal logic, while circuits and protocols are modeled as state–transition systems. An efficient search procedure is used to determine automatically if a specification is satisfied by some transition system. The technique has been used in the past to find subtle errors in a number of non-trivial examples.

During the last few years, the size of the state-transition systems that can be verified by model checking techniques has increased dramatically. By representing transition relations implicitly using Binary Decision Diagrams (BDDs), we have been able to check some examples that would have required $10^{20}$ states with the original algorithm. Various refinements of the BDD-based techniques have pushed the state count up to $10^{100}$. By combining model checking with various abstraction techniques, we have been able to handle even larger systems. For example, we have used this technique to verify the cache coherence protocol in the IEEE Futurebus+ Standard. We found several errors that had been previously undetected. Apparently, this is the first time that formal methods have been used to find nontrivial errors in an IEEE standard.

**Main Publications:**

1. E. Clarke, O. Grumberg and D. Long. Verification tools for finite-state concurrent systems. In: *A Decade of concurrency–Reflections and Perspectives* . Lecture Notes in Computer Science, 803, 1994.
2. J.R. Burch, E.M. Clarke, K.L. McMillan, D.L. Dill, and J. Hwang. Symbolic model checking: 10E20 states and beyond. In *LICS*, 1990.
3. E.M. Clarke, O. Grumberg,H. Hiraishi, S. Jha, D.E. Long, K.L. McMillan, and L.A. Ness. Verfication of the Futurebus+cache coherence protocol. In L. Claesen, editor, *Proceedings of the Eleventh International Symposium on Computer Hardware Description Languages and their Applications*. North-Holland, April 1993.
4. W. Marrero, E.M. Clarke, and S. Jha. Model Checking for Security Protocols. Technical Report CMU-SCS-97-139, Carnegie Mellon University, May 1997.
5. E.M.Clarke and E.A. Emerson and A.P. Sistla. Automatic verification of finite-state concurrent systems using temporal logic specifications. In *ACM Transactions on Programming Languages and Systems*, 8(2):244- 263, 1986.
6. E.M. Clarke and E.A. Emerson. Synthesis of synchronization skeletons for branching time temporal logic. In *Logic of Programs: Workshop, Yorktown Heights, NY, May 1981* Lecture Notes in Computer Science, vol. 131, Springer-Verlag. 1981.

For a complete list of publications, please visit `http://www.cs.cmu.edu/{\%}7Eemc/`

VITA.pdf

**Service to the Profession:** Editorial Board Member, Distributed Computing, 1986-2000; Editorial Board Member, Logic and Computation, 1990-1993; Editorial Board Member, ACM Transactions on Design Automation of Electronic Systems (TOAES), 1996-1999; Editor-in-Chief, Formal Methods in System Design, Kluwer Academic Publishers; Associate Editor, IEEE Transactions on Software Engineering, published by IEEE Computer Society; Editorial Board Member, Microelectrics Journal; Advisory Board Member, Software Tools and Technology Transfer; Organizing Committee Member, Logic in Computer Science.

**Teaching:** Previous PhD students include the following: Pankaj Chauhan, Muralidhar Talupur, Anubhav Gupta, Alex Groce, Sagar Chaki, Dong Wang, W. Marrero, Y. Lu, M. Minea, V. Hartonas-Garmhausen, S. Jha, S. Campos, X. Zhao, D.E. Long, J.R. Burch, K.L. McMillan, M.C. Browne, D.L. Dill, B. Mishra, A.P. Sistla, C.N. Nikolaou, E.A. Emerson.

**Honours and Awards:** Member, National Academy of Engineering, 2005; IEEE Harry H. Goode Memorial Award, 2004; Award Winner, ACM Kanellakis Award, 1999; Allen Newell Award for Excellence in Research, Carnegie Mellon Computer Science Department, 1999; Technical Excellence Award, Semiconductor Research Corporation, 1995; Sigma Xi; Phi Beta Kappa.

# COCCHIARELLA, Nino B.

**Specialties:** Formal ontology, tense, modal and intensional logic, higher-order logic, philosophical logic, Montague grammar.

**Born:** USA.

**Educated:** UCLA, PhD, 1966, UCLA, MA, 1962, Columbia University, BS, 1958.

**Dissertation:** *Tense Logic: A Study of Temporal Reference*; supervisor Richard M. Montague.

**Regular Academic or Research Appointments:** PROFESSOR EMERITUS, INDIANA UNIVERSITY; Professor of Philosophy, Indiana University. Assistant Professor, San Francisco State University, 1964-68.

**Research Profile:** Cocchiarella proved the first completeness theorems in tense logic and second-order modal logic. He was the first to develop several second-order logics with nominalized predicates as abstract singular terms and then to use those systems in a consistent logical reconstruction of both Frege's and Russell's early logics and in the application of those reconstructions to the semantic analysis of natural language. This work also led to Cocchiarella's development of formal theories of predication and comparative formal ontology, including especially logical reconstructions of nominalism, conceptualism, logical realism, and the logic of natural kinds. Cocchiarella also showed how logical atomism is compatible with logical necessity as a modality, and that it is the only ontology in which logical necessity, as opposed to other kinds of modalities, makes sense. Cocchiarella's own preferred ontological framework is conceptual realism, which he has been formally developing for many years, and which contains a logic of both actualism and possibilism in terms of a distinction between concepts that entail concrete existence and those that do not. It also contains a logic of classes as many as plural objects, which is the basis of Cocchiarella's semantics for plurals and mass nouns in natural language, and in which the Leonard-Goodman calculus of individuals (and therefore Leœniewski's mereology as well) is reducible. Cocchiarella has also shown that Leœniewski's ontology, which is also called a logic of names, is reducible to his theory of reference in conceptual realism, and that the medieval *suppositio* theories of Ockham, Buridan, and other medieval logicians can be logically reconstructed in terms of this theory of reference. Cocchiarella is currently continuing his work on different subsystems of conceptual realism, including in particular a logic of events as truth-makers.

**Main Publications:**

1. *Modal Logic: Introduction to its Syntax and Semantics*, co-author with Max Freund, Oxford University Press, Oxford and N.Y., 2008.
2. *Formal Ontology and Conceptual Realism*, Synthese Library vol, 339, Springer, Dordrecht, 2007.
3. *Logical Studies in Early Analytic Philosophy*, Ohio State University Press; Columbus, 1987.
4. *Logical Investigations of Predication Theory and the Problem of Universals*, Bibliopolis Press; Naples, 1986.
5. "A Completeness Theorem in Second Order Modal Logic," *Theoria*, vol. 35 (1969) pp. 81-103.
6. "On the Primary and Secondary Semantics of Logical Necessity," *Journal of Philosophical Logic*, vol. 4 (1975), pp. 13-27.
7. "Nominalism and Conceptualism as Predicative Second Order Theories of Predication," *Notre Dame Journal of Formal Logic*, vol. 21 (1980), pp. 481-500.
8. "Richard Montague and the Logical Analysis of Language," in *Contemporary Philosophy: A New Survey*, vol. 2, *Philosophy of Language/Philosophical Logic*, G.

Flostad, ed., Martinus Nijhoff, the Hague, 1981, pp. 113-155.

9. "The Development of the Theory of Logical types and the Notion of a Logical Subject in Russell's Early Philosophy," *Synthese*, vol. 45 (1980), pp. 71-115.

10. "Philosophical Perspectives on Quantification in Tense and Modal Logic," in *Handbook of Philosophical Logic*, vol. 2, eds. D. Gabbay and F. Guenthner, D. Reidel Pub. Co., Dordrecht, 1984, pp. 309-353.

11. "Frege, Russell and Logicism: A Logical Reconstruction," in *Frege Synthesized: Essays on the Philosophical and Foundational Work of Gottlob Frege*, Leila Haaparanta and Jaakko Hintikka, eds., D. Reidel Pub. Co., Dordrecht, 1986: 197-252.

12. "Two λ-Extensions of the Theory of Homogeneous Simple Types as a Second-Order Logic," *Notre Dame Journal of Formal Logic*, vol. 26, no. 4 (Oct. 1985): 377-407.

13. "Frege's Double Correlation Thesis and Quine's Set Theories NF and ML," *Journal of Philosophical Logic*, vol. 14, no. 4 (1985): 1-39.

14. "Conceptualism, Ramified Logic, and Nominalized Predicates," *Topoi*, vol. 5, no. 1 (March 1986): 75-87.

15. "Predication Versus Membership in the Distinction between Logic as Language and Logic as Calculus," *Synthese*, vol. 75 (1988): 37-72.

16. "Conceptual Realism Versus Quine on Classes and Higher-Order Logic," *Synthese*, vol. 90 (1992): 379-436.

17. "Cantor's Power-Set Theorem Versus Frege's Double-Correlation Thesis," *History and Philosophy of Logic*, vol. 13 (1992): 179-201.

18. "Formally Oriented Work in the Philosophy of Language," in *Philosophy of Meaning, Knowledge and Value in the Twentieth Century*, vol. X of *Routledge History of Philosophy*, edited by Jack V. Canfield, Routledge, London and New York, 1997, pp. 39–75.

19. "On the Logic of Classes as Many," *Studia Logica*, vol. 70 (2002): 303–338.

20. "Denoting Concepts, Reference, and the Logic of Names, Classes as Many, Groups and Plurals," *Linguistics and Philosophy*, Vol. 28 (2005): 135–179.

**Service to the Profession:** Member of the Editorial board of *Journal of Philosophical Logic*.
Member of the Advisory Editorial board of *Synthese*.
Member of the Editorial board of *Axiomathes*.
Member of the Editorial board of *Metalogicon*.
Member of International Advisory Board of *Philosophy and History of Science: A Taiwanese Journal*.
Reviewer for *Mathematical Reviews*.
Referee for *Journal of Philosophical Logic, Journal of Symbolic Logic, Notre Dame Journal of Formal Logic, Noûs, Reports on Mathematical Logic, Synthese, Studia Logica, Journal of Logic and Computation*.

**Teaching:** Cocchiarella has taught introductory, intermediate, and advanced courses in logic, semantics, set theory and Montague Grammar, as well as seminars on some of the most recent areas of research in logic. He has placed an emphasis in his teaching on the logical analysis of natural language and the ontological interpretations of both scientific and mathematical language.

**Vision Statement:** Cocchiarella sees logic as a powerful tool for the analysis of our scientific theories and the structures that underlie natural language and our commonsense understanding of the world. The study of logical categories in particular provides an important way to study the semantic and ontological categories underlying our scientific and commonsense world views.

**Honours and Awards:** Gladiatore d'Oro award by the Province of Benevento, Italy, 2004; Cittadinanza Onoraria, Keys to the City by Fragneto L'Abate, Benevento Province, Italy, 2003; "Diploma di Benemerenza" for excellence in scholarship by the city of Fragneto L'Abate, Italy, 2002; NEH Fellowship for Independent Study and Research, 1988-89 (Conceptualism, Realism, and Intensional Logic); Italian Govt. Award, Faculty Seminar Lecture Series, 1987, University of Padua, Italy; Research Grant, Institut für Linguistik, University of Stuttgart, Germany, 1981; NEH Fellowship for Independent Study and Research, 1977-78 (Logic and Ontology); NSF (Post-Doctoral) Grant, 1973-74 (Investigations on Formal Ontology); NSF (Post-Doctoral), 1971-73 (Investigations of Higher Order Modal Logics).

# CONSTABLE, Robert L.

**Specialties:** Complexity theory, programming logics, semantics, constructive tree theory, formal methods, and verification.

**Born:** 20 January 1942 in Detroit, Michigan, USA.

**Educated:** A.B., Princeton University, Mathematics, 1964; M.A., University of Wisconsin, Mathematics, 1965; Ph.D., University of Wisconsin, Mathematics, 1968.

**Dissertation:** *Extending and Refining Hierarchies of Computable Functions*, supervisor Stephen Cole Kleene

**Regular Academic or Research Appointments:** DEAN OF THE FACULTY OF COMPUTING AND INFORMATION SCIENCE, CORNELL UNIVERSITY, 1999–; Chair, Computer Science Department, Cornell University, 1993–99; Professor, Department of Computer Science, Cornell University, 1978–; Associate Professor, Department of Computer Science, Cornell University, 19732–78; Assistant Professor, Department of Computer Science, Cornell University, 1968–72; Instructor, University of Wisconsin, 1968–68.

**Visiting Academic or Research Appointments:** Warwick University, England; Edinburgh University, Scotland (3 visits); Cambridge University, England; Ben-Gurion University, Israel (2 visits); Tel Aviv University, Israel.

**Research Profile:** My early computer science research was in computational complexity theory and subrecursive hierarchies. I worked with Hartmanis on the relationship between computational complexity and program size and with Hopcroft and Borodin on families of complexity classes. I initiated the study of computational complexity for operators and higher-order functions. Kurt Mehlhorn wrote a thesis with me on PTime at type 2. This is still an active area of research with connections to the analysis of functional programming languages. This work led me to the study of programming language semantics, such as the results with Egli on denotational semantics for recursive operators.

In 1971, I introduced the notion of *proofs-as-programs* using Kleene realizability. Later Bates and I extended this to deBruijn/Martin-Löf realizability in terms of propositions-as-types. These ideas lie at the heart of my current work on Nuprl and have been widely studied and implemented in various program synthesis systems and program verification systems. In the same year, Cherniavsky and I began work on logics for reasoning about functional programs.

By 1975 O'Donnell and I began work on logics for imperative programs and started our book, *A Programming Logic*. One of the central innovations was the idea of *asserted programs* as proofs (also called proof outlines). We developed a block structured natural deduction system for these new kinds of *dynamic* proofs. These ideas became the basis for the PL/CV program verification system that Johnson and I built and wrote about in a second book with Eichenlaub.

I began to see this as a wider subject called *programming logics*. I proved some basic completeness and decidability results and supervised Clarke's thesis with his famous incompleteness theorem. During this period we were working on the implementation of PL/CV and enriching its type system. As the type system of the programming logics became richer, I was led to the study of type theory and eventually, by 1978, to constructive type theory and the work of Per Martin-Löf as a semantic basis for programming logics.

Constructive type theory has been one of my main interests since this time. I developed a theory called V3 and then, working with Bates, Allen and Howe, we began the design of the Nuprl type theory which has been evolving ever since. The Cornell work on type theory has added several new type constructors to Martin-Löf's theories: such as subset types; quotient types; inductive and coinductive types (with Mendler); bar types (with Smith); and very dependent types (Hickey). We have introduced reflection (with Allen, Howe and Aitken), and Allen has introduced new semantic methods.

I have explored the rule of constructive type theory as a foundation for computational mathematics and in some sense for computer science. We have shown it to be a good basis for programming language semantics in work with Harper, Cleaveland and Crary. We have shown it to be an excellent basis for automated reasoning in contributions by Howe, Basin, Murthy, Jackson and Hickey; these contributions have made Nuprl one of the major *theorem provers* in use today. Nuprl introduced the notion of a *tactic-tree-proof* (with Bates, Knoblock and Griffin) and the notion of a *refinement logic* due to Bates.

The Nuprl system has evolved through several versions, from one to the current five. Many people have helped with the design, principally Allen, Howe, Jackson, Hickey and Eaton (who is also the chief programmer). The system is being used in various applications in both hardware and software verification and in computer algebra system semantics (Jackson). The current application to the Ensemble group communication system was driven by the work of Hayden, Hickey and Kreitz.

Currently, we are also investigating how to express computational complexity in constructive type theory, and we are using Nuprl to help create a distributed digital library of formalized mathematics. This work is displayed on the Nuprl home page along with a list of over 80 publications about the system and its type theory. See www.cs.cornell.edu/Info/Projects/Nuprl.

**Main Publications:**

1. *Implementing Mathematics with the Nuprl Proof Development System*, pages 266–281, Prentice-Hall, 1986 (with PRL Group).
2. A Causal Logic of Events in Formalized Computational Type Theory. In *Logical Aspects of Secure Computer Systems, Proceedings of International Summer School Marktoberdorf, 2005*, to be published, 2006 (with Mark Bickford).
3. Naive Computational Type Theory . In *Proof and System-Reliability* H. Schwichtenberg and R. Stein-

bruggen (eds.), pages 213–259, 2002.

4. Computational Complexity and Induction for Partial Computable Functions in Type Theory. In *Reflections on the Foundations of Mathematics: Essays in Honor of Solomon Feferman*, editors W. Sieg, R. Sommer, and C. Talcott, Association for Symbolic Logic, 2001, pages 166–183 (with K. Crary).

5. Nuprl's Class Theory and its Applications. In *Foundations of Secure Computation*, editors F. L. Bauer and R. Steinbrueggen, NATO Science Series F, IOS Press, Amsterdam, 2000, pages 91–116.

6. Constructively Formalizing Automata. In *Proof Language and Interaction: Essays in Honour of Robin Milner*, MIT Press, Cambridge, 2000, pages 213–238 (with P. B. Jackson, P. Naumov, and J. Uribe).

7. Types in Logic, Mathematics and Programming. In *Handbook of Proof Theory*, editor S. R. Buss, Elsevier Science B.V., 1998, pages 683–786.

8. Metalogical Frameworks. In *Logical Environments*, editors G. Huet and G. Plotkin, Cambridge University Press, 1993, pages 1–29 (with David A. Basin).

9. Implementing Metamathematics as an Approach to Automatic Theorem Proving. In *A Source Book of Formal Approaches in Artificial Intelligence*, North-Holland, 1990, pages 45–75 (with D. Howe).

10. Assigning Meaning to Proofs: A Semantic Basis for Problem Solving Environments. In *Constructive Methods in Computing Science*, editor M. Broy, NATO ASI Series, Vol. F55, Springer-Verlag, 1989, pages 63–91.

11. Themes in the Development of Programming Logics Circa 1963-1987. In *Annual Review of Computer Science*, Vol. 3, 1988, pages 147–165.

12. Innovations in Computational Type theory using Nuprl. To appear in *Journal of Applied Logic*, 2006 (with PRL Group).

13. Computational Foundations of Basic Recursive Function Theory. In *Theoretical Computer Science B: Logic, Semantics, and Theory of Programming*, Vol. 120, 1993, pages 89–112 (with S. F. Smith).

14. On Writing Programs that Construct Proofs. In *Journal of Automated Reasoning*, Vol. 1, 1985, pages 285–326 (with T. Knoblock and J. Bates).

15. Proofs as Programs. In *Transactions on Programming Languages and Systems*, Vol. 7(1), 1985, pages 113–136.

16. On Computational Complexity of Scheme Equivalence. In *Proceedings of the Eighth Princeton Conference on Information Sciences and Systems*, 1974; also *SICOMP*, Vol. 9(2), 1980, pages 396–416 (with H. Hunt and S. Sahni).

17. The Operator Gap. In *Proceedings of the Ninth IEEE Symposium on Switching and Automata Theory*, 1969, pages 20–26 (expanded in *Journal of the ACM*, Vol. 19(1), 1972, pages 175–183).

18. On the Efficiency of Programs in Subrecursive Formalisms. In *Proceedings of the Tenth IEEE Symposium on Switching and Automata Theory*, 1972, pages 60–67; also as Subrecursive Programming Languages I; also, in *Journal of the ACM*, Vol. 19(3), 1972, pages 526–586 (with A. Borodin).

19. On Classes of Program Schemata. In *SIAM Journal of Computing*, Vol. 1(1), 1972, pages 66–118 (with D. Gries).

20. Extracting Programs from Constructive HOL Proofs via IZF Set-Theoretic Semantics. In *Proceedings of International Joint Conference on Automated Reasoning (IJCAR 2006)*, With Wojciech Moczydlowski.

21. Building Reliable, High-Performance Systems from Components. In *Proceedings of 17th ACM Symposium on Operating System Principles (SOSP'99)*, Operating Systems Review, vol. 34, no. 5, pages 80–92, 1999 (with Xiaoming Liu, Christoph Kreitz, Robbert van Renesse, Jason Hickey, Mark Hayden and Ken Birman).

**Service to the Profession:** Editorships: *The Computer Journal*, Oxford University Press; *Journal of Logic and Computation*, Oxford University Press; *Formal Methods in System Design*, Kluwer Academic Publishers; *Journal of Symbolic Computation*, Academic Press. Director, NATO Summer School at Marktoberdorf.
Memberships: Computing Research Association (CRA) Board (elected); General Chair, LICS (1991-1994) Association for Symbolic Logic, elected member of ASL Council (1995-1998) ACM, SIGACT, SIGART, and SIGPLAN.

**Teaching:** Professor Constable is a graduate of Princeton University where he worked with Alonzo Church, one of the pioneers of computer science, and he did his PhD with Stephen Cole Kleene, another pioneer. He joined the Cornell faculty in 1968. He has supervised over forty PhD students in computer science. Constable is known for his work connecting programs and mathematical proofs which has led to new ways of automating the production of reliable software. The method he introduced in 1971 is known by the slogan proofs-as-programs. He has written three books on this topic as well as numerous research articles. He and his colleagues at Cornell are known for creating Computational Type Theory which is the basis of the Nuprl Proof Development system that has been used since 1984 in the design and verification of software systems. Two open problems in mathematics were solved using Nuprl ("new pearl"), and many important theorems in constructive mathematics have been proved using the system. The CTT type theory is related to Martin-Lof's Intuitionistic Type Theory. He has supervised 43 PhD students to date.

**Vision Statement:** The intellectual future of logic is bright because its universality adds value to our understanding in diverse fields of study from philosophy and law to mathematics, linguistics, and computer science. Lately computer science has greatly expanded the technical scope and relevance of logic, and I see that continuing unabated. I foresee a world in which interactive and automatic theorem provers will be used to create a vast formal digital library of computer checked and computer generated mathematics. A significant part of this formal material will include algorithms that are embedded in formal explanations of what they do. Already the demonstration that computer automation of reasoning significantly extends our ability to solve hard scientific problems is one of the significant contributions of computer science to logic and to intellectual history in general. The partnership of logic and computer science has made it possible to realize the dreams of Gottfried Wilhelm Leibniz expressed in his unpublished drafts that have stimulated logicians for three centuries.

**Honours and Awards:** ACM Fellow, 1994; John Simon Guggenheim Fellowship, 1990-91; Outstanding Educator Award, 1987.

# CORCORAN, John

**Specialties:** History of logic, philosophy of logic, mathematical logic, metaphysics, epistemology, philosophy of mathematics, linguistics (syntax, semantics, and pragmatics).

**Born:** 20 March 1937 in Baltimore, Maryland, USA.

**Educated:** University of California, Mathematics, 1965; Yeshiva University, Mathematics, 1964; Johns Hopkins University, BES Mechanical Engineering, 1959, MA Philosophy, 1962; PhD Philosophy, 1963; Baltimore Polytechnic Institute A, 1959.

**Dissertation:** *Generative Structure of Two-valued Logics*; supervisor, Robert McNaughton.

**Regular Academic or Research Appointments:** PROFESSOR OF PHILOSOPHY, UNIVERSITY OF BUFFALO (SUNY), 1973–. University of Pennsylvania, Linguistics, 1965-1969; University of Buffalo, Philosophy, 1969-1973; IBM Research Center, Mathematics, 1963-1964.

**Visiting Academic or Research Appointments:** Visiting Lecturer in Philosophy, University of California, Berkeley, 1964-1965; Visiting Associate Professor of Philosophy and Research Associate, University of Michigan, 1969–1970; Associate Professor, Linguistic Institute, 1971; Visiting Scholar, Linguistic Institute 1976; Visiting Professor, University of Santiago de Compostela, 1994.

**Research Profile:** Corcoran's work in history of logic spans most of the periods with original contributions concerning Aristotle, the Stoics, Ockham, Boole, Dedekind, the American Postulate Theorists, Tarski, and Quine. His work in philosophy of logic focuses on the nature of logic, the conceptual structure of logic, the metaphysical and epistemological presuppositions of logic, the nature of mathematical logic and the gaps between logical theory and mathematical methods. His work in mathematical logic treats propositional logics, modal logics, identity logics, syllogistic logics, the logic of variable-binding term operators, second-order logics and the theory of strings, the theory which is foundational in all areas of logic and which provides essential background for all of his other work. In philosophy of mathematics Corcoran has consistently been guided by a nuanced and inclusivionary platonism which strives to do justice to all aspects of mathematical and logical experience especially those aspects most emphasized by competing philosophical perspectives such as logicism, constructivism, deductivism, and formalism. Although several of his philosophical papers presuppose little history or mathematics, his historical papers often involve original mathematics. He has referred to this aspect of his approach to history as mathematical archeology. Moreover, his philosophical papers often involve original history. He has been guided by the Aristotelian principle that the nature of a modern concept is sometimes best understood in light of its historical development, a view that he attributes to Arthur Lovejoy's History of Ideas program at Johns Hopkins University and in which he has been encouraged by the American philosopher Peter Hare.

**Service to the Profession:** Founding member of the Editorial Board, History and Philosophy of Logic, 1980-; Chair, Buffalo Logic Colloquium, 1970-; Co-chair, Conference on Gaps between Logical Theory and Mathematical Practice (Shapiro, Scanlan, Tiezsen, Kearns, et. al.), 2001; Sponsor of Alonzo Church for Doctor Honoris Causa at University of Buffalo, 1989; Reviewer, Philosophy of Science and Journal of Symbolic Logic; Conference Organizer, Buffalo New York, Church Symposium (Church, Davis, Henkin, Rogers), 1989, Nature of Logic (Tarski, Putnam, Friedman, Jech, Vesley, Goodman, et. al.), 1973, Ancient Logic (Corcoran, Kretzmann, Mueller, et, al.), 1972; Founder, Buffalo Logic Colloquium, 1970; Reviewer, Mathematical Reviews, 1969-1998; Cofounder, Philadelphia Logic Colloquium (with George Weaver), 1966.

**Teaching:** In his teaching Corcoran emphasizes

the intensely and essentially personal nature of all knowledge including logical knowledge and he also emphasizes how much each person can benefit in the personal search for truth from cooperation with other objective researchers.

**Vision Statement:** MISSING

**Honours and Awards:** Festschrift special double issue of *History and Philosophy of Logic*, 2000 (Eds. M. Scanlan, S. Shapiro); Exceptional Scholar Award from The University of Buffalo, 2002; Doctor Honoris Causa from University of Santiago de Compostela (Spain), 2003; Corcoran Symposium, University of Santiago de Compostela (Spain), 2003.

# CRAIG, William

**Specialties:** Algebraic first-order logic, proof theory, philosophy of logic.

**Born:** 13 November 1918 in Nuremberg, Germany.

**Educated:** Harvard University, PhD Philosophy, 1951; Princeton University, 1949-1950; Swiss Federal Institute of Technology, 1948-1949; Harvard University, 1946-1948; University of California, Berkeley 1940-1941; Cornell University, BA Philosophy and Physics, 1940.

**Dissertation:** *A Theorem About First Order Functional Calculus With Identity and Two Applications*; supervisor W. V. O. Quine.

**Regular Academic or Research Appointments:** PROFESSOR EMERITUS, PHILOSOPHY, UNIVERSITY OF CALIFORNIA, BERKELEY 1989– ; Professor, Philosophy, University of California, Berkeley, 1961-1989; Associate Professor, Mathematics, Pennsylvania State University, 1957-1961; Assistant Professor, Mathematics, Pennsylvania State University, 1952-1957; Mathematics Instructor, Pennsylvania State University, 1951-1952.

**Visiting Academic or Research Appointments:** Mathematics, University of California, Berkeley, 1960-1961; Mathematics, University of Notre Dame, 1957-1958.

**Research Profile:** In philosophy of science, a distinction has often been made between notions linked to observations and notions that are of a more theoretical nature. In his PhD Thesis and in [1960], predicates occurring in any given first-order axiom system are divided into two classes, auxiliary and non-auxiliary. With the help of Herbrand's Theorem or Gentzen's extended Hauptsatz, derivations from the axioms of those theorems in which there occur no auxiliary predicates are restructured so as to minimize manipulation of the auxiliary predicates. This yields an axiomatization, recursive but, in general, not very transparent, of the set of those theorems that do not contain auxiliary predicates. The Herbrand-Gentzen Theorem is used in a similar manner in [1957(a)] and [1957(b)] to obtain an interpolation theorem for first-order logic. It can be used to replace implicit by explicit definitions. An important failure of interpolability is discussed in [1965(a)]. Use of auxiliary predicates for obtaining finite axiomatization is discussed in [1958(a)], written jointly with R.L. Vaught. It turned out that recursive axiomatizability of subtheories can be obtained much more generally by "Craig's trick" described in [1953] and [1956]. Its philosophical significance, or the lack of it, has been discussed by Carl Hempel, Ernest Nagel and others.

Among operations that are logical or invariant, there are significant differences. This gives rise to subclasses with quite distinct features. One subclass is discussed in [1965(b)] and in [1978], another in [1989(a)] and [1989(b)], with emphasis on proof-theoretic aspects, and a third in [1974(b)] and [2006].

Between linear reasoning and equational reasoning there are major similarities and major differences. Craig's interest in the former led him to investigate the latter. In [1974(a)], there are given three equational systems of first-order logic (with equality) that are sound and complete. The first of these concerns operations on set of sequences that are of finite length. Its relationship to mathematical practice is closer than that of the others. One should like to obtain a system of this kind in which the role of the operations that are not Boolean, and also interaction with Boolean operations, comes out more clearly. [2006] is intended to serve as one of the preparatory steps towards this.

**Main Publications:**

1. 1953 "On Axiomatizability within a System," *JSL* 18(1): 30-2.
2. 1956 "Replacement of Auxiliary Expressions," *Phil. Rev.*, LXV (1): 38-55.
3. 1957(a) "Linear Reasoning. A New Form of the Herbrand-Gentzen Theorem," *JSL*, 22(3): 250-68.
4. 1957(b) "Three Uses of the Herbrand-Gentzen Theorem in Relating Model Theory and Proof Theory," *JSL*, 22 (3): 269-85.
5. 1958(a) "Finite Axiomatizability Using Additional Predicates," (with R.L. Vaught), *JSL*, 23 (3): 289-309.
6. 1958(b) Definitional Independence of Combinators, Section 5, pp. 179-84, in: *Combinatory Logic*, by H.B. Curry and R. Feys, Amsterdam, North-Holland Publishing Co.

7. 1960 "Bases for First-Order Theories and Subtheories," *JSL*, 25 (2): 97-142.

8. 1965(a) "Satisfaction for n-th Order Languages Defined in n-th Order Languages," *JSL*, 30 (1): 13-25.

9. 1974(a) Logic in Algebraic Form. Three Languages and Theories. North-Holland Publishing Co. 1974.

10. 1974(b) "Diagonal Relations," Proceedings of the Tarski Symposium. Association for Symbolic Logic. 1974, pp. 91-104.

11. 1978 "Boolean Logic and the Everyday Physical World", Proc. And Addresses American Philosophical Association, 52: 751-778.

12. 1989(a) Near-equational and Equational Systems of Logic for Partial Functions, Journal of Symbolic Logic, vol. 54, pp. 795-827 and 1181-1215.

13. 1989(b) Logical Partial Functions and Extensions of Equational Logic. Logic Colloquium '88. Ferro, Bonotto, Valentini and Zanard (eds.). North-Holland, 1989, pp. 319-354.

14. 2006 Semigroups Underlying First-Order Logic, Memoirs of the American Mathematical Society, November 866, XXV+ 263 pp.

*Work in Progress*

15. An article entitled "Uniform Functions That Underlie First-Order Logic" is about to be submitted. It is intended to serve as a less abstract follow-up of [2006]. Whereas, in [2006], functions on sequences and their converses are treated as elements of a semigroup with involution, in the article one considers conditions satisfied by these functions that are meaningful for any unary functions. A set of conditions is given that are necessary and sufficient for an arbitrary unary partial algebra to be isomorphic to one of the intended algebras of unary partial functions on finite sequences.

**Service to the Profession:** Editor, Journal of Symbolic Logic, 1976-1981; President, Pacific Division of the American Philosophical Association, 1978-1979; President, Association for Symbolic Logic, 1965-1967.

**Honours and Awards:** NSF Senior Post-Doctoral Fellow, 1968-1969; Member, Miller Institute for Basic Research in Science, 1965-1966; Howison Travelling Fellowship in Philosophy, UC Berkeley, 1940-1941, 1946-1949; President's Scholarship, Cornell University, 1938-1940.

# CRESSWELL, Maxwell John

**Specialties:** Modal logic, formal semantics

**Born:** 19 November, 1939 in Wellington, New Zealand.

**Educated:** Victoria University of Wellington, LitD, 1972; University of Manchester, PhD, 1964; University of New Zealand, MA (1st class honours in Philosophy) 1961, BA, 1960.

**Dissertation:** *General and Specific Logics of Functions of Propositions*; supervisor, Arthur Prior.

**Regular Academic or Research Appointments:** PROFESSOR OF PHILOSOPHY, UNIVERSITY OF AUKLAND, 2004–; Massey University, Professor of Philosophy, 2001; Victoria University of Wellington, Professor of Philosophy, 1974–2000; Reader in Philosophy, 1973; Senior Lecturer in Philosophy, 1968–1972; Lecturer in Philosophy, 1963–1967.

**Visiting Academic or Research Appointments:** Professorial Fellow in Defence Studies, Massey University, 2003, 2006, 2008; Visiting Professor, Texas A&M University, 2002–; Visiting Professor, University of California, Davis, January–March, 2000; Visiting Professor, Colby College, August–December, 1996; Visiting Professor, University of Bologna, March 1996; Visiting Professor, University of Massachusetts, September–December 1989–1992; Visiting Professor, Universität Konstanz, June–July 1983; Visiting Professor, Universität Stuttgart, May–June 1976; Visiting Professor, UCLA January–March 1970

**Research Profile:** Max Cresswell's research interests in logic are principally in modal logic. He is best known for the introductions to modal logic that he wrote with George Hughes in 1968, 1984 and 1996. Among his early papers in modal logic is perhaps the first completeness proof of a first-order logic weaker than S5 with the Barcan formula. This was received in 1965 by the *Notre Dame Journal of Formal Logic*, though it did not appear until 1969. More recently he has addressed the problem of frame completeness in modal predicate logic. In particular he has established that the predicate extensions of a wide class of logics with a 'discreteness axiom', such as the logics of discrete time, or the so-called 'provability' logics, are none of them characterizable by any class of relational frames, whether or not they contain the Barcan Formula. He has also applied logical methods to the formal semantics of natural language, concentrating in particular on the 'intensional' features of such languages in such areas as the analysis of adverbial modification. This work has resulted in the development of tests for the ontological commitments of natural languages, whose syntax may not admit explicit variable-binding. In particular he has argued that logical tests establish that natural language requires an indexical semantics which is as powerful as explicit variable-binding, including over such things as possible worlds. His research interests also include the history of philosophy and he has published articles in ancient philosophy, and on the philosophy of John Locke and F.H. Bradley.

**Main Publications:**

1. Possibility semantics for intuitionist logic. *Australasian Journal of Logic*, Vol 2, 2004, pp.11–29
2. How to complete some modal predicate logics. *Advances in Modal Logic*, Vol 2, (ed M. Zakharyaschev, K. Segerberg, M. de Rijke and H. Wansing.), Stanford, CSLI Publications, 2001, pp. 155–178
3. Some incompletable modal predicate logics. *Logique et Analyse* No 160, 1997, pp. 321-334 [Published October 2000]
4. A note on de re modalities. *Logique et Analyse* No 158, 1997, pp.147-153
5. *A New Introduction to Modal Logic*, (with G.E. Hughes) London, Routledge, 1996
6. *Semantic Indexicality*, Dordrecht, Kluwer, 1996
7. Incompleteness and the Barcan Formula. *Journal of Philosophical Logic* Vol 24, 1995, pp.379–403
8. *Language in the World*, Cambridge, Cambridge University Press, 1994
9. *Entities and Indices*, Dordrecht, Kluwer, 1990
10. Magari's theorem via the recession frame. *Journal of Philosophical Logic* Vol 16, 1987, pp.13–15
11. *Structured Meanings: The Semantics of Propositional Attitudes*, Bradford Books/MIT Press, 1985
12. (with G.E. Hughes) *A Companion to Modal Logic*, London, Methuen, 1984
13. An incomplete decidable modal logic. *The Journal of Symbolic Logic* Vol 49, 1984, pp.520–527
14. The completeness of KW and K1.1. *Logique et Analyse* No 102, 1983, pp.123-127
15. *Logics and Languages*, London, Methuen, 1973
16. (with G.E. Hughes) *An Introduction to Modal Logic*, London, Methuen, 1968
17. A Henkin completeness theorem for T. *Notre Dame Journal of Formal Logic* Vol 8, 1967, pp.186–190
18. From modal discourse to possible worlds. *Studia Logica*, vol. 82, 2006, 307-327.
19. Now is the time. *Australasian Journal of Philosophy*, vol. 84, 2006, pp. 311–332.
*Work in Progress*
20. A book with A.A. Rini on *The World Time Parallel*. This will argue that the formal logical parallel between modal and temporal discourse is too close to be ignored.

**Service to the Profession:** President, NZ Division of the Australasian Association of Philosophy, 2003; President, Australasian Association of Philosophy, 1985-1986; Member of the council of the Association for Symbolic Logic, 1977-1980; Secretary, NZ Division of the Australasian Association of Philosophy and President, 1968.

**Teaching:** Cresswell taught for 36 years at the Victoria University of Wellington. His most notable logic student is probably Rob Goldblatt, now Professor of Mathematics at VUW. Although he was not a PhD adviser to any student at the University of Massachusetts he taught many students of the likes of Ted Sider, Geoff Goddu, Hotze Rullmann, Kai von Fintel Adriane Rini and others in logic and semantics.

**Vision Statement:** For logicians who are also philosophers the importance of logic is often as a test of the consistency or coherence of a philosophical position. One particular area is that of modal predicate logic. Most modal logicians find that the demands of predicate logic do not suit the mathematical techniques that serve so well in propositional modal logic, and sometimes they try to change the rules to suit this. On the other hand most philosophers who use modal predicate logic do not have the experience in the technicalities of modal logic to be able to evaluate important philosophical positions.

**Honours and Awards:** New Zealand Government Marsden grant (2007–2009) with A. A. Rini, to study the 'world-time parallel. Victoria University of Wellington, Emeritus Professor, 2001; *Logique et Analyse*, Number 181, March 2003 [published November 2004] is a 'Festschrift for Max Cresswell on the occasion of his $65^{th}$ birthday'; Hägerström Lecturer, Uppsala University, 2000; Faculty Fellowship, Institute for Advanced Studies in the Humanities, University of Edinburgh, 2000; Visiting Scholarship, St John's College, Cambridge, 1992; Visiting Fellowship, Centre for Cognitive Science, University of Edinburgh, 1992; Residence at Rockefeller Study and Conference Center, Bellagio, 1988; Claude McCarthy Fellowship 1988; Commonwealth Universities Interchange Scheme Visitorship (United Kingdom) 1979; Commonwealth Scholarship (United Kingdom) 1961–1963

# D

## DA COSTA, Newton C. A.

**Specialties:** Non classical logics, model theory, induction and probability, foundations of physics and philosophy of science.

**Born:** 16 September 1929 in Curitiba, Brazil.

**Educated:** Federal University of Paraná, Engineering, 1952; Mathematics, 1956, PhD, Mathematics, 1961.

**Dissertation:** *Topological spaces and continuous functions*; supervisor: J. Rémy Freire.

**Regular Academic or Research Appointments:** RETIRED, 1999; Professor of Logic and Philosophy of Science, University of São Paulo, 1985; Professor of Philosophy, State University of Campinas, 1985; Professor of Mathematics, University of São Paulo, 1970; Professor of Mathematics, State University of Campinas, 1968; Professor of Mathematics, Federal University of Paraná, 1957.

**Visiting Academic or Research Appointments:** Visiting Professor of Philosophy at the Federal University of Santa Catarina, Brazil; Visiting Professor of Philosophy, University of Sienna, 1986; Visiting Professor of Philosophy, University of Turin, 1982; Visiting Professor of Philosophy, Australian University, 1976; Visiting Professor, Catholic University of Chile, 1976; Visiting Professor, University of Torun, 1975; Visiting Professor, University of Paris, 1972 and 1999; Visiting Scholar, University of California, 1972; Visiting Professor, University of Buenos Aires, 1969; Visiting Professor, University of Bahia Blanca (Argentina), 1968; Visiting Scholar, University of Paris, 1967; Visiting Professor, Federal University of Rio de Janeiro, 1963.

**Research Profile:** Newton da Costa was one of the founders of paraconsistent logic; he studied several applications of such logic in philosophy, law, computing and Artificial Intelligence. He constructed the theory of quasi-truth that constitutes a generalization of Tarski's theory of truth, and applied it to the foundations of science. He also worked in model theory through the theory of valuations and in generalized Galois theory. He investigated the foundations of physics, specially the axiomatic bases of quantum theory and relativity. Da Costa and his colleague, F. A. Doria have being working in complexity theory, obtaining, among other results, theorems of relative consistency connected with the problem P=NP. At present, he is investigating the so-called abstract logics, which constitute a species of structures that generalizes the usual logical systems; his main objective is to unify abstract Galois theory, as developed by J. Sebastião e Silva and M. Krasner, and the theory of abstract logics.

**Main Publications:**

1. Calculs propositionels pour les systèmes formels inconsistants, *C.R.Acad.Sc.Paris*, 257, 1963, 3790–3793.
2. Calculs de prédicats pour les systèmes formels inconsistants, *C.R.Acad.Sc.Paris*, 258, 1964, 27–29.
3. Sur un système inconsistant de la théorie des ensembles, *C.R.Acad.Sc.Paris*, 258, 1964, 3144–3147.
4. Opérations non-monotones dans les treillis, *C.R.Acad.Sc.Paris*, 263A, 1966, 429–432.
5. On a set theory suggested by Ehresmann and Dedecker, *Proceedings of the Japan Academy of Sciences* 45,1 1969, 880–888.
6. On the theory of inconsistent formal systems, *Notre Dame Journal of Formal Logic*, XV (4), 1974, 497–510.
7. Pragmatic Probability, *Erkenntnis*, 25, 1986, 141–162.
8. *Logique Classique et Non-Classique*, Paris, Masson, 1997
9. Pragmatic Truth and the Logic of Induction, with S. French, *The British Journal for the Philosophy of Science*, 40, 1989, 333-356.
10. The Model-Theoretic Approach in the Philosophy of Science, with S. French, *Philosophy of Science*, 57, 1990, 248–265.
11. Towards an Acceptable Theory of Acceptance: Partial Structures, Inconsistency and Correspondence, with S. French, in *Correspondence, Invariance and Heuristics*, S. French and H. Kamminga eds., Dordrecht, Kluwer Academic Publishers, 1993, 137–158.
12. A Model Theoretic Approach to 'Natural Reasoning', with S. French, *International Studies in the Philosophy of Science*, 7, 1993, 177–190.
13. *Partial Truth: A unitary approach to models and scientific reasoning*, with Steven French, Oxford Un. Press, 2003.
14. Complementarity and paraconsistency, with D. Krause, in *Logic, Epistemology, and the Unity of Science*, S. Rahman, J. Symons, D. M. Gabbay, J. -P. van Bendegen eds., Kluwer Ac. Press, 2004, 557–568.
15. Paraconsistent logics as a formalism for reasoning about inconsistent knowledge bases, with V. S. Subrahmanian, *Artificial Intelligence in Medicine*, 1, 1989, 167–174. DAVIS, Martin D. 57

16. Outlines of a paraconsistent category theory, with O. Bueno and A. G. Volkov, in *Alternative Logics: do sciences need them?*, P.Weingartner ed., Springer, 2004, 95–114.

17. The logic of pragmatic truth, with O. Bueno and S. French, *Journal of Philosophical Logic*, 27, 1998, 603–620.

18. The paraconsistent logic PT, with V. S. Subrahmanian and C. Vago, *Zeitschr. f. math. Logik und Grundlagen d. Math.*, 37, 1991, 139–148.

19. Pragmatic truth and approximation to truth, with I. Mikenberg and R. Chuaqui, *Journal of Symbolic Logic*, 51, 1986, 201–221.

20. Consequences of an exotic formulation for P=NP!, with F. A. Doria, *Applied Mathematics and Computation*, 145, 2003, 655–665.

*Work in Progress*

21. 21. A book covering the fields of paraconsistent logic and its applications, in collaboration with Décio Krause and Otávio Bueno; a book on the foundations of physics, in collaboration with Francisco A. Doria.

**Service to the Profession:** Member, Institute for Advanced Studies, University of São Paulo; President, the Paranaense Society of Mathematics; President, the Brazilian Association of Logic; Member, Committee of the Association of Symbolic Logic of South America; Director, Institute of Mathematics, University of São Paulo.

**Teaching:** da Costa has taught a variety of courses in the areas of logic, mathematics and philosophy of science, at the graduate and undergraduate levels. He supervised more than 30 Master and PhD theses, in Brazil and other countries.

**Vision Statement:** da Costa believes that the significant progress in the field of logic will give rise to new fundamental developments in computing and technology, especially in connection with non classical logics and their applications. Logic will also become more and more important in the areas of philosophy (in particular in philosophy of science), mathematics, science, technology, and computer science.

**Honours and Awards:** Member, Academy of Sciences of the State of São Paulo; Member, Academy of Sciences of Chile; Member, Institute of Philosophy of Peru; Member, Institute of Philosophy of Paris; 'Moinho Santista' Prize in Exact Sciences, 1991; 'Jabuti' Prize in Exact Sciences, 1992; Nicholas Copernic Medal, University of Torun, 1990; Doctor Honoris Causa, Federal University of Paraná, 1993; ASTEF Fellowship, France, 1967; Fulbright Fellowship, 1982; Exact Sciences Medal, State of Paraná, Brazil, 1998.

# DAVIS, Martin D.

**Specialties:** Recursion theory, Diophantine decision problems, automated deduction, history of logic.

**Born:** 8 March 1928 in New York, USA.

**Educated:** Princeton University, PhD, 1950; Princeton University, MA, 1949; City College of the College of the City of NY, BS, 1948.

**Dissertation:** *On the Theory of Recursive Unsolvability*; supervisor, Alonzo Church.

**Regular Academic or Research Appointments:** NEW YORK UNIVERSITY, PROFESSOR EMERITUS OF MATHEMATICS 1965–. Joint appointment with Computer Science, 1969–; Chair of Computer Science, 1988–1990; Yeshiva University (Belfer Graduate School of Science), Associate Professor and Professor, 1960–1965; Rensselaer Polytechnic Institute (Hartford Graduate Division), Assistant Professor and Associate Professor of Mathematics 1956–1959; Ohio State University Assistant Professor of Mathematics, 1955–1956; University of California, Davis Assistant Professor of Mathematics, 1954–1955; Institute for Advanced Study, School of Mathematics, Visiting Member 1952–1954; University of Illinois, Champaign-Urbana, Research Associate, Control Systems Laboratory 1951–1952; Research Instructor in Mathematics, 1950–1951.

**Visiting Academic or Research Appointments:** Adjunct Professor of Mathematics, Mills College, 1997; Visiting Scholar, University of California, Berkeley, 1996–; Visiting Professor of Mathematics, University of California, Santa Barbara, 1978–1979; Visiting Professor, Yeshiva University (Belfer Graduate School of Science), 1970–1971; Visiting Professor, University of London, Westfield College, 1968–1969; Adjunct Associate Professor of Mathematics, New York University, Research Scientist, 1959–1960.

**Research Profile:** In his dissertation, Martin Davis began his life-long involvement with Hilbert's tenth problem (which sought an algorithm for determining solvability in integers of polynomial Diophantine equations), obtaining a normal form for recursively enumerable sets expressed as a quantificational prefix preceding a Diophantine equation in which all the quantifiers are existential except for a single bounded universal quantifier. The work towards eliminating that universal quantifier (and thus obtaining the unsolvability of the tenth problem) proceeded with joint work with Hilary Putnam that showed how to do this by permitting variable exponents in the equation, subject however to the hypothesis (proved in 2005 by Tao and Green) that there are arbitrarily long arithmetic progressions of primes. (Julia Robinson showed how to do without that hypothesis, which in 1960 was crucial. Together with her earlier work, this reduced the problem to finding a single Diophantine equation satisfying a certain rate of growth condition which Yuri Matiyasevich famously and elegantly did in 1970.) Davis's dissertation also introduced the important hyperarithmetic hierarchy and proved some of its simpler properties. His work on automated deduction began with his computer program for Presburger arithmetic. Joint work with Hilary Putnam on proof procedures for first-order logic introduced the Davis-Putnam procedure for the satisfiability problem which (with a modification introduced by Davis in collaboration with Don Loveland and George Loveland) is still in use. Davis's later work emphasized the crucial importance of complementary matching of literals and introduced much of the framework and terminology of that field. Much of Davis's research effort was devoted to reshaping the manner of exposition of aspects of logic leading to his books on computability and nonstandard analysis. In collaboration with Elaine Weyuker, he worked on theoretical aspects of software testing. His most recent efforts have been in the history of logic, including explorations of Gödel's thought, editing E.L. Post's collected works, and publishing a "prehistory" of the computer for a popular audience.

**Main Publications:**

1. Arithmetical Problems and Recursively Enumerable Predicates, *Journal of Symbolic Logic*, vol. 18, 1953, pp. 33–41.
2. *Computability and Unsolvability*, McGraw-Hill, New York 1958; reprinted with an additional appendix, Dover 1983.
3. Reductions of Hilbert's Tenth Problem, with Hilary Putnam, *Journal of Symbolic Logic* vol.23, 1958, pp. 183-187.
4. A Program for Presburger's Algorithm, Summaries of Talks Presented at the Summer Institute for Symbolic Logic, Cornell University, 1957, Institute for Defense Analyses, 1960, pp. 215–223; reprinted in *Automation of Reasoning*, Siekmann, Jörg and Graham Wrightson eds., vol. 1, Springer Verlag, 1983, pp. 41–48.
5. A Computing Procedure for Quantification Theory, with Hilary Putnam, *Journal of the Association for Computing Machinery*, vol. 7, 1960, pp. 201–215; reprinted in *Automation of Reasoning, vol. 1*, Siekmann, Jörg and Graham Wrightson eds., Springer Verlag, 1983, pp. 125–139.
6. The Decision Problem for Exponential Diophantine Equations, with Hilary Putnam and Julia Robinson, *Annals of Mathematics*, vol.74, 1961, pp. 425–436.
7. Eliminating the Irrelevant from Mechanical Proofs, *Proceedings of Symposia in Applied Mathematics*, vol.15, 1963, pp. 15–30. reprinted in *Automation of Reasoning*, Siekmann, Jörg and Graham Wrightson eds., vol. 1, Springer Verlag, 1983, pp. 315–330.
8. Editor, *The Undecidable*, Raven Press 1965, reprinted Dover 2004.
9. One Equation to Rule Them All, *Transactions of the New York Academy of Sciences*, Sec. II, vol. 30, 1968, pp. 766–773.
10. An Explicit Diophantine Definition of the Exponential Function, *Communications on Pure and Applied Mathematics*, vol.24, 1971, pp. 137–145.
11. On the Number of Solutions of Diophantine Equations, *Proceedings of the American Mathematical Society* vol.35, 1972, pp. 552–554.
12. Hilbert's Tenth Problem is Unsolvable, *American Mathematical Monthly*, vol. 80, 1973, pp. 233–269; reprinted in Davis & Martin, *Computability and Unsolvability*, Dover 1983.
13. Hilbert's Tenth Problem: Diophantine Equations: Positive Aspects of a Negative Solution, with Yuri Matijasevic and Julia Robinson, *Proceedings of Symposia in Pure Mathematics*, vol.28, 1976, pp. 323–378; reprinted in *The Collected Works of Julia Robinson*, Feferman & Solomon eds., Amer. Math. Soc. 1996, pp.269-378.
14. *Applied Nonstandard Analysis*, Interscience-Wiley, 1977, reprinted Dover 2005.
15. A Relativity Principle in Quantum Mechanics, *International Journal of Theoretical Physics*, vol.16, 1977, pp. 867–874.
16. Obvious Logical Inferences, *Proceedings of the Seventh Joint International Congress on Artificial Intelligence*, 1981, pp. 530–531.
17. Why Gödel Didn't Have Church's Thesis, *Information and Control*, vol. 54, 1982, pp. 3–24.
18. A Formal Notion of Program-Based Test Data Adequacy, with Elaine J. Weyuker, *Information and Control*, vol. 56, 1983, pp. 52–71.
19. Solvability, Provability, Definability: The Collected Works of Emil L. Post, edited by Martin Davis and including the article: Emil L. Post: His Life and Work, pp. xi-xxviii, Birkhäuser, 1994.
20. The Universal Computer: The Road from Leibniz to Turing, W.W. Norton, 2000; Paperback edition: *Engines of Logic: Mathematicians and the Origin of the Computer*, W.W. Norton, 2001.
21. The Myth of Hypercomputation, *Alan Turing: Life and Legacy of a Great Thinker*, Christof Teuscher ed., Springer 2004, pp. 195–212.
22. What Did Gödel Believe and When Did He Believe It? *Bulletin of Symbolic Logic*, vol. 11, 2005, pp. 194–206.

**Service to the Profession:** Moderator, FOM (Foundations of Mathematics) email list; Editorial Board, Journal of Automated Reasoning; Editorial Board, Journal of Symbolic Logic; Editorial Board, Journal of the Association for Computing Machinery; MAA award committees, Chauvenet

Prize and Hedrick Lecturer; Nominations Committee, American Mathematical Society; Nominations Committee, Section A, A.A.A.S; Chairman, Nominations Committee, Association for Symbolic Logic; Chairman, Committee on Academic Freedom, Tenure, and Employment, Security of the American Mathematical Society; Program Committee of Fifth Conference on Automated Deduction; Local Arrangements Chairman for Sixth Conference on Automated Deduction; Committee Chairman, first winner of prize for a "landmark" contribution to automatic theorem proving; Program Committee Member, "Logic in Computer Science," 1989; American Mathematical Society Representative, AAAS Section T (Information, Computing and Communication); Wrote section on "Theoretical Computer Science" for "Outlook for Science and Technology - The Next Five Years," prepared by the National Research Council for the Congress of the United States, 1982.

**Teaching:** Taught varied graduate and undergraduate courses in mathematics and computer science over my long and varied career. 24 doctoral students including Donald Loveland, Robert Di Paola, Eugenio Omodeo, Donald Perlis, Roberto Policriti.

**Vision Statement:** I regard Gödel's legacy to us as the problem of determining the relevance of incompleteness to important open problems. On the level of utter fantasy: might it be proved one day that the Riemann Hypothesis is not provable in ZFC but follows from the existence of a measurable cardinal?

**Honours and Awards:** Herbrand Prize, Conference on Automated Deduction, 2005; Townsend Harris medal, City College Alumni Association, 2001; Trjitzinsky Memorial Lecturer, University of Illinois, 2001;. Elected to Gamma Chapter, Phi Beta Kappa, 1995; Guggenheim Foundation Fellow, 1983-1984; Elected Fellow of the A.A.A.S., 1982; Earle Raymond Hedrick Lecturer, Mathematical Association of America, 1976; Leroy P. Steele Prize, American Mathematical Society, 1975; Chauvenet Prize & Lester R. Ford Prize, both awarded by the Mathematical Association of America, 1975.

# DAWSON, John W., Jr.

**Specialties:** Axiomatic set theory, history of modern logic.

**Born:** 4 February 1944 in Wichita, Kansas, USA.

**Educated:** University of Michigan, PhD Mathematics, 1972; Massachusetts Institute of Technology, SB Mathematics, 1966

**Dissertation:** *Definability of ordinals in the rank hierarchy of set theory*; supervisors, David Kueker and Andreas Blass

**Regular Academic or Research Appointments:** PROFESSOR OF MATHEMATICS, PENNSYLVANIA STATE UNIVERSITY, YORK, 1975–2006. Instructor in Mathematics, Pennsylvania State University, University Park, 1972-1975.

**Visiting Academic or Research Appointments:** Member, Institute for Advanced Study, 1982-1984; Visiting Scholar, Stanford, 1985, 1986, 1988.

**Research Profile:** Following early work in axiomatic set theory, catalogued *Nachlass* of Kurt Gödel. Biographer of Gödel and co-editor of his *Collected Works*. Author of 30 articles in refereed journals, contributor to numerous handbooks and encyclopedias.

**Main Publications:**

1. *Kurt Gödel: Collected Works.* (Co-editor and translator, with Solomon Feferman, Warren Goldfarb, Stephen C. Kleene, Gregory H. Moore, Charles Parsons, Wilfried Sieg, Robert M. Solovay and Jean van Heijenoort.) Oxford University Press, New York. (Volume I, 1986; Volume II, 1990; Volume III, 1995; Volumes IV and V, 2003.)
2. *Logical Dilemmas: The Life and Work of Kurt Gödel.* A.K. Peters, Ltd. (Wellesley, Mass., 1997)
3. (With Cheryl A. Dawson) Future tasks for Gödel scholars, *Bulletin of Symbolic Logic* 11 : 2 (June 2005), 150-171.
4. The golden age of mathematical logic. *The Cambridge History of Philosophy, 1870-1945* (Cambridge University Press, 2003), 590-597.
5. The compactness of first-order logic: from Gödel to Lindström. *History and Philosophy of Logic 14* (1993), 15-37.
6. Facets of incompleteness. In *Mathematical Logic and its Applications* (ed. D.G. Skordev), Plenum Publishing Company (New York, 1988), 9-21.
7. The reception of Gödel's incompleteness theorems. In *PSA 1984: Proceedings of the Biennial Meeting of the Philosophy of Science Association (Chicago, 1984), Volume 2* (1985), 253-271.
8. (With Richard Mansfield) Boolean-valued set theory and forcing. *Synthese 33* (1976), 223-252.
9. (With Paul E. Howard) Factorials of infinite cardinals. *Fundamenta Mathematicae XCIII* (1976), 185-195.
10. Ordinal definability in the rank hierarchy. *Annals of Mathematical Logic 6* (1973), 1-39. (Corrigendum, *ibid.* 7 (1974), 325.)

11. Classical logic's coming of age. *Handbook of the Philosophy of Science*, vol. 5, ed. Dale Jacquette (North-Holland Pub. Co.), 497–522.

12. In quest of Kurt Gödel: reflections of a biographer. *Notices of the American Mathematical Society*, 53:4 (April, 2006), 440–443.

13. Why do mathematicians re-prove theorems? *Philosophia Mathematica* (III) 14:3 (2006), 269–286.

14. *Kurt Gödel:Das Album/The Album.* (Catalog of the Gödel Centenary Exhibition in Vienna; co-editor and translator, with Karl Sigmund and Kurt Mühlberger.) Vieweg, Wiesbaden, 2006.

**Service to the Profession:** Co-editor (with Volker Peckhaus), *History and Philosophy of Logic*, 2006–; Editor, *History and Philosophy of Logic*, 2001–2006.

**Honours and Awards:** National Merit Scholar, M.I.T., 1962–1966.

# DEHORNOY, Patrick

**Specialties:** Set theory, connection with algebra.

**Born:** 11 September 1952 in Rouen.

**Educated:** Ecole Normale Superieure, Paris, France, PhD Mathematics, 1975.

**Dissertation:** *Intersection d'ultrapuissances iterees*; supervisor, Kenneth McAloon.

**Regular Academic or Research Appointments:** PROFESSOR OF MATHEMATICS, UNIVERSITÉ DE CAEN, FRANCE, 1989–; Associate Professor, Mathematics, Universite de Caen, France, 1983-1989; Charge de recherche, CNRS, Universite de Paris, 1982-1983; Attache de recherche, CNRS, Universite de Paris, 1975-1981.

**Visiting Academic or Research Appointments:** Charge de cours logique, Ecole Normale Superieure, Paris, 2004-.

**Research Profile:** Patrick Dehornoy first studied intersections of iterated ultrapowers of models of ZFC in the vein of K. Kunen's work. Later he investigated iterations of elementary embeddings of a rank into itself, emphasizing the role of the self-distributivity law LD: $x(yz)=(xy)(xz)$. Together with those of R.Laver, his results gave intriguing purely algebraic statements, such as the decidability of the word problem for the LD law, whose proof relies on a very strong large cardinal axiom. Subsequently, Dehornoy found an alternative proof involving no Set Theory by attaching to the LD law a certain group reminiscent of R.Thompon's group. As this group turns out to be an extension of Artin's braid group, unexpected braid results followed, in particular an ordering that is now classical. These results are arguably applications of Set Theory, as one can doubt they would have been discovered without the motivation given by the latter. Dehornoy continued to work on braids and in group theory, introducing in particular the rather successful notion of a Garside group. In set theory, he popularized Woodin's work on CH.

**Main Publications:**

1. Iterated ultrapowers and Prikry forcing, *Ann. Math. Logic*, 15 (1978) 109-160.

2. An application of iterated ultrapowers, *J. Symb. Logic*, 48-2 (1983) 225-235.

3. Pi11-complete families of elementary sequences, *Ann. P. Appl. Logic*, 38 (1988) 257-287.

4. An alternative proof of Laver's result on the algebra generated by an elementary embedding; *Set Theory of the Continuum* (H. Judah & al. eds.), MSRI Publications 26, Springer (1992) 27-33.

5. Structural monoids associated to equational varieties, *Proc. Amer. Math. Soc.*, 117-2 (1993) 293-304.

6. Braid groups and left distributive operations, *Trans. Amer. Math. Soc.*, 345-1 (1994) 115-151.

7. From large cardinals to braids via distributive algebra, *J. Knot Theory \& Ramifications*, 4-1 (1995) 33-79.

8. Another use of set theory, *Bull. Symb. Logic*, 2-4 (1996) 379-391.

9. *Braids and Self-Distributivity*, Progress in Math. vol. 192, Birkhäuser (2000).

10. Study of an identity, *Alg. Universalis*, 48 (2002) 223-248.

11. Elementary embeddings and algebra, *Handbook of Set Theory* (M.Foreman, A.Kanamori, M.Magidor, eds.), to appear.

12. Progrès récents sur l'hypothèse du continu, d'après Woodin, Seminaire Bourbaki, *Astérisque*, 294 (2004) 147-172.

*Work in Progress*

13. A textbook on Set Theory (in French).

A full list of publications is available at Dehornoy's home page http://www.math.unicaen.fr/~dehornoy

**Service to the Profession:** Member, Agence Nationale de la Recherche, 2006-; Associated Editor, *Algebra Universalis*, 2005-; Advisory Editor, *Annals Pure and Appl. Logic*, 2001-; Advisory editor, *J. Knot Th. and Ramifications*, 2003-; Editor, *J. of Algebra*, 2002-; Chair, Groupement de Recherches Tresses, CNRS GDR 2105, 1999-2004; Chair, Laboratoire de Mathematiques Nicolas Oresme, CNRS UMR 6139, Universite de Caen, 1997-2005; Member, Comite National de la Recherche Scientifique, 1995-2000; Member, Comite National des Universites, 1990-1995; Organizer, International Workshop on Set Theory, Luminy, 1990, 1992, 1994, 1996, 1998, 2000, 2002, 2004.

**Teaching:** Dehornoy has had twelve PhD students in the fields of Set Theory, Low Dimensional Topology, and Group Theory.

**Honours and Awards:** Prix Paul Langevin, Academie des Sciences de Paris, 2005; Senior member, Institut Universitaire de France, 2002-; Ferran Sunyer I Balaguer Prize, 1999.

# DEMOPOULOS, William

**Specialties:** Philosophy of logic and mathematics, history of modern logic, philosophy of science.

**Born:** 21 February 1943 in Buffalo, New York, USA.

**Educated:** University of Minnesota, University of Pittsburgh and University of Western Ontario, PhD Philosophy, 1974; University of Buffalo and University of Minnesota, BA Philosophy, 1964.

**Dissertation:** *The Possibility Structure of Physical Systems*; supervisor, Jeffrey Bub.

**Regular Academic or Research Appointments:** PROFESSOR OF PHILOSOPHY, UNIVERSITY OF WESTERN ONTARIO, 1986–; Professor of Logic and Philosophy of Science, University of California at Irvine, 2004-2006; Associate Professor of Philosophy, University of Western Ontario, 1979-1986; Assistant Professor of Philosophy, University of Western Ontario, 1975-1986; Assistant Professor Philosophy, University of New Brunswick, 1970-1975.

**Visiting Academic or Research Appointments:** Visiting Fellow, All Souls College, Oxford University, 2004; Visiting Professor of Philosophy, Harvard University, 2000; Benjamin Meeker Visiting Professor, University of Bristol, 2000; Erskine Fellow, University of Canterbury, 1999; Visiting Scholar, University of British Columbia, 1991; Visiting Professor, University of Illinois at Chicago, 1986; Visiting Fellow, Minnesota Center for the Philosophy of Science, 1978.

**Research Profile:** William Demopoulos began his academic career with research into the role of logic in non-relativistic quantum mechanics. This work spanned philosophy of logic and the philosophy of physics; it is an area of research to which he has recently returned. Demopoulos has been at the forefront of the re-assessment of Frege's philosophy of arithmetic and its subsequent development by Russell and others; he has also made significant contributions to the application of the methods and ideas of modern logic to the reconstruction of scientific theories and to our understanding of the history of such applications over the last century. His historical studies in philosophy of logic, mathematics and science have encompassed the work of Frege, Dedekind, Ramsey, Russell and Carnap.

**Main Publications:**

1. Fundamental statistical theories, in *Logic, Probability and Quantum Mechanics*, ed. by P. Suppes (Reidel: 1975) 421-431.
2. The possibility structure of physical systems, in *Foundations of Probability Theory, Statistical Inference and Statistical Theories, Vol. III*, ed. by C.A. Hooker and W.L. Harper (Reidel: 1976) 55-80.
3. Completeness and realism in quantum mechanics, *Foundational Problems in the Special Sciences*, ed. by R. Butts & J. Hintikka (Reidel: 1977) 81-88.
4. Russell's *Analysis of Matter*: Its historical context and contemporary interest, with Michael Friedman, *Philosophy of Science* **52** (1985) 621-639.
5. On some fundamental distinctions of computationalism, *Synthese* **70** (1987) 79-96.
6. The homogeneous form of logic programs with equality, *Notre Dame Journal of Formal Logic* **31** (1990) 291-303.
7. On applying learnability theory to the rationalism-empiricism controversy, in *Learnability and Linguistic Theory*, ed. by R. J. Matthews and W. Demopoulos (Kluwer: 1990), 77-88.
8. Frege, Hilbert, and the conceptual structure of model theory, *History and Philosophy of Logic* **15** (1994) 211-225.
9. Frege and the rigorization of analysis, *Journal of Philosophical Logic* **23** (1994) 225-245.
10. *Frege's Philosophy of Mathematics*, editor, (Harvard University Press: 1995).
11. The philosophical basis of our knowledge of number, *Noûs* **32** (1998) 481-503.
12. The theory of meaning of 'On denoting,' *Noûs* **33** (1999) 439-458.
13. The origin and status of our conception of number, *Notre Dame Journal of Formal Logic*, **41** (2000) 210-26.
14. Russell's structuralism and the absolute description of the world, *The Cambridge Companion to Russell*, N. Griffin (ed.), (Cambridge University Press: 2003) 392-419.
15. On the rational reconstruction of our theoretical knowledge, *British Journal for the Philosophy of Science*, **54** (2003) 371-403.
16. Elementary propositions and essentially incomplete knowledge: A framework for the interpretation of quantum mechanics, *Noûs* **38** (2004) 86-109.
17. On our knowledge of numbers as self-subsistent objects, *Dialectica* **59** (2005) 141-159.
18. The logicism of Frege, Dedekind and Russell, with Peter Clark, *The Oxford Handbook of the Philosophy of Logic and Mathematics*, Stewart Shapiro (ed.) (Oxford University Press: 2005) 129-165.
19. Carnap on the reconstruction of scientific theories, *The Cambridge Companion to Carnap*, Richard

Creath and Michael Friedman (eds) (Cambridge University Press: 2007) 248–272.

20. *Physical Theory and Its Interpretation: Essays in Honor of Jeffrey Bub*, editor with Itamar Pitowsky (Springer: 2006).

*Work in Progress*

21. A further study of the role of logical notions in the analysis of non-relativistic quantum mechanics, one which develops a notion of completeness relevant to certain physical theories.

22. A reconstruction of the 1910 *Principia*'s theory of propositional functions and classes that dispenses with the no-classes theory of classes.

23. The nature of the a priori and the role of objectivity in Frege's philosophy of arithmetic and its bearing on the representation of his logicism.

A fuller list of publications is available at Demopoulos's home page http://www.uwo.ca/philosophy/facultyandstaff/wdemopoulos

**Service to the Profession:** Managing Editor, *Western Ontario Series in the Philosophy of Science*, 1997-; Associate and Occasional Acting Editor, *Philosophy of Science*, 1980- 1990; Director of Graduate Studies, University of Western Ontario Philosophy 1984-1989.

**Teaching:** Demopoulos has been a mainstay of the graduate program in the philosophy of science at the University of Western Ontario since his appointment as an assistant professor in 1975; he has directed eighteen PhD theses. His students' work spans a wide range of research topics in logic and the philosophy of science.

**Vision Statement:** The discovery of modern logic toward the end of the $19^{th}$ century and the beginning of the $20^{th}$ had a profound influence, first on the philosophy of mathematics, but subsequently on the philosophy of language and philosophy of science. The depth of these early logical insights and their subsequent elaboration continue to set the course for the development of philosophy.

# DE QUIEROZ, Ruy

**Specialties:** Mathematical logic, proof theory, foundations of mathematics, philosophy of mathematics.

**Born:** 11 January 1958 in Recife, Pernambuco, Brazil.

**Educated:** Imperial College, London, PhD Computing, 1990; Universidade Federal de Pernambuco, MSc Informatics, 1984; Escola Politecnica de Pernambuco, BEng Electrical Engineering, 1980.

**Dissertation:** *Proof Theory and Computer Programming. An Essay into the Logical Foundations of Computation*; supervisor, Thomas S. E. Maibaum.

**Regular Academic or Research Appointments:** ASSOCIATE PROFESSOR, INFORMATICS, UNIVERSIDADE FEDERAL DE PERNAMBUCO, RECIFE, BRAZIL, 1993–.

**Visiting Academic or Research Appointments:** Edward Larocque Tinker Visiting Professor, Philosophy, Stanford University, 2006; Research Associate, Department of Computing, Imperial College, London, 1991-1993; Research Assistant, Department of Computing, Imperial College, London, 1989-1991.

**Research Profile:** In the late 1980s, de Queiroz offered a reformulation of Martin-Lof's type theory based on a novel reading of Wittgenstein's 'meaning-is-use' where the explanation of the consequences of a given proposition gives the meaning to the logical constant dominating the proposition, amounting to a non-dialogical interpretation of logical constants via the effect of elimination rules over introduction rules, finding a parallel in Lorenzen's and Hintikka's dialogue/game-semantics. This led to a type theory called 'Meaning as Use Type Theory' published in item 10 below. In reference to the use of Wittgenstein's dictum, he has shown (in item 12 below) that the aspect concerning the explanation of the consequences of a proposition has been present for some time. In an early letter to Russell, Wittgenstein refers to the universal quantifier only having meaning when one sees what follows from it. Since later in the 1990's de Queiroz has been engaged, jointly with D. Gabbay, in a program of providing a general account of the functional interpretation of classical and non-classical logics via the notion of labeled natural deduction. As a result, novel accounts of the functional interpretation of the existential quantifier, as well as the notion of propositional equality, were put forward, the latter allowing for a recasting of Statman's notion of *direct computation*, and a novel approach to the dichotomy 'intensional versus extensional' accounts of propositional equality via the Curry-Howard interpretation. Since the early 2000's, de Queiroz has been investigating, jointly with A. de Oliveira, a geometric perspective of natural deduction based on a graph-based account of Kneale's symmetric natural deduction.

**Main Publications:**

1. (with de Oliveira, A.) Geometry of Deduction via Graphs of Proof. In *Logic for Concurrency and Synchronisation*, R. de Queiroz (ed.), volume 18 of the *Trends in*

*Logic* series, Kluwer Acad. Pub., Dordrecht, July 2003, ISBN 1-4020-1270-5, pp. 3-88.

2. Meaning, function, purpose, usefulness, *consequences* - interconnected concepts. *Logic Journal of the Interest Group in Pure and Applied Logics*, **9**(5):693-734, September 2001, Oxford Univ. Press.

3. (with Gabbay, D.) Labelled Natural Deduction. In *Logic, Language and Reasoning. Essays in Honor of Dov Gabbay*, H.J. Ohlbach and U. Reyle (eds.), volume 5 of *Trends in Logic* series, Kluwer Academic Publishers, Dordrecht, June 1999, pp. 173-250.

4. (with de Oliveira, A.) A Normalization Procedure for the Equational Fragment of Labelled Natural Deduction. *Logic Journal of the Interest Group in Pure and Applied Logics*, **7**(2):173-215, 1999, Oxford Univ. Press. Full version of a paper presented at 2nd WoLLIC'95, Recife, Brazil, July 1995. Abstract appeared in *Journal of the Interest Group in Pure and Applied Logics* **4**(2):330-332, 1996.

5. (with Gabbay, D.) The Functional Interpretation of the Existential Quantifier, in *Bulletin of the Interest Group in Pure and Applied Logics* **3**(2-3):243-290, 1995. (Special Issue on *Deduction and Language*, Guest Editor: Ruth Kempson). Full version of a paper presented at *Logic Colloquium '91*, Uppsala. Abstract in *JSL* **58**(2):753-754, 1993.

6. Normalisation and Language-Games. In *Dialectica* **48**(2):83-123, 1994. (Early version presented at *Logic Colloquium '88*, Padova. Abstract in *JSL* **55**:425, 1990.)

7. (with Gabbay, D.) Extending the Curry-Howard interpretation to linear, relevant and other resource logics, in *Journal of Symbolic Logic* **57**(4):1319-1365. Paper presented at *Logic Colloquium '90*, Helsinki. Abstract in *JSL* **56**(3):1139-1140, 1991.

8. Meaning as grammar *plus* consequences, in *Dialectica* **45**(1):83-86.

9. (with Maibaum, T.) Abstract Data Types and Type Theory: Theories as Types, in *Zeitschrift für mathematische Logik und Grundlagen der Mathematik* **37**:149-166.

10. (with Maibaum, T.) Proof Theory and Computer Programming, in *Zeitschrift für mathematische Logik und Grundlagen der Mathematik* **36**:389-414.

11. A Proof-Theoretic Account of Programming and the Rôle of Reduction Rules, in *Dialectica* **42**(4):265-282.

12. The mathematical language and its semantics: to show the consequences of a proposition is to give its meaning. In Weingartner, Paul and Schurz, Gerhard, editors, *Reports of the Thirteenth International Wittgenstein Symposium 1988*, volume 18 of *Schriftenreihe der Wittgenstein-Gesellschaft*, Vienna, 304pp. Hölder-Pichler-Tempsky, pp. 259-266. Symposium held in Kirchberg/Wechsel, Austria, August 14-21 1988.
*Work in Progress*

13. A book on the general account of labeled natural deduction, with D. Gabbay & A. de Oliveira, to be published by World Scientific.

14. A paper on the normalization of N-graphs. (with G. Alves & A. de Oliveira).

**Service to the Profession:** Council Member, Association for Symbolic Logic, 2006-2008; Executive Editor, *Logic Journal of the Interest Group in Pure and Applied Logics,* Oxford University Press, 1993-; Coordinator and Co-founder (with D. Gabbay), *Interest Group in Pure and Applied Logics (IGPL)*, the clearing house of the European Association for Logic, Language and Information (FoLLI), 1990-; Guest Editor of several volumes (in partnership with several world standing logicians and computer scientists such as John Baldwin, Sergei Artemov, Bruno Poizat, Dexter Kozen, Angus Macintyre, Grigori Mints), *Annals of Pure and Applied Logic, Theoretical Computer Science,* one volume of *Information and Computation,* several volumes of *Electronic Notes in Theoretical Computer Science*; Creator and Prime Organizer of the series of workshops *WoLLIC* `http://www.cin.ufpe.br/~wollic`; Editorial Board Member, *International Directory of Logicians*, D. Gabbay & J. Woods (eds.), Elsevier.

**Teaching:** de Queiroz has taught several disciplines related to logic and theoretical computer science, including Set Theory, Recursion Theory (as a follow-up to a course given by Solomon Feferman), Logic for Computer Science, Discrete Mathematics, Theory of Computation, Proof Theory, Model Theory, Foundations of Cryptography. He has had four PhD students in the fields of Mathematical Logic and Theoretical Computer Science.

**Vision Statement:** Studying the logical foundations of the notion of computation one is almost inevitably led to the study of the foundations of mathematics and the nature of mathematical objects. Brouwer, Wittgenstein and Godel set the tone, each one from their own perspective, for the investigation between language, meaning, and mathematical objects.

**Honours and Awards:** Edward Larocque Tinker Visiting Professorship at Stanford University, awarded by The Tinker Foundation, after the nomination given by Solomon Feferman and Grigori Mints, 2005; Overseas Research Student scholarship award, Committee of Vice-Chancelors and Principals, University of London, 1985-1987.

# DE RIJKE, Maarten

**Specialties:** Information processing, computational logic

**Born:** August 1, 1961 in Vlissingen, Zeeland, The Netherlands.

**Educated:** University of Amsterdam MSc (Philosophy, cum laude) 1989, University of Amsterdam MSc (Mathematics, cum laude) 1990, University of Amsterdam PhD Computer Science, 1993.

**Dissertation:** *Extending Modal Logic*; supervisor Johan van Benthem.

**Regular Academic or Research Appointments:** PROFESSOR OF INFORMATION PROCESSING AND INTERNET, INFORMATICS INSTITUTE, UNIVERSITY OF AMSTERDAM, 2004–. Associate/assistant professor, Institute for Logic, Language and Computation, University of Amsterdam, 2001–2003/1998-2000. Warwick Research Fellow, University of Warwick, 1996-1997. Research scientist, Center for Mathematics and Computer Science, Amsterdam, 1994–1995.

**Visiting Academic or Research Appointments:** Institute for Research in Cognitive Science, University of Pennsylvania, Spring 2001; University of Cape Town, Winter 1994, 1995, 1996

**Research Profile:** With a background in mathematical logic, De Rijke started his research career in computational modal logic, with a special interest in expressive power and its relation to computational complexity. Since the late 1990's, De Rijke become increasingly interested in research into information access, with a focus on unstructured and semi-structured information. Since 2004, De Rijke leads the Information and Language Processing Systems group. While relatively young, this group has rapidly established itself as one of the leading academic research groups in information retrieval in Europe. His current research focus is on intelligent web information access, with projects on vertical search engines, question answering, weakly or semi-structured documents, and multilingual information.

**Main Publications:**

1. The Importance of Length Normalization for XML Retrieval, (with J. Kamps and B. Sigurbjörnsson). *Information Retrieval*, 8(4):631–654, 2005
2. Boosting Web Retrieval through Query Operations, (with G. Mishne). In: D.E. Losada and J.M. Fernández-Luna, editors, *Advances in Information Retrieval: Proceedings 27th European Conference on IR Research (ECIR 2005)*, LNCS 3408, Springer, pages 502–516, 2005
3. Semantic Characterizations of Navigational XPath, (with M. Marx). *ACM SIGMOD Record* 34(2):41–46, 2005
4. Monolingual Document Retrieval for European Languages, (with V. Hollink, J. Kamps and C. Monz). *Information Retrieval*, 7:33–52, 2004
5. Enriching the Output of a Parser Using Memory-Based Learning, (with V. Jijkoun). In: *Proceedings of the 42nd Annual Meeting of the Association for Computational Linguistics (ACL 2004)*, pages 311–318, 2004
6. Deciding the Guarded Fragments by Resolution, (with H. de Nivelle). *Journal of Symbolic Computation*, 35(1):21–58, 2003
7. A Modal Perspective on Path Constraints, (with N. Alechina and S. Demri). *Journal of Logic and Computation* 13(6):939–956, 2003
8. Resolution in Modal, Description and Hybrid Logics, (with C. Areces and H. de Nivelle). *Journal of Logic and Computation* 11(5):717–736, 2001
9. Encoding two-valued non-classical logics in classical logic, (with H.-J. Ohlbach, A. Nonnengart and D.M. Gabbay). In: A. Robinson and A. Voronkov, editors, *Handbook of Automated Reasoning*, pages 1403–1486, Elsevier Science Publishers, 2001
10. *Modal Logic*, (with P. Blackburn and Y. Venema). Cambrdige University Press, 2001. Revised 2004.
11. Light-Weight Entailment Checking for Computational Semantics, (with C. Monz). In: P. Blackburn and M. Kohlhase, editors, *Proceedings ICoS-3*, 2001
12. Expressiveness of Concept Expressions in First-Order Description Logics, (with N. Kurtonina). *Artificial Intelligence*, 107(2):303–333, 1999
13. A System of Dynamic Modal Logic. *Journal of Philosophical Logic*, 27:109–142, 1998
14. Simulating without Negation, (with N. Kurtonina). *Journal of Logic and Computation*, 7:503–524, 1997
15. A Lindström Theorem for Modal Logic. In: A. Ponse, M. de Rijke and Y. Venema, editors, *Modal Logic and Process Algebra*, Lecture Notes 53, CSLI Publications, Stanford, pages 217–230, 1995
16. The Modal Logic of Inequality. *Journal of Symbolic Logic* 57:566–584, 1992

*Work in Progress*

17. Retrieving Answers from Frequently Asked Questions Pages on the Web, with V. Jijkoun
18. Query Operations for Improving Web Retrieval Effectiveness, with G. Mishne and J. Kamps
19. Data-driven Type Checking in Open Domain Question Answering, with S. Schlobach, D. Ahn, and V. Jijkoun
20. Mixing Rule-Based and Data-Driven Methods for Interpreting Temporal Information, with D. Ahn and S. Fissaha Adafre

**Service to the Profession:** *Chair*, Standing committee of the European Summer School in Logic, Language and Information, 2002–2004; *Coordinator*, evaluation efforts for multilingual web retrieval and Dutch question answering at CLEF, 2003–; *Book review editor*, Journal of Logic, Language and Information, 1996–2004; *Editor*, ACM Transactions on Computational Logic, 1999–; *Editor*, Foundations and Trends in Information Retrieval, 2005–; *Editor*, Research on Language and

Computation, 1999–; *Editor*, Studies in Logic, Language and Information, 1994–2004; *Editor-in-chief*, FoLLI series in Logic, Language and Information, 2004–; *Founder*, Advances in Modal Logic, 1996; *Founder*, Inference in Computational Semantics, 1998; *Vice-President*, European Association for Logic, Language and Information 2004–;

**Teaching:** De Rijke strongly believes in communicating his own and his group's research achievements to students at all levels, ranging from primary schools to advanced intensive courses. He has made important contributions to the European Summer School in Logic, Language and Information, both as an organiser and as a lecturer. Since 2005 he is program director for Information Studies at the University of Amsterdam.

**Vision Statement:** Computer science has been an inspiring source of research questions for logic, with databases and verification as driving forces. Information access yields new sources of research questions for logic, such as semi-structured data and the semantic web. The prime challenge here will be to blend traditional exact but very brittle high-precision rule-based approaches with far more robust data-driven methods.

**Honours and Awards:** Pionier (Persoonsgerichte Impuls voor Onderzoeksgroepen met Nieuwe Ideeën voor Excellente Research) grant, 2001–2006.

# DEVLIN, Keith J.

**Specialties:** Set theory, linguistics.

**Born:** 16 March 1947 in Birth Hull, England.

**Educated:** University of Bristol, PhD Mathematics, 1971; Kings College, London, BSc Mathematics, 1968.

**Dissertation:** *Some Weak Versions of Large Cardinal Axioms*; supervisor, Fred Rowbottom

**Regular Academic or Research Appointments:** EXECUTIVE DIRECTOR, CENTER FOR THE STUDY OF LANGUAGE AND INFORMATION, STANFORD UNIVERSITY, 2001–. Consulting Professor, Department of Mathematics, Stanford University, 2001-; Dean, School of Science and Professor of Mathematics, Saint Mary's College of California, 1993–2001; Carter Professor of Mathematics and Chair of Department, Department of Mathematics and Computer Science, Colby College, 1989–1993; Associate Professor of Mathematics and Philosophy, Stanford University, 1987–1989; Reader in Mathematics, Lancaster University, Lancaster, UK, 1977–1987; Assistant Professor of Mathematics, Bonn University, Bonn, Germany, 1974–1976; Milner Scholar, Department of Mathematics, University of Aberdeen, Scotland, 1971-1972.

**Visiting Academic or Research Appointments:** University of Siena, Italy, 1984; University of Gdansk, Poland, 1982; University of Toronto, Canada, 1981; University of Essen, Germany, 1981; University of Colorado, Boulder, 1980; University of Toronto, Canada, 1979; Pennsylvania State University, 1978; University of Toronto, Canada, 1976; University of Heidelberg, Germany, 1974; University of Oslo, Norway, 1973; Banach Center, Warsaw, Poland, 1973; University of Manchester, England, 1973; University of Oslo, Norway, 1972.

**Research Profile:** Early research work (1971-85) in set theory, followed by applications of logic to linguistics and everyday logical reasoning, together with the development of materials to disseminate mathematics to a wider audience.

**Main Publications:**

1. *Aspects of Constructibility*. Springer-Verlag, Lecture Notes in Mathematics 354 (1973)

2. *The Souslin Problem* (Joint with H. Johnsbraten). Springer-Verlag, Lecture Notes in Mathematics 405 (1974)

3. *The Axiom of Constructibility: A Guide for the Mathematician.* Springer-Verlag, Lecture Notes in Mathematics 617 (1977)

4. *Constructibility*. Springer-Verlag (1984)

5. *Logic and Information*. Cambridge University Press (1991)

6. *Language at Work: Analyzing Communication Breakdown in the Workplace to Inform Systems Design.* (Joint with Duska Rosenberg) Stanford University: CSLI Publications and Cambridge University Press (1996)

7. *Fundamentals of Contemporary Set Theory.* Springer-Verlag (1979)

8. *Sets, Functions and Logic.* Chapman and Hall (1981, 1992, 2003)

9. *The Joy of Sets*, Springer-Verlag (1993)

10. *Goodbye Descartes: The End of Logic and the Search for a New Cosmology of the Mind*, Wiley (1997)

11. Some Weak Versions of Large Cardinal Axioms. *Annals of Mathematical Logic* 5 (1973), pp.291-325.

12. More on the Free Subset Problem. (with J. B. Paris). *Annals of Mathematical Logic* 5 (1973), pp.27-30.

13. Marginalia to a Theorem of Silver (with R. B. Jensen). in *Proceedings of the Logic Conference at Kiel, 1974*, Springer-Verlag: Lecture Notes in Mathematics 499 (1975), pp.115-142.

14. à$_1$-trees. *Annals of Mathematical Logic* 13 (1978), pp.267-330.

15. A Weak Version of Diamond Which Follows From $2^{\hat{}} \aleph_0 < 2^{\hat{}} \aleph_1$. (with S. Shelah). *Israel Journal of Mathematics* 29 (1978), pp.239-247.

16. Concerning the Consistency of the Souslin Hypothesis with the Continuum Hypothesis. *Annals of Mathematical Logic* 19 (1980), pp.115-125.

*Work in Progress*

17. Current research is focused on the design of information/reasoning systems for intelligence analysis. Other current research interests include: theory of information, models of reasoning, applications of mathematical techniques in the study of communication; mathematical cognition; and the use of different media to teach and communicate mathematics to diverse audiences.

**Service to the Profession:** Fellow, American Association for the Advancement of Science (AAAS); Council, American Mathematical Society; American Mathematical Society Publications Committee; National Academy of Sciences Mathematical Sciences Education Board; Committee of Science Policy of the Joint Policy Board for Mathematics; Editor, FOCUS, Mathematical Association of America newsletter; Contributing Editor, Notices of the American Mathematical Society.

**Honours and Awards:** Pythagoras Prize for mathematics exposition, Italy, 2005; California State Assembly Certificate of Recognition for Innovative Work in the Field of Mathematics and its Relation to Logic and Linguistics, 2003; Peano Prize for mathematics exposition, Italy, 2002; USA Joint Policy Board for Mathematics: Communications Award, 2001; American Association of Publishers, "Most Outstanding Book in Computer Science and Data Processing for 1991" (for *Logic and Information*), 1991.

# DIPERT, Randall R.

**Specialties:** History of logic, philosophies of logic and mathematics, the philosophy of artifacts, aesthetics, action theory, and metaphysics.

**Educated:** Indiana University, Bloomington, PhD, 1976-1977; University of Michigan, Ann Arbor, BA, 1973.

**Dissertation:** *Development and Crisis in Late Boolean Logic: The Deductive Logics of Peirce, Schroeder and Jevons*; supervisor, J. Michael Dunn.

**Regular Academic or Research Appointments:** C. S. PEIRCE PROFESSOR OF AMERICAN PHILOSOPHY, SUNY UNIVERSITY AT BUFFALO, 2000–; Professor, U.S. Military Academy at West Point, 1995–2000; Professor, SUNY University at Fredonia, 1977–1995.

**Research Profile:** His research interests include the history and philosophy of logic, Peirce and early American pragmatism, logic and the philosophies of logic and mathematics, the philosophy of artifacts, aesthetics, action theory, and metaphysics, especially the metaphysics and logic of relations. He also has interests in ethics and political philosophy and most recently, in the philosophy of war and peace (Just War theory) and value-theoretic implications of computational results in game theory, especially of the Iterated Prisoner's Dilemma.

He has written articles for *Encyclopedia Britannica*, the *Routledge Encyclopedia of Philosophy*, as well as for several anthologies and a number of professional journals in philosophy. He has for some time written my own logic teaching software, and has also written a number of programs (in Prolog and LISP) to do what I would computer-assisted research in logic and philosophy. He has authored one book (on artifacts) and co-authored another (a logic textbook). He is currently at work on a book on metaphysics as well as a historically- and philosophically-oriented textbook in logic.

**Main Publications:**

1. Co-authored with M. Schagrin and W. Rapaport, *Logic: A Computer Approach* (McGraw Hill, 1985).
2. *Artifacts, Artworks and Agency* (Temple University Press, 1993).

Papers on "Peirce's propositional logic" (1981),

3. Set-Theoretical Representatins of Ordered Pairs and their Adequacy for the Loic of Relations (1982),
4. "Peirce, Frege, Church's Theorem and the Logic of Relations" (1984),
5. "Peirce's Underestimated Place in the History of Logic,"(1989),
6. "Individuals and Extensional Logic in Schröder's Vorlesungen über die Algebra der Logik," "The Life and Work of Ernst Schröder"(1990-1991) ,
7. "History of Modern Logic" and on "20th Century Logic" for Macropedia article "Logic" in the Encyclopedia Britannica, ed. 1993-current,
8. "The Life and Logical Contributions of O.H. Mitchell," Peirce's Gifted Student (1994),
9. The Mathematical Structure of the World: The World as Graph (1997),
10. C.S. Peirce's Philosophical Conception of Sets (1997),
11. "Logic Machines and Diagrams" and "Logic in the 19th Century," for the Routledge Encyclopedia of Philosophy (1999) ,
12. C.S. Peirce's Logic (2004),
13. "Preventive War, Game Theory, and the Epistemological Dimensionof the Morality of War" (2006).

More information can be found at www.dipert.org.

**Service to the Profession:** Co-editor, Transactions of the C.S. Peirce Society.

# DI PRISCO, Carlos Augusto

**Specialties:** Set Theory, infinitary combinatorics.

**Born:** 4 October 1949 in Caracas, Venezuela.

**Educated:** Massachusetts Institute of Technology, Universidad Central de Venezuela.

**Dissertation:** *Combinatorial properties and supercompact cardinals*; supervisor, Eugene Kleinberg.

**Regular Academic or Research Appointments:** INVESTIGADOR TITULAR, INSTITUTO VENEZOLANO DE INVESTIGACIONES CIENTIFICAS, 1987–; PROFESOR TITULAR, UNIVERSIDAD CENTRAL DE VENEZUELA, 1992–; Investigador Asociado, Instituto Venezolano de Investigaciones Cientificas, 1976/1987.

**Visiting Academic or Research Appointments:** Visiting Professor University of Paris VII, Paris, France, 2000, 2006, 2008; Investigador Visitante, Institucio Catalana de Recerca I Estudis Avancats (ICREA), Barcelona, Spain; Research Associate, University of California, Berkeley, 1992-1993; Member, Center for Logic, Epistemology and Philosophy of Science, University of Campinas, Sao Paulo, Brazil, since 1988.

**Research Profile:** Carlos Di Prisco has made contributions to the study of combinatorial structures related to large cardinal properties. He has been interested in combinatorial properties of sets of numbers, specifically sets of natural numbers and sets of real numbers. He has established results which clarify the relation of perfect set properties for sets of real numbers and the existence of non principal ultrafilters on the set of natural numbers. He has been active in the development of the Ramsey Theory of the Real Numbers, studying polarized and parametrized partition relations defined on products.

**Main Publications:**

1. Supercompact cardinals and a partition property, *Advances in Mathematics*, Vol. 25, No. 1 (1977), 46-55.
2. (with J. Henle) On the compactness of $aleph_1$, and $\aleph_2$. *Journal of Symbolic Logic*, Vol. 43, No. 3 (1978), 394-401.
3. (with W. Marek) Some properties of stationary sets, *Dissertationes Mathematicae*, Vol. CCXVIII (1982), 1-37. Premio Anual del CONICIT, Area de Matem\'aticas, 1980.
4. (with J.Barbanel and I.B. Tan) Many times huge and superhuge cardinals, Journal of Symbolic Logic, Vol. 49, No. 1 (1984), 112-122.
5. (with W. Marek) A filter on $[\lambda]^\kappa$. Proceedings of the American Mathematical Society}, Vol. 90 (1984), 591-598.
6. (with W. Marek) Some aspects of the theory of large cardinals, Mathematical Logic and Formal Systems (Ed. L.P. Alcántara) Lecture Notes in Pure and Applied Mathematics, Marcel Dekker, pp. 89-139 (1985).
7. (with J. Llopis) On some extensions of the projective hierarchy, Annals of Pure and Applied Logic. Vol. 36 (1987) pp 105-113.
8. (with J. Henle) Partitions of Products. Journal of Symbolic Logic 58 (1993) 860-871.
9. (with W.A. Carnielli) Polarized partition properties of higher dimension. Mathematical Logic Quarterly 39 (1993) 461-474.
10. Partition properties and perfect sets. Notas de Lógica Matematica, No. 38 INMABB-CONICET, Bahia Blanca, Argentina. (1993) 119-127.
11. (with M.C. Carrasco and A. Millán) Partitions of the set of finite sequences. Journal of Combinatorial Theory, Series A 71 (1995) 255-274.
12. (with O. De la Cruz) Weak forms of choice. Proceedings of the American Mathematical Society 126 (1998) 867-876.
13. (with Omar De la Cruz) Weak forms of the axiom of choice and partitions of infinite sets. in Set Theory: Techniques and Applications, Kluwer (1998) 47-70.
14. (with S. Todorcevic) Perfect-Set Properties in L(R)[U]. Advances in Mathematics 139 (1998) 240-259.
15. (with S. Todorcevic) A cardinal defined by a polarized partition property. Israel Journal of Mathematics 109 (1999) 41-52.
16. (with J. Henle) Doughnuts, floating ordinals, square brackets and ultraflitters. Journal of Symbolic Logic 65 (2000) 461-473.
17. (with J. Llopis y S. Todorcevic) Borel partitions of products of finite sets and the Ackermann function. Journal of Combinatorial Theory, Series A 93 (2001) 333-349.
18. (with Todorcevic) Souslin partitions of products of finite sets. Advances in Mathematics 176 (2003) 145-173.
19. (with J. Llopis y S. Todorcevic) Parametrized partitions of products of finite sets. Combinatorica 24 (2004) 209-232.
20. (with Todorcevic) Canonical forms of shift-invariant maps on $[N]^\infty$. Discrete Mathematics.

**Service to the Profession:** President, Asociacion Matematica Venezolana 2004–2006, 2006–2008; Associate Editor, Interciencia, 1982-1999; Asessor, International Union of History and Philosophy of Science, Division of Logic, Methodology and Philosophy of Science, 1987-1991, 1995-1998; Council Member, Association for Symbolic Logic, 1995-1998; Organizing Committee, Simposio Latinoamericano de Lógica Matemática, 1983; Organizer, Escuela Venezolana de Matematica, 1998.

**Teaching:** Carlos Di Prisco teaches at Universidad Central de Venezuela and Instituto Venezolano de Investigacionbes Cientiicas. He conducts a logic seminar in Caracas with participation of faculty and students from different institutions.

He has had thirteen students at the graduate level, including Carlos Uzcategui, Elias Tahhan, Jimena Llopis, Gisela Mendez, Omar De la Cruz, Andres Millan, Nelson Hernandez, Maria Carrasco, Jose Gregorio Mijares. Five PhD students.

**Vision Statement:** Some of the most interesting developments in set theory are those which solve or shed light on problems in analysis, topology and some other areas of mathematics.

Foundational research regarding the continuum has advanced considerably in recent years, to the point that it is now reasonable to think that the continuum problem could be settled in the near future.

**Honours and Awards:** Fellow, John Simon Guggenheim Memorial Foundation 1991-1992; Member, Academia de Ciencias Fisicas, Matematicas y Naturales, Caracas, 2001; Prize Lorenzo Mendoza Fleury, Fundacion Polar, 1983.

# DOMINICY, Marc

**Specialties:** History and philosophy of logic, philosophy of language, philosophy of mind, argumentation theory.

**Born:** 17, June, 1948 in Etterbeek, Brussels, Belgium.

**Educated:** Université Libre de Bruxelles, PhD Linguistics, 1975; Université Libre de Bruxelles, MA Romance Languages and Literatures, 1970.

**Dissertation:** *La périphrase verbale venir de plus infinitif et ses équivalents dans quelques langues romanes*, supervisor, Jacques Pohl.

**Regular Academic or Research Appointments:** PROFESSEUR ORDINAIRE, UNIVERSITÉ LIBRE DE BRUXELLES, 1979–. Part-time professor (Deeltijds Hoogleraaar), Vrije Universiteit Brussel, 1976-2008; Research Fellow of the Belgian Fonds National de la Recherche Scientifique, 1970-1979.

**Visiting Academic or Research Appointments:** Visiting professor, University of Antwerp, 1990 and 1996; Visiting professor, Université Lumière-Lyon 2, 1989-1995. Visiting researcher: University of Massachusetts, Amherst, 1978 (Barbara Hall Partee); University of Bologna, 1976; Central University, Barcelona, 1972-1973.

**Research Profile:** In order to study the conceptual foundations of the Port-Royal *Grammaire* and *Logique*, Marc Dominicy provided a formal reconstruction of the Cartesian theory of ideas. He showed that the syntactical correspondence between the calculus of classes and the calculus of ideas can be conserved if the latter is interpreted in an intensional semantics (a rather primitive possible-worlds semantics where modalities conflate with Aristotelian quantification). This reconstruction helped him to understand why Arnauld rejected Descartes's views on material falsity and was unable to refute Leibniz's metaphysics of individual notions. In the long run, Marc Dominicy found that the best strategy for reconstructing the logico-grammatical achievements of Arnauld and Nicole would consist in adopting a fragment of French as the object-language to be interpreted directly. In related work bearing on the whole tradition of French "Grammaire Générale", Marc Dominicy showed how Arnauld's logico-grammatical reflection paved the way for insightful linguistic inquiries, even if the formal limitations of the theory of ideas finally led to the failure of this scientific programme.

Marc Dominicy's work in the philosophy of logic was mainly devoted to *that's all* statements. He proved that Popperian falsificationism cannot be hold by sticking to Popper's own typology of first-order statements. If first-order *that's all* statements are no natural laws, they should belong to the set of initial conditions; yet, any verification of an empirical first-order *that's all* statement involves the verification of another statement of the same kind. One solution consists in considering such statements as *ad hoc* hypotheses. Marc Dominicy proposed another way out, which amounts to eliminating those *that's all* statements that falsifiabilize unfalsifiable statements; he showed that, under reasonable assumptions, this strategy also applies to Goodman's grue-bleen paradox. In a more recent paper, he analyzed Wittgenstein's inconclusive discussion of this question in the first texts of his second period, and in particular in the Vienna Circle sessions recorded by Waismann. Using a semantic approach to *that's all* statements, Marc Dominicy also attempted to prove that non-supervenience facts about first-order properties like consciousness are positive and follow from the physical.

As a linguist, Marc Dominicy has a crucial interest in the philosophy of language and argumentation theory. His contributions to this field deal with Speech Act Theory and Illocutionary Logic, with discourse acts and rhetoric. He showed that many apparent shortcomings of Searle & Vanderveken's approach to illocutionary acts can be avoided by assuming a pragmatic reduction of non-expressive speech acts to expressive ones. His research on argumentation theory led him to question the Aristotelian parallelism between rhetoric and dialectic, and to emphasize the non-dialectical (i.e. poetical and evocative) nature of the *epideictic* praise and blame. He claimed that, from a logico-philosophical viewpoint, the most striking features of *epideictic* speeches have to do with the conceptualization of human decision and action. More recently, Marc Dominicy wrote a synthetic article on Chaim Perelman and the Brussels School (Dupréel, Olbrechts-Tyteca) where he tried to trace Perelman's views on semantics (especially on the *vagueness* or *fuzziness* of notions) back to Dupréel's pioneering work on this topic and an ill-founded conception of logic as a universal language. Marc Dominicy suggested that Perelman's later evolution did not involve any modification of his previous views on logic, but derived from his empirical analysis of forensic argumentation.

**Main Publications:**

1. *La naissance de la grammaire moderne. Langage, logique et philosophie à Port-Royal*, Bruxelles, Mardaga, 1984.
2. "On Abstraction and the Doctrine of Terms in Eighteenth-Century Philosophy of Language", *Topoi*, 4 (1985), 201-205.
3. "L'analyse du langage à Port-Royal", *Archives et Documents de la Société d'Histoire et d'Épistémologie des Sciences du Langage*, n° 8 (1987), 28-112.
4. "Langage et logique à Port-Royal", *L'Âge de la Science*, 4 (1991), 171-192.
5. "Le programme scientifique de la grammaire générale", in S. Auroux (ed.), *Histoire des idées linguistiques*, tome 2, Bruxelles, Mardaga, 1992, 326-343.
6. "Sur la logique de Port-Royal: du calcul des idées à la sémantique formelle", in M. Dominicy (ed.), "Port-Royal" = *Revue Internationale de Philosophie*, n° 4/1994, 485-503.
7. "La grammaire générale et sa survie dans les traditions de langues romanes. Une esquisse méthodologique", in P. Schmitter (ed.), *Geschichte der Sprachtheorie*, Tübingen, Narr, vol. 5, 1996, 3-23.
8. "Falsification and Falsifiabilization. From Lakatos to Goodman", *Revue Internationale de Philosophie*, n° 144-145 (1983), 163-197.
9. "Wittgenstein et les limites du monde", *Logique et Analyse*, n° 167-168 (1999), 411-440.
10. "Conscience et réalité physique: sur une thèse de David Chalmers", in F. Beets & M.-A. Gavray (eds), *Logique et ontologie. Perspectives diachroniques et synchroniques. Liber amicorum in honorem Huberti Hubiani*, Liège, Éditions de l'Université de Liège, 2005, 133-147.
11. "Effabilité", in S. Auroux (ed.), *Les notions philosophiques*, Paris, Presses Universitaires de France, 1990, vol. I, 751-753.
12. With Nathalie Franken, "Speech Acts and Relevance Theory", in D. Vanderveken & S. Kubo (eds.), *Essays in Speech Act Theory*, Amsterdam-Philadelphia, John Benjamins (Pragmatics and Beyond Series, NS 77), 2001, 263-283.
13. "Le raisonnement pratique met-il en œuvre une nécessité logique? Une critique de Georg Henrik von Wright", in N. Zaccaï-Reyners (ed.), *Explication—Compréhension. Regards sur les sources et l'actualité d'une controverse épistémologique*, Bruxelles, Éditions de l'Université de Bruxelles, 2003, 185-202.
14. "Rhétorique et cognition: vers une théorie du genre épidictique", *Logique et Analyse*, n° 150-151-152, (1995), 159-177. Shorter version: "Le genre épidictique: une argumentation sans questionnement?", in C. Hoogaert (ed.), *Argumentation et questionnement*, Paris, Presses Universitaires de France, 1996, 1-12.
15. "L'épidictique et la théorie de la décision", in M. Dominicy & M. Frédéric (eds.), *La mise en scène des valeurs. La rhétorique de l'éloge et du blâme*, Lausanne-Paris, Delachaux et Niestlé, 2001, 49-77.
16. With Nathalie Franken, "Épidictique et discours expressif", in M. Dominicy & M. Frédéric (eds.), *La mise en scène des valeurs. La rhétorique de l'éloge et du blâme*, Lausanne-Paris, Delachaux et Niestlé, 2001, 79-106.
17. "La dimension sémantique du discours argumentatif: le travail sur les notions", in R. Amossy, R. Koren & G.-E. Sarfati (eds.), *Après Perelman: quelles politiques pour les nouvelles rhétoriques? L'argumentation dans les sciences du langage*, Paris-Montréal, L'Harmattan, 2002, 121-150.
18. "Les "topoï" du genre épidictique: du modèle au critère, et vice-versa", in E. Eggs (ed.), *Topoï, discours, arguments* (= *Zeitschrift für französische Sprache und Literatur*, Beiheft 32), Stuttgart, Franz Steiner Verlag, 2002, 49-65.
19. "Langage, interprétation, théorie. Fondements d'une épistémologie moniste et faillibiliste", in J.-M. Adam and U. Heidmann, (eds.), *Sciences du texte et analyse de discours. Enjeux d'une interdisciplinarité*, Genève, Slatkine, 2005, 231-258.
20. "Perelman und die Brüsseler Schule", in J. Kopperschmidt (ed.), *Die neue Rhetorik. Studien zu Chaim Perelman*, Munich, Wilhelm Fink Verlag, 2006, 73-134. French version available at http://www.philodroit.be/ under "Working Papers").
21. "Sémantique et philosophie de l'esprit: les rapports de perception visuelle", in F. Neveu & S. Pétillon (eds.), *Sciences du langage et sciences de l'homme. Actes du colloque 2005 de l'Association des Sciences du Lange*, Limoges, Lambert-Lucas, 2007, 65-82.
22. "Epideictic Rhetoric and the Representation of Human Decision and Choice", in K. Korta & J. Garmendia (eds.), *Meaning, Intentions, and Argumentation*, Stanford, Center for the Study of Language and Information, 2008, forthcoming.

For further details visit http://www.ulb.ac.be/rech/inventaire/unites/ULB582.html

*Work in Progress*

23. Deals with speech acts and action theory, the se-

mantic and ontological status of facts, the cognitive foundations of discourse evocation.

**Service to the Profession:** Editor of "Port-Royal", special issue of the *Revue Internationale de Philosophie*, n° 4/1994; Editorial Committee, *Logique et Analyse*; Head, "Laboratoire de Linguistique Textuelle et de Pragmatique Cognitive" (LTPC, Université Libre de Bruxelles) and the doctorate school "Théorie du Langage et de l'Esprit: structure des représentations et mécanismes interprétatifs" (Université Libre de Bruxelles); Member ("Promoteur") of the "Centre National Belge de Recherches de Logique"; Member of SCOLA ("Centre de Recherches Transdisciplinaires en Sciences Cognitives et du Langage", Université Libre de Bruxelles); Co-supervisor of the research projects "Perception, Intention, and Action. A transdisciplinary approach to cognitive representation", Université Libre de Bruxelles, 2002-2006; "Pragmatique linguistique", SPPS-SSTC, Belgium, 1990-1996; "Sémantique lexicale", ACCT, Paris, 1989-1991; "Formalisation et implémentation d'un module pragmatique pour le traitement du langage naturel", SPPS, Belgium, 1988-1992; Administrative Staff, "Société d'Histoire et d'Épistémologie des Sciences du Langage", Paris, 1984-1992; Administration, "Société Belge de Logique et de Philosophie des Sciences", 1980-1988 ; Established a curriculum in Cognitive Science and the Philosophy of Mind, École Normale Supérieure, Paris, 1997.

**Teaching:** Marc Dominicy supervised the PhD dissertations of Christian Plantin (published version: *Essais sur l'argumentation*, Paris, Kimé, 1990), Philippe Kreutz (*Les prédicats factifs: une enquête logique et linguistique*; see two papers by Kreutz: "Burton-Roberts's approach to presupposition: A critical appraisal", *Lingua*, 88 (1992), 301-330; ""Ou": la disjonction et les modalités", *Cahiers Chronos*, 4 (1999), 53-76), Emmanuelle Danblon (published version: *Rhétorique et rationalité. Essai sur l'émergence de la critique et de la persuasion*, Bruxelles, Éditions de l'Université de Bruxelles, 2002) and Mikhail Kissine (*Contexte et force illocutoire. Vers une théorie cognitive des actes de langage*; see two papers by Kissine: "Direction of fit", *Logique et Analyse*, n°198 (2007), 113-128; "Why *will* is not a modal", *Natural Language Semantics*, 16(2), forthcoming). He was also involved in the supervision of the PhD dissertations of André Leclerc, *Le traitement des aspects illocutoires de la signification dans la grammaire philosophique de l'époque classique*, Université du Québec à Trois-Rivières, 1989) and Philippe De Brabanter (*Making sense of mention, quotation and autonymy. A semantic and pragmatic survey of metalinguistic discourse*, Université Libre de Bruxelles, 2002). He has supervised Ana-Lêda de Araújo's post-doctoral research work on Speech Act Theory and the logic of relevance.

**Vision Statement:** Marc Dominicy hopes intensive research will be devoted to developing a logic of perception and action which could deal with the interplay between the event ontology and the intensionality of those descriptions of the physical that involve a mentalistic stance.

## DOŠEN, Kosta

**Specialties:** Proof theory, categorical logic, substructural logics

**Born:** 5 June 1954 in Belgrade, Serbia.

**Educated:** University of Oxford, St John's College, D.Phil. (Mathematical Logic) 1980, University of Belgrade B.A. (Philosophy) 1977

**Dissertation:** *Logical Constants: An Essay in Proof Theory*; supervisor Michael Dummett.

**Web Address:** `http://www.mi.sanu.ac.yu/~kosta`

**Regular Academic or Research Appointments:** PROFESSOR OF LOGIC, FACULTY OF PHILOSOPHY, UNIVERSITY OF BELGRADE, 2003–. Professor, Mathematical Institute, Serbian Academy of Sciences and Arts, Belgrade, 1982–; Professor, Institut de Recherche en Informatique de Toulouse, Université de Toulouse III, 1994–1998.

**Visiting Academic or Research Appointments:** Part-Time Professor, Faculty of Mathematics, University of Belgrade, 2000–2001, 1991–1992, 1985–1986; Visiting Professor, Department of Mathematics, University of Athens, 2001; Visiting Professor, Wilhelm-Schickard-Institut, Universität Tübingen, 1997; Visiting Professor, Département de Mathématiques et Informatique, Université de Montpellier III, 1992–1994; Visiting Scholar, Institut de Recherche en Informatique de Toulouse, Université de Toulouse III, 1992; Visiting Scholar, Zentrum Philosophie und Wissenschaftstheorie, Universität Konstanz, 1989–1990; Visiting Professor, Institute of Mathematics, University of Montenegro, Podgorica, 1988–1989; Visiting Professor, Department of Philosophy, University of Notre Dame, 1986–1987.

**Research Profile:** Došen's work in proof theory belongs to what, after Dag Prawitz, is called *general proof theory*. Since the late 1990s he has concentrated on the problem of identity criteria for proofs, and used as the main tool representations of proofs as arrows in freely generated categories with structure. This area of general proof theory started with Joachim Lambek's work in the 1960s and is called *categorial* (or categorical) *proof theory*. Among the not very numerous books in the field, Došen's are two, and he is preparing a third one. In the monograph [1] one finds a nontrivial notion of identity of proofs for classical propositional logic — something which was previously considered unattainable. This notion, which is in accordance with Gentzen's cut-elimination procedure for plural (multiple-conclusion) sequent systems modified by adding new principles called *union* of proofs and *zero* proofs, is inspired by taking *generality* of proofs as a criterion for identity of proofs (cf. [3]) and leads to what in category theory is called *coherence*. In his previous monograph [6] Došen showed that fundamental notions of category theory, and in particular the notion of adjoint functor, can be characterized proof-theoretically by cut, i.e. composition, elimination. This book contains a systematic presentation of various alternative equational definitions of the notions of adjunction and comonad (i.e., monad, or triple). It contains also a geometrical model of adjunction, related to developments in knot theory arising from Jones' polynomial (cf. [4]). Roughly speaking, adjunction is about "straightening a serpentine" — this notion is caught by Reidemeister moves of planar ambient isotopies. The third book [20] is a study of coherence for star-autonomous categories, which are closely tied to the proof nets of classical linear logic. The paper [7] is about isomorphism of formulae in linear logic. This categorial notion, which codifies a relation stronger than logical equivalence, is proposed as an analysis of the notion of propositional identity (see [8] and [2]).

Since his doctoral thesis (partly summarized in [19] and [15]) Došen investigated substructural logics in a unified manner. For all of them he keeps the rules for logical operations unchanged and varies only Gentzen's structural rules. Relevant logic (without distribution of additive conjunction over additive disjunction) is then simply characterized by lack of thinning. The rules for logical operations, which take the form of equivalences between premises and conclusions, are related to William Lawvere's thesis that all logical operations are tied to adjoint situations. Došen amends Lawvere's thesis by suggesting that one of the functors should be a structural functor, where "structural" is understood in the sense of Gentzen. This requires that the adjunction characterizing implication should be different from Lawvere's adjunction between functors based on product and exponentiation in cartesian closed categories. Došen proposes an adjunction underlying the *functional completeness*, in the sense of Lambek, of closed categories (see [5], [9] and [10]) from which Lawvere's adjunction is derived. In his doctoral thesis Došen proposed a criterion for the demarcation of logic (see [15]) which is tied to Lawvere's thesis in its amended form. Došen investigated linear logic *avant la lettre* (it is prefigured in his thesis). In [16] (made public in 1985) intuitionistic linear nonmodal propositional logic was called system M, after "multiset". He is not only a pioneer in the unified investigation of substructural logics: in 1990 he coined the name "substructural logics" (see the often cited collection of papers [12]; the first appearance in print of the term is in [14]). He is one of those authors who in the 1980s, before the advent of linear logic, revived interest in the Lambek calculus (using for it the name that has become standard both in logic and in theoretical linguistics; see [17]). Došen is one of the pioneers in the investigations of intuitionistic modal logic, using Kripke models with two accessibility relations, one intuitionistic and the other modal (see [18]). He applied such models to study negation understood as a modal operator. He defined for intuitionistic propositional logic most general models in the style of Kripke where the accessibility relation satisfies conditions weaker, and sometimes more complicated, than reflexivity and transitivity (see [13]). He was led to these models by embeddings of intuitionistic logic in modal systems weaker than S4, which he investigated previously.

**Main Publications:**

1. *Proof-Theoretical Coherence* (with Zoran Petrić) *Studies in Logic 1*, King's College Publications, London, 2004, xiv+369 pp.

2. "Identity of proofs based on normalization and generality", *The Bulletin of Symbolic Logic* **9** (2003) 477–503.

3. "Generality of proofs and its Brauerian representation" (with Zoran Petrić) *The Journal of Symbolic Logic* **68** (2003) 740–750.

4. "Self-adjunctions and matrices" (with Zoran Petrić) *Journal of Pure and Applied Algebra* **184** (2003) 7–39.

5. "Abstraction and application in adjunction", in Z. Kadelburg (ed.) *Proceedings of the Tenth Congress of Yugoslav Mathematicians*, Faculty of Mathematics, University of Belgrade, 2001, 33–46 (http:// arXiv. org/ math. CT/ 0111061).

6. *Cut Elimination in Categories*, Trends in Logic 6, Kluwer, Dordrecht, 1999, xii+229

7. "Isomorphic objects in symmetric monoidal closed categories" (with Zoran Petrić) *Mathematical Structures in Computer Science* **7** (1997) 639–662.

8. "Logical consequence: A turn in style", in M.L. Dalla Chiara et al. (eds.) *Logic and Scientific Methods, Volume One of the Tenth International Congress of Logic, Methodology and Philosophy of Science*, Florence, August 1995, Kluwer, Dordrecht, 1997, 289–311.

9. "Deductive completeness", *The Bulletin of Symbolic Logic* **2** (1996) 243–283, p. 523.

10. "Modal functional completeness" (with Zoran Petrić) in H. Wansing (ed.) *Proof Theory of Modal Logic*, Kluwer, Dordrecht, 1996, 167–211.

11. "Equality in substructural logics, Substructural predicates", in W. Hodges et al. (eds.) *Logic, from Foundations to Applications: European Logic Colloquium*, Oxford University Press, Oxford, 1996, 71–101.

12. *Substructural Logics* (ed.) (with Peter Schröder-Heister) Oxford University Press, Oxford, 1993.

13. "Rudimentary Kripke models for the intuitionistic propositional calculus", *Annals of Pure and Applied Logic* **62** (1993) 21–49.

14. "Modal translations in substructural logics", *Journal of Philosophical Logic* **21** (1992) 283–336.

15. "Logical constants as punctuation marks", *Notre Dame Journal of Formal Logic* **30** (1989) 362–381; reprinted in a slightly amended version in D.M. Gabbay (ed.) *What is a Logical System?*. Oxford: Oxford University Press, 1994; 273–296.

16. "Sequent-systems and groupoid models", *Studia Logica* **47** (1988) 353–385, **48** (1989) 41–65, **49** (1990) 614.

17. "A completeness theorem for the Lambek calculus of syntactic categories", *Zeitschrift für mathematische Logik und Grundlagen der Mathematik* **31** (1985) 235–241. (Cf. A brief survey of frames for the Lambek calculus, *ibid.* **38** (1992) 179–187.)

18. "Models for normal intuitionistic modal logics" (with Milan Božić) *Studia Logica* **43** (1984) 217–245, **44** (1985) 39–70.

19. "Sequent-systems for modal logic", *The Journal of Symbolic Logic* **50** (1985) 149–168.

*Work in Progress*

20. *Proof-Net Categories* (with Zoran Petrić) preprint, Mathematical Institute, Belgrade, March 2005, viii+135

**Service to the Profession:** Editor, *Notre Dame Journal of Formal Logic* 1991–; Editor, *Studia Logica* 1993–1997; Reviewer, *Mathematical Reviews* 1985–1997; Chairman of the Logic Seminar, Mathematical Institute, Belgrade, 1985–1989, 2001–; Leader of the Study Group for Quantum Groups, Belgrade, 1999–2000.

**Teaching:** There is in Belgrade a school of categorial proof theory, of which Došen is the founder. His student and closest collaborator in the last ten years is Zoran Petrić, who wrote an award-winning doctoral thesis in 1997 (partly published in papers of Petrić in the Annals of Pure and Applied Logic and Studia Logica). Došen worked together for some time with Djordje Čubrić, who also made significant contributions to the field. Došen's other students are Silvia Ghilezan, Mirjana Borisavljević and Branislav Boričić, who made contributions to the typed lambda calculus and general proof theory. He teaches an undergraduate course in mathematical logic, and has taught at various places advanced courses in proof theory, categorical logic, the lambda calculus, intuitionistic logic, modal logic and formal grammars.

**Vision Statement:** If in the order of explanation deducing precedes asserting, as asserting precedes naming, the theory of proofs should have precedence over the rest of logic. The difference between this theory and the rest of logic is that in it one studies a consequence *graph* rather than a consequence *relation*.

**Honours and Awards:** Alexander von Humboldt Fellow, 1989–1990.

# DOWNEY, Rodney Graham

**Specialties:** Computability theory, computable structure and model theory, complexity theory and algorithmic information theory.

**Born:** 20 September, 1957 in Toowoomba, Queensland, Australia.

**Educated:** University of Queensland BSc (Honours Mathematics) 1979, Monash University PhD Mathematics, 1982.

**Dissertation:** *Abstract Dependence, Recursion Theory, and the Lattice of Recursively Enumerable Filters*; supervisor John Crossley.

**Regular Academic or Research Appointments:** PROFESSOR PERSONAL CHAIR IN MATHEMATICS, VICTORIA UNIVERSITY OF WELLINGTON, 1995–. Victoria University of Wellington School of Mathematical and Computing Sciences, 1986–; Maclaurin Fellow 2003, based at Victoria University; Inaugural New Zealand Institute for Mathematics and its Applications, University of Illinois at Urbana-Champaign Department of Mathematics, 1985–1986; National University of Singapore Department of Mathematics, 1983–1985; Chisholm Institute of Technology, Department of Mathematics 1982.

**Visiting Academic or Research Appointments:** Visiting Scholar, University of Chicago, 2003; Visiting Scholar, University of Chicago, 2001; Visiting Professor, University of Notre Dame, 2000; Visiting Scholar, University of Wisconsin, Madison, 1999; Visiting Professor, National University of Singapore, 1999; Visiting Scholar, University of Siena, 1997; Visiting Professor, Mathematics Department, Cornell University, 1995; Lee Kong Chang Visiting Fellow, National University of Singapore, 1993; Member, Mathematical Sciences Institute, Ithaca, 1992; Visiting Scholar Cornell University, 1992; Member, Mathematical Sciences Research Institute, Berkeley, 1989.

**Research Profile:** Downey is well known for the breadth of his research. His early work was in classical computability theory, particularly the structure of the computably enumerable sets and degrees. Of note is the results on automorphisms of the lattice of computably enumerable sets, such as the proof that there is no property of supersets alone that can guarantee incompleteness (Cholak, Downey, Stob,[6]) and the introduction of the new orbits in Downey and Stob [5], which pointed at the importance of splittings in this area; these ideas culminating in the recent work of Harrington and Cholak, and the Cholak-Downey, Harrington solution to the Slaman-Woodin conjecture. Downey is also known for his results on definability in the computably enumerable degrees, and in strong reducibilities such as the natural definition of contiguity, and low$_2$-ness.

In the early 1990's, together with Mike Fellows, Downey introduced a new paradigm for measuring complexity. The idea is to design a complexity theory to understand the situation in, for instance, database theory, where the size or shape of a query is small or well-behaved, yet the size of the database is large. The result is known as *parameterized complexity*, and resulted in the monograph Downey and Fellows [1]. The applicability of the theory often stems from the fact that in real applications of algorithms, data is often presented in a much more ordered way than it would appear from the theory.

Since 2000, Downey has become fascinated with algorithmic information theory and has begun a program of calibrating randomness by using, say, relative initial segment Kolmogorov complexity. Part of Downey's work with co-authors such as LaForte, Hirschfeldt, Nies and Stephan has been a new requirement free solution to Post's problem based on Kolmogorov complexity. A monograph on this material will appear soon. (Downey-Hirschfeldt [15]).

Downey is especially known for his work with postdoctoral fellows which include Michael Moses, Reed Solomon, Denis Hirschfeldt, Peter Cholak, Walker White, Richard Coles, Evan Griffiths, Geoff LaForte, Yu Liang, Joe Miller, and Wu Guohua.

**Main Publications:**

1. *Parameterized Complexity*, (with M. Fellows) Monographs in Computer Science, Springer-Verlag, 1999, xiii+533.
2. "Maximal theories", *Annals Pure and A Logic.*, **33** (1987) 245–282.
3. "$T$-degrees, jump classes and strong reducibilities", (with C. Jockusch) *Trans. Amer. Math. Soc.*, **301** (1987) 103–136.
4. "Lattice nonembeddings and initial segments of the recursively enumerable degrees", *Annals Pure and Appl. Logic*, **49** (1990) 97–119.
5. "Automorphisms of the lattice of recursively enumerable sets: Orbits", (with M. Stob) *Advances in Math.*, **92** (1992) 237–265.
6. "Automorphisms of the lattice of recursively enumerable sets: promptly simple sets", (with P. Cholak and M. Stob) *Trans. American Math. Society*, **332** (1992) 555–570.
7. "Computability Theory and Linear Orderings", in Ershov, Goncharov, Nerode and Remmel (eds.), *Handbook of Recursive Mathematics*, Vol 2. North-Holland, 1998; 823–977.
8. "Fixed-parameter tractability and completeness II: on completeness for W[1]", (with M. Fellows) *Theoretical Comput. Sci.* **141** (1995) 109–131.
9. "Parameterized Complexity Analysis in Computational Biology", (with H. Bodlaender, M. Fellows, M. Hallett, and H. Todd Wareham) *Computer Applications in the Biosciences*, **11** (1995) 49–57
10. "There is no fat orbit", (with L. Harrington) *Annals of Pure and Applied Logic*, **80** (1996) 227–289.
11. "Undecidability results for low complexity time classes", (with A. Nies) *Journal of Computing and Sys. Sci.*, **60** (1999) 465–479.
12. "Every set has a least jump enumeration", (with R. Coles and T. Slaman) *Journal London Math. Soc.*, **62**(2) (2000) 641–649.
13. "Automorphisms of the lattice of $\Pi_0^1$ classes: perfect thin classes and anr degrees", (with P. Cholak, R. Coles and E. Herrmann) *Trans. Amer. Math. Soc.* **353** (2001) 4899–4924.
14. "Randomness, computability, and density", (with D. Hirschfeldt and A. Nies) *SIAM J. Comput.* **31** (2002) 1169–1183.

*Work in Progress*

15. *Algorithmic Randomness and Complexity*, (with Denis Hirschfeldt). Monographs in Computer Science, Springer-Verlag. To appear.
16. "Automorphisms of the lattice of computably enumerable sets: the Slaman-Woodin conjecture", (with P. Cholak and L. Harrington).
17. "Randomness and reducibility", (with G. Laforte and D. Hirschfeldt) extended abstract appeared in *Mathematical Foundations of Computer Science*, J. Sgall, A. Pultr, and P. Kolman (eds.) 2000; *Mathematical Foundations of Computer Science* 2001; *Lecture Notes in Computer Science* 2136 Springer, 2001, 316–327. Final version accepted for publication in *Journal of Computing and System Sciences*.

18. "Computability-theoretical and proof-theoretical aspects of partial and linear orderings", (with D. Hirschfeldt, S. Lempp, and R. Solomon) *Israel J. Math.*. To appear.

19. "Some Computability-Theoretical Aspects of Reals and Randomness", in a vol. from the *Lecture Notes in Logic* Series, Symbolic Logic Association. To appear.

20. *Asian Logic Conferences*, (edited) World Scientific. To appear.

21. On the orbits of computably enumerable sets, (with P. Cholak and L. Harrington) to appear, *Journal of the American Mathematical Society*

22. (with A. Montabán) The isomorphism problem for torsion-free Abelian groups is analytic complete, to appear *Journal of Algebra*.

**Service to the Profession:** Editor *Journal of Symbolic Logic*, 1999–, and Coordinating editor, 2000–; Co-director, New Zealand Mathematical Sciences Research Institute; Exectutive, Governing Board, New Zealand Institute for Mathematics and its Applications; 1997, 2002– Marsden panel for Mathematical and Information Sciences; 1999– New Zealand Mathematical Sciences Advisory Group; 2001– New Zealand representative on the International Mathematics Union; 2001– President, New Zealand Mathematical Society; Prizes committee, council, and Australasian committee Association for Symbolic Logic; 2000–2001 Vice-President, New Zealand mathematics Society; 1996–2001 Fellows' Committee Royal Society New Zealand.

**Teaching:** Downey's students include Catherine McCartin, now at Massey University, and Wu Gouhua, at Nangyang Technological University in Singapore.

**Vision Statement:** Downey believes that logic had an intense growth period where it was very much concerned with often very difficult technical questions driven by internal considerations. He believes that we are seeing logic re-applied to the foundations of mathematics and as an engine in computer science and this trend will continue.

**Honours and Awards:** New Zealand Royal Society Hamilton Award for Science, 1990; New Zealand Mathematical Society Award for Research, 1992; New Zealand Association of Scientists Research Medal for the best New Zealand based scientist under 40, 1994; Personal Chair in Mathematics, Victoria University, 1995; Elected Fellow of the Royal Society (NZ), 1996; Vice-Chancellor's Award for Research Excellence, 2000; Elected Fellow of the New Zealand Mathematical Society, 2003; Inaugural MacLaurin Fellow, New Zealand Institute for Mathematics and its Applications (Center of Research Excellence), 2003; Royal Society James Cook Fellowship (2008–2010); Elected Fellow of the Association for Computing Machinery, 2008.

# DUMMETT, Michael Anthony Eardley

**Specialties:** Intuitionism, quantum logic, intermediate logics, intermediate modal logics, proof-theoretic justifications of logical laws, philosophy of mathematics.

**Born:** 27 June 1925 in London, England, United Kingdom.

**Educated:** Oxford University, MA, 1954; Oxford University, DLitt, 1989; Oxford University, BA First Class, Philosophy, Politics and Economics, 1950.

**Regular Academic or Research Appointments:** HONORARY FELLOW, NEW COLLEGE, OXFORD, 1999–; EMERITUS WYKEHAM PROFESSOR OF LOGIC, OXFORD UNIVERSITY, 1992–; EMERITUS FELLOW, ALL SOUL'S COLLEGE, OXFORD, 1979–; Emeritus Fellow, New College, Oxford, 1992–1999; Wykcham Professor of Logic, Oxford University, 1979-1992; Fellow, New College, Oxford, 1979–1992; Senior Research Fellow, All Souls' College, Oxford, 1975–1979; Reader, Philosophy of Mathematics, Oxford University, 1962–1975; Extraordinary Research Fellow, All Souls' College, Oxford, 1962–1975; Research Fellow, All Souls' College, Oxford, 1957–1962; Junior Research Fellow, All Souls' College, Oxford, 1950–1957; Assistant Lecturer, Philosophy, Birmingham University, 1950–1951.

**Visiting Academic or Research Appointments:** Dewey Lecturer, Columbia University, New York, 2001; Gifford Lecturer, University of St. Andrews, 1996; Fellow, Center for Advanced Study in the Behavioral Sciences, Stanford, 1988–1989; William James Lecturer, Harvard University, 1976; Visiting Professor, Rockefeller University, New York, 1974, 1978; Visiting Professor, Princeton University, 1972; Visiting Professor, University of Minnesota, 1970; Visiting Professor, Stanford University, 1960, 1962, 1966; Visiting Lecturer, University of Ghana, Legon, 1958; Alexander von Humboldt Prize, University of Münster, 1957; Visiting Researcher, University of Münster, 1954.

**Research Profile:** Dummett's academic work has lain principally in philosophy, including the philosophy of mathematics, together with social choice theory, which he values as comprising the theory of voting, rather than in mathematical logic. Within mathematical logic, his main interest is in intuitionism, which is one of the areas of logic most relevant to philosophical reflection, because the motivation for the intuitionist rejection of classical logic is itself philosophical. Intuitionistic logic and mathematics represent the most sophisticated development of an anti-realist view of any subject-metter, far deeper than one based on a straightforward opposition to bivalence as exemplified by three-valued logic, and hence a model for anti-realist treatments of other subject-metters. Dummett approached the notion of Kripke trees by observing that distributive lattices, known to characterize intuitionistic sentential logic, could be represented by open subsets of partially ordered sets (those containing all elements $\leq$ to any element); but he did not use the notion of variable domains to extend this to predicate logic.

**Main Publications:**

1. 'A Propositional Calculus with Denumerable Matrix'; *Journal of Symbolic Logic,* Vol. 24, 1959, pp. 97-106.
2. 'Modal Logics between S4 and S5' (with John Lemmon), *Zeitschrift fur mathematische Logik und Grundlagen der Mathematik,* Vol. 5, 1959, pp. 250-64.
3. 'Wittgenstein's Philosophy of Mathematics', *Philosophical Review,* Vol. 68, 1959, pp. 324-48.
4. 'The Epistemological Significance of Gödel's Theorem', *Proceedings of the First International Congress on Logic, Methodology and Philosophy of Science,* Amsterdam, 1961.
5. 'The Philosophical Significance of Gödel's Theorem', *Ratio,* Vol. 5, 1963, pp. 140-55.
6. 'The Philosophical Basis of Intuitionistic Logic', *Logic Colloquium 1973,* edited by H.E. Rose & J. C. Shepherdson, Amsterdam, 1975, pp. 5-40.
7. 'Frege on the Consistency of Mathematical Theories', *Studien zu Frege,* edited by M. Schirn, volume 1, 1976, pp. 229-42.
8. *Elements of Intuitionism,* Oxford, 1977, second revised edition 2000.
9. 'On a Question of Frege's about Right-Ordered Groups' (with Peter Neumann and Samson Adeleke), *Bulletin of the London Mathematical Society,* Vol. 19, 1987, pp. 513-21.
10. *The Logical Basis of Metaphysics,* Cambridge, Mass., 1991.
11. *Frege: Philosophy of Mathematics,* London, 1991.

*Work in Progress*

12. Having reached the age of 81, Dummett is reluctant to undertake any more creative work. There is one theorem he would like to prove, in social choice theory (which has not here been included in logic). This is that no system making essential use of quotas is monotonic. A quota is a number $n$ such that, for any number $k \geq 1$ and any set $A$ of candidates in a vote to elect some number $> k$ representatives, if a set of $nk$ voters prefer every candidate in $A$ to every candidate not in $A$, then at least $k$ members of $A$ are guaranteed to be elected. A system is monotonic if a voter cannot adversely affect candidate $I$'s chance of being elected by ranking him higher on his preference scale, the other preference scales remaining as before. This aside, it is unlikely that Dummett will attempt any more academic work.

**Service to the Profession:** Organizer, With John Crossley, Eighth Logic Colloquium (in conjunction with a meeting for the ASL), Oxford, having raised money for it from NATO and other sources, 1963; Proceedings of this colloquium, edited by the two organizers, North Holland, *Formal Systems and Recursive Functions,* published 1965.

**Teaching:** With the help of John Crossley, in 1962 Dummett instigated the creation at Oxford University of a new Honours School (undergraduate course leading to the BA) in Mathematics and Philosophy; it has been very successful, and continues to flourish. The syllabus includes a very large component of mathematical logic, intended as a bridge subject. For many years after the creation of the School, Dummett undertook twice the number of lectures and tutorials formally required of him as Reader in Philosophy of Mathematics, assisted first by Crossley and then by Robin Gandy, who came as Reader in Mathematical Logic and also did far more than his stint; the lectures were principally in logic. After a time, an increase in the number of those competent to teach logic allowed some relaxation. Dummett feels that he has had some wonderful students, but that it would be invidious to single out any by name.

**Vision Statement:** Mathematical logic was invented as a tool for the philosophy of mathematics. In so far as it bears on deductive argument of all kinds, rather than just on mathematical proof – which of course it does – it is relevant to all branches of philosophy. By its nature, however, it not only admits but demands reasoning of a mathematical character to establish its results. A consequence is that its origins tend to be forgotten, so that it comes to be thought of simply as a branch of mathematics. Modern logic should bear in mind its origins. Those who work in it ought to be sensitive to the relevance of results in logic to philosophical questions, whether in the philosophy of mathematics or in other areas of the subject. Because, in several of its branches, mathematical logic has become too technical for many philosophers to follow easily, its results, though not necessarily its proofs, ought to be expounded in an accurate but accessible manner for the sake of philosophers who can reflect on them and perhaps make use of them. Such exposition is not easy; it is similar to the problem faced (not as yet very successfully) by philosophers of physics; but it is important.

**Honours and Awards:** Hon. D. Litt., University of Athens, 2005; Hon. D. Univ, Stirling University, 2002; Hon. D. Litt., Aberdeen University, 1995; Rolf Schock Prize in Logic and Philosophy, 1994; Hon. D. Litt., University of Caen, 1993; Imre Lakatos Prize for Philosophy of Science, 1993; Foreign Honorary Member, American Academy of Arts and Sciences, 1975-; Hon. D. Phil., University of Nijmegen, 1983; Alexander van Humboldt Prize, 1982; Fellow, British Academy, 1970-1981 (resigned), re-elected Senior Fellow, 1995-.

# DUNN, Jon Michael

**Specialties:** Non-classical logics, especially computation and information based logics, including relevance logic, other substructural logics, quantum logic, algebraic and proof-theoretical approaches to logic.

**Born:** 19 June 1941 in Fort Wayne, Indiana, USA.

**Educated:** University of Pittsburgh, PhD Philosophy (Logic), 1966; Oberlin College, AB Philosophy, 1963.

**Dissertation:** *The Algebra of Intensional Logics*; supervisor, Nuel D. Belnap.

**Regular Academic or Research Appointments:** PROFESSOR EMERITUS OF INFORMATICS, PROFESSOR EMERITUS OF COMPUTER SCIENCE, AND OSCAR EWING PROFESSOR EMERITUS OF PHILOSOPHY, INDIANA UNIVERSITY, 2007–; Professor, Informatics, Indiana University, 2000–2007; Dean, School of Informatics, Indiana University, 1999–2007; Oscar Ewing Professor of Philosophy, Indiana University, 1989–2008; Professor, Computer Science, Indiana University, 1989; Core Faculty, Cognitive Science Program, Indiana University, 1989; Professor , Philosophy, Indiana University, 1976; Associate Professor, Philosophy, Indiana University, 1969; Assistant Professor, Philosophy, Wayne State University, 1966.

**Visiting Academic or Research Appointments:** Visitor, Moscow State University and the Russian Academy of Sciences, 1996; Visitor, Academia Sinica, Taipei, Taiwan, 1994; Adjunct Professor, Department of Philosophy, University of Massachusetts-Amherst, 1985; Senior Fellow, University of Pittsburgh Center for the Philosophy of Science, 1984; Distinguished Faculty Visiting Scholar, University of Melbourne, 1983; Short-term Visitor, Wolfson College, Oxford, 1982; Senior Visitor, Mathematical Institute, Attached to Wadham College, Oxford, 1978; Visiting Fellow, Australian National University Institute for Advanced Studies, 1975-1976; Visiting Assistant Professor, Philosophy, Yale University, 1968-1969.

**Research Profile:** Dunn's research focuses on information based logics and relations between logic and computer science. His early work was mainly on relevance logic but his work broadened to include other so-called "sub-structural logics" including intuitionistic logic, relevance logic, linear logic, BCK-logic, and the Lambek Calculus. He has also worked on modal logic. He has developed an algebraic approach to these and many other logics under the heading of "gaggle theory" (ggl, for generalized galois logics), which is contained in a series of papers and in his book with Gary Hardegree *Algebraic Methods in Philosophical Logic* (Oxford, 2001). He has over 80 publications and has just finished writing a book with Katalin Bimbo titled *Generalized Galois Logics: Relational Semantics of Nonclassical Logical Calculi*. He has done recent work on the relationship of quantum logic to quantum computation, and has a general interest in cognitive science and the philosophy of mind.

**Main Publications:**

1. 1969, "E, R, and Gamma" (with R. K. Meyer), *Journal of Symbolic Logic*, 34, pp. 460–474. Reprinted with alterations by A. R. Anderson and N. D. Belnap, Jr. in their book *Entailment: The Logic of Relevance and Necessity*, vol. 1, Princeton (Princeton University Press), 1975.
2. 1970 "Algebraic Completeness Results for R-Mingle and its Extensions," *Journal of Symbolic Logic*, vol. 35, pp. 1–13.
3. 1973 "A TruthValue Semantics for Modal Logic," in *Truth, Syntax and Modality*, Amsterdam (North Holland Publishing Co.), pp. 87–100.
4. 1975 "Intensional Algebras" and various other sections of *Entailment: The Logic of Relevance and Necessity*, vol. 1, principal authors A. R. Anderson and N. D. Belnap, Jr., contributions by J. M. Dunn and R. K. Meyer, with further contributions by others, Princeton (Princeton University Press), 18, pp. 180–206.
5. 1976 "Intuitive Semantics for First-Degree Entailments and Coupled Trees," *Philosophical Studies*, vol. 29, pp. 149–168.
6. 1976 "Quantum Mathematics": *PSA 80*, 1982, vol. 2, eds. P. Asquith and R. Gierce, Philosophy of Science Association, East Lansing, Michigan.
7. 1985 "Relevance Logic and Entailment," in *Handbook of Philosophical Logic*, vol. 3, eds. D. Gabbay and F. Guenthner, D. Reidel, Dordrecht, Holland, pp. 117-224. Called just "Relevance Logic" (with G. Restall), *Handbook of Philosophical Logic*, $2^{nd}$ edition, 2002,

vol. 6, eds. D. Gabbay and F. Guenthner, Kluwer Academic Publishers, pp. 1-136.

8. 1987 "Relevant Predication 1: The Formal Theory," *Journal of Philosophical Logic*, vol. 16, pp. 347-381.

9. 1990 "Gaggle Theory, an Abstraction of Galois Connections and Residuation, with Applications to Negation, Implication, and Various Logical Operators," in *Logics in AI* (European Workshop JELIA 1990, Amsterdam), ed. J. Van Eijck, Lecture Notes in AI, Springer Verlag, pp. 31-51.

10. 1992. (Book) *Entailment: The Logic of Relevance and Necessity*, vol. 2 (with A. R. Anderson and N. D. Belnap, Jr.), Princeton University Press, 749 + xxvii pp.

11. 1992 "Stone Duality for Lattices" (with C. Hartonas), *Algebra Universalis*, vol. 37, pp. 391-401. Preliminary version, "Duality Theorems for Partial Orders, Semilattices, Galois Connections and Lattices" in Indiana University Logic Group Preprint Series IULG-93-26, 1993, 27 pp.

12. 1993 "Perp and Star: Two Treatments of Negation," in *Philosophical Perspectives* vol.7: *Language and Logic*, ed. James Tomberlin, pp. 331-357.

13. 1995 "Positive Modal Logic," *Studia Logica*, vol. 55, pp. 301-317.

14. 1997 "Combinatory Logic and Structurally Free Logic" (with R. K. Meyer), *Journal of the Interest Group in Pure and Applied Logic*, Oxford University Press, vol. 5, no. 4, 1997, pp. 505-537.

15. 2001 "A Representation of Relation Algebras Using Routley-Meyer Frames," *Logic, Meaning and Computation: Essays in Memory of Alonzo Church*, eds. C. A. Anderson and M. Zeleny, pp. 77-108. Preliminary version in Indiana University Logic Group Preprint Series, IULG-93-28, 1993.

16. 2001 (Book) *Algebraic Methods in Philosophical Logic* (with G. Hardgree), Oxford University Press, 470 + xv pp.

17. 2003 "Four-valued Logic" (with K. Bimbo), *Notre Dame Journal of Formal Logic*, vol. 42, pp. 171-192.

18. 2005 "Canonical Extensions of Ordered Algebraic Structures and Relational Completeness of Some Substructural Logics" *Journal of Symbolic Logic* (with Mai Gehrke and Alessandra Palmigiano), vol. 70, no. 3, pp.713-740.

19. 2005 "Quantum Logic as Motivated by Quantum Computing," *Journal of Symbolic Logic* (with Tobias J. Hagge, Lawrence S. Moss, and Zhenghan Wang), vol. 70, pp. 353-359.

*Work in Progress*

20. *Generalized Galois Logics. Relational Semantics of Nonclassical Logical Calculi* (with K. Bimbo), Center for the Study of Language and Information Lecture Notes, 394pp.

**Service to the Profession:** Chair, Iota Phi Nu, National Informatics Honorary Society 2005-2007; Vice-Chair, Computing Research Association's IT Deans Group 2004-2006; North American Collecting Editor, *Bulletin of the Section of Logic of the Polish Academy of Sciences*, 1998-; President, Society for Exact Philosophy 1988-1990; Coordinating Editor, *Journal of Philosophical Logic* 1987-1996; Editor, *Journal of Symbolic Logic*, with membership on Council of Association for Symbolic Logic, 1982-1988; Executive Committee Member, Association for Symbolic Logic, 1978-1981.

**Teaching:** Dunn has directed/co-directed 17 dissertations in logic and cognitive science, mostly in Philosophy, but also in Computer Science and Mathematics. PhD students: Dolph Ulrich, James Freeman, Randal Dipert, Daniel Cohen (co-directed with Nuel Belnap), Edwin Mares, Yu-Houng Houng, Gordon Beavers (with William Wheeler), Gerard Allwein, David Chalmers (with Douglas Hofstadter), Chrysafis Hartonas (with Jon Barwise), Paul Syverson, Raymundo Morado, Gregg Rosenberg, Katalin Bimbo, Pragati Jain, Yuko Murakami (with Larry Moss), Chunlai Zhou (with Larry Moss).

**Vision Statement:** Logic is important for computer science (but regrettably growing less influential in mathematics and philosophy). Automated reasoning techniques (including statistics) become important with the "Semantic Web." Alternative logics are a good thing in my "engineering" view of logics as tools.

**Honours and Awards:** Provost's Medal, 2007; Sagamore of the Wabash, 2007 (highest honor by the state of Indiana); i-School Caucus "Bookends Award" for vision and pioneering leadership in the formation of the i-schools community," 2006; *Who's Who in the World*, 2005- ; Elected Honorary Member, Phi Beta Kappa, 2004; First Finalist INITA Mira Award for IT Educator of Year, 2003; *Who's Who in America*, 1985–; Fellow, American Council of Learned Societies, 1984; Fulbright Research Senior Scholar, 1975; Visiting Fellow, Australian National University Institute for Advanced Studies, 1975; National Endowment for the Humanities Fellowship (declined), 1974; Woodrow Wilson Dissertation Fellow, 1965-1966; Andrew Mellon Fellow, 1964-1965 ; Woodrow Wilson Fellow, 1963-1964; National Merit Scholar, 1959-1963.

## DŽAMONJA, Mirna

**Specialties:** Set theory, connections with other subjects, such as analysis, topology, and model theory.

**Born:** 12 December 1965 in Sarajevo, Yugoslavia (Bosnia and Herzegovina).

**Educated:** University of Wisconsin-Madison, USA, PhD Mathematics, 1993; Mathematics, University of Wisconsin-Madison, USA, MSc, 1990; Mathematics, University of Sarajevo, Yugoslavia (Bosnia and Herzegovina), BSc, 1988.

**Dissertation:** *A Set Theoretic Approach to Some Problems in Measure Theory*; supervisor, Kenneth Kunen.

**Regular Academic or Research Appointments:** READER IN MATHEMATICS, UNIVERSITY OF EAST ANGLIA, 2000–; EPSRC ADVANCED FELLOW IN MATHEMATICS, 2002–2007; Lecturer in Mathematics, University of East Anglia, 1998–2002.

**Visiting Academic or Research Appointments:** Senior Visiting Researcher, Kurt Gödel Centre, Vienna, 2005; Visiting Professor in Mathematics, University Paris VII, 2004; Visiting Logic Professor, University of Wisconsin-Madison, 2003; Visiting Associate Professor, Baruch College, CUNY, 2001- 2002; Van Vleck Visiting Assistant Professor, University of Wisconsin-Madison, 1995-1998; Forchheimer Postdoctoral Fellow in Mathematics, Hebrew University of Jerusalem, 1994; Postdoctoral Fellow in Mathematics, Hebrew University of Jerusalem, 1993.

**Research Profile:** Mirna Džamonja is a set-theorist with a strong interest in the connections between set theory and other areas of mathematics. She has worked in many areas within set theory itself, such as infinitary combinatorics, forcing, pcf and large cardinals. On the other hand, she has worked in set-theoretic topology, measure theory, functional analysis and model theory. She is noted for her work on the question of the existence of universal models. In her papers joint with Saharon Shelah she developed a method for showing the consistency of the existence of a small number of models universal in a certain class of models, independently of the behaviour of the continuum function. In further work she applied this and other methods from universality to settle a long-standing open question about uniform Eberlein compacta and she developed further applications to the theory of trees and partial orders. In measure theory, starting from her work joint with Kenneth Kunen and through her later work with Grzegorz Plebanek she developed a number of insightful examples of curious spaces with measures of certain kind, and then worked on a characterisation of Boolean algebras which carry certain types of measures. In model theory, jointly with Saharon Shelah, she developed the theory of properties that prevent a model-theoretic class from having universal models.

In combinatorial set theory, with Moore and Hrusak, she was one of the authors who discovered the theory of parametrized diamond principles.

**Main Publications:**

1. Measures on Compact HS Spaces, Fundamenta Mathematicae 143-1 (1993), pp. 41-54. (with Kenneth Kunen)
2. Similar but not the same: various versions of the club principle do not coincide, Journal of Symbolic Logic, vol. 64 (1), pp. 180-198 (03/1999). (with Saharon Shelah)
3. A Note on the Splitting Property in Strongly Dense Posets of Size $\aleph_0$, Radovi Matematički, Vol. 8, No 1 (1998), pg. 321-326.
4. On $\wp_\kappa \lambda$ combinatorics using a third cardinal, Radovi Matematicki, Vol. 9, No 2 (2000), pg. 141-155.
5. On the existence of universal models, Archive for Mathematical Logic, Vol. 43 (2004), pg. 901-936. (with Saharon Shelah)
6. On universal Eberlein compacta and c-algebras, Topology Proceeedings, Vol. 23, pg. 143-150, (1998).
7. Universality of uniform Eberlein compacta, to appear in the Proceedings of the American Mathematical Society.
8. On D-spaces and Discrete Families of Sets, in AMS, DIMACS: Series in Discrete Mathematics and Theoretical Computer Sciences, ed. by S. Thomas 58 (2002), 45-63.
9. Universal graphs at the successor of a singular cardinal, Journal of Symbolic Logic, Vol. 68, No 2 (June 2003), pg. 366-387. (with Saharon Shelah)
10. Wild Edge Colourings of Graphs, Journal of Symbolic Logic, vol. 69, no.1, (2004), pg. 255-264. (with Peter Komjáth and Charles Morgan)
11. Some remarks on a question of D.H. Fremlin regarding $\varepsilon$-density, Archive for Mathematical Logic, Vol. 40 (7), (2001), pg. 531-540. (with Arthur Apter)
12. Generalisations of epsilon-density, Acta Universitatis Carolinae Mathematica et Physica, vol. 44 No. 2, (2003), pp. 57-64. (with Grzegorz Plebanek)
13. A family of trees with no uncountable branches, Volume 28, no. 1, (2004), Topology Proceedings. (with Jouko Väänänen)
14. On properties of theories which preclude the existence of universal models, Annals of Pure and Applied Logic, vol. 139, (1-3), (2006), pg. 280-302. (with Saharon Shelah)
15. On Efimov spaces and Radon measures, *Topology and its Applications*, vol. 154 (2007), pp 2063–2072. (with Grzegorz Plebanek)
16. An Expansion of a Poset Hierarchy, Central European Journal of Mathematics, vol. 4, no. 2, (2006), pg. 1-18. (with Katherine Thompson)
17. Strictly Positive Measures on Boolean Algebras, to appear in *Journal of Symbolic Logic*. (with Grzegorz Plebanek)
18. A partition theorem for large dense linear orders,

to appear in *The Israel Journal of Mathematics*. (with Jean Larson and Bill Mitchell)

19. Measure Recognition Problem, em Philos. Trans. Royal Soc. A, vol. 364 (2006), pp. 3171-3182.

20. Global complexity, in *Quaderni di Matematica: Set Theory and Applications*, ed. by A. Andretta vol. 17 (2007), pp. 25–47. (with Sy-David Friedman and Katherine Thompson).

A full list of publications is available at M. Džamonja's home page http://www.mth.uea.ac.uk/~h020/papers.html

**Service to the Profession:** Member of the Executive Committee Association of Symbolic Logic, 2008–2011; Programme Chair, Association for Symbolic Logic Annual Meeting, 2007; Member of the Board of Jurors, Young Scholar's Competition, Kurt Gödel Centenary Conference, Vienna, 2006; Editorial Board Member, "Sarajevo Journal of Mathematics," 2005-; Secretary of the British Logic Colloquium, 2002; Member, EPSRC Peer Review College, 2000-; Co-organiser LMS and BLC sponsored series of conferences on Set Theory and its Neighbours, 1998-; Council Member, Association of Symbolic Logic, 2003- 2005.

**Teaching:** M. Džamonja has had a varied teaching career with teaching experience in the United States and in the United Kingdom. She taught many mathematics courses on all levels, starting from basics of higher mathematics up to graduate courses in various subjects of mathematical logic. She has had quite a number of graduate students, both on the PhD and MSc level. Her students include Katherine Thompson (PhD 2003). She has also collaborated as an external adviser with universities such as the University of Ljubljana in Slovenia (advised an MSc student and advising a PhD student). Her teaching philosophy is that the most important skill to impart on students is the skill of critical thinking.

**Vision Statement:** Set theory is part both of mathematics and of logic. We have logical techniques that show the limitations of the axioms to solve certain mathematical questions. Now we should use these techniques to understand exactly which questions we CAN solve.

**Honours and Awards:** Advanced Research Fellow, Engineering and Physical Sciences Research Council UK, 2002-; Forchhheimer Postdoctoral Fellows, Lady Davis Foundation at the Hebrew University of Jerusalem 1994-1995.

# E

## ENGELER, Erwin

**Specialties:** Model theory, computational logic, foundations and philosophy of mathematics.

**Born:** 13 February 1930 in Schaffhausen, Switzerland.

**Educated:** ETH Zurich, Dr.sc.math, 1958.

**Dissertation:** *Untersuchungen zur Modelltheorie*; supervisor, Paul Bernays.

**Regular Academic or Research Appointments:** PROFESSOR OF LOGIC AND COMPUTER SCIENCE, TTH ZURICH 1972–1997, EMERITUS 1997–. Professor, Mathematics, 1967–1972, Associate Professor, Mathematics, 1963-1967; Assistant Professor, Mathematics, University of Minnesota, 1958-1962.

**Visiting Academic or Research Appointments:** Visiting Professor, University of Zurich 1970-1971; Forschungsinstitut für Mathematik ETH Zurich, 1967-1968; IBM Research Institute Zurich, 1963-1964; Visiting Assistant Professor, University of California Berkeley 1961-1963.

**Research Profile:** Erwin Engeler contributed to the early development of model theory by shifting interest to combinatorial model constructions, to interaction with infinitary logics and its expressive power, e.g. categoricity, and by a sustained effort to cast model theory into a frame of category theory, using filters of mappings between models (Key Publications 5,6,7). Engeler was one of the pioneers of the logic of computations, in particular by characterizing structures up to computational equivalence, e.g. the real numbers and various construction geometries (8,9). By creating rich models of combinatory logic (16), Engeler initiated a broad research program in combinatory algebra reaching from universal algebra to computer algebra to process-modeling (20), e.g. by creating a theory of the differential ring of programmable functions over the real and complex numbers (18). Early influences by his teachers Bernays and H.Weyl, formed Engeler's interest in the foundations and philosophy of mathematics and resulted in investigations into the approximation by computable theories of a sequence of finite set-objects to a limit theory of infinite sets (13,19), in following up foundational aspects of combinatory algebra into the foundations of set theory and of the basic concepts of logical languages (14,15).

**Main Publications:**

1. Eine Konstruktion von Modellerweiterungen. Z. Math. Logik Grundl. Math. 5 (1959) 126-131.
2. Unendliche Formeln in der Modelltheorie. Z. Math. Logik Grundl. Math. 7 (1961) 154-160.
3. Zur Beweistheorie von Sprachen mit unendlich langen Formeln. Z. Math. Logik Grundl. Math. 7 (1961) 213-218.
4. A reduction principle for infinite formulas. Mathem. Annalen 151 (1963) 296-303.
5. Categories of mapping filters. In: Conf. Categorical Algebra, eds S. Eilenberg et al. (Springer: New York, 1966) pp. 247-253.
6. On structures defined by mapping filters. Math. Annalen 167 (1966) 105-112.
7. On the structure of elementary maps. Z. Math. Logik u. Grundl. d. Math. 13 (1967) 323-328.
8. Algorithmic properties of structures. Math. Systems Theory 1 (1967) 183-195.
9. Formal Languages: Automata and Structures. (Markham: Chicago 1968) 81 pp.
10. Introduction to the theory of computation. (Academic Press: New York, 1973), 231 pp. (Transl. into Japanese 1975).
11. Generalized Galois theory and its application to complexity. Theoret. Comput. Sci. 13 (1981) 271-293.
12. Metamathematik der Elementarmathematik (Springer: New York, 1983) 132 pp. English Translation: Foundations of Mathematics (Springer, 1993) 100 pp. Transl. into Russian 1987, Transl. into Chinese 1995.
13. An algorithmic model of strict finitism. Colloq. Math. J. Bolyai (1978) 345-357.
14. Zur wissenschaftstheoretischen Bedeutung der kombinatorischen Algebra. In: Jahrbuch der Kurt Goedel Gesellschaft, (Wien 1990) pp. 45-53.
15. Existenz und Negation. In: Jahrbuch der Kurt Goedel Gesellschaft, (Wien 1994), pp. 37-53.
16. Algebras and combinators. Algebra Universals 13 (1981) 389-392.
17. Representation of varieties in combinatory algebras. Algebra Universalis 25 (1988) 85-95.
18. Combinatory differential fields. Theoretical Computer Science 72 (1990) 119-131.
19. Algorithmic Properties of Structures Selected papers of E. Engeler. World Publ. Co., 1994, 257 pp.
20. The Combinatory Programme (with K. Aberer et al.) Birkhäuser Boston, 1995, 143 pp.
21. see also homepage: www.math.ethz.ch/$\sim$engeler

*Work in Progress*

22. Neural nets quite naturally embody subalgebras of rich combinatory algebras. Application to interrelating the structure and function in neurology,

**Service to the Profession:** Editor, Dialectica,

1966-2004, Honorary Editor 2004-; Editor, Theoretical Computer Science from founding to 1997, Honorary Editor, 1997-; Member, National Research Council of Switzerland, Mathematics and Computer Science, 1984-1996; Member, Scientific Board of ICSI, Berkeley, 1989-1997; Founding President of the Swiss Association for Research in Computer Science (SARIT), 1989-1996.

**Teaching:** Engeler has been instrumental in introducing logic and computer algebra into curricula, including corresponding laboratories. He had 37 PhD students in diverse fields of logic and computer science, number theory, computer algebra, universal algebra, etc. of which about one third now hold university appointments, others are in industry and in finance. Four mathematicians obtained their Habilitation with Engeler.

**Vision Statement:** I would love to see that logic attracts again the best minds in mathematics, as it did before my time.

**Honours and Awards:** Member, Academia Europea, 1990; Fellow, Association for Computing Machinery, 1992.

## ERNST, Zachary

**Specialties:** Automated reasoning, game theory.

**Born:** 18 October 1972 in Detroit, Michigan, USA.

**Dissertation:** *Evolutionary Game Theory and the Evolution of Fairness*; supervisor Elliott Sober.

**Regular Academic or Research Appointments:** ASSISTANT PROFESSOR, UNIVERSITY OF MISSOURI-COLUMBIA, 2006–; Assistant Professor, Florida State University, 2003-2006; Special Term Appointee, Argonne National Laboratory, 2001-2003.

**Visiting Academic or Research Appointments:** Visiting Research Faculty, Argonne National Laboratory, 2004.

**Research Profile:** Zachary Ernst's work in logic is concerned with automated theorem proving and the application of automated theorem proving techniques to the solution of open problems in logic. He has published results (with Branden Fitelson, Kenneth Harris, and Larry Wos) on the axiomatization of the implicational fragments of modal and substructural logics. Currently, he is concerned with extending automated theorem proving techniques to develop tools for the automatic discovery of infinite models of logics lacking the finite model property.

**Main Publications:**

1. Ernst, Zachary; Fitelson, Branden; Harris, Kenneth; Wos, Larry Shortest axiomatizations of implicational S4 and S5. Notre Dame J. Formal Logic 43 (2002), no. 3, 169–179 (2003).
2. Ernst, Zachary Completions of $TV\sb {\to}$ from $H\sb {\to}$. Bull. Sect. Logic Univ. \Lód\'z 31 (2002), no. 1, 7–14.

*Work in Progress*

3. "Tree Automata as Infinite Models of Propositional Calculi"

**Service to the Profession:** Associate Editor, *Studia Logica*, 2007-.

**Teaching:** Zachary Ernst teaches in the Department of Philosophy at the University of Missouri-Columbia.

**Vision Statement:** "There is need for development of a theory concerning the finite presentation of infinite models of systems lacking the finite model property. Such a theory could help develop tools for automating the search for such models using automate reasoning techniques."

## ERSHOV, Yuri L.

**Specialties:** Computability, model theory.

**Born:** 1 May 1940 in Novosibirsk, USSR.

**Educated:** Institute of Mathematics (IM), SB AS USSR Candidate Degree (PhD), 1964; Novosibirsk State University, Diploma, 1963; Novosibirsk State University, 1961-1963; Tomsk State University, 1958-1961.

**Dissertation:** *Decidable and Undecidable Theories*; supervisor, A.I.Malcev; IM, Doctor Degree, 1966, *Elementary Theories of Fields*.

**Regular Academic or Research Appointments:** DIRECTOR, SOBOLEV INSTITUTE OF MATHEMATICS, SB RAS, 2002–; Director, Institute of Discrete Mathematics and Informatics,1993-2003; Rector, Novosibirsk State University (NSU), 1985-1993; Head, Mathematical Logic Division of IM,1967-2002; Professor, NSU,1966-2002; Researcher, Sobolev IM, Senior, 1964-1967, Junior, 1963-1964.

**Visiting Academic or Research Appointments:** Technical University Darmstadt, Germany, 2006, 2004,...,1994; University of San Paulo, Brazil, 2006; Isaac Newton Institute, Cambridge, UK, 2005; Helsinki University, Finland, 2001; Mittag-Leffler Institute, Djursholm, Sweden, 2001; Konstanz University, Germany, 1999; Institute for Studies in Theoretical Physics and Mathematics, Tehran, Iran, 1997; Uppsala University, Sweden, 1995; Paris University VII, France, 1992; UCLA, USA, 1980.

**Research Profile:** The main achievements of Ershov belong to Mathematical Logic and concern the decidability problems of elementary theories, model-theoretical algebra, and computability. He proposed a very powerful method of demonstrating the undecidability of elementary theories: the method of relative elementary definability, and used it to prove the undecidability of a great variety of interesting algebraic theories (for example, the undecidability of the elementary theory of finite simple groups). He proved decidability of many important elementary theories: distributive lattices with relative complements, the field of p-adic numbers (Tarski's problem proved independently by J.Ax and S.Kochen), the field of totally p-adic algebraic numbers, etc. He carefully studied the model-theoretical properties of (multi-)valued fields. He discovered the importance of the local-global principle for rational points of varieties and continuity of the local elementary properties for the nice model theory of multi-valued fields, found wonderful extensions for fields of algebraic numbers with decidable theory. These extensions were used for new formulation of global class field theory. Regarding computability, Ershov made fundamental contributions in numbering NUMBER? theory, many-one reducibility, the theory of constructive models and computability in (special) admissible sets. He developed a general theory of numberings and used it for introducing the class of partial computable functionals of finite types. Investigations of these functionals leaded him to defining new very useful topological notions, including domains (proposed independently by D.Scott) and A-spaces. Ershov found a "right" definition for the m-jump operation in m-degrees, and introduced and investigated the difference hierarchy (known now as Ershov's hierarchy) closely connected with m-jump. His most important result is a full algebraic characterization of the upper semilattice of m-degrees by a property of extensions of embeddings on ideals. In constructive model theory, Ershov proved several important general theorems (existence of decidable models for decidable theories; a theorem on the core model; usefulness of Skolem functions for existence of constructive models). Most unexpected among the general computability (sigma-definability) results of Ershov's in special admissible sets is the characterization of the theories having uncountable models sigma-definable in the hereditarily finite superstructure (HF(L)) over a dense liner order L. In philosophy of mathematics, he proposed a modification of Hilbert's program, division into 'finite (real)' or 'ideal' statements (objects) depending a class of initial problems. He, jointly with S.Goncharov and D.Sviridenko, suggested a concept of semantic programming.

**Main Publications:**

1. Decidability of the elementary theory of distributive lattices with relative complements and theory of filters (in Russian), Algebra Logic, 3, 3(1964), 17-38
2. Unsolvability of the theories of symmetric and simple finite groups, Soviet Math.Dokl. 5 (1964), 1309-1311
3. On the elementary theory of maximal normed fields, Soviet Math.Dokl. 6 (1965),1390-1393
4. (with I.Lavrov,A.Tajmanov,M.Taitslin) Elementary theories, Russian Math.Surv. 20, 4 (1965), 35-105
5. A hierarchy of sets I,II,III.Algebra Logic 7 (1968), 25-43; 212-232;9 (1970), 20-31
6. Computable functionals of finite types, Algebra Logic, 11 (1972), 203-242
7. Existence of constructivizations, Soviet Math.Dokl. 13 (1972), 779-783
8. The theory of A-spaces,Algebra Logic, 12 (1973), 209-232
9. Theories of nonabelian varieties of groups, Proc Tarski Symp., Berkeley 1971, Proc.Symp. Pure Math., v.25, AMS, 1974, 255-264
10. The upper semilattice of numerations of a finite set, Algebra Logic,14(1975),159-175
11. Theory of Numberings (in Russian), Moscow, Nauka, 1977, 416 p
12. Model **C** of continuous partial functionals, in: Logic Coll.76, Studies in Logic , vol. 87, North.Holl.1977,455-467
13. (with E.A.Palyutin) Mathematical Logic, 1979, 1987, 2004, 2005 (in Russian), 1984 (in English), 1990 (in Spanish)
14. Problems of decidability and constructive models (in Russian), Moscow, Nauka, 1980, 415 p
15. (with S.S.Goncharov and K.F.Samokhvalov) Introduction to the Logic and Methodology of Science (in Russian), Interprax, Moscow, 1994, 255 p.
16. Definability and Computability, Plenum, 1996, 264 p.
17. (with S.S.Goncharov) Constructive Models, Kluwer Academic/Plenum Publ. 2000, 360 p
18. Multi-Valued Fields, Kluwer Acad./Plenum Publ. 2001, 270 p.
19. Nice extensions and global class field theory, Dokl.Mathematics, 67, #1 (2003), 21-23
20. Rogers semilattices of finite partially ordered sets, Algebra Logic, 45, #1 (2006), 26-48

*Work in Progress*

21. Survey on HF-computability (with V.Puzarenko and A.Stukachev)
22. Subfields of the adele rings

**Service to the Profession:** Editor in Chief, Siberian Math. Journ., 2003-; Editor in Chief, Algebra and Logic, 1987-; Editor, Algebra and Logic, 1967-1987; Editor, SMJ, 1969-2003; Editor, Handbook of Recursive Mathematics, Elsevier, 1998; Dean of Math. Faculty, NSU, 1963-1975.

**Teaching:** Ershov taught the primary courses in Mathematical Logic and the Theory of Algorithms as well as several special courses at Math. Faculty of NSU, 1964-2002. He had more than 40 PhD students in Mathematics, including S.Goncharov, M.Peretyt'kin, S.Badaev, A.Pinus, K.Kudaibergenov, A.Khutoretski, and S.Denisov.

**Vision Statement:** Mathematical Logic began as a technical tool in the Foundations of Mathematics, but has now become an unavoidable part of Modern Mathematics. It has provided new techniques for mathematical proofs and served as an important source for new visions and concepts in practically any part of Mathematics, including classical parts such as Number Theory. Mathematical Logic has also a decisive impact on Theoretical Computer Science.

**Honours and Awards:** State Prize, Science and Technology, 2002; Malcev Prize, Russian Academy of Sciences (RAS), 1993; Full Member, RAS, 1991; Corresponding Member of USSR Academy of Science, 1970.

# ESAKIA, Leo

**Specialties:** Intuitionistic logic, modal logics (including Provability Logic), algebraic logic.

**Born:** 14 November 1934 in Tbilisi, Georgia.

**Educated:** Moscow State University, Moscow, PhD,1980; Tbilisi State University, Tbilisi, BA and MS, Mathematics, 1958.

**Dissertation:** *Semantical analysis of Intermediate Logics and normal extensions of the Modal System S4*; supervisor, none.

**Regular Academic or Research Appointments:** HEAD OF THE LOGIC SECTOR, THE RAZMADZE MATHEMATICAL INSTITUTE, GEORGIAN ACADEMY OF SCIENCES, 2003–; Department Head, Mathematical Logic, Institute of Cybernetics, Georgian Academy of Sciences, 1974-2002; Senior Researcher, Institute of Cybernetics, Georgian Academy of Sciences, 1961-1974; Junior Researcher, Institute of Cybernetics, Georgian Academy of Sciences, 1959-1961.

**Visiting Academic or Research Appointments:** Professor, Joint Appointment, Department of Foundations of Mathematics, Tbilisi State University, 1989-; Associate Professor, Department of Foundations of Mathematics, Tbilisi State University, 1985-1989; Lecturer, Department of Algebra and Geometry, Tbilisi State University, 1976-1985.

**Research Profile:** The core of Esakia's reaserch can be summarized as follows: semantical analysis of Intermediate Logics and Modal Systems (including Provability Logic). One common tendency in this research is development of new modellings, by employing the dualities between the topological Kripke frames and Heyting, Closure and Derivative algebras.

The duality theory has proved its usefulness in many areas. The intermediate logics and normal extensions of the classical Lewis system S4 is one such area where the duality approach has provided an illuminating and clarifying framework. The study of dual objects of Heyting lattices, Closure and Derivative algebras provides a tool which permits the resolution in a uniform way of a considerable number of otherwise disconnected problems.

Esakia has founded and developed the Duality theory for Heyting lattices, Closure and Derivative algebras. Based on this duality, Esakia obtained Representation Theorem for Heyting lattices, Closure and Derivative algebras in terms of topological Kripke frames and established that the category of Heyting algebras is equivalent to the category of Closure algebras in which every element is a finite Boolean combination of closed elements. He showed that the Grzegorczyk's modal system is the largest modal system in which the Intuitionistic Logic can be embedded by the Gödel translation. Moreover he established that the lattice of Intermediate Logics is isomorphic to the lattice of all normal extensions of the Grzegorczyk modal system. In terms of perfect Kripke models he constructed weak and almost-direct product Heyting algebras and presented a complete description of the lattice of congruences and subalgebras, subdirectly irreducible and finitely approximable Heyting, Closure and Derivative algebras.

The link between Diagonalizable algebras and Cantor scattered spaces is pointed out by Esakia for the first time in explicit form. He found an "equational" characterization of Cantor scattered spaces and an axiomatic definition of Hausdorff reducible spaces. He has established topological completeness of the Grzegorczyk modal system with respect to the Hausdorff reducible spaces, and of the Gödel-Löb modal system (alias, Provability Logic) with respect to the Cantor scattered spaces. Esakia has developed the idea of interpreting possibility modality as the topological derivative (alias, limit); in this direction, he discovered a simple equational system of postulates for Derivative algebras, which captures algebraic properties of the topological derivative operation and established topological completeness of a new modal system wK4. Esakia has investigated three modal logical systems of finite trees (*reflexive, unreflexive* and *irreflexive* ones), establishing various algebraic and topological properties of their semantical counterparts.

He has introduced an Amended Intuitionistic Predicate Logic, admitting a provability interpretation (via Gödel's translation and Solovay's arithmetical completeness theorem). Esakia has established some important features of the Amended Intuitionistic Logic: 1) Quantifier models of the Amended Calculus include all Kripke models with well-founded base, and hence the ones with finite base. 2). Among the sheaf models, this Calculus admits the sheaf toposes only over scattered Cantor spaces.

Esakia has introduced a conservative enrichment of Intuitionistic Logic by temporal (Always-Before) modalities which have the following striking features: 1) in them, one can speak about not only temporal aspects of reasoning, but also about creative and critical states of knowledge as well, 2) their models have a tree structure; 3) this tree model property is important for obtaining good complexity results and for the design of practical algorithms; 4) they are decidable fragments of undecidable constructive protothetics (that is Second Order Intuitionistic propositional Logic). Esakia has developed representation and duality theory of modalized Heyting algebras – algebraic counterpart of the conservative enrichment of the intuitionistic logic with temporal modalities.

**Main Publications:**

1. (with G.Bezhanishvili, D.Gabelaia) *Some Results on Modal Axiomatization and Definability for Topological Spaces*, Studia Logica, 2005, vol.81, 325-356
2. *Intuitionistic Logic and Modality via Topology*, Annals of Pure and Appled Logic, 2004, *vol 127, n 1-3, 155-170*
3. *Gödel's Embedding the Intuitionistic Calculus into Modal Logic: Recent and New Observations*", IV Intern. Logic Conference, Moscow, Russ.Acad.Sci., 2003, 72-75
4. *A modal version of Godel's second Theorem and McKinsey system*, Logical Investigations, Moscow, "Nauka", 2002, vol. 9, 292-300 (in Russian)
5. *Gödel-Löb modal system – addendum*, III Intern. Logic Conference, Moscow, Russ.Acad.Sci. 2001, 77-99 (in Russian)
6. *Weak Transitivity – restitution*, Logical Investigations, Moscow, "Nauka", 2001, vol.8, 244-254 (in Russian)
7. *Synopsis of Fronton Theory*,Logical Investigations,Moscow,"Nauka", 2000, vol. 7, 137-147 (in Russian)
8. (with M.Jibladze, D.Pataraia*) Scattered Toposes*, Annals of Pure and Appled Logic, 2000, vol.103, 97—107
9. *Quantification in Intuitionistic Logic with provability smack*, Bull.Sect. Logic, 1998, vol.27, n 1/2, 26-28
10. *Provability Interpretation of the Intuitionistic Logic*, Logical Investigations, Moscow, "Nauka", 1998, vol. 5, 19-24 (in Russian)
11. *Heyting algebras. Duality Theory.* Tbilisi, Acad.Press "Mecniereba", 1985, 104 pp.
12. *On the variety of Grzegorczyk algebras*, Sel.Math. Sov. 1984, vol.3, n .4, 343-366
13. *Algebraic Logic. Ordered sets and Lattices*, Saratov, Saratov State University, 1983, 15-26 (in Russian)
14. *Weak decomposition of Heyting and Boolean algebras*, VII Internat. Congress for Logic, Methodology and Philosophy of Science, Austria, Salzburg, 1983, 71-73
15. *Diagonal Constructions, Lob's formula and Cantor scattered spaces*, Logic and Semantic investigations, Tbilisi, "Mecniereba", 1981, 128-143(in Russian)
16. *A Theory of Modal and Superintuitionistic systems*, Logical Inference, Moscow, "Nauka", 1979, 147-172 (in Russian)
17. (with R.Grigolia*) The criterion of Brouwerian and Closure algebras to be finitely generated*, Bull.Sect.Log. 1977, vol.6, n 2, 46-52
18. (with V.Meskhi) *Five critical modal systems*, Theoria, 1977, vol.43, n 1, 52-60
19. (with R.Grigolia) *Christmas tree. On the free cyclic algebras in some varieties of closure algebras*, Bull.Sect.Logic, 1975, vol.43, n 1, 52-60
20. *On topological Kripke models,* Dokl.AN USSR, 1974, vol 214, n 2, 298-301 (in Russian)

*Work in Progress*

21. Study of certain "intrinsic" reincarnations of the standard Provability Predicate in Peano Arithmetic *PA* that are of special interest in connection with the study of Provability Logic. These special *reincarnations* (reflexive *distortion*, modest *enrichment* and *iteration*) of standard provability predicate are internally definable metamathematical predicates distinct from the standard provability, yet strong enough to satisfy Hilbert-Bernays Derivability conditions. The Gödel-Löb modal system (alias, Provability logic) is a modal logic that is used to investigate standard provability predicates in a propositional language. The purpose is to establish modal systems arising from a study of such reincarnations of the standard provability predicate. Among such modal systems can be found ones refuting modal versions of both Gödel incompleteness theorems while retaining certain weakening of the modal form of the Löb axiom.

22. Investigation of a new class of topological semantics, for which restrictions are imposed not on the class of spaces but rather on *the admissible valuations* of propositional variables; in particular, study of the modal logic of the Euclidean space under the restriction that all involved valuations are *pointwise discontinuous* functions in the sense of Baire.

**Service to the Profession:** Co-organizer, Conferences on *Algebraic and Topological Methods* in *Non-Classical* Logics, Tbilisi, Georgia, 2003, Spain, 2005; Co-organizer, Series of Conferences: *Logic, Language and Computation*, Georgia, 1995, 1997, 1999, 2001, 2003; Guest Editor, *Studia Logica* issue *Provability Logic*, 50 (1991), n 1; Co-organizer, with A. Preller, Marseille University, Series of Workshops: *Logic and Computer Science around the $42^{nd}$ Parallel*, France, Marseille, Centre International de Recontres Mathematicques, Luminy, 1988, 1990, 1992, 1994; Advisory Editor, *Studia Logica*, 1975-; Collecting Editor, *Bulletin of the Section of Logic*, 1975-1986.

**Teaching:** Leo Esakia was reading regularly courses in mathematical logic at the Tbilisi State University; typically they included a regular seminar in *Algebraic Semantics of Intuitionistic Logic and Basic Modal Systems*, and a course concerning *Duality Theory for Heyting and Modal algebras*. Esakia had sixteen MSc and PhD students. Among them are Slava Meskhi, PhD Thesis: *Theory of Models for Logical Calculus with Temporal Operators*, Moscow Pedagogical Institute, 1976; Revaz Grigolia, PhD Tesis *Algebraic Analysis of n-valued Lukasiewicz-Tarski Systems*, Moscow Pedagogical Institute, 1976; Guram Dardjania, PhD Thesis *Complexity of Deduction and Complexity of Tree Countermodels for Some Logical Calculus'*, Steklov Mathematics Institute (Leningrad Division), 1986; Merab Abashidze, PhD Thesis *Algebraic Analysis of the Godel-Lob Modal System*, Moscow Pedagogical Institute, 1987; Guram Bezhanishvili, PhD Thesis *An Algebraic approach to Intuitionistic Modal Logics over MIPC*, Tokyo Institute of Technology, 1998 (co-Advisor: Hiroakira Ono, Japan, JAIST)

**Vision Statement:** "Among the main trends which are in my opinion welcome is the continuing increasing penetration of the mathematical apparatus of universal algebra, topology and category theory in the semantical investigation of various kinds of logical systems. One particularly important still unaccomplished goal that I would like to see progress in is creation of a more or less flexible Meaning Theory. Another related one is development of a theory of intensional invariants of translations between Natural Languages"

**Honours and Awards:** INTAS, Proposal Nr 1-04-77-7080 "Algebraic and Deduction methods in Non-Classical Logic and their Applications to Computer Science", 2005-2007; Georgian-U.S. Bilateral Grants Program (CRDF-GRDF), *Applications of Topology and Universal Algebra to Modal logic*, N 3303, 2003-2004; Georgian Academy of Sciences Grants: *a) Topological frames and Tree models of Godel Modal system and Intuitionistic Logic*, 1997-1999, N 1.25; *b) Theory of Modalized Heyting Lattices*, 2000-2001, N 1.26; *c)The Proof-Intuitionistic Logic and its Modal companions*, N 1.27, 2002-2003; Fellow, Japan Society for Promotion of Science, 1998.

# ETCHEMENDY, John William

**Specialties:** Philosophy of logic, philosophical logic, semantics, model theory

**Born:** 16 May 1952 in Reno, Nevada, USA.

**Educated:** Stanford University, PhD, 1982; University of Nevada/Reno, MA, 1976; University of Nevada/Reno, BA, 1973.

**Dissertation:** *Tarski, Model Theory and Logical Truth*; supervisor, John Perry.

**Regular Academic or Research Appointments:** PROVOST, STANFORD UNIVERSITY, 2000–; PATRICK SUPPES FAMILY PROFESSOR OF HUMANITIES AND SCIENCES, STANFORD, 2003–; Professor, Philosophy, Stanford, 1993-; Senior Associate Dean, Humanities and Sciences, Stanford, 1993-1997; Director, Center for the Study of Language and Information, Stanford, 1990-1993; Associate Professor, Philosophy, Stanford, 1988-1993; Assistant Professor, Philosophy, Stanford, 1983-1988; Assistant Professor, Philosophy, Princeton University, 1981-1983.

**Research Profile:** Etchemendy is the author of a frequently discussed criticism of the model-theoretic analysis of logical consequence, originally advocated by Alfred Tarski and, in an earlier form, by Bernard Bolzano. He has also made contributions to our understanding of the Paradox of the Liar and to the formal definition of truth. With his co-author Jon Barwise, he pioneered the study of heterogeneous logic, that is, logic involving multiple forms of representation, such as diagrams and sentences. He has also developed widely used textbooks and software for teaching logic.

**Main Publications:**

1. *The Liar: An Essay on Truth and Circularity,* with Jon Barwise, (1987) New York and Oxford: Oxford University Press, 185+xii. Second edition with postscript (1988), 194+xii.

2. *The Concept of Logical Consequence,* (1990) Cambridge and London: Harvard University Press, 174+xii. (1999) Stanford: CSLI Publications.

3. *Tarski's World,* with Jon Barwise, (1991) Stanford: CSLI Publications, 122+xviii.

4. *Turing's World: An Introduction to Computability Theory,* with Jon Barwise, (1993) Stanford: CSLI Publications, 123+ix.

5. *Hyperproof,* with Jon Barwise, (1994) Stanford: CSLI Publications, 255+xvii.

6. *Language, Proof and Logic,* with Jon Barwise, in collaboration with Gerard Allwein, Dave Barker-Plummer and Albert Liu, (2000) Stanford: CSLI Publications, 587+x.

7. "Representing Visual Decision Making," with Dave Barker-Plummer, in *Visual and Spatial Analysis: Advances in Data Mining, Reasoning, and Problem Solving,* Boris Kovalerchuk and James Schwing, ed., London: Springer, 2004, 79-109.

8. "Applications of Heterogeneous Reasoning in Design," with Dave Barker-Plummer, in *Machine Graphics and Vision,* Volume 12, Number 1, 2003, 39-54.

9. "A Computational Architecture for Heterogeneous Reasoning," in Theoretical Aspects of Rationality and Knowledge, I. Gilboa, ed., San Francisco: Morgan Kaufmann, 1998, 1-27.

10. "Computers, Visualization, and the Nature of Reasoning," with Jon Barwise, in *The Digital Phoenix: How Computers are Changing Philosophy.* T. W. Bynum and James H. Moor, eds. London: Blackwell, 1998, 93-116.

11. "Heterogeneous Logic," with Jon Barwise, in *Diagrammatic Reasoning: Cognitive and Computational Perspectives,* Janice Glasgow, N. Hari Narayanan and B. Chandrasekaran, eds., Cambridge, Mass: The MIT Press, 1995, 211-234.

12. "Hyperproof: Logical Reasoning with Diagrams," with Jon Barwise, in *Proceedings of the 1992 AAAI Spring Symposium on Diagrammatic Reasoning,* Stanford: AAAI, 1992, 80-84. Reprinted in *Reasoning with Diagrammatic Representations,* Menlo Park: AAAI Press, 1994.

13. "Visual Information and Valid Reasoning," with Jon Barwise, in *Visualization in Mathematics,* MAA Notes No. 19, Walter Zimmermann and Stephen Cunningham, eds., Washington, D.C.: Mathematical Association of America, 1991, 9-24. Reprinted in *Philosophy and the Computer,* Leslie Burkholder, ed., Boulder: Westview Press, 1992, 160-182.

14. "Model-theoretic Semantics," with Jon Barwise, in *Foundations of Cognitive Science,* Michael Posner, ed., Cambridge: MIT Press, 1989, 207-243.

15. "Models, Semantics and Logical Truth," *Linguistics and Philosophy,* Volume 11, 1988, 91-106.

16. "Tarski on Truth and Logical Consequence," *Journal of Symbolic Logic,* Volume 53, March 1988, 51-79.

17. "The Doctrine of Logic as Form," *Linguistics and Philosophy,* Volume 6, 1983, 319-334.

**Service to the Profession:** Executive Council Member, Association for Symbolic Logic, 1999-2005; Editor, *Journal of Symbolic Logic,* 1999-2005; Editorial Board Member, *Philosophia Mathematica,* 1994-; Editorial Board Member, *Synthese,* 1990-.

**Teaching:** Etchemendy has supervised nine PhD students in logic and philosophy, including Genoveva Martí (Barcelona), Sun-Joo Shin (Yale), and Patricia Blanchette (Notre Dame). He has received several teaching awards, including the Bing Award for Excellence in Teaching and the Educom Medal for pioneering contributions to logic instruction.

**Vision Statement:** Over the past 100 years, logicians have made great strides understanding the logic of formal languages, developing tools that allow us to study these languages with great precision. I would like to see more attention paid to reasoning using other forms of representation, such as diagrams, graphs and maps. Many of the same tools pioneered for the study of formal languages can be applied to these representations, but new tools will also have to be developed. Too often, an overly restrictive conception of logic has prevented logicians from addressing these phenomena with the seriousness they deserve.

# F

## FAGIN, Ronald

**Specialties:** Applications of logic to computer science, finite model theory, reasoning about knowledge, database theory.

**Born:** 1 May 1945 in Oklahoma City, Oklahoma.

**Educated:** Dartmouth College, Bachelor's Degree in Mathematics (*Summa Cum Laude* and With Highest Distinction in Mathematics) 1967; University of California at Berkeley, Ph.D. in Mathematics 1973.

**Dissertation:** *Contributions to the Model Theory of Finite Structures*; supervisor, Robert Vaught.

**Regular Academic or Research Appointments:** MANAGER, FOUNDATIONS OF COMPUTER SCIENCE, IBM SAN JOSE RESEARCH LABORATORY (LATER IBM ALMADEN RESEARCH CENTER), 1979–. Research Staff Member, IBM San Jose Research Laboratory (later IBM Almaden Research Center), 1975–; Research Staff Member, IBM Watson Research Center, 1973–1975.

**Visiting Academic or Research Appointments:** Research Fellow, IBM Haifa Research Laboratory, 1996–1997; Visiting Professor, Pontifícia Universidade Católica do Rio de Janeiro, Summer 1981.

**Research Profile:** Ronald Fagin is generally considered to be the father of finite model theory, based on the work in his Ph.D. thesis. The main result was "Fagin's Theorem", which gives an equivalence between the important complexity class NP and existential second-order logic over finite models. He also proved the 0-1 law for first-order logic over finite models; this law says that, surprisingly, every property expressible in first-order logic is either almost surely true or almost surely false. His main focus has been on applications of logic to computer science. In particular, he has done a great deal of research in database theory (a database can be viewed as a dynamically changing finite model). Another of his research areas has been epistemic logic and its applications to distributed computing; this led to his co-authored 1995 book, "Reasoning about Knowledge" (paperback, 2003).

**Main Publications:**

1. "Generalized first-order spectra and polynomial-time recognizable sets". *Complexity of Computation*, ed. R. Karp, SIAM-AMS Proceedings 7, 1974, pp. 43–73.
2. "Probabilities on finite models". *J. Symbolic Logic* 41(1)(March 1976): 50–58.
3. "Multivalued dependencies and a new normal form for relational databases". *ACM Trans. on Database Systems* 2(3)(Sept. 1977): 262–278.
4. "Extendible hashing—a fast access method for dynamic files", with Jurg Nievergelt, Nicholas J. Pippenger, and H. Raymond Strong. *ACM Trans. on Database Systems* 4(3)(Sept. 1979): 315–344.
5. "Horn clauses and database dependencies". *J. ACM* 29(4)(Oct. 1982): 952–985. Preliminary version appeared in *Proc. 12th ACM Symposium on the Theory of Computing*, 1980, pp. 123–134.
6. "On the semantics of updates for databases", with Jeffrey D. Ullman and Moshe Y. Vardi. *Proc. 2nd ACM Symposium on Principles of Database Systems*, Atlanta, 1983, pp. 352–365.
7. "On the desirability of acyclic database schemes", with Catriel Beeri, David Maier, and Mihalis Yannakakis. *J. ACM* 30(3)(July 1983): 479–513.
8. "Degrees of acyclicity for hypergraphs and relational database schemes". *J. ACM*, 30(3)(July 1983): 514–550.
9. "Inclusion dependencies and their interaction with functional dependencies", with Marco Casanova and Christos Papadimitriou. *J. Computer and System Sciences* 28(1)(Feb. 1984): 29–59. Preliminary version appeared in *Proc. 1st ACM Symposium on Principles of Database Systems*, Los Angeles, March 1982, pp. 171–176.
10. "Belief, awareness, and limited reasoning", with Joseph Y. Halpern. *Artificial Intelligence* 34, 1988, pp. 39–76. Preliminary version appeared in *International Joint Conference on Artificial Intelligence* (IJCAI-85), Aug. 1985, pp. 491–501.
11. "Reachability is harder for directed than for undirected finite graphs", with Miklos Ajtai. *J. Symbolic Logic* 55(1)(March 1990): 113–150. Preliminary version appeared in *Proc. 29th IEEE Symposium on Foundations of Computer Science*, 1988, pp. 358–367.
12. "A logic for reasoning about probabilities", with Joseph Y. Halpern and Nimrod Megiddo. *Information and Computation* 87(July/Aug 1990): 78–128. (Special issue for selected papers from the 1988 IEEE Symposium on Logic in Computer Science Conference).
13. "Finite model theory—a personal perspective". *Theoretical Computer Science* 116, 1993, pp. 3–31. Preliminary version appeared as Invited paper, *3rd International Conference on Database Theory*, Dec. 1990, Springer-Verlag *Lecture Notes in Computer Science 470*,
14. "Reasoning about knowledge and probability", with Joseph Y. Halpern. *J. ACM* 41, 2, 1994, pp. 340–367. Preliminary version appeared in *Second Conf. on Theoretical Aspects of Reasoning about Knowledge*, ed.

M. Y. Vardi, Morgan Kaufmann, 1988, pp. 277–293. Corrigendum: *J. ACM* 45(1)(Jan. 1998): 214.

15. *Reasoning about Knowledge*, with Joseph Y. Halpern, Yoram Moses, and Moshe Y. Vardi. MIT Press, 1995. Paperback edition, 2003.

16. "On monadic NP vs. monadic co–NP", with Larry Stockmeyer and Moshe Vardi. *Information and Computation* 120(1)(July 1995): 78–92. Preliminary version appeared in *1993 IEEE Structure in Complexity Theory Conference*, pp. 19–30.

17. "Combining fuzzy information from multiple systems". *J. Computer and System Sciences* 58(1999): 83–99. (Special issue for selected papers from the 1996 ACM Symposium on Principles of Database Systems).

18. "A formula for incorporating weights into scoring rules", with Edward L. Wimmers. *Theoretical Computer Science* 239(2000): 309–338. (Special issue for selected papers from the 1997 International Conference on Database Theory). Preliminary version appeared under the title "Incorporating user preferences in multimedia queries", in *Proc. 6th International Conference on Database Theory*, Jan. 1997, Springer–Verlag *Lecture Notes in Computer Science 1186*, ed. F. Afrati and Ph. Kolaitis, Delphi, pp. 247–261.

19. "Optimal aggregation algorithms for middleware", with Amnon Lotem and Moni Naor. *J. Computer and System Sciences* 66(2003): 614–656. (Special issue for selected papers from the 2001 ACM Symposium on Principles of Database Systems).

20. "Data exchange: semantics and query answering", with Phokion Kolaitis, Renee J. Miller, and Lucian Popa. *Theoretical Computer Science* 336(2005): 89–124. (Special issue for selected papers from the 2003 International Conference on Database Theory).

**Service to the Profession:** Program Committee Chair, International Conference on Database Theory, 2009. Program committee chairman, ACM Symposium on Theory of Computing, 2005; Program committee chairman, Theoretical Aspects of Reasoning about Knowledge, 1994; Program committee chairman, ACM Symposium on Principles of Database Systems, 1984; member of 30 Program Committees; Conference chairman, Theoretical Aspects of Reasoning about Knowledge, 1992; General chairman, ACM Symposium on Principles of Database Systems, 1983; Associate Editor, *Journal of Computer and System Sciences*, 1984–2008; Editor, *Journal of Computer and System Sciences*, 2007–; Editor, *Chicago Journal of Theoretical Computer Science*, 1994–; Member, Editorial Board, *Foundations and Trends in Databases*, 2006–.

**Vision Statement:** A very important role logic can play is in its applications to computer science. Fagin has never given up on trying to resolve the P versus NP question through logic.

**Honours and Awards:** 2004 ACM SIGMOD Edgar F. Codd Innovations Award for "fundamental contributions to database theory"; Highly Cited Researcher, ISI, 2002; Docteur Honoris Causa, University of Paris, 2001; Fellow of the ACM (Association for Computing Machinery), 2000, for "creating the field of finite model theory, and for fundamental research in relational database theory and in reasoning about knowledge"; Fellow of the IEEE (Institute of Electrical and Electronic Engineers), 1997, for "contributions to finite-model theory and to relational database theory"; Fellow of the AAAS (American Association for the Advancement of Science), for "fundamental contributions to computational complexity theory, database theory, and the theory of multi-agent systems"; Best Paper Award, ACM Symposium on Principles of Database Systems, 2001; Best Paper Award, International Joint Conference on Artificial Intelligence, 1985; 7 IBM Outstanding Innovation Awards; 2 IBM supplemental Patent Issue Awards, given for key IBM patents; Elected to the IBM Academy of Technology for "fundamental contributions to computer science theory and its application to IBM products", 2007.

## FARIÑAS DEL CERRO, Luis

**Specialties:** Modal logic, automated deduction, knowledge representation.

**Born:** 24 October 1949 in Tolede, Spain.

**Educated:** University Paul Sabatier, Toulouse, Habilitation à Diriger des Recherches, 1985; Complutense University, Madrid, Spain, Doctoral Thesis in Mathematics, 1983; University of Paris VII, "Thèse d'Etat" in Computer Science, 1981; Central University of Madrid, Diploma in Mathematical Sciences (Computer Sciences).

**Dissertation:** Doctoral Thesis: *Conceptos Temporales para el estudio del comportamiento de programas*; "Thèse d'Etat": *Automatic Deduction and Modal Logic*.

**Regular Academic or Research Appointments:** ASSISTANT DIRECTOR FOR EUROPEANS AND INTERNATIOAL RELATIONS OF THE STIC DEPARTMENT OF CNRS, 2004–2006; DEPUTY DIRECTOR OF THE STIC DEPARTMENT OF CNRS, 2001–2004; DIRECTOR OF IRIT, INSTITUT DE RECHERCHE EN INFORMATIQUE DE TOULOUSE, 1999–. Director of Research, Computer Science Research IRIT, UMR 5505 - CNRS/INPT/UPS, 1991; Creator, "Journal of Applied Non-Classical Logics" (JANCL), published by Hermès International, 1990; Researcher, Laboratory of Informatics for Human Sciences, CNRS, Marseille, 1980.

**Research Profile:** His works concern the applications of Computer Logic, a field of research between theoretical informatics and artificial intelligence. Luis Fariñas del Cerro and his students are authors of several articles which have allowed to extend methods of automatic deduction of classical logic to the modal logic, as for example the principle of resolution and its application in logical programming for modal logics. They also defined indirect methods of proof, via a translation of modal logics to first order specific theories. More recently they introduced a generic method of automatic deduction for the modal logic of array type.

He defined and applied, in the field of computer science, multimodal systems allowing us to represent a very wide spectrum of concepts. These systems also enable us to represent the uncertainty as well as allow us to represent geometrical notions. In particular, for the manipulation of these last concepts, methods of automatic deduction using the algebra of relations were defined.

**Main Publications:**

1. AUDUREAU E., ENJALBERT P., FARIÑAS DEL CERRO L. Logique Temporelle Sémantique et Validation de Programmes parallèles. Masson, 1990
2. CROCCO G, FARIÑAS DEL CERRO L. and HERZIG A. (Eds) Conditionals: from philosophy to Computer Sciences. (Eds.) Oxford University Press, 1995
3. FARIÑAS DEL CERRO L. and RAGGIO A. Some results in intuitionistic modal logic. *Logique et Analyse*, 102, 219-224, 1983.
4. FARIÑAS DEL CERRO L. Modalities for total correctness. *Fundamenta Informaticae*, 7(3), 301-311, 1984.
5. FARIÑAS DEL CERRO L. and ORLOWSKA E. Dal, a logic for data analysis. *Theoretical Computer Science*, 36, 251-264, 1985.
6. CIALDEA M. and FARIÑAS DEL CERRO L..A Modal Herbrand Property. *Zeitschrift für Mathematische Logik und Grundlagen der Mathematik*, 32, 523-530, 1986.
7. FARIÑAS DEL CERRO L. and MOLOG : a system that extends PROLOG with modal logic. *New Generation Computing*, 4, 35-50, 1986.
8. FARIÑAS DEL CERRO L. and PENTTONEN M. A note on the complexity of the satisfiability of modal Horn clauses. *The Journal of Logic Programming*, 4, 1-10, 1987.
9. DEMOLOMBE R. and FARIÑAS DEL CERRO L. An algebraic evaluation method for deduction in incomplete database. *The Journal of Logic Programming*, 5(3), 183-207, 1988.
10. FARIÑAS DEL CERRO L. and PENTTONEN M. Grammar logics. *Logique et Analyse*, Nauwelaerts Printings S.A., 121-122, 1988
11. ENJALBERT P. and FARIÑAS DEL CERRO L. Modal resolution in clausal form. *Theoretical Computer Science*, 65, 1-65, 1989.
12. FARIÑAS DEL CERRO L., HERZIG A. and LANG J. Ordering-based nonmonotonic reasoning. *Journal of Artificial Intelligence*, 66, 375-393, 1994.
13. BALBIANI P., FARIÑAS DEL CERRO L., TINCHEV T. and VAKARELOV D., A modal logic of incidence space, *Journal of Logic and Computation*, 7(1), 59-78, 1997.
14. BALBIANI P. and FARIÑAS DEL CERRO L., *Complete axiomatization of a relative modal logic with composition and intersection. JANCL* vol. 8, pp 325-335, 1998.
15. FARIÑAS DEL CERRO L. and GASQUET O. Tableaux Based Decision Procedures for Modal Logics of Confluence and Density. Fundamenta Informaticae, V.41 N.1, p.1-17, janvier 2000.
16. CHETCUTI-SPERANDIO N. and L. FARIÑAS DEL CERRO " A Mixed Decision Method for Duration Calculus. " *Journal of Logic and Computation,* Oxford University Press. Vol. 10 numéro 6, pages 877-895, 2000.
17. BALBIANI P., CONDOTTA J.-F. and FARIÑAS DEL CERRO L. Tractability results in the block algebra. Journal of Logic and Computation, vol.12 (2002), 885-909.
18. BIDOIT N., FARINAS DEL CERRO L., FDIDA S. et VALLEE B. (Eds). Paradigmes et enjeux de l'informatique. Série informatique et systèmes d'information. Hermès, Lavoisier, Paris, 2005.

**Vision Statement:** Given the variety of tools of representation which are used to represent knowledge : texts, figures, diagrams, he was interested in the extension of the traditional frame of logic, that may be qualified as linguistic, to the one that allows to integrate visual aspects. In this sense, having defined the methods of proof such as rewriting for geometry (that is the mathematical base of visual modality) he took charge of the definition of logical systems allowing the manipulation of the visual aspects of representations. His purpose is to define a visual logical system for the specifications of complex systems and to use the associated methods of proof as interpreters that can manipulate figures.

**Honours and Awards:** Elected Fellow of the ECCAI Society (2005).

# FEFERMAN, Solomon

**Specialties:** Mathematical logic, foundations of mathematics, philosophy of mathematics, history of modern logic.

**Born:** 13 December 1928 in New York City, New York, USA.

**Educated:** University of California at Berkeley, PhD Mathematics, 1957; California Institute of

Technology, BS Mathematics, 1948.

**Dissertation:** *Formal Consistency Proofs and Interpretability of Theories*; supervisor, Alfred Tarski.

**Regular Academic or Research Appointments:** PROFESSOR OF MATHEMATICS AND PHILOSOPHY, EMERITUS AND PATRICK SUPPES FAMILY PROFESSOR OF HUMANITIES AND SCIENCES, EMERITUS, STANFORD UNIVERSITY, 2004–. Patrick Suppes Family Professor of Humanities and Sciences, Stanford University, 1993–2003; Professor of Mathematics and Philosophy, Stanford University, 1968–2003; Associate Professor of Mathematics and Philosophy, Stanford University, 1962–1968; Assistant Professor of Mathematics and Philosophy, Stanford University, 1958–1962; Instructor of Mathematics and Philosophy, Stanford University, 1956–1958.

**Visiting Academic or Research Appointments:** Visiting Professor of Philosophy, University of California at Berkeley, Spring 2003; Fellow, Mittag-Leffler Institute, Djursholm Sweden, April-May 2001; Fellow, Center for Advanced Study in the Behavioral Sciences, Stanford, 1995-1996; Visiting Professor of Philosophy, Ecole Normale Supérieure, March 1992; Fellow, Stanford Humanities Center, Stanford University, 1989-1990; Guggenheim Fellow, University of Rome, Spring 1987; Guggenheim Fellow, ETH Zürich, Winter 1987; Visiting Fellow, Wolfson College, Oxford University, Spring 1980; Visiting Fellow, All Souls College, Oxford University, 1979-1980; Guggenheim Fellow, University of Paris, 1973; Guggenheim Fellow, Mathematics Institute, Oxford University, Autumn 1972; Visiting Associate Professor of Mathematics, Massachusetts Institute of Technology, 1967-1968; National Science Foundation Senior Post-doctoral Fellow, University of Paris and University of Amsterdam, 1964-1965; National Science Foundation Post-doctoral Fellow, Institute for Advanced Study, Princeton, 1959-1960.

**Research Profile:** Solomon Feferman has made fundamental contributions to proof theory and constructive and semi-constructive foundations of mathematics, and has made significant contributions to the other three principal areas of mathematical logic, namely model theory, set theory, and recursion theory. In proof theory and foundations of mathematics, he is noted for his fundamental work on the arithmetization of metamathematics, transfinite progressions of theories, the limits and extent of predicative mathematics, the proof theory of subsystems of classical and constructive analysis, proof-theoretic ordinals, and systems of explicit mathematics that bridge constructive, predicative and classical systems and also have applications to computer science. In model theory he is noted for his seminal work with Robert Vaught on properties of generalized products of relational systems as well as for his contributions to abstract model theory; in set theory his work on applications of the notions of forcing and generic sets directly following that of Cohen's was very influential; and in recursion theory he has contributed to classifications of recursive functions in transfinite hierarchies and to computation on abstract structures. Feferman has also worked on type-free theories of truth and on the foundations of category theory with self-membered categories. Finally he has contributed to the history of modern logic, especially in his analyses of the logical work of Hilbert, Weyl, Turing, Gödel, Tarski, and others, as well as through his work as Editor-in-Chief of the five volume edition of the Collected Works of Kurt Gödel. The collection *In the Light of Logic* (item 14 in the list below) contains the material on Hilbert, Weyl and Gödel together with essays spanning the period 1979-1993 on foundational problems, foundational ways, proof theory, and countably reducible mathematics.

**Main Publications:**

1. (with R. L. Vaught), The first order properties of products of algebraic systems, *Fundamenta Mathematicae* 47 (1959), 57-103.
2. Arithmetization of metamathematics in a general setting, *Fundamenta Mathematicae* 49 (1960), 35-92.
3. Transfinite recursive progressions of axiomatic theories, *J. Symbolic Logic* 27 (1962), 259-316.
4. Systems of predicative analysis, *J. Symbolic Logic* 29 (1964), 1-30.
5. Some applications of the notions of forcing and generic sets, *Fundamenta Mathematicae* 56 (1965), 325-345.
6. A language and axioms for explicit mathematics, in *Algebra and Logic* (J. N. Crossley, ed.), Lecture Notes in Mathematics 450 (1975), 87-139.
7. Categorical foundations and foundations of category theory, in *Logic, Foundations of Mathematics and Computability Theory* (R. E. Butts and J. Hintikka, eds.) vol. 1, Reidel, Dordecht (1977), 149-169.
8. Constructive theories of functions and classes, in *Logic Colloquium '78* (M. Boffa, et al., eds.), North-Holland, Amsterdam (1979), 159-224.
9. (with W. Buchholz, W. Pohlers, and W. Sieg*)*, *Iterated Inductive Definitions and Subsystems of Analysis: Recent Proof-theoretical Studies*, Lecture Notes in Mathematics 897 (1981).
10. Toward useful type-free theories I, *J. Symbolic Logic* 49 (1984), 75-111.
11. (with J. Barwise, eds.), *Model-theoretic Logics*, Springer-Verlag, Berlin, 1985.

12. Proof theory: a personal report, Appendix to *Proof Theory*, 2nd edn., by G. Takeuti, North-Holland, Amsterdam (1987), 447-485.

13. Computation on abstract data types. The extensional approach, with an application to streams, *Annals of Pure and Applied Logic* 81 (1996), 75-113.

14. *In the Light of Logic*, Oxford University Press (1998).

15. Does mathematics need new axioms?, *American Mathematical Monthly* 106 (1999), 99-111.

16. (with J. Avigad) Gödel's functional ("Dialectica") interpretation, in *The Handbook of Proof Theory* (S. Buss, ed.), North-Holland, Amsterdam (1998), 337-405.

17. Logic, logics, and logicism, *Notre Dame J. of Formal Logic* 40 (1999), 31-54.

18. (as Editor-in-Chief), Kurt Gödel. Collected Works, Vol. I. Publications 1929-1936. Vol. II, Publications 1938-1974. Vol. III, Unpublished essays and lectures. Vol. IV. Correspondence A-G. Vol. V. Correspondence H-Z. All, Oxford University Press, 1986-2003.

19. Predicativity, in *The Oxford Handbook of Philosophy of Mathematics and Logic* (S. Shapiro ed.), Oxford University Press, Oxford (2005), 590-624.

20. (with Anita Burdman Feferman), *Alfred Tarski: Life and Logic*, Cambridge University Press, New York (2004).

A full list of publications is available at Feferman's home page http://math.stanford.edu/~feferman/

*Work in Progress*

21. A book on systems of explicit mathematics and their applications, with Gerhard Jäger and Thomas Strahm.

22. A successor volume to *In the Light of Logic*, to include essays on Gödel's program for new axioms, as well as essays on Turing and Tarski.

23. Further development of operational set theory as a framework for common generalizations of large cardinal axioms in classical and admissible set theory and related systems of ordinal notations.

24. Elaboration of the use of open-ended interactive axiom systems to model mathematics in practice.

**Service to the Profession:** Editor-in-Chief, Kurt Gödel, *Collected Works*, 1982-2003; Editor, *Perspectives in Mathematical Logic*, 1986-2003; Editor, *Ergebnisse der Mathematik*, 1986-2003; Chair, Department of Mathematics, Stanford University, 1985-1992; Member, Steering Committee, International Congress of Mathematics, 1983-1986; President, Association for Symbolic Logic, 1980-1982; Member, American Mathematical Society Committee on Translations, 1980-1982; Editor, *Transactions* and *Memoirs* of the American Mathematical Society, 1976-1979; Member, Executive Committee and Council, Association for Symbolic Logic, 1964-1967.

**Teaching:** Feferman has been a mainstay of the interdepartmental logic program at Stanford since his appointment to the faculty in 1956. He has had eighteen PhD students in the fields of Mathematics, Philosophy and Computer Science, including Jon Barwise, Wilfried Sieg, Carolyn Talcott, and Paolo Mancosu. He has also influenced the PhD work of Jeremy Avigad, Gerhard Jäger, Wolfram Pohlers, and Thomas Strahm. "Reflections", a symposium honoring Feferman on the occasion of his $70^{th}$ birthday (a.k.a. the "Feferfest") was held at Stanford University, Dec. 11-13, 1998. The symposium volume, *Reflections on the Foundations of Mathematics. Essays in honor of Solomon Feferman* (W. Sieg, R. Sommer, C. Talcott, eds.), Lecture Notes in Logic 15, ASL (2002), testifies to the influence of Feferman's teaching and research over the years.

**Vision Statement:** Foundational programs were one of the main forces behind the flowering of logic in the $20^{th}$ century. The subsequent technical development of this subject both internally and through a variety of applications has been remarkable, but the motivating programs were battered under critical examination and largely left behind. This does not mean that foundational concerns must be abandoned. Rather, foundationally directed work retains its prime importance, only now to be approached with a greater clarity of aims and sophisticated use of a variety of tools of modern logic.

**Honours and Awards:** Rolf Schock Prize in Logic and Philosophy, 2003; University of California at Irvine Chancellor's Distinguished Fellow, 1999; Patrick Suppes Family Professor of Humanities and Sciences, Stanford University, 1993-2003 (Emeritus, 2004); Fellow, American Academy of Arts and Sciences, 1990- .

# FEJER, Peter Andrew

**Specialties:** Computability theory.

**Born:** 29 July 1952 in Chicago, Illinois, USA.

**Educated:** University of Chicago, PhD Mathematics, 1980, MS Mathematics, 1976; Reed College, BS Mathematics, 1974.

**Dissertation:** The Structure of Definable Subclasses of the Recursively Enumerable Degrees; supervisor, Robert Soare.

**Regular Academic or Research Appointments:** PROFESSOR, COMPUTER SCIENCE, UNIVERSITY OF MASSACHUSETTS BOSTON, 2001–; Professor, Mathematics and Computer Science, University of Massachusetts Boston, 1995–2001;

Associate Professor, Mathematics and Computer Science, University of Massachusetts Boston, 1988–1995; Assistant Professor, Mathematics and Computer Science, University of Massachusetts Boston, 1984–1988; H. C. Wang Assistant Professor of Mathematics, Cornell University, 1980-1984.

**Visiting Academic or Research Appointments:** Guest Professor, University of Heidelberg, 1998, 1990-1991; National Science Foundation Post-Doctoral Research Fellow, Cornell University, 1980–1981.

**Research Profile:** Peter Fejer has made contributions to computability theory, in particular to the study of algebraic properties of the computably enumerable (c.e.) degrees under various reducibilities. In his thesis, he showed that the nonbranching (i.e., meet-irreducible) computably enumerable degrees are dense in the computably enumerable degrees. This was the first theorem after the Density Theorem of Sacks that showed that a class of c.e. degrees is dense. Also in his thesis, he invented a fundamental technique for constructing branching (i.e., meet reducible) c.e. degrees. In later work, he showed the decidability of the two-quantifier theory of the computably enumerable weak truth-table degrees and other distributive degree structures (with K. Ambos-Spies, S. Lempp, and M. Lerman), showed that every computably enumerable truth-table degree is branching (with R. Shore), and showed that local nondistributivity coincides with local nonmodularity in the c.e. degrees (with K. Ambos-Spies, extending work of R. Downey and S. Lempp). Much of Fejer's work has involved lattice embeddings and he has written an expository article on the use of lattice representations in this endeavor. He has contributed to the history of logic through his article with K. Ambos-Spies on the history of degree theory to appear in the *Handbook of the History of Logic*. Fejer has also published Volume 1 of a two volume series on Mathematical Foundations of Computer Science (with D. Simovici) in the Springer Verlag Texts and Monographs in Computer Science series.

**Main Publications:**

1. The plus-cupping theorem for the recursively enumerable degrees (with R.I. Soare), in M. Lerman, J.H. Schmerl and R.I. Soare, eds., *Proceedings of the Logic Year 1979-1980*, Lecture Notes in Mathematics, Vol 859, Springer-Verlag, Berlin, 1981, 49-62.
2. Branching degrees above low degrees, *Transactions of the American Mathematical Society*, 273, 1(1982), 157-180.
3. The density of the nonbranching degrees, *Annals of Pure and Applied Logic*, 24, 2(1983), 113-130.
4. Embeddings and extensions of embeddings in the r.e. tt- and wtt- degrees (with R.A. Shore), in H.D. Ebbinghaus, G.H. Muller and G.E. Sacks, eds., *Recursion Theory Week (Proceedings of a Conference held in Oberwolfach, West Germany, April 15-21, 1984)*, Lecture Notes in Mathematics, Vol. 1141, Springer-Verlag, Berlin, 1985, 121-140.
5. Infima of recursively enumerable tt-degrees (with R.A. Shore), *Notre Dame Journal of Formal Logic*, 29, 3(1988), 420-437.
6. Degree theoretical splitting properties of recursively enumerable sets (with K. Ambos-Spies), *Journal of Symbolic Logic*, 53, 4(1988), 1110-1137.
7. Embedding lattices with top preserved below non-GL2 degrees, *Zeitschrift fuer Mathematische Logik und Grundlagen der Mathematik*, 35, 1(1989), 3-14.
8. A direct construction of a minimal recursively enumerable truth-table degree (with R. Shore), in K. Ambos-Spies, G. Mueller and G. E. Sacks, eds., *Recursion Theory Week (Proceedings of a Conference held in Oberwolfach, FRG, March 19-25, 1989)*, Lecture Notes in Mathematics, Vol. 1432, Springer-Verlag, Berlin, 1990, 187-204.
9. *Mathematical Foundations of Computer Science, Volume 1: Sets, Relations, and Induction* (with D. Simovici), Texts and Monographs in Computer Science, Springer-Verlag, New York, 1991.
10. Embedding distributive lattices preserving 1 below a nonzero recursively enumerable Turing degree (with K. Ambos-Spies and D. Decheng), in J. Crossley et al., eds., *Logical Methods: In Honor of Anil Nerode's Sixtieth Birthday*, Progress in Computer Science and Applied Logic, Volume 12, Birkhauser, Boston, 1993, 92-129.
11. Decidability of the two-quantifier theory of the recursively enumerable weak truth-table degrees and other distributive upper semi-lattices (with K. Ambos-Spies, S. Lempp and M. Lerman), *Journal of Symbolic Logic*, 61, 3(1996), 880-905.
12. Lattice representations for computability theory, *Annals of Pure and Applied Logic*, 94, 1-3(1998), 53-74.
13. Collapsing polynomial-time degrees (with K. Ambos-Spies, L. Bentzien, W. Merkle, F. Stephan), in S. Buss et al., eds., *Logic Colloquium '98*, Lecture Notes in Logic, Volume 13, Association for Symbolic Logic, Natick, MA, 2000, 1-24.
14. Every incomplete computably enumerable truth-table degree is branching (with R. Shore), *Archive for Mathematical Logic*, 40, 2(2001), 113-123.
15. Embeddings of N5 and the contiguous degrees (with K. Ambos-Spies), *Annals of Pure and Applied Logic*, 112, (2001), 151-188.
16. Enumerations of the Kolmogorov function (with R. Beigel, H. Buhrman, L. Fortnow, P. Grabowski, L. Longpre, A. Muchnik, F. Stephan, L. Torenvliet), *Journal of Symbolic Logic*, 71, 2(2006), 501-528.

*Work in Progress*

17. *Mathematical Foundations of Computer Science, Volume 2: Logical Foundations* (with D. Simovici)

18. Degrees of unsolvability (with K. Ambos-Spies), to appear in D. Gabbay and J. Woods, eds., *Handbook of the History of Logic,* Volume 9, *Logic and Computation,* J. Siekmann, co-ed.

19. Work with K. Ambos-Spies on generating sets for the computably enumerable degrees.

**Service to the Profession:** Chair, Computer Science Department, University of Massachusetts Boston, 2001-.

**Teaching:** Peter Fejer devotes most of his teaching efforts to instructing Computer Science students on the mathematical foundations of the subject.

**Vision Statement:** Mathematical logic in the last half of the twentieth century made great progress on internally-generated problems. As with any area of mathematics, it is perfectly appropriate for logicians to study problems local to their own concerns, but, especially in view of its origins, logic in the twenty first century needs to look also to problems involving connections to other areas of mathematics and to computer science.

**Honours and Awards:** National Science Foundation Research Grant, 1981-1983.

# FENSTAD, Jens Erik

**Specialties:** Recursion theory, nonstandard analysis, logic and natural language semantics, logic and probability.

**Born:** 15 April 1935 in Trondheim, Norway.

**Educated:** University of Oslo, Mag.Scient. in Mathematical Logic, 1959.

**Regular Academic or Research Appointments:** PROFESSOR OF MATHEMATICAL LOGIC, UNIVESITY OF OSLO, 1968–2003; Vice-Rector, University of Oslo, 1988-1992; University Research Fellow, University of Oslo, 1961-1967.

**Visiting Academic or Research Appointments:** Visiting Scientist, XEROX PARC, Palo Alto, 1994; Visiting Professor, Stanford University, 1983-1984, 1994; Visiting Professorial Fellow, Wolfson College, Oxford University, 1975; Visiting Research fellow, Stanford University, 1965-1966; University Fellow, University of California, Berkeley, 1959-1960.

**Research Profile:** Jens Erik Fenstad has led three large research programmes at the University of Oslo and most of his research activities have been carried out in the context of these programmes. The first was a programme in general recursion theory (see item 10 in *Key Publications*), with the main activity taking place in the period from 1970 to 1980; the second was a programme in nonstandard analysis with applications to stochastic analysis and mathematical physics (see item 12), running from 1975 to 1990; the third was a programme in logic and natural language technology, partly in collaboration with CSLI of Stanford University (see item 13). The last programme started in the late 1970s and is now being continued as a formal university programme in "Language, Logic and Information". In addition, Fenstad has also made several contributions to logic and probability theory, in particular, to representation theory with applications to the foundation of statistics.

**Main Publications:**

1. Representations of probabilities defined on first order languages, in *Sets, Models and Recursion Theory* (J.N.Crossley, ed.), North-Holland, Amsterdam (1967), 156-172.

2. The structure of logical probabilities, *Synthese* 18 (1968), 1-23.

3. (ed.), *Th.Skolem. Selected Works in Logic*, Universitetsforlaget, Oslo (1970).

4. (ed.), *Proceedings of the Second Scandinavian Logic Symposium*, North-Holland, Amsterdam (1971).

5. (with P.G.Hinman, eds.), *Generalized Recursion Theory*, North-Holland, Amsterdam (1974).

6. Between recursion theory and set theory, in *Logic Colloquium 76* (R.O.Gandy, J.M.E.Hyland eds.), North-Holland, Amsterdam (1977), 393-406.

7. Models for natural languages, in *Essays on Mathematical and Philosophical Logic* (J.Hintikka, I.Niiniluoto, E.Saarinen, eds.), D.Reidel, Dordecht (1978), 315-340.

8. (with R.O.Gandy, G.Sacks, eds.), *Generalized Recursion Theory II*, North-Holland, Amsterdam (1978).

9. (with S.Albeverio, R.Høegh-Krohn), Singular perturbations and nonstandard analysis, *Trans.Amer.Math.Soc.* 252 (1979), 275-295.

10. *Generalized Recursion Theory*, Springer-Verlag, Berlin (1980).

11. (with S.Albeverio, R.Høegh-Krohn, W.Karwowski, T.Lindstrøm) Perturbations of the Laplacian supported by null sets, with applications to polymer measures and quantum fields, *Physics Letters* 104 (1984), 396-400.

12. (with S.Albeverio, R.Høegh-Krohn, T.Lindstrøm), *Nonstandard Methods in Stochastic Analysis and Mathematical Physics*, Academic Press, Orlando (1986).

13. (with P-K. Halvorsen, T.Langholm, J. van Benthem), *Situations, Language and Logic*, D.Reidel, Dordrecht (1987).

14. The discrete and the continuous in mathematics and the natural sciences, in *Infinity in Science* (G.T. di Francia, ed.), Enciclopedia Italiana, Rome (1987).

15. (with I.T.Frolov, R.Hilpinen, eds.), *Logic, Methodology and Philosophy of Science VIII*, North-Holland, Amsterdam (1989).

16. Remarks on the science and technology of language, *European Review* 4 (1996), 107-120.

17. Partiality, in *Handbook of Logic and Language* (J. van Benthem, A. ter Meulen, eds.), Elsevier, Amsterdam (1997), 649-682.

18. Formal semantics, geometry, and mind, in *Discourse, Interaction and Communication* (X.Arrazola et al, eds.), Kluwer, Dordrecht (1998), 85-103.

19. Computability theory: structure or algorithms, in *Reflections on the Foundation of Mathematics* (W.Sieg, R.Sommer, C.Talcott, eds.), Lecture Notes in Logic 15, ASL (2002), 182-207.

20. Tarski, truth and natural languages, *Annals of Pure and Applied Logic* 126 (2004), 15-26.

*Work in Progress*

21. Future efforts will be concentrated on a continued study of the interactions between logic, language and the cognitive sciences. In addition to several forthcoming papers, a research monograph is planned on these topics. Fenstad has also published many studies on the methodology, history and ethics of science; this work will be continued.

**Service to the Profession:** Chair of the Board, Niels Henrik Abel Memorial Fund, 2002-2004; President, UNESCO World Commission on the Ethics of Scientific Knowledge and Technology, 2002-2005; Member of the Executive Board, International Council of Science, 1996-1999; Chairman of the Physical and Engineering Science Committee, European Science Foundation, 1995-1999; Scientific Advisor, Norwegian Foreign Office, 1994-1998; Member, NATO Science Committee, 1992-2004; President, International Union of History and Philosophy of Science, 1991-1995, Chair, General Program Committee, International Congress on Logic, Methodology and Philosophy of Science in Moscow, 1987; Chair, Natural Science Research Council of Norway, 1985-1989.

For a number of years, starting in the mid 1970s, he was chair of the ASL European Committee.

He has been editor/member of the editorial board of Annals of Mathematical Logic, Journal of Symbolic Logic, Journal of Philosophical Logic, Zeitschrift f. Mathematische Logik und Grundlagen der Mathematik, Archive for Mathematical Logic, and Synthese.

**Teaching:** Fenstad's teaching activities in Oslo have to a large extent been connected to the three major research programmes mentioned under the section on research above. Some PhD students connected to these programmes are: D.Normann, J.Moldestad in recursion theory; T.Lindstrøm in nonstandard analysis; and T.Langholm, H.F.Sem, E.Colban, J.T.Lønning in logic and natural languages.

**Vision Statement:** A main motivating force for the development of logic was the need to understand such "natural phenomena" as the preformal notions of proofs and algorithms. Another motivating force was the need to understand the geometric continuum, i.e. to resolve the tension between "points and fields". In both of these endeavors logic has had great success, and from my own experience I see interesting possibilities for future (see e.g. item 19 above). Today we also see a revival of the links between logic, grammar and meaning. This is still an unfolding story. Logic has had some successes, but the logician should remember that logic inspired by natural languages is not necessarily logic relevant for the study of language. The logician, as an applied scientist, needs to cross traditional disciplinary boundaries.

**Honours and Awards:** Member, Norwegian Academy of Letters and Science, 1976-; Member, Academia Europaea, 1989-; Sierpinski Medal, University of Warsaw, 1988; Honorary Doctor, Uppsala University, Sweden, 1998.

# FERMÜLLER, Christian Georg

**Specialties:** Automated deduction, proof theory, fuzzy logic, many-valued logic.

**Born:** 13 April 1963 in Graz, Austria.

**Educated:** Vienna University of Technology: Diplom Ingenieur, 1987; Dr.techn., 1991.

**Dissertation:** *Deciding Classes of Clause Sets by Resolution*; s*upervisor,* Alexander Leitsch.

**Regular Academic or Research Appointments:** ASSOCIATE PROFESSOR OF COMPUTER SCIENCE, VIENNA UNIVERSITY OF TECHNOLOGY, 1996–. Research assistant and lecturer *(Universitätsassistent)*, Vienna University of Technology, 1989-1995.

**Visiting Academic or Research Appointments:** Visiting Scholar, CSLI, Stanford University, 1995-1996

**Research Profile:** Christian Fermüller's earlier contributions have explored new uses of the resolution calculus and other methods of automated deduction; in particular, for a problem dual to classical theorem proving, detecting satisfiability and algorithmic extraction of models from proof search. He has also contributed to the development of proof theory for many valued logics, t-norm based fuzzy logics and related non-classical logics. More recently he has been exploring the connection between formal dialogue games and analytic proof systems.

**Main Publications:**

1. C. Fermüller, A. Leitsch, T. Tammet, N. Zamov: Resolution Methods for the Decision Problem (Monograph). LNAI 679, Springer-Verlag, 1993.
2. G. Gottlob, C. Fermüller: Removing Redundancy from a Clause. *Artificial Intelligence* 61 (1993), 263-289.
3. M. Baaz, C. Fermüller, A. Leitsch, A Non-Elementary Speed-Up in Proof Length. Proceedings of the 9th Annual IEEE Symposium of Logic in Computer Science, LICS'94, IEEE Computer Society Press, 213-219.
4. M. Baaz, C. Fermüller: Resolution-Based Theorem Proving for Many-valued Logics. Journal of Symbolic Computation, 19 (1995), 353-391.
5. M. Baaz, C. Fermüller, R. Zach, G. Salzer: Labeled Calculi and Finite-Valued Logics. Studia Logica 61, 1998, 7-33.
6. M. Baaz, C. Fermüller: Analytic Calculi for Projective Logics. In: Automated Reasoning with Analytic Tableaux and Related Methods, Neil V.\ Murray (Ed.), TABLEAUX'99, Saratoga Springs, NY, USA, June 1999, LNAI~1617, Springer-Verlag, 1999, 36-50.
7. C. Fermüller, A. Leitsch, U. Hustadt, T. Tammet: Resolution Decision Procedures. In: Handbook of Automated Reasoning. Editors: A. Robinson and A. Vorkonov, Elsevier, 2001, 1791-1849.
8. M. Baaz, C. Fermüller, G. Salzer: Automated Deduction for Many-Valued Logic. In: Handbook of Automated Reasoning. Editors: A. Robinson and A. Vorkonov, Elsevier, 2001, 1355-1402.
9. M. Baaz, A. Ciabattoni, C. Fermüller: Sequents of Relations Calculi: A Framework for Analytic Deduction in Many-Valued Logics. In: Beyond Two: Theory and Applications of Multiple-Valued Logic, eds:Melvin Fitting and Ewa Orlowska. Studies in Fuzziness and Soft Computing, Physica-Verlag, 2002, 157-180.
10. C. Fermüller, A. Ciabattoni: From Intuitionistic Logic to Gödel-Dummett Logic via Parallel Dialogue Games. In: ISMVL 2003, 33rd IEEE International Symposium on Multiple-valued Logic, May 16-19, 2003, IEEE Computer Society, Los Alamitos, 88-95.
11. C. Fermüller: Parallel Dialogue Games and Hypersequents for Intermediate Logics. In: Automated Reasoning with Analytic Tableaux and Related Methods. International Conference, TABLEAUX 2003, Rome, Italy, September 2003, Proceedings, Marta Cialdea Mayer, Fiora Pirri (Eds.), 48-64.
12. C. Fermüller: Theories of Vagueness Versus Fuzzy Logic: Can Logicians Learn from Philosophers? Neural Network World Journal 13(5), 2003, 455-466.
13. M. Baaz, A. Ciabattoni, C. Fermüller: Hypersequent Calculi for Gödel Logics — a Survey. Journal of Logic and Computation, Vol. 13 No.6, Oxford UP, 2003, 835-861
14. A. Ciabattoni, C. Fermüller, G. Metcalfe: Uniform Rules and Dialogue Games for Fuzzy Logics. In: Franz Baader, Andrei Voronkov (Eds.): Logic for Programming, Artificial Intelligence, and Reasoning, 11th International Conference, LPAR 2004, Montevideo, Uruguay, March 14-18, 2005, Proceedings. LNCS 3452, Springer 2005, 496-510.

*Work in Progress*

15. (with A. Ciabattoni and G. Metcalfe:) Deriving analytic calculi for fuzzy logics from dialogue games and betting schemes.
16. Exploration of the use of Lorenzen style dialogue games (and corresponding analytic systems) as semantic foundation of reasoning with vague propositions.

**Service to the Profession:** Various functions, *Kurt Gödel Society* 1989-; Steering committee, *TABLEAUX*, 2001–2003, 2008–; PC and OC memberships for various conferences and workshops in Automated Deduction and related fields, 1993-.

**Teaching:** Lecturing regularly on Theoretical Computer Science, Automated Theorem Proving, Recursion Theory and Nonclassical Logics at Vienna University of Technology since 1991.

**Vision Statement:** Mathematical logic will continue to be invigorated by new problems and topics emerging from computer science. In particular interfaces between complexity theory, game theory, linguistics and proof theory provide new methodological challenges that might change the very nature of the field.

**Honours and Awards:** Erwin-Schrödinger-Scholarship of the Austrian Science Foundation 1995-1996.

# FERNANDES, António Marques

**Specialties:** Mathematical logic, foundations of mathematics.

**Born:** 16 January 1965 in Lisboa, Portugal.

**Educated:** University of Lisbon, PhD Mathematics, 2001; University of Lisbon, MS Mathematics, 1993; University of Lisbon, BS Mathematics, 1988.

**Dissertation:** *Investigações em sistemas de análise exequível (Investigations in feasible systems of analysis)*; supervisor, Fernando Ferreira.

**Regular Academic or Research Appointments:** ASSISTANT PROFESSOR, TECHNICAL UNIVERSITY OF LISBON, 2001–.

**Research Profile:** António Fernandes has made contributions to the investigation of the role played by infinitary principles (weak König's lemma, Baire category theorem, etc.) in the formalization of analysis in weak systems of arithmetic.

**Main Publications:**

1. (with Fernando Ferreira) On extracting algorithms from intuitionistic proofs, in Mathematical Logic Quarterly, 44, pp. 143-160 (1998).
2. A new conservation result of WKL over $RCA_-0$, Archive for Mathematical Logic, vol. 41, 1, pp 55–63 (Jan. 2002).
3. (with Fernando Ferreira) Groundwork for weak analysis, Journal of Symbolic Logic 67, pp. 557-578 (2002). The Baire category theorem over a feasible base theory, in Reverse Mathematics 2001, Stephen Simpson (org.), Lecture Notes in Logic, Association for Symbolic Logic, pp. 164–174 (2005).
4. (with Fernando Ferreira) Basic applications of weak König's lemma in feasible analysis, in Reverse Mathematics 2001, Stephen Simpson (org.), Lecture Notes in Logic, Association for Symbolic Logic, pp. 175–188 (2005).
5. Strict $\Pi_1^1$-refection in bounded arithmetic, submitted to the Archive for Mathematical Logic.

*Work in Progress*

6. Papers on the demonstrability strength of the bounded collection principle over theories with bounded induction, and Cantini's conjecture about the equivalence of weak König's lemma and strict $\Pi_1^1$-reflection over some weak theory of arithmetic are in preparation.
7. There is also some ongoing research on the consistency of Quine's NF.

**Teaching:** Fernandes is currently teaching Mathematical Analysis in the Mathematics Department of Instituto Superior Técnico–Technical University of Lisbon.

# FIELD, Hartry

**Specialties:** Semantic paradoxes and paradoxes of naive property theory, logic of vagueness, indeterminacy of truth value in mathematics.

**Born:** 30 November 1946 in Boston, Massachusetts, USA.

**Educated:** Harvard University, PhD Philosophy, 1972; Harvard University, MA Philosophy, 1968; University of Wisconsin, BA Mathematics, 1967.

**Dissertation:** *Reference Truth and Meaning*; supervisor, Hilary Putnam.

**Regular Academic or Research Appointments:** SILVER PROFESSOR OF PHILOSOPHY, NEW YORK UNIVERSITY, 1997–. Kornblith Distinguished Professor, Philosophy, City University of New York Graduate Center, 1991-1997; University of Southern California, Philosophy, 1976-1991; Princeton University, Philosophy, 1970-1976.

**Visiting Academic or Research Appointments:** Visiting Research Social Scientist, University of Arizona Cognitive Science Program, 1987; Visiting Professor, Massachusetts Institute of Technology, 1984; Visiting Fellow, All Souls College, Oxford, 1979.

**Research Profile:** My work has not been primarily in logic, but in the last few years, after becoming deeply dissatisfied with treatments of truth and properties within classical logic, I have developed an account which allows us to consistently keep the "naive schemas" (e.g. "$\langle A \rangle$ is true if and only if $A$" and "For all $x$, $x$ instantiates the property of being $F$ if and only if $Fx$"). Paradox is avoided by restrictions on the law of excluded middle, and by the development of a new conditional that reduces to the old one when excluded middle is assumed. This logic turns out to handle paradoxes of vagueness as well, and to have other ramifications which I have been exploring. I've also been investigating ways in which one might alter the treatment of the conditional and still avoid the paradoxes. The work connects with more traditional issues in logic, especially in set theory, and I hope to soon connect it up with work on subsystems of second order arithmetic, as well as with more philosophical work on how to formally treat the idea that certain questions in mathematics are not fully objective.

**Main Publications:**

1. *Saving Truth from Paradox*, Oxford, 2008.
2. Solving the Paradoxes, Escaping Revenge. In JC Beall, *Revenge of the Liar*, Oxford, 2007.
3. Truth and the Unprovability of Consistency, *Mind*, 2006.
4. The Consistency of the Naive Theory of Properties, in G. Link, *One Hundred Years of Russell's Paradox* (de-Gruyter 2004).
5. The Semantic Paradoxes and the Paradoxes of Vagueness, in JC Beall, *Liars and Heaps* (Oxford 2004).
6. A Revenge-Immune Solution to the Semantic Paradoxes, *Journal of Philosophical Logic* 2003.
7. Which Undecidable Mathematical Sentences Have Determinate Truth Values, in H.G. Dales and G. Olivieri, *Truth in Mathematics* (Oxford 1998).
8. A Nominalistic Proof of the Conservativeness of Set Theory, *Journal of Philosophical Logic* (1992).
9. Metalogic and Modality, *Philosophical Studies* (1991).
10. *Science Without Numbers* (Blackwell Publishers and Princeton University Press, 1980).
11. Logic, Meaning and Conceptual Role, *Journal of Philosophy* 1977.

**Service to the Profession:** Member of Program Committee for Association of Symbolic Logic *Logic Colloquium '05*, Athens, 2005.

**Vision Statement:** I believe work in the foundations of mathematics to be paramount, and that it can be enriched by studying non-classical extensions of set theory that contain their own truth predicate, and also certain non-classical weakenings that allow for indeterminacy in how far the ordinals extend and/or indeterminacy in the extent of power sets.

**Honours and Awards:** Elected to American Academy of Arts and Sciences, 2003; Lakatos Prize in Philosophy of Science, 1986.

## FINE, Kit

**Specialties:** Modal logic, relevance logic, philosophical logic, foundations of mathematics.

**Born:** 26 March 1946 in Farnborough, Hants, England.

**Educated:** University of Warwick, PhD; University of Oxford, BA (PPE), 1967.

**Dissertation:** *For Some Proposition and So Many Possible Worlds*; supervisor, Arthur Prior.

**Regular Academic or Research Appointments:** SILVER CHAIR IN PHILOSOPHY, NEW YORK UNIVERSITY, 2003–; AFFILIATE PROFESSOR, MATHEMATICS, COURANT INSTITUTE, NEW YORK UNIVERSITY, 2002–. Professor, Philosophy, New York University, 1997-; Flint Chair in Philosophy, University of California, Los Angeles, 1993-1997; Professor, Philosophy, University of California Los Angeles, 1988-1997; Professor, University of Michigan, Ann Arbor, 1978-1988; Professor, Philosophy, University of California Irvine, 1977-1978; Associate Professor, University of California Irvine, 1975-1977; Lecturer, Philosophy, University of Edinburgh, 1971-1973; Junior Research Fellow, St. John's College, Oxford, 1969-1971; Assistant Lecturer, Philosophy, University of Warwick, 1967-1969.

**Visiting Academic or Research Appointments:** Santayana Visiting Professor, Philosophy, Harvard University, 2003-2004; Visiting Fellow, All Souls College, Oxford, 2003; Visiting Professor, Philosophy, Princeton University, 1999-2001; Visiting Fellow, Automatic Reasoning Project, Australian National University, 1990; Visiting Professor, Philosophy, UCLA, 1987; Visiting Professor, Philosophy, University of Melbourne, 1985; Visiting Fellow, Australian National University, 1985; Visiting Professor, Philosophy, UCLA, 1983; Honorary Fellow, Center for Cognitive Science, University of Edinburgh, 1983–; Visiting Professor, Philosophy, University of Arizona, 1977; Visiting Associate Professor, Philosophy, University of Toronto, 1974-1975; Visiting Assistant Professor, Philosophy, Stanford University, 1974.

**Research Profile:** Kit Fine has worked in the areas of modal and relevance logic, the theory of arbitrary objects, the semantics for prolog, formal language theory and the foundations of mathematics.

**Main Publications:**

1. Propositional Quantifiers in Modal Logic, *Theoria* 36:3 (1970), 336346. Translation reprinted in Polish collection (1981).
2. The Logics Containing S4.3, *Zeitschrift fur Mathematische Logik und Grundlagen der Mathematik* 17 (1971), 371376.
3. Logics Containing S4 Without the Finite Model Property, *Conference in Mathematical Logic '70* (New York: SpringerVerlag, 1972), 98102.
4. In So Many Possible Worlds, *Notre Dame Journal of Formal Logic* 13:4 (October 1972), 516520.
5. An Incomplete Logic Containing S4, *Theoria* 40 (1974), 2329.
6. Models for Entailment, *Journal of Philosophical Logic* 3 (October 1974), 347372. Substantially reprinted in *Entailment II* (ed. Belnap et al), Princeton Univ. Press, 1992.
7. Logics Containing K4, Parts I-II: Part I, *The Journal of Symbolic Logic* 39:1, (March 1974) 3142; Part II, *The Journal of Symbolic Logic* 50:3 (September 1985), 619651.
8. Some Connections Between Elementary and Modal Logic, *Proceedings of the Third Scandinavian Logic Symposium* (ed. Stig Kanger), Amsterdam: North Holland, 1975, 1531.
9. Normal Forms in Modal Logic, *Notre Dame Journal of Formal Logic* 16:2 (April 1975), 229237.
10. Model Theory for Modal Logic, Parts I–III: Part I, The De Re/De Dicto Distinction, *Journal of Philosophical Logic* 7 (May 1978), 125156; Part II, The Elimination of *De Re* Modality, *Journal of Philosophical Logic* 7 (August 1978), 277306; Part III, Existence and Predication, *Journal of Philosophical Logic* 10 (August 1981), 293307.
11. Failures of the Interpolation Lemma in Quantified Modal Logic, *The Journal of Symbolic Logic* 44:2 (June 1979), 201206.
12. FirstOrder Modal Theories, Parts I–III: Part I, Sets, *Nôus* 15 (May 1981), 177205; Part II, Propositions *Studia Logica* 39:2/3 (1980), 159202; Part III, Facts, *Synthese* 53 (October 1982), 43122.
13. Reasoning with Arbitrary Objects, Blackwell: Oxford (1985)
14. Semantics for Quantified Relevance Logic, *Journal of Philosophical Logic*, 17 (1988), 27-59. Substantially reprinted in *Entailment II*.
15. The Justification of Negation as Failure, *Proceedings of the Congress on Logic, Methodology and the Philosophy of Science*, VIII, (ed. Fenstad et al.), Elsner Science Publishers B.V. (1989), 263-301.
16. Incompleteness for Quantified Relevance Logics, *Directions in Relevant Logics*, (ed. Sylvan, Norman),

Kluwer: Dardrecht (1989). Substantially reprinted in *Entailment II*.

17. Transparency, in *Proceedings of the Conference on Logic in Computer Science, 89*, in MSRI Series (1992), Springer-Verlag.

18. Semantics for the Logic of Essence, Journal of Philosophical Logic, vol. 29, pp. 543-584, 2000.

19. The Limits of Abstraction, Clarendon Press: Oxford (2002)

20. Class and Membership, Journal of Philosophy v. 102, no. 11, 547-72, 2005.

*Work in Progress*

21. The Method of Postulation, a book in which I develop a new 'procedural' approach to the foundations of mathematics.

**Service to the Profession:** Member of the consulting editorial board, *Studia Logica*, 2002–; Member of Program subcommittee, Joint Meeting of the American Philosophical Association and the Association of Symbolic Logic, 1999-200; Editor of *Lecture Notes on Logic*, published under the auspices of the ASL, 1994-1997; Editor of *Reports on Logic*, 1990-; Member of the Editorial Board, *The Journal of Applied Non-classical Logics*, 1990-; Associate member of the UCLA. Cognitive Science Group, 1990-97; Chair of the Oversight Committee for the Journal of Philosophical Logic, 1988; Editor of *Notre Dame Journal of Formal Logic*, 1984-87; Member of the Executive Committee of the Association of Symbolic Logic, 1983-1987; Chair of the Ad Hoc Committee on the Future of the JSL/ASL, 1983-1985; Editor of *The Journal of Symbolic Logic*, 1979-1987; coordinator for the editors, 1983-1985.

**Teaching:** Kit Fine has taught undergraduate courses on classical mathematical logic (at all levels), modal logic and set theory; and he has taught graduate courses on philosophical logic, modal and relevance logic, model theory and the foundations of mathematics. His students include Steve Kuhn (Georgetown) and Roy Benton (Columbia Union College), Fabrice Correia (University of Geneva), and Marcus Kracht (UCLA).

**Vision Statement:** Logic is losing its traditional connection with philosophy. I would like to see more philosophers informed about logic and more logicians, especially from mathematics and computer science, informed about philosophy.

**Honours and Awards:** Honorary Doctorate, University of Bucharest, 2006; Corresponding Member of the British Academy, 2005; John Locke Lecturer, Oxford, 2003; Inaugural Blackwell/Brown Lecturer, 2002; Visiting Fellow, All Souls College, Oxford, 1995-1996; Fellow, Institute for Advanced Studies in the Humanities, University of Edinburgh, 1981-1982; Fellow, American Council of Learned Societies, 1981-1982; Rackham Fellow, University of Michigan, Ann Arbor, 1981; Guggenheim Fellow, 1978-79; Recipient of Regents' Award in the Humanities, University of California, 1977.

## FINN, Victor

**Specialties:** Many-valued logics, logics for Intelligent Systems, logics for sociology and life sciences, formal epistemology, history of logic.

**Born:** 15 July 1933 in Moscow, former USSR.

**Educated:** Moscow State University, MS Mathematics, 1966, Moscow State University, MS Philosophy, 1957, USSR Academy of Sciences, PhD Mathematics, 1975.

**Dissertation:** MS dissertation, *Logical Problems of Informational Search*; supervisor, Dmitry Bochvar; DSc Dissertation, *Plausible Reasoning in Expert Systems with Incomplete Information*.

**Regular Academic or Research Appointments:** PROFESSOR OF LOGICS AND COMPUTER SCIENCE, RUSSIAN STATE UNIVERSITY FOR HUMANITIES, 1994–. Professor of Logics, High School of Economics (State University), 2005–; Associated Professor of Logics, Moscow State University, 1967–1968; Head of the Department of Intelligent Systems, All-Russian Institute for Scientific and Technical Information of Russian Academy of Science, 1992–; Researcher, All-USSR Institute for Scientific and Technical Information of USSR Academy of Science, 1957–1991.

**Research Profile:** Victor Finn has made significant contribution to many-valued logics, namely has developed the method of axiomatization of many-valued logics with Boolean and non-Boolean variables. He has investigated class of logics with quasi-lattices being their algebras. The examples of such logics are n-valued generalizations of 3-valued Bochvar logic (these algebras classes form quasi-varieties). Classification of 3-valued logics that has been suggested by Victor Finn consists of logics formalizing uncertainty and nonsense correspondingly. He has formulated a criterion of functional completeness for 3-valued Bochvar logic. In Łukasiewicz logics he is noted for two important results: he has proved that these n-valued logics are functionally precomplete if and only if n − 1 is a primitive number; if n − 1 is a power of primitive number, then the disjunctive normal form does exist (function representation theorem). He has also suggested the class of non-Postian logics being extensions of

Łukasiewicz logics, which preserve Boolean restriction of logical connectives. In logics for social sciences he also investigated 4-valued argumentation logics with non-associative logical connectives. He has created the method of automatic generation of hypotheses (JSM-method) using special class of plausible reasoning. These reasoning are formed by interaction of induction, analogy and abduction. JSM-plausible reasoning formalization is realized by infinite-valued argumentation logics with counting set of one-placed J-operators. Truth-values of these logics are degrees of plausibility of generated hypotheses. JSM reasoning contains formalizations of J.S. Mill's inductive methods and C.S. Peirce's abduction. JSM-method logical means are used in Intelligent Systems for knowledge discovery in pharmacology, medicine, biochemistry, sociology and linguistics.

**Main Publications:**

1. The Precompleteness of a class of functions that corresponds to the three-valued logic of J. Łukasiewicz (in Russian), *Nauchno-Technicheskaya Informatsiya, ser.2,* 10 (1969), 35–38.
2. (with D.A. Bochvar) On many-valued logic useful for antinomies analysis (in Russian), In: *Researches on mathematical linguistic, mathematical logic and informational languages,* Moscow, Nauka (1972).
3. Axiomatization of some three-valued calculus and their algebras (in Russian), In: *Philosophy and Logic,* Moscow, Nauka (1974).
4. A Criterion of Functional Completeness for $B_3$, *Studia Logica,* Vol.XXXIII, N 2, (1974), 121–125.
5. Some remarks on no-Postian logics, 5th *International Congress on Logic, Methodology and Philosophy of Science, Contributed papers, Section 1,* London – Ontario (1975).
6. Logical Problems of Informational Search (in Russian), Moscow, Nauka (1976).
7. (with R. S. Grigolia) $B_n$-algebras and the Corresponding Propositional Calculus, *Bull. of the Section of Logic,* Vol.9, N 1 (1980), 39–46.
8. (with O.M. Anshakov, R. S. Grigolia, M.I. Zabezhailo) Many-valued logics as Fragments of formalized Semantics, *Acta Filosofica Fennica,* 35 (1982), 239–272.
9. (with O.M. Anshakov, D.P. Skvortsov) On Axiomatization of Many-Valued Logics Associated with Formalization of Plausible Reasoning, *Studia Logica,* Vol. XLVIII, 4 (1989), 423–447.
10. Plausible Inferences and Plausible Reasoning, *Journal of Soviet Mathematics,* vol. 56, N 1 (1991), 2201–2248.
11. (with O.M. Anshakov, D.P. Skvortsov) On logical construction of JSM-method of automatic hypotheses generation, *Doklady Mathematics,* Vol. 44, 4 (1991).
12. Plausible reasoning in JSM-type intelligent systems (in Russian), *Itogi Nauki i tekhniki, ser. "Informatika",* 15 (1991), 54–101.
13. (with R. S. Grigolia) Nonsense Logics and Their Algebraic Properties, *Theoria,* Vol. LIX, Pt. 1 – 3 (1993), 207-273.
14. A form of argumentation logic, *Automatic Documentation and Mathematical Linguistics,* 30, N 3 (1996), 3–27.
15. The Synthesis of Cognitive Procedures and the Problem of Induction (in Russian), *Nauchno-Technicheskaya Informatsiya, ser.2,* 1 – 2 (1999), 8–45.
16. (with S.M. Gusakova and M.A. Mikheyenkova) On logical means of automated opinion analysis (in Russian), *Nauchno-Technicheskaya Informatsiya, ser.2,* 5 (2001), 4–24.
17. Intelligent Systems and Society (in Russian), Moscow, URSS (2006).

*Work in Progress*

18. A book "Cognitive reasoning: formal approach", with O.M. Anshakov, T. Gergely and S.O. Kuznetsov.
19. A book "Logical means for social data analysis", with M.A. Mikheyenkova.
20. Further development of argumentation logics semantics.

**Service to the Profession:** Scientific Editor, C.S. Peirce, *Reasoning and the Logic of Things,* translation into Russian, 2006; Member, Editorial Council, Russian State University for Humanities, 1995–; Member, Council of Russian Association of Artificial Intelligence, 1988–; Member, Editorial Board, Studia Logica, 2005, 1991; Head, Department of Philosophy, Russian Fund for Basic Research, 2003–2006, 1995–2001; Member, Editorial Board, Foundation of Science, 1995–1996;Scientific Editor, N.D. Belnap, T.B. Steel, *The Logic of Questions and Answers,* translation into Russian, 1981; Scientific Editor, H.E. Kyburg, *Probability and Inductive Logic,* translation into Russian, 1978; Editor-in-Chief, *"Mathematical linguistics",* 1964.

**Teaching:** Finn is the author of the program for the "Intelligent Systems for Humanities" degree educational course (for Russian Universities). He is a Head of the Department of Intelligent Systems for Humanities, Russian State University for Humanities. He has had seventeen PhD Students in the fields of Mathematics, Computer Science and Philosophy.

**Vision Statement:** A development of formal theories for plausible reasoning, which contains interaction of induction, analogy, abduction and deduction, is considered to be fundamental problem. These theories correctness would be investigated in their deductive imitation. Further development of such theories and their applications to open empirical fields will generate a possibility for experimental justification of logical methods in Computer Science and Robotics.

**Honours and Awards:** Member, Russian Academy of Natural Science, 1991–; Member, Polish Association of Logic and Philosophy of Science, 2005–; Russian Academy of Sciences Award, 2000; Honoured Science Worker, 2007.

# FITELSON, Branden

**Specialties:** Inductive logic, philosophical logic, formal philosophy, automated reasoning.

**Born:** 17 August 1969 in Syracuse, New York, USA

**Educated:** University of Wisconson, PhD Philosophy, 2001; MA Philosophy, 1997; BS Mathematics and Physics, 1992.

**Dissertation:** *Studies in Bayesian Confirmation Theory*; supervisor, Malcolm Forster.

**Regular Academic or Research Appointments:** ASSISTANT PROFESSOR, DEPARTMENT OF PHILOSOPHY (ALSO GROUP IN LOGIC AND THE METHODOLOGY OF SCIENCE, AND INSTITUTE FOR COGNITIVE AND BRAIN SCIENCES), UNIVERSITY OF CALIFORNIA, BERKELEY, 2003–; Assistant Professor, Department of Philosophy, San José State University, 2002–2003.

**Visiting Academic or Research Appointments:** Visiting Assistant Professor, Department of Philosophy, Stanford University, 2001–2002.

**Research Profile:** Four major areas of research: philosophy of science (specifically, confirmation theory, and the foundations of statistical inference), inductive logic (historical, philosophical, and logical aspects thereof), formal philosophy (specifically, formal epistemology), and automated reasoning (broadly construed, so as to include mechanical theorem proving, as well as decision methods, simulation, *etc.*)

**Main Publications:**

1. Likelihoodism, Bayesianism, and Relational Confirmation, *Synthese*
2. Symmetries and Asymmetries in Evidential Support (with E. Eells) *Philosophical Studies*
3. A Bayesian Account of Independent Evidence with Applications *Philosophy of Science*
4. The Plurality of Bayesian Measures of Confirmation and the Problem of Measure Sensitivity, *Philosophy of Science*
5. How Not to Detect Design: A Review of William Dembski's *The Design Inference*, (with C. Stephens and E. Sober), *Philosophy of Science*
6. Shortest Axiomatizations of Implicational S4 and S5, (with Z. Ernst, K. Harris, and L. Wos), *Notre Dame Journal of Formal Logic*
7. Short Single Axioms for Boolean Algebra (with W. McCune, R. Veroff, K. Harris, A. Feist, L. Wos), *Journal of Automated Reasoning*
8. Vanquishing the XCB Question: The Methodological Discovery of the Last Shortest Single Axiom for the Equivalential Calculus, (with L. Wos and D. Ulrich), *Journal of Automated Reasoning*
9. Using *Mathematica* to Understand the Computer Proof of the Robbins Conjecture, *Mathematica in Education and Research*
10. More information can be found at http://www.fitelson.org/.

*Work in Progress*

11. Currently writing a book on inductive logic entitled *Logical Foundations of Inductive Support*. Also in the process of perfecting a user-friendly decision procedure for the probability calculus called PrSAT (a prototype of which is already available for download from Professor Fitelson's website).

**Service to the Profession:** Editorial Boards: *Studia Logica*, *Formal Epistemology*. Referee for many journals, including *Studia Logica*, *Synthese*, *Erkenntnis*, *Philosophy of Science*, *British Journal for the Philosophy of Science*, *Journal of Automated Reasoning*. Co-organizer of annual *Formal Epistemology Workshops* (FEW), and organizing committees for various conferences, including the Pacific Division of the American Philosophical Association, the Philosophy of Science Association, and Philosophy and Computing (member of APA committee on philosophy and computers)

**Teaching:** Has taught (and continues to teach) courses on deductive logic (formal and philosophical), inductive logic, philosophy of science, epistemology, and metaphysics.

**Vision Statement:** I see logic (and, more generally, formal methods) playing a more central role in philosophy in the future. I'd like to see more emphasis on logical precision and exactness in philosophical discourse. I'd also like to see more emphasis on the role of automated reasoning (and, more generally, computational methods, including computational simulation, *etc.*) as a tool for problem solving.

**Honours and Awards:** Best Essay by a Graduate Student: Philosophy of Science Association Contest, 2000; Wisconsin Alumni Research Foundation Fellowship, 1998; Oliver Prize for best essay by a graduate student, 1998.

# FITTING, Melvin Chris

**Specialties:** Modal logic, philosophical logic, logic in computer science, general proof theory.

**Born:** 24 January 1942 in Troy, New York, USA

**Educated:** Rensselaer Polytechnic Institute, Troy, New York, BA in Mathematics, 1963. Yeshiva University, MA 1965, PhD 1968, in Mathematics.

**Dissertation:** *Intuitionistic Logic, Set Theory, and Forcing*; supervisor Raymond Smullyan.

**Web Address:** `comet.lehman.cuny.edu/fitting`

**Regular Academic or Research Appointments:** DEPARTMENT OF MATHEMATICS AND COMPUTER SCIENCE, LEHMAN COLLEGE, CUNY, 1968–. Department of Computer Science, Graduate Center, CUNY, 1986–present; Department of Philosophy, Graduate Center, CUNY, 1987–present; Department of Mathematics, Graduate Center, CUNY; 1988–present.

**Research Profile:** Fitting's work has touched on computer science, philosophy, and mathematics. This should not be surprising since much, but not all, of Fitting's research has to do with modal logic and its close relatives. His dissertation carried out in detail the application of intuitionistic logic as a tool for proving independence results in classical set theory. As an adjunct to this, it also examined relationships between algebraic semantics, Kripke semantics, and tableau proofs for intuitionistic logic. The dissertation itself was published as a book, and had some influence, especially in Eastern Europe. The tableau theme was further developed in journal papers and a subsequent book, presenting two families of tableaus, one using 'prefixes,' one not. Both families have become well-known tools, especially for automated theorem proving with modal logics, an area to which Fitting has contributed directly. And as a by-product, Fitting also wrote a book on classical logic, organized around tableaus and with applications to automated theorem proving, which has had some success.

First-order modal logics became an interest of Fitting, and a series of papers examined the utility of adding what has come to be called predicate abstraction, to the standard machinery. This device was introduced by R. Thomason and R. Stalnaker and has turned out to be quite useful. Besides purely technical results (a modal version of Herbrand's theorem, for instance) it allows one to address several philosophical problems having to do with non-rigidity in a natural, direct manner. In recent years this work has resulted in a joint book, with Richard Mendelsohn, on first-order modal logics and their philosophical applications, and also a book on higher-order intensional logic, specifically applied to ontological arguments. There is also an earlier joint book with Raymond Smullyan, in which modal logic is used to establish independence results in classical set theory. One can look at this as a kind of intensional set theory, and it is a topic that Fitting is currently exploring.

Fitting's primary area of research apart from modal logic has been in semantics for logic programming languages, and related topics. The primary motivation here was Saul Kripke's work on a theory of truth which, as it happens, introduced mathematical methods that were as applicable to programming languages as they were to natural languages. Fitting wrote a series of papers in this area, with the so-called Kripke/Kleene semantics being introduced early on. Eventually Matt Ginsberg's notion of bilattices was also found to be a useful tool, and the combination of it and Kripke's ideas led to work of purely lattice-theoretic interest. The result was applied by Fitting both to stable model/answer set program semantics, and back to Kripke's theory of truth.

There has also been some work of mixed character, applying modal logic to database theory, and investigating multi-valued modal logics. There has also been some work examining the relationships between modal semantics and modules over boolean algebras. Fitting's primary research over the past few years has concerned the logic LP (Logic of Proofs) and its relatives, all of which may be looked at as logics of knowledge with explicit evidence.

**Main Publications:**

1. *Intuitionistic Logic Model Theory and Forcing*. North-Holland Publishing Co., Amsterdam, 1969.
2. "Tableau methods of proof for modal logics". *Notre Dame Journal of Formal Logic*, 13(1972): 237–247.
3. *Fundamentals of Generalized Recursion Theory*. North-Holland Publishing Co., Amsterdam, 1981.
4. *Proof Methods for Modal and Intuitionistic Logics*. D. Reidel Publishing Co., Dordrecht, 1983.
5. "A Kripke/Kleene semantics for logic programs". *Journal of Logic Programming*, 2(1985): 295–312.
6. *Computability Theory, Semantics and Logic Programming*. Oxford University Press, 1987.
7. "First-order modal tableaux". *Journal of Automated Reasoning*, 4(1988): 191–213.
8. "Bilattices and the theory of truth". *Journal of Philosophical Logic*, 18(1989): 225–256.
9. "Bilattices and the semantics of logic programming". *Journal of Logic Programming*, 11(1991): 91–116.
10. "The family of stable models". *Journal of Logic Programming*, 17(1993): 197–225.
11. "A modal Herbrand theorem". *Fundamenta Informaticae*, 28(1996): 101–122.
12. *First-Order Logic and Automated Theorem Proving*. Springer-Verlag, Second Edition 1996.

13. *Set Theory and the Continuum Problem*, with Raymond Smullyan. Oxford University Press, 1996.
14. "A theory of truth that prefers falsehood". *Journal of Philosophical Logic*, 26(1997): 477–500.
15. *First-Order Modal Logic*, with Richard Mendelsohn. Kluwer, Dordrecht and New York, 1998.
16. "Fixpoint semantics for logic programming-a survey". *Theoretical Computer Science*, 278(2002): 25–31.
17. *Types, Tableaus, and Gï¿½del's God*. Kluwer, Dordrecht and New York, 2002.
18. "The logic of proofs, semantically". *Annals of Pure and Applied Logic*, 132(2005): 1–25.

**Service to the Profession:** Editor *Studia Logica* for over 20 years, contributor to *A Companion to Philosophical Logic*, Blackwell, to *Handbook of Tableau Methods*, Oxford, to *Handbook of Logic in Artificial Intelligence and Logic Programming*, Oxford.

**Teaching:** Fitting has been teaching for almost 40 years, undergraduate and graduate, with courses ranging over computer science, mathematics, and philosophy. Like the prospect of execution, it concentrates the mind wonderfully; unlike that, the effects have a beneficial long-term effect.

# FØLLESDAL, Dagfinn

**Specialties:** Philosophy of logic, Philosophy of Language

**Born:** 22 June 1932 in Askim, Norway.

**Educated:** Harvard University, PhD Philosophy, 1961; University of Oslo, Mathematics 1955-1957; University of Oslo, MA Philosophy, 1956; University of Göttingen, Mathematics 1954-1955; University of Oslo Cand.mag. Mathematics, Astronomy, Mechanics 1953.

**Dissertation:** MA; supervisor, Thoralf Skolem; PhD, *Referential Opacity and Modal Logic*; supervisor, W. V. Quine.

**Regular Academic or Research Appointments:** C.I. LEWIS PROFESSOR OF PHILOSOPHY, STANFORD UNIVERSITY, 1976-; PROFESSOR, PHILOSOPHY, STANFORD UNIVERSITY, 1968-; Professor, Philosophy, University of Oslo, 1967-1999; Instructor, Assistant Professor, Philosophy, Harvard University, 1961-1964; Research Fellow, Ionospheric Physics, Norwegian Research Council, 1955-1957.

**Visiting Academic or Research Appointments:** ´Ecole normale supérieure, Paris, 2003; London School of Economics, 2002; University of Salzburg, 1992, 2001; University of Frankfurt, 1997-1998; University of Auckland,1982; Collège de France, 1977; University of California, Berkeley, 1977.

**Research Profile:** Dagfinn Føllesdal's early interests were in mathematics and science rather than in logic, but he did have a strong side interest in philosophy and wrote the little book Husserl und Frege (1956).

After having come across Quine's *From a Logical Point of View* he applied for a fellowship to study with Quine and wrote his dissertation with him on *Referential Opacity and Modal Logic* (1961). A main result in this dissertation is that names and other genuinely referring expressions have to refer to the same object in all possible worlds. This view was supported by two kinds of arguments: it results from a semantic analysis of how reference works and in addition it blocks Quine's argument in *Word and Object* that in quantified modal logic modal distinctions collapse.

Some of Føllesdal's later work has been devoted to studying the tie between referring expressions and their objects. He has proposed a "normative theory of reference," where the tie is not causal, but results from a commitment to preserve the reference of singular terms through the vicissitudes of changes in the world and changes in our views of the world.

Føllesdal has also been working on the theory of meaning, exploring the consequences of taking seriously the view that language and communication are social phenomena.

A major part of Føllesdal's work has been devoted to the study of contemporary continental philosophy, especially the views of Edmund Husserl and his followers. From 1956 on he has interpreted Husserl in a manner that makes him comprehensible to philosophers coming from other traditions and showed how Husserl's ideas are pertinent to contemporary issues in cognitive science, the philosophy of mind, perception, the role of the body, intersubjectivity and also psychotherapy and interpretation of literature. In particular, he has proposed new ways of looking at two central topics in Husserl: intentionality and ultimate justification. Føllesdal and his students have long argued that the schism between continental philosophy and "analytic philosophy" is both unfortunate and unjustified.

**Main Publications:**

Books relating to logic:
1. *Husserl und Frege: Ein Beitrag zur Beleuchtung der Entstehung der phänomenologischen Philosophie*. Thesis for the degree of Magister artium, Oslo 1956. (Avhandlinger utgitt av Det Norske Videnskapsakademi i Oslo. Hist.-Filos. Klasse. II. 1958. No 2.) Oslo: Aschehoug, 1958. 60 pp. Translation: English: "Husserl

and Frege: a contribution to elucidating the origins of phenomenological philosophy." Translation, by Claire Hill, in Leila Haaparanta, ed., *Mind, Meaning and Mathematics*, Dordrecht: Kluwer, 1994, pp. 3-47.

2. *Referential opacity and modal logic*. Thesis for the Ph.D. degree, Harvard 1961. Mimeographed, slightly expanded version, Oslo University Press, 1966. Reprint of §§ 16-19 in P.W. Humphreys and J.H. Fetzer, eds., *The New Theory of Reference*, Dordrecht: Kluwer, 1998, pp. 181-202. Printed, with an Introduction and Addenda, published in the series *Dissertations in Philosophy*, edited by Robert Nozick. London: Routledge, 2004, ISBN 0415938511

3. *Argumentasjonsteori og vitenskapsfilosofi*. [*Theory of Argumentation and Philosophy of Science*.] Written together with Lars Walløe and Jon Elster. Oslo: Oslo University Press 1976 and seven later editions. German translation, by Matthias Kaiser: *Rationale Argumentation* (Grundlagen der Kommunikation) Berlin: De Gruyter, 1986. ISBN 3 11 011075 X. Paperback edition. Danish translation, Karsten Klint Jensen: *Politikens Introduktion til moderne filosofi og videnskabsteori*. Copenhagen: Politikens Forlag, 1992. Swedish translation, by Mats Söderlind: *Argumentationsteori, språk och vetenskapsfilosofi*. Stockholm: Bokförlaget Thales, 1993, Second edition 1995, Third edition 2001.

Volumes edited relating to logic:

4. *The Philosophy of Mathematics*. Special issue of *The Monist*, January, 1984, Volume 67, Number 1.

5. *Phenomenology and the Formal Sciences*. Thomas E. Seebohm, Dagfinn Føllesdal, Jittendra Nath Mohanty, eds., (Contributions to phenomenology, vol. 8), Dordrecht: Kluwer, 1991. 262 pp. ISBN 0-7923-1499-9

6. *The Philosophy of W. V. Quine*. Editor and Introductions, five volumes of articles on Quine for Garland Publishers, New York, 2001:
Vol. 1 General, Reviews, Analytic/Synthetic ISBN 0815337388
Vol. 2 Naturalism and Ethics ISBN 0815337396
Vol. 3 Indeterminacy of Translation ISBN 081533740X
Vol. 4 Ontology ISBN 0815337418
Vol. 5 Logic, Modality and Philosophy of Mathematics ISBN 0815337426

7. Two volumes of papers by W. V. Quine, some previously unpublished, edited together with Douglas Quine: *Confessions of a Confirmed Extensionalist* and *Quine in Dialogue*. Cambridge: Harvard University Press, 2008. ISBN: 978-0-674-03084-8 and 978-0-674-03083-1. Articles relating to logic:

8. "Quantification into causal contexts." *Boston Studies in the Philosophy of Science*. A volume in honor of Philipp Frank. (Proceedings of the Boston colloquium for the philosophy of science.) Dordrecht: Reidel, 1965, pp. 263-74. ed in L. Linsky, ed., *Reference and Modality*. (Oxford Readings in Philosophy.) Oxford: Clarendon Press, 1971, pp. 52-62. Italian translation in Italian edition of Leonard Linsky, ed., *Reference and Modality*.

9. *A model theoretic approach to causal logic*. (Det Kgl. Norske Videnskabers Selskab. Skrifter. 1966. No. 2) Trondheim: Bruns Bokhandel, 1966. Italian translation in Claudio Pizzi, ed., *Leggi di natura, modalita, ipotesi: La logica del ragiona-mento controfattuale*. Milano: Feltrinelli, 1978, pp. 204-214.

10. "Knowledge, identity and existence." *Theoria* 33 (1967), pp. 1-27

11. Interpretation of quantifiers." In: B. van Rootselaar and J. F. Staal, eds., *Logic, Methodology and Philosophy of Science*. (Proceedings of the Third International Congress for Logic, Methodology and the Philosophy of Science, Amsterdam 1967.) Amsterdam: North-Holland, 1968, pp. 271-81. Reprinted in Dale Jacquette, ed., *Philosophy of Logic* (Blackwell Philosophy Anthologies), Oxford: Blackwell, 2002.

12. "Quine on modality." *Synthese* 19 (1968), pp. 147-57. (Reply by Quine: page 306.) [Not the same as my 2004 d) article with the same title.] Reprinted in: Donald Davidson and Jaakko Hintikka, eds., *Words and Objections: Essays on the Work of W. V. Quine*. (Synthese Library.) Dordrecht: Reidel, 1968, pp. 175-85, and in Farangh Zabech, E. D. Klemke, and Arthur Jacobson, eds., *Readings in Semantics*. Urbana, Chicago, London: Univ. of Illinois Press, 1974.

13. "Deontic logic: An introduction." By D.F. and Risto Hilpinen. In: Risto Hilpinen, ed., *Deontic logic: Introductory and Systematic Readings*. (Synthese Library.) Dordrecht: Reidel, 1970, pp. 1-35.

14. "Reference and sense." Symposium on Reference, together with Saul Kripke and Peter F. Strawson, Chair: W.V. Quine, at the XVIIth World Congress of Philosophy, Montreal, August 21 - 27, 1983. In Venant Cauchy (ed.), *Philosophy and Culture: Proceedings of the XVIIth World Congress of Philosophy*. Montreal: Editions du Beffroi, Editions Montmorency, 1986, pp. 229-239.

15. "Essentialism and reference." In Lewis E. Hahn and Paul Arthur Schilpp, eds., *The Philosophy of W.V. Quine* (The Library of Living Philosophers). La Salle, Ill.: Open Court, 1986, pp. 97-113. (Reply by Quine: pp. 114-115).

16. "Von Wright's modal logic." In Paul Arthur Schilpp and Lewis Edwin Hahn, eds., *The Philosophy of Georg Henrik von Wright* (The Library of Living Philosophers). La Salle, Ill.: Open Court, 1989, pp. 539-556. (Reply by von Wright: pp. 848-854.)

17. "Gödel and Husserl." In Jaakko Hintikka, ed., *From Dedekind to Gödel: Essays on the Development of the Foundations of Mathematics*, Dordrecht: Kluwer, 1995, pp. 427-446. Reprinted in Jean Petitot, Francisco J. Varela, Bernard Pachoud and Jean-Michel Roy, eds., *Naturalizing Phenomenology: Issues in Contemporary Phenomenology and Cognitive Science*. Stanford University Press, 1999, pp. 385-400. French translation: "Gödel et Husserl." In Jean Petitot, Francisco J. Varela, Bernard Pachoud and Jean-Michel Roy, eds., *Naturaliser la phénoménologie: Essais sur la phénoménologie contemporaine et les sciences cognitives*. Paris: CNRS Éditions, 2002, pp. 503-523.

18. "Introductory note to Kurt Gödel, 'The mod-

ern development of the foundations of mathematics in the light of philosophy'." In Kurt Gödel, *Collected Works*, Volume III: Unpublished Essays and Lectures, ed. Solomon Feferman et.al., Oxford: Oxford University Press, 1995, pp. 364-373.

19. "In Memoriam: Willard Van Orman Quine 1908-2000." Together with Charles Parsons, *The Bulletin of Symbolic Logic* Vol. 8, Number 1, March 2002, pp. 105-110.

20. "Quine on Modality." In Roger Gibson, ed., *The Cambridge Companion to Quine*. Cambridge: Cambridge University Press, 2004, pp. 200-213. [Not the same as my 1968 article by the same title.]

21. "Answers to five questions." In Vincent F. Hendricks and John Symons, eds., *Formal Philosophy: Aim, Scope, Direction*. Automatic Press, 2005, pp. 35-51. (ISBN 10-87-991013-1-9 hardback, ISBN-10-87-991013-0-0 paperback)

**Service to the Profession:** President, Norwegian Academy of Science, 1993, 1995, 1997; Chair of the Board, Center for Advanced Study, Norwegian Academy of Science and Letters, 1991–1992; Chair, Norwegian Government's Council for Information Technology, 1984–1987; Editorial Board Member, *Philosophy of Science*, 1981–1984; Editorial Board Member, *Studia Logica*, 1976–1993; Editorial Board Member, *Journal of Philosophical Logic*, 1972–2003; Editor, *Journal of Symbolic Logic*, 1970–1982; Council Member, Association for Symbolic Logic, 1966–1968 and 1970–1982.

**Teaching:** Since 1966 Føllesdal has divided his time between Oslo and Stanford. The quality of teaching, measured in number of students per teacher, the amount of written work and the feedback on the work, and also the broadness of the curriculum, is so much better in the United States that he has recommended good philosophy students in Norway to go abroad for their PhD. Among these are Jon Elster, Olav Gjelsvik, Bjørn Ramberg, Øystein Linnebo and many others. Among PhD students at Stanford are: David Smith, Ronald McIntyre, Izchak Miller and John Lad, all in phenomenology. Among Harvard students are Terry Malick and Daniel Dennett.

**Vision Statement:** Logic is more and more becoming a mathematical subject. Very few are able to do top work in logic as well as in philosophy. Still, logic remains an important tool in philosophy. Logic is, like mathematics and many other intellectual fields, concerned with structures. One reason for this is that philosophy, at least the way I conceive of it, is very holistic. One is concerned with how one's view on one issue fits in with one's view on other issues, and how apparently small modifications on one point may have far-reaching repercussions elsewhere.

Logic is a good tool for studying such structures; I look upon logic as an instrument on a par with telescopes and microscopes. Our eye is a very good all-round instrument which enables us to get the general picture and note what is worthy of our attention. However, when we start attending to matters that we regard as important and want to see them more clearly and study their consequences for other issues, it is important to focus on what matters and disregard features that seem irrelevant. This is just what we do when we use logic.

For this reason, I think that philosophers benefit from the study of logic. The study should make us aware of the importance of apparently small differences in philosophical views. This, in turn should make us skeptical towards "-ism" debates, where philosophers with widely different views are lumped together under one label. The study of logic should generally train us to stress arguments and evidence and not be taken in by persuasion and rhetorical devises.

**Honours and Awards:** Lauener Prize, 2006; Dr. honoris causa, University of Stockholm, 2003; Nansen Prize for Research, Norwegian Academy of Science and Letters, 2003; Alexander von Humboldt Research Prize 1997, University of Oslo Research Prize 1995.

Member, American Academy of Arts and Sciences, Academia Europaea, Akademie der Wissenschaften zu Göttingen, Palestinian Academy for Science and Technology, and the scientific academies of Denmark, Finland, Sweden and Norway.

Fellow, Center for Advanced Study, Norwegian Academy of Science and Letters, 1995-1996 and 2003-2004; Fellow, Wissenschaftskolleg, Berlin, 1989-1990; Member, Institute for Advanced Study, Princeton, 1985-1986; Fellow, American Council of Learned Societies, 1983-1984; Fellow, Center for Advanced Study in the Behavioral Sciences, Stanford, 1981-1982; Guggenheim Fellow, 1978.1979; Santayana Fellow, Harvard University, 1964-1965.

Gunnerus lecture, Royal Norwegian Society of Sciences and Letters, 1995; Tanner lectures, Buenos Aires, Commentator, 1988; Hägerström lectures, Uppsala University, 1987; Gurwitsch lecture, Society for Phenomenology and Existential Philosophy (in Toronto), 1986; Alfred Schutz lecture, American Philosophical Association, California Institute of Technology, 1975; Gurwitsch lecture, New School for Social Research 1975; Wolfson College Lecture, Oxford, 1974.

# FRIEDMAN, Sy David

**Specialties:** Mathematical logic, set theory, higher recursion theory

**Born:** 23 May 1953 in Chicago, Illinois, USA.

**Educated:** Massachusetts Institute of Technology, PhD Mathematics 1976.

**Dissertation:** *Recursion on inadmissible ordinals*; supervisor, Gerald E. Sacks

**Regular Academic or Research Appointments:** ORDENTLICHE UNIVERSITÄTS-PROFESSOR FÜR MATHEMATISCHE LOGIK, UNIVERSITY OF VIENNA, 1999–; DIRECTOR, KURT GÖDEL RESEARCH CENTER FOR MATHEMATICAL LOGIC, 2004–; Professor of Mathematics, Massachusetts Institute of Technology (MIT), 1990–2004; Associate Professor of Mathematics, MIT, 1983–1990; Assistant Professor of Mathematics, MIT, 1978–1983; National Science Foundation Postdoc, University of California at Berkeley, 1978–1979; L. E. Dickson Instructor, University of Chicago, 1976–1978.

**Visiting Academic or Research Appointments:** ICREA Visiting Research Professor, Centre de Recerca Matematica, Bellaterra, January, February, September 2006 and February, March, September 2005; Visitor, Centre de Recerca Matematica, Bellaterra, September 2003; Visitor, Mittag-Leffler Institute, Djursholm, September 2000; Visiting Fellow, Balliol College, Oxford, February 2000; Visiting Professor, Universities of Paris 6 and 7, February-May 1996; Visiting Professor, University of Paris 7, February-March 1995; Visitor, Centre de Recerca Matematica, Bellaterra, January 1995; Resident Scholar, Rockefeller Center, Bellagio, December 1994; Visiting Professor, University of Siena, October-November 1994; Visiting Professor, University of Paris 7, May 1994; Visiting Associate Professor, University of California at Berkeley, January-June 1990; Visiting Fellow, All Souls College, Oxford, April-June 1988; Visiting Fellow, Wolfson College, Oxford, September-June 1986.

**Research Profile:** Sy-David Friedman is recognized for his important contributions to higher recursion theory and to set theory. In the former area, he made applications of set-theoretic methods to study the analogue of the Turing degrees for admissible and inadmissible ordinals. One of his results in this area uses Silver's work on the singular cardinal problem to show that for singular cardinals of uncountable cofinality in GÄ¶del's universe of constructible sets, the degrees above some fixed degree are well-ordered with successor given by the jump. He also obtained a generalization of the Friedberg-Muchnik theorem (establishing the existence of incomparable Turing degrees of recursively enumerable sets) to certain inadmissible ordinals using Jensen's Diamond principle. Another significant result in higher recursion theory is his use of forcing and infinitary model theory to obtain a generalization to the uncountable of Sacks' theorem that each countable admissible ordinal is the least admissible relative to some real. Friedman's best-known work in set theory is in the area of coding. By combining Jensen's coding method with other techniques, he succeeded in resolving a number of long-standing conjectures: The admissibility spectrum conjecture asserts that there is a real relative to which the class of admissible ordinals is nontrivial but well-behaved, for example equal to the admissible limits of admissible ordinals. Friedman proved this conjecture using his elaborate strong coding method. The Pi-1-2 singleton conjecture asserts that Silver's real number 0-sharp is the least Pi-1-2 singleton in terms of constructibility degree. Friedman refuted this conjecture through a combination of coding methods and a version of Kleene's recursion theorem. Friedman also unexpectedly lifted the coding method into the context of Woodin cardinals, circumventing obstacles posed by Woodin's extender algebra. Friedman's more recent work characterizes theories which are classifiable in the sense of model theory in terms of their set-theoretic behavior (joint work with Hytinnen and Rautila) and, together with his young colleagues at the Kurt Gödel Research Center, develops the internal consistency program, which aims to determine the large cardinal hypotheses needed to establish the truth of set-theoretic statements in inner models. This program has led to his inner model hypothesis, a natural principle of strong absoluteness which has striking implications for the nature of the set-theoretic universe.

**Main Publications:**

1. Beta-Recursion Theory, Transactions AMS. Volume 255, 1979, pp. 173–200.
2. Post's Problem Without Admissibility, Advances in Mathematics, Vol. 35, No. 1, 1980. pp. 30–49.
3. Negative Solutions to Post's Problem I, Generalized Recursion Theory II, North Holland, 1978, pp. 127–133.
4. Negative Solutions to Post's Problem II, Annals of Mathematics, Vol. 113, 1981. pp. 25–43.
5. Uncountable Admissibles I: Forcing, Transactions AMS, Vol. 270, No.1, 1982, pp. 61–73.
6. Uncountable Admissibles II: Compactness, Israel Journal of mathematics, Vol. 40, No. 2, 1981, pp. 129–149.

7. Strong Coding, Annals of Pure and Applied Logic, 1987, pp. 1–98 and A Guide to Strong Coding, Annals of Pure and Applied Logic, 1987, pp. 99–122.
8. The Pi-1-2 Singleton Conjecture, Vol.3, Journal of the American Mathematical Society, 1990, pp. 771–791.
9. The Genericity Conjecture, Journal of Symbolic Logic, Vol.59, No.2, 1994, pp. 606–614.
10. Iterated Class Forcing, Mathematical Research Letters, Vol.1, No.4, 1994, pp. 427–436.
11. (with Peter Koepke) An Elementary Approach to the Fine Structure of L, Bulletin of Symbolic Logic, Vol.3, No.4, 1997, pp. 453–468.
12. Generic Saturation, Journal of Symbolic Logic, Vol.63, No.1, 1998, pp. 158–162.
13. Fine Structure and Class Forcing, book, De Gruyter series in Logic and its Applications, 2000.
14. Cantor's set theory from a modern point of view, Jahresbericht der Deutsche Mathematiker-Vereinigung 104, Heft 4, pp. 165–170, 2002.
15. (with Tapani Hytinnen and Mika Rautila) Classification theory and 0-sharp, Journal of Symbolic Logic, Vol. 68, No. 2, pp. 580–588, 2003.
16. Completeness and iteration in modern set theory, One Hundred Years of Russell's Paradox, Godehard Link, ed., Walter de Gruyter, 2004, pp. 85–92.
17. Genericity and large cardinals, Journal of Mathematical Logic, Vol. 5, No. 2, pp. 149–166, 2005.
18. Stable axioms of set theory, in Set Theory: Centre de recerca Matemà tica, Barcelona, 2003-2004, Trends in Mathematics. Birkhäuser Verlag, pp.275–283, 2006.
19. Internal consistency and the inner model hypothesis, Bulletin of Symbolic Logic, Vol.12, No.4, December 2006, pp. 591–600.
20. Large cardinals and L-like universes, to appear, in Set theory: recent trends and applications, Quaderni di Matematica, Alessandro Andretta, editor.

A full list of publications is available at http://logic.univie.ac.at/~sdf/
*Work in Progress*
21. The internal consistency program: Joint work with Dobrinen, Futas, Ondrejovic and Thompson concerning the internal consistency strength of costationarity for P-kappa-lambda, failures of the singular cardinal hypothesis, realization of Easton functions and generalized dominating numbers.
22. The inner model hypothesis: Upper and lower bounds on the consistency strength of the inner model hypothesis in its weak and strong forms.
23. The outer model program: What L-like properties can be forced while preserving large cardinals? Is the fine structure theory consistent with superstrong cardinals?
24. Projective regularity: Joint work with Schrittesser, using coding methods to separate regularity properties at the projective level.

**Service to the Profession:** Chair, Program Committee for Set Theory at Oberwolfach, 2002–; Editor, Monatshefte für Mathematik, 2002 –; Chair, ASL Committee on Logic in Europe, 2000–2003; Chair, Committee on the Sacks Prize, 1994–1999; Editor, Journal of Symbolic Logic, 1990–1996.

**Teaching:** Friedman directed the MIT logic program for 20 years, in collaboration with Gerald Sacks. During his years at MIT he supervised 6 doctoral students and 3 postdocs. Since he took up the Logic Chair at the University of Vienna in 1999, he has directed an additional 5 doctoral theses with 3 more in progress. His 10 postdocs to date at the University of Vienna include Riccardo Camerlo (Associate Professor at Torino), David Aspero (ICREA Junior Professor at Barcelona), Heike Mildenberger, (Assistentin at the Gödel Center), Andres Caicedo (Bateman Instructor at Caltech), John Krueger (Morrey Assistant Professor at Berkeley) and Matteo Viale (currently at the Gödel Center, winner of the 2006 Sacks prize)

**Vision Statement:** I see two important tasks for set theory in the coming years. The first is to fully justify the need for large cardinal axioms. This cannot be achieved by examining their implications or by deriving them from principles of absoluteness for set-generic extensions of the universe. It will be achieved through compelling principles of strong absoluteness which take arbitrary, and not just generic, extensions of the universe into account. I expect such principles to also lead to a strong negative solution to the continuum problem: the cardinality of the reals is not the alpha-th cardinal for any absolutely definable ordinal alpha. The second task is to provide model theory with new techniques for analyzing theories in terms of absoluteness properties of their models. This may serve to bring these subjects close together, as they were in the 1970's.

**Honours and Awards:** ICREA Visiting Research Professor, 2005-2006; Visiting Fellow, Balliol College, Oxford, 2000; Visiting Professor, University of Bonn, 1996; Resident Scholar, Rockefeller Center, Bellagio, 1994; Visiting Fellow, All Souls College, Oxford, 1988; Visiting Fellow, Wolfson College, Oxford, 1985-1986.

# FUHRMANN, André Theodor

**Specialties:** Modal logic, relevance logic, belief revision, philosophical logic, philosophy of logic.

**Born:** 30 October 1958 in Essen, Germany.

**Educated:** University of Konstanz, Habilitation 1993; Australian National University, PhD, 1988;

University of St. Andrews, MPhil, 1984; University of Marburg.

**Dissertation:** *Relevant Logic, Modal Logic, and Theory Change*; supervisor, Richard Sylvan.

**Regular Academic or Research Appointments:** FULL PROFESSOR (LOGIC AND PHILOSOPHY OF SCIENCE), UNIVERSITY OF FRANKFURT, 2006–; Full Professor, Universidade S. Judas Tadeu, SÃ£u Paulo, 2002–2005; Assistent, Dozent, Heisenberg-Fellow, University of Konstanz, 1989–2002.

**Visiting Academic or Research Appointments:** Visiting Professor, Federal University of Rio de Janeiro, 1998-2002, Visiting Professor, Columbia University, New York, 1994; Visiting Professor, Indiana University, Bloomington, 1993.

**Main Publications:**

1. Ultrafilter and normality logic, *Studia Logica (Festschrift issue for Sven Ove Hansson)*, 73:197–207, m2003.
2. Russell's way out of the paradox of propositions, *History and Philosophy of Logic*, 3: 197–213, 2002.
3. When hyperpropositions meet ..., *Journal of Philosophical Logic*, 28: 559–74, 1999.
4. *An Essay on Contraction*, Studies in Logic, Language and Information, CSLI Publications / Cambridge University Press, Stanford, 1997.
5. A relevant theory of conditionals, *Journal of Philosophical Logic*, 24: 645–665, 1995.
6. On S, *Studia Logica*, 53: 75–91, 1994.
7. (with Sven Ove Hansson), A survey of multiple contractions, *Journal of Logic, Language and Information*, 3: 39–76, 1994.
8. Theory contraction through base contraction, *Journal of Philosophical Logic*, 20: 256–281, 1991.
9. Models for relevant modal logics, *Studia Logica*, 49: 301–315, 1991.
10. Reflective modalities and theory change, *Synthese*, 81: 115–134, 1989.
11. *Relevant Logic, Modal Logic and Theory Change*, PhD Thesis, Department of Philosophy and Automated Reasoning Project, Institute of Advanced Studies, Australian National University, 1988.

**Service to the Profession:** Managing Editor, Editorial Board, *Studia Logica*; Reviews Editor, Editorial Board, *Erkenntnis*; Editorial Board, *Manuscrito*; Editorial Board, *Theoria* (Sweden).

**Honours and Awards:** Heinz Meier-Leibnitz Prize for work in cognitive science, 1993.

# G

## GABBAY, Dov

**Specialties:** Nonclassical logics, modal logics, temporal logis, intuitionistic logic, quantum logics, nonmonotonic logics, labelled deductive systems, logic of practical reasoning, goal directed systms, combining logics, metalevel object=level systmes.

**Born:** 16 October, 1945 in Baghdad, Iraq.

**Educated:** Hebrew University of Jerusalem, BSc mathematics and physics 1966; MSc mathematics, 1967; PhD mathematics and logic 1969. Science, 1976.

**Dissertation:** *Non-classical Logics*; supervisors, A. Levy, M. O. Rabin and Y. Bar-Hillel.

**Regular Academic or Research Appointments:** AUGUSTUS DE MORGAN PROFESSOR OF LOGIC, KING'S COLLEGE LONDON, 1998–. Professor of Computing, Imperial College, London, 1983-1998; Lady Davis Professor of Logic, Bar-Ilan University, Israel, 1977–1983; Associate Professor of Mathematics and Computer Science, Bar-Ilan University, Israel, 1975–1977; Associate Professor of Philosophy, Stanford University, USA, 1970–1973; Instructor, Hebrew University of Jerusalem, 1968–1970.

**Visiting Academic or Research Appointments:** Leverhulme major research fellowship, 2007-2010; Alexander von Humbolt research fellow, 2001; SERC senior research fellow, 1992–1997; Adjunct Professor, University of Georgia, 1985–1996; Research Professor, Max-Planck Institute, Saarbrucken, 1992–1995; Visiting Professor, University of Stuttgart, 1992; Visiting Professor, University of Munich, 1990.

**Research Profile:** Dov Gabbay has made major contributions to the evolution of logic in the 20th century by developing the foundations of logical systems arising from the applications of logic in computer science, artificial intelligence, and in language. Espeically significant are his contributions to temporal logics, nonmonotonic logics, labelled deductive systems, goal directed proof theory, dynamical syntax and practical reasoning.

**Main Publications:**

1. *Investigations in Modal and Tense Logics with Applications*, D. Reidel, 1976.
2. *Semantical Investigations in Heyting's Intuitionistic Logic*, D. Reidel, 1981.
3. *Temporal Logic: Mathematical Foundations and Computational Aspects*, with I. Hodkinson and M. Reynolds, Oxford University Press, 1994.
4. *Labelled Deductive Systems: Principles and Applications. VOl 1: Basic Principles*, Oxford University Press, 1996.
5. *Elementary Logic: A Procedural Perspective*, Prentice-Hall, 1998.
6. *Fibring Logics*, Oxford University Press, 1998.
7. *The Imperative Future*, with H. Barringer, M. Fisher, R. Owens and M. Reynolds, Research Studies Press, 1996.
8. *Temporal Logic: Mathematical Foundations and Computational Aspects. Vol. 2: Computational Aspects*, with M. Reynolds and M. Finger, Oxford University Press, 2000.
9. *Goal Directed Algorithmic Proof Theory*, with N. Olvietti. Kluwer Academic Publishers, 2000.
10. *Dyanamic Syntax: The Flow of Language Understanding* with R. Kempson and W. Meyer-Viol, Blackwell, 2000.
11. *Many Dimensional Modal Logics*, with A. Kuruc F. Wolter and M. Zakharyaschev, Elsevier, 2003.
12. *Agenda Relevance*, with J. Woods, Elsevier, 2003.
13. *The Reach of Abduction*, with J. Woods, Elsevier, 2005.
14. *The Imperative Future*, with H. Barringer, M. Fisher, R. Owens and M. Reynolds. Research Studies Press/J. Wiley, 1996.
15. *Neural-Symbolic Learning Systems: Foundations and Applications* with A. S. D'Avila Garcez and K. Broda. Springer-Verlag, 2002.
16. *Compiled Labelled Deductive Systems for Modal and Conditional Logics* with K. Broda, L. C. Lamb and A. Russo. Research Studies Press/J. Wiley, 2004.
17. *Interpolation and Definability, Volume 1: Modal and Intuitionistic Logic*, with L. Maksimova. Oxford University Press, 2005.
18. *Connectinist Non-classical Logics: Sistributed Reasoning and Learning in Neural Networks*, with A. S. D'Avila Garcez and L. C. Lamb. Springer-Verlag.
19. *Analysis and Synthesis of Logics*, with W. Carnielli, M. Coniglio, P. Gouveia and C. Sernadas. Draft available.

A full list of publications is available at www.dcs.kcl.ac.uk/staff/dg

*Work in Progress*

20. *Proof Theory for Fuzzy Logics*, with G. Metcalfe and N. Olivetti. Partialdraft available. To appear with Research Studies Press/J. Wiley.
21. *Quantum Logic*, with K. Engesser and D. Lehmann.
22. *Second-order Quantifier Elimination* with R. Schmidt and A. Szalas.

23. *Quatntification in Non-classical Logics*, with V. Shehtman and D. Skvortsof.
24. *Revision by Translation*, with O. Rodrigues and A. Russo.
25. *Seductions and Shortcuts, Errors in Cognitive Economy*, with John Woods. Volume 3 of Gabbay and Woods, *A Practical Logic of Cognitive Systems*.

**Service to the Profession:** Editor of multiple Handbooks, including *Handbook of Philosophical Logic*, Kluwer/Springer; *Handbook of Logic in Computer Science*, OUP, *Handbook of Logic in Artificial Intelligence and Logic Programming*, OUP, *Handbook of Defeasible Reasoning and Uncertainty*, Kluwer, *Handbook of Tableaux*, Kluwer, *Handbook of Logic of Argument and Inference: The Turn Toward the Practical*, Elsevier; *Handbook of the History of Logic*, Elsevier; *Handbook of the Philosophy of Science*, Elsevier.

Editor of major journals, book series, treasury of logic, Internatioanl Federation of Computational Logic.

**Teaching:** Professor Gabbay is a gifted teacher and has won several teaching citations for his courses. He is also a devoted PhD supervisor and many of his sudents are now senior academics all over the world. Many graduate students from other institutions come to spend a year with him.

**Vision Statement:** We take the point of view that when the Almighty created man he sprinkled from some lump of logic into his creation. What flows from this belief is hat ther is coherence and overall unity in the reasoning (formal or commonsense) and daily behaviour of the human agent.

Current research areas and communities involved in the study of the human agent are diverse and non-integrated. they range from mathematical logic and formal philosophy, formal linguistics and formal computation through araes such as common sense reasoning, artificial intelligence, psychology, social decision theory, all the way to neural nets and Bayesian networks, and argumenation structures. Our vision is to develop methodologies and general principles which can bring together these diverse areas and their research communities and give us some idea of the nature of the original "lump" of "logic" sprinkled into humanity.

There are several steps to be taken which we will exemplify

1. Find a point of view which can bring together several diverse disciplines, e.g. one natural point of view for neural nets, argumentation theory and, e.g. discrete logic (such as classical or intuitionistic logic).

2. Apply this point of view to internally restructure the participating disciplines themselves. So if for example, a subarea of discrete logic may be deemed marginal by its community, it may emerge as more central when the interaction with other disciplines and communities is taken into account.

3. Seek and identify methodologies common or unique to different areas and apply the methods of unique to one area to other areas. For example, the idea of combining systmes is current in many areas but the details of combination are different. We can apply the method of one area in another to see what we get.

To sum up, my vision is that there is a body of methods an ideas which can be combined and specialised to give the various "logics" manifested in the widely diveres areas of human action and behaviour.

**Honours and Awards:** Fellow of the British Computer Society, 2004; Fellow of the Royal Society of Arts, UK, 2002, External Fellow of the Royal Society of Canada, 2000, Doctor Honoris Causa (Science), Univ. Paul Sabatier, France, 1996.

# GAIFMAN, Haim

**Specialties:** Mathematical logic, foundations of probability, theoretical computer science, philosophy of language, philosophy of mathematics, early analytic philosophy.

**Born:** 26 September 1934 in Rovno Ukraine, emigrated to Israel (then Palestine) in 1936.

**Educated:** University of California, Berkeley, PhD Mathematics, 1960-1962; Hebrew University, MS Mathematics, 1954-1958.

**Dissertation:** *Contributions to The Theory of Boolean Algebras*; supervisor, Alfred Tarski.

**Regular Academic or Research Appointments:** PROFESSOR OF PHILOSOPHY, COLUMBIA UNIVERSITY, 1990–; Chairman of the Philosophy Department, Columbia University, 1999–2002; Eleanor Roosevelt Professor of Logic and Philosophy of Science, Hebrew University, 1976–1992; Professor of Mathematics, Hebrew University, 1974–1992; Associate Professor of Mathematics, Hebrew University, 1968–1974; Assistant Professor of Mathematics, Hebrew University, 1963–1968; Ritt Instructor of Mathematics, Columbia University, 1962–1963.

**Visiting Academic or Research Appointments:** Visiting Researcher, IBM Hawthorn, 1989; Visiting Professor, Computer Science, Stanford University, 1988-1989; Visiting Researcher, IBM Almaden, 1988; International Fellow, Stanford Research Institute, Stanford, 1987-1988; Visiting Professor, Computer Science, Université de Paris VI, 1982-1983; Visiting Professor, Mathematics, Université de Paris VII, 1977-1978; Visiting Professor, Philosophy, Stanford University, 1972-1973; Visiting Professor, Mathematics, University of California Los Angeles, 1971; Visiting Professor, Mathematics, Aahrus Universitet, Denmark, 1969, 1970; Visiting Associate Professor, Mathematics, UCLA, 1967-1968; Visiting Associate Professor, Mathematics, University of California, Berkeley, 1968, Instructor, Mathematics, University of California Berkeley, 1962; Researcher, Linguistics, Rand Corporation, 1960; Research Assistant to Carnap, University of California Los Angeles, 1959-1960.

**Research Profile:** : Haim Gaifman has made fundamental contributions in set theory, models of arithmetic, foundations of probability and inductive logic, and the modeling of self reference. He has also contributed significant results in widely different areas: formal languages, Boolean algebras, finite model theory and query languages. The following sketch and the *key publications* entry do not cover his philosophical works in philosophy of language, philosophy of mathematics and metaphysics.

In 1963 he invented the iteration technique for internally defined elementary-extensions of models of set theory and proved that if the extending model is well founded, so are all models obtained from well-ordered iterations; the images of the critical point under such iterations form a closed unbounded class of ordinals, and the corresponding $L_\alpha$'s(the constructible universe up to $\alpha$) form an elementary chain. In 1964, using the functorial properties of the extension operator, Gaifman derived, from the existence of measurable cardinals, strong claims concerning $L$. The functorial properties imply immediately that the above class consists of indiscernible in the resulting universe, hence also the existence of $O^\#$ - a fact not noted in Gaifman's 1964 work, but derived later by Silver from Ramsey cardinals following Rowbottom's work. Iterated ultraproducts were subsequently given a direct combinatorial definition by Kunen. Iterations are now a staple device of large cardinal theory. In 1965 Gaifman generalized internally defined operators to uniform extension operators; these act as functors in some category of models and elementary embeddings, and also associate with every model in the category an elementary embedding of it in the same category, such that the resulting diagram commutes. Such an operator admits, for any order type, a corresponding natural iteration. Gaifman introduced definable types and minimal end-extension types for models of arithmetic, which give rise to uniform extension operators and using the iteration techniques got many results concerning the lattices of elementary submodels and the order types of models of Peano. He has also worked in other areas of Peano models. His results led to natural questions and to lines of research that were considerably developed by other researchers. In the foundations of probability and induction Gaifman proposed in 1960 a general definition of a probability defined over sentences of any first order language, establishing its basic properties. Later, with Snir, he investigated probabilities over rich languages containing arithmetic, proved convergence theorems and theorems that establish limits on inductive confirmation in terms of the logical complexity of the definition of the probability. That work also contains for the first time a general approach to randomness. He also proposed a framework that unifies objective and subjective probabilities, as well as a rigorous treatment of higher order probabilities. In the treatment of self-reference he proposed and developed the pointer semantics, which associates truth values with pointers - an abstract generalization of tokens. Among important theorems in various domains are: the equivalence of phrase structure and categorial grammars, proved in 1959 (which easily implies Greibach's result on hardest context free language), the existence of a Boolean algebra with the countable chain condition without a strictly positive finitely additive measure, the Gaifman-Hales theorem (proved independently by Hales) that the free Boolean algebra, on infinitely many generators with infinitary operations (subject to certain distributive laws) is a proper class, and results concerning the expressive powers of various extensions of first order logic. In particular, he has defined a metric, with respect to which first-order definitions have a locality property, which turned out to be a very useful tool in many researches by logicians and computer scientists. Gaifman also worked and published results, in various versions of Prolog, parallel computation, and decidability of problems arising in query languages.

**Main Publications:**

1. On categorial and phrase structure grammars (with Y. Bar-Hillel and E. Shamir) *The Bulletin of the Research Council of Israel*, vol. 9F, 1960, pp. 1 – 6.
2. Concerning measures in first order calculi, *Israel*

*Journal of Mathematics*, vol. 2, 1964, pp. 1 – 18.

3. Infinite Boolean Polynomials I, *Fundamenta Mathematicae*, vol. 54, 1964, pp. 229 – 250.

4. Uniform extension operators for models and their applications, in *Sets, Models and Recursion Theory*, Crossley J. ed, North-Holland, 1967, pp. 122 – 155.

5. On local functions and their applications for constructing types of Peano's arithmetic, in *Mathematical Logic and Foundations of Set Theory*, Bar-Hillel Y. editor, Proceedings of the International Colloquium Held Under the Auspices of the Israeli Academy of Science and Humanities, North-Holland, 1970, pp. 105 – 121. (Results announced and lectured on in 1965: Leicester Logic Colloquium, AMS abstract, 65T-195 and technical report 21, for the U.S. Office of Naval research).

6. Operations on relational structures, functors and classes I *Tarski Symposium, UC Berkeley 1971* L. Henkin editor, Symp. in Pure Mathematics 25, AMS - for the Assoc. of Symb. Logic, 1974, pp. 20 -39 (Postscript with new results added in the second edition of this volume, 1979).

7. Elementary embeddings of models of set theory and certain subtheories *Axiomatic Set Theory*, Scott D. and T. Jech editors, Symp. in Pure Mathematics 13, AMS 1974, vol II, pp. 33 – 101. (Results announced and lectured on in 1964: AMS abstract 64T-505, contributed paper at the 1964 International Congress for Logic Philosophy and Methodology of Science. Mimeographed handwritten version of a long paper circulated in the same year.)

8. Models and types of Peano's arithmetic, *Annals of Mathematical Logic*, vol. 9, 1976, pp. 223 – 306.

9. On local and non-local properties, *Logic Colloquium 1981*, J. Stern editor, North-Holland, 1982, pp. 105 – 132.

10. Probabilities over rich languages, testing and randomness (with M. Snir) *Journal of Symbolic Logic*, vol. 43, No. 3, 1982, pp. 495 – 548.

11. Towards a unified concept of probability, *Proceedings of the 1983 International Congress for Logic Methodology and Philosophy of Science*, R. Barcan Marcus, G.J.W. Dorn and P. Weingartner editors, North Holland 1986, pp. 319-350.

12. A theory of higher order probabilities, *Theoretical Aspects of Reasoning About Knowledge\/}*, J. Halpern editor, Morgan Kaufman, 1986, pp. 275 - 292. A fuller version under this title appeared also in *Causation Chance and Credence*, Skyrms B. and W.L. Harper editors, Reidel 1988, pp. 191-219.

13. Decidable optimization problems for database logic programs (with Cosmadacis, Kannelakis and Varzi) *STOC (Annual ACM Symposium on Theory of Computing)* May 1988, IEEE Computer Society Press, 1988, pp. 477 – 490.

14. Operational Pointer Semantics: Solution to Self-Referential Puzzles *Theoretical Aspects of Reasoning About Knowledge* M. Vardi editor, Morgan Kaufman 1988 pp. 43 - 59.

15. Pointers to Truth, *Journal of Philosophy*, LXXXIX, 5 (1992), pp. 223-261. Appeared also in the *Philosopher's Annual* as one of the best ten philosophical papers of 1992.

16. Pointers to Propositions, in *Circularity, Definition, and Truth*, edited by Chapuis A. and Gupta A., Indian Council of Philosophical Research, New Delhi (2000), pp. 79 -122.

17. Contextual Logic and Its Applications to Vagueness, *The Bulletin of Symbolic Logic* Abstracts of Invited Talks, ASL Meeting, 2001, March 10-13.

18. Reasoning with Limited Resources and Assigning Probabilities to Arithmetical Statements *Synthese*, May 2004, vol. 140, 1-2, pp. 97-119.

*Work in Progress*

19. I am preparing "Vagueness, Tolerance and Contextual Logic" (41 pages) that sums up works presented at various conferences in the period 1997 – 2001, and can be found on my website, for publication.

20. "Naming and Diagonalization, from Cantor to Gödel to Kleene" is due to appear in the *Logic Journal of the IGPL*. I am working on a sequel that develops further the theory of Naming Systems.

21. I am preparing "Russellian Substitutional Algebras", a philosophical paper with a strong technical component, offering formally rigorous reconstruction of Russell's theory of propositions and the way it can serve as a framework for the *Principia*. I presented a sketch of a preliminary draft (29 pages) in a conference held in January 2005 in Bombay.

22. I am working on a paper, "Frege's logical perspectives" that focuses on Frege's conception of logic and his metatheoretic views, especially in the context of his dispute with Hilbert.

23. A longer term project on the foundations of probability, aimed at a monograph that will elaborate the conception of a unified probability and the inevitable limits of Bayesian approaches.

The articles mentioned in 1 and 2, as well as works from 1992 on, can be found on my website
`http://www.columbia.edu/~hg17/`

**Service to the Profession:** Vice President, International Congress for Logic Methodology and the Philosophy of Science, 2003-; Program Committee Chairman, ASL, New York, 2005; Executive Committee Member, Association for Symbolic Logic 1994–1998; Chairman, Program of History and Philosophy of Science, Hebrew University, 1973-1977.

**Teaching:** As a young teacher Gaifman was involved with a group of extremely bright students, including Avron, Gabbay, Magidor, Shelah and Stavi. He supervised Shelah's master thesis and Avron's and Stavi's PhD's. He also organized various courses in philosophy of science. At Columbia he was central in organizing the logic program, both philosophical and mathematical (including set theory and recursive functions). His text, written specifically for the main logic course has been successfully used in the last 12 years. He also taught courses in philosophy of language, philosophy of logic, philosophy of mathematics, realism, vagueness, Frege, Russell, Wittgenstein, inductive logic and rational choice. And he has supervised dissertations on Frege, Wittgenstein, in metaphysics and in the philosophy of mind.

**Vision Statement:** Logic emerged from great foundational projects that have lost some of their vitality. Also the sociology of academia is not always kind to logicians. But as far as the subject is concerned logic should remain flourishing, albeit in different modes. Besides the purely mathematical developments, it remains the major tool of conceptual analysis, whenever this can be done at a high level of rigor, e.g., in analyzing the notion of algorithm. Foundational questions remain. I would like to get a better grip on the possibilities of ultra fintism. And there is always the possibility of an earth shaking independence result, say for P = NP, in a way that would not give us any indication about its truth...

# GALVIN, Fred

**Specialties:** Model theory, classical set theory.

**Born:** 10 November 1936 in Saint Paul, Minnesota, USA.

**Educated:** University of Minnesota, PhD. 1967; University of Minnesota, MA, 1961; University of Minnesota, BA, 1958.

**Dissertation:** *Horn Sentences*; supervisor, Bjarni Jónsson.

**Regular Academic or Research Appointments:** PROFESSOR, UNIVERSITY OF KANSAS, 1978–; Associate Professor, University of Kansas, 1975-1978; Assistant Professor, University of California at Los Angeles, 1968–1975; Lecturer, University of California at Berkeley, 1967–1968; Acting Instructor, University of California at Berkeley, 1965–1967.

**Visiting Academic or Research Appointments:** Exchange Scientist, Mathematical Institute of the Hungarian Academy of Sciences, 1980-1981; Visiting Professor, University of Colorado, 1978-1979; Visiting Research Mathematician, University of Calgary, 1976; Exchange Scientist, Mathematical Institute of the Hungarian Academy of Sciences, 1972–1973; Postdoctoral Fellow, University of Alberta, 1969–1970.

**Research Profile:** Fred Galvin made several contributions to logic and set theory. In logic, he proved that every first-order sentence is equivalent to a Boolean combination of Horn sentences; also, he showed that H. J. Keisler's characterization of Horn classes, as the only elementary classes closed under the reduced product operation, does not require any set-theoretic assumption beyond ZFC. His most noteworthy achievements in set theory were his work with Karel Prikry on Ramsey's theorem for Borel partitions, and his work with András Hajnal on cardinal exponentiation.

**Main Publications:**

1. Distributive sublattices of a free lattice, with B. Jónsson, *Canad. J. Math.*, 13, 1961, 265–272.
2. Reduced products, Horn sentences, and decision problems, *Bull. Amer. Math. Soc.*, 73, 1967, 59–64.
3. Horn sentences, *Ann. Math. Logic*, 1, 1970, 389–422.
4. Borel sets and Ramsey's theorem, with Karel Prikry, *J. Symbolic Logic*, 38 1973, 193–198.
5. On set-systems having large chromatic number and not containing prescribed subsystems, with P. Erdős and A. Hajnal, *Infinite and Finite Sets* (Colloq. Keszthely, 1973; dedicated to P. Erdős on his $60^{th}$ birthday), Vol. I, pp. 425–513. *Colloq. Math. Soc.*, János Bolyai, Vol. 10, North-Holland, Amsterdam, 1975.
6. On a partition theorem of Baumgartner and Hajnal, *Infinite and Finite Sets* (Colloq. Keszthely, 1973; dedicated to P. Erdős on his $60^{th}$ birthday), Vol. II, pp. 711–729. *Colloq. Math. Soc. János Bolyai*, Vol. 10, North-Holland, Amsterdam, 1975.
7. Inequalities for cardinal powers, with András Hajnal, *Ann. of Math.*, 101(2), 1975, 491–498.
8. Infinitary Jonsson algebras and partition relations, *Algebra Universalis*, 6, 1976, 367–376.
9. An ideal game, with T. Jech and M. Magidor, *J. Symbolic Logic*, 43, 1978, 284–292.
10. Indeterminacy of point-open games, *Bull. Acad. Polon. Sci. Sér. Sci. Math. Astronom. Phys.*, 26, 1978, 445–449.
11. Generalized Erdős cardinals and $0^{\#}$, with James E. Baumgartner, *Ann. Math. Logic*, 15, 1978, 289–313.
12. Chain conditions and products, *Fund. Math.*, 108, 1980, 33–48.
13. A Ramsey-type theorem for traceable graphs, with I. Rival and B. Sands, *J. Combinatorial Theory Ser.*, 33(B), 1982, 7–16.
14. $\gamma$-sets and other singular sets of real numbers, with Arnold W. Miller, *Topology and Appl.*, 17, 1984, 145–155.
15. Stationary strategies in topological games, with Rastislav Telgársky), *Topology Appl.*, 22, 1986, 51–69.
16. Cylinder problem, with K. Ciesielski, *Fund. Math.*, 127, 1987, 171–176.
17. Cardinal representations for closures and preclosures, with E. C. Milner and M. Pouzet, *Trans. Amer. Math. Soc.*, 328, 1991, 667–693.
18. Graph colorings and the axiom of choice, with P. Komjáth, *Period. Math. Hungar.*, 22, 1991, 71–75.

19. Fonctions aux differences f(x)-f(a+x) continues, with Gilbert Muraz and Pawel Szeptycki, *C. R. Acad. Sci. Paris Sér. I Math.*, 315, 1992, 397–400.

20. Generating countable sets of permutations, *J. London Math. Soc.* 51(2), 1995, 230–242.

**Teaching:** Fred Galvin's one doctoral student, Marion Scheepers, is a very active researcher in set theory.

## GÄRDENFORS, Peter

**Specialties:** Decision theory, concept formation, cognitive semantics, models of knowledge and information, evolution of cognition.

**Born:** 1949 in Forsakar, Sweden.

**Educated:** Lund University, PhD Philosophy, 1974; Lund University, BSc Mathematics, philosophy, computer science, 1970.

**Dissertation:** *Group Decision Theory.*

**Regular Academic or Research Appointments:** PROFESSOR, COGNITIVE SCIENCE, LUND UNIVERSITY, 1994–; Associate Professor, Philosophy, Lund University, 1980-1988; Associate Professor, Philosophy of Science, Umeå University, 1975-1977; Various, Philosophy, Lund University, 1970-1980.

**Visiting Academic or Research Appointments:** Visiting Professor, Cà Foscari University of Venice, 2005; Visiting Professor, UCSD, 2005; Scholar, British Academy, 2005; Visiting Scholar, CREA, Paris, 1999; Visiting Professor, Rome University La Sapienza, 1995; Visiting Professor, École Normale Superieur, Cachan, 1992; Visiting Professor, Universidad de Buenos Aires, 1990; Visiting Researcher, Cognitive Science, Swedish Council for Research in Humanities and Social Sciences, 1988-1994; Visiting Fellow, Australian National University, 1986-1987; Visiting Professor, Auckland University, 1986; Visiting Scholar, Stanford University, 1983-1984; Visiting Fellow, Princeton University, 1973-1974.

**Research Profile:** Previous research focused on philosophy of science, decision theory, belief revision and nonmonotonic reasoning. Main current research interests are concept formation (using conceptual spaces based on geometrical and topological models), cognitive semantics, models of knowledge and information (including the relation between natural human information codes and computer codes), and the evolution of cognition.

**Main Publications:**

1. **Knowledge in Flux: Modeling the Dynamics of Epistemic States**, Bradford Books, MIT Press, 1988.

2. **Blotta Tanken**, Nya Doxa, 1992.

3. **Fängslande information**, Natur och Kultur, 1996 (second enlarged edition 2003).

4. **Conceptual Spaces**, Bradford Books, MIT Press, 2000.

5. **Hur Homo blev sapiens: Om tänkandets evolution**, Nya Doxa, 2000. Revised English version published as **How Homo Became Sapiens: On the Evolution of Thinking**, Oxford University Press, 2003.

6. **The Dynamics of Thought**, Springer Verlag, 2005.

*Work in Progress*

7. 2007j "De farligaste orden" (The most dangerous words), manuscript.

8. 2007i (together with Massimo Warglien) "Semantics, conceptual spaces and the meeting of minds", manuscript.

9. 2007h "'We and they", to appear in **Frameworks**.

10. 2007g "'Grottan och den andra världen" (The cave and the other world), manuscript.

A full list of publications is available at his home page http://www.lucs.lu.se/People/Peter.Gardenfors/

**Service to the Profession:** Editor, *Journal of Logic, Language and Information*, 1992-1995; Editor, *Theoria* 1978-1986.

**Honours and Awards:** Member, Leopoldina Deutsche Akademie für Naturforscher, 2004-; Electronic Festschrift, Spinning Ideas, on my 50th birthday; Member, Academia Europaea, 1999-; Senior Individual Grant, Swedish Foundation for Strategic Research, 1997; Member, Royal Swedish Academy of Letters, History and Antiquities, 1996-; Rausing Prize, Humanities, 1996.

## GELFOND, Michael

**Specialties:** Computer science, knowledge representation and reasoning, logic programming, mathematical logic.

**Born:** November 1945 in St. Petersburg, Russia.

**Educated:** Institute of Mathematics of the Academy of Sciences, PhD Mathematics, St. Petersburg, 1975; St. Petersburg University, Russia, MS Mathematics, 1968.

**Dissertation:** *Classes Of Formulae Of Classical Analysis Compatible With Constructive Interpretation* (in Russian); supervisor, Nikolai Alexandrovich Shanin.

**Regular Academic or Research Appointments:** PROFESSOR, COMPUTER SCIENCE, TEXAS TECH UNIVERSITY, 2000–; Professor, Computer Science, University of Texas at El Paso, 1992–1999; Associate Professor, Computer Science,

University of Texas at El Paso, 1987–1992; Assistant Professor, Computer Science, University of Texas at El Paso, 1984–1987; Assistant Professor, Mathematics, University of Texas at El Paso, 1981–1984; Lecturer, Department of Mathematics, St. Petersburg School of Business Administration and Economics, Russia, 1974-1976.

**Visiting Academic or Research Appointments:** Visiting Assistant Professor of Mathematics, University of Texas at El Paso, 1980–1981.

**Research Profile:** Michael Gelfond has made substantial contributions to work on formal knowledge representation and inference related to commonsense reasoning. He is the co-inventor (with Vladimir Lifshitz) of the Stable Model Semantics of logic programs with default negation, and contributed to the design of Answer Set Prolog (the language of logic programs with default and explicit negations and epistemic disjunction) and to the Answer Set Programming Paradigm, which is founded on this language. His work also significantly influenced developments in the theory of action and change, and helped to explain the relationship between various non-monotonic logics. Gelfond is noted for applications of his theoretical work to building software systems. This includes a decision support system for the space shuttle controllers, known as USA-Advisor. The system, written mostly in Answer Set Prolog, is capable of performing non-trivial and useful planning and diagnostics tasks. In his earlier work Gelfond used mathematical logic to investigate the relationship between classical and constructive analysis.

**Main Publications:**

1. M. Gelfond. "On Stratified Autoepistemic Theories". Proceedings of Sixth National Conference on Artificial Intelligence, 207 - 212, Seattle, July 1987.
2. M. Gelfond, V. Lifschitz, "The Stable Model Semantics for Logic Programs", Proceedings of the Fifth International Conference and Symposium on Logic Programming, 1070-1080, Seattle, August 1988.
3. M. Gelfond, V. Lifschitz, "Classical Negation in Logic Programs and Deductive Databases", New Generation Computing, vol 9, Nos. 3,4 pp. 365 – 387, 1991.
4. M. Gelfond, V. Lifschitz, A. Rabinov, "What are the Limitations of the Situation Calculus?", Automated Reasoning, Essays in Honor of Woody Bledsoe, Edited by S. Boyer, Kluwer Academic Publishers, 1991, pp. 167-181.
5. M. Gelfond and V. Lifschitz, "Representing Actions and Change by Logic Programs", the Journal of Logic Programming, vol. 17, Num. 2,3,4, pp. 301–323, 1993.
6. C. Baral, M. Gelfond, "Reasoning about effects of concurrent actions", Journal of Logic Programming, vol. 31, Num. 1,2 and 3, pp. 85-119, 1997.
7. M. Gelfond. Representing Knowledge in A-Prolog, volume 2408 of Computational Logic: Logic Programming and Beyond, Essays in Honour of Robert A. Kowalski, Part II, pages 413-451. Springer-Verlag, Berlin, 2002.
8. M. Balduccini and M. Gelfond. Diagnostic reasoning with A-Prolog. Theory and Practice of Logic Programming, 3(4-5):425-461, Jul 2003.
9. C. Baral, M. Gelfond, and N. Rushton. Probabilistic reasoning with answer sets. The Proceedings of the 7th International Conference on Logic Programming and Nonmonotonic Reasoning, pp 21-33, 2004.
10. M. Balduccini, M. Gelfond, and M. Nogueira. Answer Set Based Design of Knowledge Systems. Annals of Mathematics and Artificial Intelligence, 47(1-2): 183–219 (2006).
11. M. Gelfond, A class of theorems with valid constructive counterparts, Lecture Notes in Mathematics, vol 873, Constructive Mathematics, 314 - 321, 1981.

*Work in Progress*

12. Continuing work on combining non-monotonic logical and probabilistic reasoning.
13. Design and implementation of inference engines combining the satisfiability and constraint solving methods.
14. Axiomatization of simple commonsense domains and their applications to question answering from natural language texts.

More publications can be found on Gelfond's home page http://www.cs.ttu.edu/~mgelfond/

**Service to the Profession:** Area Editor, International Journal of Logic Programming, 1994-2000; Area Editor, International Journal of Theory and Practice of Logic Programming, 2001-; Executive Editor, Journal of Logic and Computation.

**Teaching:** Gelfond helped to develop successful research groups in the area of logic programming and non-monotonic reasoning at the University of Texas at El Paso and at Texas Tech University. He has graduated five PhD students.

**Vision Statement:** My primary research objective is to help create mathematical models of commonsense reasoning.

The advent of Computer Science and Artificial Intelligence has brought new and important challenges to traditional fields of Logic. Some of the most interesting ones are related to a logic based approach to programming, in which knowledge relevant to a given domain is captured in a set of formal axioms in an appropriate logical language. Computational problems about the domain can then be formulated as sentences in the same language, and be analyzed by general purpose inference algorithms.

In order to succeed in non-mathematical domains, this approach requires further research in several related directions. First, it requires careful analysis of our every day commonsense knowledge and reasoning. This analysis, which began taking shape in the 1970's, led to ground-breaking research on nonmonotonic logics and logic programming. The second direction is the development and careful mathematical investigation of existing logics and reasoning algorithms. Finally, this must be accompanied by an effort to axiomatize various non-mathematical commonsense domains, and by the use of these axiomatizations for building novel software systems. The later will help us to discover new questions and to better evaluate existing theories.

**Honours and Awards:** Fellow, American Association for Artificial Intelligence; Elected

Member, European Academy of Sciences; International Association of Logic Programming, recognized as the writer of the most influential paper in twenty years, together with Vladimir Lifschitz, 2004; Distinguished Achievement Award in Research, University of Texas at El Paso, 1994.

The symposium honoring Michael Gelfond on the occasion of his 50th birthday was held at the University of Texas in El Paso in 1995. The selected papers from the symposium, "Logic Programming, Nonmonotonic Reasoning, and Reasoning about Actions" were published in the Annals of Mathematics and artificial Intelligence.

# GHILARDI, Silvio

**Specialties:** Mathematical logic, modal logic, categorical and algebraic logic, automated reasoning.

**Born:** 12 May 1958 in Alzano Lombardo, Bergamo, Italy.

**Educated:** Università degli Studi di Milano, Degree in Philosophy 1982; Università degli Studi di Milano, PhD in Mathematics 1990.

**Dissertation:** *Dissertation: Modalities and Categories*; supervisor: Gian Carlo Meloni.

**Regular Academic or Research Appointments:** FULL PROFESSOR IN LOGIC AND PHILOSOPHY OF SCIENCE, UNIVERSITA' DEGLI STUDI DI MILANO, 2002–; Associate Professor in Mathematical Logic, Università degli Studi di Milano, 1998-2002; Researcher in Algebra and Geometry, Università degli Studi di Milano, 1991-1998.

**Visiting Academic or Research Appointments:** Research Fellow, Mc Gill University, Montreal, Canada, 1992-1993; Visiting Fellow, JAIST Kanazawa, Spring 1994.

**Research Profile:** Silvio Ghilardi worked in different fields of mathematical logic.

He obtained a series of incompleteness results concerning Kripke semantics of quantified modal logics and studied more powerful alternative semantics, mostly suggested by category-theoretic frameworks like presheaves and sheaves. He found (in a series of joint papers with G. Meloni) a simple axiomatization for counterparts and for topological semantics, by carefully analyzing the interplay of intensional operators and substitutions.

In algebraic logic, he analyzed the structure of finitely presented algebras, in connection both to normal forms of formulae and to differentiatedness of finite models. He gave important contributions (in a series of joint papers and in a joint book with M. Zawa-dowski) to uniform interpolation theory, by analyzing both category- and model-theoretic implications of the existence of uniform interpolants.

In unification theory, he proposed a new conceptual approach and proved finitarity of unification type for many common propositional logics (among them: intuitionistic logic, modal logics K4, S4, GL) : armed by this deep result, he gave a new neat solution to the admissibility problem for inference rules in these logics.

Since 2002, S. Ghilardi has been interested also in automated reasoning, in particular in techniques for modular combinination of decision procedures. His main contribution in this area is an extension of Nelson-Oppen combination method to the case of non disjoint signatures. A similar extension, involving model-theoretic techniques, was used (in a joint paper with F. Baader and C. Tinelli) to produce a strong decidability transfer result for combined word problems: such a result entails a general decidability transfer result for fusion of (non necessarily normal) modal logics. A further powerful extension of Nelson-Oppen method beyond the first-order case has been recently proposed in his joint work with E. Nicolini and D. Zucchelli.

**Main Publications:**

1. (with G.Meloni) *Modal and tense predicate logic: models in presheaves and categorical conceptualization*, in F.Borceux (ed.) "Categorical algebra and its applications, Proceedings Louvain-la-Neuve 1987", Lecture Notes in Mathe-matics, 1248, Springer, pp.130-142, (1988).

2. *Incompleteness results in Kripke semantics*, Journal of Symbolic Logic, 56, pp. 517-538, (1991).

3. *An algebraic theory of normal forms*, Annals of Pure and Applied Logic, 71, pp.189-245, (1995).

4. *Irreducible models and definable embeddings*, in Gabbay D. - Csirmaz L. - De Rijke M. (eds.) "Logic Colloquium '92 (Proceedings)", Studies in Logic, Language and Information, CSLI Publications, pp. 95-213, (1995).

5. (with G. Meloni) *Relational and partial variable sets and basic predicate logic*, Journal of Symbolic Logic, 61, pp.843-872, (1996).

6. (with G.Meloni) *Constructive canonicity in non-classical logics*, Annals of Pure and Applied Logic, 86, pp.1-32, (1997).

7. *Unification in intuitionistic logic*, Journal of Symbolic Logic, 64, pp.859-880, (1999).

8. *Best solving modal equations*, Annals of Pure and Applied Logic, 102, pp.183-198, (2000).

9. *Substitution, quantifiers and identity in modal logic*, in Hieke A. - Morscher E. (eds.) "New Essays in Free Logic (in honour of K. Lambert)", Applied Logic Series, vol.23, pp.87-115, Kluwer, (2001).

10. (with M. Zawadowski) *Sheaves, Games and Model Completions* (a categorical approach to non classical propositional logics), Trends in Logic Series, Kluwer, (2002).

11. (with C. Fiorentini) *Combining Word Problems Through Rewriting in Categories with Products*, Theoretical Computer Science, 294, pp. 103-149 (2003).

12. *Unification, Finite Duality and Projectivity in Varieties of Heyting Algebras*, Annals of Pure and Applied Logic, 127, 1-3, pp.99-115, (2004).

13. *Model Theoretic Methods in Combined Constraint Satisfiability*, Journal of Automated Reasoning, 33, pp.221-249, (2004).

14. (with L. Sacchetti) *Filtering Unification and Most General Unifiers in Modal Logic*, Journal of Symbolic Logic, 69, pp.879-906 (2004).

15. (with F. Baader) *Connecting many-sorted theories*, in R. Nieuwenhuis (ed.) "Proceedings of the 20-th Conference on Automated Deduction (CADE-20)", Lecture Notes inArtificial Intelligence, Springer, 3632, pp. 278-294, (2005).

16. (with C. Lutz and F. Wolter ) *Did I damage my Ontology ?*, Proceedings of Principles of Knowledge Representation and Reasoning, KR 06, (2006).

17. (with T. Brauner) *First-Order Modal Logic*, in P. Blackburn, J. van Benthem, F. Wolter (eds.) "Handbook of Modal Logic", Elsevier, (2006).

18. (with F. Baader and C. Tinelli) *A New Combination Procedure for the Word Problem that generalizes Fusion Decidability in Modal Logic*, Information and Computation, 204:10, 1413–1452, 2006.

19. (with G. Bezhanishvili) *An Algebraic Approach to Subframe Logics. Intuitionistic Case*, Annals of Pure and Applied Logic, 147, 84–100, 2007.

20. (with E. Nicolini and D. Zucchelli) *A Comprehensive Combination Framework*, ACM Transactions on Computational Logic, 9:2, March 2008.

A full list of publications is available at Ghilardi's home page http://homes.dsi.unimi.it/~ghilardi/

*Work in Progress*

21. Further combination results involving intensional logics, higher order fragments, software verification theories, and model-checking (this is part of a large joint project with E. Nicolini, S. Ranise, and D. Zucchelli).

22. Algebraic analysis of results concerning subframe logics, by means of nuclei on finitary algebraic structures (joint project, partially already published, with G. Bezhanishvili).

23. Applications of uniform interpolants to conservativity problems in modal and description logics (this project continues joint recently published work with C. Lutz, F. Wolter, M. Zakharyaschev).

24. Unification theory for modal logic K (this project involves a series of people including, among possibly others, V. Goranko and M. Zakharyaschev).

25. Axiomatization issues for counterpart and topological frames in quantified modal logic.

**Teaching:** Silvio Ghilardi has been teaching, since 1996, regular courses in Logic for both undergraduate, master and PhD students in mathematics and in computer science, at the Università degli Studi di Milano. He also gave undergraduate courses in mathematics and in natural language semantics. He is currently supervising two PhD students.

**Vision Statement:** I feel that logic should pursue its intrinsic multidisciplinary connections, by contributing to the development of mathematics, of linguistics and of computer science. The contribution of logic should concern, besides foundational aspects, concrete emerging applications, where our discipline can bring new insight, new theoretically sophisticated conceptualizations, and powerful tools. Recent and current developments of logic in computer science (for instance, in formal methods for hardware and software verification) can be considered as a concrete paradigm for the kind of advance I wish our discipline is going to attain in the future.

# GILMORE, Paul C.

**Specialties:** Logical foundations of mathematics and recursive function theory, nominalist interpretations of higher order logic.

**Born:** 5 December 1925 in Lethbridge, Alberta, Canada.

**Educated:** University of Amsterdam, Dr of Mathematics, 1953; Cambridge University, MA, 1955; Cambridge University, BA Honours Mathematics, 1951; University of British Columbia, BA Honours Mathematics and Physics, 1949.

**Dissertation:** *Griss' criticism of intuitionistic mathematics and its effect upon theories formalized within intuitionistic logic;* Supervisors, Evert W. Beth & Arend Heyting.

**Regular Academic or Research Appointments:** PROFESSOR EMERITUS, DEPARTMENT OF COMPUTER SCIENCE, UNIVERSITY OF BRITISH

COLUMBIA, CANADA, 1989–; Professor, Dept. Computer Science, Univ. B.C., 1984-1989; Head, Dept. Computer Science, Univ. B.C., 1977-1984; Staff Mathematician & Manager, Thomas J. Watson Research Center, Yorktown Heights N.Y., 1958-1971, 1972-1974, 1975-1977; Assistant to IBM Vice-President & Chief Scientist, IBM Corporate HQ, 1974-1975; Visiting Professor, Dept. Mathematics, Univ. B.C., 1971-1972; Adjunct Professor, Dept. of Math. & Statistics, Columbia University, 1972-1977; Adjunct Professor, School of Eng. & Ap.Sci., Columbia University, 1966-1968; Assistant Professor Mathematics, Penn. State Univ., 1955-1958; National Research Council Post-doctoral Fellow, Dept. Mathematics, Univ. Toronto, 1953-1955.

**Research Profile:** Alone and in collaborations Gilmore has made contributions to pure and applied mathematics. He is a co-author with Abraham Robinson of what some regard as the first paper in non-standard analysis. He designed and implemented one of the first computer programs for first order theorem-proving and described how computer supported models of a theory can be used to assist in computer based theorem proving for the theory. With Ralph Gomory and Alan Hoffman he co-authored papers that contributed to both combinatorics and industrial applications of mathematics. As Chair of the Steering Committee of CDNnet, 1984-88, a Canadian based international e-mail network, he was instrumental in the testing of the EAN e-mail software of Gerald Neufeld. He has revived an interest in Carnap's distinction between the intensions and extensions of predicates by describing their roles in the design of databases for computers and in the logical foundations of analysis and recursive function theory. At the same time he has substantiated Sellar's explanation of the paradoxes of set theory and higher order logics as resulting from a confusion of the use and mention of predicate names.

**Main Publications:**

1. *Metamathematical Considerations on the Relative Irreducability of Polynomials*, Canadian Journal of Mathematics, 7, (1955), pp. 483-489, with Abraham Robinson.
2. *A Linear Programming Approach to the Cutting Stock Problem, I & II*, Journal of Operations Research, 9, No. 6 (Nov.-Dec. 1961), pp. 849-859; 11, No. 6, (Nov.-Dec. 1963), pp. 863-888, with Ralph Gomory,
3. *A Characterization of Comparability and Interval Graphs*, Canadian Journal of Mathematics, 15, (1964), pp. 539-548, with Alan Hoffman.
4. *A Solvable Case of the Traveling Salesman Problem*, Proc. National Academy of Science, 51, (Feb. 1964), pp. 178-181, with Ralph Gomory.
5. *An Examination of the Geometry Theory Machine*, Artificial Intelligence, Vol. 1, (1970), pp. 171-187.
6. *Combining unrestricted abstraction with universal quantification*, pp. 99-123 of "To H.B. Curry: Essays on combinatorial logic, lambda calculus and formalization", eds. J.P. Seldin & J.R. Hindley, Academic Press, 1980.
7. *An intensional type theory: Motivation and Cut-elimination*, Journal of Symbolic logic, vol. 66 (2001), pp. 383-400.
8. *Logicism Renewed: Logical Foundations for Mathematics and Computer Science*, Lecture Notes in Logic 23, the Association for Symbolic Logic, A K Peters, Ltd., Wellesley, Mass.

*Work in Progress*

9. *A Nominalist Motivated Intuitionist Type Theory*, work in progress.

**Service to the Profession:** Reviewer, Mathematical Reviews, 1980-; Associate Editor, INFOR, Canadian Journal of Operational Research and Information Processing, 1978-1984; Council Member, Canadian Mathematical Congress, 1977-1979; Membership Chair, Association for Symbolic Logic, 1965-1966; Representative, Division of Mathematical Sciences, National Research Council of USA, 1970-1975; Chair Program Committee, International Symposium on the Theory of Graphs, 1966.

**Teaching:** As Adjunct Professor at Columbia University he taught logic to undergraduates and graduates, emphasizing how mathematical logic can be used to guide database design and to define recursive functions; he supervised one PhD student. His teaching as Professor of Computer Science at UBC continued a similar emphasis on applications of mathematical logic; he supervised one PhD student and numerous MSc students.

**Vision Statement:** Abraham Robinson's non-standard analysis provided insight into the apparently conflicting views of Newton and Leibniz on the nature of real numbers: The two views could be seen to arise from two different, but equally acceptable, models of the recursively defined natural numbers. Since every recursively defined predicate may have non-standard models in the same sense that the natural numbers do, it may be worthwhile to develop a better understanding of these models, since it is ultimately models that guide applications, not formal theories. This study may be particularly rewarding since in a nominalist view of logic the truths of logic are analytic, resulting from natural language specifications for the correct use of predicate names.

**Honours and Awards:** Recognized by the IBM Center for Advanced Study as a "Canadian Pioneer in Computing" (2005); Lanchester Prize of the Operations Research Society of America for the outstanding paper on Operations Research, with Ralph Gomory, 1963.

# GIVANT, Steven Roger

**Specialties:** Mathematical logic, algebraic logic, universal algebra, model theory.

**Born:** 1 September 1943 in Berkeley, California, United States.

**Educated:** Dartmouth College, Ludwig-Maximilians Universität München, University of California at Berkeley, BA (Magna Cum Laude) 1967, MA 1969, PhD 1975.

**Dissertation:** *Universal classes categorical or free in power*; supervisor, Robert Vaught. (Unofficial advisor: Alfred Tarski).

**Regular Academic or Research Appointments:** PROFESSOR, MATHEMATICS AND COMPUTER SCIENCE, MILLS COLELGE, 1987–; Associate Professor, Mathematics and Computer Science, Mills College, 1981-1987; Assistant Professor, Mathematics and Computer Science, Mills College, 1975-1981; Assistant and Associate Research Mathematician, University of California at Berkeley, 1975-1981.

**Research Profile:** Givant has made important contributions to universal algebra/model theory, to algebraic logic, and to the foundations of mathematics. In universal algebra/model theory he gave a complete description (up to definitional equivalence) of all universal Horn classes categorical in some infinite power, and he showed that the assumption of a weak form of categoricity in one infinite power implies that form of categoricity in all infinite powers. This work eventually led to the description (by others) of equationally axiomatized theories with less than continuum many countable models (the solution of the Vaught conjecture for equational theories) and to the description of equationally axiomatized theories in which the number of finite models in a give power grows at a polynomially-bounded rate. Givant's main work in algebraic logic has been in the domain of relation algebras. One aspect has been the structural analysis and the construction of the basic algebraic building blocks of the models, the so-called "simple" relation algebras; the goal of this research is the reduction of the description of simple relation algebras to that of a very special subclass, the socalled "integral" relation algebras. Another aspect has been the development of general, easy-to-use methods for showing that equational theories of relation algebras are, or are not, decidable; these methods were applied by Givant and others to solve a number of open decision problems posed by Tarski. In a seminal contribution to the foundations of mathematics, Givant and Tarski showed that a very elegant variable-free equational formalism (based on the calculus of relations), with ten simple axioms and with the high school rule of replacing of equals by equals as the only rule of inference, provides an adequate framework for the development of all of classical mathematics. One consequence of this work is that there are variable-free subsystems of propositional logic that are adequate for the development of mathematics; another is that, in standard first-order developments of set theory, three-variables suffice for doing all of mathematics. In related work, Givant and Tarski proved that Peano arithmetic is definitionally equivalent to a Zermelo-like theory of finite sets of finite rank. They also gave an alternative axiomatization of this finite set theory that is very close in spirit to Peano's axiomatization of arithmetic.

**Main Publications:**

1. *Universal Horn classes categorical or free in power*, Annals of Mathematical Logic 15 (1978), 1-53.
2. *A representation theorem for universal Horn classes categorical in power*, Annals of Mathematical Logic 17 (1979), 91-116.
3. (with Blum, L.) *Increasing the participation of college women in fields that use mathematics*, American Mathematical Monthly 87 (1980), 785-793. Also appeared as: *Increasing the participation of college women in mathematics-related fields*, in: Women and minorities in science, Strategies for increasing participation (Sheila Humphreys, ed.), AAAS Selected Symposia Series, vol. 66, Praeger, 1982, pp. 119-137.
4. *The number of non-isomorphic denumerable models of certain universal Horn classes*, Algebra Universalis 13 (1981), 56-68.
5. (co-editor with McKenzie, R.N.) *Alfred Tarski: Collected papers*. Volumes I-IV, Contemporary Mathematicians, Birkhäuser Verlag, Basel, 1986.
6. *Bibliography of Alfred Tarski*, Journal of Symbolic Logic 51 (1986), 913-941.
7. *A portrait of Alfred Tarski*, Mathematical Intelligencer 13 (1991), 16-32. Czech translation: *Portrét Alfréda Tarského*, Pokroky Mathmatiky Fyziky & Astronomie 37 (1992), 185-205. Polish translation of an extended version of the article: *Alfred Tarski w kalejdoskopie impresji osobistych*, (coauthored with V. Huber-Dyson), Roczniki Polskiego Towarzystwa Matematycznego, Seria II: Wiadmosoci Matematyczne 32 (1996), 95-127.
8. (with Tarski, A.) *A formalization of set theory without variables*, Colloquium Publications, vol. 41, American Mathematical Society, Providence, R.I., 1987, xxii + 318 pp.

9. *The structure of relation algebras generated by relativizations*, Contemporary Mathematics, vol. 156, American Mathematical Society, Providence, R.I., 1994, xv + 134 pp.

10. (with Andréka, H. and Németi, I.) *The lattice of varieties of representable relation algebras*, Journal of Symbolic Logic 59 (1994), 631-661.

11. (with Shelah, S.) *Universal theories categorical in power and k-generated models*, Annals of Pure and Applied Logic 69 (1994), 27-51.

12. (with Andréka, H. and Németi, I.) *Decision problems for equational theories of relation algebras*, Memoirs of the American Mathematical Society, vol. 126, no. 604, American Mathematical Society, Providence, R.I., 1997, xiv + 126 pp.

13. (with Andréka, H., Mikulá, Sz., Németi, I., and Simon, A.) *Notions of density that imply representability in algebraic logic*, Annals of Pure and Applied Logic 91 (1998), 93-190

14. (with Halmos, P.) *Logic as algebra*, Dolciani Mathematical Expositions, no. 21, Mathematical Association of America, Washington, D.C., 1998, x + 141 pp.

15. *Unifying threads in Alfred Tarski's work*, Mathematical Intelligencer 21 (1999), 47-58.

16. (with Venema, Y.) *The preservation of Sahlqvist equations in completions of Boolean algebras with operators*, Algebra Universalis 41 (1999), 47-84.

17. *Universal classes of simple relation algebras*, Journal of Symbolic Logic 64 (1999), 575-589.

18. (with Tarski, A.) *Tarski's system of geometry*, Bulletin of Symbolic Logic 5 (1999), 175-214.

19. (with Andréka, H.) *Groups and algebras of binary relations*, Bulletin of Symbolic Logic 8 (2001), 38-64.

20. *Inequivalent representations of geometric relation algebras*, Journal of Symbolic Logic 68 (2002), 267-310.

21. (with Halmos, P) *Introduction to Boolean Algebras*. Undergraduate Texts in Mathematics, Springer Verlag, New York, 2008.

22. (with Andréka, H.) *Simple Relation Algebras*, Springer Verlag, to appear 2008.

Work in Progress

23. (with Tarski, A.) *The definitional equivalence of Peano arithmetic and a Zermelo-like theory of sets of finite rank* (paper).

24. (with Andréka, H.) *Groups and algebras of relations* (monograph).

25. (with Andréka, H. and Németi, I.) *The structure of cylindric algebras generated by relativizations* (monograph).

26. (with Andréka, H. and Németi, I.) *Complete and incomplete representations of algebras of logic* (paper).

27. (with Andréka, H.) *Homomorphisms between weak cylindric set algebras* (paper).

28. (with Andréka, H.) *A representation theorem for rectangularly dense algebras of logic* (paper).

29. (with Andréka, H.) *Funtionally dense relation algebras* (paper).

**Service to the Profession:** Co-editor (with R. McKenzie) of *Alfred Tarski: Collected Papers, Volumes I-IV*; Alfred Tarski Committee Founding Member, University of California at Berkeley; Co-founder (with L. Henkin) and Director, Summer Mathematics Institute for Undergraduate Women, Mills College.

**Teaching:** Givant is a gifted teacher at an undergraduate liberal arts college for women. He has the ability to reach out to students at very different levels of preparation, and to inspire their interest and develop their abilities in mathematics. Together with Lenore Blum, he created an innovative, nationally recognized pre-calculus program that aimed to prepare students for calculus in one semester, no matter what their backgrounds. Together with Leon Henkin, he founded and directed the Mills Summer Mathematics Institute, a summer mathematics program aimed at motivating and preparing talented undergraduate women to pursue advanced degrees in mathematics. The program, which was national in scope, became a model for several other collegiate-level mathematics programs for undergraduate women, including the Carleton and St. Olaf College's Summer Mathematics Program for Women, and the George Washington University Summer Program for Women in Mathematics.

**Vision Statement:** Most of logic has focused on the infinite, but I believe logical investigations involving finite structures will find the greatest applicability to the real world. In particular, the theory of finite relations, as it develops, will find many important applications, just as finite field theory and Boolean algebra have already found important applications in computer science.

**Honours and Awards:** Danforth Foundation Associateship (in recognition of outstanding contributions to the teaching profession); International Research and Exchange Scholar (Poland and Hungary).

# GOCHET, Paul

**Specialties:** History of contemporary logic, modal logic, epistemic logic, relevant logic.

**Born:** 21 March 1932 in Bressoux, Belgium.

**Educated:** University of Liège, PhD Philosophy, 1968; University of Brussels, BA, Philology, 1954; BA, Philosophy 1959.

**Dissertation:** *Esquisse d'une théorie nominaliste de la proposition*; supervisor, Philippe Devaux.

**Regular Academic or Research Appointments:**
PROFESSOR OF LOGIC AND EPISTEMOLOGY,

EMERITUS, UNIVERISTY OF LIEGE; Promoter, National Centre for Research in Logic, 1972-.

**Visiting Academic or Research Appointments:** Invited lecturer (course of logic for graduate students) University of Lille 3, Spring 2008; Guest speaker, 'Conférences Pierre Duhem', Paris, Opening Session, 2006; Visiting Professor, University of Tsukuba and Shizuoka, 2003; Visiting Professor, University of Puerto-Rico, 2001; Research Fellow, Australian National University, Canberra, Automated Reasoning Project, 1995; Visiting Francqui Leerstoel Professor, Ghent University, 1988; Research Fellow, Research School of Social Sciences, Australian National University, Canberra, 1984; Invited Lecturer, Collège de France, 1981; A.C.L.S. Research Fellow, Stanford and Berkeley, 1974-1975; Visit to the Department of Philosophy, Harvard University, National Foundation for Scientific Research (Belgium) Grant, 1971.

**Research Profile:** Following in the footsteps of his late professor Philippe Devaux, Paul Gochet has contributed to the introduction of Anglo-American philosophy into the French-speaking world. During the first part of his career, he devoted several books and papers to central issues in analytic philosophy with emphasis on the philosophy of logic. He translated three books: J. L. Austin's *Sense and Sensibilia*, W. V. Quine's *Word and Object* (with Prof. Joseph Dopp) and L. Linski's *Referring* (with Prof. Philippe Devaux and Prof. Suzanne Stern Gillet). During the second part of his career, he focused on formal approaches to natural language and wrote extensively on formal semantics and formal pragmatics within Richard Montague's theoretical framework, mainly with Prof. André Thayse (Philips Laboratory and Université of Louvain) and his collaborators. Concurrently, he moved to logic applied to Artificial Intelligence and worked in an international team chaired by Prof. Philippe Smets (University of Brussels). Since 1990, he has been working continuously on logic with Professor Pascal Gribomont (University of Liège). With him and with Didier Rossetto, research engineer, he developed a new algorithm for relevant logic B. With Prof. Gribomont, he wrote a detailed systematic presentation of formal epistemic logic from 1960 to 2000. With him, he is now expanding the monograph to cover new developments.

**Main Publications:**

1. *Esquisse d'une théorie nominaliste de la proposition*, Paris, Armand Colin, 1972.
2. *Quine en perspective*, Paris, Flammarion, 1978.
3. *Outline of a Nominalist Theory of Propositions*, Dordrecht, Reidel, 1980 (trans. of (1)).
4. *Quine zur Diskussion*, Berlin, Ullstein, 1984 (trans. of (2)).
5. *Ascent to Truth*, Munich, 1986.
6. Contributions to Philippe Smets, et al.(eds.), *Non-Standard Logics for Automated Reasoning*, London, Academic Press, 1988.
7. with André Thayse, "Intensional Logic and natural language", "Montague Semantics", in A. Thayse, et al.(eds.), *From Modal Logic to Deductive Databases. Introducing a Logic Based Approach to Artificial Intelligence*, Chichester, Wiley, 1989, 55-163. French version, Paris, Bordas,1989, Russian translation, Moscow, Mir, 1998.
8. with Pascal Gribomont, *Logique, Méthodes pour l'informatique fondamentale*, vol. 1, Paris, Hermès, 1990, 1991, 1998.
9. with Pascal Gribomont, *Logique, Méthodes pour la vérification des programmes*, vol. 2, Paris, Hermès,1984.
10. with Pascal Gribomont and André Thayse, *Logique, Méthodes pour l'intelligence artificielle*, vol. 3, Paris, Hermès-Lavoisier, 2000.
11. with Eric Gillet, "On Professor Weingartner's Contribution to Epistemic Logic", in *Advances on Scientific Philosophy*, G. Schurz and G. Dorn (eds.), Amsterdam, Rodopi, 1991, 97-115.
12. "On Sir Alfred Ayer's Theory of Truth", in *The Philosophy of Sir Alfred Ayer*, E. Hahn (ed.) (The Library of Living Philosophers), La Salle Open Court, 1992, 201-20.
13. with Pascal Gribomont and Didier Rossetto, "Algorithms for Relevant Logic", *Logique et Analyse*, 150-51-152, 1995, reprinted in *Logic, Thought and Action*, D. Vanderveken (ed.), Dordrecht, Springer, 2005, 479-96.
14. with E. Gillet, "A new approach to Logical Omniscience", in *Calculemos... Matematicas y Libertad*, J. Etchevarria, et al. (eds), Madrid, Editorial Trotta, 1996, 321-32.
15. with E. Gillet, "Quantified Modal Logic, Dynamic Semantics and S5", *Dialectica* 53, 1999, 243-52.
16. "Quantifiers, Being, and Canonical Notation", in *A Companion to Philosophical Logic*, D. Jacquette (ed.), Oxford, Blackwell, 2002, 265-80.
17. "The Dynamic Turn in Epistemic Logic", in *Knowledge and Belief. Wissen und Glauben* W. Loffler and P. Weingartner, (eds.), Vienna, öbvethpt, 2004, 129-34.
18. "Hybrid Logic, its theoretical and practical significance", in *Proceedings of the 2003 MLG meeting at Shizuoka*, Japan, 2004, 6-9.
19. "Formal Philosophy', in *Masses of Formal Philosophy*, V. Hendricks & John Symons, (eds.), Automatic Press, 2006,15-25.
20. with Pascal Gribomont, "Epistemic Logic", in D. Gabbay & J. Woods, (eds.), *The Handbook of History of Logic* vol.7, Amsterdam, Nord Holland, Elsevier, 2006, 99-195.
21. "Un problème ouvert en épistémologie: la for-

malisation du savoir-faire" in Paul Gochet & Philippe de Rouilhan, *Logique épistémique & Philosophie des Mathématiques*, Thierry Martin (ed.), Paris, Vuibert Sciences, 2007, 3-37.

22. Foreward to the posthumous publication of Philippe Devaux, *La cosmologie de Whitehead*, vol.1, M.Weber (ed.), Louvain-la-Neuve, Editions Chromatika, 2007, I-VIII.

**Service to the Profession:** Président d'honneur, Société de Philosophie analytique (SOPHA), Paris, 1997-2003; President, European Association for Logic, Language and Information (FOLLI), Amsterdam, 1999-2002; Member, 19th Selection Committee of the National Foundation for Scientific Research, Brussels, 1983-1993; Co-director, Institut des Hautes Etudes de Belgique, Brussels, 1982-; Secretary, National Centre for Research in Logic, 1972-1988; Editorial Board Member, *Knowledge, Rationality and Action*; Editorial Board Member, *Logique et Analyse*; Editorial Board Member, *Dialectica*; Editorial Board Member, *Grazer Philosophische Studien*; Editorial Board Member, *Philosophiques*; Editorial Board Member, *Revue internationale de Philosophie*.

**Teaching:** Paul Gochet promoted the teaching of logic both in the Department of Philosophy (Faculty of Philosophy and Letters) and in the Department of Computer Science (Faculty of Applied Sciences) at the University of Liège. With philosophers from the Universities of Hull and Murcia, he launched an Erasmus exchange program among the Departments of Philosophy of the three institutions.

A *Festschrift* entitled *Logique en perspective, Mélanges offerts à Paul Gochet* prepared by 27 of his colleagues and pupils appeared in 2000 (François Beets and Eric Gillet (eds.), Brussels, Ousia).

**Vision Statement:** The same problems are sometimes investigated in several areas (logic, linguistics, A. I.). When the solutions are compared, stresses and strains appear. Questions, e.g. 'How can indefeasible semantics be combined with defeasible pragmatics?' call for an integrated approach. This gives rise to a bottom-up interdisciplinarity which is often productive.

**Honours and Awards:** Doctor Honoris Causa Univesity of Nancy 2, 2007; Member, Royal Academy of Belgium; Member, International Institute of Philosophy; Member, Académie internationale de la Philosophie des Sciences.

## GOLDBLATT, Robert Ian

**Specialties:** Algebraic logic, model theory of modal and other intentional logics, conceptual foundations of mathematics.

**Born:** 21 April 1949 in Wellington, New Zealand.

**Educated:** Victoria University of Wellington, BA Honours (1st Class), 1970; PhD, 1974.

**Dissertation:** *Metamathematics of Modal Logic*. Supervisor: M. J. Cresswell.

**Regular Academic or Research Appointments:** PROFESSOR OF PURE MATHEMATICS, VICTORIA UNIVERSITY OF WELLINGTON, 1985–; Personal Chair, Victoria University of Wellington, 1981-; Reader, Victoria University of Wellington, 1979-1981; Lecturer, Victoria University of Wellington, 1975-1981; Junior Lecturer, Victoria University of Wellington, 1971-1974; Chairperson, Department of Mathematics 1987-1991; Head, School of Mathematical and Computing Sciences, 1996-1998.

**Visiting Academic or Research Appointments:** Visiting Fellow, Institute For Advanced Study, Bologna University, 2002; Visiting Fellow, Japan Advanced Institute for Science and Technology, 2001; Visiting Fellow, Australian National University, 1992; Fulbright Senior Scholar and Visiting Professor, Stanford University, 1985-1987; Exchange Professor, University of Auckland, 1984; Nuffield Foundation Fellow, Oxford University, 1977-1978; Visiting Scientist, Simon Fraser University, 1976–1977.

**Research Profile:** Rob Goldblatt constructed a category-theoretic duality between relational models (Kripke frames) and algebraic models of modal logics, providing the notion of a "descriptive" frame as one isomorphic to its double dual. He generalized this to a theory of duality between general relational structures and Boolean algebras with operators, identifying intimate relationships between the first-order logic of structures and the equational logic of algebras. This methodology has been widely adopted in studies of definability and axiomatisability in non-classical logic.

He has written about the historical origins of mathematical semantics for modal logics, and has developed model theories and axiomatisations for various systems, including: the modal interpretation of Grothendieck topologies, relational semantics for the logic of ortholattices, the temporal logic of special relativity theory, the provability interpretation of intuitionistic logic, aspects of the dynamic logic of programs, and infinitary propositional proof theories. He has also resolved a number of technical questions of interest, including:

failure of the McKinsey axiom to be canonically valid or determined by any elementary class of frames; existence of certain canonically valid logics that are not elementarily determined; and failure of the orthomodular law of quantum logic to be first-order definable. More recently he has applied the mathematics of modality to the logic of coalgebras - structures that are used to model data structures and state-based transition systems in theoretical computer science. His current work concerns the semantics of quantification in relevant logic and other substructural logics.

His interest in the notion of orthogonality as a geometric primitive led him to develop complete axiomatisations of the first-order theories of Minkowski spacetime and other metric affine geometries over real-closed ordered fields.

Goldblatt has a strong interest in the explication and pedagogy of mathematical ideas. His text on category-theoretic logic (topos theory) continues to be republished after three decades. He has written a widely-used book on logics of time and computation, as well as books on nonstandard analysis, the geometry of orthogonality, and the axiomatisation of quantified dynamic logic.

**Main Publications:**

1. *Topoi : The Categorial Analysis of Logic.* Studies in Logic 98, North-Holland, 1979. Russian translation, 'Mir', Moscow, 1983. Revised expanded edition 1984. Internet edition, *Cornell University Library Historical Mathematical Monographs Collection*, http://historical.library.cornell.edu/math.html, 2002. Dover Publications paperback edition, 2006.
2. *Logics of Time and Computation.* Lecture Notes 7, Center for the Study of Language and Information, Stanford University, distributed by University of Chicago Press, 1987. Revised expanded edition 1992.
3. *Orthogonality and Spacetime Geometry.* Universitext, Springer-Verlag, 1987.
4. *Mathematics of Modality.* Lecture Notes 43, Center for the Study of Language and Information, Stanford University, distributed by University of Chicago Press, 1993.
5. *Lectures on the Hyperreals,* Graduate Texts in Mathematics 188, Springer-Verlag, 1998.
6. Semantic analysis of orthologic. *Journal of Philosophical Logic* 3, 1974, 19-35.
7. Axiomatic classes in propositional modal logic (with S.K. Thomason). In *Algebra and Logic*, ed. by J.N. Crossley, Lecture Notes in Mathematics 450, Springer-Verlag, 1975, 163-173.
8. Metamathematics of modal logic. *Reports on Mathematical Logic* 6, 41-78 (Part I) and 7, 21-52 (Part II), 1976.
9. Diodorean modality in Minkowski spacetime. *Studia Logica* 39, 1980, 219-236.
10. Orthomodularity is not elementary. *The Journal of Symbolic Logic* 49, 1984, 401-404.
11. Varieties of complex algebras. *Annals of Pure and Applied Logic* 44, 1989, 173-242.
12. First-Order Spacetime Geometry. In *Logic, Methodology, and Philosophy of Science VIII*, J.E. Fenstad et al. (eds.), Studies in Logic 126, North-Holland, 1989.
13. The McKinsey axiom is not canonical. *The Journal of Symbolic Logic* 56, 1991, 554-562.
14. Elementary generation and canonicity for varieties of Boolean algebras with operators. *Algebra Universalis* 34, 1995, 551-607.
15. Mathematical Modal Logic: A View of its Evolution. *Journal of Applied Logic* 1, nos. 5-6, 2003, pp 309-392. Revised version to appear in *The Handbook of the History of Logic*, vol, 7, Elsevier.
16. Equational Logic of Polynomial Coalgebras. In *Advances in Modal Logic* 4, Philippe Balbiani, Nobu-Yuki Suzuki, Frank Wolter, and Michael Zakharyaschev, editors. King's College Publications, King's College London, 2003, 149-184.
17. Observational Ultrapowers for Polynomial Coalgebras. *Annals of Pure and Applied Logic* 123, 2003, 235-290.
18. . Erdos Graphs Resolve Fine's Canonicity Problem (with Ian Hodkinson and Yde Venema). *The Bulletin of Symbolic Logic* 10, no. 2, 2004, pp 186-208.
19. On Canonical Modal Logics That Are Not Elementarily Determined (with Ian Hodkinson and Yde Venema). *Logique et Analyse* 181, 2003, 77-101.
20. An Alternative Semantics for Quantified Relevant Logic (with Edwin D. Mares). *The Journal of Symbolic Logic* 71, no. 1, 2006, 163-187.

Recent papers can be downloaded from Goldblatt's webpage at www.mcs.vuw.acc.nz/~rob

**Service to the Profession:** Editor, Journal of Symbolic Logic, 2001–2006, Coordinating Editor, Journal of Symbolic Logic, 2004-2005; Managing Editor, Studia Logica, 1993-; Advisory Board Member, Studia Logica, 1981-1993; Council Member, Association for Symbolic Logic, 2000-2006; Executive Committee Member, Association for Symbolic Logic, 2000-2002; Committee on Logic, Australasia, 1988–1994; President, New Zealand Mathematical Society, 1997-1999, Council member 1987-1993, 1996-2000; Council Member, Royal Society of New Zealand, 1991-1995, Marsden Fund research committee, 1998-2001; New Zealand representative to the International Mathematical Union 1997-2005.

**Teaching:** Goldblatt's undergradute course in logic is one of the most popular in the mathematics curriculum at Victoria University. He has also taught many Honours courses in mathematical logic, topology, classical and universal algebra, and category theory. His books on logics of time and computation and on nonstandard analysis evolved directly from notes for courses he developed on these subjects at Stanford and at Wellington.

Together with colleagues in philosophy and computer science he introduced a Logic and Computation major that has attracted a number of students to advanced study. Over the years he has supervised a dozen Masters dissertations and one PhD. Several of his students have gone on to doctoral study in the northern hemisphere.

**Vision Statement:** Model theory now blends geometry with algebra. The study of proofs is informed by category theory, and used in computer science and linguistics. Our view of computability and complexity is influenced by ideas about quantum physics, biology and randomness. There will be more such applications and interactions as the progress of science throws up other symbolic formalisms, as yet unheard of, for the logician to contemplate.

It would be a great triumph for the logical enterprise if the answer to some deep question (worth a million dollars?) was found by resolving it in some alternative mathematical universe, perhaps based on a nonclassical logic, and then logically transferring the solution to the standard world.

**Honours and Awards:** Honorary Life Member, NZ Mathematical Society, 2004; New Zealand Mathematical Society Award for Research, 1991; Fellow of the Royal Society of New Zealand, 1990; Research Medal of the New Zealand Association of Scientists, 1985.

# GOLDFARB, Warren

**Specialties:** Mathematical logic, history of modern logic, philosophy of logic and mathematics

**Born:** 25 August 1949 in New York City, New York, USA.

**Educated:** Harvard University, AB Philosophy and Mathematics, 1969; Harvard University, AM Philosophy, 1971; Harvard University, PhD Philosophy, 1975.

**Dissertation:** *On Decision Problems for Quantification Theory*; supervisor, Burton Dreben.

**Regular Academic or Research Appointments:** WALTER BEVERLY PEARSON PROFESSOR OF MODERN MATHEMATICS AND MATHEMATICAL LOGIC, HARVARD UNIVERSITY, 1995–; Professor, Philosophy, Harvard University, 1982–1995; Associate Professor, Philosophy, Harvard University, 1979–1982; Assistant Professor, Philosophy, Harvard University, 1975–1979; Junior Fellow, Harvard University Society of Fellows, 1971–1975.

**Visiting Academic or Research Appointments:** Visiting Professor, Philosophy, University of California, 1984.

**Research Profile:** Goldfarb's work in mathematical logic has focused on proof theory and decision problems. In the former, historical interest in Herbrand's writings were joined to projects on applications of Herbrand's Theorem in proof-theoretic contexts, leading, for example, to detailed analyses (for Peano Arithmetic) of omega-consistency arguments, and to results connecting Herbrand's Theorem with incompleteness and rates of growth of provably recursive functions. Goldfarb has also contributed to the area of provability logic. But most of his mathematical work has been on decision problems. In 1980, he showed the undecidability of the second-order unification problem. Primarily, he has concentrated on decision problems for fragments of first-order logic. Positive solvability results, for classes specified by fine-grained syntactical features, were collected in his 1979 book with Dreben (item 3 below); since then, his aim has been more to analyze the known results through newer techniques, in aid of understanding why the line between solvable and unsolvable falls where it does. Goldfarb has also proved a number of unsolvability results, the most significant of which was that of the (minimal) Gödel Class with Identity, i.e., prenex formulas of first-order predicate logic with identity with prefixes AAE. (This refutes a claim Gödel made in 1933, and is the final result in settling the decidability of all classes of quantification theory specified by quantifier prefix and vocabulary.)

Goldfarb's work on history of logic includes a critical edition of Herbrand's logical writings and co-editorship of volumes III - V of Gödel's *Collected Works*. He is currently editing a volume of Rudolf Carnap's writings on logic and philosophy of mathematics.

Starting in the late 1980s Goldfarb turned to more philosophical pursuits, and has written extensively on the philosophy of logic and mathematics, especially as it figures in the most important philosophers in the analytic tradition: Frege, Russell, Wittgenstein, Carnap, and Quine.

**Main Publications:**

1. (as editor) Jacques Herbrand, *Logical Writings*, Reidel and Harvard University Press (1971).

2. (with Thomas M. Scanlon) The omega-consistency of number theory via Herbrand's theorem, *J. Symbolic Logic* 39 (1974), 678-692.

3. (with Burton Dreben) *The Decision Problem: Solvable Classes of Quantificational Formulas* (with Burton Dreben), Addison-Wesley (1979).

4. Logic in the twenties: the nature of the quantifier, *J. Symbolic Logic* 44 (1979), 351-368.

5. The undecidability of the second-order unification problem, *Theoretical Computer Science* 13 (1981), 224-230.

6. The unsolvability of the Gödel class with identity, *J. Symbolic Logic* 49 (1984), 1237-1252.

7. Poincaré against the logicists, in *History and Philosophy of Modern Mathematics* (W. Aspray and P. Kitcher, eds.), University of Minnesota Press (1988), 61-81.

8. Russell's reasons for ramification, in *Rereading Russell: Essays on Bertrand Russell's Metaphysics and Epistemology* (C. W. Savage and C. A. Anderson, eds.), University of Minnesota Press (1989), 24-40.

9. Herbrand's Theorem and the incompleteness of arithmetic, *Iyyun* 39 (1990), 45-64.

10. (with Thomas Ricketts) Carnap and the philosophy of mathematics, in *Science and Subjectivity* (D. Bell and W. Vossenkuhl, eds.), Akademie Verlag Berlin (1992), 61-78.

11. Random models and solvable Skolem classes, *J. Symbolic Logic* 58 (1993), 908-914.

12. The philosophy of mathematics in early positivism, in *Origins of Logical Empiricism* (R.N. Giere and A.W. Richardson, eds.), University of Minnesota Press (1996), 213–230.

13. (as co-editor) Kurt Gödel. *Collected Works*,Vol. III, Unpublished essays and lectures. Vol. IV. Correspondence A-G. Vol. V. Correspondence H-Z, Oxford University Press, 1995-2003.

14. Frege's Conception of Logic, in *Futures Past: Reflections on the History and Nature of Analytic Philosophy* (J. Floyd and S. Shieh, eds), Oxford University Press, 25-41

15. *Deductive Logic*, Hackett Publishing, Indianapolis (2003).

16. On Gödel's Way In: The Influence of Rudolf Carnap, *Bull. of Symbolic Logic* 11 (2005), 185-193.

*Work in Progress*

17. (as editor) Rudolf Carnap, *Collected Works*, vol. III: logic and philosophy of mathematics, 1927-32.

**Service to the Profession:** Editorial Board Member, *Notre Dame Journal of Formal Logic*, 1992-; Editorial Board Member, Rudolf Carnap *Collected Works,* 2002-; Editorial Board Member, Kurt Gödel *Collected Works*, vols. III - V, 1990-2003; Executive Committee Member, Association for Symbolic Logic, 1982-1985.

**Teaching:** Goldfarb's main teaching in logic has been at the undergraduate level. He has had the primary responsibility for teaching introductory logic (for non-mathematicians) at Harvard since 1979, and in 2003 published a textbook for this purpose (item 15 above). He has taught an intermediate course focusing on Gödel's Theorems every other year since 1975. The latter course has seen a large number of students who have gone in to academic careers in philosophy, mathematics, and computer science. He has supervised eighteen senior honors theses in logic, two of which eventuated in publication. As professor in a philosophy department, Goldfarb has had no PhD's in mathematical logic, although he has had several in areas of philosophy that are more-or-less adjacent.

## GOLDSTERN, Martin

**Specialties:** Forcing, cardinal characteristics of the continuum, applications of set theory.

**Born:** 7 May 1963 in Wien (Vienna), Austria.

**Educated:** University of California, Berkeley, PhD, 1991; Vienna University of Technology, Dr.techn. 1986; Vienna University of Technology, Dipl.Ing. (M.A.) 1985.

**Dissertation:** Asymptotische Verteilung spezieller Folgen (Asymptotic distribution of special sequences), Vienna 1986. *Sets of reals and countable support iteration*, Berkely 1991; supervisor Robert F. Tichy (Vienna 1986). Jack Silver and Haim Judah (Berkeley 1991)

**Regular Academic or Research Appointments:** A. O. UNIVERSITY PREOFESSOR (ASSOCIATE PROFESSOR), VIENNA UNIVERSITY OF TECHNOLOGY, 1997–; Assistant professor, Vienna University of Technology, 1993-1997; Research Assistant, Free University of Berlin, 1992-1993.

**Visiting Academic or Research Appointments:** Visiting scholar, Free University of Berlin, 1996-1997; Visiting scholar, Carnegie Mellon University, 1995-1996; Postdoc, Bar Ilan University, Israel, 1991-1992.

**Research Profile:**
In set theory, Martin Goldstern is perhaps best known for his joint paper with Saharon Shelah introducing the "Bounded Proper Forcing Axiom" and the related notion of "Sigma-1-reflecting cardinals", and for his survey paper "Tools for your forcing construction" in which he reviews Shelah's "preservation theorems" for forcing iterations over countable supports. He uses and develops such preservation theorems also in his investigations of cardinal characteristics of the continuum (such as the smallest size of a set that does not have strong measure zero). In his more recent

work, Goldstern (again with Shelah as a frequent coauthor) has found applications of set-theoretical methods to problems originating in lattice theory and universal algebra.

**Main Publications:**

1. Martin Goldstern and Saharon Shelah: Clones from creatures. *Transactions AMS 357 (2005) 3525-3551.*
2. Martin Goldstern, Lattices, interpolation, and set theory. *Contributions to general algebra, 12 (Vienna, 1999),* 23–36, *Heyn, Klagenfurt,* 2000.
3. Martin Goldstern and Saharon Shelah, There are no infinite order polynomially complete lattices, after all. *Algebra universalis 42 (1999), 49-57.*
4. Martin Goldstern and Saharon Shelah, Order polynomially complete lattices must be large. *Algebra universalis 39 (1998), 197-209.*
5. Martin Goldstern and Haim Judah,The incompleteness phenomenon. A new course in mathematical logic. *A K Peters, Ltd., Natick, MA, 1995*
6. Martin Goldstern and Saharon Shelah: The bounded proper forcing axiom. *Journal of symbolic logic, 60 (1995), 58-73.*
7. Martin Goldstern, An application of Shoenfield's absoluteness theorem to the theory of uniform distribution. *Monatshefte für Mathematik 116 (1993), 237-243.*
8. Martin Goldstern and Haim Judah, Iteration of Souslin forcing, projective measurability and the Borel conjecture. *Israel Journal of Mathematics 78 (1992), 335-362.*
9. Martin Goldstern, Tools for your forcing construction. *Set theory of the reals, 305-360. Israel Math. Conference Proceedings 6, Bar-Ilan Univ., 1993.*
10. Martin Goldstern and Saharon Shelah, Ramsey ultrafilters and the reaping number – Con(r<u). *Annals of pure and applied logic 49 (1990), 121-142.*
*Work in Progress*
11. Martin Goldstern: Applications of logic in algebra: Clones. (A survey of results and open problems concerning questions from universal algebra that seem to need methods from logic, in particular set theory and model theory.)
12. Martin Goldstern and Saharon Shelah: All creatures large and small. (An extension of a previous result to uncountable cardinals.)

**Service to the Profession:**

Editorial Board Member, Contributions to Discrete Mathematics, 2005-; Conference Organizer, "Cantor's set theory" and "Cantor's set theory 2", Free University of Berlin, 1993, 1997; Local Organizing Committee Member, Logic Colloquium 2001, Vienna.

**Vision Statement:** I see a potential danger to the development of mathematical logic in the expanding rift between its two main directions, that can be vaguely described as "finitary/syntactic/proof theoretic" and "infinitary/semantic/model theoretic". On the other hand, the breadth of existing and ongoing research is also an opportunity for finding new connections between branches of logic, as well as connections or applications to other fields such as analysis, algebra and computer science.

## GONCHAROV, Sergei Savostyanovich.

**Specialties:** Mathematical logic, theory of algorithms, theory of models, constructive models, algebra and their applications in informatics.

**Born:** 24, September, 1951, Novosibirsk city, Russia.

**Educated:** Department of Mechanics and Mathematics of Novosibirsk State University 1973, corresponding member of the Russian Academy of Sciences 1997, Doctor degree of physics and mathematics 1981, Ph.D. Physics and Mathematics 1974.

**Dissertation:** thesis for Doctor degree is ¡¡Nonautoequivalent constructivisations¿¿ 1981, thesis for Ph. D. degree is ¡¡Constructive Boolean Algebras¿¿ 1974; supervisor, Yuri Leonidovich Ershov.

**Regular Academic or Research Appointments:** HEAD OF THE DEPARTMENNT OF MECHANICS AND MATHEMATICS OF NOVOSIBIRSK STATE UNIVERSITY, 1996–; HEAD OF THE LABORATORY OF INSTITUTE OF MATHEMATICS OF SB RAS 1992–; PROFESSOR OF THE CHAIR OF ALGEBRA AND LOGIC AT NOVOSIBIRSK STATE UNIVERSITY, 1983–; VICE-DIRECTOR OF THE INSTITUTE OF DISCRETE MATHEMATICS AND INFORMATICS; 1991–; research probationer, junior research worker, senior research worker, Institute of Mathematics (Siberian Branch of the Russian Academy of Sciences), 1973–1996; assistant, assistant professor, Novosibirsk State University, 1973–1983; student of the Department of Mechanics and Mathematics, Novosibirsk State University, 1968 – 1973.

**Visiting Academic or Research Appointments:** Visiting Researcher in Cornell University; 1988–1989, 1994. Visiting Professor in Alma-Ata University; 1980, 1990. Visiting Professor in Tehran University; 1997. Visiting Researcher in Notre Dame University; 1999, 2000, 2001, 2002, 2004, 2006. Visiting Professor in Wisconsin University; 1998. Visiting Researcher in Monash University, Australia, 1984. Visiting Researcher in Auckland University (New Zealand); 1999, 2001, 2003, 2004.

**Research Profile:** S. S. Goncharov is a famous specialist in the field of the theory of algorithms, the theory of models, algebra and their applications in informatics. He has contributed significantly to various aspects of algebra, logic and informatics having obtained outstanding results making him internationally recognized. He has published more than 100 research works.

The most important results were gained by S. Goncharov in the theory of algorithms and model theory. He has developed the theory of algorithmical dimensionality based on his own fundamental assumption about the existence of unstable models of finite algorithmical dimensionality. S. Goncharov has elaborated new strong methods of proof of the infinity of algorithmical dimensionality allowing to solve the problem of spectrum characterisation of algorithmical dimensionality for particular classes of models and algebraic systems; different types of reductions and their relationships were also investigated.

Significant contribution was made to the theory of decidable models with establishing the fundamental criterion of the decidability of homogeneous models. He solved the Morley problem and the problem of Peretyat'kin – Denisov on the base of this criterion. He also solved the problem of characterizing the axioms of the classes with strong epimorphisms and homomorphisms set up by academician A. I. Mal'tsev in 1961. The theory of constructive Boolean algebras was developed by S. Goncharov as well. Recent works under his supervision include the elaboration of such problems as group structure and their automorphism, subalgebra lattices, enrichment with ideals and subalgebras and many others. S. Goncharov studied nilpotent groups of finite algorithmical dimensionality and arrived at characterization of autostability of finite-rank nilpotent groups without torsion, Abel r-groups. In the field of the classical theory of algorithms he contributed fundamentally to the theory of computable enumerations; he developed a new method of constructing computable enumerations enabling to solve a series of problems about the number of Friedburg enumerations and about families with only positive as well as many others. In cooperation with American scientists R. Soare, P. Cholak and B. Khusainov, S. Goncharov obtained the solution of the old problem about autostability of finite constant enrichments of autostable models in 1995; together with B. Khusainov they solved the problem of 2-element spectrum with recursive T-rank; together with S. Badaev they solved the problem about the family with one-element Rogers semilattice with an untrivial embedding in 1996; he also solved the problem of existence of very constructive homogeneous extensions in 1996; and in cooperation with an Italian logician A. Sorbi he investigated Rogers semilattice of computable enumerations of arithmetic sets.

In 2003, together with S. Lempp and R. Solomon, he has obtained a characterization of autostable ordered Abelian groups. In 2004 – 2005, together with S. Badaev and A. Sorbi he has solved a series of questions on the countability of elementary theories of Rogers semilattices on each level of arithmetical hierarchy and on the triviality of intersections of Rogers semilattices for different levels of arithmetical hierarchy. Together with B. Khoussainov, he has constructed examples of countably categorical and of uncountably categorical theories for all levels of arithmetical hierarchy. In 2004, together with R. Shore, J. Khight, and some other american logicians he has studied algorithmic properties of Harrisson's structures in Kleene's notation system. In 2005, together with J. Knight, V. Harizanov, Ch. McCoy, and R. Miller, he has developed a general method to construct computable structures based upon the theories of computable, arithmetical, and hyperarithmetical numberings. He has also solved a series of questions on the relationship between definability and syntactical properties of computable models.

**Main Publications:**

1. Yu. Ershov, S. Goncharov, Constructive models, Consultants Burean, xii, New-York, 2000, 293 pp.
2. S. S. Goncharov, Countable Boolean algebras and Decidability, Handbook of Recursive Mathematics, Plenum Publishers, New York, 1997, 373 pp.
3. S. Goncharov, S. Badaev, On Rogers semilattices of families of arithmetic sets, Algebra and Logic, Plenum Publ.Corp., v. 40, N 5, 2001, 283-291.
4. S. Goncharov, S. A. Badaev, A. Sorbi , Completeness and universality of arithmetical numberings, Computability and models, Kluwer Academic/Plenum Publishers, 2003, 11-44.
5. P. Cholak, S. Goncharov, B. Khoussainov, R. Shore, Computably categorical structures and expansions by constans, Journal of Symbolic Logic, v.64, N 1, 1999, 13-37.
6. S. Goncharov, B. Khoussainov, Open problems in the theory of constructive algebraic systems, Contemporary Mathematics (AMS), v.257, 2000, 145-170.
7. S. Goncharov, V. Harizanov, J. F. Knight, R. A. Shore, $\Pi_1^1$-relations and paths through, The Journal of Symbolic Logic, v.69, N 2, 2004, 585-611.
8. S. Goncharov, V. Harizanov, J. F. Knight, Ch. F. D. McCoy, Relatively Hyperimmune Relations on Structures, Algebra and Logic, Plenum Publ.Corp, v.43, N 2, 2004, 94-101.
9. S. Goncharov, S. Lempp, R. Solomon, The computable dimension of ordered Abelian groups, Advances in Math., v. 175, N 1, 2003, 102-143.
10. S. Goncharov, V. Harizanov, M. Laskowski, S. Lempp, Ch. F. D. McCoy, Trivial, strongly minimal

theories are model complete after naming constants, Proceedings of AMS, v.195, 2003, 12 pp.

Full list of publication http://mmfd.nsu.ru

**Service to the Profession:** Corresponding Member, Russian Academy of Sciences,1997 – present; Member of Council of Association for Symbolic Logic, 2004-2006. Member European Academy of Sciences, 2002 – present; Full Member, International Academy of Sciences of Higher School, 1995 – present; Editor-in-Chief, journal ¡¡Vestnik NGU. Series: mathematics, mechanics, and theoretical computer science¿¿; Vice Editor-in-Chief, journal ¡¡Algebra and Logic¿¿; Associate editor, series of monographs ¡¡Siberian School of Algebra and Logic¿¿; Vice-chairman, Siberian Fund of Algebra and Logic; Member of Editorial Boards, journal ¡¡Siberian Journal of Mathematics¿¿, Member of Editorial Boards, ¡¡Matematichsekie Trudy¿¿, Member of Editorial Boards, ¡¡Siberian Advances in Mathematics¿¿; Member of Editorial Boards, ¡¡Computational Systems¿¿; Member of Editorial Boards, ¡¡Problems of specialized education¿¿; Supervisor, Leading Scientific School NSh-2112.2003.1, 2003 – 2006; Supervisor, Leading Scientific School NSh-4413.2006.1, 2006 – present; Supervisor and participant of Russian and international projects (grants) by Russian Foundation for Basic Research, INTAS, NSF (USA), NSF of China, and of Royal Scientific Society of New Zealand.

**Teaching:** S. Goncharov also actively participates in organization and evaluation of research as a member of Scientific Councils. He gives much attention to supervising his disciples. Among them there are 23 candidates (Ph.D.) and 8 doctors of science and professors. Some of them were awarded with medals and prizes and hold academic positions. His followers work at Institute of Mathematics, Novosibirsk State University, other colleges of Russia and CIN, as well as in the USA, Australia, New Zealand, Sweden.

**Honours and Awards:** Mal'sev prize of Russian Academy of Sciences, 1997; elector, Russian Academy of Sciences, 1991; Siberian Branch of Russian Academy of Sciences Prize, 1983; Silver badge of Algebra and Logic Seminar, 1982; Lenin Komsomol Prize on science and technology, 1976; Graduated with honours, Department of Mechanics and Mathematics of Novosibirsk State University, 1973.

S. Goncharov repetedly visits universities of the USA, Australia, Iran, Italy and other countries giving lectures and participating in research programs. He was the head of scientific groups working on the problems supported by well-known Russian and international funds (grants by the Russian Fund for Fundamental Research, PECO etc.) S. Goncharov was invited with plenary papers to many Russian and international conferences. Only in 1996 he participated in international conferences in Germany and Spain, in a seminar honoring A. Kurosh and a UNESCO congress. In 1997 he was invited as a plenary speaker to a Logic Colloquium in Lids (UK). S. Goncharov was a member of program committees of many all-union and international conferences. In 1997 he was a co-chairman of the International Conference on recursion theory in Kazan and a member of the program committee of LFCS-97. In 2002 S. Goncharov was invited with tutorial talk to Mathematicians Congress in San-Diego (USA), to the 8th Asian Conference in Logic, to Satellite Conference of International Mathematical Congress in Chuncin (China), to International INTAS Conference ¡¡Models and Computability¿¿ in Alma-Ata (Kazakhstan).In 2005 S. Goncharov was invited as a plenary speaker to a Logic Colloquium in Athens (Greece). Chairman of Programme Committee of the 9th Asian Conference 2005 in Logic.

## GOODMAN, Nicolas Daniels

**Specialties:** Mathematical logic, philosophy of mathematics, intuitionism and constructive mathematics, epistemic logic.

**Born:** 23 June 1940 in Berlin, Germany.

**Educated:** Stanford University, PhD Mathematics, 1968; Stanford University, MS Mathematics, 1963; Harvard College, BA Mathematics 1961.

**Dissertation:** *Intuitionistic Arithmetic as a Theory of Constructions*; supervisor, Dana Scott.

**Regular Academic or Research Appointments:** PROFESSOR, MATHEMATICS EMERITUS, STATE UNIVERSITY OF NEW YORK AT BUFFALO, 2005–; Professor, Mathematics, State University of New York at Buffalo, 1989-2005; Vice Provost for Undergraduate Education, State University of New York at Buffalo, 1993-2000; Associate Professor, Mathematics, State University of New York at Buffalo, 1973-1989; Assistant Professor, Mathematics, State University of New York at Buffalo, 1969-1973; Assistant Professor, Mathematics, University of Santa Clara, 1968-1969; Instructor, Mathematics, University of Illinois at Chicago Circle, 1965-1966.

**Research Profile:** Nicolas Goodman has made fundamental contributions to the foundations of intuitionistic logic and to the proof-theoretic analysis of theories formulated in epistemic logic. He has also written widely on the philosophy of mathematics, and especially on the metaphysical issues raised by constructive mathematics.

**Main Publications:**

1. *A theory of constructions equivalent to arithmetic,* in Intuitionism and Proof Theory, ed. A. Kino, J. Myhill, and R. E. Vesley, North-Holland Pub. Co, Amsterdam (1970), pp. 101 – 120.
2. (With John Myhill) *The formalization of Bishop's constructive mathematics,* in Toposes, Algebraic Geometry and Logic, ed. F. W. Lawvere, Springer-Verlag, Berlin (1972), pp. 83 – 96.
3. *A simplification of combinatory logic,* Journal of Symbolic Logic, v. 37(1972), pp. 225 – 246.
4. *The arithmetic theory of constructions,* in Cambridge Summer School in Mathematical Logic, Proceedings (1971), ed. A. R. D. Mathias and H. Rogers, Springer-Verlag, Berlin (1973), pp. 274 – 298.
5. *The faithfulness of the interpretation of arithmetic in the theory of constructions,* Journal of Symbolic Logic, v. 38(1973), pp. 453-459.
6. *The theory of the Goedel functionals,* Journal of Symbolic Logic, v. 41(1976), pp. 574-580.
7. *Relativized realizability in intuitionistic arithmetic of all finite types,* Journal of Symbolic Logic, v. 43(1978), pp. 23 – 44.
8. *The nonconstructive content of sentences of arithmetic,* Journal of Symbolic Logic, v. 43(1978), pp. 497 – 501.
9. (With John Myhill) *Choice implies excluded middle,* Zeitschrift fur Mathematische Logik and Grundlagen der Mathematik, v. 24(1978), p. 4611.
10. *Mathematics as an objective science,* American Mathematical Monthly, v. 86(1979), pp. 540 – 551. Reprinted in New Directions in the Philosophy of Mathematics, ed. T. Tymoczko, Birkhauser, 1985, pp. 79 – 94.
11. *The logic of contradiction,* Zeitschrift fur Mathematische Logik and Grundlagen der Mathematik, v. 27(1981), pp. 119 – 126.
12. *The experiential foundations of mathematical knowledge,* History and Philosophy of Logic, v. 2(1981), pp. 55 – 65.
13. *Reflections on Bishop's philosophy of mathematics,* in Constructive Mathematics, ed. F. Richman, Lecture Notes in Mathematics 873, Springer-Verlag, Berlin (1981), pp. 135 – 145. Reprinted in The Mathematical Intelligencer, v. 5(1983), pp. 61 – 68.
14. *Epistemic arithmetic is a conservative extension of intuitionistic arithmetic,* Journal of Symbolic Logic, v. 49(1984), pp. 192 – 203.
15. *A genuinely intensional set theory,* in Intensional Mathematics, ed. S. Shapiro, North Holland Pub. Co., 1985, pp. 63 – 80.
16. *The knowing mathematician,* Synthese, v. 60 (1984), pp. 21 – 38.
17. *Flagg realizability in arithmetic,* Journal of Symbolic Logic, v. 51(1986), pp. 387 – 392.
18. *Mathematics as natural science,* Journal of Symbolic Logic, v. 55(1990), pp. 182 – 193.
19. *Topological models of epistemic set theory,* Annals of Pure and Applied Logic, v. 46(1990), pp. 147 – 167.
20. *Modernizing the philosophy of mathematics,* Synthese, v. 88(1991), pp. 119 – 126. Reprinted in Essays in Humanistic Mathematics, ed. Alvin White, MAA (1993), pp. 63 – 66.

**Service to the Profession:** Member, U. S. National Committee for the International Union of the History and Philosophy of Science, 1986-1993.

**Teaching:** Goodman's PhD students included Julius Barbanel, Robert Flagg, Norollah Talebi, and Mai Tong. He also played an important role on the doctoral committees of a number of students both in Mathematics and in Philosophy, including Andrej Scedrov, and Stewart Shapiro, among others.

# GOTTLOB, Georg

**Specialties:** Finite model theory, complexity, nonmonotonic logic, Logic in AI, database theory.

**Born:** 30 June 1956 in Vienna, Austria.

**Educated:** TU Vienna, PhD, 1981; TU Vienna, MSc 1979 (Diplom-Ingenieur).

**Dissertation:** *Multi-Valued Logic – Structure and Application in Computer Science* (in German); supervisor Curt Christian.

**Regular Academic or Research Appointments:** PROFESSOR, COMPUTING SCIENCE, OXFORD UNIVERSITY, 2006–; Professor, Computer Science, TU Vienna, 1988-2005; Staff Researcher, Institute of Applied Mathematics CNR, Genoa, Italy, 1985-1988, Research Associate, Politecnico di Milano, Milan, Italy, 1982-1985; Research Scholar and Lecturer, Stanford University, 1985-1987; University Assistant, TU Vienna, 1981-1982.

**Visiting Academic or Research Appointments:** Visiting Professor, Université Paris VII, Denis Diderot, 2002; Visiting Professor and McKay Lecturer, University of California, Berkeley, 1999; Visiting Scientist, ETH Zurich, 1993; short term appointments as Visiting Professor (Professore a Contratto) at several universities in Italy, among which: University of Rome La Sapienza, University of Pisa, University of Calabria, University of Bari.

**Research Profile:** Gottlob has made numerous contributions to the logical and algorithmic foundations of non-monotonic reasoning, finite model theory, database theory, Web data processing, logic programming, constraint satisfaction, modal logic, automated theorem proving, and Artificial Intelligence at large. A few highlights of his research are given in the following.

Gottlob proved that reasoning with non-monotonic propositional logics, such as Default Logic and Autoepistemic Logic, is hard for complexity classes at the second level of the Polynomial-Time Hierarchy and thus presumably harder than reasoning in classical propositional logic (*J. of Logic and Computation*, 1992). With Th. Eiter, similar complexity results were found for almost all other forms of non-monotonic reasoning.

Gottlob also investigated the complexity and expressive power of many other non-classical logics or formalisms. For example, he showed that reasoning in the propositional modal logic proposed by Carnap in *Meaning and Necessity (1949)*, is complete for the complexity class NP[O(log n)] consisting of all problems that can be solved in polynomial time with a logarithmic number of queries to an oracle in NP (*J. of the ACM* 42:2, 1995). Moreover, Gottlob proved that first order logic with Henkin Quantifiers captures NP[O(log n)] over ordered finite structures, and proved a meta-theorem for analyzing the complexity and expressive power for other extensions of first order logic with generalized quantifiers (*Journal of Symbolic Logic* 62:2, 1997).

Gottlob was the initiator of the ongoing long-term research programme *Second Order Logic over Finite Structures* whose aim is to determine the complexity of evaluating formulas of quantificational prefix classes of second order logic (SO) over different types of finite structures, e.g., graphs, trees, strings. A complete characterization of existential SO over strings was obtained by Eiter, Gottlob, and Gurevich (*J. of the ACM*, 47:1, 2000). Existential SO over graphs was fully characterized by Gottlob, Kolaitis, and Schwentick (*J. of the ACM* 51:2, 2004).

Gottlob worked on finding tractable subclasses of NP-hard problems. The structure of the instances of problems, such as, conjunctive database queries (CQs), or constraint satisfaction can often be described via hypergraphs. Gottlob and his colleagues Leone and Scarcello introduced the concepts of *hypertree decomposition* and *hypertree width*, the latter being a new cyclicity measure of hypergraphs (*J. of Computer and System Sciences*, vol. 64:3, 2002). They showed that queries and constraint satisfaction problems of bounded hypertree width are polynomially solvable and actually highly parallelizable, since their evaluation is complete for the complexity class LOGCFL (*J. of the ACM* 48:3, 2001). The class of acyclic conjunctive database queries expresses precisely the guarded fragment of positive existential disjunctive first-order logic. Similar characterizations were obtained for CQs of hypertree-width k, and a game theoretic characterization was given (*J. of Computer and System Sciences*, 66:4, 2003).

Together with Christoph Koch and Reinhard Pichler, Gottlob proved that the well-known XPATH query language can be evaluated in polynomial time (*ACM Transactions on Database Systems* 30:2, 2005) and studied the precise complexity of various fragments of XPATH (*J. of the ACM* 52:2, 2005). With Koch, he laid the logical and algorithmic foundations of Web data extraction based on monadic second order logic over trees and proved that monadic datalog is exactly as expressive as monadic SO (*J. of the ACM* 51:1, 2004). With students and post-docs, Gottlob built the Web data extraction system *Lixto* that essentially uses monadic datalog as a core extraction language (*Intl. Conference on Very Large Databases* VLDB 2001, pp. 119-128). This system is now successfully used with many customers by the Lixto Software company (www.lixto.com).

**Main Publications:**

1. G. Gottlob. Complexity results for nonmonotonic logics. Journal of Logic and Computation, 2(3):397–425, June 1992.

2. Eiter, T; Gottlob, G. "The complexity of logic-based abduction." *Journal of the ACM* 42:1 pp. 3-42, 1995.

3. Gottlob, G. "Translating default logic into standard autoepistemic logic." *Journal of the ACM* 42:4 pp.711-740, 1995.

4. Gottlob, G. "Relativized logspace and generalized quantifiers over finite ordered structures." *Journal of Symbolic Logic* 62:2 pp.545-574, 1997.

5. Eiter, T; Gottlob, G. "On the expressiveness of frame satisfiability and fragments of second-order logic." *Journal of Symbolic Logic* 63:1 pp.73-82, 1998.

6. Gottlob, G. "NP trees and Carnap's modal logic." Journal of the ACM 42:2 pp.421-457, 1995.

7. Eiter, T; Gottlob, G; Gurevich, Y. "Existential second-order logic over strings." *Journal of the ACM* 47:1 pp.77-131, 2000.

8. Gottlob, G; Kolaitis, PG; Schwentick, T. "Existential second-order logic over graphs: Charting the tractability frontier", *Journal of the ACM* 51:12, pp.312-362, 2004

9. Gottlob, G; Leone, N; Scarcello, F. "Hypertree decompositions and tractable queries." *Journal of Computer and System Sciences* 64:3 pp.579-627, 2002.

10. Eiter, T; Gottlob, G; Makino, K. "New results

on monotone dualization and generating hypergraph transversals." *SIAM Journal on Computing* 32:2, pp. 514-537, 2003.

11. Gottlob, G; Koch, C; Schulz, KU. "Conjunctive queries over trees." *Journal of the ACM 53:2* pp.238-272, 2006.

12. Gottlob, G; Greco, G; Scarcello, F. "Pure Nash equilibria: Hard and easy games." *Journal of Artificial Intelligence Research* 24: 357-406, 2005.

13. Gottlob, G; Koch, C; Pichler, R., Segoufin, L. "The complexity of XPath query evaluation and XML typing." *Journal of the ACM* 52:2, pp. 284-335, 2005.

14. Gottlob, G; Koch, C; Pichler, R. "Efficient algorithms for processing XPath queries." *ACM Transactions on Database Systems* 30:2: pp.444-491, 2005.

15. Gottlob, G; Koch, C. "Monadic datalog and the expressive power of languages for Web information extraction." *Journal of the ACM* 51:1 pp. 74-113, 2004.

16. Gottlob, G; Leone, N; Scarcello, F. "Robbers, marshals, and guards: game theoretic and logical characterizations of hypertree width." Journal of Computer and System Sciences 66:4 pp.775-808, 2003.

17. Gottlob, G; Leone, N; Scarcello, F. "The complexity of acyclic conjunctive queries." *Journal of the ACM* 48:3 431-498, 2001.

18. Gottlob, G; Leone, N; Scarcello, F. "A comparison of structural CSP decomposition methods." *Artificial Intelligence* 124:2, pp. 243-282, 2000.

19. R. Baumgartner, S. Flesca,G. Gottlob. Visual Web Information Extraction with Lixto. Proceedings of the 27th International Conference on Very Large Data Bases (VLDB'01), Rome, September 2001, pp. 119-128, 2001.]

*Work in Progress*

20. Gottlob is currently working on logical data exchange problems:

21. G. Gottlob and A. Nash. Data Exchange – Computing Cores in Polynomial Time. Full version of the homonymous conference paper published at the 25. ACM SIGACT SIGMOD SIGART Symposium on Principles of Database Systems (PODS-06), June 26-28, 2006, Chicago, IL, pp. 40-49, 2006.

**Service to the Profession:** General Chair, ACM PODS 2005, 2006; Program Chair, IJCAI 2003; Program Chair, ACM PODS 2000; Area Editor, *Theory and Practice of Logic Programming*, 2000-2003; Editorial Board Member, *IEEE Transactions on Knowledge and Data Engineering*, 1999-2003; Co-Program Chair, CSL, 1998; Editorial Board Member, *Journal of Logic Programming*, 1997-2000; Editorial Board Member, *Journal of Artificial Intelligence Research*, 1996-1998; Co-Program Chair, ICDT '95; Editorial Board Member, *Journal on Information Processing and Cybernetics*, 1994-1996; Editorial Board Member, *Very Large Databases*, 1993-1998; Editorial Board Member, *Computing*, 1992-1996; Editorial Board Member, *J. of Computer and System Sciences*; Editorial Board Member, *Artificial Intelligence*; Editorial Board Member, *J. of Applied Logic*; Editorial Board Member, *Annals of Mathematics and Artificial Intelligence*; Editorial Board Member, *Web Intelligence and Agent Systems*; Editorial Board Member, *Journal of Discrete Algorithms*.

**Teaching:** Gottlob has taught courses at all levels on a wide variety of subjects. He has supervised over 70 Master theses and over 20 PhD theses. Many of his former students have successfully embarked in an academic career and are Professors by now. Among these are Wolfgang Nejdl (Hannover), Michael Schrefl (Linz), Markus Stumptner (Adelaide), Tom Fruehwirth (Ulm), Thomas Eiter (Vienna), Helmut Veith (Munich), Christoph Koch (Saarbruecken; Cornell), Reinhard Pichler (Vienna).

**Vision Statement:** Some time ago I cast the motto: "*Computer Science is the Continuation of Logic by Other Means.*" In fact, almost all aspects of computing are deeply rooted in logic. I wish this link becomes yet more visible in education and research. In particular, I hope that the *Semantic Web* will be built on solid logical foundations.

**Honours and Awards:** Royal Society Wolfson Research Merit Award, Royal Society, London, 2006; Member, European Academy of Sciences *Academia Europaea*, London,2006; Member, German Academy of Sciences "Leopoldina", 2006; ACM Recognition of Service Award, 2005; Member, Austrian Academy of Sciences 2004 (Corresponding Member 1999); Fellow, ECCAI, European Artificial Intelligence Society, 2002; Best Paper Award (with Ch. Koch), ACM PODS Conference, 2002; Honorary Scientist, Guizhou Acad. of Sciences, Guyang, China, 2000; Wittgenstein Award, 1998; Senior Fellow, Christian Doppler Society, 1996.

# GRATTAN-GUINNESS, Ivor Owen

**Specialties:** History of algebraic and mathematical logic and related topics, 1800-1950, relationship between logic and mathematics.

**Born:** 23 June 1941 in Bakewell, Derbyshire, England.

**Educated:** London, DSc History and Philosophy of Science, 1978; London, PhD History of Mathematics, 1969; London, MSc (Econ) Philosophy of Science, 1966; Oxford, MA, 1967; Oxford, BA ($2^{nd}$) Mathematics, 1962.

**Dissertation:** The historical development of mathematical analysis from Euler to Riemann; supervisor, Cyril Offord, J.R. Ravetz.

**Regular Academic or Research Appointments:**
RETIRED, Professor, History of Mathematics and Logic, Middlesex University, England 1993–2002.

**Visiting Academic or Research Appointments:** Lecturer, Brazil, 1997; Lecturer, Australia, 1997, 1977, 1976; Lecturer, Portugal, 1990; Lecturer, South Africa, 1987; Lecturer, Italy, 1980; Insitute for Advanced Study, Princeton, 1979; Lecturer, New Zealand, 1977; Lecturer, Australia, 1977.

**Research Profile:** Grattan-Guinness has focused on the history of mathematics and logic, especially 1750–1950 and with especial reference to mathematical analysis and mathematical physics. Philosophy of mathematics, especially regarding the formation of theories. Fuzzy set theory. History of numerology. History of scientific institutions and education. The careers and works of particular logicians and set theorists, especially Cantor, Russell, Peirce: also Whitehead, Post, Gödel, MacColl, Peano, Carnap, Pieri.

**Main Publications:**

1. *Dear Russell - Dear Jourdain. A Commentary on Russell's Logic, Based on his Correspondence with Philip Jourdain* (1977, London: Duckworth; New York: Columbia University Press).
   Editor. *From the Calculus to Set Theory, 1630-1910: An Introductory History* (1980, London: Duckworth). Reprinted (2000, Princeton: Princeton University Press).Esp. chs. 4–6.
   Editor. *Companion Encyclopaedia of the History and Philosophy of the Mathematical Sciences* (1994, London: Routledge), 2 volumes. Repr. (Baltimore: Johns Hopkins University Press, 2003). Esp. Parts 3–6.
2. *The Fontana History of the Mathematical Sciences. The Rainbow of mathematics* (1997, London: Fontana). Also as *The Norton History of the Mathematical Sciences. The Rainbow of Mathematics* (1998, New York and London: Norton). Esp. chs. 8, 12, 16.
3. *The Search for Mathematical Roots, 1870-1940. Logics, Set Theories and the Foundations of Mathematics from Cantor Through Russell to Gödel* (2000, Princeton: Princeton University Press).
   Editor. *Landmark Writings in Western Mathematics 1640–1940* (2005, Amsterdam: Elsevier). Esp. chs. 36, 43, 46, 55, 61, 71, 77.
4. Philip E. B. Jourdain, *Selected Essays on the History of Set Theory and Logics (1906–1918)* (1991, Bologna: CLUEB).
5. George Boole, *Selected Manuscripts on Logic and its Philosophy* (ed. with G. Bornet; 1997, Basel: Birkhäuser).
6. 'An unpublished paper by Georg Cantor: *Principien einer Theorie der Ordnungstypen. Erste Mittheilung* ', *Acta mathematica*, 124 (1970), 65–107.
7. 'Towards a biography of Georg Cantor', *Annals of science*, 27 (1971), 345-391 and plates xxv–xxviii.
8. 'The rediscovery of the Cantor-Dedekind correspondence', *Jahresbericht der Deutschen Mathematiker-Vereinigung*, 76 (1974–1975), part 1, 104–139.
9. 'Wiener on the logics of Russell and Schröder. An account of his doctoral thesis, and of his subsequent discussion of it with Russell', *Annals of Science*, 32 (1975), 103–132.
10. 'Preliminary notes on the historical significance of quantification and of the axioms of choice in the development of mathematical analysis', *Historia Mathematica*, 2 (1975), 475–488.
11. 'Fuzzy membership mapped onto intervals and many-valued quantities', *Zeitschrift für mathematischen Logik und Grundlagen der Mathematik*, 22 (1976), 149–160.
12. 'In memoriam Kurt Gödel: his 1931 correspondence with Zermelo on his incompletability theorem', *Historia Mathematica*, 6 (1979), 294–304.
13. 'Forays into the meta-theory of fuzzy set theory', *Logique et Analyse*, 22 (1979), 321–337.
14. 'Georg Cantor's influence on Bertrand Russell', *History and Philosophy of Logic*, 1 (1980), 61–93.
15. 'Russell's logicism versus Oxbridge logics, 1890-1925. A contribution to the real history of twentieth-century English philosophy', *Russell*, new ser., 5 (1985–1986), 101–131.
16. 'Living together and living apart: on the interactions between mathematics and logics from the French Revolution to the First World War', *South African Journal of Philosophy*, 7 (1988), no. 2, 73–82.
17. 'The manuscripts of Emil L. Post', *History and Philosophy of Logic*, 11 (1990), 77–83.
18. 'Structure-similarity as a cornerstone of the philosophy of mathematics', in J. Echeverria, A. Ibarra and T. Mormann (eds.), *The Space of Mathematics. Philosophical, Epistemological, and Historical Explorations* (1992, Berlin and New York: de Gruyter), 91–111.
19. 'Peirce between logic and mathematics', in N. Houser, D. Roberts and J. van Evra (eds.), *Studies in the Logic of Charles S. Peirce* (1997, Bloomington: Indiana: Indiana University Press), 23–42.
20. 'A retreat from holisms: Carnap's logical course, 1921–1943', *Annals of Science*, 54 (1997), 407–421.
21. 'Forms in algebras, and their interpretations: some historical and philosophical features', in L. Albertazzi (ed.), *Shapes of Forms. From Gestalt Psychology and Phenomenology to Ontology and Mathematics* (1999, Dordrecht: Kluwer), 177–190.
22. 'Are other logics possible? MacColl's logic and some English reactions, 1905-1912', *Nordic Journal for Philosophical Logic*, 2(1999), 1–16.
23. 'Mathematics and symbolic logics: some notes on an uneasy relationship', *History and Philosophy of Logic*, 20 (1999: publ. 2000), 159–167.

24. 'Re-interpreting '()': Kempe on multisets and Peirce on graphs, 1886–1905', *Transactions of the C.S. Peirce Society*, 38 (2002), 327–350.

25. 'Algebras, projective geometry, mathematical logic, and constructing the world: intersections in the philosophy of mathematics of A.N. Whitehead', *Historia Mathematica*, 29 (2002), 427–462. [Printing correction: 30 (2003), 96.]

26. 'The mathematical turns in logic', in D. Gabbay and J. Woods (eds.), *The Rise of Modern Logic from Leibniz to Frege* (Amsterdam: Elsevier, 2004), 545–556.

27. Contributions to the section on logic, in J.W. Dauben (ed.), *Bibliography of the History of Mathematics* (1985, New York: Garland), 286–296. Revised in A.C. Lewis (ed.), *The History of Mathematics; A Selective Bibliography* (2000, Providence R.I.: American Mathematical Society), CD-ROM.

28. 'Cumulative indexes. Volumes 1 to 10, 1980 to 1989', *History and Philosophy of Logic*, 11 (1990), 193–202.

29. 'Notes on the fate of logicism from Principia mathematica to Gödel's incompletability theorem', *History and philosophy of logic*, 5 (1984), 57–78.

30. 'The correspondence between George Boole and Stanley Jevons, 1863–1864', *History and philosophy of logic*, 12 (1991), 15–35.

31. 'Where does Grassmann fit in the history of logic?', in G. Schubring (ed.), *Hermann Gunther Grassmann (1809–1877) — visionary scientist and neohumanist scholar* (1996, Dordrecht: Kluwer), 187–191.

32. 'Forms in algebras, and their interpretations: some historical and philosophical features', in L. Albertazzi (ed.), *Shapes of forms. From Gestalt psychology and phenomenology to ontology and mathematics* (1999, Dordrecht: Kluwer), 177–190.

33. 'Forms in algebras, and their interpretations: some historical and philosophical features', in L. Albertazzi (ed.), *Shapes of forms. From Gestalt psychology and phenomenology to ontology and mathematics* (1999, Dordrecht: Kluwer), 177–190.

34. 'Structural similarity or structuralism? Comments on Priest's analysis of the paradoxes of self-reference', *Mind*, new ser., 107 (1998), 823–834.

35. 'Boole's algebraic logic after the mathematical analysis of logic ', in J. Gasser (ed.), *A Boole anthology. Recent and classical studies in the logic of George Boole*, Dordrecht (Kluwer), 213–216.

36. 'Are other logics possible? MacColl's logic and some English reactions, 1905-1912', *Nordic journal for philosophical logic*, 2(1999), 1–16.

37. 'The duo from Trinity: A.N. Whitehead and Bertrand Russell on the foundations of mathematics, 1895-1925', in S. Mitten and P. Harman (eds.), *Cambridge scientific minds* (2002, Cambridge: Cambridge University Press), 141–154.

38. 'Foundational studies and logics during the 1930s: Gonseth's Entretiens (1941) and its background', in J. Gasser and H. Volken (eds.), *Twentieth-century logic: the Swiss connection* (2000, Bern: Schweizerische Philosophische Gesellschaft), 10–18.

39. 'Mathematics in and behind Russell's logicism, and its reception', in N. Griffin (ed.), *The Cambridge companion to Russell* (2003, Cambridge: Cambridge University Press), 51–83.

40. 'Mathematics and philosophy', in M. Weber and others (eds.), *Alfred North Whitehead's science and the modern world* (2006, Frankfurt and Lancaster: Ontos Verlag), 77–80.

41. 'A.N. Whitehead: foundations of mathematics and logicism', in M. Weber and W. Desmond, Jr. (eds.), *Handbook of Whiteheadian process thought*, 2 vols. (2008, Frankfurt and Lancaster: ontos verlag), vol. 1, 33–39.

42. 'Set theory, symbolic logics and foundations for mathematics: principal interests of the major figures, 1890-1940', in V. Peckhaus (ed.), *Methodisches Denken im Kontext. Festschrift für C. Thiel* (2008, Paderborn: mentis Verlag), 177–183.

43. 'The reception of Gödel's 1931 incompletability theorems by mathematicians', in C. Binder (ed.), *Von den Tontafel zum Internet* (2006, Vienna: TU Vienna), 229–239.

**Service to the Profession:** Associate Editor, Mathematicians and Statistians, *Oxford Dictionary of National Biography* (2004), 1994–2004; President, British Society for the History of Mathematics, 1985–1988; Collaborating Editor, *Handbook of Ontology and Metaphysics*, Philosophia Verlag, Munich, 1987–1991; Editorial Board Member, *Reason and Argument* series, Martinus Nijhoff, The Hague, Netherlands, 1982–1994; Contributing Editor, *C.S. Peirce Chronological Edition*, Indiana University Press, Bloomington, Indiana, 1978–; Advisory Editorial Board Member, *Collected Paper of Bertrand Russell*, Routledge, London, 1979–; Consultant, *Russell. The Journal of the Bertrand Russell Archives*, 1979–; *Annals of Science,* Editor, 1974-1981, Book Reviews Editor, 1974–1987, Board Member, 1974-; Founder-Editor, *History and Philosophy of Logic*, Book Review Editor, 1979–2008 and Board Member, 1979–; Reviewing Panel Member, *Mathematical Reviews*, 1977-; Reviewing Panel Member, *Zentralblatt für Mathematik*, 1972–.

**Teaching:** Maria Panteki PhD, 1984-1992, on relationships between algebra, logics and differential equations in England, 1800-1860. Adrian Rice PhD, 1994-1997, on Augustus de Morgan and the development of mathematics in London in the 19th century. Alison Walsh PhD, 1988-2000, on relationships between algebra and logics, 1860-1900s in Benjamin and C. S. Peirce.

# GROENENDIJK, Jeroen

**Specialties:** Formal semantics and pragmatics of natural language, logic of conversation

**Born:** 20 July 1949 in Amsterdam, The Netherlands.

**Educated:** Universiteit van Amsterdam, PhD Philosophy, 1984; Universiteit van Amsterdam, MA Philosophy, 1974.

**Dissertation:** (joint with Martin Stokhof): *Studies on the Semantics of Questions and the Pragmatics of Answers*; supervisors: R. Bartsch and J. van Benthem.

**Regular Academic or Research Appointments:** FULL PROFESSOR, PHILOSOPHY OF LANGUAGE, UNIVERSITEIT VAN AMSTERDAM, 1998–; Associate Professor, Philosophy of Language, Department of Philosophy, Universiteit van Amsterdam, 1988-1998; Assistant Professor, Logic and Semantics, Department of Computational Linguistics, Universiteit van Amsterdam, 1986-1992; Assistant Professor, Philosophy of Language, Universiteit van Amsterdam, 1976-1988.

**Visiting Academic or Research Appointments:** Visiting Professor, Institute for Advanced Studies at the Hebrew University in Jerusalem, 1997, 1998; Senior Researcher and Advisor, Rosetta Machine Translation Project, Philips Research Laboratories, 1986-1988; Visiting Researcher, Department of Computational Lingusitics, Katholieke Universiteit Brabant, 1985; Researcher, Dutch Organizaiton for the Advancement of Pure Researcher (Z.W.O.), 1974-1978.

**Research Profile:** Jeroen Groenendijk has worked on descriptive and theoretical issues in the formal semantics and pragmatics of natural language.

Together with Martin Stokhof he developed the so-called 'partition approach' to the semantics of questions, in the early 1980's. This approach makes a purely semantic analysis of questions possible, one that satisfies the same strict requirements as other branches of logical semantics. Such a semantic analysis can be embedded in a pragmatic, information-based analysis of various relations of answerhood.

Also with Martin Stokhof, and later with Frank Veltman and others, he initiated and explored the so-called 'dynamic approach' to meaning in natural language in the 1990s. This dynamic approach abandons the common reference and truth based analysis of natural language meaning that semantics inherited from classical logic, and treats meaning as 'information change potential' (making use of concepts from dynamic logic). This approach allows for a conceptual integration of semantics and pragmatics, and extends naturally to the analysis of larger discourses and of linguistic interactions.

Around the turn of the century Jeroen Groenendijk returned to the study of the semantics and pragmatics of questions, but now in the setting of dynamic semantics, and with a special interest for questions which are difficult to handle in a partition semantics, because they allow for complete answers which do not exclude each other. Examples of such questions are conditional questions, and so-called alternative questions. This led to a widening of the dynamic notion of meaning in terms of information change, into an alternative semantic notion of meaning directly related to information exchange. A key-feature of the alternative semantics is that disjunction does more than providing information; it also raises an issue; it gives rise to two alternatives. Due to that hybrid nature of disjunction, questions can be analyzed in terms of disjunction.

**Main Publications:**

1. (With M. Stokhof) A pragmatic analysis of specificity, in: F. Heny (ed), *Ambiguities in Intensional Contexts*, Dordrecht, Reidel, 1981, pp. 98-123
2. (With M. Stokhof) Semantic analysis of wh-complements, *Linguistics and Philosophy*, 5(2), 1982, pp. 175-235
3. (With M. Stokhof) Type-shifting rules and the semantics of interrogatives, G. Chierchia, B. Partee & R. Turner (eds), *Properties, Types and Meaning. Vol. II: Semantic Issues*, Dordrecht, Reidel, 1988, pp. 21-69 [Reprinted in P. Portner & B.H. Partee (eds), *Formal Semantics. The Essential Readings*, Oxford: Blackwell, 2002, pp. 421-456]
4. (With J. van Benthem, D. de Jongh, M. Stokhof, H. Verkuyl) *Logic, Language and Meaning. Vol. I: Introduction to Logic*, Chicago, The University of Chicago Press, 1990 [Spanish translation: *Introducción a la Lógica*, Buenos Aires: Eudeba, 2002]
5. (With J. van Benthem, D. de Jongh, M. Stokhof, H. Verkuyl) *Logic, Language and Meaning. Vol. II: Intensional Logic and Logical Grammar*, Chicago, The University of Chicago Press, 1990
6. (With M. Stokhof) Dynamic Montague grammar, L. Kálmán & L. Pólos (eds), *Proceedings of the Second Symposion on Logic and Language*, Hajdúszoboszló, September 5-9, 1989, Budapest, Eötvös Loránd University, 1990, pp. 3-48
7. (With M. Stokhof) Dynamic predicate logic, *Linguistics and Philosophy*, 14(1), 1991, pp 39-100 [Reprinted in: P. Grim, P. Ludlow, & G. Mar (eds), *The Philosopher's Annual*, Vol. XIV,1991, Atascadero, California: Ridgeview, 1993, pp. 67-128; and in: S. Davis & B. Gillon (eds), *Semantics: A Reader*, Oxford: Oxford

University Press, 2004, pp. 263-305]

8. (With M. Stokhof, F. Veltman) Coreference and Modality, in: S. Lappin (ed), *Handbook of Contemporary Semantic Theory*, Oxford, Blackwell, 1996, pp. 179-216

9. (With M. Stokhof), Questions, in: van Benthem & A. ter Meulen (eds), *Handbook of Logic and Language*, Amsterdam/Cambridge, Mass., Elsevier/MIT Press, 1997, pp. 1055-1124

10. Questions in update semantics, in: J.H. Hulstijn & A. Nijholt, *Formal Semantics and Pragmatics of Dialogue: Proceedings of the Thirteenth Twente Workshop on Language Technology*, Universiteit Twente, Faculteit Informatica, Enschede, 1998, pp. 125-137

11. (With M. Stokhof) Meaning in motion, in: K. von Heusinger & U. Egli (eds), *Reference and Anaphora*, Kluwer, Dordrecht, 1999, pp. 47-76

12. The logic of interrogation, in: T. Matthews and D.L. Strolovitch (eds) *The Proceedings of the Ninth Conference on Semantics and Linguistic Theory*, CLC Publications, Ithaca, NY, 1999, pp. 109-126 [Reprinted in: M. Aloni, A. Butler & P. Dekker, *Questions in Dynamic Semantics*, Elsevier, Oxford, 2007, pp. 43-62].

*Work in Progress*

13. Alternative Logical Semantics and the Logic of Discourse

14. Alternative Dynamic Epistemic Semantics and the Dynamics of Inquiry

A full list of publications is available at `http://dare.uva.nl/auteur/groenendijk`

**Service to the Profession:** Associate Editor, Semantics, *Linguistics and Philosophy*, 1990-; Editorial Board Member, *Natural Language Semantics*, 1992-; Editorial Board Member, *Current Research in the Semantics/Pragmatics Interface* (Elsevier), 1997-; Editorial Board Member, *Journal for Research on Language and Computation*, 1997-; Organization, several of the bi-annual *Amsterdam Colloquium*, 1976-.

**Teaching:** Groenendijk has taught logic and formal semantics for many years, to philosophy students as well as to students in computational linguistics. As director of the Teaching Institute of Philosophy, Groenendijk co-organized the introduction of the Bachelor/Master structure at the Faculty of Humanities of the University of Amsterdam. Together with J. van Benthem, D. de Jongh, M. Stokhof and H. Verkuyl he wrote the two Gamut textbooks on logic and logical grammar. He supervised 10 PhD students, among whom were Paul Dekker, Jelle Gerbrandy, and Maria Aloni.

**Vision Statement:** Logical semantics and pragmatics for natural language should not just involve the application of existing logical tools; it should lead to innovations in logic. Traditionally, the central logical notion is validity of arguments. Argumentation is one function of natural language, but by far not the most central one. An argument, in the logical sense, is basically a monologue, but the most central uses of language concern communication in dialogue, the exchange of information in a cooperative dynamic process of raising and resolving issues. Of course, that may also lead to an argument in the more common sense.

To get at a logical analysis of what is at the heart of the use of natural language, we should develop logics which do not have entailment and validity as their core logical notions, but new notions which rule logically coherent cooperative dialogue.

The dominant logical semantical notion of meaning as propositional content is rooted in a logic that was not designed to deal with communication. If we redesign logic such that it does directly deal with that, we may arrive at a new notion of meaning that directly reflects what we use language for.

And once the notion of meaning is directly linked to what is at the heart of language use, it is likely to turn out that the whole semantics-pragmatics distinction was an epiphenomenon of basing natural language semantics on a logic which was dedicated to a peripheral instance of language use.

Such an enterprise may not only turn out to be of philosophical and linguistic interest, but in the interest of pure logic as well.

## GROHE, Martin

**Specialties:** Logic in computer science, complexity theory.

**Born:** 10 July 1967 in Blankenstein, Germany.

**Educated:** University of Freiburg, Habilitation, 1998; University of Freiburg, PhD, 1994, University of Freiburg, Diplom, 1992.

**Dissertation:** *The structure of fixed-point logics*; supervisor, Heinz-Dieter Ebbinghaus.

**Regular Academic or Research Appointments:** PROFESSOR, THEORETICAL COMPUTER SCIENCE, HUMBOLDT UNIVERSITY BERLIN, 2003–; Reader, Computer Science, University of Edinburgh, 2001-2003; Assistant Professor, Mathematics and Computer Science, University of Illinois at Chicago, 2000-2001; Hochschulassistent, Mathematics, University of Freiburg, 1996-2000.

**Visiting Academic or Research Appointments:** Postdoctoral Fellow, Department of Computer Science, University of California at Santa Cruz; Visiting Scholar, Department of Mathematics, Stanford University, 1995-1996.

**Research Profile:** Martin Grohe's research is centred on the connections between logic, complexity theory, and discrete mathematics. Important results are algorithmic meta theorems, for example, the linear time decidability of first-order properties of planar graphs. Grohe also characterized the polynomial time decidable properties of several natural classes of structures. In finite model theory, Grohe has investigated the expressive power of extensions of first-order logic on finite structures. More recently, he has worked in parameterized complexity, a relatively new branch of computational complexity theory; his work (most of it joint work with Jörg Flum) emphasized the fundamental role logic plays in this area. Grohe has also been interested in applications of logic in other areas of computer science, for example, database theory.

**Main Publications:**

1. A. Bulatov and M. Grohe. The complexity of partition functions. Theoretical Computer Science, 348:148–186, 2005.
2. J. Flum, M. Frick, and M. Grohe. Query evaluation via tree decompositions. Journal of the ACM, 49(6):716–752, 2002.
3. J. Flum and M. Grohe. The parameterized complexity of counting problems. SIAM Journal on Computing, 33(4):892–922, 2004.
4. J. Flum and M. Grohe. Parameterized Complexity Theory. Springer-Verlag, 2006.
5. M. Frick and M. Grohe. Deciding first-order properties of locally tree decomposable structures. Journal of the ACM, 48:1184–1206, 2001.
6. M. Frick and M. Grohe. The complexity of first-order and monadic second order logic revisited. Annals of Pure and Applied Logic, 130:3–31, 2004.
7. M. Grohe. Arity hierarchies. Annals of Pure and Applied Logic, 82:103–163, 1996.
8. M. Grohe. Fixed-point logics on planar graphs. In Proceedings of the 13th IEEE Symposium on Logic in Computer Science, pages 6–15, 1998.
9. M. Grohe. Large finite structures with few $L^k$-types. Information and Computation, 179(2):250–278, 2002.
10. M. Grohe. The complexity of homomorphism and constraint satisfaction problems seen from the other side. In Proceedings of the 43rd Annual IEEE Symposium on Foundations of Computer Science, pages 552–561, 2003.
11. M. Grohe. Local tree-width, excluded minors, and approximation algorithms. Combinatorica, 23(4):613–632, 2003.
12. M. Grohe. Computing crossing numbers in quadratic time. Journal of Computer and System Sciences, 68(2):285–302, 2004.
13. M. Grohe and N. Schweikardt. Lower bounds for sorting with few random accesses to external memory. In Proceedings of the 24th ACM Symposium on Principles of Database Systems, pages 238–249, 2005.
14. M. Grohe and T. Schwentick. Locality of order-invariant first-order formulas. ACM Transactions on Computational Logic, 1:112–130, 2000.

**Service to the Profession:** Journal of Symbolic Logic, Editor, 2002–2007, Coordinating Editor, 2006–2007; Editorial Board Member, Journal of Discrete Algorithms, 2002–2007; Member, ASL council, 2001–.

**Honours and Awards:** Heinz-Maier-Leibnitz Preis, German National Young Investigators Award, 1999.

# GRZEGORCZYK, Andrzej

**Specialties:** Mathematical logic, metamathematics, philosophy of mathematics, decision problems, applications of logic and logical analysis to philosophical problems.

**Born:** 22 August 1922 in Warsaw Poland.

**Educated:** Warsaw University, PhD Mathematics, 1950; Krakow Jagiellonian University, 1945; Underground University, Warsaw during German Occupation, Philosophy, 1940-1944.

**Dissertation:** *Topological Spaces in Pointless Topological Algebras* (in Polish), *Supervisor*: Andrzej Mostowski.

**Regular Academic or Research Appointments:** EMERITUS PROFESSOR, PHILOSOPHY, INSTITUTE OF PHILOSOPHY AND SOCIOLOGY OF POLISH ACADEMY OF SCIENCE, 1992–; Professor, Logic and Philosophy, Institute of Philosophy and Sociology of Polish Ac. of Sc., 1972-1992; Professor, Mathematics, Inst. of Math. Polish Ac. of Sc., 1960-1972; Assistant and Docent, Math, Inst. of Math. Polish Ac. of Sc., 1948-1960.

**Visiting Academic or Research Appointments:** Beth's Institute Amsterdam, 1964; Istituto per il Calcolo Rome, 1970.

**Research Profile:** Andrzej Grzegorczyk has made some contributions to three areas of logic: 1) Recursion, hierarchies and decidability; 2) Constructive reasoning and intuitionism; 3) Application of logic to philosophical problems.

In the area 1) he made some observations on recursive analysis, undecidability of some elementary theories (algebraic and topological), and on sub-recursive hierarchy [in1] (the paper 1). inspired some investigations on complexity). In the area 2) he proposed a version of intuitionistic logic and an application of type-stratified $\lambda$-combinators. In the area 3) he gave a veridical interpretation of semantic paradoxes [in 10. and

11.], stressed the use of logic in the foundation of psychological distinctions, and initiated a psycho-philosophical cognitive interpretation of Gödel's discovery by taking [in 9.] a theory of texts as appropriate medium for the proof of undecidability, (without any arithmetical construction).

**Main Publications:**

1. *Some Classes of Recursive Functions*, Rozprawy Matematyczne Nr IV, Instytut Matematyczny PAN, Warszawa 1953, 45 s.
2. *An Outline of Mathematical Logic*, Reidel-Holland 1974, 596 s.
3. Undecidability of some topological theories, *Fundamenta Mathematicae* 38/1951, 137-152.
4. (with K. Kuratowski) On Janiszewski's property of topological spaces, Annales de la Société Polonaise de Mathématique 25/1952, 69-82.
5. Some proofs of undecidability of arithmetic, Fundamenta Mathematicae 43/1956, 166-177.
6. (with A. Mostowskim and C. Ryll-Nardzewskim) The classical and the omega-complete arithmetic, Journal of Symbolic Logic 23/1958, 188-206.
7. An example of two weak essentially undecidable theories F and F*, Bulletin of Académie Polonaise des Sciences, Série Math. 10/1962, 5-9.
8. Recursive objects in all finite types, Fundamenta Mathematicae 54/1964, 73-93.
9. Undecidability without Arithmetization, Studia Logica 79 (2005, 2). 163-230.
10. *Logic – a Human Affair*, Scholar, Warszawa 1997, 147 s.
11. The paradox of Grelling and Nelson presented as a veridical observation concerning naming, in: *The Lvov-Warsaw School and Contemporary Philosophy* (ed. K. Kijania-Placek), Kluwer 1998, 183-190.
12. Is Antipsychologism Still Tenable? in: *Alfred Tarski and the Vienna Circle*, (ed. J. Woleński and E. Köhler), Kluwer 1999, 109-114.
*Work in Progress*
13. An elementary geometry without point.

**Service to the Profession:** [Grzegorczyk only claims small contributions to a positive atmosphere for logic in Poland.] Assessor, ASL, c 1970.

**Teaching:** He has taught mathematical logic at Warsaw University in 1960-1968 and 1956-1960 at the University Curie-Sklodowska in Lublin. But his textbook (item 2) (and in French *Fonctions Recursives*) has been well received as it is not difficult for weaker students.

**Vision Statement:** Logic is *homo sapiens'* essential mental implementation. It should be used to improve the human condition. In converting to logical thinking, we may hope to transgress the clash of civilizations, and the accomplishment of a durable peace.

# GUPTA, Anil K.

**Specialties:** Philosophical logic, epistemology, metaphysics.

**Born:** 5 February 1949 in Ambala, Haryana, India.

**Educated:** University of Pittsburgh, MA Philosophy 1973, PhD Philosophy 1977; Derby and District College of Technology, University of London, BSc Mechanical Engineering 1969.

**Dissertation:** *The Logic of Common Nouns: An Investigation in Quantified Modal Logic*; supervisor, Nuel Belnap.

**Regular Academic or Research Appointments:** DISTINGUISHED PROFESSOR OF PHILOSOPHY AND PROFESSOR OF HISTORY AND PHILOSOPHY OF SCIENCE, UNIVERSITY OF PITTSBURGH, 2001–; Rudy Professor of Philosophy, Indiana University, 1995-2000; Professor of Philosophy, Indiana University, 1989–1995; Associate Professor of Philosophy, University of Illinois at Chicago, 1982–1989; Associate Professor of Philosophy, McGill University, 1980–1982; Assistant Professor of Philosophy, McGill University, 1975–1980.

**Visiting Academic or Research Appointments:** Visiting Professor of Philosophy, University of Valencia, June 2006; Fellow, Center for Advanced Study in the Behavioral Sciences, Stanford, 1998–1999; Fellow, Institute for the Humanities, University of Illinois at Chicago, 1985-1986; Visiting Professor, University of Padua, June 1985; Visiting Fellow, Princeton University, Fall 1982; Visiting Assistant Professor of Philosophy, University of Pittsburgh, 1979–1980.

**Research Profile:** Anil Gupta is best known for his work in the logic of definitions and the theory of truth. He has developed a general theory of definitions that sustains the logical legitimacy of interdependent definitions and, indeed, of circular definitions. He has argued that some of our ordinary notions—in particular, our semantic notions—are circular. Gupta is one of the originators (with Hans Herzberger and Nuel Belnap) of the Revision Theory of Truth. More recently, Gupta has used the idea of interdependence to develop a novel account of experience.

Gupta has contributed also to quantified modal logic, the theory of conditionals, the logic of relevance, and the theory of rational choice. Also notable is his critique of deflationism, which has been influential.

**Main Publications:**

1. *The Logic of Common Nouns*, Yale University Press, 1980, 142 pp.
2. A Theory of Conditionals in the Context of Branching Time, with Richmond H. Thomason, *The Philosophical Review*, 89, 1980, pp. 65–90. Reprinted in William Harper, Robert Stalnaker, and Glenn Pearce eds., *Ifs: Conditionals, Belief, Decision, Chance, and Time*, D. Reidel Publishing Company, 1981.
3. A Consecution Calculus for Positive Relevant Implication with Necessity, with Nuel Belnap and J. Michael Dunn, *Journal of Philosophical Logic*, 9, 1980, pp. 343–362. Revised version is reprinted in Alan Anderson, Nuel Belnap, and J. Michael Dunn, *Entailment* (Vol. II), Princeton University Press, 1992.
4. Truth and Paradox, *Journal of Philosophical Logic*, 11, 1982, pp. 1–60. Revised version with a brief "Postscript 1983" is reprinted in Robert L. Martin ed., *Recent Essays on Truth and the Liar Paradox*, Oxford University Press, 1984.
5. A Fixed Point Theorem for the Weak Kleene Valuation Scheme, with Robert L. Martin, *Journal of Philosophical Logic*, 13, 1984, pp. 131–135. A correction to this paper appears in *Journal of Philosophical Logic*, 14, 1985, p. 229.
6. Remarks on Definitions and the Concept of Truth, *Proceedings of the Aristotelian Society* 89, 1988–89, pp. 227–246. Reprinted in Patrick Grim, Gary Mar, and Peter Williams eds., *The Philosopher's Annual*, 12, Ridgeview Publishing Company, 1991, pp. 67–86.
7. *The Revision Theory of Truth*, with Nuel Belnap, MIT Press, 1993, 299 pp.
8. A Critique of Deflationism, *Philosophical Topics*, 21, 1993, pp. 57–81. Reprinted in Simon Blackburn and Keith Simmons eds., *Truth*, Oxford Readings in Philosophy, Oxford University Press, 1999, pp. 282–307. Also reprinted in Michael Lynch ed., *The Nature of Truth*, MIT Press, 2001, pp. 527–557; and, with a postscript, in Bradley Armour-Garb and J. C. Beall eds., *Deflationary Truth*, Open Court, 2005, pp. 199–226.
9. Meaning and Misconceptions, in Ray Jackendoff, Paul Bloom, and Karen Wynn eds., *Language, Logic, and Concepts: Essays in Memory of John Macnamara*, The MIT Press, 1999, pp. 15–41.
10. On Circular Concepts, in André Chapuis and Anil Gupta (eds.), *Circularity, Definition and Truth*, Indian Council of Philosophical Research, 2000, pp. 123–153.
11. An Argument Against Tarski's Convention T, in Richard Schantz ed., *What is Truth?*, Walter de Guyter, 2002, pp. 225–237.
12. Experience and Knowledge, in Tamar Szabó Gendler and John Hawthorne eds., *Perceptual Experience*, Oxford University Press, 2006, pp. 181–204.
13. *Empiricism and Experience*, Oxford University Press, 2006, 255 pp.
*Work in Progress*
14. A collection of essays on truth and definitions.

**Service to the Profession:** Editorial Board, *Notre Dame Philosophical Reviews*, 2001– ; Editorial Board, *Journal of Indian Council of Philosophical Research*, 1991–; Board of Consultants, *The Philosophical Gourmet Report*, 2002– ; Editor, *Journal of Philosophical Logic*, 1991–1994; Program Committee, 1995 Annual Meeting of the Central Division of the American Philosophical Association; Member, Executive Committee, Association for Symbolic Logic, 1991-1994.

**Teaching:** Gupta teaches logic, epistemology, and metaphysics at all levels, graduate and undergraduate. He has directed or co-directed fourteen doctoral dissertations, including those of Ingo Brigandt, André Chapuis, Adam Kovach, Byeongdeok Lee, and Caleb Liang. He is currently directing the work of six Ph.D. students. Gupta received a Teaching Excellence Recognition Award from Indiana University in 1998.

**Vision Statement:** Logic and philosophy have been, and remain, of critical importance to each other. The fundamental philosophical problems—some of them, if not all—are rooted in logical misconceptions. These problems can be resolved only through a careful attention to logical issues and, indeed, only through an enrichment of our logical ideas.

**Honours and Awards:** Fellowship for University Teachers, National Endowment for the Humanities, 1988–1989, 1995–1996, 2003–2004; Teaching Excellence Recognition Award, Indiana University, 1998; American Council for Learned Society Fellowship, 2003–2004; Fellow, Center for Advanced Study in the Behavioral Sciences, Stanford, 1998–99; Fellow, American Academy of Arts and Sciences, 2006– .

# GUREVICH, Yuri

**Specialties:** Algebra, logic, computation theory, foundations of software engineering.

**Born:** 7 May 1940 in Nikolayev, USSR.

**Educated:** Urals University, USSR, MSc in Math 1962, PhD in Math 1964, Dr of Math 1968.

**Dissertation:** *Elementary Properties of Ordered Abelian Groups*, supervisor P. G. Kontorovich. *Dr of Math Dissertation:* The Decision Problem for Some Algebraic Theories.

**Regular Academic or Research Appointments:** SR. RESEARCHER, MICROSOFT RESEARCH, REDMOND, WA, USA, FROM 1998; Prof. of Computer Science, University of Michigan, Ann Arbor, MI, USA, 1982–98; Prof. (1978–82) and Assoc. Prof. (1974–78), Math, Ben Gurion University, Beer Sheva, Israel; Prof. and Chair, Algebra and Computer Science Dept., Kuban

University, Krasnodar, USSR, 1971–72; Prof. and Chair, Math Dept., Institute for National Economy, Sverdlovsk, USSR, 1969–71; Assoc. Prof. (1966–69) and Assistant Prof. (1965–66), Math, Urals University, Sverdlovsk, USSR; Assistant Prof., Krasnoyarsk University, Krasnoyarsk, USSR, 1964–65.

**Visiting Academic or Research Appointments:** Centre National de la Recherche Scientifique, Paris, France, 1995–96; Stanford University and IBM Almaden Research Center, California, 1988–89; Simon Fraser University, British Columbia, Canada, 1978–79; Bowling Green State University, Ohio, 1981–82; Hebrew University, Israel, 1980–81.

**Research Profile:** Gurevich pioneered behavioral computation theory. He introduced abstract state machines (ASMs) and put forward the ASM thesis: every algorithm is behaviorally identical to an appropriate ASM [13]. In [15], he axiomatized sequential algorithms and used the axiomatization to prove the sequential version of the ASM thesis. This has been extended to parallel algorithms [16] and to intra-step interactive parallel algorithms. At the time of this writing (July 2005), the work in intra-step interactive algorithms has been published only partially [17, 18, 19, 20]. In 1998, upon invitation of Microsoft Research, Gurevich founded a group on Foundations of Software Engineering there. ASM-based tools, developed by the group, are widely used within Microsoft. See also the ASM academic website http://www.eecs.umich.edu/gasm.

Gurevich started his career in the Soviet Union as an algebraist and self-taught logician. In his PhD thesis, solving a problem of Alfred Tarski, he classified ordered abelian groups (OAGs) by first-order (FO) properties and proved the decidability of the FO theory of OAGs [1]. But much of the interesting OAG algebra is not FO. Fortunately there is a fragment of monadic second-order logic that is a good match for OAGs. It gives rise to a rich theory of OAGs that expresses much of the relevant algebra but is nevertheless decidable [2].

Gurevich made a prominent contribution to the classical decision problem, that is the problem of classifying standard fragments of first-order logic as decidable or undecidable. This work is covered in the book [11].

In 1973, Gurevich moved to Israel where he succeeded to prove or disprove a number of conjectures in Saharon Shelah's seminal paper on the monadic second order theory of order in Annals of Mathematics 102 (1975). This gave rise to a most fruitful collaboration between the two on monadic second-order theories [5]. The Gurevich-Harrington theorem on forgetful determinacy of games [3] is related to that body of work.

In 1982, Gurevich moved to Michigan and to computer science. He noticed that classical logic, developed to confront the infinite, is ill prepared to deal with finite structures [4, 8]. He became one of the pioneers of finite model theory (FMT). He authored or co-authored more papers in the area of FMT than in any other area. His earlier results are covered by the survey [6]. Other notable FMT articles include [7] and [10]. Several papers bridge FMT and behavioral computation theory, most notably [14].

Gurevich made significant contributions to computational complexity theory, especially to average-case complexity theory [9, 12].

Additional information on Gurevich's research is found at his webpage http://research.microsoft.com/~gurevich/.

**Main Publications:**

1. "Elementary properties of ordered abelian groups", Algebra and Logic 3:1 (1964), 5–39 (Russian, PhD Thesis); AMS Translations 46 (1965), 165–192.
2. "Expanded theory of ordered abelian groups", Annals of Mathematical Logic 12 (1977), 193–228.
3. "Automata, trees, and games" (with Leo Harrington), 14th Annual Symposium on Theory of Computing, ACM, 1982, 60–65.
4. "Toward logic tailored for computational complexity", Springer Lecture Notes in Math. 1104 (1984), 175–216.
5. "Monadic second-order theories", In *Model-Theoretical Logics* (eds. J. Barwise and S. Feferman), Springer-Verlag, 1985, 479–506.
6. "Logic and the challenge of computer science", In *Current Trends in Theoretical Computer Science* (ed. E. Börger), Computer Science Press, 1988, 1–57.
7. "Fixed-point extensions of first-order logic" (with Saharon Shelah), Annals of Pure and Applied Logic 32 (1986), 265–280.
8. "Monotone versus positive" (with Miklos Ajtai), J. of ACM, 34, 1987, 1004–1015.
9. "Average case completeness", J. of Computer and System Sciences 42:3, June 1991, 346–398.
10. "Datalog vs. first-order logic", J. of Computer and System Sciences 49: 3, Dec 1994, 562–588.
11. "Classical Decison Problem" (with Egon Börger and Erich Grädel), Springer Verlag, Perspectives in Mathematical Logic, 1997.
12. "Matrix transformation is complete for the average case" (with Andreas Blass), SIAM J. on Computing 24:1, 1995, 3–29.
13. "Evolving algebra 1993: Lipari guide", in *Specification and Validation Methods* (ed. E. Börger), Oxford University Press, 1995, 9–36.
14. "Choiceless polynomial time" (with Andreas Blass and Saharon Shelah), Annals of Pure and Applied

Logic 100 (1999), 141–187.

15. "Sequential abstract state machines capture sequential algorithms", ACM Transactions on Computational Logic, 1:1 (2000), 77–111.

16. "Abstract state machines capture parallel algorithms" (with Andreas Blass), ACM Transactions on Computation Logic, 4:4 (2003), 578-651.

17. "Ordinary interactive small-step algorithms", parts I, II and III (with A. Blass), ACM Transactions on Computation Logic, to appear.

18. "Interactive algorithms 2005", Springer Lecture Notes in Computer Science 3618 (2005), 26–38.

19. "General interactive small-step algorithms" (with Andreas Blass, Dean Rosenzweig and Benjamin Rossman), in preparation.

20. "Interactive wide-step algorithms" (with Andreas Blass, Dean Rosenzweig and Benjamin Rossman), in preparation.

**Service to the Profession:** Gurevich served on numerous editorial boards and various steering and programming committees. He is a popular lecturer. His most notable contribution is the long-running continuing column on Logic in Computer Science in the Bulletin of the European Association for Theoretical Computer Science.

**Vision Statement:** Computer science has become a much larger consumer of logic than mathematics. Logic is growing more pragmatic as a result. Practical feasibility concerns supersede those of computability or even polynomial time computability. Fortunately, for those of us attracted to logic because of its role in foundations of mathematics, new foundational issues arise. "What's an algorithm?" is the most notable of them.

**Honours and Awards:** Fellow of John Simon Guggenheim Memorial Foundation, 1995; Fellow of the Association for Computing Machinery, 1996; Dr. Honoris Causa from University of Limburg, Belgium, 1998, and from Urals State University, Ekaterinburg, Russia, 2005.

# H

## HAACK, Susan Wendy

**Specialties:** Philosophy of logic, non-standard logics, theories of truth, pragmatism, epistemology, metaphysics, philosophy of science; legal theories of evidence, logic in the law.

**Born:** 23 July 1945 in Burnham, Bucks, UK.

**Educated:** Cambridge, Ph.D. 1972, Oxford, B.Phil.1968; Oxford, B.A. 1966.

**Dissertation:** Ph.D. dissertation: *Deviant Logic;* supervisor Timothy Smiley, B.Phil. dissertation: *Ambiguity*; supervisor; David Pears.

**Regular Academic or Research Appointments:** DISTINGUISHED PROFESSOR IN THE HUMANITIES, 2007–; COOPER SENIOR SCHOLAR IN ARTS AND SCIENCES, 1997–; PROFESSOR OF PHILOSOPHY, 1990– PROFESSOR OF LAW, 2000–, UNIVERSITY OF MIAMI; Professor of Philosophy, University of Warwick, UK, 1982–90; Reader in Philosophy, University of Warwick, 1976–82; Lecturer in Philosophy, University of Warwick, 1971–6; Fellow, Lecturer, New Hall, Cambridge, 1968–71.

**Visiting Academic or Research Appointments:** Stanisław Kamiński Professor, John Paul II Catholic University of Lublin (Poland), May 2007; Visiting Professor, Faculty of Laws, University of Bologna (Italy), March 2005; Joseph Wunsch Lecturer, The Technion (Israel), June 2004; Landsdowne Professor, University of Victoria (Canada), March 2002; Spencer-Leavitt Proessor, Union College, April 2000; Cowling Professor, Carleton College, January 2000; Visiting Professor, Dept. of Philosophy, Aarhus University (Denmark), June 1999; Visiting Professor, Faculty of Philosophy, University of Santiago de Compostela (Spain), May 1997; Visiting Fellow, School of Social Sciences, Australian National University, December 1981-September 1982; Visiting Professor, Dept. of Philosophy, University of Virginia, January–May 1980; Visiting Professor, Dept. of Philosophy, University of Guelph (Canada), September 1978; Harkness Fellow, Dept. of Philosophy, Princeton University, academic year 1975-6.

**Research Profile:** In philosophy of logic, Haack is well-known for her work on deviant systems and on theories of truth. Her *Philosophy of Logics* has been continuously in print in English since 1978, and introduced contemporary work in the area to readers in China, where it circulated in illegal photocopy for many years before it was published in Chinese. In epistemology, Haack is well known for her new account of epistemic justification ("foundherentism"), and her analogy of the structure of evidence as like a crossword puzzle. In philosophy of science she has proposed a distinctive account of evidence and of method ("Critical Common-Sensism") which is distinctively "worldly," i.e., takes into account the relation of scientific vocabulary to things and events in the world, and of scientists' interactions with the world and each other, as well as of narrowly logical matters. (The scope and limits of formal methods in philosophy has been a long-standing preoccupation.) In legal scholarship she is best known for work on the rules governing scientific testimony, and has recently begun to publish in jurisprudence, including questions about the role of logic in the law. She is a respected scholar of pragmatism, especially of Peirce.

**Main Publications:**

1. *Deviant Logic*, Cambridge: Cambridge University Press, 1974, also published in Spanish, Madrid: Paraninfo: 1980. Second, expanded edition, *Deviant Logic, Fuzzy Logic: Beyond the Formalism*, Chicago: University of Chicago Press, 1996, including 8, 9, 10, 11.
2. *Philosophy of Logics*, Cambridge: Cambridge University Press, 1978; also published in Spanish, Madrid: Cátedra, 1982; in Italian Milan: Franco Angelli, 1984); in Korean (Seoul: 1984); in Portuguese Saõ Paulo: Editoria UNESP, 2002; in Chinese Beijing: Commercial Press, 2003; and in Croatian Zagreb: Scopus, 2005.
3. *Evidence and Inquiry: Towards Reconstruction in Epistemology*, Oxford: Blackwell, 1993; also published in Spanish Madrid: Tecnos, 1997; and in Chinese Beijing: Renmin University Press, 2005.
4. *Manifesto of a Passionate Moderate: Unfashionable Essays*, Chicago: University of Chicago Press, 1998; including 13, 14, 15 16, and an expanded version of 17.
5. *Defending Science – Within Reason: Between Scientism and Cynicism*, Amherst, NY: Prometheus Books, 2003. A Chinese edition is under contract with Renmin University Press, Beijing.
6. *Pragmatism, Old and New: Selected Writings*, with Associate editor Robert Lane, Amherst, NY: Prometheus Books, 2006.
7. *Putting Philosophy to Work: Inquiry and Its Place in Culture*, Amherst, NY: Prometheus Books, 2008.
8. Mentioning Expressions, *Logique et Analyse* 17, 1974: 277–94.

9. The Justification of Deduction, *Mind*, 85, 1976: 11209. Reprinted in Irving M. Copi and James A. Gould, *Contemporary Philosophical Logic* New York: St. Martin's Press, 1978, 54–62; in R.I.G. Hughes ed., *A Philosophical Companion to First Order Logic*, Indianapolis: Hackett, 1993, 76–84.

10. Analyticity and Logical Truth in The Roots of Reference, *Theoria*, XLIII(2), 1977, 129–43.

11. Do We Need 'Fuzzy Logic'?, *International Journal of Man-Machine Studies*, 11, 1979, 432–45.

12. Dummett's Justification of Deduction, *Mind*, 95, 1982: 216–39.

13. Peirce and Logicism: Notes Towards an Exposition, *Transactions of the Charles S. Peirce Society*, XXIX(1), 1993, 35–56.

14. Knowledge and Propaganda: Reflections of an Old Feminist, *Partisan Review*, LX, 4, Fall 1993, 556–63. Reprinted in *Measure*, 124, September/October, 1994, 1 and 7-10; in Kurzweil and William Phillips eds., *Our Country, Our Culture* Boston: Partisan Review Press, 1994, 57-65; in Haack, *Manifesto of a Passionate Moderate*, 1998, 123–36; and in Pinnick C. Koertge N. and Almeder R. eds., *Scrutinizing Feminist Epistemology*, Piscataway, NJ: Rutgers University Press, 2003, 7–19.

15. 'We pragmatists...': Peirce and Rorty in Conversation, *Agora*, 15(1), 1996, 53–68. Reprinted in *Partisan Review*, LXIV(1), 1997, 91–107; and in Haack, *Manifesto of a Passionate Moderate* (1998), 31–47. In Portuguese in *Filosofia e Filosofia de Educacão*, Portal Brasileiro de Filosofia, December 2001. Forthcoming in Chinese in *World Philosophy*, Beijing.

16. As for that phrase 'studying in a literary spirit . . . , *Proceedings and Addresses of the American Philosophical Association*, 70, 2, 1996, 57–75. Reprinted in Haack, *Manifesto of a Passionate Moderate*, 1998, 31–47. In Spanish in *Analogia Filsófica; Revista de Filosofía*, XII(1), 1998, 157–87. In Portuguese in *Filosofia Analitica, Pragmatismo e Ciência*, Paulo Margutti et al. eds., Minas Gerais, Brazil: Editoria UFMG, 1998, 40–70.

17. Preposterism and Its Consequences, in *Social Philosophy and Policy*, 13(2), 1996, 296–315; reprinted in *Scientific Innovation, Philosophy and Public Policy*, eds. Ellen Frankel Paul, Fred D. Miller, and Jeffery Paul (New York: Cambridge University Press, 1996), 296–315.

18. Concern for Truth: What It Means, Why It Matters, in Paul R. Gross, Norman Levitt, and Marin W. Lewis, eds, *The Flight from Science and Reason*, Annals of the New York Academy of Sciences, 775, 1996, 57–62 (reprinted Baltimore: Johns Hopkins University Press, 1997. In Spanish in María-José Frápolli and Juán A. Nicolas, eds, *Teorías de la Verdad en el Siglo XX*, Madrid: Tecnos, 1988, 53–62. In French in *Logique en Perspective: Mélanges offerts au Professeur Gochet*, eds. Francois Beets and Eric Gilet Brussels: Ousia, 2000, 289–302.

19. A Foundherentist Theory of Empirical Justification, in Louis Pojman, ed., *The Theory of Knowledge: Classic and Contemporary Readings* Belmont, CA: Wadsworth, 1998, 283–93. Reprinted in Ernest Sosa and Jaegwon Kim, eds., *Epistemology: An Anthology*, Oxford: Blackwell, 2000, 226–36; in Huemer, M. ed. *Epistemology: Contemporary Readings*, New York: Routledge, 2002, 417–34; and in Luper, S. ed., *Essential Knowledge*, New York: Longman's, 2004, 157–67. In Spanish in *Agora*, 18(1), 1999, 35–53. In French in *Carrefour*. Ottawa, October, 2001, 39–60.

20. Staying for an answer: the untidy process of groping for truth, *Times Literary Supplement*, July $9^{th}$, 1999, 12–14; reprinted in Daphne Patai and Will Corrall, eds. *Theory's Empire: An Anthology of Dissent*, New York: Columbia University Press, 2005, 85–108. In Portuguese in *Best of*, Lisbon, February 2000, 5–14. In Spanish in *El Malpensante: Lecturas Paradójicas*, Colombia, 28, February-March 2001, 28–41.

21. Formal Philosophy: A Plea for Pluralism, in Victor Hendricks and John Symons, eds., *Formal Philosophy*, New York: Automatic Press/VIP, 2005, 77–98.

22. On Legal Pragmatism: Where Does 'The Path of the Law' Lead Us?, *The American Journal of Jurisprudence*, 50, 2005, 71–105.

23. Scientific Secrecy and 'Spin': The Sad, Sleazy Story of the Trials of Remune. *Law and Contemporary Problems*, 69.5, 2006, 47–67.

24. On Logic in the Law: 'Something, but not all,' *Ratio Juris*, 20.1, March 2007, 1–31.

25. Of Truth, in Science and in Law, *Brooklyn Law Review*, 73.2, 2008.

For a complete list, see Haack's webpage: http://www.miami.edu/phi./haa

**Service to the Profession:** Editorial Board Member, *Theoria* (Sweden), 2007–; Consultant to the Thomas B. Fordham Foundation, reporting on science standards across the 50 states, 2005; Editor in Chief, With Chen Bo of Book Series, Western Philosophy in Translation, 2005-; Editorial Board Member, *Syntaxis*, Uruguay, 2005–; Editorial Board Member, *Contrastes*, Spain, 2005–; Advisory Board Member, *Ratio Juris*, Italy, 2004–; Member, *SSRN Abstracting Journal of Evidence and Evidentiary Procedure*, 2004–; Editorial Board Member, *Episteme*, 2002–; Editorial Board Member, *Journal of Philosophy, Science, and Law*, 2001–; Editorial Board Member, *Teorema*, Spain, 1998–; Editorial Board Member, *Metaphilosophy*, 1996–; Advisory Board Member, Leverhulme Evidence Project; President, Charles S. Peirce Society, 1994–1995; Advisory Board Member, Shannon Center for Advanced Studies, University of Virginia, 1994–2004; Editor, issue of the American Philosophical Association *Newsletter on Philosophy and Law* on scientific testimony, 2003; External appraiser, College of Liberal Arts, Concordia University, Canada, 2003; External appraiser, Philosophy, University of Western

Ontario, Canada, 2003; Advisory Board Member, Peirce Edition Project, 1998–; Member of the Advisory Board of the Peirce Edition Project, 1998–; Editor, special issue of *The Monist* on Feminist Epistemology, For and Against, 1994; Editorial Board Member, *The Monist*, 1994–; Editorial Board Member, *History and Philosophy of Logic*, 1979–.

**Teaching:** Award for Outstanding Graduate Mentor, University of Miami, 1997; Award for Excellence in Teaching, American Philosophical Association, 1995; Award for Excellence in Teaching, University of Miami, 1994. Former students include: Cambridge: Graham Priest (undergraduate: now at the University of Melbourne); Warwick: Mark Migotti (M.Phil; now at the University of Calgary); Luciano Floridi (Ph.D.; now at Wolfson College, Oxford). Oxford: Cheryl Misak (D. Phil., jointly supervised with with David Wiggins; now at the University of Toronto). Miami: Cornelis de Waal (Ph.D., now at Indiana University/Purdue University); Robert Lane (Ph.D.; now at University of West Georgia); Jason Borenstein (Ph. D.; now at Georgia Tech); Rosa Mayorga (Ph.D.; now at Virginia Tech); Kiriake Xerohemona (Ph.D.; now at Florida International University); Cheng His Heng (Ph.D.; now at National Tsing Hua University of Taiwan).

**Vision Statement:** As I make clear in "Preposterism and Its Consequences" (17 above), I am allergic to this concept; however, in "Formal Philosophy: A Plea for Pluralism" I supply a list of some important philosophical questions which I believe (to borrow Peirce's phrase) presently "press for industrious and solid investigation".

**Honours and Awards:** *Susan Haack – A Lady of Distinctions: A Philosopher Responds to Critics*, ed. Cornelis de Waal, presently in press; Selma V. Forkosch Award for Excellence in Writing, 2006; listed in the *Sunday Independent* London as one of the ten most important women philosophers of all time, 2005; included in Peter J. King, *One Hundred Philosophers: The Life and Work of the Worlds's Greatest Thinkers* (New York: Barrons, 2004); Faculty Senate Distinguished Scholar Award, University of Miami, 2002; British delegate to the Institut Internationale de Philosophie, 1999-; (National) Romanell Phi Beta Kappa Professor, 1997–1998; Provost's Award for Scholarly Activity, University of Miami, 1997; Honorary member of Phi Beta Kappa from 1997; Included in Mary Warnock, ed. *Women Philosophers*, 1995; Honorary member of Phi Kappa Phi from 1994; Harkness Fellow, 1975–1976.

*Philosophy of Logics* cited in the *Oxford English Dictionary* on the meaning of "variable."
Entry on foundherentism in the Fontana/Norton *Dictionary of Modern Thought*.

## HÁJEK, Petr

**Specialties:** Mathematical logic, logical foundations of computer science.

**Born:** 6 February 1940 in Prague, Czechoslovakia (now Czech Republic).

**Educated:** Charles University Prague, Faculty of Mathematics and Physics 1957-1963, master thesis from algebra; CSc (equivalent of PhD) in the Czechoslovak Academy of Sciences 1965.

**Dissertation:** from set theory, informal supervisor Petr Vopěnka. DrSc ("big doctorate") 1990 (Academy of sciences).

**Regular Academic or Research Appointments:** INSTITUTE OF COMPUTER SCIENCE, ACADEMY OF SCIENCES OF THE CZECH REPUBLIC: from 2000 senior researcher, 1992-2000 director of the institute. 1963-1992 Mathematical Institute of Czechoslovak academy of sciences (aspirant, junior researcher, senior researcher).

**Visiting Academic or Research Appointments:** University Heidelberg (Germany) 1968-69 (two semesters), Cuban Academy of Sciences La Habana 1983 (3 months), University Siena (Italy) 1989 (two months).

**Research Profile:** 1963-1970 work in set theory (independence proofs by interpretations); cca 1969-1992 metamathematics of arithmetic - the joint monograph with P. Pudlák has become one of basic references in the domain. In parallel since 1963 foundations of mechanized hypothesis formation (the GUHA method), culminating by the joint monograph with T. Havránek. Since about 1995 till present mathematical fuzzy logic — his monograph Metamathematics of fuzzy logic has become a central source for fuzzy logic as a kind of fully fledged symbolic (formal) many-valued logic, both propositional and predicate logic, with double semantics — standard and algebraic. This logic is developed further by an international group of logicians.

**Main Publications:**

1. What is Mathematical Fuzzy Logic. Fuzzy Sets and Systems, Vol. 157, 2006, pp. 597-603

2. On Arithmetic in the Cantor-Lukasiewicz Fuzzy Set Theory. Archive for Mathematical Logic, Vol. 44, 2005, pp. 763-782

3. Arithmetical Complexity of Fuzzy Predicate Logics - A Survey. Soft Computing, Vol. 9, 2005, pp. 935-941

4. (with Holeňa M.) Formal Logics of Discovery and Hypothesis Formation by Machine. Theoretical Computer Science, Vol. 292, 2003, pp. 345-357

5. A New Small Emendation of Gödels Ontological Proof. Studia Logica, Vol. 71, 2002, pp. 149-164

6. (with Shepherdson J.) A Note on the Notion of Truth in Fuzzy Logic. Annals of Pure and Applied Logic, Vol. 109, 2001, pp. 65-69

7. (with Paris J., Shepherdson J.) The Liar Paradox and Fuzzy Logics. Journal of Symbolic Logic, Vol. 65, 2000, No. 1, pp. 339-346

8. Metamathematics of Fuzzy Logic. - Dordrecht, Kluwer Academic Publ. 1998, 297p. (Trends in Logic Vol. 4)

9. Interpretability and Fragments of Arithmetic. In: Arithmetic, Proof Theory and Computational Complexity. (Ed.: Clote P., Krajíček J.) - Oxford, Clarendon Press 1993, pp. 185-196

10. (with Pudlák P.) Metamathematics of First-Order Arithmetic. Berlin, Springer-Verlag 1993, 460p. Perspectives in Mathematical Logic

11. (with Havránek T., Jiroušek R.) Uncertain Information Processing in Expert Systems. Boca Raton, CRC Press 1992, 285p.

12. (with Clote P.,Paris J.) On Some Formalized Conservation Results in Arithmetic. Archive for Mathematical Logic, Vol. 31, 1991, pp. 201-218

13. (with Montagna F.) The Logic of Pi-1 Conservativity. Archive for Mathematical Logic, Vol. 30, 1990, pp. 113-123

14. (with Kučera A.) On Recursion Theory in IE1. The Journal Symbolic of Logic, Vol. 54, 1989, pp. 576-589

15. Arithmetical Interpretations of Dynamic Logic. Journal of Symbolic Logic, Vol. 48, 1983, pp. 704-71

16. (with Havránek T.) Mechanizing Hypothesis Formation - Mathematical Foundations of a General Theory. Berlin, Springer-Verlag 1978, 396p.

17. (with Vopěnka P.) The Theory of Semisets. Prague, Academia 1972, 332p.

18. (with Hájková M.) On Interpretability in Theories Containing Arithmetic. Fundamenta Mathematicae, Vol. 76, 1972, pp. 131-137

19. (with Havel I., Chytil M.) The GUHA-Method of Automatic Hypotheses Determination. Computing, Vol. 1, 1966, pp. 293-308

20. The Consistency of Church's Alternatives. Bull. Acad. Polon. Sci., Vol. 14, 1966, pp. 31-47

A full list of publications available on the address http://www.cs.cas.cz/people/php.list.php automatically generated.

*Work in Progress*

21. Foundations of fuzzy logic. Seminar on applied mathematical logic (existing from 1965).

**Service to the Profession:** Member of editorial board of several journals (Archive for mathematical logic, Soft computing, Czechoslovak Mathematical Journal, Kybernetika, Fundamenta Informaticae, earlier also Studia Logica). Member of the Association for Symbolic logic (1992-1994 member of the Council). 1995-1999 first vice-president of the International Union of History and Philosophy of Science (Division Logic and methodology of science). 1996-2003 president of Kurt Gödel Society. Docent 1993, (full) professor 1997 (Charles University Prague). Member of program committees of many conferences.

**Teaching:** Even if in the Academy of sciences the main activity is research, most fellows do some teaching on universities. Hájek has lectured for years on the Faculty of mathematics and physics of the Charles University (mathematical logic), faculty of nuclear engineering of the Czech Technical University Prague (Foundations of fuzzy logic) and also on the Technical University Vienna. Advisor of some PhD theses.

**Vision Statement:** The term "fuzzy logic" has been frequently used in a very broad sense, meaning just any use of the notion of fuzzy sets. In the narrow sense, (mathematical) fuzzy logic is a sort of many- valued mathematical logic with a comparative notion of truth, most elaborated fuzzy propositional and predicate logics being t-norm based (see e.g. my Metamathematics of fuzzy logic, Kluwer 1998). Problems of continuing importance include various kinds of semantics, (in)completeness, computational and arithmetical complexity etc., fuzzy logics in the context of substructural logics, fuzzy logic as logics of resoning under vagueness, development of important parts of mathematics as axiomatic theories over fuzzy logic. Mathematical fuzzy logic is expected to be generally recognized as a fully fledged part of mathematical logic sui generis with deep results and interesting problems which can, in the full recongition of classical (two-valued, Boolean) logic as the queen of logic, contribute to the building a bridge between formal logical systems and aspects of logical reasoning in natural language.

**Honours and Awards:** Medal for the credit of the state of the state in the domain of science from the President of the Czech Republic, 2006; Medal of the minister of education of the Czech Republic 1996; honorary professor of the Technical University Vienna 1999; medal "De scientiae et humanitate optime meritis" of the Academy of Sciences of the Czech Republic 2006. Member of

## HALBACH, Volker

**Specialties:** Philosophical logic, philosophy of mathematics, theories of truth,.

**Born:** 21 October 1965 in Ingolstadt, Germany.

**Educated:** University of Munich, MA; Constance, Habilitation, Universities of Florence and Munich, Dr Phil.

**Dissertation:** *Tarski-hierarchies*; supervisor, Matthias Varga von Kibed.

**Regular Academic or Research Appointments:** READER, UNIVERSITY OF OXFORD, 2004–; TUTORIAL FELLOW, NEW COLLEGE, OXFORD, 2004–; Assistant, University of Constance, 2000-2004; Wissenschaftlicher Mitarbeiter, University of Constance, 1997-2000.

**Visiting Academic or Research Appointments:** Vertretungsprofessor Marburg, 2003-2004; Vertretungsprofessor, 2002-2003; Visiting Fellow, University of California at Irvine, 2001; Visiting Scholar University of Notre Dame, 1996.

**Research Profile:** Volker Halbach has worked extensively on formal theories of truth. In particular, he has focused on the proof theory of axiomatic theories truth on obtained various result on the strength of various axiomatic systems of truth.

**Main Publications:**

1. "A system of complete and consistent truth", *Notre Dame Journal of Formal Logic* 35 (1994), 311-327
2. *Axiomatische Wahrheitstheorien*, Akademie Verlag, Berlin, 1996
3. "Tarskian and Kripkean truth", *Journal of Philosophical Logic* 26 (1997), 69-80
4. "Conservative theories of classical truth", *Studia Logica* 62 (1999), 353-370
5. "Disquotationalism and Infinite Conjunctions", *Mind* 108 (1999), 1-22
6. "Two proof-theoretic remarks on EA+ECT" (with Leon Horsten), *Mathematical Logic Quarterly* 46 (2000), 461-466
7. "How Innocent is Deflationism?", *Synthese* 126 (2001), 167-194
8. "Disquotational Truth and Analyticity", *Journal of Symbolic Logic* 66 (2001), 1959-1973
9. "Possible Worlds Semantics for Modal Notions Conceived as Predicates" (with Hannes Leitgeb and Philip Welch), *Journal of Philosophical Logic* 32 (2003), 179-223
10. "Bealers Masterargument: ein Lehrstück zum Verhältnis von Metaphysik und Semantik" (with Holger Sturm), *Facta Philosophica* 6 (2004), 97-110
11. "Computational Structuralism" (with Leon Horsten), *Philosophia Mathematica* 13 (2005), 174-186
12. "Axiomatizing Kripke's Theory of Truth" (with Leon Horsten), to appear in the *Journal of Symbolic Logic*

For further publications see http://users.ox.ac.uk/~sfop0114/fsp.htm

*Work in Progress*

13. A book on axiomatic theories of truth summarizing the development in this field over the last 30 years
14. Various articles concerning the formal properties of predicates of necessity as opposed to sentential operators of necessity
15. An article on the axiomatisation of stable truth (revision semantics)

**Service to the Profession:** Principal Overseas Partner, research project Logical Methods in Epistemology, Semantics, and Philosophy of Mathematics, 2004–; Vice President, Gesellschaft für Analytische Philosophie, 2003–.

**Vision Statement:** "Philosophy without logic is empty, logic without philosophy is blind."

**Honours and Awards:** Wolfgang-Stegmüller-Preis Award, Gesellschaft für Analytische Philosophie, 2003; Feodor-Lynen Scholarship, Alexander-von-Humboldt Foundation, 1996.

## HAMKINS, Joel David

**Specialties:** Mathematical logic, set theory, infinitary computability theory.

**Born:** 1 February 1966 in Racine, Wisconsin, USA.

**Educated:** University of California at Berkeley, PhD Mathematics, 1994; University of California at Berkeley, CPhil in Mathematics, 1991; California Institute of Technology, BS (with honors), 1988.

**Dissertation:** *Lifting and Extending Measures; Fragile Measurability*; supervisor, W. Hugh Woodin.

**Web Address:** http://jdh.hamkins.org

**Regular Academic or Research Appointments:** PROFESSOR OF MATHEMATICS, THE CITY UNIVERSITY OF NEW YORK (CUNY), THE COLLEGE OF STATEN ISLAND, 2003–; DOCTORAL FACULTY IN MATHEMATICS, THE CUNY GRADUATE CENTER, 1997–; DOCTORAL FACULTY IN COMPUTER SCIENCE, THE CUNY GRADUATE CENTER, 2002–; Associate Professor of Mathematics, Georgia State University, 2002–2003; Associate Professor of Mathematics, The College of Staten Island of CUNY, 1999–2002; Assistant

Professor of Mathematics, The College of Staten Island of CUNY, 1995–1998.

**Visiting Academic or Research Appointments:** Visiting Researcher, NWO Bezoekersbeurs, Universiteit van Amsterdam, Institute for Logic, Language and Computation, June–August 2005; Mercator-Gastprofessor, The Deutsche Forschungsgemeinschaft, Universität Münster, Institut für mathematische Logik, 2004; Visiting Associate Professor of Mathematics, Carnegie Mellon University, 2000–2001; Research Fellow, Japan Society for the Promotion of Science, Kobe University Graduate School of Science and Technology, Japan, 1998; Visiting Assistant Professor, Mathematics, University of California at Berkeley, 1994–1995.

**Research Profile:** Joel David Hamkins conducts research in mathematical logic, particularly in set theory and its connections to other areas, broadly emphasizing the mathematics and philosophy of the infinite and the abstract relations between models of set theory. He has focused on the fundamental interaction of forcing and large cardinals, investigating how large cardinals are affected by the move to a forcing extension and how the large cardinal embeddings of a forcing extension relate to those in the ground model. His work on the approximation and covering properties showed that the large cardinal embeddings of a forcing extension, in very general circumstances, are amenable to the ground model. These ideas underlie Richard Laver's recent proof that every model of set theory is a definable class in all its forcing extensions. With Jonas Reitz, he introduced the Ground Axiom, which asserts that the universe is not a forcing extension of any inner model, and despite the prima facie second order nature of this axiom, it is actually first order expressible in the language of set theory. He introduced several new forcing axioms, the Maximality Principles, which express directly the abstract idea behind all forcing axioms, that any statement that could become true in all forcing extensions is already true. He introduced the corresponding forcing interpretation of modal logic in set theory and in joint work with Benedikt Loewe, determined the exact modal power of this interpretation. He provided some principal forcing techniques for several new large cardinal axioms, such as the unfoldable cardinals, tall cardinals, and the Wholeness Axiom. With Andy Lewis, he introduced the notion of infinite time Turing machines, which provide a natural model of infinitary computability by extending the operation of ordinary Turing machines into transfinite ordinal time. This work in infinite time computability theory has now led to infinite time complexity theory, where in joint work with Ralf Schindler and Vinay Deolalikar he proved the infinitary analogue of $P \neq NP \cap coNP$, and to infinite time computable model theory, which extends the classical finite time theory to uncountable structures and theories. His work on the automorphism tower problem in group theory includes joint work with Simon Thomas on the malleability of the automorphism tower of a group by forcing, as well as the fundamental group theoretic fact that every group has a terminating automorphism tower. His most recent work includes the application of set theoretical methods, such as the Proper Forcing Axiom, to long open problems in models of arithmetic, as well as an investigation of the generalization of PFA to kappa-proper forcing. His co-authors include many of the world's leading researchers.

**Main Publications:**

1. Fragile measurability, *Journal of Symbolic Logic*, 59(1), pp. 262–282, 1994.
2. Canonical seeds and Prikry trees, *Journal of Symbolic Logic*, 62(2), pp. 373–396, 1997.
3. Every group has a terminating transfinite automorphism tower, *Proceedings of the American Mathematics Society*, 126(11), pp. 3223–3226, 1998.
4. Small forcing makes any cardinal superdestructible, *Journal of Symbolic Logic*, 63(1), pp. 51–58, 1998.
5. Superdestructibility: a dual to Laver's indestructibility, with S. Shelah, *Journal of Symbolic Logic*, 63(2), pp. 549–554, 1998, [HmSh:618].
6. Universal indestructibility, with A. W. Apter, *Kobe Journal of Mathematics*, 16(2), pp. 119–130, 1999.
7. Infinite time Turing machines, with A. Lewis, *Journal of Symbolic Logic*, 65(2), pp. 567–604, 2000.
8. The lottery preparation, *Annals of Pure and Applied Logic*, 101(2-3), pp. 103–146, 2000.
9. Changing the heights of automorphism towers, with S. Thomas, *Annals of Pure and Applied Logic*, 102(1–2), pp. 139–157, 2000.
10. Small forcing creates neither strong nor Woodin cardinals, with W. H. Woodin, *Proceedings of the American Mathematics Society*, 128(10), pp. 3025–3029, 2000.
11. Unfoldable cardinals and the GCH, *Journal of Symbolic Logic*, 66(3), pp. 1186–1198, 2001.
12. The Wholeness Axiom and V=HOD, *Archive for Mathematical Logic*, 40(1), pp. 1–8, 2001.
13. Post's problem for supertasks has both positive and negative solutions, A. Lewis, *Archive for Mathematical Logic*, 41(6), pp. 507–523, 2002.
14. A simple maximality principle, *Journal of Symbolic Logic*, 68, pp. 527–550, June 2003.
15. Extensions with the approximation and cover properties have no new large cardinals, *Fundamenta Mathematicae*, 180(3), pp. 257–277, 2003.
16. $P \neq NP \cap coNP$ for infinite time Turing machines, with R.-D. Schindler, *Journal of Logic and Com-*

*putation*, 15, pp. 577–592, October 2005.

17. The necessary maximality principle for c.c.c. forcing is equiconsistent with a weakly compact cardinal, with W. H. Woodin, *Mathematical Logic Quarterly*, 51(5), pp. 493–498, 2005.

*Work in Progress*

18. Diamond (on the regulars) can fail at any strongly unfoldable cardinal, with M. Džamonja, to appear in the *Annals of Pure and Applied Logic*.

19. *Forcing and Large Cardinals*, monograph in progress.

20. The Ground Axiom and V≠HOD, with J. Reitz and W. H. Woodin, in progress.

A full list of publications is available at http://jdh.hamkins.org

**Service to the Profession:** Editor, *Notre Dame Journal of Formal Logic*, 2004–; Advisory Board Member, MidAtlantic Mathematical Logic Seminar, 1997–; Referee, *Annals of Pure and Applied Logic*; Referee, *Journal of Symbolic Logic*; Referee, *Journal of Mathematical Logic*; Referee, Journal *of Philosophical Logic*; Referee, *Archive for Mathematical Logic*; Referee, *Notre Dame Journal of Formal Logic*; Referee, *Studia Logica*; Referee, *Integers*; Referee, National Science Foundation and the Nederlandse Organisatie voor Wetenschappelijk; Organizer, New York City Logic Conferences 1999, 2002, 2005.

**Teaching:**
Professor Hamkins has been recognized for his effective, lively teaching at both the undergraduate and graduate level. He regularly earns among the highest teaching evaluations in his department, and is often asked to speak at special events, such as those recruiting students to mathematics. He was awarded the Distinguished Undergraduate Teaching Award ("Teacher of the Year"), at the University of California at Berkeley in 1995. He is active in all aspects of graduate education at the CUNY Graduate Center, and has served as dissertation supervisor for George Leibman, Victoria Gitman, Thomas Johnstone and Jonas Reitz, and on the dissertation committees of Sidney Raffer and Erez Shochat.

**Vision Statement:**
Set theory has matured to become the study of its fundamental objects, the models of set theory, and now exhibits a category theoretic nature, in which these models are connected by the forcing relation and diverse large cardinal embeddings.

**Honours and Awards:** Visiting Researcher Grant, Nederlandse Organisatie voor Wetenschappelijk, Bezoekersbeurs, Universiteit van Amsterdam, Institute for Logic, Language and Computation, 2005; Mercator-Gastprofessor, Deutsche Forschungsgemeinschaft, Universität Münster, Institut für mathematische Logik, 2004; Performance Excellence Award, CUNY, College of Staten Island, 2000; National Science Foundation Research Grant, 1999–2002; Research Fellow, Japan Society for the Promotion of Science, 1998; PSC-CUNY Grant Awards, The City University of New York, 1995–2006; Collaborative Incentive PSC-CUNY grant awards, 1996–1998, 1999–2001, 2005–2007.

## HANSSON, Sven Ove

**Specialties:** Belief revision, formal epistemology, formal value theory, logic of preferences, deontic logic, decision theory, logical foundations of economics and social science.

**Born:** 6 December 1951 in Kävlinge, Sweden.

**Educated:** Lund University, PhD Practical Philosophy, 1999; Uppsala University, PhD Theoretical Philosophy, 1991; Uppsala University, BA, 1981; Lund University, Bachelor of Medicine, 1972.

**Dissertations:** PhD Practical Philosophy, *Structures of Value*; supervisor, Wlodek Rabinowicz; PhD Theoretical Philosophy, *Belief Base Dynamics*; supervisor, Stig Kanger.

**Regular Academic or Research Appointments:** DEPARTMENT HEAD, PHILOSOPHY AND THE HISTORY OF TECYNOLOGY, 2005–; PROFESSOR, ROYAL INSTITUTE OF TECHNOLOGY (KTH), STOCKHOLM, 2000–; Researcher, Royal Institute of Technology (KTH), Stockholm 1999–2000; Associate Professor, Uppsala University 1993–1999.

**Research Profile:** In belief revision theory, Sven Ove Hansson has focused on constructing cognitively realistic models. He has developed the use of finite belief bases and provided an axiomatic characterization of base-generated changes on belief sets, i.e. operations on a (logically closed) belief set that are generated from operations on some finite base for that set. [See 6] He has introduced a new type of contraction, kernel contraction. This is a generalization of partial meet contraction that is the standard (AGM) operator of belief contraction. [9] He has also developed representations of non-prioritized belief revision, i.e. belief revision in which only some inputs are accepted whereas others are rejected. [14, 16] He has studied the interpretation of belief revision theories and the connections between formal and informal approaches to epistemology. [17] Recently, he has investigated the relationship between coherentism and foundationalism. Based on studies of how mutual support can be distributed in a (coherent) set of objects, such as sentences, he argues that we need to modify conventional views on the structure of coherentist theories. [20] Hansson is currently developing a new model for belief revision (specified meet contraction) in which he introduces cognitive limitations by requiring that belief sets be axiomatizable, but without the introduction of belief bases.

In preference logic, Hansson has studied rationality criteria for preferences and investigated the formal relations between preferences on wholes and preferences on their parts. [2] He has investigated how monadic predicates representing "good" and "bad" can be introduced into preference logic. [3] He has also provided a formal representation of changes in preferences. [11]

In deontic logic Hansson has criticized standard deontic logic (SDL). SDL identifies our obligations with how we would act in an ideal world. Hansson claims that this is a fundamental error, since to act as if one lived in an ideal world is bad moral advice, associated with wishful thinking rather than well-considered moral deliberation. [19] He has proposed another semantic approach that is based on preference logic, according to the simple principle that "What is worse than something forbidden is itself forbidden." In a deontic logic built on preference logic according to this principle, the usual deontic paradoxes do not arise. [2, 18] Hansson has also developed a formal framework for the representation of various types of legal relations such as rights. This framework contains a general mechanism for expressing how legal rights can be activated by performative actions such as making a claim and granting a permission. [12]

In his studies of the logic of social choice, Hansson has extended the standard decision-theoretical framework so that major types of procedural preferences that influence an individual's choices can be represented. Procedural preferences, such as preferences for concensus, preferences for being part of the winning majority, and preferences against ties, induce participants to take part in compromises that would not be motivated by their strictly consequentialist preferences. His formal analysis shows how such preferences provide mechanisms for decisional stability in cases when preferences restricted to outcomes cannot do this. This seems to be a major reason why decision instability plays a smaller role in actual decision-making than what is predicted by Arrovian social choice theory. [4]

In his studies of the logical foundations of economics, Hansson has investigated ways to express how one person's well-being may depend on the material resources of other persons. He has shown that Pareto efficiency on the level of well-being may require "non-Paretian" reduction of inequality on the level of material resources. He has also investigated foundational issues in cost-benefit analysis.

In his methodological writings, Hansson has defended the application of logic and other formal methods to problems in philosophy, but he has also warned of some of the pitfalls of formalization. [15] In addition to his work in logic Hansson publishes extensively in other areas of philosophy, such as decision theory, moral and political philosophy, philosophy of risk, medical ethics, and the philosophy of science and technology.

**Main Publications:**

1. Sven Ove Hansson, A Textbook of Belief Dynamics. Theory Change and Database Updating. Kluwer 1999.
2. Sven Ove Hansson, The Structure of Values and Norms, Cambridge University Press, 2001.
3. Sven Ove Hansson, "Defining 'good' and 'bad' in terms of 'better'", Notre Dame Journal of Formal Logic 31:136-149, 1990.
4. Sven Ove Hansson, "A Procedural Model of Voting", Theory and Decision 32:269-301, 1992.
5. Sven Ove Hansson, "In Defense of the Ramsey Test", Journal of Philosophy 89:522-540, 1992.
6. Sven Ove Hansson, "Theory Contraction and Base Contraction Unified", Journal of Symbolic Logic 58:602-625, 1993.
7. Sven Ove Hansson, "Reversing the Levi Identity", Journal of Philosophical Logic 22:637-669, 1993.
8. Sven Ove Hansson, "A Note on Anti-Cyclic Properties of Complete Binary Relations", Reports on Mathematical Logic 27:41-44, 1993.
9. Sven Ove Hansson, "Kernel Contraction", Journal of Symbolic Logic 59:845-859, 1994.
10. Sven Ove Hansson, "Some Solved and Unsolved Remainder Equations", Mathematical Logic Quarterly 41:362-368, 1995.
11. Sven Ove Hansson, "Changes in Preference", Theory and Decision 38:1-28, 1995.
12. Sven Ove Hansson, "Legal Relations and Potestative Rules", Archiv für Rechts- und Sozialphilosophie 82:266-274, 1996.
13. Sven Ove Hansson and David Makinson, "Applying Normative Rules with Restraint", pp. 313-332 in ML Dalla Chiara et al (eds) Logic and Scientific Method, Kluwer 1997.
14. Sven Ove Hansson, "Semi-revision" Journal of Applied Non-Classical Logic 7:151-175, 1997.
15. Sven Ove Hansson, "Formalization in philosophy", Bulletin of Symbolic Logic 6:162-175, 2000.
16. Sven Ove Hansson, Eduardo Fermé, John Cantwell, and Marcelo Falappa, "Credibility-Limited Revision", Journal of Symbolic Logic 66:1581-1596, 2001.
17. Sven Ove Hansson, "Ten Philosophical Problems

in Belief Revision", Journal of Logic and Computation 13:37-49, 2003.

18. Sven Ove Hansson, "A new representation theorem for contranegative deontic logic", *Studia Logica* 77:1-7, 2004.

19. Sven Ove Hansson, "Ideal Worlds – Wishful Thinking in Deontic Logic", Studia Logica 82:329-336, 2006.

20. Sven Ove Hansson, "Coherence in Epistemology and Belief Revision", Philosophical Studies 128:93-108, 2006.

21. Replacement – a Sheffer stroke for belief revision, Journal of Philosophical Logic, in press.

22. Specified Meet Contraction, Erkenntnis, in press.

23. Contraction Based on Sentential Selection, Journal of Logic and Computation, 17: 479-498, 2007.

24. The false dichotomy between coherentism and foundationalism, Journal of Philosophy, 104:(6):290-300, 2007.

An extensive list of publications is available at Hansson's homepage: http://www.infra.kth.se/~soh

**Service to the Profession:** Member, Swedish National Committee for Logic, Methodology, and Philosophy, Royal Swedish Academy of Sciences, 2003–; Member, Swedish Government's Research Advisory Board, 2000–2005; Editor, Theoria, 1999–; Board Member, Swedish Natural Science Foundation, 1989–1992; Committee Member, Physics and Mathematics, Natural Science Foundation, 1987–1992; Founding Chairperson, and Board Member, Swedish Skeptics.

**Teaching:** Hansson is the author of *A Textbook of Belief Dynamics. Theory change and database updating).* [1] PhD theses in logic that he has supervised are: Erik Olsson, PhD in theoretical philosophy 1997, Eduardo Fermé, PhD in computer science 1999, and Martin Peterson, PhD in philosophy 2003. Hansson is also a writer of popular science.

**Vision Statement:** The use of formal methods, including logic, can increase clarity and precision in many areas of philosophy. Formalized philosophy can also contribute to the construction of new and more sophisticated representations of human beliefs, preferences, norms, and interactions that are useful in economics and other branches of the social sciences. However, in spite of its great potential formalized philosophy is currently an endangered specialty. There is an urgent need to revitalize it and to strengthen its connections with other parts of philosophy.

# HAREL, David

**Specialties:** Dynamic logic, program specification, computability, finite model theory, visual languages.

**Born:** 12 April 1950 in London, UK.

**Educated:** Mass. Inst. Technology, PhD Computer Science; Tel-Aviv University, MSc Computer Science, 1976; Bar-Ilan University, BSc Mathematics and Computer Science, 1974.

**Dissertation:** MSc: *Completeness Issues for Inductive Assertions and Hoare's Method*", supervisor, Amier Puneli; Doctoral: *Logics of Programs: Axiomatics and Descriptive Power*; supervisor, Vaughan Pratt.

**Regular Academic or Research Appointments:** PROFESSOR, DEPARTMENT OF COMPUTER SCIENCE AND APPLIED MATHEMATICS, WEIZMANN INSTITUTE OF SCIENCE, REHOVOT, ISRAEL, 1989–; Assoc.Professor; 1983-1989; Senior Scientist, 1980-1983.

**Visiting Academic or Research Appointments:** Visiting Professor, The University of Singapore, 2004; Visiting Professor, Computer Science Dept., University of Birmingham, UK, 2003; Visiting Professor, VERIMAG, University of Grenoble, 2001-2002; Founder and Chief Scientist, SenseIT Technologies, Inc. (later DigiScents, Inc), 1998-2001; Founder and Chief Scientist, I-Logix, Inc., Andover, MA, and I-Logix Israel, Ltd, 1984-; Adjunct Professor, Open University, Israel, 1990-1999; Visiting Professor, Computer Science Department, Cornell University, 1994-1995; Visiting Professor, Computer Science Department, Carnegie-Mellon University, 1986-1987.

**Research Profile:** In the past David Harel has worked in several areas of theoretical computer science, including computability theory (especially high levels of undecidablility), logics of programs (especially dynamic logic), database and finite model theory theory (the power of queries on structured data), and automata theory. Over the years, his activity in these areas diminished, and in more recnet years he has become involved in several other areas, including software and systems engineering, object-oriented analysis and design, visual languages, layout of diagrams, modeling and analysis of biological systems, and the synthesis and communication of smell. He has published widely on these topics, including several books. He is the inventor of the language of statecharts (1983), and co-inventor of live sequence charts (1999), and was part of the team that designed the tools Statemate (1984-1987), Rhapsody (1997) and the Play-Engine (2003), and the idea of reactive animation (2002). His work is central to the behavioral aspects of the UML. He devotes part of his time to educational and expository work: In

1984 he delivered a lecture series on Israeli radio (see the book version), and in 1998 he hosted a series of programs on Israeli television. Some of his writing is intended for a general audience.

**Main Publications:**

1. D. Harel, *First-Order Dynamic Logic, Lecture Notes in Computer Science*, Vol. 68, Springer-Verlag, New York (133 pp.), 1979.
2. D. Harel, *Algorithmics: The Spirit of Computing*, Addison-Wesley, Reading, MA, (425 pp.) 1987. 2nd edition, 1992; 3rd edition, 2004 (with Y. Feldman). (Dutch, 1989; Hebrew (Open University Press), 1991; Polish, 1992.)
3. D. Harel, D. Kozen and J. Tiuryn, *Dynamic Logic*, MIT Press, 2000.
4. D. Harel and R. Marelly, *Come, Let's Play: Scenario-Based Programming Using LSCs and the Play-Engine*, Springer-Verlag, 2003.
5. D. Harel, A. R. Meyer and V. R. Pratt, "Computability and Completeness in Logics of Programs", *Proc. 9th ACM Symp. on Theory of Computing*, pp. 261-268, Boulder, Colorado, May 1977.
6. A. K. Chandra and D. Harel, "Computable Queries for Relational Data Bases", *J. Comput. System Sciences* 21 (1980), 156-178. (Also, *Proc. ACM 11th Symp. on Theory of Computing*, pp. 309-318, Atlanta, Georgia, April 1979.)
7. D. Harel and A. Pnueli, "On the Development of Reactive Systems", in *Logics and Models of Concurrent Systems* (K. R. Apt, ed.), NATO ASI Series, Vol. F-13, Springer-Verlag, New York, 1985, pp. 477-498.
8. D. Harel, "Effective Transformations on Infinite Trees, with Applications to High Undecidability, Dominoes and Fairness", *J. Assoc. Comput. Mach.* 33 (1986), 224-248.
9. D. Harel, "Statecharts: A Visual Formalism for Complex Systems", *Sci. Comput. Programming* 8 (1987), 231-274.
10. D. Harel, "Hamiltonian Paths in Infinite Graphs", *Israel J. Math.* 76:3 (1991), 317-336.
11. D. Harel, "Towards a Theory of Recursive Structures", *11th Ann. Symp. on Theoretical Aspects of Computer Science* (invited paper), Lecture Notes in Computer Science, Vol. 775, Springer-Verlag, 1994, pp. 633-645.
12. D. Harel and E. Gery, "Executable Object Modeling with Statecharts", *Computer* 30:7 (July 1997), IEEE Press, 31-42.
13. W. Damm and D. Harel, "LSCs: Breathing Life into Message Sequence Charts", *Formal Methods in System Design* 19:1 (2001), 45-80. (Preliminary version in *Proc. 3rd IFIP Int. Conf. on Formal Methods for Open Object-Based Distributed Systems* (FMOODS'99), (P. Ciancarini, A. Fantechi and R. Gorrieri, eds.), Kluwer Academic Publishers, 1999, pp. 293-312.)
14. D. Harel and R. Marelly, "Specifying and Executing Behavioral Requirements: The Play In/Play-Out Approach", *Software and System Modeling* (SoSyM) 2 (2003), 82-107.
15. D. Harel, L. Carmel and D. Lancet, "Towards an Odor Communication System", *Computational Biology and Chemistry* (formerly *Computers & Chemistry*) 27 (2003), 121-133.
16. N. Kam, D. Harel, H. Kugler, R. Marelly, A. Pnueli, E.J.A. Hubbard and M.J. Stern, "Formal Modeling of C. elegans Development: A Scenario-Based Approach", *Proc. 1st Int. Workshop on Computational Methods in Systems Biology* (ICMSB 2003), Lecture Notes in Computer Science, Vol. 2602, Springer-Verlag, pp. 4-20, Feb. 2003. (Revised version in *Modeling in Molecular Biology* (G. Ciobanu and G. Rozenberg, eds.), Springer, Berlin, 2004, pp. 151-173.)
17. D. Harel, "A Grand Challenge for Computing: Full Reactive Modeling of a Multi-Cellular Animal", *Bulletin of the EATCS*, European Association for Theoretical Computer Science, no. 81, 2003, pp. 226-235. (Reprinted in *Current Trends in Theoretical Computer Science: The Challenge of the New Century*, Algorithms and Complexity, Vol I (Paun, Rozenberg and Salomaa, eds.), World Scientific, pp. 559-568, 2004.)
18. S. Efroni, D. Harel and I.R. Cohen, "Towards Rigorous Comprehension of Biological Complexity: Modeling, Execution and Visualization of Thymic T Cell Maturation", *Genome Research* 13 (2003), 2485-2497.
19. D. Harel, "A Turing-Like Test for Biological Modeling", *Nature Biotechnology* 23 (2005), 495-496.

**Honours and Awards:** Doctor Honoris Causa, University of Rennes, 2005; Israel Prize, 2004; ACM Karlstrom Outstanding Educator Award, 1992; Book Prize, Main Selection of the Macmillan Library of Science, 1988.

# HÅSTAD, Johan

**Specialties:** Complexity theory, cryptography.

**Born:** 19 November 1960 in Danderyd Sweden.

**Educated:** Massachusets Institute of Technology, PhD Mathematics, 1986; Uppsala University, Master of Science in Mathematics, 1984; Stockholm University, Bachelor of Science in Mathematics, 1981.

**Dissertation:** *Computational Limitations of Small-Depth Circuits*; supervisor, Shafrira Goldwasser.

**Regular Academic or Research Appointments:** PROFESSOR, THEORETICAL COMPUTER SCIENCE, ROYAL INSTITUTE OF TECHNOLOGY, 1992–; Associate Professor, Computer Science, Royal Institute of Technology, 1988-1992; Massachusetts Institute of Technology, Post-Doc, 1986-1987.

**Visiting Academic or Research Appointments:** Institute for Advanced Study, 2000-2001.

**Research Profile:** Johan Håstad started out in the area of lower bounds for various limited computational models, in particular Boolean circuits and formulas. He has also done work in cryptography and a famous result in this area, obtained jointly with Impagliazzo, Luby and Levin, is that the existence of one-way functions is a necessary and sufficient condition for the existence of pseudorandom generators. Lately the focus of Håstad's research has shifted towards approximability of NP-hard optimization problem.

**Main Publications:**

1. J. Håstad, Almost Optimal Lower Bounds for Small Depth Circuits, in Randomness and Computation, Advances in Computing Research, Vol 5, ed. S. Micali, 1989, JAI Press Inc, pp 143–170.
2. P. Beame and J. Håstad, Optimal Bounds for Decision Problems on the CRCW PRAM, Journal of ACM, 1989, Vol 36, No 3, pp 643–670.
3. J. Håstad, The Shrinkage Exponent is 2, SIAM Journal on Computing, 1998, Vol 27, pp 48–64.
4. J. Håstad, R. Impagliazzo, L. Levin, and M. Luby, A Pseudorandom Generator from any one-way function, SIAM Journal on Computing, Vol, 28:4, 1999, pp 1364–1396.
5. J. Håstad, Clique is Hard to Approximate within $n^{1-\epsilon}$, Acta Mathematica, Vol. 182, 1999, pp 105–142.
6. J. Håstad, Some optimal inapproximability results, Journal of ACM, Vol 48, 2001, pp 798-859.

**Service to the Profession:** Editor, Theory of Computing, 2004-; Editor, Journal of the ACM, 1997-2003; Editor, SIAM Journal of Computing 1991-1999; Editor, Computational Complexity 1991-; Editor, Information Processing Letters, 1990-1993.

**Teaching:** Johan Håstad have supervised 8 PhD students.

**Vision Statement:** We have only scratched the surface in the study of efficient computation. Maybe computation is too difficult to understand completely but I believe that much remains to be discovered and the reward will be well worth the effort.

**Honours and Awards:** Plenary speaker at the ECM, Stockholm, 2004; Member, Royal Swedish Academy of Sciences, 2001; Winner, Göran Gustafsson prize in mathematics, 1999; Gödel prize, 1994; Chester Carlson's research prize, 1990; ACM Doctoral Dissertation Award, 1986.

# HAUSER, Kai

**Specialties:** Mathematical logic and foundations of mathematics, philosophy of mathematics, epistemology, philosophy of science.

**Educated:** California Institute of Technology, PhD Mathematics, 1989; Ruprecht- Karls-Universität, Heidelberg, Diplom in Mathematik und Philosophie, 1985.

**Dissertation:** *Independence Results for Indescribable Cadinals*; supervisor, W. Hugh Woodin.

**Regular Academic or Research Appointments:** ICREA RESEARCH PROFESSOR, PHILOSOPHY, UNIVERSITAT DE BARCELONA, 2005–; PROFESSOR, MATHEMATICS (APL.), TECHNISCHE UNIVERSITÄT BERLIN, 2000–; Assistant Professor, Mathematics, University of Michigan Ann Arbor, 1998–1992; Member, Scientific Staff, Institut für Mathematik, Ruprecht- Karls-Universität, Heidelberg 1985–1988.

**Visiting Academic or Research Appointments:** Visiting Professor, Center For Pure and Applied Mathematics, University of California Berkeley, 2005–2006; Guest Professor, German Cancer Research Center, 2005-; Visiting Professor, Philosophy, Bergische Universität Wuppertal, 2003–2004; Guest Professor, Interdisciplinary Studies, Bergische Universität Wuppertal, 2002–2003, Postdoctoral Research Fellow, Group in Logic and the Methodology of Sciences, University of California Berkeley, 1991–1997

**Research Profile:** Kai Hauser's mathematical work is mainly in set theory (infinitary combinatorics, forcing, large cardinals, core model theory and descriptive set theory). In philosophy he has been studying the objective existence of ideal objects and the ways in which they can be known.

**Main Publications:**

1. Gödel's Program Revisited. Part I: The Turn to Phenomenology, *Bulletin of Symbolic Logic,* 12 (2006), 529–590.
2. Is Choice Self-Evident?, *American Philosophical Quarterly* 42 (4), 237–261 (2005). (Abstract in Review of Metaphysics 59 (2005), 466.)
3. The Consistency Strength of Projective Absoluteness, *Annals of Pure and Applied Logic* 74 (1995), 245–295.
4. Indescribable Cardinals and Elementary Embeddings, *Journal of Symbolic Logic* 56 (1991), 439–457.

*Work in Progress*

5. The Axiom of Determinacy, (with Adrian R. D. Mathias and W. Hugh Woodin).

**Service to the Profession:** Guest Professor, University of Kabul and University of Herat, 2006; Editorial Advisory Board Member, Iranian Journal of Science and Technology, 1996-; Academic Mentorship Program for Minority Students, University of Michigan, Ann Arbor, USA, 1990–1991.

**Teaching:** Kai Hauser has taught a wide range of couses in mathematics and in philosophy at several institutions.

**Vision Statement:** Modern developments in mathematical logic should be incorporated, among other things, into philosophical discussions about truth, knowledge and consciousness. At the same time, there are fundamental questions about the status of new axioms in mathematics which call for a philosophical elucidation.

**Honours and Awards:** Heisenberg Fellow Deutsche Forschungsgemeinschaft (DFG), 1994–1999; Group in Logic and the Methodology of Sciences and DFG, University of California Berkeley, 1991–1997.

# HENLE, James Marston

**Specialties:** Combinatorial set theory, infinite-exponent partition relations, the structure of the real line without choice, nonstandard models, the philosophy of mathematics.

**Born:** 13 November 1946 in Washington, D.C., U.S.A.

**Educated:** Massachusetts Institute of Technology, Ph.D. Mathematics, 1976; Dartmouth College, AB Mathematics, 1968.

**Dissertation:** *Aspects of choiceless combinatorial set theory*; supervisor, Eugene M. Kleinberg.

**Regular Academic or Research Appointments:** PROFESSOR, SMITH COLLEGE, 1988–; Associate Professor, Smith College, 1983–1988, Assistant Professor, Smith College, 1976–1983.

**Visiting Academic or Research Appointments:** Visiting Fellow, Wolfson College, 1996; Professor, Graduate Faculty of Mathematics, University of Massachusetts, 1984–; Visiting Professor, University of the Philippines, 1980; Visiting Instructor, University of the Philippines College in Baguio, 1968–1970.

**Research Profile:** The bulk of James Henle's work is located in the school of E. M. Kleinberg and his students. This is characterized by an interest in powerful axioms contradicting the Axiom of Choice and a preference for combinatorial techniques. Henle's early contributions (see *Key Publications*, 1-7, 10) center on infinite-exponent partition relations. Several (see 2 and 4 in particular), notably devise combinatorial equivalents to model-theoretic and recursion-theoretic methods.

A second theme of Henle's work draws inspiration from the real line. In one direction, he has explored the theoretical possibilities of the line (see 8, 9, 11.). In another, he has found analogs among large cardinals (see 2 and 18). In yet another, he has investigated the varieties of nonstandard approaches (see 6 and 12).

Henle maintains a strong interest in the philosophical foundations of mathematics. Paper 13 took a clear position. Paper 14 and book 16 made points more obliquely and the projected book, 19, will move further in this direction.

Perhaps Henle's most important contribution lies in his books which bring the sometimes arcane world of logic to the mathematical public and the general public. Infinitesimal Calculus, with E. M. Kleinberg, was the first to offer nonstandard analysis to undergraduates. An Outline of Set Theory shared with undergraduates the excitement of set-theoretic possibility. Sweet Reason, with T. Tymoczko, persuasively placed logic at the center of every student's intellectual life.

**Main Publications:**

1. Researches into the world of $\kappa \rightarrow (\kappa)^{\kappa}$, *Annals of Mathematical Logic*, 17, 151–169, 1980.
2. Supercontinuity, with A. R. D. Mathias, *Mathematical Proceedings of the Cambridge Philosophical Society*, 92(1), 1–16, 1982.
3. Magidor-like and Radin-like forcing, *Annals of Pure and Applied Logic*, 25, 59–72, 1983.
4. Infinite subscripts from infinite exponents, with J. Baumgartner, *Journal of Symbolic Logic*, 49(2), 558–562, 1984.
5. Spector forcing, *Journal of Symbolic Logic*, 49(2), 542–554, 1984.
6. Weak strong partition cardinals, *J. Symbolic Logic*, 49(2), 555–557, 1984.
7. An extravagant partition relation for a model of arithmetic, *Contemporary Mathematics*, 31, 109–113, 1984.
8. A barren extension, with A. R. D. Mathias and H. Woodin, *Methods in Mathematical Logic*, Proceedings, Caracas 1983. Lecture Notes in Mathematics No. 1130, Springer-Verlag, 1985, 195–207.
9. The consistency of one fixed omega, *Journal of Symbolic Logic* 60(1), 172–177, 1995.
10. Partitions of the Reals and Choice, with C. A. Di Prisco, *Models, Algebras, and Proofs*, Xavier Caicedo and Carlos Monenegro, eds., Marcel Dekker, Inc., 1999.
11. The calculus of partition sequences, changing cofinalities, and a question of Woodin, with A. W. Apter and S. C. Jackson, *Transactions of the A.M.S.*, 352(3), 969–1003, 2000.
12. Doughnuts, floating '2's, and ultraflitters, with C. A. Di Prisco, *Journal of Symbolic Logic*, 65(1), 461–473, 2000.
13. Second-order Non-nonstandard Analysis, *Studia Logica*, 74(3), 399–426, 2003.

14. The happy formalist, *The Mathematical Intelligencer*, 13(1), 12–18, 1991.

15. Classical Mathematics. Baroque Mathematics. Romantic Mathematics? Also Atonal, New Age, Minimalist, and Punk Mathematics, *The American Mathematical Monthly*, 103(1), 18–29, 1996.

16. *Infinitesimal Calculus*, with E. M. Kleinberg, M. I. T. Press, 1979; Dover, 2003.

17. *An Outline of Set Theory*, Springer-Verlag, 1986; Japanese translation, 1988.

18. *Sweet Reason: A Field Guide to Modern Logic*, with Thomas Tymoczko, W. H. Freeman & Co., 1995, second printing, Springer-Verlag, 1999, Spanish translation (Razon, Dulce Razon), 2002.

*Work in Progress*

19. Calculus on strong partition cardinals
20. The Proof and the Pudding

**Teaching:** With T. Tymoczko, James Henle founded the Logic program at Smith College. With J. Garfield, he developed the Logic major at Smith, the annual Alice Ambrose/Tom Tymoczko/Logic Lecture, and the Five College Logic Certificate program. With both Tymoczko and Garfield, he developed the introductory logic course to its current status as the largest undergraduate course at the college. The role of logic at Smith College is unique among liberal arts colleges and a model for the synthesis of analytic and writing skills. Henle has had one doctoral student. Several undergraduate students have become logicians.

**Vision Statement:** Logic both gains and suffers from its philosophical position vis-a-vis mathematics. On the one hand, researchers are drawn to the subject for the significance of its theorems. On the other hand, even modest results are freighted with ideological concerns that jeopardize their intellectual innocence. In this sense, I am as ambivalent as any. To me, the most exciting development in logic is the recent work of Hugh Woodin toward a resolution of the Continuum Hypothesis. The scope, the ambition, the power, and the beauty of the attempt takes my breath away. On the other hand, I firmly believe, on philosophical grounds, that the attempt will fail.

I also believe that philosophical fundamentalism will remain a minority position and that researchers in and out of logic will continue to appreciate logical results as much for their beauty, cleverness, and charm as for their signal importance to mathematics.

# HIGGINBOTHAM, James T.

**Specialties:** Philosophical logic, modal and intensional logic, model theory.

**Born:** 17 August 1941 in Chattanooga, Tennessee, USA.

**Educated:** Columbia University, PhD, 1973; Columbia University, BS, 1967.

**Dissertation:** *Some Problems in Semantics and Radical Translation*. Dissertation Supervisor Charles Parsons.

**Regular Academic or Research Appointments:** PROFESSOR, PHILOSOPHY AND LINGUISTICS, UNIVERSITY OF SOUTHERN CALIFORNIA, 1999–; Professor, General Linguistics, University of Oxford, 1993-2000; Professor, Philosophy, Massachusetts Institute of Technology, 1987-1993, Associate Professor, 1982-1987; Senior Fellow in the Humanities, Columbia University, 1979-1980; Assistant Professor, Philosophy, Columbia University, 1978-1979; Lecturer and Fellow in the Humanities, Columbia University, 1976-1978; Assistant Professor, Philosophy, Columbia University, 1973-1976, Instructor, 1970-1973.

**Visiting Academic or Research Appointments:** Enseignant Étranger, Département d'Études Cognitives, École Normale Supérieure, Paris, France, 2004; Nelson Philosopher in Residence, University of Michigan at Ann Arbor, 1999; Visiting Professor, Cognitive Science, Rutgers University, 1998; Visiting Researcher, Institute for Computational Linguistics, University of Stuttgart, 1997; Visiting Professor, Philosophy, Massachusetts Institute of Technology, 1996; Distinguished Visitor, Philosophy, University of Kansas, Lawrence, Kansas, 1996; Visiting Professor, Philosophy, Princeton University, 1990.

**Research Profile:** Higginbotham has long been interested in using the resources of logical theory to probe meaning in natural language, and in the logical problems that come to light through the investigation of language. His current work is focused on several specific topics in linguistic semantics, and on the general issue of compositionality.

**Main Publications:**

1. "Remarks on Compositionality." Gillian Ramchand and Charles Reiss (eds.), *The Oxford Handbook of Linguistic Interfaces*. Oxford: Oxford University Press, 2007. pp. 425-444.

2. "Sententialism: The Thesis that Complement Clauses Refer to Themselves." Ernest Sosa and Enrique Villanueva (eds.), *Philosophical Issues 16: Philosophy of Language* (a supplementary volume to *Noûs*). Oxford: Blackwell Publishing, 2006. pp. 101-119.

3. "The English Progressive." Jacqueline Gueron and Jacqueline Lecarme (eds.), *The Syntax of Time*. Cam-

bridge, Massachusetts: The MIT Press, 2004. pp. 329-358.

4. "Conditionals and Compositionality." John Hawthorne and Dean Zimmerman (eds.), *Philosophical Perspectives 17* (2003). pp. 181-194. Malden, Massachusetts: Blackwell Publishing.

5. "On Second-Order Logic and Natural Language." Gila Sher and Richard Tieszen (eds.), *Between Logic and Intuition: Essays in Honor of Charles Parsons*. Cambridge: Cambridge University Press, 2000. pp. 79-99.

6. "Tense, Indexicality, and Consequence." Jeremy Butterfield (ed.), *The Arguments of Time*. A British Academy 'Centenary' Monograph. Published for The British Academy by Oxford University Press, Oxford 1999. pp. 197-215.

7. "On Higher Order Logic and Natural Language." T.J. Smiley (ed.), *Proceedings of the British Academy 95: Philosophical Logic*. Published for The British Academy by Oxford University Press, Oxford, 1998. pp. 1-27. Reprinted in T.R. Baldwin and T.J. Smiley (eds.), *Studies in the Philosophy of Logic and Knowledge*, Oxford, 2004, pp. 249-275.

8. "Tensed Thoughts." *Mind and Language* **10**, 3 (1995). pp. 226-249. Reprinted in Wolfgang Künne, Albert Newen, and Martin Anduschus (eds.), *Direct Reference, Indexicality, and Propositional Attitudes*. Stanford, California: CSLI Publications. pp. 21-48.

9. "The Semantics of Questions." Shalom Lappin (ed.), *The Handbook of Contemporary Semantic Theory*. Oxford: Basil Blackwell, 1995. pp. 361-383.

10. "Mass and Count Quantifiers." *Linguistics and Philosophy* **17**, 5 (1994). pp. 447-480. Reprinted in Emmon Bach *et al.* (eds.), *Quantification in Natural Language*, Kluwer Academic Publishers, Dordrecht, Holland, 1995, pp. 383-419.

11. "Interrogatives." Kenneth Hale and Samuel Jay Keyser (eds.), *The View From Building 20: Essays in Linguistics in Honor of Sylvain Bromberger*. Cambridge, Massachusetts: The MIT Press, 1993. pp. 195-227.

12. "Frege, Concepts, and the Design of Language." Enrique Villanueva (ed.), *Information, Semantics and Epistemology*. Oxford: Basil Blackwell, 1990. pp. 153-171.

13. "Plurals" (with Barry Schein). *NELS XIX Proceedings*. GLSA, University of Massachusetts, Amherst, 1989. pp. 161-175.

14. "English Is Not a Context-Free Language." *Linguistic Inquiry* 15, 2 (1984). Reprinted in Walter Savitch *et al.* (eds.), *The Formal Complexity of Natural Language*. Dordrecht, Holland: D. Reidel, 1987. pp. 335-348.

15. "The Logic of Perceptual Reports: An Extensional Alternative to Situation Semantics." *The Journal of Philosophy* **80**, 2 (1983). pp. 100-127.

16. "Questions, Quantifiers, and Crossing" (with Robert May). *The Linguistic Review* **1**, 1 (1981). pp. 41-80.

*Work in Progress*

17. "The English Perfect and the Metaphysics of Events." For a volume growing out of the Paris Conference on Time, Tense, and Modality, December 2005. Jacqueline Lecarme and Jacqueline Guéron, eds. Berlin: Springer Verlag.

18. "Compositionality and Opacity." For Markus Werning *et al.* (eds.), *The Oxford Handbook of Compositionality*.

**Service to the Profession:** Series Editor, Martin Davies, John O'Keefe, Christopher Peacocke, and Kim Plunkett, *Oxford Cognitive Science*, Oxford: Oxford University Press; Representative to the Linguistics Section of the American Academy of Arts and Sciences, Association for Symbolic Logic, 1996-2002; Consulting Editor, *Journal of Philosophy*, 1979-; Associate Editor, *Natural Language and Linguistic Theory*, 1990-1993; Associate Editor, *Linguistic Inquiry*; Associate Editor, *Mind and Language*; Associate Editor, *Rivista di Linguistica*; Associate Editor, *Natural Language Semantics*; Associate Editor, *Pragmatics and Cognition*; Associate Editor, *Linguistics and Philosophy*; Associate Editor, *Research in Language* (Poland).

**Teaching:** Since 1980 he has divided his teaching between Philosophy and Generative Linguistics, concentrating on Semantics.

**Vision Statement:** The involvement of logic in problems of computation and natural language parsing has brought great dividends. Logical theory has certainly clarified many properties of language, as in accounts of modality, generalized and polyadic quantification, and the use of choice functions, dynamic logic, and game-theoretic techniques in the understanding of cross-reference in discourse. Conversely, I believe that there is still much to be learned in logic proper from the careful probing of the lexical and combinatorial semantics of historically given natural languages, as illustrated by the ongoing discussion of plurals, tenses, conditionals, and aspectual markers, the scope and boundaries of higher-order systems, and the characterization of truth and logical consequence for languages with indexicals and other context-dependent expressions.

**Honours and Awards:** Linda Hilf Chair in Philosophy, USC, 2004-; Fulbright Distinguished Professor of the Philosophy of Language, University of Venice Ca'Foscari, Italy, 2003; Fellow, British Academy, 1995-; Visiting Fellow, All Souls College Oxford, 1990; Visiting Scholar, Scuola Normale Superiore, Pisa, Italy, 1987.

# HILPINEN, Risto Juhani

**Specialties:** Inductive logic, deontic logic, the logic of action and practical reasoning, epistemic logic and epistemology, semiotics and the philosophy of language, philosophy of science, the philosophy of C. S. Peirce, the history of logic, and the metaphysics of artifacts, works, and cultural entities.

**Born:** : 9 March, 1943, in Lahti, Finland.

**Educated:** : University of Helsinki, Finland. Cand. Phil. 1967, Lic. Phil. 1967, Ph.D. 1969, University of Helsinki.

**Dissertation:** : *Rules of Acceptance and Inductive Logic*, 1968; supervisor Jaakko Hintikka.

**Regular Academic or Research Appointments:** PROFESSOR OF PHILOSOPHY, UNIVERSITY OF MIAMI, CORAL GABLES, FLORIDA, 1998–; UST Distinguished Professor of Ethics, University of Miami, 1995–1998; Research Professor of Philosophy, University of Miami, 1987–1998; University of Queensland Research Fellow, University of Queensland, Brisbane, Australia, 1973-1974; Professor of Theoretical Philosophy, University of Turku, Turku, Finland, 1972–1999; Professor of Philosophy, University of Jyväskylä, Jyväskylä, Finland, 1970–1971; Junior Fellow of the Finnish National Research Council for the Humanities, 1970.

**Visiting Academic or Research Appointments:** Visiting Professor of Philosophy, University of Graz (Karl-Franzens-Universität Graz), Austria, 1995; Visiting Fellow, Center for Philosophy of Science, University of Pittsburgh, 1986; Distinguished Visiting Professor of Philosophy, University of Miami, 1985; Visiting Professor of Philosophy and Visiting Research Associate, Florida State University, 1985, 1980; Visiting Assistant Professor of Philosophy, Stanford University, 1971; Visiting Professor of Philosophy, University of Rochester, 1969.

**Research Profile:** In his early work in the 1960's Risto Hilpinen studied epistemic decision theory and probabilistic rules of inductive acceptance, and developed a measure of evidential support based on the shared information content of a hypothesis and the evidential data. Since the 1970's he has worked in several areas of what is often called "philosophical logic", especially in deontic logic and the logic of action, and applied the tools of intensional semantics to the analysis of the concept of verisimilitude (truthlikeness) and other methodological concepts. He has been interested in the philosophy of C. S. Peirce, especially in Peirce's pioneering work in logic and semiotics and in his pragmatic theory of meaning and knowledge. His studies in the metaphysics of artifacts, works, and other products of culture reflect his long-standing interest in art.

**Main Publications:**

1. Rules of Acceptance and Inductive Logic. Acta Philosophica Fennica 22, Amsterdam: North-Holland, 1968. 134 pp.
2. 'On the Information Provided by Observations', in Information and Inference, ed. by J. Hintikka and P. Suppes, Dordrecht: D. Reidel, 1970, pp. 97-122.
3. 'Decision Theoretic Approaches to Rules of Acceptance', in Contemporary Philosophy in Scandinavia, ed. by R. E. Olson and A. Paul, Baltimore: The Johns Hopkins Press, 1972, pp. 147-168.
4. 'Approximate Truth and Truthlikeness', in Formal Methods in the Methodology of Empirical Sciences, ed. by M. Przeecki, K. Szaniawski and R. Wójcicki, Wrocaw: Ossolineum, Dordrecht: D. Reidel, 1976, pp. 19-42.
5. 'On C. S. Peirce's Theory of the Proposition: Peirce as a Precursor of Game-theoretical Semantics', The Monist 65 (1982), 182-188.
6. 'Knowledge and Conditionals', in Philosophical Pespectives, 2: Epistemology, 1988, ed. James Tomberlin, Atascadero, Calif.: Ridgeview Press Publishing Company, 1988, pp. 157-182.
7. 'On Peirce's Philosophical Logic: Propositions and Their Objects', Transactions of the Charles S. Peirce Society 28 (1992), 467-488.
8. 'Deontic Logic, Pragmatics and Modality', in Pragmatik: Handbuch des pragmatischen Denkens, Vol. 4, ed. by H. Stachowiak, Hamburg: Felix Meiner Verlag, 1993, pp. 295-319.
9. 'Actions in Deontic Logic', in Deontic Logic in Computer Science: Normative System Specification, ed. by J.-J. Ch. Meyer and R. Wieringa, New York: John Wiley & Sons, 1993, pp. 85-100.
10. 'Authors and Artifacts', Proceedings of the Aristotelian Society 93 (1993), 155-178.
11. 'Peirce on Language and Reference', in Peirce and Contemporary Thought, ed. by Kenneth L. Ketner, Bronx, N.Y: Fordham University Press, 1995, pp. 272-303.
12. 'States, Actions, Omissions and Norms', in Contemporary Action Theory, Vol. 1, ed. by G. Holmström-Hintikka and R. Tuomela, Dordrecht: Kluwer Academic Publishers, 1997, pp. 83-107.
13. 'Belief Systems as Artifacts', The Monist 78 (1995), 136-155.
14. 'Deontic Logic', in Blackwell Guide to Philosophical Logic, ed. by Lou Goble, Oxford, UK, and Malden, Mass.: Blackwell Publishers, 2001, pp. 159-182.
15. 'Peirce's Logic', in Handbook of the History of Logic, Vol. 3: The Rise of Modern Logic: From Leibniz to Frege, ed. by D. M. Gabbay and J. Woods, Amster-

dam: Elsevier BV — North-Holland, 2004, pp. 611-58.

16. 'On a Pragmatic Theory of Knowledge and Meaning', Cognitio: Revista de Filosofia 5:2 (2004), 150-167

17. 'Hintikka on Epistemic Logic and Epistemology', in The Philosophy of Jaakko Hintikka, ed. by R. E. Auxier and L. E. Hahn, Chicago: Open Court Publishing Company, 2006, pp. 783-818.

18. 'Norms, Normative Utterances, and Normative Propositions', Análisis Filosófico 26:2 (2006): Homenaje a Carlos E. Alchourrón, ed. by E. Bulygin and G. Palau, pp. 229-241.

19. 'On Practical Abduction', Theoria: Swedish Journal of Philosophy 73 (2007), 207-220.

20. 'Conditionals and Possible Worlds: On C. S. Peirce's Conception of Conditionals and Modalities', forthcoming in The Development of Modern Logic: A Philosophical Perspective, ed. by L. Haaparanta. New York: Oxford University Press.

21. 'Remarks on the Iconicity and Interpretation of Existential Graphs' (forthcoming).

For a more inclusive list of publications, see http://www.as.miami.edu/phi/hilpinen/hilpinenPUBLIST.pdf

**Service to the Profession:** Editor or co-editor of several books in deontic logic and the philosophy of science: Deontic Logic: Introductory and Systematic Readings, 1971, Deontische Logik und Semantik (with A. Conte and G. H. von Wright), 1977, Rationality in Science: Studies in the Foundations of Science and Ethics, 1980, New Studies in Deontic Logic: Norms, Actions, and the Foundations of Ethics, 1981, Juristische Logik, Rationalität und Irrationalität im Recht (with A. Arnaud, J. Wróblewski, and R. J. Vernengo), 1985, Logic, Methodology and Philosophy of Science VIII: Proceedings of the Eighth International Congress of Logic, Methodology and Philosophy of Science, Moscow, 1987 (with J. E. Fenstad and I. T. Frolov), 1989, Realism and Anti-Realism in the Philosophy of Science: Beijing International Conference, 1992 (with R. S. Cohen and Qiu Renzong), 1996. Editor, Synthese, 1977-1979, 1982-1999, Associate Editor, 1971-1976. Member of the editorial boards of several scholarly journals. Secretary of the International Union of History and Philosophy of Science, Division of Logic, Methodology and Philosophy of Science (IUHPS/DLMPS), 1983-1991, Secretary General of IUHPS, 1986, 1988-1990. Member of the General Committee of ICSU (International Council of Scientific Unions), 1985-1986, 1988-1990. Member of the ICSU Committee on Structure and Statutes, 1988-1990, Member of the Joint Commission for History and Philosophy of Science, IUHPS, 1985-1986. Association for Symbolic Logic, Member of the European Committee, 1986-1993. Member of the Program Committee of several international workshops, conferences, and summer schools: Deontic Logic in Computer Science (DEON), 2008 (Luxembourg), 2006 (Utrecht), 2000 (Toulouse, Co-Chair), 1996 (Sesimbra) 1994 (Oslo); International conference "Applying Peirce", 2007 (Helsinki), Third International Summer School in the Theory of Knowledge, 2000 (Madralin, Poland), International conference "Vienna Circle and Contemporary Science and Philosophy in Memory of Tscha Hung", 1994 (Beijing). Organizer of an international conference "The Background of Contemporary Philosophical Logic", University of Miami, 1987.

**Teaching:** Professor Hilpinen has taught courses in logic, the history of logic, philosophy of language, action theory, philosophy of science, and other areas of philosophy at several universities, mainly at the University of Miami and the University of Turku, and has served as the dissertation advisor or a member of the dissertation committee of some 35 students. He has given more than 100 guest lectures at various universities and institutes in Europe, North and South America, Asia, and Australia.

**Vision Statement:** According to C. S. Peirce's method of "logical magnifying-glass", a logical study of a phenomenon should be based on cases in which the main features of the phenomenon are displayed in an exaggerated form, as clearly as possible. This recipe guides the construction of conceptual models in philosophy.

**Honours and Awards:** Member of the Finnish Academy of Science and Letters, 1973; Member of Institut International de Philosophie, 1977; Cooper Fellowship in Arts and Sciences, University of Miami, 2008-2011, Provost's Award for Scholarly Activity, University of Miami, 2006-2007; A.C.L.S. American Studies Fellowship 1975-1976, U.S. Department of State Grant, 1970-1971.

# HINDLEY, J. Roger

**Specialties:** Lambda-calculus, combinatory logic, type theory

**Born:** 1939 in Belfast, Northern Ireland, UK

**Educated:** University of Newcastle upon Tyne, England, PhD, 1964; Queen's University, Belfast, BSc, 1960, MSc, 1961, DSc, 1991.

**Dissertation:** *The Church-Rosser Property and a Result in Combinatory Logic*: supervisor, Ronald Harrop

**Regular Academic or Research Appointments:** HONORARY RESEARCH FELLOW, MATHEMATICS DEPARTMENT, SWANSEA UNIVERSITY, UK, 1999–; Reader, Mathematics, University of Wales Swansea, 1989–1998; Lecturer, Mathematics, University of Wales Swansea, UK, 1968–1989; Temporary Lecturer, Mathematics, University of Bristol, UK, 1966—1968; Visiting Assistant Professor, Mathematics, Pennsylvania State University, USA, 1964–1966.

**Main Publications:**

1. The principal type-scheme of an object in combinatory logic, *Trans. American Math. Soc.* 146 (1969), 29–60.
2. *Combinatory Logic Volume* II (with H. B. Curry, J. P. Seldin), North-Holland Co., Amsterdam 1972.
3. *Introduction to Combinatory Logic*, (with B. Lercher, J. P. Seldin), Cambridge University Press, England 1972.
4. Standard and normal reductions, *Trans. American Math. Soc.* 241 (1978), 253–271.
5. Lambda-calculus models and extensionality, (with G. Longo), *Zeit. Math. Logik* 26 (1980), 289–310.
6. *To H. B. Curry, Essays on Combinatory Logic, Lambda Calculus and Formalism* (edited, with J. P. Seldin), Academic Press, London 1980.
7. The completeness theorem for typing lambda terms, *Theoretical Computer Sci.* 22 (1983), 1–17.
8. *Introduction to Combinators and Lambda-calculus* (with J. P. Seldin), Cambridge University Press, England 1986.
9. Principal type-schemes and condensed detachment (with D. Meredith), *J. Symbolic Logic* 55 (1990), 90–105.
10. Types with intersection, an introduction, *Formal Aspects of Computing* 4 (1992), 470–486.
11. *Basic Simple Type Theory*, Cambridge University Press, England 1997.
12. *Typed Lambda Calculi and Applications TLCA'97* (edited, with P. de Groote), Springer Verlag, Berlin 1997.
13. Lambda-calculus and Combinators in the 20th Century, (with F. Cardone), in *Handbook of the History of Logic* (Editors D. Gabbay, J. Woods), Elsevier. 2008.
14. *Lambda-calculus and Combinators, an Introduction* (with J. P. Seldin), Cambridge University Press, England 2008.

# HINTIKKA, Jaakko J.

**Specialties:** Epistemic logic, free logics, inductive logics, game theory, set theory and other modal logics.

**Born:** 12 January 1929 in Vantaa, Finland.

**Educated:** University of Helsinki, PhD Mathematics, 1956, Lic. Ph., 1952, Cand. Ph., 1952.

**Dissertation:** *Distributive Normal Forms in the Calculus of Predicates*; supervisor, G.H. von Wright.

**Regular Academic or Research Appointments:** PROFESSOR OF PHILOSOPHY, BOSTON UNIVERSITY, 1990–. Professor of Philosophy, Florida State University, 1979–1990; Research Professor, Academy of Finland, 1970–81; Professor of Philosophy, Stanford University, 1965–1981; Professor of Philosophy, University of Helsinki, 1959–70; Junior Fellow, Harvard University, 1956–59.

**Visiting Academic or Research Appointments:** Visiting Professor, Hebrew University, 1974; U.C. Berkeley, 1963; Brown University, 1962.

**Research Profile:** In his early work, Jaakko Hintikka contributed significantly to the theory of distributive normal forms, free logics, the tree method, epistemic logic, including the logic of questions and answers, inductive logic, especially the theory of inductive generalization, to possible world semantics, to the clarification of the notion of information and infinitely deep logics. Later, he has been the main architect of game-theoretical semantics and independence-friendly logic. Hintikka has also worked in the foundations of mathematics, among other things criticizing axiomatic set theory and vindicating Hilbert's program.

**Main Publications:**

1. *Distributive Normal Forms in the Calculus of Predicates*, Acta Philosophica Fennica, vol. 6, 1953.
2. *Knowledge and Belief: An introduction to the Logic of the Two Notions*, Cornell U.P., 1962. (Second ed., 2005)
3. Depth Information and Surface Information in *Information and Inference*, ed. by Jaakko Hintikka and Patrick Suppes, D. Reidel, 1970, pp. 263-297.
4. *Logic, Language Games and Information*, Oxford U.P., 1973.
5. Quantifiers vs. Quantification Theory, *Dialectica*, vol. 27 (1974), pp. 329-358.
6. "Is", Semantical Games and Semantical Relativity, *Journal of Philosophical Logic*, vol. 8 (1979), pp. 433-468.
7. (with Jack Kulas) *Anaphora and Definite Descriptions; Two Applications of Game-theoretical Semantics*, D. Reidel, 1985.
8. 'Standard vs. Nonstandard Distinction: A Watershed in the Foundations of Mathematics, in *From Dedekind to Gödel*, ed. by Jaakko Hintikka, Kluwer Academic, 1995, pp. 21-44.
9. *The Principles of Mathematics Revisited*, Cambridge U.P., 1996.
10. (with Gabriel Sandu) "Game-theoretical Semantics" in *Handbook of Logic and Language*, ed. by J. van

Benthem and Alice ter Meulen, Elsevier, 1996, pp. 361-410.
11. (with Gabriel Sandu) "A Revolution in Logic?", *Nordic Journal of Philosophical Logic*, vol. 1 (1996), pp 169-183.
12. No Scope for Scope?, *Linguistics and Philosophy*, vol. 20 (1997), pp. 515-544.
13. Truth Definitions, Skolem Functions and Axiomatic Set Theory, *Bulletin of Symbolic Logic*, vol. 4 (1998), pp. 303-337.
14. *Language, Truth and Logic in Mathematics: Selected Papers III*, Kluwer Academic, 1998.
15. Post-Tarskian Truth, *Synthese*, vol. 126 (2001), pp. 17-36.
16. Hyperclassical Logic (a.k.a. IF logic) and its Implications for Logical Theory, *Bulletin of Symbolic Logic*, vol. 8 (2002), pp. 404-423.
17. A Second Generation Epistemic Logic and its General Significance, in *Knowledge Contributors*, ed. by Vincent Hendricks et al, Kluwer Academic, 2003, pp. 33-55.
18. On the Different Identities of Identity, in *Language, Meaning, Interpretation*, ed. by Guttorm Fløistad, Kluwer Academic, 2004, pp. 117-139.
19. Independence-friendly Logic and Axiomatic Set Theory, *Annals of Pure and Applied Logic*, vol. 126 (2004), pp. 313-333.
20. What is the True Algebra of Logic?'in *First-Order Logic Revisited*, ed. by Vincent Hendricks et al, Logos Verlag, 2004, pp. 117-128.
21. Truth, Negation and Other Basic Notions of Logic. In *The Age of Alternative Logics: Assessing Philosophy of Logic and Mathematics Today*, edited by Johan van Benthem *et al.*, Springer, pp 195–220
22. (with Besim Karakadilar) How to Prove the Consistency of Arithmetic. In *Truth and Games: Essays in Honour of Gabriel Sandu*, Acta Philosophica Fennica 78, ed. by Tuomo Aho and Ahti-Veikka Pietarinen, Societas Philosophica Fennica, 2006, 1–15.
23. Who Has Kidnapped the Notion of Information? In *Socratic Epistemology; Explorations of Knowledge-Seeking by Questioning*, Cambridge University Press, 2007, pp. 189–210.
24. (with Gabriel Sandu) What Is Logic? In *Handbook of the Philosophy of Science. Philosophy of Logic*, ed. by Dov M. Gabbay, Paul Thagard and John Woods, Elsevier, 2007, 13–39.

*Work in Progress*
Hintikka's papers in progress include the following:
25. Truth, Axiom of Choice and Axiomatic Set Theory: What Tarski Never Told Us.
26. The Notion of Probability in IF Logic.
27. From c-Functions to q-Functions: A New Approach to Quantum Theory.
28. Is Nonlocality Merely a Mathematical Illusion?

**Service to the Profession:** Editor-in-chief, *Synthese*, 1965-76, 1982-2003; Editor, *Synthese Library*, 1965-2003; President, Institut International de Philosophie, 1999-2002; Co-Chair, American Organizing Committee of the Twentieth World Congress of Philosophy, 1998; Vice-President, Federation Internationale des Societies Philosophique, 1993-98; President, American Philosophical Association (Pacific Division), 1975-76; President, of DLMPS of the International Union of History and Philosophy of Science, 1975, Vice-President 1972-75.

**Teaching:** Hintikka has argued that introductory logic and reasoning is almost universally taught in a wrong way. The focus is on so-called rules of inference. These rules do not guide inferences, however. They are merely permissive in that they tell which conclusions may be drawn from which premises. They do not tell anyone which conclusions one should draw in which circumstances. Instead of such rules the pedagogical emphasis should be focused on strategies of deduction. Hintikka has tried to correct these mistakes through his teaching and through his introductory textbook *What If....? Toward Excellence in Reasoning* (with James Bachman), Mayfield, 1991.

Several of Hintikka's former doctoral students have reached an international status, among them Simo Knuuttila, Risto Hilpinen, Raimo Tuomela, Juha Manninen, Ilkka Niiniuluoto, Gabriel Sandu and John Symons.

**Vision Statement:** Quantifiers do not express only the emptiness or nonemptyness of certain predicates. By their formal dependencies on each other, they express material dependencies between the corresponding variables. Once this is realized, the indispensability of independence-friendly (IF) logic is also realized. In IF logic the law of excluded middle does not hold; hence it is elementary in a perfectly good sense. If unlimited use of contradictory negation is allowed, we obtain the strength of second-order logic on the first-order level. This strength suffices for the entire classical mathematics. This strength cannot be captured by any first-order axiomatic set theory. Such a set theory must be rejected for other reasons as well; for one thing, false theorems can be proved in it. An implementation of the idea of irreducible mutual dependence turns out to require an extension of the notion of function. This extension remains to be carried out, but it looks as if the resulting mathematics would help to solve the conceptual problems of quantum theory.

IF logic also brings new algebraic and geometric structure to logical space. The strong (dual) negation characteristic of IF logic can be thought of as a generalization of the notion of orthogonality, which makes it possible to introduce such notions as dimension, coordinate representation etc.

These results pose new problems to, and open entire new fields for, research in logic.

**Honours and Awards:** Member of seven academies of science in five countries. Honorary doctorates from Turku, 2003; Oulu, 2002; Uppsala, 2000; Krakow, 1995; and Liège, 1989. Prizes: Schock Prize, 2005; E.J. Nyström Prize, Societas Scientarum Fennica, 1988; Wihuri International Prize, 1976. Immanuel Kant Lectures (Stanford), 1985; Axel Hägerström Lectures, (Uppsala) 1983; Guggenheim Fellow, 1980-81; W.T. Jones Lectures (Pomona College), 1975; John Locke Lectures (Oxford), 1964. A Library of Living Philosophers volume devoted to Jaakko Hintikka: *The Philosophy of Jaakko Hintikka* in The Library of Living Philosophers, vol 30 (Open Court) 2006.

# HIRSCHFELDT, Denis Roman

**Specialties:** Computability theory, algorithmic information theory, reverse mathematics.

**Born:** 13 November 1971 in Rio de Janeiro, Rio de Janeiro, Brazil.

**Educated:** Cornell University, PhD Mathematics 1999; Cornell University, MS Computer Science 1998; University of Pennsylvania, BA Mathematics 1993.

**Dissertation:** *Degree Spectra of Relations on Computable Structures*; supervisor, Richard A. Shore.

**Regular Academic or Research Appointments:** ASSOCIATE PROFESSOR OF MATHEMATICS, UNIVERSITY OF CHICAGO, 2005–; Assistant Professor of Mathematics, University of Chicago, 2002-2005; Dickson Instructor of Mathematics, University of Chicago, 2000-2002; Postdoctoral Fellow, Victoria University of Wellington, 1999-2000.

**Visiting Academic or Research Appointments:** Visiting Scholar, University of Notre Dame, 2005; Visiting Assistant Professor, University of Wisconsin-Madison, 2003; Honorary Fellow, University of Wisconsin-Madison, 1999.

**Research Profile:** Hirschfeldt's research is in computability theory and its applications. In computable structure theory, he has worked on issues connected with computable dimension, degree spectra of relations, and related notions. In computable model theory, he has worked on the effective content of some of the fundamental model-theoretic notions and constructions introduced by Vaught. He has also worked on the proof-theoretic and computability-theoretic strength of combinatorial principles, especially those related to Ramsey's Theorem for pairs and Weak König's Lemma. Finally, he has worked on effective randomness, including the study of measures of relative randomness and classes of sets that can be defined using algorithmic information theory, such as the Martin-Löf random sets and the $K$-trivial sets.

**Main Publications:**

1. Degree Spectra of Relations on Computable Structures, *Bulletin of Symbolic Logic* 6 (2000), 197 - 212.
2. Undecidability and 1-types in Intervals of the Computably Enumerable Degrees (with K. Ambos-Spies and R. A. Shore), *Annals of Pure and Applied Logic* 106 (2000), 1 - 47.
3. Degree Spectra of Intrinsically C.E. Relations, *Journal of Symbolic Logic* 66 (2001), 441 - 469.
4. A $\Delta_2^0$ Set with no Infinite Low Subset in Either It or Its Complement (with R. G. Downey, S. Lempp, and R. Solomon), *Journal of Symbolic Logic* 66 (2001), 1371 - 1381.
5. (with R. G. Downey, eds.) *Aspects of Complexity: Minicourses in Algorithmics, Complexity, and Computational Algebra, NZMRI Summer Meeting, Kaikoura, New Zealand, January 7 - 15, 2000, de Gruyter Series in Logic and its Applications* 4 (de Gruyter, 2001).
6. Degree Spectra and Computable Dimensions in Algebraic Structures (with B. Khoussainov, R. A. Shore, and A. M. Slinko), *Annals of Pure and Applied Logic* 115 (2002) 71 - 113.
7. Degree Spectra of Relations on Structures of Finite Computable Dimension, *Annals of Pure and Applied Logic* 115 (2002) 233 - 277.
8. Randomness, Computability, and Density (with R. G. Downey and A. Nies), *SIAM Journal on Computing* 31 (2002) 1169 - 1183.
9. Trivial Reals (with R. G. Downey, A. Nies, and F. Stephan), in R. Downey, D. Decheng, T. S. Ping, Q. Y. Hui, and M. Yasugi (eds.), Proceedings of the 7th and 8th Asian Logic Conferences (Singapore University Press and World Scientific, 2003).
10. Computability-Theoretic and Proof-Theoretic Aspects of Partial and Linear Orderings (with R. G. Downey, S. Lempp, and R. Solomon), *Israel Journal of Mathematics* 138 (2003) 271 - 290.
11. Randomness and Reducibility (with R. G. Downey and G. LaForte), *Journal of Computer and System Sciences* 68 (2004), 96 - 114.
12. Bounding Prime Models (with B. F. Csima, J. F. Knight, and R. I. Soare), *Journal of Symbolic Logic* 69 (2004) 1117 - 1142.
13. Relativizing Chaitin's Halting Probability (with R. Downey, J. S. Miller, and A. Nies), *Journal of Mathematical Logic* 5 (2005) 167 - 192.
14. Computable Trees, Prime Models, and Relative Decidability, *Proceedings of the American Mathematical Society* 134 (2006) 1495 - 1498.
15. Calibrating Randomness (with R. Downey, A.

Nies, and S. A. Terwijn), to appear in the *Bulletin of Symbolic Logic*.

16. Using Random Sets as Oracles (with André Nies and Frank Stephan), to appear.

17. Combinatorial Principles Weaker than Ramsey's Theorem for Pairs (with R. A. Shore), to appear.

18. Bounding Homogeneous Models (with B. F. Csima, V. S. Harizanov, and R. I. Soare), to appear.

*Work in Progress*

19. *Algorithmic Randomness and Complexity* (with R. G. Downey), to be published by Springer-Verlag New York.

20. The Atomic Model Theorem (with R. A. Shore and T. A. Slaman), in preparation.

A full list of publications is available at Hirschfeldt's home page http://www.math.uchicago.edu/~drh

**Service to the Profession:** Member, Program Committee, Conference on Logic, Computability, and Randomness, Buenos Aires, Argentina, 2007; Co-organizer, Workshop on Effective Randomness, American Institute of Mathematics, 2006; Member, Association for Symbolic Logic Committee on Meetings in North America, 2006-; Member, Association for Symbolic Logic Nominating Committee, 2005; Reviews editor, *Bulletin of Symbolic Logic*, 2004-; Member, Program Committee, Conference on Logic, Computability, and Randomness, Córdoba, Argentina, 2004; Member, Program Committee, Association for Symbolic Logic Annual Meeting, 2004; Member, Program Committee, Association for Symbolic Logic Annual Meeting, 2003; Member, Organizing Committee, New Zealand Mathematics Research Institute Summer Meeting, Kaikoura 2000.

**Teaching:** Hirschfeldt has taught at the University of Chicago since 2001. He has served as secondary advisor for Barbara Csima and five current graduate students.

**Honours and Awards:** Sacks Prize of the Association for Symbolic Logic, 1999; Alfred P. Sloan Doctoral Dissertation Fellowship, 1998-1999; Robert John Battig Prize of the Cornell Department of Mathematics, 1998.

# HJORTH, Greg

**Specialties:** Descriptive set theory, model theory.

**Born:** June 14 1963 in Melbourne, Australia.

**Educated:** University of California at Berkeley, PhD, 1963.

**Dissertation:** *The influence of u2*; supervisor, W. Hugh Woodin.

**Regular Academic or Research Appointments:** PROFESSOR OF MATHEMATICS, UNIVERSITY OF CALIFORNIA AT LOS ANGELES, 2001–; PROFESSORIAL FELLOW, DEPARTMENT OF MATHEMATICS AND STATISTICS, UNIVERSITY OF MELBOUNE, 2006–; Associate Professor of Mathematics, University of California at Los Angeles, 1997-2001; Assistant Professor of Mathematics, University of California at Los Angeles, 1995-1997; Bateman Research-Instructor, Caifornia Institute of Technology, 1993-1995.

**Research Profile:** Greg Hjorth works in descriptive set theory, though has also written a number of papers in model theory. Most of his work has involved equivalence relations, especially those induced by the continuous action of a Polish group.

**Main Publications:**

1. A converse to Dye's theorem. Transactions of the American Mathematical Society 357 (2005), no. 8, 3083-3103.

2. Vaught's conjecture on analytic sets. Journal of the American Mathematical Society 14 (2001), no. 1, 125-143.

3. Classification and orbit equivalence relations. Mathematical Surveys and Monographs, 75. American Mathematical Society, Providence, RI, 2000.

**Teaching:** Hjorth has taught a wide range of mathematics courses at UCLA and has several former graduate students.

**Honours and Awards:** Karp Prize in Mathematical Logic, shared with Alexander Kechris, 1998-2003; Sacks Prize, outstanding thesis in Mathematical Logic, 1993-1994.

# HOARE, Charles Antony Richard (Tony)

**Specialties:** Theory of programming, proofs of program correctness.

**Born:** 11 January 1934 in Colombo, Sri Lanka.

**Educated:** Oxford University, Certificate in Statistics, MA (Literae Humaniories).

**Regular Academic or Research Appointments:** EMERITUS PROFESSOR, OXFORD UNIVERSITY, 1999–; James Martin Professor of Information Engineering, Oxford University, 1993-1999; Professor, Computation, Oxford University, 1977-1993; Professor, Computer Science, Queen's University, Belfast, Northern Ireland, 1968-1977.

**Visiting Academic or Research Appointments:** PRINCIPAL RESEARCHER, MICROSOFT RESEARCH LTD., CAMBRIDGE, ENGLAND,

1999-; Einstein Professorship, Chinese Academy of Sciences, 2006; Lee Kuan Yew Distinguished Visitor, Singapore, 1992; Admiral R. Inman Centennial Chair in Computing Theory, University of Texas at Austin, 1986-1987; Visiting Professor, Stanford University, 1972-1973.

**Research Profile:** In Tony Hoare's first employment as a computer programmer (1960), his first major responsibility was the implementation of a mechanical translator for the new computer programming language ALGOL 60. Its syntax had been formally specified by a technique due originally to Chomsky, but its semantics were only informally described. He suggested that the most appropriate style of semantic presentation would be axiomatic, because axioms give greatest freedom of choice between a variety of efficient implementation methods, while providing enough information to the programmer to write programs and prove their correctness.

Realizing that this topic required long-term research, he left industry and entered academic life at the Queen's University, Belfast (1968). He adopted Floyd's assertional method of program proof, and applied it (with adaptation) to many features of standard programming languages. (Under the title of 'Hoare logic' this excited some interest among logicians.) More controversially, he suggested that more complicated programming features, whose correctness could not be readily formalized for purposes of proof, should be avoided by programmers and by programming language designers. Clarity of logical foundation is therefore an objective scientific criterion of good programming language design.

In his later years at Oxford, with his long-term research colleague He Jifeng, he embarked on an effort to classify and categorise theories of programming into families with well understood connections. Tarski's calculus of relations provided the basic conceptual framework.

On retirement from Oxford in 1999, he joined Microsoft Research Ltd. in Cambridge, where he pursues his interests in program verification and concurrency. There are increasing signs that these interests will bring practical benefit in the design and implementation of widely used commercial software.

**Main Publications:**

1. An Axiomatic Basis for Computer Programming. Comm. ACM 12(10):576-580, 583, 1969.
2. Proof of a program FIND. Comm ACM 14(1):39-45, Jan 1971.
3. Proof of Correctness of Data Representations. Acta Informatica 1(4):271-281, 1972.
4. (with Niklaus Wirth) An Axiomatic Definition of the Programming Language PASCAL. Acta Informatica 2(4):335-355, 1973.
5. (with Peter Lauer) Consistent and Complementary Theories of the Semantics of Programming Languages. Acta Informatica 3(2):135-153, 1974.
6. Parallel Programming: an Axiomatic Approach. Computer Languages 1(2):151-160, Jun 1975.
7. Some Properties of Predicate Transformers, JACM 25(3):461-480, Jul 1978.
8. A Calculus for Total Correctness of Communicating Processes. Science of Computer Programming 1(1):49-72, Oct 1981.
9. (with A.W.Roscoe) Programs as Executable Predicates. Proc. Int. Conf. on Fifth
10. Generation Computer Systems, ICOT, Tokyo 1984.
11. A Couple of Novelties in the Propositional Calculus. Z. Math. Logik Grundlag. Math. 31(2):173-178, 1985.
12. (with J.C.Shepherdson, ed) Mathematical Logic and Programming Languages. Phil. Trans. Royal Soc., Series A, 312, 1984.
13. (with A.W.Roscoe) Laws of occam Programming. Technical Monograph PRG-53, Oxford Univ. Comp. Lab. Feb 1986.
14. (with others) The Laws of Programming. Comm ACM 30(8) 672-687 1987, with corrections in Comm. ACM 30(9) 770.
15. (with A.P.Ravn and Zhou Chao Chen) A Calculus of Durations. Inf. Proc. Lett. 40(5):269-276,1992.
16. Guest Editorial. J. Logic and Computation 4(3):215-216 1994.
17. (with He Jifeng) Unifying Theories of Programming Prentice Hall, 1998.
18. The Verifying Compiler: a Grand Challenge for Computing Research. J. ACM 50(1) 63-69 2003.

*Work in Progress*

19. Recent developments in multi-core computer architecture, needed to achieve continuing performance improvements, have significantly increased problems of correctness of concurrent programs. Earlier research results in process algebra must be adapted to meet these new needs.

Recent advances in mechanical model checking, constraint satisfaction and automatic proof search offer the prospect of application to the design, development and evolution computer components and systems; specifications of their desired properties and behaviour will ideally be proved correct with a high degree of automation. Realization of this ideal may require a long-term international collaboration of scientists working in different specialist areas of Computer Science. A Grand Challenge is envisaged, something like the Human Genome project. Work is in progress to identify the leading teams, and inspire them to participate in the project.

**Service to the Profession:** Director, Technical Excellence Week, Pune, India 2003-; Director, Smith Institute, 1996-; Director, UT Year of Programming, 1987; Director, Marktoberdorf Summer School, 1982-; Editor, Prentice Hall International Series in Computer Science (over 100 volumes), 1977-1998; Organizer, Royal Society Discussion Meetings: Mathematical Logic and Programming Languages, Scientific application of Multiprocessors, Mechanised reasoning and Hardware Design, 1982; Editorial Board Member, Acta Informatica; Editorial Board Member, IEEE Transactions on Software Engineering; Editorial Board Member, Computer Programming Languages; Editorial Board Member, British Computer Journal; Editorial Board Member, Fifth Generation Computing; Editorial Board Member, Mathematical Structures in Computing Science.

**Teaching:** Tony Hoare established the first undergraduate courses in Computer Science at the Queen's University Belfast, and at Oxford University. He established graduate programmes at Oxford, including a professional Master's Degree in Software Engineering in the Department of Continuing Education at Oxford. He has been Director and Lecturer at Summers Schools in Villard-de-Lans, Le Breau sans Nappe, Marktoberdorf (repeatedly), Austin, Wollongong, Jindabyne, Santa Cruz (twice), Moscow, Beijing, and Pune. Bill Roscoe and Steve Brookes are among his distinguished students.

**Vision Statement:** Computer programming is an applied branch of logic. The ideal of logical reasoning by machine is close to practical realization. Even in the study of pure logic, emphasis is shifting from what can't be done to what can.

**Honours and Awards:** Member, US National Academy of Engineering, 2006; Fellow, Computer History Museum, 2006; Fellow, Royal Academy of Engineering, 2006; Kyoto Prize, 2000; Knight Bachelor, 2000; corr. mem., Bavarian Academy of Sciences, 1997; Member, Academia Europaea, 1989; Foreign Member, Accademia dei Lincei, 1988; IEE Faraday Medal, 1985; Fellow, Royal Society, 1982; ACM Turing Award, 1980; Distinguished Fellow, British Computer Society, 1978.

# HODGES, Wilfrid Augustine

**Specialties:** Model theory, semantics.

**Born:** 27 May 1941 in Reading, Berks, England.

**Educated:** Oxford, BA Literae Humaniores, 1963; Oxford, BA Theology, Oxford, 1965; DPhil Literae Humaniores, 1970.

**Dissertation:** *Some questions on the structure of models*; supervisor, John Crossley.

**Regular Academic or Research Appointments:** RETIRED; Professor of Mathematics, Queen Mary, University of London, 1987–2006; Reader, Mathematical Logic, Queen Mary, University of London, 1984–1987; Reader, Mathematical Logic, Bedford College, University of London 1981–1984; Lecturer, Mathematics, Bedford College, University of London 1974–1981; Lecturer, Philosophy and Mathematics, Bedford College, University of London 1968–1974.

**Visiting Academic or Research Appointments:** Professor, Queen Mary, University of London, 2006-2008; Visiting Associate Professor, Mathematics, University of Colorado at Boulder, 1979-1980; Acting Assistant Professor, Philosophy, University of California at Los Angeles, 1967–1968.

**Research Profile:** Wilfrid Hodges' mathematical research has largely been concerned with ways in which structures of one type can be systematically extended or expanded to structures of another type; for example each field has an algebraic closure unique up to isomorphism over the field. He has described algebraic sufficient conditions and necessary conditions (the latter in work with Saharon Shelah) for such a construction to be uniformly definable in set theory. A byproduct was some work on the axiom of choice in algebra, for example the result that the existence of maximal ideals in commutative rings implies the axiom of choice. Within model theory such constructions can be studied through relative categoricity; with Ian Hodkinson, Anand Pillay and others he has contributed both to the general (cohomological) theory of relative categoricity (which has the theory of covers as a special case), and to the identification of examples of relative categoricity, for example in abelian groups and in linear orderings. His book 'Building Models by Games' brought together Martin Ziegler's game-theoretic interpretation of forcing and some constructions of Shelah, to give an integrated game-theoretic approach to a range of model-theoretic constructions, particularly of existentially closed structures; this approach was generalised and applied widely by Hodkinson and his associates. More recently Hodges began work on the semantic assumptions of model theory. Here he gave compositional truth definitions (built up by recursion on the complexity of formulas) for several languages which had been claimed to have no such truth definition; they include Jaakko Hintikka's adaptation of the branching quantifier languages of Leon Henkin. These compositional truth definitions are the basis for work of several authors, for example the PhD theses of Francine

Dechesne and Allen Mann, and above all Jouko Väänänen's team logic.) Hodges showed that the existence — though not the exact form — of such compositional truth definitions follows from some very general principles that relate semantics to syntax and that apply also to natural languages with a well-defined syntax; linguists and philosophers who have discussed this work include Dag Westerståhl, Markus Werning and Jeff Pelletier. Other areas in which Hodges has published, all related to semantics, are: (1) The logical basis of some specification languages; this work applies the sufficient conditions for a construction to be set-theoretically definable, and has encouraged recent work by several computer scientists. (2) The history of semantics, ranging from translation of writings of Ibn Sina and medieval Arabic semanticists to several studies of the development of Alfred Tarski's early views on model theory. (3) The formalisation of informal arguments, and in particular some analyses of why students and logical amateurs find certain forms of argument difficult.

**Main Publications:**

1. Logic, Penguin Books, London 1977, 331pp.; 2nd edition 2001.
2. Building Models by Games, London Math. Soc. Student Texts no. 2, Cambridge University Press 1985, 311pp. Reprinted by Dover Books 2006.
3. Model Theory, Cambridge University Press, Cambridge 1993, 772pp.
4. A Shorter Model Theory, Cambridge University Press, Cambridge 1997.
5. with A. H. Lachlan and Saharon Shelah: Possible orderings of an indiscernible sequence, Bull. London Math. Soc. 9 (1977) 212-215.
6. Krull implies Zorn, J. London Math. Soc. 19 (1979) 285-287.
7. with Ian Hodkinson, Daniel Lascar and Saharon Shelah: The small index property for $\omega$-stable $\omega$-categorical structures and for the random graph, J. London Math. Soc. 48 (1993) 204-218.
8. with Anand Pillay: Cohomology of structures and some problems of Ahlbrandt and Ziegler, J. London Math. Soc. 50 (1994) 1-16.
9. The meaning of specifications, I: Initial models, Theoretical Computer Science 152 (1995) 67-89.
10. Compositional semantics for a language of imperfect information, Logic Journal of the IGPL 5 (1997) 539-563.
11. 'Some strange quantifiers', in Structures in Logic and Computer Science, ed. Jan Mycielski et al., Lecture Notes in Computer Science 1261, Springer, Berlin 1997, pp. 51-65.
12. An editor recalls some hopeless papers, Bulletin of Symbolic Logic 4 (1998) 1-16.
13. Formal features of compositionality, Journal of Logic, Language and Information 10 (2001) 7-28.
14. What languages have Tarski truth definitions?, Annals of Pure and Applied Logic 126 (2004) 93-113.
15. 'Definability and automorphism groups', in Proceedings of International Congress in Logic, Methodology and Philosophy of Science, Oviedo 2003, ed. Petr Hajek et al., King's College Publications, London 2005, pp. 107-120.
16. with Ian Chiswell: Mathematical Logic, Oxford University Press 2007.
17. Tarski's theory of definition. In *New Essays on Tarski and Philosophy*, ed. Douglas Patterson, Oxford University Press 2008.
18. with Anatoly Yakovlev: Relative categoricity in abelian groups II, in Proceedings of Mathematics Conference in St Petersburg 2005, ed. Yi Zhang (to appear).
*Work in Progress*
19. with Saharon Shelah: Naturality and definability II.
20. Commented translation of Ibn Sina, I'bara (draft available).

**Service to the Profession:** President, Division of Logic, Methodology and Philosophy of Science of the International Union of History and Philosophy of Science 2008–2011; President of the International Union 2010–2011; Editorial Board Member, Annals of Applied Logic, 2003–; Scientific Advisory Committee, ILLC, University of Amsterdam, 1999-; Editorial Board Member, Logic and its Applications, 1998–; Committee Member, Electronic Publishing and Communication (CEIC) of International Mathematical Union, 1998–2002; Vice-President, London Mathematical Society, 1996–1997; President, European Association for Logic, Language and Information (FoLLI), 1995–1996; Executive Committee, Association for Symbolic Logic, 1994–1996; Editorial Board Member, Logic Journal of the IGPL, 1994–; Editorial Board Member, Mathematical Logic Quarterly, 1992–2002; Editorial Board Member, Perspectives in Logic, 1991–2005; President, British Logic Colloquium, 1990–1995; Editorial Board Member, Journal of Logic and Computation, 1989–; Advisory Editor, Logic and Set Theory, London Mathematical Society, 1988–1996; With Otto Kegel and Peter Neumann, Organiser of Durham Symposium on Model Theory and Groups, 1988; Editorial Board Member, Journal of Symbolic Logic, 1979–1987; Editorial Board Member, Journal of Philosophical Logic, 1974–1994.

**Teaching:** PhD students (13 in all) include Alex Wilkie, Michael Mortimer, Anand Pillay, Patricia K. Rogers, Kornelia Kalfa, Simon Thomas, Ian Hodkinson, and Jeremy Clark. Four logic textbooks in print (and a fifth in press) at levels ranging from popular to research. Educational lectures: London Mathematical Society Popular Lecture 1987; British Association for the Advancement of Science 1987, 1998; Coulter McDowell Annual Lecture 2001; Third Annual John Venn Lecture 2003. Undergraduate teaching includes an innovative course using logic as a framework for teaching mathematical writing.

**Vision Statement:** In the allowed forty words I can only quote three pieces of sound advice: 'Only connect' (E. M. Forster); 'Go round about' (the Bøjg to Peer Gynt); 'Think of something new and then prove it' (Jerome Keisler in conversation, 1967).

# HOMER, Steven

**Specialties:** Computability theory, complexity theory, mathematical logic, quantum computation.

**Born:** 27 March 1952 in Chicago, Illinois, USA.

**Educated:** MIT, PhD Mathematics 1978, University of California, Berkeley, BA Mathematics, 1973.

**Dissertation:** *Priority Arguments in Beta-Recursion Theory*; supervisor, Gerald Sacks.

**Regular Academic or Research Appointments:** CO-DIRECTOR OF BOSTON UNIVERSITY CENTER FOR RELIABLE INFORMATION SYSTEMS AND CYBER SECURITY, 2004–; PROFESSOR OF COMPUTER SCIENCE, BOSTON UNIVERSITY, 1992–. Associate Professor of Computer Science, Boston University, 1986-1992; Computer Science Department Chairman, 1986-1990, 1993; Assistant Professor of Computer Science, Boston University, 1982-1986; Assistant Professor of Mathematics, DePaul University, 1978-1982.

**Visiting Academic or Research Appointments:** Visiting Professor, Mathematical Institute, Oxford University, 1996-1997; Guest Professor and Fulbright Fellow, Mathematical Institute, Heidelberg University, 1988-1989.

**Research Profile:** Trained in mathematical logic, Dr. Homer wrote his PhD on ordinal recursion theory under Gerald Sacks at MIT (1978). His research in recursion theory continued until 1982 when his interests turned toward theoretical computer science, and to complexity theory in particular. Research in complexity theory has centered on the computational complexity of NP, exponential time problems and reductions between combinatorial problems. Homer has also done research in quantum computation, security, error correcting codes, parallel and randomized algorithms, mathematical logic and computational learning theory. He has published over 60 research papers in these fields.

**Main Publications:**

1. Two Splitting Theorems for Beta-Recursion Theory, Annals of Mathematical Logic (18) 1980, pp. 137-151.
2. Quadratic Automata, (with Jerry Goldman), Journal of Computer and Systems Sciences, (24) 1982, pp. 180-196.
3. Inverting the 1/2-Jump, (with Gerald Sacks), Transactions of the AMS, (278) 1983, pp. 317-331.
4. Oracle Dependent Properties of the Lattice of NP Sets, (with W. Maass), Theoretical Computer Science (24) 1983, pp. 279-289.
5. Minimal Degrees for Polynomial Reducibilities, Journal of the Association for Computing Machinery, (34), 1987, pp. 480-491.
6. Oracles for Structural Properties: The Isomorphism Conjecture and Public-Key Cryptography (with Alan Selman), Journal of Computer and Systems Sciences, (44), 1992, pp. 287-301.
7. Almost Everywhere Complexity Hierarchies for Nondeterministic Time, (with E. Allender, R. Beigel and U. Hertrampf), Theoretical Computer Science, 115, 1993, pp. 225-241.
8. Minimal Pairs and Complete Problems, (with K. Ambos-Spies and R. Soare), Theoretical Computer Science, 132, 1994, pp. 229-241.
9. On Reductions of NP Sets to Sparse Sets, (with Luc Longpre), Siam Journal on Computing, 1994, pp. 324 – 336.
10. Learning Discretized Geometric Concepts, (with Zhixiang Chen and Nader Bshouty), Annual Conferences on Foundation of Computer Science, 1994, pp. 54-63.
11. On the Performance of Polynomial-Time CLIQUE-Approximation Algorithms on very large Graphs (with Marcus Peinado), In Cliques, Coloring and Satisfiability: Second DIMACS Implementation Challenge 1993, David Johnson and Michael Trick, editors, AMS Press, DIMACS Series 26, 1996, pp. 147-168.
12. Quantum NP is Hard for PH (with Stephen Fenner, Fred Green and Randall Pruim), Royal Society of London A (1999) 455, pp 3953 – 3966.
13. *Computability and Complexity Theory*, (with Alan Selman), Springer Texts in Computer Science, Springer-Verlag, 2001.
14. Hyper-Polynomial Hierarchies and the NP-Jump (with Stephen Fenner, Randall Pruim and Marcus Schaefer), Theoretical Computer Science 262, 2001, pp. 241-256.
15. Quantum Lower Bounds for Fanout, (with Steve Fenner, Fred Green, Maosen Fang and Y. Zhang), Journal of Quantum Information and Computation, 2006, Vol. 6, no 1, pp. 46–57, and in lanl.arXiv.org, quant-ph/0312208.

*Work in Progress*

16. A second, expanded edition of *Computability and Complexity Theory*, (with Alan Selman), Springer Texts in Computer Science.
17. An examination of the role and necessity of ancil-

lae (extra work bits) in small, efficient quantum circuits.

18. A study of the complexity-theoretic relationships between (resource-bounded) Kolmogorov random sets, and their NP-jumps, and complete sets.

**Service to the Profession:** Member of External Review Committee for B.S. and M.S. Programs in Computer Science, SUNY at Buffalo, February, 2004; Co-organizer, Biennial MIT Mathematical Logic Meeting, 2003; Program Committee Member, Annual IEEE Conference on Computational Complexity, 2003; Program Committee Member, International Computing and Combinatorics Conference, 2001; Editorial Board Member, Journal of Universal Computer Science, 1999-present; Member of the IFIP (Int'l Federation for Information Processing) committee on descriptional complexity, 1998-present; Steering Committee member, annual IEEE Conference on Computational Complexity, 1998-1999; NSF Panel Member, typically 1-2 panels per year, 1995-present; Conference Chair, Annual Structure in Complexity Theory Conference (currently IEEE Conference on Computational Complexity), 1994 – 1997; Committee Chair, Fulbright-Hays Area Committee on Computer Science Fellowship, 1994-95; Editorial Board Member, Chicago Journal of Theoretical Computer Science, 1993-2002; Program Committee Chair of the annual Structure in Complexity Theory conference, 1993; Member of the Fulbright-Hays Area Committee on Computer Science Fellowship, 1992-1995; Local arrangements Co-Chair, Annual Structure in Complexity Theory Conference, 1992.

**Teaching:** Since joining Boston University, Professor Homer has taught most of the classes in the Computer Science curriculum. He has specialized in teaching the theory classes, algorithms and complexity theory, teaching each of them many times and at all levels. He also regularly teaches classes in programming and data structures, and advanced topics such as randomized algorithms and quantum computing. In 1995 he won a computational science teaching award from the Department of Energy for co-developing (with R. Giles) and teaching an undergraduate course in programming supercomputers (The CM-5). He has supervised 12 PhD students and 3 postdoctoral fellows.

**Honours and Awards:** Fulbright Senior Research Professor, Mathematisches Institut, Heidelberg University, 1988-1989; Computational Science Department of Energy Undergraduate Teaching Award, Ames Laboratory, September 1994; Phi Beta Kappa; Sigma Chi.

# HORTY, John F

**Specialties:** Philosophical logic, logic in artificial intelligence, decision theory, formal theories of practical reasoning.

**Born:** 23 November, 1954 in Johnstown, Pennsylvania, USA.

**Educated:** University of Pittsburgh, PhD, 1986; Oberlin College, BA, 1977.

**Dissertation:** *Some aspects of meaning in mathematical language*; supervisor, Nuel Belnap.

**Regular Academic or Research Appointments:** PROFESSOR, PHILOSOPHY DEPARTMENT AND INSTITUTE FOR ADVANCED COMPUTER STUDIES; AFFILIATE PROFESSOR, COMPUTER SCIENCE DEPARTMENT, UNIVERISTY OF MARYLAND, 1999–; Assistant Professor, University of Maryland, 1989-1994; Associate Professor, University of Maryland, 1994-1998.

**Research Profile:** Horty has worked in nonmonotonic logic, knowledge representation, deontic logic, tense logic, the logic of action, formal philosophy of language, decision theory, and formal theories of practical reasoning.

**Main Publications:**

1. *Agency and Deontic Logic*. Oxford University Press (2001), xiv+192 pp. Also included in Oxford Scholarship Online.
2. em Frege on Definitions: A Case Study of Semantic Content. Oxford University Press, forthcoming.
3. A clash of intuitions: the current state of nonmonotonic multiple inheritance systems (D. Touretzky, J. Horty, and R. Thomason). In *Proceedings of the Tenth International Joint Conference on Artificial Intelligence* (IJCAI-87), Morgan Kaufmann Publishers (1987), pp. 476–482.
4. A skeptical theory of inheritance in nonmonotonic semantic networks (J. Horty, R. Thomason, and D. Touretzky). *Artificial Intelligence*, vol. 42 (1990), pp. 311–348.
5. Frege on the psychological significance of definitions. *Philosophical Studies* (Special Issue on Definitions), vol. 69 (1993), pp. 113–153.
6. . Moral dilemmas and nonmonotonic logic. *Journal of Philosophical Logic*, vol. 23 (1994), pp. 35–65.
7. Some direct theories of nonmonotonic inheritance. In *Handbook of Logic in Artificial Intelligence and Logic Programming, Volume 3: Nonmonotonic Reasoning and Uncertain Reasoning*, D. Gabbay, C. Hogger, and J. Robinson (eds.), Oxford University Press (1994), pp. 111–187.
8. The deliberative stit: a study of action, omission, ability, and obligation (J. Horty and N. Belnap). *Journal of Philosophical Logic*, vol. 24 (1995), pp. 583–644.
9. Agency and obligation. *Synthese*, vol. 108 (1996), pp. 269–307.

10. Nonmonotonic foundations for deontic logic. In *Defeasible Deontic Logic*, D. Nute (ed.), Kluwer Academic Publishers (1997), pp. 17–44.

11. Evaluating new options in the context of existing plans (J. Horty and M. Pollack). *Artificial Intelligence*, vol. 127 (2001), pp. 199–220.

12. Argument construction and reinstatement in logics for defeasible reasoning. *Artificial Intelligence and Law*, vol. 9 (2001), pp. 1–28.

13. Skepticism and floating conclusions. *Artificial Intelligence*, vol. 135 (2002), pp. 55–72.

14. Reasoning with moral conflicts. *Nous*, vol. 37 (2003), pp. 557–605.

15. Defaults with priorities. Forthcoming in the *Journal of Philosophical Logic*

Work in Progress

16. *Reasons as Defaults*, Oxford University Press, under contract.

**Service to the Profession:** Editor, *Journal of Philosophical Logic*, 2008–.

**Honours and Awards:** Paper selected for *Philosopher's Annual* ("Ten Best Papers"), 1995; National Endowment for Humanities Fellowships, 1993-94, 2005-2006; National Science Foundation Grants, 1990-1991, 1990-1993, 1997-2000; Andrew Mellon Fellowship, University of Pittsburgh, 1980; Florence Frew Prize in Classics, Oberlin College, 1977; Phi Beta Kappa, Oberlin College, 1977; Graduation with Highest Honors, Oberlin College, 1977.

# I

## IMMERMAN, Neil

**Specialties:** Descriptive complexity, Model checking, Static analysis, Finite Model Theory.

**Born:** 24 November, 1953 in Manhasset, New York, USA.

**Educated:** Yale University BS 1974, MS 1974, Cornell University, Ph.D., 1980

**Dissertation:** *First Order Expressibility as a New Complexity Measure*; supervisors, Juris Hartmanis and Anil Nerode.

**Regular Academic or Research Appointments:** PROFESSOR, COMPUTER SCIENCE DEPT., UNIVERSITY OF MASSACHUSETTS, AMHERST, 1995–. Associate Professor, Computer Science Dept., University of Massachusetts, Amherst, 1989–1995; Associate Professor, Computer Science Dept., Yale University, 1986–1989; Assistant Professor, Mathematics and Computer Science Depts., Yale University, 1983–1986; Assistant Professor of Computer Science, Dept. of Mathematics, Tufts University, 1980–1983.

**Visiting Academic or Research Appointments:** Visiting Professor, Computer Science Dept., University of Wisconsin, Madison, 2003–2004; Visiting Professor, Computer Science Dept., Cornell University, 1995–1996; Visitor, Mathematical Sciences Research Institute, Berkeley, CA, Fall, 1985; Visiting Scientist, Laboratory for Computer Science, M.I.T., 1980–1983.

**Research Profile:** Neil Immerman is one of the principal founders of the field of Descriptive Complexity which has characterized all important computational complexity classes in terms of the richness of logical languages needed to express the computational problems. This has led to fundamental new insights in computational complexity and database theory. In particular, Immerman used these methods to prove that the nondeterministic space complexity classes are closed under complementation. For this result, which shocked the complexity community, Immerman and Robert Szelepcsényi shared the Gödel prize in theoretical computer science. Immerman's book, *Descriptive Complexity*, is an in-depth introduction to the field. While continuing to work on complexity, Immerman has more recently been studying formal methods whose ultimate goal is the computer-aided construction of reliable systems.

**Main Publications:**

1. *Descriptive Complexity*. In *Springer Graduate Texts in Computer Science*. Springer, Berlin, 1999.
2. Simulating Reachability using First-Order Logic with Applications to Verification of Linked Data Structures, with T. Lev-Ami, T. Reps, M. Sagiv, S. Srivastava, and G. Yorsh. *20th Int'l. Conf. Automated Deduction (CADE)*, ? (2005): ??–??.
3. The Boundary Between Decidability and Undecidability for Transitive-Closure Logics, with A. Rabinovich, T. Reps, M. Sagiv, and G. Yorsh. *Computer Science Logic (CSL)*, ? (2004): 160–174.
4. Complete Problems for Dynamic Complexity Classes, with W. Hesse. *IEEE Symp. Logic In Comput. Sci.*, ? (2002): 313–322.
5. An n! Lower Bound On Formula Size, with M. Adler. *ACM Transactions on Computational Logic*, 4(3) (2003): 296–314.
6. Reachability Logic: An Efficient Fragment of Transitive Closure Logic, with N. Alechina. *Logic Journal of the IGPL*, 8(3) (2000): 325–338.
7. Number of Variables Is Equivalent To Space, with J. Buss, and D.M. Barrington. *J. Symbolic Logic*, 66(3) (2001): 1217–1230.
8. Model Checking and Transitive Closure Logic, with M. Vardi. *Symposium on Computer-Aided Verification*, ? (1997): 291–302.
9. Tree Canonization and Transitive Closure, with K. Etessami. *Information and Computation* 157(1,2) (2000): 2–24.
10. Dyn-FO: A Parallel, Dynamic Complexity Class, with S. Patnaik. *J. Comput. Sys. Sci.*, 55(2) (1997): 199–209.
11. A First-Order Isomorphism Theorem, with E. Allender and J. Balcázar. *SIAM J. Comput.*, 26(2) (1997): 557–567.
12. An Optimal Lower Bound on the Number of Variables for Graph Identification, with J. Cai and M. Fürer. *Combinatorica*, 12(4) (1992): 389–410.
13. On Uniformity Within $NC^1$, with D.M. Barrington and H. Straubing. *J. Comput. Sys. Sci.*, 41(3) (1990): 274–306.
14. Definability (with Bounded Number of Bound Variables, with D. Kozen. *Information and Computation*, 83(1989): 121–139.
15. Expressibility and Parallel Complexity. *SIAM J. of Comput.*, 18(1989): 625–638.
16. Nondeterministic Space is Closed Under Complementation. *SIAM J. Comput.*, 17(5) (1988): 935–938.
17. Languages That Capture Complexity Classes. *SIAM J. of Computing*, 16(4) (1987): 760–778.
18. Relational Queries Computable in Polynomial Time, *Information and Control*, 68(1986): 86–104.
19. Upper and Lower Bounds for First Order Express-

ibility. *J. Comput. Sys. Sci.*, 25(1982): 76–98.

20. Number of Quantifiers is Better Than Number of Tape Cells. *J. Comput. Sys. Sci.*, 22(1981): 384–406.

**Service to the Profession:** Editor, *SIAM Journal on Computing*, 2003–; *Information and Computation*, 1987–2004; *Chicago Journal of Theoretical Computer Science*, 1994–2003; *Journal of Symbolic Logic*, 1996–1999.

Program Committee Chair, *Finite Model Theory Workshop*, Bedlewo, Poland, 2003, *Structure in Complexity Theory*, 1991.

Program Committee Member, *Int'l. Conf. Logic Programming, AI, and Reasoning*, 2001; *Principles of Database Systems*, 2001, *Computer Science Logic*, 1999; *Logic in Computer Science*, 1998, 2005; *ACM Symp. on Theory of Computation*, 1995, 1986; *Principles of Database Systems*, 1995; *Structure in Complexity Theory*, 1994, 1989, 1987.

**Teaching:** Ph.D. Students: Ruben Michel, Sushant Patnaik, Kousha Etessami, Jose Antonio Medina, and William Hesse.

**Vision Statement:** Descriptive complexity has taught us that the fundamental questions in computational complexity can be understood as problems in logic. Logic is also key to the creation of environments for building correct, flexible, and secure software.

**Honours and Awards:** Guggenheim Fellow, 2003–2004; Fellow of the Association of Computing Machinery, 2002–; Gödel Prize in Theoretical Computer Science, 1995

# ISRAEL, David

**Specialties:** Philosophy of rational agency, theory of knowledge representation in artificial intelligence, philosophy of logic, semantics of natural languages.

**Born:** 27 September 1943 in Boston, MA, USA.

**Educated:** University of California, Berkeley, PhD Philosophy, 1973; Harvard College, BA Philosophy, 1965.

**Dissertation:** *??*; supervisor, Barry Stroud.

**Regular Academic or Research Appointments:** PROGRAM DIRECTOR, ARTIFICIAL INTELLIGENCE CENTER, SRI INTERNATIONAL, 2001–; Senior Computer Scientist, AIC, SRI International, 1984-; Computer Scientist, Artificial Intelligence Department, Bolt Beranek and Newman, 1979-1984; Assistant Professor, Philosophy, Tufts University, 1970-1977.

**Visiting Academic or Research Appointments:** Assistant Professor, Philosophy, Hampshire College, 1978-1979; Assistant Professor, Philosophy, University of Connecticut, 1977-1978.

**Research Profile:** For much of the last 30 years David Israel has been working on projects that involve the application of logical concepts and techniques to the design and/or analysis of artificial intelligence systems. In many cases, and especially with regard to work on rational agency, this has meant theorizing about distinctively, and perhaps uniquely, human phenomena, but his focus has always been on what can be learned from the human case applied to the design of artificial systems. Starting about 8 years ago, Dr. Israel has undergone a not-so-gradual, and by no quite complete transubstantiation into a manager of others' research.

**Main Publications:**

1. What's Wrong with Non-Monotonic Logic. *Proceedings of the First Annual Conference of the American Association for Artificial Intelligence*, Stanford University, August 1980.

2. Distinctions and Confusions: A Catalogue Raisonne, (with R. J. Brachman). *Proceedings of the Seventh International Joint Conference on Artificial Intelligence*, August 1981. Reprinted in *Context-Directed Pattern Recognition and Machine Intelligence Techniques for Information Processing*, Y. Pao and G. Ernst (eds.), 1982, IEEE Computer Society Press.

3. Interpreting Network Formalisms. In a special issue on Computational Linguistics of the *International Journal of Computers and Mathematics*, Vol. 9, No. 1, 1983, N. Cercone (ed.). This issue was reprinted by Pergamon Press, Oxford, 1983.

4. Some Remarks on the Semantics of Representation Languages (with R.J. Brachman). *On Conceptual Modeling: Perspectives from Artificial Intelligence, Databases and Programming Languages*, M. Brodie, J. Mylopoulos, and J. Schmidt (eds.), Springer Verlag, NY, 1984.

5. A Prolegomenon to Situation Semantics. *Proceedings of the Association for Computational Linguistics Conference*, MIT, Cambridge, MA, June 1983.

6. Some Remarks on the Place of Logic in Knowledge Representation. In a special issue of *IEEE Computer on Knowledge Representation in Artificial Intelligence*, October 1983, Vol. 16, No. 10. Reprinted in *The Knowledge Frontier: Essays in the Representation of Knowledge*, N. Cercone and G. McCalla (eds.), Springer-Verlag, New York, 1987.

7. A Short Companion to the Naive Physics Manifesto. *Formal Theories of the Common Sense World*, J. R. Hobbs and R.C. Moore (eds.), Ablex Publishing, Norwood, NJ, 1985.

8. The Role of Propositional Objects of Belief in Action. *CSLI Report No. CSLI-87-72*, April 1987.

9. Plans and Resource-Bounded Practical Reasoning, (with Bratman, M.E., and Pollack, M.E.). *Computational Intelligence*, Vol. 4, No. 4, 1988. Reprinted in Philosophy and AI: Essays at the Interface, R. Cummins and J. Pollock, (eds.), MIT Press, Cambridge, 1991.

10. What is Information? (with John R. Perry). *Information, Language and Cognition: Vancouver Studies in Cognitive Science*, Vol. I, ed. P. Hanson, University of British Columbia Press, 1990.

11. Fodor and Psychological Explanations (with John Perry). *Meaning in Mind*, B. Loewer and G. Rey (eds.), Basil Blackwell, Oxford, pages 165-180, 1991.

12. Actions and Movements (with John Perry and Syun Tutiya). *Proceedings of IJCAI-91*, Sydney, Australia, August, 1991.

13. Information and Architecture (with John Perry). *Situation Theory and its Applications, II*, K.J. Barwise, J. M. Gawron, G. Plotkin, and S. Tutiya, (eds.), CSLI Lecture Notes, 1991.

14. Lucid and Intensional Logic. *Proceedings of the Fifth International Symposium on Lucid and Intensional Programming Languages*, San Francisco, April, 1992.

15. The Role(s) of Logic in Artificial Intelligence. *The Handbook of Logic in Artificial Intelligence and Logic Programming, Volume I*, D. M. Gabbay, C. J. Hogger, and J. A. Robinson (eds.), Oxford University Press, 1993.

16. Executions, Motivations, and Accomplishments, (with J. Perry and S. Tutiya). *Philosophical Review*, Vol. 102, No. 4, October, 1993.

17. The Very Idea of Dynamic Semantics. *Proceedings of the Ninth Amsterdam Colloquium*, Amsterdam, December, 1993.

18. FASTUS: A Cascaded Finite-State Transducer for Extracting Information from Natural-Language Text,. (with D.Appelt, J.Bear, J.Hobbs, M.Kameyama, M.Stickel and M.Tyson). *Finite State Devices for Natural Language Processing*, E.Roche and Y.Schabes, (eds.), MIT Press, Cambridge, MA, 1996.

19. Where Monsters Dwell (with J.R. Perry). *Logic, Language and Computation*, D.Westerstahl and J.Seligman (eds.), CSLI Publications, Stanford, 1996.

20. Interpretation of Discourse and Proof Theory: Questions and Directions'. *JFAK. Essays Dedicated to Johan van Benthem on the Occasion of his 50th Birthday*, J. Gerbrandy, M. Marx, M. de Rijke and Y. Venema (eds.), Amsterdam University Press, 1999.

**Vision Statement:** While I maintain some of my earlier interest in aspects of pure logic, the main focus of what vision I have left is on the intersection of logic with Computer Science generally and Artificial Intelligence, more particularly.

# J

## JACQUETTE, Dale

**Specialties:** Philosophical logic, philosophy of logic, history of logic.

**Born:** 19 April 1953 in Sheboygan, WI, USA.

**Educated:** Brown University, PhD Philosophy, 1983; Brown University, MA Philosophy, 1981; Oberlin College, BA Philosophy, 1975.

**Dissertation:** *The Object Theory Logic of Intention*; supervisor, Roderick M. Chisholm.

**Regular Academic or Research Appointments:** PROFESSOR, PHILOSOPHY, PENNSYLVANIA STATE UNIVERSITY, 1986–.

**Visiting Academic or Research Appointments:** Visiting Assistant Professor, Philosophy, University of Nebraska, Lincoln, 1985-1986; Visiting Assistant Professor, Philosophy, Franklin and Marshall College, 1983-1985.

**Research Profile:** Jacquette has made philosophical contributions to contemporary logic both as system builder and critic of logical systems. His projects in logic have largely centered on efforts to do justice, first, to the semantics of the true or false sentences of natural science, and in the second instance on extrascientific uses of formal and colloquial languages. Jacquette's first logical system-building involved intensional systems intended to support the reference and true predication of constitutive properties to any and all objects regardless of their ontic status. The purpose of such logics is to provide a more general intuitive semantics for the truth values of propositions ostensibly about nonexistent entities, including law-governed idealizations, in the language of natural science, that can also do duty for predications in fiction, practical reasoning, and discourse outside of science, than was available in standard extensionalist logics and semantics, inspired by Bertrand Russell's theory of definite descriptions. The task is to adapt the notation of standard logic, in part to determine how far and in what directions it could be stretched, to suit the purposes of an ontically agnostic logic to which a semantics with both an extensional and extra-extensional component could productively be attached. The ease with which such formalisms can be developed, despite the need to overcome certain technical obstacles, and the satisfying fit of the interpretation, satisfaction and truth conditions that the logic made possible, provide a clue to deeper philosophical questions about the nature and meaning of logical relations, which has characteristically been Jacquette's focus. Jacquette also been engaged over the years in criticisms of received logical systems and more especially of key assumptions about the nature of widely discussed paradoxes, puzzles, and limitations in logic. These prominently include the liar, Russell, Curry, and other paradoxes, but additionally paradoxes originating in the medieval tradition, such as the problem of Buridan's bridge and Walter Burleigh's challenges to hypothetical syllogism. Jacquette has been more concerned about less respected problems like Grelling's paradox, and the Pseudo-Scotus or validity paradox, including his own variation of the inference, which he has labeled the soundness paradox. From a philosophical perspective, Jacquette finds it natural to regard paradoxes as conceptual challenges to be overcome, rather than as ultimate threats to logical consistency. Jacquette is suspicious of, although he offers no conclusive reason to deny the proposal that all paradoxes might surrender to the same general style of solution. Instead, he considers each according to its own terms and the special problems it raises. His approaches to the logical paradoxes typically involve specific but usually pedestrian methods of identifying equivocations; imposing intuitively plausible restrictions on certain distinctions; or by uncovering other errors of reasoning that have contributed to the mistaken judgment that a genuine paradox needs to be solved. In his work to date, Jacquette has gestured toward a branching synthesis in which several kinds of solutions can be knit together into distinct families, suggesting in most cases that the apparent paradoxes themselves belong to corresponding categories in which logical or conceptual blunders of different sorts are being made. In this regard, Jacquette's policy toward the paradoxes has been most profoundly influenced by Wittgenstein in both his early and later periods. The methods he have found most expedient for deflating the standard paradoxes has nevertheless virtually nothing to do with Wittgenstein's logical atomism, picture theory of meaning, or general form of proposition, nor, except in the most general terms that most Wittgensteinians would probably disown. In particular, Jacquette is out of sympathy with Wittgenstein's later appeal to philosophical grammar as a kind of therapy in which paradoxes among other traditional philosophical problems can be psychologically overcome in the sense of being quieted at least in their urgency. Jacquette insists whenever possible on using the most widely accepted methods of logic to uncover errors in ordinary logic. Ideally, Jacquette expects logic to cure itself, to take care of itself, as Wittgenstein maintains in both the *Tractatus Logico-Philosophicus* and *Notebooks 1914-1916*. This is the limited sense, especially in confronting the paradoxes, in which Jacquette's approach to logic remains Wittgensteinian.

**Main Publications:**

1. Translation with Introduction and Critical Commentary of Gottlob Frege, *Foundations of Arithmetic: A Logical-Mathematical Investigation into the Concept of Number* (Longman), 2007.
2. Critical Introduction to Bertrand Russell, *An Introduction to Mathematical Philosophy* (Longman), 2007.
3. *Philosophy of Logic* (edited) Handbook of the Philosophy of Science series, edited by Dov Gabbay, John Woods and Paul Thagard (North-Holland Press (Elsevier), 2006) (xiv + 1218 pp.) (forthcoming also in Russian and Chinese translations).
4. *A Companion to Philosophical Logic* (edited) (Blackwell Publishers, 2002), (xiii + 816 pp.). (Electronic e-book NetLibrary version published to licensed distributors by Baker & Taylor, Inc.; http://www.netLibrary.com). Paperback edition with corrections 2006.
5. *Symbolic Logic* (with instructor's and student solutions manuals prepared in collaboration with Andrew R. Martinez and interactive logic exercises on CD–ROM designed in collaboration with Nelson Pole) (Wadsworth Publishing, 2001) (xix + 488 pp.).
6. *Ontology* (Acumen Books, McGill–Queen's University Press / Central Problems of Philosophy Series, 2002) (xv + 348 pp.).
7. *On Boole* (Wadsworth Publishing / Wadsworth Philosophers Series, 2002), (vi + 97 pp.).
8. *Meinongian Logic: The Semantics of Existence and Nonexistence* (Walter de Gruyter & Co., 1996) (xiii + 297 pp.).
9. "Bochenski on Property Identity and the Refutation of Universals", *Journal of Philosophical Logic*, 35, 2006, 293–316.
10. "Propositions, Sets, and Worlds", *Studia Logica: An International Journal for Symbolic Logic*, special double issue 'Ways of Worlds' on 40 Years of Possible Worlds Semantics, Vol. 1: On Possible Worlds and Related Notions, guest edited by Vincent F. Hendricks and Stig Andur Pedersen, 82, 2006, 337–343.
11. "Grelling's Revenge", *Analysis*, 64, 2004, 251–256.
12. "The Soundness Paradox", *Logic Journal of the Interest Group in Pure and Applied Logics* (IGPL), 11, 2003, 547–556.
13. "Diagonalization in Logic and Mathematics", *Handbook of Philosophical Logic*, 2nd Edition, Volume 11, edited by Dov M. Gabbay and Franz Guenthner (Dordrecht: Kluwer Academic Publishing, 2002), 55–147.
14. "Analysis of Quantifiers in Wittgenstein's *Tractatus*: A Critical Survey", *Logical Analysis and History of Philosophy*, 4, 2001, 191–202.
15. "An Internal Determinacy Metatheorem for Lukasiewicz's *Aussagenkalküls*", *Bulletin of the Section of Logic*, 29, 2000, 115–124.
16. "Paraconsistent Logical Consequence", *Journal of Applied Non-Classical Logics*, 8, 1998, 337–351.
17. "The Validity Paradox in Modal $S_5$", *Synthese*, 109, 1996, 47–62.
18. "Tarski's Quantificational Semantics and Meinongian Object Theory Domains", *Pacific Philosophical Quarterly*, 75, 1994, 88–107.
19. "On the Completeness of a Certain System of Arithmetic of Whole Numbers in which Addition Occurs as the Only Operation", translation of and commentary on Mojzesz Presburger, "Über die Vollständigkeit eines gewissen Systems der Arithmetik ganzer Zahlen, in welchem die Addition als einzige Operation hervortritt", *History and Philo-sophy of Logic*, 12, 1991, 225–233.
20. "Metamathematical Criteria for Minds and Machines", *Erkenntnis*, 27, 1987, 1–16.

*Work in Progress*

21. *Logic, Philosophy, Analysis* (Acumen Books)
22. *Frege: A Philosophical Biography* (Cambridge University Press)
23. *Mathematical Entity*
24. *Early Wittgentein on Identity*

**Service to the Profession:** Editor, *American Philosophical Quarterly*, 2002-2005; Series General Editor, *New Dialogues in Philosophy*, Rowman & Littlefield; Guest editor, *Journal of Value Inquiry*, 35, 2001, special issue on 'Aristotle's Theory of Value'; Guest editor, *Philosophy and Rhetoric*, 30, 1997, special issue on 'The Dialectics of Psychologism'; Guest co-editor (with Nicholas Griffin), *Russell: Journal of the Bertrand Russell Society*, special issue commemorating publication of Russell's 'On Denoting' (forthcoming).

**Teaching:** Supervision of numerous PhD and MA theses in philosophy, especially on Wittgenstein, Pennsylvania State University.

J. William Fulbright Distinguished Lecture Chair in Contemporary Philosophy of Language, University of Venice, Italy, 1996

SE Oberseminar — Philosophische Logik, 5-day pro-seminar lecture course, in collaboration with Professor Dr. Edgar Morscher, Fachbereich Philosophie, Universität *? Salzburg, Salzburg, Austria — 1. Grenzwerten für logische Formulismuse; 2. Referenz, Bedeutung und Wahrheit; 3. Intensionale und extensionale Logik und Semantik; 4. Paradoxien und Diagonalisierung; 5. Verschwommenheit und Allgemeinheit. May 15-19,

2006

Logik und Ontologie, 8-week pro-seminar lecture course, Bayerische–Julius–Maximilians–Universität–Würzburg, Würzburg, Germany, Spring 2001

Philosophical Problems of Artificial Intelligence: Intentionality, Mechanism, and Connectionism (5-day lecture course) – 'Searle's Causal Theory of Intentionality', 'Syntax and Semantics in the Turing Test and Chinese Room', 'Assumption and Mechanical Simulation of Hypothetical Reasoning', 'Artificial Intelligence and Counterfactual Inference', 'Gödel's Proof, Connectionism, and Metamathematical Criteria for Minds and Machines', International Summer Schools for Philosophy and Artificial Intelligence, Castle Mareccio, Bolzano, Italy, July 9–13, 1990

**Vision Statement:** Logic will continue to develop a wide array of nonstandard formalisms dedicated to concepts and inference structures that do not lend themselves to classical symbolisms. Intensional logics in particular will increasingly take precedence over traditional extensional systems.

**Honours and Awards:** Netherlands Institute for Advanced Study (NIAS), 2005-2006; Alexander von Humboldt–Stiftung, 1989-1990, 2000-2001; Associate, Center for Philosophy of Science, University of Pittsburgh, 1999-; Term Fellowship, Institute for the Arts and Humanistic Studies, Pennsylvania State University, 1997-1999; Institute for the Arts and Humanistic Studies, Pennsylvania State University, 1997; Melvin and Rosalind Jacobs Research Fellowship, Pennsylvania State University, 1993; National Endowment for the Humanities (NEH) 1984.

## JECH, Thomas

**Born:** 29 January 1944 in Prague, Czech Republic.

**Educated:** Doctorate: Charles University, Prague, 1966.

**Regular Academic or Research Appointments:** PROFESSOR, MATHEMATICAL INSTITUTE OF C.A.S. AND CENTER FOR THEORETICAL STUDY, PRAGUE, 2000–PRESENT. Research Associate, Charles University, 1966-68. Junior Fellow, University of Bristol, 1968-69. Associate Professor, S.U.N.Y. at Buffalo, 1969-74. Professor, The Pennsylvania State University, 1974-2000.

**Visiting Academic or Research Appointments:** Visiting Associate Professor, U.C.L.A., 1970-71. Visiting Associate Professor, Princeton University, Fall 1972. Member, Institute for Advanced Study, Spring 1973. Visiting Professor, Stanford University, Fall 1974. Visit. Professor, U.C.L.A., Winter and Spring 1981. Visiting Professor, University of Hawaii, Fall 1984. Visiting Fellow, All Souls College, Oxford, Michaelmas term 1988. Fulbright Professor, Hebrew University, Spring 1989. Visiting Professor, California Institute of Technology, Spring 1991. Center for Theoretical Studies, Prague, Fall 1995. Fulbright Professor, Université de Caen, Spring 1996. Visiting Professor, Université Paris VII, May–June 1975. National Academy of Sciences Exchange Scientist, Budapest, July 1977. Honorary Professor, Beijing Normal University, 1985. National Academy of Sciences Visiting Scholar, China, May–June 1987. Visitor, Mathematical Institute, ETH Zürich, December 1988–January 1989. Invited Professor, Université Pierre et Marie Curie, Paris, June 1989. Member, Mathematical Sciences Research Institute, Berkeley, January 1990. Invited Professor, Université Paris VII, June 1990. Visitor, Center for Theoretical Study, Prague, January 1991, July–August 1991, June–August 1992, June–July 1993, July–August 1994. Visitor, Argonne National Laboratory, May 1994. Visitor, C.R.M. Barcelona, November 1995, March–May 2001, April–May 2004. Resident scholar, Bellagio Center, Rockefeller Foundation, June 1998. Visiting Professor, National University of Singapore, November 2003.

**Service to the Profession:** Axiomatic Set Theory, Proc. Symp. Pure Math. XIII (2), 1974. Proceedings Amer. Math. Society, 1980–1988. Managing Editor, Annals of Pure and Applied Logic, 1986–present. Mathematics Advisory Committee, Council for International Exchange of Scholars, 1989–1992. American Czech-and-Slovak Education Fund, 1991–1994. Academic Advisory Board, Center for Theoretical Study, Prague, 1991–present.

**Teaching:** Ten graduated since 1972. Over 170 colloquium talks and other invited lectures in 20 countries.

## JOCKUSCH, JR, Carl G.

**Specialties:** Mathematical logic, especially computability theory (recursion theory) and its connections with other areas, in particular combinatorics, model theory, and reverse mathematics.

**Born:** 13 July 1941 in San Antonio, Texas, USA.

**Educated:** Vanderbilt University, 1959-1960; Swarthmore College, BA with Highest Honors, 1963; Massachusetts Institute of Technology, PhD Mathematics, 1966.

**Dissertation:** *Reducibilities in Recursive Function*; supervisor, Hartley Rogers, Jr.

**Regular Academic or Research Appointments:** PROFESSOR EMERITUS, MATHEMATICS, UNIVERSITY OF ILLINOIS AT URBANA-CHAMPAIGN, 2004–; Professor, Mathematics, University of Illinois at Urbana-Champaign, 1975-2004; Associate Professor, Mathematics, University of Illinois at Urbana-Champaign, 1971-1975; Assistant Professor, Mathematics, University of Illinois at Urbana-Champaign 1967-1971; Instructor, Mathematics, Northeastern University, 1966-1967.

**Visiting Academic or Research Appointments:** Member, Mathematical Sciences Research Institute, 1989; Visiting Senior Fellow, University of Leeds, 1982-1983; Visiting Associate Professor, Duke University, 1973; Numerous one-month visiting appointments at the University of Chicago, 1980-1998.

**Research Profile:** Carl Jockusch has worked in computability theory and its connections with combinatorics and model theory. Many of his papers address the effective content and logical strength of various forms of Ramsey's theorem from combinatorics. Together with Robert Soare, he studied the Turing degrees of paths through computable trees, and in particular proved the low basis theorem, which asserts that every infinite computable tree of binary strings has a path of low degree, i.e. the halting problem relative to the path is Turing equivalent to the usual halting problem. Since the solutions to many problems in combinatorics, algebra, and logic can be represented as paths through computable trees, his work in this area has found many applications. Together with Richard Shore, he developed the theory of pseudojump operators, which are natural extensions of the Turing jump operation. This has been a powerful tool in the study of Turing degrees. In separate joint work with Robert Soare and Rod Downey he has studied the Turing degrees of Boolean algebras. In joint work with Theodore Slaman he showed the decidability of a fragment of the first-order theory of the Turing degrees. Together with Paul Bateman and Alan Woods he studied the decidability of various theories with a predicate for the primes, assuming Schinzel's hypothesis from number theory.

**Main Publications:**

1. Semirecursive sets and positive reducibility, *Transactions of the American Mathematical Society* 131 (1968), 420-436.
2. Ramsey's theorem and recursion theory, *J. Symbolic Logic* 37 (1972), 268-280.
3. (with Robert I. Soare), $\Pi_1^0$ classes and degrees of theories, *Trans. Amer. Math. Soc.* 173 (1972), 33-56.
4. (with S. G. Simpson), A degree theoretic definition of the ramified analytical hierarchy, *Ann. Math. Logic* 10 (1975), 1-32.
5. (with D. B. Posner), Double jumps of minimal degrees *J. Symbolic Logic* 43 (1978), 715-724.
6. Degrees of generic sets, in *Recursion Theory: its Generalisations and Applications*, edited by F. R. Drake and S. S. Wainer, Cambridge University Press, 1980, 110-139.
7. (with R. A. Shore), Pseudo jump operators I: the r.e. case, *Trans. Amer. Math. Math. Soc.* 275 (1983), 599-609.
8. (with K. Ambos-Spies, R. Shore, and R. Soare), An algebraic decomposition of the recursively enumerable degrees and the coincidence of several degree classes with the promptly simple degrees, *Trans. Amer. Math. Soc.*, 281 (1984), 143-153.
9. (with R. A. Shore), Pseudo-jump operators II: Transfinite iterations, hierarchies, and minimal covers *J. Symbolic Logic* 49 (1984), 1205-1236.
10. (with R. G. Downey and M. Stob), Array nonrecursive sets and multiple permitting arguments, in *Recursion Theory Week, Proceedings*, 1989, edited by K. Ambos-Spies, G. H. Müller, and G. E. Sacks, Lecture Notes in Mathematics, Vol. 1432, Springer-Verlag, Berlin, Heidelberg, Tokyo, New York, 1990, 141-173.
11. (with T. Slaman), On the $\Sigma_2$ – theory of the upper semilattice of Turing degrees *J. Symbolic Logic* 58 (1993), 193-204.
12. (with P. T. Bateman and A. Woods), Decidability and undecidability of theories with a predicate for the primes, *J. Symbolic Logic* 58 (1993), 672-687.
13. (with Frank Stephan), A cohesive set which is not high, *Mathematical Logic Quarterly* 39 (1993), 515-530. (A corrective note, keeping the main result intact, appeared in the same journal, vol. 43 (1997), page 569.)
14. (with R. Soare), Boolean algebras, Stone spaces, and the iterated Turing jump, *J. Symbolic Logic* 59 (1994), 1121-1138.
15. (with R. Downey and M. Stob), Array nonrecursive sets and genericity, pp. 93-104 in *Computability, Enumerability, Unsolvability: Directions in Recursion Theory*, eds. S. B. Cooper, T. A. Slaman, S. S. Wainer, London Mathematical Society Lecture Notes Series # 224, Cambridge University Press, 1996.
16. (with P. Cholak and T. Slaman), On the strength of Ramsey's theorem for pairs, *J. Symbolic Logic*, 66 (2001), 1-55.
17. (with T. Hummel), Ramsey's theorem for computably enumerable colorings, *J. Symbolic Logic*, 66 (2001), 873-880.
18. (with A. Li and Y. Yang), A join theorem for the computably enumerable degrees, *Trans. Amer. Math. Soc.*, 356 (2004), 2557-2568.
19. (with R. Downey and J. Miller), On self-embeddings of computable linear orderings, *Annals of Pure and Applied Logic*, 138 (2006), 52-76.

20. (with D. Hirschfeldt, B. Kjos-Hanssen, S. Lempp, and T. Slaman), The strength of some combinatorial principles related to Ramsey's Theorem for pairs, to appear in *Proceedings of the Institute for Mathematical Sciences*

*Work in Progress*

21. A joint paper with Valentina Harizanov and Julia Knight on the Turing degrees of chains and antichains in computable partial orderings.

22. An attempt to better understand the strength of Ramsey's Theorem for pairs. In particular, Jockusch wants to know whether this combinatorial principle is implied by its "stable" version, and whether it implies Weak König's Lemma in $RCA_0$, the usual base system for Reverse Mathematics.

A full list of publications is available at Jockusch's home page http://math.uiuc.edu/~jockusch/

**Service to the Profession:** Editorial Board Member, Mathematical Logic Quarterly, 1998-; Editorial Series Board Member, Perspectives in Mathematical Logic, 2001-; Editor, Proceedings of the American Mathematical Society, 1997-2005; Chair, Program Committee for the year 2000 annual meeting of the Association for Symbolic Logic held in Urbana, 1998-2000; Chair, Committee on Meetings in North America of the Association for Symbolic Logic, 1993-1996; Member, Executive Committee of the Association for Symbolic Logic, 1989-1992; Editor, Journal of Symbolic Logic, 1974-1976.

**Teaching:** Jockusch taught many graduate courses in various areas of logic in his thirty-seven year career at the University of Illinois. He supervised the PhD theses of thirteen students. His final student, Joseph Mileti, was awarded the Sacks Prize by the Association for Symbolic Logic for the most outstanding doctoral dissertation in mathematical logic for 2004, jointly with Nathan Segerlind.

**Vision Statement:** Mathematical logic has at least three major connections with mathematics. First, it has provided strong foundations for mathematics. Second, it has been a useful tool for investigators working in several areas of mathematics, including algebra, analysis, number theory, and differential geometry. Finally, it is a thriving branch of mathematics in its own right. All of these roles are valuable, and none deserve to be denigrated. Furthermore, mathematical logic has had many fruitful interactions with computer science. The development of Reverse Mathematics has shown that the technical tools of computability theory (among other methods) are useful for obtaining precise foundational information on the strength of results in ordinary mathematics.

# K

## KANAMORI, Akihiro

**Specialties:** Set theory, history of set theory, philosophy of mathematics.

**Born:** 23 October 1948 in Tokyo, Japan.

**Educated:** California Institute of Technology, BS Mathematics 1970; Cambridge University, PhD Mathematics 1975.

**Dissertation:** *Ultrafilters over Uncountable Cardinals*; supervisor, Adrian Mathias.

**Regular Academic or Research Appointments:** PROFESSOR OF MATHEMATICS, BOSTON UNIVERSITY, 1992–. Associate Professor, Boston University, 1982-1991. Assistant Professor, Baruch College of the City University of New York, 1981-1982. Benjamin Pierce Assistant Professor, Harvard University, 1977-1981. Lecturer, University of California at Berkeley, 1975-1975.

**Visiting Academic or Research Appointments:** Senior Fellow, Dibner Institute for the History of Science and Technology, 2002-2003. Visiting Professor, Institute of Mathematics, Hebrew Universiy of Jerusalem, 1995. Berman Visiting Professor, Institute of Mathematics, Hebrew University of Jerusalem, 1988-1989.

**Main Publications:**

1. The Siege of Chitral as an Imperial Factor, *Journal of Indian History* 46(3)(1968), 387-404.
2. Weakly Normal Filters and Irregular Ultrafilters, *Transactions of the American Mathematical Society* 220(1976), 393-399.
3. Ultrafilters over a Measurable Cardinal, *Annals of Mathematical Logic* 11(1976), 315-356.
4. Strong Axioms of Infinity and Elementary Embeddings, *with William N. Reinhardt and Robert M. Solovay*, *Annals of Mathematical Logic* 13(1978), 73-116.
5. The Evolution of Large Cardinal Axioms in Set Theory, *with Menachem Magidor*, in: G.H. Müller and D.S. Scott (editors), Higher Set Theory (Proceedings, Oberwolfach, Germany 1977), Lecture Notes in Mathematics #669, (Springer-Verlag, 1978), pp.99-275.
6. Perfect-Set Forcing for Uncountable Cardinals, *Annals of Mathematical Logic* 19(1980), 97-114.
7. Morasses in Combinatorial Set Theory, in: A.R.D. Mathias (ed.) *Surveys in Set Theory* (Cambridge: Cambridge University Press, 1983), 167-196.
8. Partition Relations for Successor Cardinals, *Advances in Mathematics* 59(1986), 152-169.
9. Finest Partitions for Ultrafilters *The Journal of Symbolic Logic* 51(1986), 327-332.
10. On Gödel Incompleteness and Finite Combinatorics, *with Kenneth McAloon*, *Annals of Pure and Applied Logic* 33(1987), 23-41.
11. Regressive Partition Relations for Infinite Cardinals, *with Andras Hajnal and Saharon Shelah*, *Transactions of the American Mathematical Society* 299(1987), 145-154.
12. Regressive Partitions, Borel Diagonalization, and $n$-subtle cardinals, *Annals of Pure and Applied Logic* 52(1991), 65-77.
13. *The Higher Infinite*, Perspectives in Mathematical Logic, Springer-Verlag), 536 pages, 1994. Corrected second printing, 1997. Japanese translation, 1998. Second edition, April 2003.
14. Complete quotient Boolean algebras, *with Saharon Shelah*, *Transactions of the American Mathematical Society,* 347(1995), 1963-1979.
15. The Mathematical Development of Set Theory from Cantor to Cohen, *The Bulletin of Symbolic Logic* 2(1996), 1–71.
16. Hilbert and Set Theory, *with Burton Dreben*, *Synthese* 110(1997), 77–125.
17. The Mathematical Import of Zermelo's Well-Ordering Theorem, *The Bulletin of Symbolic Logic* 3(1997), 281–311.
18. Does GCH imply AC locally? *with David Pincus*, in: Gábor Halász, László Lovász, Miklós Simonivits and Vera T. Sós (editors), *Paul Erdös and His Mathematics*, Bolyai Society Mathematical Studies, (Berlin: Springer 2002), volume II, 413–426.
19. The Empty Set, the Singleton, and the Ordered Pair, *The Bulletin for Symbolic Logic* 9(2003), 273–298.
20. Zermelo and Set Theory, *The Bulletin for Symbolic Logic* 10(2004), 487–553.

*Work in Progress*

21. Set-Theoretic and Mathematical Knowledge: Beyond True and False, 30pp.
22. Gödel and Set Theory, 18pp.
23. Editing *Handbook of Set Theory*.
24. Editing *Collected Works of Ernst Zermelo*.

**Service to the Profession:** Association for Symbolic Logic: Member of the Executive Council, 1981-1990; Editor for set theory reviews for *The Journal of Symbolic Logic*, 1981-1990; Representative to the American Association for the Advancement of Science, 1995-8; on program committees for 1998 Winter Meeting (Baltimore, Maryland), 2000 Annual Meeting (Urbana, Illinois), 2000 Summer European Meeting (Paris, France), 2005 Winter Meeting, (Atlanta, Georgia, program chair); Member of the Executive Council, 1999–2002; Editor for articles for the *The Bulletin of Symbolic Logic* 2001–. Editor, Volume 6,

Analytic Philosophy and Logic, *The Proceedings of the Twentieth World Congress of Philosophy*, held at Boston University 1998, (Philosophy Documentation Center: Bowling Green State University 2000).

**Teaching:** 2005 Metcalf Award for Excellence in Teaching, university-wide for Boston University.

**Honours and Awards:** New England Open Co-Champion of Chess, 1984; Marshall Scholarship (British Government), 1970-1972; Danforth Foundation Fellowship, 1970-1975.

## KAPLAN, David Benjamin

**Specialties:** Logic, philosophical logic, modality, philosophy of langauge, metaphysics and epistemology.

**Born:** 1933

**Educated:** PhD in Philosophy, UCLA, 1964.

**Dissertation:** *Foundations of Intensional Logic*' supervisor: Rudolf Carnap.

**Regular Academic or Research Appointments:** PROFESSOR OF PHILOSOPHY, UCLA.

**Research Profile:** Kaplan's work is primarily focused on issues in the philosophy of language and logic. Although, these ventures sometimes take him into related issues in other fields, such as the philosophy of mind. Kaplan's most influential contribution to the philosophy of language is his semantic analysis of indexicals and demonstratives which is advanced in his articles "DThat", "On the Logic of Demonstrative,and Demonstrtives and afterthoughts. Also important is Kaplan's article "Quantifying-in" 1968 which disucsses issues in intentional and indirect discourse, subh as substitution failure, existential generalisation failure, and the distinction between *de re* and *de dicto* propositional attitude attribution.

Kaplan provides an apparatus which allows one to quantify into intensional contexts even if they permit a certain kind of substitution failure.

**Main Publications:**

1. "Quantifying In," *Synthese*, XIX 1968.
2. "On the Logic of Demonstratives," *Journal of Philosophical Logic*, VIII 1978: 81-98; and reprinted in French et al. (eds.), Contemporary Perspectives in the Philosophy of Language (Minneapolis: University of Minnesota Press, 1979): 401-412.
3. "Dthat," *Syntax and Semantics*, vol. 9, ed. P. Cole (New York: Academic Press, 1978); and reprinted in The Philosophy of Language, ed. A. P. Martinich (Oxford: Oxford University Press, 1985).
4. "Bob and Carol and Ted and Alice," in *Approaches to Natural Language* (J.Hintikka et al., eds.), Reidel, 1973.
5. "How to Russell a Frege-Church," *The Journal of Philosophy*, LXXII 1975.
6. "Opacity," in *W.V. Quine* (L. Hahn, ed.) Open Court, 1986.
7. "Demonstratives" and "Afterthoughts" in *Themes From Kaplan* (Almog, et al., eds.), Oxford 1989. ISBN 978-0195052176
8. "Words," *The Aristotelian Society*, Supplementary Volume, LXIV 1990
9. "A Problem in Possible World Semantics," in *Modality, Morality, and Belief* (W. Sinnott-Armstrong et al.,eds.) Cambridge, 1995.
10. "Reading 'On Denoting' on its Centenary", *Mind*, 114 2005: 934-1003.

**Teaching:** In most years, Kaplan teaches an upper division course on philosophy of language, focusing on the work of either Gottlob Frege, Bertrand Russell, or P.F. Strawson. His lively lectures often focus on selected paragraphs from Russell's "On Denoting" as well as Frege's "On Sense and Reference."

## KASHER, Asa

**Specialties:** Pragmatics, philosophy of language.

**Born:** 6 June 1940 in Jerusalem, Israel.

**Educated:** Hebrew University of Jerusalem, PhD Philosophy, 1971; Hebrew University of Jerusalem, MSc Mathematics 1963.

**Dissertation:** PhD Thesis: *The Logical Status of Indexical Sentences*; supervisor, Yehoshua Bar-Hillel.

**Dissertation:** MSc Thesis: *Constructive Analysis*. Supervisor: Michael Rabin.

**Regular Academic or Research Appointments:** EMERITUS PROFESSOR, PHILOSOPHY, TEL AVIV UNIVERSITY, 2004–; Laura Schwarz-Kipp Professor of Professional Ethics and Philosophy of Practice, Tel Aviv University, 1993-2006; Avraham Horodisch Professor of Philosophy of Language, Tel Aviv University, 1986-1993; Professor, Philosophy, Tel Aviv University, 1979-2004; Associate Professor, Philosophy, Tel Aviv University, 1974-1979.

**Visiting Academic or Research Appointments:** Fellow, University of California Humanities Research Institute, Irvine, 1997; Visiting Fellow, University of California, Los Angeles, 1991, 1990, 1988-1989; Visiting Scholar, University of California, Los Angeles, 1986-1987; Senior Exchange

Scholar, University of Pavia, 1985; Visiting Fellow, Wolfson College, Oxford, 1979; Research Associate, Rijksuniversiteit Gent and Vriej Brussel, 1978; Research Associate, Ruhr Universitat Bochum, 1977; Research Associate, University of Amsterdam, 1976; Research Associate, Technische Universitat Berlin, 1973, 1974, 1975; Assistant Professor and Research Associate, University of Texas at Austin, 1971-1972; Member, Applied Logic Branch, Hebrew University of Jerusalem, 1962-1964, 1969-1970.

**Research Profile:** Kasher has made major contributions to a theory of natural language use (pragmatics of natural language), some parts of which have logical components. At the background of Kasher's theory is a unique combination of insights: From Wittgenstein's philosophy of language he has taken the idea of language games that are rule-governed practices, while from Chomsky's linguistic and philosophy of language he has taken the modular approach as well as the methodology of research of natural language. On the level of basic insights, Kasher has added to those taken from Wittgenstein and Chomsky the idea that there is a pragmatic competence that consists, mainly, (a) of basic speech act types, each related to a certain cognitive function, such as forming beliefs, solving problems, or invoking norms, and of a modular nature, (b) of operations that produce new types of speech act from given types of speech act by adding elements to some components of the generic structure of speech act types, and (c) of a central component that processes output of modular speech act systems. Here is where implicatures are produced.

The first component involves a generic definition of a speech act type. Kasher's work on "pragmemes" has shown how such types can be analyzed in terms of preference relations. Logic systems of preference are here applicable.

The third component includes a theory of conversational and other implicatures analyzed by Kasher in terms of rationality principles as applied to speech acts. Theories of (instrumental) rationality are here applicable.

As is clear from Kasher's publications, he has applied formal methods of different types to a variety of philosophical issues, from rules of induction, through formal semantics of some terms and phenomena, to collective identity.

**Main Publications:**

1. *Philosophical Linguistics:* An Introduction (with S. Lappin), Scriptor Verlag, Kronberg, 1977.
2. *Linguistics and Logic*: Conspectus and Prospects, Scriptor Verlag, Kronberg, 1975.
3. "Measures of syntactic complexity" (with Y. Bar-Hillel and E. Shamir), in: *Machine Translation* (A.D. Booth, ed.) North-Holland, Amsterdam 1967, 29-50; Russian translation of chapter 1: Kiberneticheskij Sbornik 4 (1967) 219-227, under the title "Degrees of nesting and depth of postponed symbols".
4. "On the puzzle of self-supporting inductive arguments", *Mind* 322 (1972) 277-279.
5. "Linguistik und Mathematik", in: *Linguistik und Nachbarwissenschaften* (R. Bartsch and T. Venneman, eds.) Scriptor, Kronberg, 1974, 59-74; English translation, in English edition of the same book, Scriptor, Kronberg 1976.
6. "Mood implicatures, A logical way of doing generative pragmatics", *Theoretical Linguistics* 1 (1974) 6-38.
7. "The proper treatment of Montague grammar in natural logic and linguistics", *Theoretical Linguistics* 2 (1975) 133-145.
8. "Pragmatical representations and language games: Beyond intensions and extensions", in: *Rudolf Carnap, Logical Empiricist* (J. Hintikka, ed.) Reidel, Dordrecht 1975, 271-292.
9. "Conversational maxims and rationality", in: *Language in Focus: Foundations*, Methods and Systems (A. Kasher, ed.), Reidel, Dordrecht 1976, 197-216.
10. "On the semantics and the pragmatics of specific and non-specific indefinite expressions" (with D.M. Gabbay), *Theoretical Linguistics* 3 (1976) 145-190.
11. "What is a theory of use?", *Journal of Pragmatics* 1 (1977) 105-120.; republished in: *Meaning and Use* (A. Margalit, ed.) Reidel, Dordrecht 1979, 37-55.
12. "On the quantifier 'There is a certain x'" (with D.M. Gabbay), *Communication and Cognition* 10 (1977) 71-76.
13. "A note on the breadth and depth of terms" (with R. Manor), *Theory and Decision* 11 (1979) 71-79; corrigendum: 14 (1982) 109.
14. "Simple present tense" (with R. Manor), in: *Time, Tense and Quantifiers* (Ch. Rohrer, ed.) Niemeyer, Tubingen 1980, 315-328.
15. "Gricean inference reconsidered", *Philosophica* (Gent) 29 (1982) 25-44.
16. "On the Psychological Reality of Pragmatics", *Journal of Pragmatics* 8 (1984) 539-557; a revised version, published in a *Reader in Pragmatics* (S. Davis, ed) Oxford University Press, Oxford, 1991.
17. "Justification of Speech, Acts, and Speech Acts", in: *New Directions in Semantics* (E. LePore, ed.) Academic Press, London 1987, 281-303.
18. "Pragmatics and Chomsky's research program", in: *The Chomskyan Turn* (A. Kasher, ed.), Blackwell, Oxford 1991, 122-149; republished in: *Noam Chomsky: Critical Assessments* (C. Otero, ed.) Routledge, London 1994, Volume II, pp. 677-706.
19. "On the question "Who is a J?", A Social Choice Approach" (with A. Rubinstein) *Logique et Analyse* 160 (1997)[2000] 385-395.

*Other Publications:*

Book reviews in various journals, including *Journal of*

*Symbolic Logic* and entries in various encyclopedias and handbooks on pragmatics, philosophy, language and linguistics.

*Work in Progress*  A book on modular pragmatics. It develops a theory that answers the question "What is a natural speech act type?", in terms of modules that combine certain general cognitive functions with particular linguistic features and of operations that use basic speech act types in order to define new types of speech act which are governed by linguistic and non-linguistic rules. It also develops a theory of implicature on grounds of rationality principles, in an attempt to incorporate into speech acts theory insights from action theory. Both parts of the book are related to what is, in a sense, informal logic, on the one hand, and to empirical studies of language use, on the other hand.

**Service to the Profession:** Editor, *Pragmatics, Vol. 1-6*, Routledge, London, 1997; Head, Cognitive Studies of Language and its Uses Graduate Program, Tel Aviv University, 1995-2003; Member, National Council of Research and Development, 1994-1997; Editor, *The Chomskyan Turn*, Blackwell, Oxford, 1991; paper back edition, 1993; Guest Editor, *Journal of Pragmatics*, issue 16:5, 1991; Editor, *Cognitive Aspects of Language Use*, North Holland, Amsterdam, 1989; Guest Editor, *Journal of Pragmatics*, issues 12:5/6, 1988; President, Israel Philosophical Association, 1986-1991; Advisory Editor, *Linguistics and Philosophy*, 1982-1990; Guest Editor, *Theoretical Linguistics*, issues 9:1, 1982, 12:2/3, 1985; Guest Editor, *Philosophica* (Gent), issues 27-29, 1981-1982; Advisory Editor, *Theoretical Linguistics*, 1978-2000; Advisory Editor, *Journal of Pragmatics*, 1976-; Member, Executive Committee, *Linguistics and Philosophy*, 1976-1982; Editor, *Language in Focus: Foundations, Methods and Systems*, Dordrecht, Reidel, 1976 [Festschrift for Yehoshua Bar-Hillel}; Editor, *Philosophia, Philosophical Quarterly of Israel*, 1973-, Co-Editor, 1971-1973.

**Teaching:** Kasher has had more than twenty PhD students in various fields of philosophy and more than forty MA level theses.

**Vision Statement:** Logic was born as a study of practical reasoning. It used informal methods of analysis and formal depictions of inference. During the last centuries it met mathematics and computer sciences, which changed its nature and resulted in deep, precise and thorough understanding of mathematics and computation. The application of mathematics-oriented formal methods to issues of natural language has not been as successful. Attempts to develop utterly new methods have been even less so. One would like to see an intermediate way that would shed light on natural language.

**Honours and Awards:** Honorary Degree, Netania Academic College, 2004; Fellow, European Academy of Sciences and Humanities; Prize of Israel, General Philosophy, 2000.

## KEARNS, John T.

**Specialties:** Illocutionary logic, modal logic, logic of natural language, combinatory logic.

**Born:** 28 October 1936 in Elgin, Illinois, USA.

**Educated:** Yale University, PhD, 1962, MA, 1960; University of Notre Dame, BA 1958.

**Dissertation:** *Lesniewski, Language, and Logic*; supervisor, Alan Ross Anderson.

**Regular Academic or Research Appointments:** DEPARTMENT OF PHILOSOPHY, UNIVERSITY AT BUFFALO, THE STATE UNIVERSITY OF NEW YORK, 1964–.

**Research Profile:** John Kearns has published in many areas of logic, including history of logic, free logic, modal logic, epistemic logic, intuitionist logic, combinatory logic, the logic of natural language, and, most recently and now nearly exclusively, illocutionary logic or the logic of speech acts. Except for the historical writing and the work on combinatory logic, the focus of most of the research has been on using logical systems to capture ordinary concepts, such as the many ordinary concepts of possibility and necessity, and on representing the actual linguistic and deductive practices of human beings. His current work in illocutionary logic aims at recapturing the epistemic dimension of logic, and is intended to expand the field of logic so that logic will more clearly accommodate both the ontological and the epistemic.

**Main Publications:**

1. *Using Language: The Structures of Speech Acts*, SUNY Press, 1984.
2. *Reconceiving Experience, A Solution to a Problem Inherited from Descartes*, SUNY, 1996.
3. Combinatory Logic with Discriminators, *The Journal of Symbolic Logic* 34 (1969), 561575.
4. The Completeness of Combinatory Logic with Discriminators, *Notre Dame Journal of Formal Logic* 14 (1973), 323330.
5. The Logic of Calculation, *Zeitschrift fur Mathematische Logik und Grundlagen der Mathematik*, Bd 23 (1977), 4558.
6. A Strong Completeness Theorem for Kleene's ThreeValued Logic, *Zeitschrift fur Mathematische Logik und Grundlagen der Mathematik*, Bd 25 (1979), 6168.
7. Intuitionist Logic, a Logic of Justification, *Studia Logica* 37 (1978), 243260.
8. A Little More Like English, *Logique et Analyse*, n.s. 87 (1979), 353368.

9. Fully Explicit Deductive Systems, in *To H. B. Curry: Essays in Combinatory Logic, LambdaCalculus, and Formalism*, R. Hindley and J. P. Seldin eds., Academic Press; London, 1980.
10. Modal Semantics Without Possible Worlds, *The Journal of Symbolic Logic*, 46 (1981), 7786.
11. A More Satisfactory Description of the Semantics of Justification, *Notre Dame Journal of Formal Logic*, XXII (1981), 109119.
12. Lesniewski's Strategy and Modal Logic, *Notre Dame Journal of Formal Logic*, 30 (1989), 291307.
13. Propositional Logic of Supposition and Assertion, *Notre Dame Journal of Formal Logic* 38(1997), 325-349.
14. An Illocutionary Logical Explanation of the Surprise Execution, *History and Philosophy of Logic* 20 (2000), 195-214.
15. La Intencionalidad Irreductible del Calcular, *Mentes reales, La ciencia cognitiva y la naturalización de la mente*, J. Botero, J. Ramos, and A. Rosas eds., Biblioteca Universitaria, Universidad Nacional de Colombia, Bogotá, Colombia, 2000, 137-153.
16. Logic is the Study of a Human Activity, *The Logica Yearbook 2001*, T. Childers and O. Majer eds., Filosofia, Prague, 2002, 101-110.
17. The Logic of Coherent Fiction, *The Logica Yearbook 2002*, T. Childers and O. Majer eds., Filosofia, Prague, 2003, 133-146.
18. An Enlarged Conception of the Subject Matter of Logic, *Ideas y Valores* 126 (2004), 57-74.
19. Russell's Epistemic Understanding of Logic, *Teorema* 24 (2005), 115-131.

*Work in Progress*
20. A book, *Restoring the Epistemic Dimension to Logic*, which develops several systems of illocutionary logic.

**Service to the Profession:** Editor, Logic and Language, State University of New York Press, 1985–2000.

**Teaching:** John Kearns has found his many years of teaching logic to be particularly useful for developing and refining his own understanding of logic as an intellectual discipline. He may derive more benefit from his teaching of logic than his students do.

**Vision Statement:** I look for the development of logical systems that accommodate truth conditions and concepts involving truth conditions, as well as conditions of rational commitment generated by illocutionary acts and concepts involving these commitment conditions.

# KECHRIS, Alexander

**Specialties:** Foundations of mathematics, mathematical logic and set theory; their interactions with analysis and dynamical systems.

**Born:** 23 March 1946 in Athens, Greece.

**Educated:** UCLA, PhD Mathematics, 1972; National Technical University of Athens, Diploma Mechanical and Electrical Engineering, 1969.

**Dissertation:** *Projective ordinals and countable analytical sets*; supervisor, Yiannis Moschovakis.

**Regular Academic or Research Appointments:** PROFESSOR OF MATHEMATICS, CALIFORNIA INSTITUTE OF TECHNOLOGY, 1981–. Associate Professor of Mathematics, California Institute of Technology, 1976-1981; Assistant Professor of Mathematics, California Institute of Technology, 1974-1976; C.L.E. Moore Instructor in Mathematics, M.I.T., 1972-1974.

**Visiting Academic or Research Appointments:** Program on Set Theory and its Applications, CRM, Bellaterra Spain, 2003; Semester in Set Theory and Analysis, Fields Institute, Toronto, Canada, 2002; Logic Year, Mittag-Leffler Institute, Djursholm Sweden, 2000; Visiting Miller Research Professor, U.C. Berkeley, 1998; Visiting Professor, University of Paris VII, 1978-1979.

**Research Profile:** Alexander Kechris has made contributions to the theory of determinacy and the definability theory of the continuum, particularly the structure theory of the projective and more complex definable sets; he has also contributed to generalized recursion theory. More recently, he has been interested in the connections and applications of descriptive set theory to other areas of mathematics, including classical and harmonic analysis, the theory of topological groups, various aspects of dynamical systems, including ergodic theory and topological dynamics, and combinatorics. He is also currently involved in the development of a theory of complexity of classification problems in mathematics and the related study of definable equivalence relations.

**Main Publications:**

1. Measure and category in effective descriptive set theory, *Ann. Math. Logic*, 5, (1973), 337-384.
2. The theory of countable analytical sets, *Trans. Amer. Math. Soc.*, **202**, (1975), 259-297.
3. (With D. A. Martin) On the theory of $\Pi^1_3$ sets of reals, *Bull. Amer. Math. Soc.*, 84, (1978), 149-151.
4. (With L. A. Harrington) On the determinacy of games on ordinals, *Ann. Math. Logic*, 20, (1981), 109-154.
5. (With D. A. Martin and R. M. Solovay) Introduction to Q-theory, *Cabal Seminar 7-81, Proc. Caltech-UCLA Logic Seminar 1979-1981*, eds. A. S. Kechris, D. A. Martin, and Y. N. Moschovakis, Lecture Notes in Math., **1019**, (1983), Springer-Verlag, 207-289.

6. (With W.H. Woodin) Equivalence of partition properties and determinacy, *Proc. Nat. Acad. Sci. USA*, **80**, (1983), 1783-86.

7. (With H. Becker) Sets of ordinals constructible from trees and the Third Victoria Delfino Problem, *Axiomatic Set Theory*, J. E. Baumgartner, D. A. Martin and S. Shelah, eds., Contemporary Mathematics, **31**, (1984), 13-29.

8. (With A. Louveau) *Descriptive Set Theory and the Structure of Sets of Uniqueness*, London Math. Society Lecture Note Series, **128**, Cambridge Univ. Press, Cambridge, 1989.

9. (With A. Louveau) A classification of Baire class 1 functions, *Trans. Amer. Math. Soc.*, **318 (1)**, (1990), 209-236.

10. (With L.A. Harrington and A. Louveau) A Glimm-Effros dichotomy for Borel equivalence relations, *J. Amer. Math. Soc.*, 3(**4**), (1990), 903-928.

11. (With R. Dougherty) The complexity of antidifferentiation, *Adv. in Math.*, **88(2)**, (1991), 145-169.

12. *Classical Descriptive Set Theory*, Graduate Texts in Mathematics, **156**, Springer-Verlag, New York, 1995.

13. (With H. Becker) *The Descriptive Set Theory of Polish Group Actions*, London Math. Society Lecture Note Series, **232**, Cambridge University Press, Cambridge, 1996.

14. (With A. Louveau) The classification of hypersmooth Borel equivalence relations, *J. Amer. Math. Soc.*, **10**, (1997), 215-242.

15. (With S. Adams) Linear algebraic groups and countable Borel equivalence relations, *J. Amer. Math. Soc.*, **13(4)**, (2000), 909-943.

16. (With S. Jackson and A. Louveau) Countable Borel equivalence relations, *J. Math. Logic*, **2(1)**, (2002), 1-80.

17. (With S. Gao) On the classification of Polish metric spaces up to isometry, *Memoirs of the Amer. Math. Soc.*, Vol. **161**, No. 766, 2003.

18. (With B.D. Miller) *Topics in Orbit Equivalence*, Lecture Notes in Mathematics, **1852**, Springer, 2004.

19. (With G. Hjorth) Rigidity theorems for actions of product groups and countable Borel equivalence relations, *Memoirs of the Amer. Math. Soc.*, Vol. **177**, No. 833, 2005.

20. (With V. Pestov and S. Todorcevic) Fraïssé limits, Ramsey theory and topological dynamics of automorphism groups, *Geometric and Functional Analysis*, **15(1)**, 2005, 106-189.

*Work in Progress*

21. Work with C. Rosendal on genericity for sequences of elements of a topological group and its implications to the theory of Polish groups and in particular automorphism groups of countable structures.

22. Research on the global structure theory of ergodic actions and equivalence relations and its applications to the study of complexity of classification problems in ergodic theory.

23. Work with S. Todorcevic on the relation between finite Ramsey theory and topological dynamics of automorphism groups of homogeneous structures.

**Service to the Profession:** President, Association for Symbolic Logic, 2004-; Member, Executive Committee, Association for Symbolic Logic, 2001-2004; Associate Editor, Journal of the American Mathematical Society, 2003-; Managing Editor, Annals of Pure and Applied Logic, 2003-; Chair, Program Committee of the European Summer Meeting of the Association for Symbolic Logic, 2002; Editor, Studies in Logic and the Foundations of Mathematics, 2001-; Associate Editor, Journal of Mathematical Logic, 1999-; Editor, Electronic Research Announcements of the American Mathematical Society, 1997-; Editor, Fundamenta Mathematica, 1994-; Executive Officer (Chair), Department of Mathematics, California Institute of Technology, 1994-1997; Editor, Mathematical Logic Quarterly, 1992-; Editor, Bulletin of Symbolic Logic, 1993-1999; Coordinator of Editors, Journal of Symbolic Logic, 1985-1987; Chair, Program Committee of the Annual Meetings of the Association for Symbolic Logic, 1985,1990; Editor, Journal of Symbolic Logic, 1983-1989; Member, Council of the Association for Symbolic Logic, 1983-1989.

**Teaching:** Kechris has been teaching a variety of logic courses at Caltech for over 30 years, and he has been actively involved, particularly through supervising research projects, in the training of many talented undergraduates, who have since become active researchers in logic or other fields of mathematics. Kechris has had fifteen PhD students and has sponsored at least as many postdocs at Caltech over this period. He is one of the main organizers of the Cabal Seminar, the joint Caltech-UCLA Logic Seminar, which has been running almost continuously since 1976, and along with colleagues at UCLA has edited four volumes of proceedings of this seminar. Work on a new updated edition of these proceedings is now in progress, edited by Alexander Kechris, Benedikt Löwe, and John Steel. He has published several research monographs in descriptive set theory and its interactions with other areas.

**Honours and Awards:** Invited address at the German Mathematical Society annual meeting, 2006; Invited address at the Canadian Mathematical Society summer meeting, 2006; Alfred Tarski Lecturer, U.C. Berkeley, 2004; Invited address at the British Mathematics Colloquium, 2004; Carp Prize of the Association for Symbolic Logic, 2003; J.S. Guggenheim Memorial Foundation Fellow, 2003; Evelyn Nelson Lecturer, McMaster University, 2002; Gödel Lecturer, Association for Symbolic Logic, 1998; Distinguished Lecture Series, Indiana University, 1990; Honorary Doctoral Degree, University of Athens, Greece, 1987; 45-minute invited address at the International Congress of Mathematicians, 1986; Invited address at the annual meeting of the American Mathematical Society, 1986; Invited addresses at the International Congresses of Logic, Methodology and Philosophy of Science, 1979, 2007; A.P. Sloan Foundation Fellow, 1978-1982; Invited addresses or series of lectures at the annual meetings and European summer meetings of the Association for Symbolic Logic, 1975,1977, 1981, 1988, 1989, 1991, 1994, 2004; Academic Distinction Doctoral Award, Division of Physical Sciences, UCLA, 1972.

# KEENAN, Edward Louis

**Specialties:** Logical properties of natural language, formal semantics of natural language

**Born:** 10 December 1937 in Somerset, Pennsylvania USA.

**Educated:** University of Pennsylvania, PhD Linguistics; Computer Science Minor Field, 1969; MA French Language, 1966; Certificat de Français Littéraire, Université de Paris, 1962; Diplôme d'Etudes Littéraires, Université de Paris, 1961; Swarthmore College, BA Philosophy and Religion, 1959.

**Dissertation:** *A Logical Base for a Transformational Grammar of English*; supervisor, John Corcoran.

**Regular Academic or Research Appointments:** DISTINGUISHED PROFESSOR OF LINGUISTICS, UNIVERSITY OF CALIFORNIA AT LOS ANGELES, 1995–. Professor, Department of Linguistics, UCLA 1976-1995; Associate Professor of Linguistics, UCLA 1974-1976; Senior Fellow, King's College, University of Cambridge, 1970-1974; NSF, Postdoctoral Fellow, Madagascar, 1969-1970.

**Visiting Academic or Research Appointments:** Erskine Professor, University of Canterbury, New Zealand, 1998; Visiting Professor, University of Paris 7, 1996; Guest Professor, Institut für Maschinelle Sprachverarbeitung-Computerlinguistik, University of Stuttgart, 1994; Visiting Scientist, Dept of Mathematics and Statistics, McGill University, 1993; Visiting Professor, Tilburg University Institute for Language Technology and Artificial Intelligence, 1992; Visiting Professor, Dept of English Linguistics, Osaka University, Japan, 1989; Visiting Fellow, Max Planck Institute, Nijmegan, Holland, 1984; Visiting Professor, Dept of Linguistics, Tel Aviv University, 1977-1979.

**Research Profile:** Edward Keenan is a linguist by profession, but one with an abiding interest the logical properties of natural language. His early work in this area concerned presupposition logic for natural language. His later and more extensive work concerns the study of boolean properties of natural language and the logical expressive power of natural language quantifiers. Often working jointly with logicians he has provided a classification of quantifier types in natural language, offered one constraint that they all satisfy, and discovered non-Fregean patterns in natural language quantifiers whereby multiple quantification is provably not representable Fregeanly by iterated application of unary quantifiers.

**Main Publications:**

1. (with L.M. Faltz) *Boolean Semantics for Natural Language* D. Reidel, Dordrecht (1985).
2. (with J. Stavi) A Semantic Characterization of Natural Language Determiners *Linguistics and Philosophy* 9:253-326. (1986).
3. (with L.S. Moss) Generalized Quantifiers and the Expressive Power of Natural Language, in
4. *Generalized Quantifiers in Natural Language* J. van Benthem and A. ter Meulen (eds.), Foris, (1985): 73-127.
5. Unreducible nary Quantifiers in Natural Language, in *Generalized Quantifiers: Linguistic and Logical Approaches* P. Gärdenfors (ed.), Reidel (1987): 109-151.
6. Beyond the Frege Boundary, in *Linguistics and Philosophy* 15: 199-221, (1992)
7. Natural Language, Sortal Reducibility and Generalized Quantifiers *J. Symbolic Logic* 58.1: 314-325. (March 1993).
8. The Semantics of Determiners, in *The Handbook of Contemporary Semantic Theory* S. Lappin (ed) Blackwell (1996):41-63.
9. (with D. Westerståhl) Generalized Quantifiers in Linguistics and Logic in *The Handbook of Language and Logic* J. van Benthem and A. ter Meulen (eds) Elsevier. (1997): 837-893.
10. Quantification in English is Inherently Sortal. in *The History of Philosophy and Logic* 20: 251-265. S. Shapiro editor. (2000).
11. Logical Objects in *Logic, Meaning and Computation* C.A. Anderson & M. Zelëny (eds) Kluwer (2001):149-180.
12. (with E.P. Stabler) Syntactic Invariants, in *Algebras, Diagrams and Decisions in Language, Logic and Information* A. Copestake and K. Vermeulen (eds). CSLI (2001):1-37.
13. (with A. Altman and Y. Winter) Monotonicity and relative scope entailments. in *Proc. of the 13$^{th}$ Amsterdam Colloquium*, R. Van Rooy and M. Stokhof (eds),

ILLC, University of Amsterdam. (2001):25-30.
14. Some Properties of Natural Language Quantifiers in *Linguistics & Philosophy* 25: 627-654. G. Carlson, F.J. Pelletier, R. Thomason editors. (2002)
15. (with E.P. Stabler) *Bare Grammar: Lectures on Linguistic Invariants*. CSLI (2003)
16. (with E.P. Stabler) Structural similarity within and among languages. In *Theoretical Computer Science* 293: 345-363. (2003)
17. (with E. P. Stabler). Linguistic invariants and language variation. In *Logic, Methodology and Philosophy of Science* P. Hajek, L. M. Valdas Villanueva and D. Westerstahl (eds). King's College Publications, London (2005):395-411
18. Excursions in Natural Logic. in *Language and Grammar: Studies in Mathematical Linguistics and Natural Language*. C. Casadio, P. Scott, R. Seely (eds). CSLI, (2005): 3-24.

*Work in Progress*

The Mid-Point Theorems: article on proportionality quantifiers in English; *In situ* interpretation without type mismatches: article on the polymorphic interpretation of generalized quantifier denoting expressions in English; Directly interpreting anaphors: article on the extension of generalized quantifier theory to referentially dependent expressions. To appear in *Journal of Language and Computation*.

**Service to the Profession:** Little besides the odd review of an article at the Logic/Linguistics interface.

**Teaching:** Supervised 11 PhD dissertations, all in Linguistics.

**Vision Statement:** I see mathematical logic as a linguistic enterprise: soundness, completeness, compactness, interpolation, definability, model theoretic logics. Would that linguists come to apply the tools and rigor of logic to their problems, and that logicians come to find these problems challenging.

**Honours and Awards:** Fellow, American Academy of Arts and Sciences 1998-; Fulbright Scholar, University of Antananarivo, Madagascar, 1995.

# KNUUTTILA, Simo J. I.

**Specialties:** History of ancient and medieval logic, philosophy of logic.

**Born:** 8 May 1946 in Peräseinäjoki, Finland.

**Educated:** University of Helsinki, Finland; University of Kiel, Germany; University of Uppsala, Sweden.

**Dissertation:** *Truth and Possibility in Scholasticism*; supervisor, Jaakko Hintikka.

**Regular Academic or Research Appointments:** ACADEMY PROFESSOR, FINNISH NATIONAL RESEARCH COUNCIL 'ACADEMY OF FINALND', 1994–; Professor, Theological Ethics and Philosophy of Religion, University of Helsinki, 1981-; Professor, Practical Philosophy, University of Helsinki, 1981; Academy of Finland, Junior Research Fellow, 1979-1980, Research Associate 1976-1978.

**Visiting Academic or Research Appointments:** Visiting Professor, École des Hautes Études en Sciences Sociales, Paris, 2003; Lecturer, Department of Philosophy, University of Uppsala, 1993, 1994.

**Research Profile:** Simo Knuuttila has studied ancient and medieval modal theories and modal logic as well as the history of the theories of identity and universal validity of logic. He has identified four basic modal paradigms in ancient philosophy: the 'statistical' temporal frequency interpretation of modality, the model of possibility as a potency, the model of diachronic modalities (antecedent necessities and possibilities), and the model of possibility as non-contradictoriness. None of these conceptions, which were well known to early medieval thinkers, was associated with the idea of modality as referential multiplicity with respect to synchronic alternatives. This new idea was introduced into Western thought in early twelfth-century discussions influenced by Augustine's theological conception of God as acting by choice between alternative histories. Ancient habits of thinking continued to play an important role in scholasticism, however, and the theoretical significance of the new conception was not fully realized before the works of John Duns Scotus and some other early fourteenth-century thinkers. In the new theory, modal notions were treated in a way which shows similarities to the basic tenets of the possible world semantics of the twentieth century. On the basis of the new 'model theoretical' semantics, it was possible to evaluate critically various philosophical arguments based on the traditional modal views and to improve modal logic. Questions of modal logic were discussed separately with respect to modal propositions *de dicto* and *de re*, and *de re* modal propositions were further divided into two groups depending on whether the subject terms referred to actual or possible beings. Aristotle's modal syllogistics was regarded as a fragmentary theory in which the distinctions between these different types of the fine structures of modal propositions were not explicated. One of the best achievements of late medieval logic was John Buridan's influential modal logic. Knuuttila has also studied the discussions of late medieval modal ideas in early modern philosophy (Descartes, Leibniz), the history of expository syllogism and the emergence of the distinction between intensional and extensional identity in late medieval logic.

**Main Publications:**

1. *Modalities in Medieval Philosophy* (London, New York: Routledge, 1993).
2. *Reforging the Great Chain of Being* (editor), Synthese Historical Library 20 (Dordrecht: Reidel, 1981).
3. *The Logic of Being* (ed. with Jaakko Hintikka), Synthese Historical Library 28 (Dordrecht: Reidel, 1986).
4. *Modern Modalities: Studies of the History of Modal Theories from Medieval Nominalism to Logical Positivism* (editor), Synthese Historical Library 33 (Dordrecht: Kluwer, 1988).
5. 'Modal logic' in A. Kenny, N. Kretzmann, J. Pinborg (eds.), *The Cambridge History of Later Medieval Philosophy* (Cambridge: Cambridge University Press, 1982), 342-357.
6. 'The Foundations of Modality and Conceivability in Descartes and His Predecessors' (with Lilli Alanen) in: *Modern Modalities* (4. above), 1-69.
7. 'Norm and Action in Obligational Disputations' (with M. Yrjönsuuri) in: O. Pluta (ed.), *Die Philosophie in 14. und 15. Jahrhundert*, Bochumer Studien zur Philosophie (Amsterdam: Gruener, 1988), 191-202.
8. 'Roger Roseth and Medieval Deontic Logic' (with Olli Hallamaa), *Logique et analyse*, 149 (1995), 75-87.
9. 'Duns Scotus and the Foundations of Logical Modalities' in: L. Honnefelder, R. Wood and M. Dreyer (eds.), *John Duns Scotus: Metaphysics and Ethics* (Leiden: Brill, 1996), 127-143
10. 'Naissance de la logique de la volonté dans la pensée médiévale', *Études philosophiques*, 3/ 1996, 291-306
11. 'Modalität und Semantik möglicher Welten' in: C. Hubig (ed.), *Cognitio humana. Dynamik des Wisens und der Werte* (Berlin: Akademie Verlag, 1997), 466-476.
12. '*Positio impossibilis* in Medieval Discussions of the Trinity' in: C. Marmo (ed.), *Vestigia, Imagines Verba: Semiotics and Logic in Medieval Theological Texts* (Turnhout: Brepols, 1997), 277-88.
13. 'Les bases médiévales des conceptions modales modernes' in: R. Salais et al. (eds.), *Institutions et conventions*, 9 (Paris: École des Hautes Études en Sciences Sociales, 1998), 73-87.
14. 'Luther's View of Logic and the Revelation', *Medioevo*, 24 (1998), 219-234.
15. 'Medieval Theories of Modality', *Stanford Encyclopaedia of Philosophy*, ed. E.N. Zalta, `http://plato.stanford.edu`
16. 'The Medieval Background of Modern Modal Conceptions', *Theoria*, 66 (2000), 185-204.
17. . 'On the History of the Modality as Alternativeness' in: T. Buchheim, C.H. Kneepkens and K. Lorenz (eds.), *Potentialität und Possibilität. Modalaussagen in der Geschichte der Metaphysik* (Stuttgart-Bad Cannstatt: Frommann-Holzboog, 2001), 219-36.
18. . 'Anselmian Modalities' in: B. Davies and B. Leftow (eds.), *The Cambridge Companion to Anselm* (Cambridge: Cambridge University Press, 2004), 111-131.
19. 'The Question of the Validity of Logic in Late Medieval Thought' in: R. Friedmann and L. Nielsen (eds.), *The Medieval Heritage in Early Modern Metaphysics and Modal Logic*, The New Synthese Historical Library 53 (Dordecht: Kluwer, 2003), 121-142.
20. 'The Reception of Aristotle and Modal Conceptions', in: L. Honnefelder et al. (eds.) *Albertus Magnus und die Anfänge der Aristoteles-Rezeption im lateinischen Mittelalter*, Subsidia Albertina 1 (Münster: Aschendorff, 2005), 705-25.

*Work in Progress*
21. How scholastic problems influenced the development of medieval logic (future contingents, modalities, expository syllogism, theories of identity)
22. The history of the logic of being and identity

**Service to the Profession:** Steering Committee Member, ESF program 'From Natural Philosophy to Science), 2003–; Vice Chairman and Chairman, Finnish Academy of Science and Letters 2002–2006; Chairman of the European Science Foundation (ESF) network 'Early Modern Thought: Reconsidering the Borderline between Late Medieval and Early Modern Times' 1999–2001; Managing Editor, New Synthese Historical Library 1994–; Co-Editor, Ashgate Studies in Medieval Philosophy; Member, Institut International de Philosophie (I.I.P.), 1990–; Administrative Board Member, Société Internationale pour l'Étude de la Philosophie Médiévale (S.I.E.P.M.), 1987–1997; Member, Academia Europea, 2005–.

**Teaching:** Knuuttila has supervised several PhD students in the history of philosophy and logic, including Henrik Lagerlund (medieval modal logic), Mikko Yrjönsuuri (medieval logic of obligations), Taneli Kukkonen (modal theories in medieval Arabic philosophy) and Toivo Holopainen (logic and theology in the eleventh century).

**Vision Statement:** Late medieval logicians developed theories which were next dealt with to the same extent only in the nineteenth and twentieth century. Many parts of these and their later influence are still unexplored. Increasing knowledge of this tradition provides the possibility to investigate influential ideas in their original context as well as the varieties of logical insights in discussions different from the contemporary ones.

**Honours and Awards:** Annual Award of the Finnish Cultural Foundation, 1998; Annual Award of the Finnish Union of University Professors, 2003; Gad Rausings prize of the Royal Swedish

Academy of Letters, History and Antiquities, 2008.

## KOEPKE, Peter

**Specialties:** Mathematical logic: set theory, history of modern logic, formal proving.

**Born:** 31 May 1954 in Herford, Westf., Germany.

**Educated:** University of Freiburg, Oxford University, Dr.rer.nat. Mathematics, 1984; University of California at Berkeley, MSc Mathematics, 1979; Bielefeld University, Technical University Berlin, Free University Berlin, University of Bonn, Diplom Mathematiker 1978.

**Dissertation:** *A Theory of short core models and some applications*, supervisor Ronald Jensen; University of Freiburg, Habilitation Mathematics, 1990, *Habilitationsschrift: Finestructure for inner models with strong cardinals*.

**Regular Academic or Research Appointments:** PROFESSOR, MATHEMATICS, UNIVERSITY OF BONN, GERMANY, 1990–; Assistant Professor, Mathematics, University of Freiburg, 1987–1990.

**Visiting Academic or Research Appointments:** Visitor, Centre de Recerca Matematica, Barcelona, 2001; Visiting Fellow, Oxford University, 1994; Junior Research Fellow, Oxford University, 1983–1987.

**Research Profile:** Peter Koepke has made significant contributions to axiomatic set theory, and he has contributed to higher computability theory and general logic. In set theory he emphasizes the viewpoint that a wealth of inner models corresponds to combinatorial strength. His doctoral dissertation on *short core models* was the first completely worked out presentation of a higher core model theory including applications to consistency strength questions. In his *Habilitationsschrift* he went on towards larger core models containing strong cardinals, collaborating with Ronald Jensen. In descriptive set theory he carried out a proof of the Martin-Steel theorem on projective determinacy solely on the basis of elementary embeddings between transitive models of set theory. Koepke has made several proposals for the simplification of inner model theory. Together with Sy Friedman he introduced hyperfine structure theory which allows easier proofs of Jensen-style combinatorial principles in constructible models of set theory. This work also lead to notions of generalized computations on ordinals which are able to "compute" constructible sets. In general logic, Koepke is working on proof checking with natural language interfaces. The Bonn logic group gave the first computer-checked proof of the Gödel completeness theorem. Koepke is a member of the editorial team of the Hausdorff edition, editing and commenting Felix Hausdorff's writings on descriptive set theory.

**Main Publications:**

1. (with D. Donder), On the consistency strength of 'accessible' Jonsson cardinals and of the weak Chang conjecture, *Annals of Pure and Applied Logic* 25 (1983), 233-261.
2. The consistency strength of the free-subset property for $\omega_\omega$, *Journal of Symbolic Logic* 49 (1984), 1198-1204.
3. Some applications of short core models, *Annals of Pure and Applied Logic* 37 (1988), 179-204.
4. An introduction to extenders and core models for extender sequences, in *Logic Colloquium '87* (H.-D. Ebbinghaus et al., ed.), North-Holland, Amsterdam (1989), 137-182.
5. Metamathematische Aspekte der Hausdorffschen Mengenlehre, in *Felix Hausdorff zum Gedächtnis* (E. Brieskorn, ed.), Vieweg, Braunschweig(1996), 71-106.
6. (with S. D. Friedman) An elementary approach to the fine structure of L, *Bulletin of Symbolic Logic* 4 (1997), 453-468.
7. Extenders, embedding normal forms, and the Martin-Steel-theorem, *Journal of Symbolic Logic* 63 (1998), 1137-1176.
8. (with V. Kanovei) Deskriptive Mengenlehre in Hausdorffs Grundzügen der Mengenlehre, in *Felix Hausdorff – Gesammelte Werke Band II* (E. Brieskorn et al., ed.), Springer-Verlag, Heidelberg (2002), 773-787.
9. The category of inner models, *Synthese* 133 (2002), 275-303.
10. (with P. Braselmann) A formal proof of Gödel's completeness theorem, a series of 7 articles, *Formalized Mathematics* 13 (2005), 5-53.
11. Computing a model of set theory, in *New Computational Paradigms* (B. Cooper et al., ed.), *Lecture Notes in Computer Science* 3526 (2005), 223-232.
12. Turing computations on ordinals, *Bulletin of Symbolic Logic* 11 (2005), 377-397.
13. (with R. Schindler) Homogeneously Souslin sets in small inner models, *Archive for Mathematical Logic* 45 (2006), 53-61.
14. (with P. Welch) On the strength of mutual stationarity, in *Set Theory, Centre de Recerca Matematica Barcelona, 2003-2004* (J. Bagaria et al., ed.), Birkhäuser (2006), 309-320.
15. Infinite time register machines, in *Logical Approaches to Computational Barriers* (A. Beckmann et al., ed.), *Lecture Notes in Computer Science* 3988 (2006), 257-266.
16. (with A. Apter) The consistency strength of $\aleph_\omega$ and $\aleph_{\omega_1}$ being Rowbottom cardinals without the axiom of choice, *Archive for Mathematical Logic* 45 (2006), 721-737.
17. Forcing a mutual stationarity property in cofinal-

ity $\omega_1$, *Proceedings of the American Mathematical Society*, 135 (2007), 1523–1533.

*Work in Progress*

18. A new fine structure theory for higher core models.

19. Lecture notes on infinitary combinatorics without the axiom of choice, in collaboration with Arthur Apter.

20. Development of a controlled mathematical language to carry out formal proofs in natural language, together with associated software systems.

**Service to the Profession:** Editor, Felix Hausdorff, Collected Works, 1998–; Editor, *Mathematical Logic Quarterly*, 1993–2006; Editor and Principal Organizer, *Logic Colloquium* and *Colloquium Logicum*, 2002; President, *Deutsche Vereinigung für Mathematische Logik und Grundlagen der Exakten Wissenschaften* (German Logic Society), 2002–2008; Member, European Council, Association of Symbolic Logic, 1999–2002.

**Teaching:** Koepke has supervised more than 50 diploma students in mathematical logic. He has had four PhD students, including Ralf Schindler, Professor of Logic, Münster. Currently Koepke is supervising three PhD students.

**Vision Statement:** Mathematics will gradually shift its emphasis from numbers and geometric objects towards information and formal systems. Logic and logicians must take an active and visible role in this process.

**Honours and Awards:** Feodor-Lynen Fellow, Alexander von Humboldt foundation, 1984; Gödecke research prize, 1983.

# KOSLOW, Arnold

**Specialties:** Mathematical logic, philosophy of mathematics, philosophy of logic, philosophy and history of science.

**Born:** 28 March 1933 in Brooklyn, New York City, USA.

**Educated:** Columbia College, Columbia University, BA Mathematical Physics and Philosophy, 1954, King's College, Cambridge Research Certificate, 1955; Columbia University, PhD Philosophy, 1965.

**Dissertation:** *Changes in the Concept of Mass, from Newton to Einstein*; supervisor, Ernest Nagel.

**Regular Academic or Research Appointments:** PROFESSOR OF PHILOSOPHY, EMERITUS, THE GRADUATE CENTER, CUNY, 2002–; Professor of Philosophy, Brooklyn College, 1971–2002; Associate Professor of Philosophy, Brooklyn College, 1968–1971; Assistant Professor of Philosophy, Brooklyn College, 1965–1968.

**Visiting Academic or Research Appointments:** Faculty, The Summer Linguistics Institute of the Linguistics Society of America, 1986; Visiting Professor, Western Electric Graduate Center, 1968; Visiting Assistant Professor, Columbia University, 1967-1968; Visiting Assistant Professor, Columbia University, 1965; Visiting Lecturer, The John Hopkins University, 1960.

**Research Profile:** Koslow's logical philosophical work over the last decade or so has returned to the roots of the G.Gentzen/ P.Hertz program and the use of its abstract concept of implication relations (and implication structures) for the study of various key logical concepts such as the logical operators, modality, and truth. This program has been explored in several different directions: (1) a uniform definition of the logical operators as a special kind of function on structures, a generalized account of Introduction and Elimination Rules, and the deep connection with Intuitionism, (2) the study of non-standard consequence relations, and (3) the study of families of truth operators on implication structures and the special status which Tarskian truth operators have in such families.

**Main Publications:**

1. Translation from the Chinese (Chou Pei Suan Ching) of a proof of the Chinese Pythagorean Theorem. Published in J.Needham, *Science and Civilization in China*, vol. III, pp.2223, Cambridge University Press, 1959.

2. *The Changeless Order, The Physics of Space, Time, and Motion*, (edited), pp.328, Braziller, New York, 1967.

3. Mach's Concept of Mass: Program and Definition, Synthese, vol.18 (1968), pp.216233.

4. The Law of Inertia: Some Remarks on Its Structure and its Significance, in *Philosophy, Science, and Method: Essays in Honor of Ernest Nagel*, Eds.S.Morgenbesser, P.Suppes, and M.White, St. Martin's Press, 1969, pp.549567.

5. Ontological and Ideological Issues of the Classical Theory of Space and Time, in *Motion and Time, Space and Matter, Interrelations in the History and Philosophy of Science*, Eds. P.K.Machamer and R.G.Turnbull, Ohio State University Press, 1976, pp.224263.

6. Quantity and Quality: Some Aspects of Measurement," in Proceedings *of the Philosophy of Science Association, PSA, 1982*, Volume 1, pp.183198.

7. Quantity and Supervenience, in *How Many Questions? Essays in Honor of Sidney Morgenbesser*, Eds. L.S.Cauman, I.Levi, C.D.Parsons, and R.Schwartz, Hackett Publishing Company, Indianapolis, 1983, pp.80-104.

8. Quantitative, but NonNumerical Relations in Scientific Theory: Eudoxus, Newton, and Maxwell, Minnessota *Studies in Philosophical and Foundational Is-*

*sues in Measurement Theory*, Eds. C.W.Savage and P. Ehrlich (1992). L. Erlbaum, Publishers, Hillsdale, N.J.

9. *A Structuralist Theory of Logic*. Cambridge University Press, Dec.1992, Paperback 2006.

10. The Implicational Nature of Logic: A Structuralist Account in A.Varzi (ed.), *The European Philosophical Review (The Nature of Logic), volume 4*, Stanford U. Press, 1999, pp.111-155.

11. Truthlike and Truthful Operators, in *Between Logic and Intuition, Essays in Honor of Charles Parsons*, (eds.) G.Sher and R.Thiesen, Cambridge University Press, 2000, pp.27- 53.

12. Ontology and the Laws of Nature, *Proceedings of III International Congress of Ontology and Nature*, San Sebastian (1998), Spain. V.G.Pin (Ed.) Number 1, 2000, Bilbao.

13. Laws, explanation and the reduction of possibilities, in *Real Metaphysics, Essays in Honour of Hugh Mellor*, eds. G. Rodriguez- Pereyra, and H.Lillehammer, Routledge Press, 2003, pp.169–183.

14. Laws and Possibilities, (Invited symposium paper) *The Philosophy of Science*, December, pp. 719–729, 2004.

15. Ramsey on Simplicity and Truth, *Metaphysica, International Journal for Ontology & Metaphysics*, Nils-Eric Sahlin (ed.), 2005, pp. 89–108.

16. The Representational Inadequacy of Ramsey Sentences, *Theoria*, vol. 72, 2006, part 2, pp. 100–125. Press).

17. Structuralist Logic: Implications, Inferences, and Consequences, in *Logica Universalis vol. 1*, pp. 167–181, Springer-Verlag, 2007. 2005. (In press).

*Work in Progress*

18. Tarskian Finitary Truth and Ramseyan Belief States: A Tale of Two Schemata. In *The Logica Yearbook 2006*, pp. 137—156, O. Tomola and R. Honzik, eds. Filofofia, Prague, 2007.

19. Truth and Simplicity, F. P. Ramsey, edited by A. Koslow. *Brit.J. Phil.Sci.*, 2007, pp. 1–7.

20. Does Modal Theory need to be introduced to Intuitionistic, Epistemic, and Conditional Logic, or have they already met? forthcoming in IGPL.

*Work in Progress*

21. Explanation and Modality (in final draft);

22. Book manuscript: Scientific Laws.

**Service to the Profession:** Chairman Columbia University Seminar in the History and Philosophy of Science, 1980–1981, 1983–1984, 1987, 1991; Executive Committee, The Conference on Methods, Chairman-Elect, Conference on Methods, 1984, Chairman, 1985-1986; Deputy Executive Officer, PhD Program in Philosophy, The Graduate Center, CUNY, 1971–1973; Executive Officer 1973–1975, Deputy Executive Officer 1985-1986.

**Teaching:** Koslow has had a considerable impact on the promotion of a rigorous training and attitude for his many students at the Graduate Center, CUNY, in logic, the philosophy of mathematics, the history and philosophy of science (including theories of knowledge and belief revision). The number of doctoral dissertations completed under his supervisions is presently sixteen with four to be completed. Some students have gone on to become well known in their fields, and some have satisfying careers as teachers with modest but very worthy contributions of their own.

**Vision Statement:** Given the vitality and variety of contemporary logics, one desideratum would be the development of some unifying theories of those concepts that have had important roles in past and present developments (e.g. the logical and modal operators, various notions of implication, and consequence, etc.) I had in mind something intellectually coherent, not lists or catalogues.

**Honours and Awards:** Ethyle Wolfe Humanities Fellow, National Endowment for the Humanities, 1998; CUNY Faculty Research Grant, 1990, 1992, 1994, 1996, 1998; National Academy of Arts and Science, Traveling, 1969; Council for Research in the Humanities, 1964; William Bayard Cutting Fellow of Columbia, Stanford, Harvard, Cambridge University, Ford Foundation Fellow, ,Columbia, and Cambridge University, 1954–1957.

# KOWALSKI, Robert Anthony

**Specialties:** Computational logic, including knowledge representation and problem solving, in artificial intelligence and cognitive science.

**Born:** 15 May 1941 in Bridgeport, Connecticut, USA.

**Educated:** University of Edinburgh, PhD Computer Science, 1970; Stanford University and University of Warsaw, MA Mathematics, 1966; University of Chicago and University of Bridgeport, BA Mathematics, 1963.

**Dissertation:** *Studies in the Completeness and Efficiency of Theorem–proving by Resolution*; supervisor, Bernard Meltzer.

**Regular Academic or Research Appointments:** EMERITUS PROFESSOR, SENIOR RESEARCH FELLOW, SENIOR RESEARCH INVESTIGATOR, IMPERIAL COLLEGE LONDON, 1999–; Professor of Computational Logic, Imperial College London, 1982–1999; Reader in Theory of Computing, Imperial College London, 1975–1982; Research Fellow, Department of Computational Logic, University of Edinburgh, 1970–1975; Research Assistant, Meta-Mathematics Unit, University of Edinburgh, 1967–1970; Assistant Professor and Acting Head, Mathematics Department,

Inter-American University, San Juan, Puerto Rico, 1966–1967.

**Visiting Academic or Research Appointments:** National Institute of Informatics, Tokyo, 2002, 2006, 2008; Centro de Inteligência Artificial, Universidade Nova de Lisboa, 2007, 2008; Universidad de Los Andes, Venezuela, 2005; Swiss Federal Institute of Technology at Lausanne (EPFL), 2001; Meme Media Laboratory, Hokkaido University, 2001, 2000; Miegunyah Distinguished Fellow, University of Melbourne, 1999; University of Syracuse, 1972, 1978, 1981; Ricerche Instituto per le Applicazioni del Calcolo, Rome, 1974; University of Marseille, Luminy, 1972, 1974.

**Research Profile:** Kowalski's PhD research was in the field of Automated Theorem Proving. His research in this field included work on semantic trees with Pat Hayes, on SL-resolution with Donald Kuehner, and on the connection-graph proof-procedure. The SL-resolution proof-procedure, with its goal-reduction proof strategy and its last-in-first-out selection of sub-goals contributed to his later work on the development of logic programming. The connection-graph proof-procedure, on the other hand, by combining selection with search, eliminates sufficiently many redundant and irrelevant steps towards a proof that its completeness is still an open research issue.

Kowalski collaborated with Alain Colmerauer on the development of logic programming (LP), the main idea of which is to apply a more flexible form of SL-resolution to execute Horn clauses as goal-reduction procedures. Kowalski contributed the theoretical insight and Colmerauer developed the programming language Prolog, based on this idea. Kowalski's book, *Logic for Problem Solving*, is often cited as an introduction to logic programming, but in fact it places logic programming in the wider context of clausal logic as a general problem solving formalism. In the late 1970s and early 1980s, he developed and led a number of projects to teach logic for problem solving to children.

Following early work on the foundations of LP, Kowalski focused on developing extensions of LP, with a view to improving its use for knowledge representation and problem solving. The first extension, developed with Kenneth Bowen, was an amalgamation of object-language and meta-language for such applications as the representation of knowledge and belief, as well as for implementing meta-interpreters in which one logic is simulated by another. Together with Marek Sergot, he developed the event calculus, a logic programming representation of causal reasoning, which has been used for such applications as database updates and the logical formalization of tense and aspect in natural language.

Kowalski was one of the early developers of abductive logic programming, in which ordinary logic programs are augmented with integrity constraints and with undefined, abducible predicates. This work led to the demonstration with Phan Minh Dung and Francesca Toni that most logics for default reasoning can be regarded as special cases of assumption-based argumentation.

The two main application areas to which Kowalski has made important contributions are legal reasoning and integrity checking in deductive databases. Working with Marek Sergot, he showed how LP and its extensions can be used to formalize legal rules and regulations, distinguishing between clear concepts, which are defined by logic programs, and vague concepts, which are undefined or abducible. Working with Fariba Sadri, he developed theorem-proving techniques, which reason forward from updates, to check for violation of integrity constraints.

Recently, Kowalski's main area of research has been the development of abductive logic programming as the thinking component of an intelligent agent interacting with the changing world. Working mainly with Fariba Sadri, he has developed an agent model in which beliefs are represented by logic programs and goals are represented by integrity constraints. Integrity constraints are used to represent different kinds of goals, including maintenance goals, prohibitions, and condition–action rules. Abducible predicates are used to represent observations and actions.

The resulting agent model includes *reactive* thinking, which uses forward reasoning to derive achievement goals from maintenance goals and observations. It also includes *proactive* thinking, which uses backward reasoning to reduce achievement goals to action sub-goals.

Kowalski's current research focuses on the application of computational logic to cognitive science. This work aims to link logic's traditional role as a normative model of thinking with a more radical proposal that logic can also serve as a descriptive model of human thinking. The work extends the combination of reactive and proactive thinking to include a kind of *pre-active* thinking, which uses forward reasoning to simulate alternative candidate actions, to derive their possible consequences, to help in choosing between them.

**Main Publications:**

1. Semantic Trees in Automatic Theorem–Proving, with Hayes P. J., in *Machine Intelligence*, 4, (eds. B. Meltzer and D. Michie), Edinburgh University Press, 1969, pp. 181–201. Reprinted in *Anthology of Automated*

*Theorem-Proving Papers*, 2, Springer–Verlag, 1983, pp. 217–232.

2. Linear Resolution with Selection Function, with Kuehner D., in *Artificial Intelligence*, 2, 1971, pp. 227–60. Reprinted in *Anthology of Automated Theorem-Proving Papers*, 2, Springer–Verlag, 1983, pp. 542–577.

3. Predicate Logic as Programming Language, in *Proceedings IFIP Congress*, Stockholm, North Holland Publishing Co., 1974, pp. 569–574. Reprinted in *Computers for Artificial Intelligence Applications*, Wah B. and Li G.-J. eds., IEEE Computer Society Press, Los Angeles, 1986, pp. 68–73.

4. A Proof Procedure Using Connection Graphs, in *JACM*, 22(4), 1975, pp. 572–595.

5. The Semantics of Predicate Logic as a Programming Language, with van Emden M., in *JACM*, 23(4), 1976, pp. 733–742.

6. *Logic for Problem Solving*, North Holland Elsevier, 1979, 287 pages.

7. Amalgamating Language and Meta–language in Logic Programming, with Bowen K., in *Logic Programming*, K. Clark and S–A. Tarnlund eds., Academic Press, 1982, pp. 153–172.

8. A Logic–based Calculus of Events, with Sergot, M., in *New Generation Computing*, 4(1), February 1986, pp. 67–95. Also in *Knowledge Base Management-Systems*, C. Thanos and J. W. Schmidt eds., Springer–Verlag, pp. 23–51. Also in *The Language of Time: A Reader*, Inderjeet Mani, J. Pustejovsky, and R. Gaizauskas eds., Oxford University Press, 2005.

9. The British Nationality Act as a Logic Program, with Sergot M., Sadri F., Kriwaczek F., Hammond P., and Cory T., in *CACM*, 29(5), 1986, pp. 370–386.

10. A Theorem-Proving approach to Database Integrity, with Sadri, F., in *Deductive Databases and Logic Programming*, J. Minker ed., Morgan Kaufman, Los Altos, Ca., 1988, pp. 313–362.

11. Abduction Compared with Negation by Failure, with Eshghi K., in *Sixth International Conference on Logic Programming*, G. Levi and M. Martelli eds., MIT Press, 1989, pp. 234–254.

12. Legislation as Logic Programs, in *Logic Programming in Action*, G. Comyn, N. E. Fuchs, M. J. Ratcliffe eds., Springer–Verlag, 1992, pp. 203–230.

13. Database Updates in the Event Calculus, in *Journal of Logic Programming*, 1992, 12(162), pp. 121–146.

14. Using Metalogic to Reconcile Reactive with Rational Agents, in *Meta–Logics and Logic Programming*, K. Apt and F. Turini eds., MIT Press, 1995. (Revised version in Proc. PAAM96).

15. The Role of Logic Programming in Abduction, with Kakas T. and Toni F., *Handbook of Logic in Artificial Intelligence and Programming 5*, 1998 D. Gabbay, C.J. Hogger, J.A. Robinson eds. Oxford University Press, pp. 235–324.

16. An Abstract Argumentation-theoretic Approach to Default Reasoning, with Bondarenko A., Dung P. M., and Toni F., *Journal of Artificial Intelligence*, 93(1–2), 1997, pp. 63–101.

17. From Logic Programming towards Multi-agent Systems, with Sadri F., *Annals of Mathematics and Artificial Intelligence*, 25, 1999, pp. 391–419.

18. Artificial intelligence and the natural world, *Cognitive Processing*, 4, 2001, pp. 547–573.

19. Dialectic proof procedures for assumption-based, admissible argumentation, with Dung P. M. and Toni F., *Journal of Artificial Intelligence*, 170(2), February 2006, pp. 114-159.

20. The Logical Way to be Artificially Intelligent, *Proceedings of CLIMA VI*, F. Toni and P. Torroni eds, Springer Verlag, LNAI, 2006, pp. 1–22.

*Work in Progress*

21. A book, *The Logical Way to be Artificially Intelligent*, which presents computational logic informally, so that it can be used by ordinary people in everyday life.

22. An investigation of thinking as the activation of links in a connection graph of sentences in clausal form. The hypothesis that the mind is organized as a collection of modules is explained by the possibility that such a graph may contain implicit or explicit sub-graphs, with a high degree of connectivity within sub-graphs and a low degree between sub-graphs.

23. An investigation of the relationships between object-orientation (OO) and abductive logic programming multi-agent systems (ALP systems).

A full list of references, including links to some of these papers and to the work in progress can be found on the author's homepage: http://www.doc.ic.ac.uk/~rak/

**Service to the Profession:** Editorial Advisory Board, Progress in Informatics, 2005-; Editorial Board, Theory and Practice of Logic Programming, 2001– ; Editorial Board, Interest Group in Propositional and Predicate Logics, 1993–; Founding Co–ordinator, European Compulog Network of Excellence, 1991–1992; Editorial board, Journal of Logic and Computation, 1990–; Editorial Board, Journal of Artificial Intelligence and Law, 1990–; Editorial Advisory Board, IEICE transactions on Information and Systems, 1990–; Scientific Advisory Board, Deutsches Forschungszentrum für Künstliche Intelligenz, 1989–1998; Coordinator, European Community Basic Research Project, Compulog, 1989–1991; Chairman, Program Committee, International Joint Logic Programming Conference, 1988; Scientific Advisor, UNDP Knowledge-Based Computer Systems Project in India, 1987–1991; Honorary Secretary, Association for Logic Programming, 1986–1998; Editorial Board, Mind and Language, 1986–; Head, Logic Programming Group, Department of Computing, Imperial College, 1985–1987, 1990–1996; Editorial Board, New Generation Computing Journal, 1983– ; Editorial Board, Logic Programming Journal, 1983–2000; Joint Chairman, Program Committee, Conference on Automated Deduction, 1980; Member of Science and Engineering Research Council Subcommittee for Computing and Communications, 1980–1983.

**Teaching:** Kowalski has supervised seventeen PhD students at Imperial College in the field of Computational Logic, including Christopher Hogger, Keith Clark, Marek Sergot, Fariba Sadri, Kave Eshghi, Suryanarayana Sripada, Francis McCabe, Francesca Toni, Tze Ho Fung, Christopher Preist, Jacinto Davila, and Yongyuth Perpoontanalarp. He also contributed to the supervision of a number of PhD students at the University of Edinburgh, including David H. Warren.

**Vision Statement:** Attempts to use traditional logic in artificial intelligence have led to the development of various improvements and extensions, including default reasoning, abduction and argumentation. These developments can also be applied to the original purpose of logic, to improve the quality of human thinking.

**Honours and Awards:** "Essays in Honour of Robert Kowalski" Computational Logic: Logic Programming and Beyond, (eds. A C Kakas and F Sadri) Springer Verlag, 2002; Special issue of ACM Transactions on Computational Logic "Dedicated to Robert Kowalski", Vol. 2, No. 4, 2001; Fellow of the Association for Computing Machinery, 2000; Fellow of the European Coordinating Committee for Artificial Intelligence, 1999; Miegunyah Distinguished Fellow, University of Melbourne, 1999; Fellow, Deutsches Forschungszentrum für Künstliche Intelligenz, 1998; Fellow, City and Guilds of London Institute, 1997; Fellow, American Association for Artificial Intelligence, 1991; "Docente a titolo individuale" Giuridica "H Kelsen", dell'Università degli Studi di Bologna, 1990; Insight Award for Contributions to Fifth Generation Computing, 1984.

# KRACHT, Marcus Andreas

**Specialties:** Modal logic; algebraic logic.

**Born:** 4 February 1964 in Reinbek/Stormarn, Schleswig-Holstein, Germany.

**Educated:** Freie Universität Berlin, Diplom; University of Edinburgh, MSc (Cognitive Science); Freie Universität Berlin, PhD; Freie Universität Berlin, Habilitation; Universität Potsdam, Habilitation (Linguistics).

**Dissertation:** *Internal Definability and Completeness in Modal Logic*; supervisor Wolfgang Rautenberg.

**Regular Academic or Research Appointments:** FULL PROFESSOR, UNIVERSITÄT BIELEFELD, 2008–; ASSOCIATE PROFESSOR, UCLA, 2006–; Assistant Professor, UCLA, 2002–2006. Research Assistant, Freie Universität Berlin, 1994–2000. Teaching Assistant, Freie Universität Berlin, 1992–1994, Researcher for NWO, Utrecht, 1991–1992. Teaching Assistant, Freie Universität Berlin, 1988–1991.

**Visiting Academic or Research Appointments:** Visiting Professor, BTU Cottbus, Summer 2001.

**Research Profile:** My research areas include all areas of modal logic, pure and applied. My ambition has been to provide a general theory of modal logic, for example general results concerning completeness. The biggest step in this direction was the result that the lattices of polymodal logics with $n$ operators is isomorphic to an interval in the lattice of monomodal logics and that the isomorphism is faithful with respect to many properties. Another step was the preservation of properties under independent fusion. Also, I have promoted the use and study of the global consequence relation in addition to the usual local relation.

I have applied modal logic in the study of sentence structure, and part of the work is now being used in the study of XML-queries, though the original motivation has been quite different. I was and still am interested in the decidability of generative grammar, and I have recently shown that the dynamic logic of multidominance structures is decidable. This moves us away from trees. Whereas in trees one may use full monadic second order logic, Rabin's theorem might not be applicable to multidominance structures.

I have been positively surprised over the years about the applications that have been found for modal logic. It has very often provided expressive and yet decidable languages. In this respect I expect that it will become even more attractive for other disciplines in the future and I am also working towards this goal.

Recently, I have started to work on modal predicate logic, especially on counterpart semantics. My interest is in providing semantics for which every modal predicate logic is complete. This is of great interest in philosophy and linguistics, where traditional semantics is often less flexible than needed.

**Main Publications:**

1. *Properties of Independently Axiomatizable Bimodal Logics*. The Journal of Symbolic Logic, 56(1991), 1469–1485. (With Frank Wolter)

2. *Splittings and the finite model property*. The Journal of Symbolic Logic, 58(1993), 139–157.

3. *How Completeness and Correspondence Theory Got Married.* In: Maarten de Rijke (Ed.): *Diamonds and Defaults*, Synthese Library vol. 229, Kluwer Academic Publishers, 1993, 175–214.

4. *Is there a genuine modal perspective on feature structures?*, Linguistics and Philosophy, 18(1995), 401–458.

5. *Syntactic Codes and Grammar Refinement.* Journal of Logic, Language and Information, 4(1995), 41–60.

6. *Power and Weakness of the Modal Display Calculus.* In: Heinrich Wansing (Ed.): *Proof Theory of Modal Logic*, Studies in Applied Logic Vol. 2, Kluwer, Dordrecht, 1996, 95–122.

7. *Inessential Features.* In: Christian Retoré (Ed.): *Logical Aspects of Computational Linguistics*, Springer Lecture Notes in Artificial Intelligence No. 1328, 1997, 43–62.

8. *Normal monomodal logics can simulate all others.* The Journal for Symbolic Logic, 64(1999), 99–138. (With Frank Wolter)

9. *Lattices of Modal Logics and Their Groups of Automorphisms.* Annals of Pure and Applied Logic, 100(1999), 99–139.

10. *Tools and Techniques in Modal Logic.* Studies in Logic Nr. 142, Elsevier, 1999.

11. *Invariant Logics.* Mathematical Logic Quarterly, 48(2002), 29–50.

12. *Elementary Models for Modal Predicate Logic. Part I: Completeness.* In: Frank Wolter et al. (Eds.): *Advances in Modal Logic 3*, World Scientific, 2002. 299–320. (With Oliver Kutz)

13. *The Mathematics of Language.* Berlin: Mouton de Gruyter, 2003.

14. *Notes on the Space Requirement for Checking Satisfiability in Modal Logics*, In: Philippe Balbiani, Nobo–Yuki Suzuki, Frank Wolter and Michael Zakharyaschev (eds:): *Advances in Modal Logic 4*, King's College Publications, 2003, 243–264.

Work in Progress *Modal Consequence Relations*, in: Johan van Benthem, Patrick Blackburn and Frank Wolter, *Handbook of Modal Logic*, Elsevier.

15. *Logically Possible Worlds and Counterpart Theory*, in: Dale Jacquette, *Handbook of the Philosophy of Logic*, Elsevier. (With Oliver Kutz)

16. *The Decidability of LGB-Type Structures. Part I: Multidominance Structures*, Given at a symposium in honour of Uwe Mönnich, Freudenstadt, November 2004.

17. *Semisimple Varieties of Boolean Algebras with Operators*, Manuscript, 2004. (With Tomasz Kowalski)

18. *Gnosis*, Manuscript, UCLA.

**Service to the Profession:** 2002– : Editor of the Journal of Logic, Language and Information.

**Teaching:** In the nineties we had a rather strong group of students interested in modal logic, which included Frank Wolter, Carsten Grefe and later Oliver Kutz and Sebastian Bauer. Unfortunately, all of us had to leave for lack of jobs. Myself, I am now teaching mathematical and computational linguistics at UCLA.

**Vision Statement:** Logic has produced many applications, not only in computer science. Like number theory, it evolved from pure theory into more and more of an applied science. As much as this is a favourable trend, one needs to be careful not to lose the original motivations, or one will get stuck just caring about applications. So often pure theory has supplied stunning applications that we should continue to devote our energy to it. I am saying this against a noticeable trend to commit universities and researchers to do more applied research. We should never concern ourselves just with applications, as we should never do only theory.

**Honours and Awards:** Heisenberg-Fellowship 1999.

# KRAJÍČEK, Jan

**Specialties:** Proof complexity, bounded arithmetic.

**Born:** 18 June 1960 in Prague, Czech Republic.

**Educated:** Charles University, RNDr. Mathematics 1985; Czechoslovak Academy of Sciences, CSc. 1990; Academy of Sciences of the Czech Republic, DrSc. 1993;

**Dissertation:** *Non-classical Foundations of Mathematics*; supervisor, Pavel Pudlák.

**Regular Academic or Research Appointments:** RESEARCHER, MATHEMATICAL INSTITUTE, ACADEMY OF SCIENCES OF THE CZECH REPUBLIC, 1985– AND PROFESSOR OF MATHEMATICAL LOGIC, CHARLES UNIVERSITY, 2004–.

**Visiting Academic or Research Appointments:** University of Illinois at Urbana-Champaign 1988–1989 and 1990–1991; University of Toronto 1993; University of Oxford and visiting scholar at Wolfson college 1997–1999; Institute for Advanced Study, Princeton, member in 2004.

**Research Profile:** Jan Krajíček is a leading figure in proof complexity and bounded arithmetic, an area connecting mathematical logic and computational complexity theory. His work has been instrumental in bringing the field from its infant years in mid eighties to the current status of a rich mathematical area with many deep results and with a variety genuine connections to other areas of mathematics and computer science.

He proved (sometimes with coauthors) key witnessing theorems and separation results in bounded arithmetic and lower bounds in proof complexity, shaping the whole field. His research often lead from a solution of a particular problem to a formulation of a general method or of a research program. His 1995 monograph has been indispensable for anybody in the field, underlying the single most important universal paradigm in the area: Theories can be viewed, in many nontrivial aspects, as uniform versions of proof systems.

Among his best known achievements is the invention of the method of feasible interpolation. In a subsequent work with P. Pudlák they demonstrated links between feasible interpolation and cryptography. This work has been extraordinarily stimulating to many researchers, leading to a variety of new results.

**Main Publications:**

1. "On the Number of Steps in Proofs". *Annals of Pure and Applied Logic*. 41(2) (1989): 153–178.
2. "Propositional Proof Systems, the Consistency of First Order Theories and the Complexity of Computations", with P. Pudlák. *J. Symbolic Logic*, 54(3) (1989): 1063–1079.
3. "Bounded Arithmetic and the Polynomial Hierarchy", with P. Pudlák and G. Takeuti. *Annals of Pure and Applied Logic*, 52(1991): 143–153.
4. "Fragments of Bounded Arithmetic and Bounded Query Classes". *Transactions of the AMS*, 338(2) (1993): 587–598.
5. 'An Application of Boolean Complexity to Separation Problems in Bounded Arithmetic", with S. Buss. *Proceedings of the London Mathematical Society*, 69(3) (1994): 1–21.
6. "Lower Bounds to the Size of Constant-Depth Propositional Proofs". *J. of Symbolic Logic*, 59(1) (1994): 73–86.
7. "An Exponential Lower Bound to the Size of Bounded Depth Frege Proofs of the Pigeonhole principle", with P. Pudlák and A. Woods. *Random Structures and Algorithms*, 7(1) (1995): 15–39.
8. "Lower bounds on Hilbert's Nullstellensatz and propositional proofs", with P. Beame, R. Impagliazzo, T. Pitassi and P.Pudlák. *Proceedings of the London Mathematical Society*, 73(3) (1996): 1–26.
9. "Interpolation theorems, lower bounds for proof systems, and independence results for bounded arithmetic". *J. of Symbolic Logic*, 62(2) (1997): 457–486.
10. "Some consequences of cryptographical conjectures for $S_2^1$ and $EF$", with P. Pudlák. *Information and Computation*, 140(1) (1998): 82–94.
11. "Proof complexity in algebraic systems and bounded depth Frege systems with modular counting", with S. Buss, R. Impagliazzo, P. Pudlák, A. A. Razborov and J. Sgall. *Computational Complexity*, 6(3) (1996/1997): 256–298.
12. "On the degree of ideal membership proofs from uniform families of polynomials over a finite field". *Illinois J. of Mathematics*, 45(1) (2001): 41–73.
13. "Uniform families of polynomial equations over a finite field and structures admitting an Euler characteristic of definable sets". *Proceedings of the London Mathematical Society*, 3(81) (2000), 257–284.
14. "On the weak pigeonhole principle". *Fundamenta Mathematicae*, 170(1–3) (2001): 123–140.
15. "Combinatorics with definable sets: Euler characteristics and Grothendieck rings", with T. Scanlon). *Bulletin of Symbolic Logic*, 3(3) (2000): 311–330.
16. "Dehn function and length of proofs". *International Journal of Algebra and Computation*, 13(5) (2003): 527–542.
17. "Dual weak pigeonhole principle, pseudo-surjective functions, and provability of circuit lower bounds", *J. of Symbolic Logic*, 69(1) (2004), 265–286.
18. "Diagonalization in proof complexity, *Fundamenta Mathematicae*, 182 (2004), 181–192.
19. *Bounded Arithmetic, Propositional Logic, and Complexity Theory*. In ???? (eds.) *Encyclopedia of Mathematics and Its Applications, Vol. 60*. Cambridge University Press, Cambridge, New York, and Melbourne, 1995, pp. ??–??.

*Work in Progress*

20. *Proof Complexity*. Monograph in preparation.

**Service to the Profession:** Editor, *Annals of Pure and Applied Logic*, 1994–; editor, *Notre Dame Journal of Formal Logic*, 2003–; editor, *Logical Methods in Computer Science*, 2004–; editor or co-editor of three refereed volumes published by Oxford Press, by ASL and A.K.Peters, and by Seconda Universitá di Napoli (Quaderni di Matematica ser.); member or chairman of around thirty program and organizing committees; member of various ASL committees, 1995–.

**Teaching:** Advanced courses in mathematical logic and complexity theory. PhD students: J. Hanika (2004) and E. Jeřábek (2005).

**Vision Statement:** Krajíček expects that logic, keeping a strictly mathematical format, will contribute to the understanding and an eventual solution of fundamental problems of computational complexity theory and of theoretical computer science in general.

**Honours and Awards:** Invited speaker, Fourth European Congress of Mathematics in Stockholm 2004; Fellow, Learned Society of the Czech Republic 2004; Prize of the Education Ministry of the Czech Republic for Research 1998; Plenary speaker, Logic Colloquia, 1995 and 1998; Invited speaker, Tenth International Congress of Logic, Methodology and Philosophy of Science in Florence, 1995; Award of the Academy of Sciences of the Czech Republic for Young Researchers, 1994; ASL Invited Plenary Address at an annual joint ASL/AMS meeting in San Antonio, 1993.

# KRAUSE, Décio

**Specialties:** Non classical logics, abstract logics, quantum logic, philosophy of science.

**Born:** 01 June 1953 in Rio de Janeiro, Brazil.

**Educated:** Catholic University of Paraná, Mathematics, 1976; Federal University of Paraná, MSc Education, 1983; University of São Paulo, PhD Logic, 1990.

**Dissertation:** *Non-Reflexivity, Indistinguishability, and Weyl's Aggregates*; supervisor, Newton C. A. da Costa.

**Regular Academic or Research Appointments:** PROFESSOR OF LOGIC AND PHILOSOPHY OF SCIENCE, FEDERAL UNIVERSITY OF SANTA CATARINA, 2000–; Professor of Mathematics, Retired, Federal University of Paraná, 1977-1999; Professor of Mathematics, Federal Center for Technological Education of Paraná, 1978-1991; Elementary School Teacher.

**Visiting Academic or Research Appointments:** Visiting Scholar, University of Florence, 1992-1993; Visiting Scholar, University of Leeds, 1995-1996.

**Research Profile:** D. Krause has investigated the logical and philosophical foundations of quantum physics. Taking for granted that there is not a *just one way* to approach this field, he studies the existence of an adequate mathematical language (logic involved) which would enable us to directly speak of the 'absolutely indiscernible *objects*' which form part of the standard discourse of physical theories, even quantum field theories, and whose objective existence cannot be simply discharged. His approach follows Heinz Post, who in 1963 said that the indistinguishability of quanta should be taken "right at the start". Krause intends to avoid the usual techniques that use standard logic and mathematics, where labels or coordinates are necessarily attached to the basic quantum objects and then symmetry conditions are postulated to overcome these *ab ovo* identifications. In this vein, he has developed a class of logics termed *non-reflexive*, where the standard notion of identity is weakened; in particular he presented a *quasi-set* theory, aiming at to cope with some of the most basic traits of quantum discourse. This case study serves to justify a general philosophical view sketched in the *Vision Statement* below. His interests go also to abstract logic, the very general approach to logic as developed by Newton da Costa, and in general topics of philosophy of science, like truth in physics, the ontology of physical theories, and the logical structure of scientific theories.

**Main Publications:**

1. *Identity in Physics: A historical, philosophical and formal analysis*, with Steven French, Oxford Un. Press, 2006.
2. 'Paraconsistent logic and paraconsistency', with Newton C.A. da Costa and O. Bueno, forthcoming in Dale Jacquette (ed.), *Handbook of the Philosophy of Science*, Volume 5: Philosophy of Logic. Elsevier, 2006.
3. 'Structures for structural realism', *The Logic Journal of the IGPL* 13 (1), 2005, 113-126.
4. 'Complementarity and paraconsistency' (with N. C. A. da Costa), in S. Rahman, J. Symons, D.M. Gabbay and J. -P. van Bendegem (eds), *Logic, epistemology, and the unity of science*, Vol. 1, Kluwer Ac. Pu., 2004, pp. 557-568.
5. 'Quantum vagueness' (with S. French), *Erkenntnis* 59 (1), 2003, 97-124.
6. "Suppes predicate for genetics and natural selection" (with J. C. M. Magalhães), *Journal of Theoretical Biology* 209 (2) 2001, 141-153.
7. "Bibel's Matrix Connection Method in Paraconsistent Logic: General Ideas and Implementation", (with E. F. Nobre and M. A. Musicante), *Proceedings of the XXI Internacional Conference Chilean Computer Society*, Punta Arenas, Chile, 5-9 Nov. 2001 (IEEE Computer Society Press).
8. "Remarks on quantum ontology", *Synthese* 125 (1/2), 2000, 155-167.
9. 'The logic of quanta' (with S. French), in Cao, T. Y. (ed.), *Conceptual foundations of quantum field theory*, Cambridge University Press, 1999, 324-342.
10. "Quasi set theory for bosons and fermions" (with A. S. Sant'Anna and A. G. Volkov), *Foundations of Physics Letters* 12 (1), 1999, 67-79.
11. "Quasi set theories for microobjects: a comparision" (with M. L. Dalla Chiara and R. Giuntini), in E. Castellani (ed), *Interpreting bodies: classical and quantum objects in modern physics*, Princeton University Press, 1998, 142-152.
12. "An intensional Schrödinger logic" (with N. C. A. da Costa), *Notre Dame Journal of Formal Logic* 38 (2), 1997, 179-194.
13. 'A formal framework for quantum non-individuality' (with S. French), *Synhese* 102, 1995, 195-214.
14. "Vague identity and quantum non-individuality" (com S. French), *Analysis* 55 (1), 1995, 20-26.
15. 'Schrödinger logics' (with N. C. A. da Costa), *Studia Logica* 53 (4), 1994, 533-550.
16. 'On a quasi-set theory', *Notre Dame Journal of Formal Logic* 33 (3), 1992, 402-411.

*Work in Progress*

17. *Paraconsistent logics and paraconsistency,* book covering the fields of paraconsistent logic and its applications, in collaboration with Newton C. A. da Costa and Otávio Bueno.
18. Editor (with Steven French and Itala D'Ottaviano), *Selected Papers of Newton da Costa*, to appear in the series Contemporary Logic, Polimerica,

Italy.

19. Editor (with Jean-Yves Béziau) of a special issue of Synthese in honour of Patrick Suppes' $80^{th}$ birthday and celebrating his stay in Florianópolis, Brasil, in 2003.

**Service to the Profession:** Coordinator of the Research Group on Logic and the Foundations of Science (Brazilian Council for Research and Scientific Development, and Federal University of Santa Catarina), 2000-; Coordinator of the Epistemology and Logic Group of the Federal University of Santa Catarina, 2001-2004; General Coordinator of Research, Federal University of Paraná, 1998-2000; Editor of the Boletim da Sociedade Paranaense de Matemática, 1995-2000.

**Teaching:** D. Krause has though various courses in mathematics, logic and philosophy of science to both undergraduate and graduate levels. He has supervised or is supervising nine Master and three PhD theses on subjects dealing with the foundations of science and mathematical education.

**Vision Statement:** Krause thinks that a wider domain of knowledge, usually roughly delineated by informal 'pre-theories', can be approached from distinct perspectives. Due to the richness of the field, these approaches may originate different and even incompatible *theories* of that domain, each one serving to illuminate different aspects of the subject. So, the 'general logic' of science, if there is some, might be non-classical, perhaps a paraconsistent one.

**Honours and Awards:** CNPq Postdoctoral Fellowship, University of Leeds, 1995-1996; CNPq Postdoctoral Fellowship, University of Florence, 1992-1993; CNPq Research Grant, 1992-; CAPES Doctoral Fellowship, University of São Paulo, 1987-1990.

# KRIPKE, Saul

**Specialties:** Kripke Semantics for modal and related logics, theory of truth, philosophy of Wittgenstein, philosophy of language.

**Born:** 13 November 1940 in Bay Shore New York, USA.

**Educated:** Harvard University.

**Regular Academic or Research Appointments:** DISTINGUISHED PROFESSOR OF PHILOSOPHY AT CUNY GRADUATE CENTER, 2003–; Princeton University, 1997–2002; Rockefeller University 1967–1997.

**Research Profile:** Kripke is best known for four contributions to philosophy:

1. Kripke semantics for modal and related logics, published in several essays beginning while he was still in his teens.

2. The 1970 Princeton lectures *Naming and Necessity* (published in 1972 and 1980), that significantly restructured the philosophy of language and, as some have put it, "made metaphysics respectable again".

3. An interpretation of the philosophy of Wittgenstein.

4. A theory of truth.

He has also contributed to set-theory

**Main Publications:**

1. 1959. "A Completeness Theorem in Modal Logic", *Journal of Symbolic Logic* 24(1):1–14.
2. 1962. "The Undecidability of Monadic Modal Quantification Theory", *Zeitschrift für Mathematische Logik und Grundlagen der Mathematik* 8:113–116
3. 1963. "Semantical Considerations on Modal Logic", *Acta Philosophica Fennica* 16:83–94
4. 1963. "Semantical Analysis of Modal Logic I: Normal Modal Propositional Calculi", *Zeitschrift fÃ¼r Mathematische Logik und Grundlagen der Mathematik* 9:67–96
5. 1964. "Transfinite Recursions on Admissible Ordinals, I" (abstract), *The Journal of Symbolic Logic*, Vol. 29, No. 3, p. 162.
6. 1964. "Transfinite Recursions on Admissible Ordinals, II" (abstract), *The Journal of Symbolic Logic*, Vol. 29, No. 3, p. 162.
7. 1964. "Admissible Ordinals and the Analytic Hierarchy" (abstract), *The Journal of Symbolic Logic*, Vol. 29, No. 3, p. 162.
8. 1965. "Semantical Analysis of Intuitionistic Logic I", In *Formal Systems and Recursive Functions*, edited by M. Dummett and J. N. Crossley. Amsterdam: North-Holland Publishing Co.
9. 1965. "Semantical Analysis of Modal Logic II: Non-Normal Modal Propositional Calculi", In *The Theory of Models*, edited by J. W. Addison, L. Henkin and A. Tarski. Amsterdam: North-Holland Publishing Co.
10. 1967. "An Extension of a Theorem of Gaifman-Hales-Solovay," *Fundamenta Mathematicae*, Vol. 61, pp. 29-32.
11. 1971. "Identity and Necessity", In *Identity and Individuation*, edited by M. K. Munitz. New York: New York University Press.
12. 1972 (1980). "Naming and Necessity", In *Semantics of Natural Language*, edited by D. Davidson and G. Harman. Dordrecht; Boston: Reidel. Sets out the causal theory of reference.
13. 1975. "Outline of a Theory of Truth", *Journal of Philosophy* 72:690–716. Sets his theory of truth (against Alfred Tarski), where an object language can contain its own truth predicate.

14. 1976. "Is There a Problem about Substitutional Quantification?", In *Truth and Meaning: Essays in Semantics*, edited by Gareth Evans and John McDowell. Oxford: Oxford University Press.
15. 1977. "Speaker's Reference and Semantic Reference", *Midwest Studies in Philosophy* 2:255–276
16. 1979. "A Puzzle about Belief", In *Meaning and Use*, edited by A. Margalit. Dordrecht and Boston: Reidel.
17. 1980. *Naming and Necessity*. Cambridge, Mass.: Harvard University Press. ISBN 0-674-59845-8 and reprints 1972.
18. 1982. *Wittgenstein on Rules and Private Language: an Elementary Exposition*. Cambridge, Mass.: Harvard University Press. ISBN 0-674-95401-7. Sets out his interpretation of Wittgenstein aka Kripkenstein.
19. 1986. "A Problem in the Theory of Reference: the Linguistic Division of Labor and the Social Character of Naming," *Philosophy and Culture (Proceedings of the XVIIth World Congress of Philosophy)*, Montreal, Editions Montmorency: 241-247.
20. 1992. "Summary: Individual Concepts: Their Logic, Philosophy, and Some of Their Uses." *Proceedings and Addresses of the American Philosophical Association* 66: 70-73
21. 2005. "Russell's Notion of Scope", *Mind* 114:1005–1037
22. 2008. "Frege's Theory of Sense and Reference: Some Exegetical Notes," *Theoria* 74:181-218

**Vision Statement:** Kripke is a devoutly religious Jew. Additionally, in an interview with Andreas Saugstad, he stated "I don't have the prejudices many have today, I don't believe in a naturalist world view. I don't base my thinking on prejudices or a world view and do not believe in materialism."

**Honours and Awards:** Kripke was the recipient of the 2001 Schock Prize in Logic and Philosophy. He has received honorary degrees from the University of Nebraska, Omaha (1977), Johns Hopkins University (1997), Unifity of Haifa, Israel (1998) and the University of Pennsylvania (2005). He is a member of the American Philosophical Society. In 1963 he was appointed to the Society of Fellows.

# KRÖGER, Fred

**Specialties:** Temporal logic in computer science.

**Born:** 12 January 1945 in Schroda, Poland.

**Educated:** Ludwig-Maximilians-University Munich, Dr.rer.nat. Mathematics, 1971; Ludwig-Maximilians-University Munich, Diplom Mathematics, 1969.

**Dissertation:** *Über die Konstruktion höherer arithmetischer Ordinalzahloperationen nach SAARNIO* (in German); supervisor, Kurt Schütte.

**Regular Academic or Research Appointments:** PROFESSOR OF COMPUTER SCIENCE, LUDWIG-MAXIMILIANS-UNIVERSITY MUNICH, 1986-; Professor of Computer Science, Technical University Munich, 1980-1986; Scientific Assistant, Technical University Munich, 1971-1980.

**Research Profile:** Fred Kröger has introduced and developed — at the same time as and independently of Amir Pnueli — temporal logic which nowadays is acknowledged and widely used in computer science as a formal tool in the field of specification and verification of state based systems. Publications cover purely logical investigations of temporal logic as well as contributions to its applications. His textbook *Temporal logic of programs* was the first comprehensive monograph of the field. Besides this main research topic, Kröger worked also on Hoare logic.

**Main Publications:**
1. Logical rules of natural reasoning about programs. In: S. Michaelson, R. Milner (eds.): *Automata, Languages and Programming*, Third International Colloquium, Edinburgh, July 20-23, 1976. Edinburgh University Press: Edinburgh 1976.
2. LAR: A logic of algorithmic reasoning. *Acta Informatica* 8 (1977), 243-266.
3. A uniform logical basis for the description, specification and verification of programs. In: E.J. Neuhold (ed.): *Formal Description of Programming Concepts*. IFIP Working Conference, St. Andrews, N.B., Canada, August 1-5, 1977. North-Holland: Amsterdam-New York-Oxford 1978.
4. Infinite proof rules for loops. *Acta Informatica* 14 (1980), 371-389.
5. A generalized nexttime operator in temporal logic. *J. Comp. Syst. Sci.* 29 (1984), 80-98.
6. On temporal program verification rules. *R.A.I.R.O Informatique théorique / Theoretical Informatics* 19 (1985), 261-280.
7. Temporal logic of programs. Springer: Berlin-Heidelberg-New York 1987.
8. Abstract modules: Combining algebraic and temporal logic specification means. *Technique et Science Informatiques* 6 (1987), 559-573
9. On the interpretability of arithmetic in temporal logic. *Theoretical Computer Science* 73 (1990), 47-60.
10. Temporal logic and state systems (with Stephan Merz). Springer: Berlin-Heidelberg, 2008.

**Service to the Profession:** Collecting editor, *Journal of Applied Non-Classical Logics*, 1990- .

**Teaching:** For many years, Kröger is giving regular courses and lectures on applications of logics in computer science. He has had 7 doctoral students.

**Vision Statement:** Computer science is a wide field for applications of logics. Realizing this more could be of great benefit for the whole field of logic.

## KUNEN, Kenneth

**Specialties:** Set theory, automated reasoning.

**Born:** 2 August 1943 in New York City, New York, USA.

**Educated:** California Institute of Technology, BS Mathematics 1965; Stanford University, PhD Mathematics 1968.

**Dissertation:** *Inaccessibility Properties of Cardinals*; supervisor, Dana Scott.

**Regular Academic or Research Appointments:** PROFESSOR OF MATHEMATICS, UNIVERSITY OF WISCONSIN AT MADSION, 1972–; Associate Professor, UW Madison 1970–1972; Research Assistant Professor, UW Madison 1968–1970.

**Visiting Academic or Research Appointments:** Visiting Professor, University of Texas at Austin, 1979-1981; Visiting Associate Professor, University of California at Berkeley, 1971–1972.

**Research Profile:** Kenneth Kunen works in applications of logic to pure mathematics. There are two different aspects to this work.

First, in areas such as general topology and measure theory, results are frequently independent of the usual axioms of set theory, ZFC, and independence proofs use logical methods such as forcing and constructibility. Kunen's work combines these logical methods with the classical methods of topology and analysis.

Second, Kunen works with tools of automated deduction and their application to solve problems in algebra. Unlike the situation in most areas of mathematical reasoning, the current computer tools can actually outperform a human in problems involving primarily equations. Kunen works on non-associative algebra, such as the theory of loops. Here, Kunen's work combines these logical methods with the classical methods of algebra.

**Main Publications:**

1. Some applications of iterated ultrapowers in set theory, Annals Math. Log., Vol. 1 (1970), pp. 179-227.
2. Ultrafilters and independent sets, Trans. AMS, Vol. 172 (1972), pp. 299-306.
3. Elementary embeddings and infinitary combinatorics J. Symbolic Logic, Vol. 36 (1971), pp. 407-413.
4. Paracompactness of box products on compact spaces, AMS Transactions 240 (1978), pp. 307-316.
5. Two more hereditarily separable non-Lindelöf spaces (with I. Juhász and M. E. Rudin), Canadian J. Math. Vol. 28 (1976), pp. 998-1005.
6. Strong S and L spaces under MA, Set Theoretic Topology (Proc. of Athens Meeting, Spring 1976), pp. 265-268.
7. Weak P-points in N*, Colloq. Math. Sco. Janos Bolyai 23 (1980), pp. 741-749.
8. A Ramsey theorem in Boyer-Moore logic, Journal of Automated Reasoning 15 (1995) 217-235.
9. Quasigroups, loops, and associative laws, J. Algebra 185 (1996) 194-204.
10. Bohr topologies and partition theorems for vector spaces, Topology and Applications 90 (1998) 97-107.
11. Bohr compactifications of discrete structures (with J. Hart), Fund. Math. 160 (1999) 101-151.
12. G-loops and permutation groups, J. Algebra 220 (1999) 694-708.
13. Compact L-spaces and right topological groups, Topology Proceedings, Vol. 24 (Summer, 1999) 295-327.
14. Every diassociative A-loop is Moufang (with M. K. Kinyon and J. D. Phillips), AMS Proceedings, 30 (2002) 619-624.
15. Matrices and ultrafilters (with J. Baker), in Recent Progress in General Topology II, Elsevier - North-Holland, 2002, pp. 59-81.
16. Chromatic numbers and Bohr topologies (with B. N. Givens), Topology and Applications 131 (2003) 189 - 202.
17. Diassociativity in conjugacy closed loops (with M. K. Kinyon and J. D. Phillips), Communications in Algebra 32 (2004) 767 - 786.
18. The complex Stone-Weierstrass property, Fund. Math. 182 (2004) 151 - 167.
19. The structure of extra loops (with M. K. Kinyon), Quasigroups and Related Systems 12 (2004) 39 - 60.
20. Small locally compact linearly Lindelöf spaces, Topology Proceedings, Volume 29, No. 1 (2005), pp. 193-198.

**Service to the Profession:** Editorial Adviser, LMS Journal of Computation and Mathematics, 1996 - present; Associate Editor, AMS Transactions, 1984-1988; Associate Editor, Journal of Symbolic Logic, 1981-1983; Editor of Annals of Math. Logic, 1976-1985;

**Teaching:** Kunen teaches the undergraduate and graduate logic courses at the University of Wisconsin. He is also active in supervising graduate students, and has had a total of 23 Ph.D. students.

**Vision Statement:** Mathematical logic arose historically out of attempts to understand philosophical questions about the foundations of mathematics. However, modern logic has become an important tool in mainstream mathematics itself, and that tool will undoubtedly become more important as time goes on. Modern mathematics has become the study of how classical mathematics behaves in various models of set theory.

**Honours and Awards:** H. I. Romnes Fellowship, 1978-1981; Alfred P. Sloan Fellowship, 1969-1971; National Science Foundation Graduate Fellowship, 1965-1968.

# L

## LADNER, Richard E.

**Specialties:** Computability theory, computational complexity, complexity of logic.

**Born:** 22, August 1943 in Berkeley, California, USA

**Educated:** University of California, Berkeley, Ph.D., Mathematics, 1971; St. Mary's College of California, B.S., Mathematics, 1965

**Dissertation:** *Mitotic Recursively Enumerable Sets*; supervisor Robert W. Robinson (and with guidance from Alistair Lachlan)

**Regular Academic or Research Appointments:** BOEING PROFESSOR IN COMPUTER SCIENCE AND ENGINEERING UNIVERSITY OF WASHINGTON, 2004–; PROFESSOR, UNIVERSITY OF WASHINGTON, 1981–; Associate Professor, University of Washington, 1976–1981; Assistant Professor, University of Washington, 1972–1976; Acting Assistant Professor, University of Washington, 1971–1972.

**Visiting Academic or Research Appointments:** AT&T-Labs Research, Florham Park, NJ, September 1999–June 2000; Victoria University of Wellington, New Zealand, January–June, 1993; Mathematical Sciences Research Institute, Berkeley, January–June, 1986; Gallaudet University, August–December, 1985; Yale University, January–June, 1978; University of Toronto, August–December.

**Research Profile:** Richard E. Ladner has made fundamental contributions to computability theory and computational complexity theory. In his dissertation he proved that mitotic recursively enumerable sets and the same as autoreducible sets. Mitotic sets are those that can be partitioned into two sets each of which is of the same Turing degree as the whole. Autoreducible sets are those that can be reduced to themselves by a reduction that never queries about its own input. In addition, he proved that there are nonrecursive, recursively enumerable degrees in which all its sets are mitotic. The techniques used in this result were extended in work with Leonard Sasso to show that there are nonrecursive, recursively enumerable degrees in which all the recursively enumerable sets have the same weak truth-table degree. After his dissertation work, he moved to the area of computational complexity. In a major result he proved that there are sets in NP − P that are not NP-complete, provided P ≠ NP. In work with Nancy Lynch and Alan Selman, he showed how different notions of polynomial time reducibility are actually distinct. For example, they showed that there are computable sets A and B such that A is polynomial truth-table reducible to B but A is not polynomial many-one reducible to B. He established the computational complexity of various modal logics. For example, he proved that the satisfiability in modal logics T, K, and S4 are PSPACE-complete, while S5-satisfiability is NP-complete. With Michael J. Fischer he proved that satisfiability in propositional dynamic logic requires deterministic exponential time. He no longer works in computability theory and computational complexity. In recent years has worked on parallel, distributed, and network algorithms, memory efficient algorithms, algorithms for media delivery, and data compression.

**Main Publications:**

1. A. Bar-Noy and R.E. Ladner. Efficient Algorithms for Optimal Stream Merging for Media-on-Demand. *SIAM Journal on Computing*, Vol. 33, No. 5, 2004, 1011-1034.
2. A.E. Mohr, E.A. Riskin, and R.E. Ladner. Unequal Loss Protection: Graceful Degradation of Image Quality Over Packet Erasure Channels Through Forward Error Correction. *IEEE Journal of Selected Areas in Communications Special Issue on Error-Resilient Image and Video Transmission*, Vol. 18, No. 6, 2000, 819-828.
3. A. LaMarca and R.E. Ladner. The Influence of Caches on the Performance of Sorting. *Journal of Algorithms* Vol. 31, 1999, 66-104.
4. R.E. Ladner, Polynomial Space Counting Problems. *SIAM Journal on Computing*, Vol. 18, 1989, 1087-1097.
5. A. Condon and R.E. Ladner, Probabilistic Game Automata. *Journal of Computer and System Sciences*, Vol. 36. No. 3, 1988, 452-489
6. R.E. Ladner and J. Reif, The Logic of Distributed Protocols. *Theoretical Aspects of Reasoning about Knowledge, Proceedings of the 1986 Conference.* Edited by Joseph Y. Halpern. Morgan Kaufmann
7. Publishers, Inc. March 1986, 207-222
8. R.E. Ladner and J. K. Norman, Solitaire Automata, *Journal of Computer and System Sciences,* Vol. 30, No. 1, 1985, 116-129.
9. R. E. Ladner, L. J. Stockmeyer, and R. J. Lipton, Alternation Bounded Auxiliary Pushdown Automata. *Information and Control*, Vol. 62, Nos. 2/3, 1984, 93-108.
10. R. E. Ladner, R. J. Lipton, L. J. Stockmeyer, Alternating Pushdown and Stack Automata. *SIAM Journal*

*on Computing*, Vol. 13, No. 1, February 1984, 135-155.

11. R. E. Ladner, The Complexity of Problems in Systems of Communicating Sequential Processes. *Journal of Computer and System Sciences*, Vol. 21, No.2, 1980, 179-194.

12. M. J. Fischer and R. E. Ladner, Propositional Dynamic Logic of Regular Programs. *Journal of Computer and System Sciences*, Vol. 18, No. 2, April 1979, 194-211

13. R. E. Ladner, The Computational Complexity of Provability in Systems of Modal Propositional Logic. *SIAM Journal on Computing*, Vol. 6, No. 3, 1977, 467-480

14. R. E. Ladner, Application of Model Theoretic Games to Discrete Linear Orders and Finite Automata. *Information and Control*, Vol. 33, No. 4, April 1977, 281-303.

15. R. E. Ladner and N. A. Lynch, Relativization of Questions about Log Space Computability. *Mathematical Systems Theory*, Vol. 10, No. 1, 1976, 19-32.

16. R. E. Ladner, N. A. Lynch and A. L. Selman, A Comparison of Polynomial Time Reducibilities. *Theoretical Computer Science*, Vol. 1, 1975, 103-123.

17. R. E. Ladner and L. P. Sasso, Jr.,The Weak Truth Table Degrees of Recursively Enumerable Sets. *Annals of Mathematical Logic*, Vol. 8, 1975, 429-448

18. R. E. Ladner, On the Structure of Polynomial Time Reducibility. Journal of the ACM, Vol. 22, No. 1, 1975, 155-171.

19. R. E. Ladner, A Completely Mitotic Nonrecursive R.E. Degree. *Transactions of the AMS*, vol. 184, 1973, 479-507

20. R. E. Ladner, Mitotic Recursively Enumerable Sets, *Journal of Symbolic Logic*, Vol. 38, No. 2, June, 1973, 199-211

A complete list of his publications can be found at http://www.cs.washington.edu/homes/ladner/papers.html

**Service to the Profession:** Chair of the ACM Special Interest Group on Algorithms and Computation Theory (SIGACT), 2005–present; Editorial Board for *ACM Transactions on Accessible Computing*, 2006–present; Editorial Board for *Theory of Computing Systems*, 2005–present; Associate Editor for *Journal of Computer and System Sciences*, 1989–present; Area Editor for *Journal of the Association of Computing Machinery*, 1986–1992; Editor for *SIAM Journal on Computing*, 1980–83.

**Teaching:** Richard E. Ladner has developed courses in computational complexity, algorithms, data structures, computer networks and data compression. He has supervised 17 Ph.D. student including: Tammy VanDeGrift (2005), Justin Goshi (2004), James Fix (2002), Ed Hong (2001), Suzanne Bunton (1996), Anthony LaMarca (1996), David Cohn (1992), H.K. Dai (1991), Soma Chaudhuri (1990), Ewan Tempero (1990), Anne Condon (1987), Robert J. Fowler (1985), Garret Swart (1985), Albert Greenberg (1983), Glenn Brooks (Goodrich) (1983), Udi Manber (1982), Paul Frank (1979). He has also supervised six master theses.

**Vision Statement:** "Mathematical logic has had a huge influence on computer science. Methods for hardware and software verification have their roots in logic. The mutual interchange of ideas between computer science and logic should be fostered to the benefit of both disciplines."

**Honours and Awards:** CRA A. Nico Habermann Award, 2008; Japan Society for the Promotion of Science Fellowship, 2006; Presidential Award for Excellence in Science, Mathematics and Engineering Mentoring, 2004; Fellow of the ACM, 1994; Fulbright Travel Grant to New Zealand, 1992–1993; Guggenheim Fellowship, 1985–86.

# LAMBERT, J. Karel

**Specialties:** Logic, applications of logic, philosophical logic.

**Born:** 10 April 1928 in Chicago, Illinois, USA.

**Educated:** Willamette University, BS in Experimental Psychology, 1950; University of Oregon, MS Experimental Psychology, 1953, Michigan State University, PhD Philosophy, 1957.

**Dissertation:** *A Logical-Mathematical Analysis of Tolman's Theory of Learning*; supervisors, Henry Leonard and M. Ray Denny.

**Regular Academic or Research Appointments:** RESEARCH PROFESSOR EMERITUS OF LOGIC AND THE PHILOSOPHY OF SCIENCE, UNIVERSITY OF CLAIRFORNIA, IRVINE, 1996-; Research Professor of Logic and the Philosophy of Science, University of California, Irvine, 1994-1996; Professor of Philosophy, University of California, Irvine, 1967-1994; Professor of Philosophy and Chairman of the Department of Philosophy, West Virginia University, 1963-1967; Associate Professor of Philosophy and Experimental Psychology, University of Alberta, Edmonton, 1959-1963; Assistant Professor of Experimental Psychology, University of Alberta, Edmonton, 1956-1959.

**Visiting Academic or Research Appointments:** Visiting Distinguished Professor of Philosophy, Salzburg Universitaet, 1993; Visiting Professor of Logic and Philosophy of Science, Bielefeld Universitaet, 1992; Visiting Professor of Philosophy and Philosophy of science, Ulm Universitaet, 1984; Fulbright-Hays Senior Fellow,

Salzburg Universitaet, 1980; NEH Fellow, College de France, 1979; NEH Senior Fellow, Internationales Forschungszentrum Fuer Grundfragen der Wissenschaften, Salzburg, 1973; Graz Universitaet, Graz, 1973; Visiting Professor of Philosophy, University of Oslo, 1970.

**Research Profile:** Karel Lambert is one of the founders of free logic, a nonstandard quantification theory, and it's most prolific exponent. (He coined the name "free logic" during an international conference at Stanford in 1959.) Distinctions between kinds of free logic (positive, negative, etc.), the first axiomatic development of a free logic, and several of the semantical approaches to the subject are due to him. With others (notably, Bas van Fraassen and Robert K. Meyer) he put free logic on a sound mathematical footing. He invented free definite description theory, the fundamental property of which is widely known as "Lambert's Law". He has shown that perhaps the most well known semantical basis for free logic, that in which the denotation function is partial, yields a breakdown in the classical substitution principle that coextensive predicates substitute everywhere *salve veritate*. He has demonstrated that when applied to Quine's pre-regimented theory of predication (with identity as the only predicate) there is a breakdown in extensionality. With Meyer and Bencivenga he has shown that an existence predicate cannot be defined in a positive free logic without identity, and with van Frassen he suggested that free logics, in which the denotation function is partial, provides a very natural foundation for the theory of partial functions. This suggestion has been exploited in various ways by mathematicians (and, also, by him and the computer scientist, R. Gumb). He has used positive free definite description theory to show that unrestricted comprehensionality cannot be derived in elementary set theory. Other contributions include various translation theorems between (positive) free logics and other logics, for example, the formal system called "first order ontology" in the Polish tradition, and various applications to science and philosophy.

**Main Publications:**

1. A Study of Latent Inference Learning, *Canadian Journal of Psychology* 14 (1960), 45-50.
2. Existential import revisited, *Notre Dame Journal of Formal Logic* 4 (1963), 288-292.
3. (with Thomas Scharle) A translation theorem for two systems of free Logic, *Logique et Analyse* 10 (1967), 328-341.
4. (with Bas C. Van Fraassen) On Free Description Theory 13 *Zeitschrift f. Mathematische Logic und Grundlagen der Mathematik* (1967), 225-240.
5. (with Robert K. Meyer.) Universally free logic and standard quantification theory, *Journal of Symbolic Logic* (1968), 33, pp. 8-26.
6. (Editor) *Philosophical Problems in Logic: Some Recent Developments*, Reidel, Dordrecht, Holland (1970): reprinted by Reidel (1980).
7. Predication and extensionality, *Journal of Philosophical Logic* 3 (1974), 255-264.
8. (with Robert K. Meyer and Ermanno Bencivenga) The ineliminability of E! in a free quantification theory without identity, *Journal of Philosophical Logic* 11(1982), 229-231.
9. (with Ermanno Bencivenga.) A free logic with simple and complex predicates, *Notre Dame Journal of Formal Logic* 27 (1986), 247-256.
10. A theory about logical theories of expressions of the form 'the so and so' where 'the' is in the singular, *Erkenntnis*, 35 (1991), 337-346.
11. (Editor) *Philosophical Applications of Free Logic*, Oxford, (1991).
12. Russell's version of the theory of definite descriptions, *Philosophical Studies* 65 (1992), 153-167.
13. A theory of definite descriptions, in *Definite Descriptions*,(Ed. G. Ostertag), M.I.T. Press (1997). A slightly revised compilation of two papers published in *The Notre Dame Journal of Formal Logic* in 1962 and 1963
14. (with Brian Skyrms) Resiliency and laws in the web of belief *Laws of Nature*, (F. Weinert ed), de Gruyter, Berlin (1995), 159- 176.
15. On the reduction of two paradoxes and the philosophical significance thereof,in *Physik, Philosophie und die Einheit der Wissenschaften* (Hrsg. L. Krüger u. B. Falkenburg), Spektrum, Heidelberg (1995), pp. 21-32.
16. (with Raymond Gumb) Definitions in nonstrict positive free logic 7 *Modern Logic* (1997), 25-55
17. *Free Logics: Their Character, Genesis and Some Applications Thereof,* Akademia Verlag, Sankt Augustin bei Bonn (1997).
18. Set theory and definite descriptions, *Grazer Philosophische Studien* 60 (2000), 1 - 11. This paper is a reprint of a Plenary Address to the ASL Section of the AMS-ASL meetings in 2000.
19. Free logics, in *The Blackwell Guide to Philosophical Logic* (L. Goble. Ed.) (2001), Blackwell, London 258-280.
20. *Free Logic: Selected Essays*, Cambridge, The University Press (2003)
*Work in Progress*
21. An investigation of the failure of the substitution of identicals on Frege's pre-analytic principles.
22. A set of lectures on the logical notion of predication.

**Service to the Profession:** Organizer, a conference on predication in science and in philosophical logic, University of California, Irvine, 1991; Advisory Committee Member, Program in Logic and Philosophy of Science at the International Center in Dubrovnik, Jugoslavia, Croatia, 1985-

1989; Member, Chair, Committee on Academic Personnel, University of California, Irvine, 1981-1984; Developer (along with Paul Weingartner), exchange program in Philosophy and in Logic and Philosophy of Science between University of California, Irvine and University of Salzburg, 1975–; Advisor, Internationales Forschungszentrum fuer Grundfragen der Wissenschaften, Austria, 1970-1996; Organizer, Colloquium on free and modal logic, University of California, 1969; Member, ASL 1956-.

**Teaching:** Lambert has taught at the University of Alberta, 1956–1963, at West Virginia University, 1963-1967, where he won a distinguished teaching award, and at the University of California, Irvine 1967–1996. In Europe he has taught virtually continuously from 1975 until retirement in 1996 at the University of California, Irvine. He has been a visiting professor of logic and/or philosophy (and a frequent colloquium member) at many universities in Austria, France, Germany, Italy, Spain, and Norway. He has been the recipient of two Festschriften: (1) *Existence and Explanation. Essays Presented in Honor of Karel Lambert*, (W. Spohn, B.Skyrms and B. van Fraassen, eds.) Kluwer, Dordrecht (1991), and (2) New *Essays in Free Logic. In Honour of Karel Lambert*, (E. Morscher and A. Hieke, eds.) Kluwer, Dordrecht, (2001). Among his best known students are Bas van Fraassen (undergraduate), David Thompson (MA), Robin Dwyer (MA), Ronald Scales, Daniel Hunter and Michael Byrd (PhDs). He has also served on many Doctoral committees in both the U.S. and in Europe.

**Vision Statement:** Philosophical applications, and implications, of free logics will endure. Since the mathematical foundations of the subject are rather well established, the question arises: where will its most fruitful applications be sought? The answer is: probably in computer science. Applications there have hardly scratched the surface of possibilities. He also believes that logic and its applications could profit from more joint work, an attitude encouraged by his early work in the sciences.

**Honours and Awards:** A workshop honoring Lambert (the "Lambertfest") on his $75^{th}$ birthday was held at the University of California on April $12^{th}$, 2003. Among the speakers were Kit Fine, Wolfgang Spohn and Bas van Fraassen. University of California Presidential Fellowship in the Humanities, 1993; Inaugural Lectures in Logic and Philosophy of Science, Bielefeld, 1992; Honorary Professorship, Universitaet Salzburg, 1984; Fulbright-Hays Senior Fellow, 1980; Medal of the College de France, 1980; NEH senior Fellow, 1979.

# LAMPORT, Leslie

**Specialties:** Specification and verification of concurrent systems.

**Born:** 7 February 1941 in New York, New York, USA.

**Educated:** PhD Mathematics, 1972; Brandeis University, MA Mathematics, 1963; MIT, BS Mathematics, 1960.

**Dissertation:** *The Analytic Cauchy Problem with Singular Data*; supervisor, Richard Palais.

**Regular Academic or Research Appointments:** SENIOR RESEARCHER, MICROSOFT RESEARCH, 2001–; Software Engineer, Digital Equipment Corporation, Compaq, 1985-2001; Researcher, SRI International, 1977-1985; Researcher, Massachusetts Computer Associates, 1970-1977; Faculty Member, Marlboro College, 1965-1969.

**Visiting Academic or Research Appointments:** Visiting Professor, Ecole Normale Superieur, Paris, 1995, 2001.

**Research Profile:** Leslie Lamport's primary research has been in the field of concurrent and distributed algorithms. The problem of ensuring the correctness of these algorithms led him to develop methods for reasoning about them based on invariance and temporal logic. These led to research into the formal specification of concurrent systems. He invented the temporal logic of actions (TLA), a form of temporal logic that makes practical the use of formal mathematics to specify concurrent systems. He designed TLA+, a complete specification language based on TLA. In attempting to reason formally about concurrent systems, he developed a method of hierarchically structuring proofs that can be used to make ordinary informal mathematical proofs easier to read and less error-prone.

**Main Publications:**

1. Proving the Correctness of Multiprocess Programs. IEEE Transactions on Software Engineering SE-3, 2 (March 1977), 125-143.

2. 'Sometime' is Sometimes 'Not Never'. Proceedings of the Seventh ACM Symposium on Principles of Programming Languages, ACM SIGACT-SIGPLAN (January 1980).

3. Proving Liveness Properties of Concurrent Programs (with Susan Owicki). ACM Transactions on Programming Languages and Systems 4, 3 (July 1982), 455-495.

4. What Good Is Temporal Logic? Information Processing 83, R. E. A. Mason, ed., Elsevier Publishers (1983), 657-668.

5. On Interprocess Communication–Part I: Basic Formalism. Distributed Computing 1, 2 (1986), 77-85.

6. The Existence of Refinement Mappings (with Martín Abadi). Theoretical Computer Science 82, 2 (May 1991), 253-284.

7. How to Write a Proof. American Mathematical Monthly 102, 7 (August-September 1993) 600-608.

8. The Temporal Logic of Actions. ACM Transactions on Programming Languages and Systems 16, 3 (May 1994), 872-923.

9. Specifying Systems: The TLA+ Language and Tools for Hardware and Software Engineers. Addison-Wesley (2002).

A full list of publications can be found through a link from Lamport's home page: http://lamport.org

**Service to the Profession:** Editorial Board Member, Applied Mathematics Letters; Associate Editor, Distributed Computing; Associate Editor, Science of Computer Programming; Associate Editor, ACM Transactions on Programming Languages and Systems.

**Vision Statement:** I would like to see computer scientists use simple first-order logic and set theory instead of inventing needlessly complicated, esoteric formalisms.

**Honours and Awards:** Honorary Doctorate, University of Lugano, 2006; Edsger W. Dijkstra Prize in Distributed Computing, 2005; IEEE Piore Award, 2004; Honorary Doctorate, Ecole Polytechnique Fédérale de Lausanne, 2004; Honorary Doctorate, Christian Albrechts University, Kiel, 2003; Honorary Doctorate, University of Rennes, 2003; PODC Influential Paper Award, 2000; Member, National Academy of Engineering, 1991-.

# LAVER, Richard Joseph

**Specialties:** Set theory.

**Born:** 20 October 1942.

**Educated:** University of California at Berkeley, PhD Mathematics, 1969; University of California, Los Angeles, BA, 1964.

**Dissertation:** *Well-quasi-orderings and Order Types Theories*; supervisor, Ralph McKenzie.

**Regular Academic or Research Appointments:** PROFESSOR, UNIVERSITY OF COLORADO, PROFESSOR, 1980–; Associate Professor, 1977-1980, Assistant Professor, 1974-1977; Acting Assistant Professor, University of California, Los Angeles, 1971-1973; Lecturer, University of Bristol, England, 1969-1970.

**Visiting Academic or Research Appointments:** NATO Research Project, Paris, Toronto and Boulder, 1996; Research Fellow, California Institute of Technology, 1989; University of California, Berkely, 1974; University of California, Los Angeles, 1973; Postdoctoral Fellow, University of Bristol, 1970-1971.

**Research Profile:** Richard Laver's results (in the linked fields of large cardinals, forcing and combinatorics) include the following.

The proof of Fraïssé's conjecture on linear orderings; a theorem on decomposing members of that class of linear orderings; the consistency of Borel's conjecture on strong measure zero sets; the consistency that there is an indestructibly supercompact cardinal; the consistency of the $\lambda_2^1$-Souslin hypothesis with the CH (with S. Shelah); the consistent solution of Erdos* and Hajnal's spolarized partition problem on $\lambda_1^1 \times \lambda_2^1$; the proof of the infinite Halpenn-Lauchli* conjecture on products of perfect trees; work showing that random reals need not add Souslinn* trees; consistent solutions of Erdos* and Hajnal's downwards transfer problem for $\lambda_2^1$ (with M. Foreman); theorems about Saks forcing (with J. Baumgartner, T. Carlson, and M. Groszek); theorems about nonregular ultrafilters; theorems about precipitous ideals; the left distributive law $[a(bc) = (ab)(ac)]$ and its relation to elementary embeddings witnessing very large cardinal axioms; a division algorithm for the free one-generated left distributive algebra; Dehornoy's connection between left-distributive algebras and the braid groups and a proof that his resulting linear ordering of the braid group, when restricted to the puritive* braids, is a well ordering; a theorem, involving a result of J. Steel, on the abovementioned large cardianal LD * algebras and a consequence of this theorem for finite combinatorics; results on strong implications between vary large cardinal axioms, and a result on noncreation of very large cardinals in small forcing extensions.

In 2001, Laver discovered a new class of large cardinal axioms, stronger than the standard ones — they postulate self-elementary embeddings on models satisfying the $\lambda$-version of $AD_{1R}$, thus generating W. Woodin's well known results about models satisfying the $\lambda$-version of AD. Laver has proved that one of the stronger of these new axioms is not preserved under small Cohen extensions; this is an example of a step which is needed if one is to solve the continuum hypothesis using large cardinals.

**Main Publications:**

1. On Fraïssé's order type conjecture, *Annals of Math* 93 (1971), 89-111.
2. An order type decomposition theorem, *Annals of Math* 98 (1973), 96-119.
3. Partition relations for uncountable cardinals $\leq 2^{\aleph_0}$, *Colloquia Mathematica Societatis Janos Bolgei 10*, North Holland (1975), 1029-1042.
4. On the consistency of Borel's conjecture, *Acta Mathematica* 137 (1976), 151-169.
5. Making the supercompatness of $k$ indestructible under $k$-directed-closed forcing, *Israel Journal of Math* 29 (1978), 385-388.
6. Better-quasi-orderings and a class of trees, *Advances in Math Supplementary Studies I*, Academic Press (1978), 31-48.
7. The $\lambda_2^1$-Souslin hypothesis, *Trans. Amer. Math. Soc.* 264 (1981), 411-417 (with S. Shelah).
8. Saturated ideals and nonregular ultrafilters, *Patras Logic Symposium*, North Holland (1982), 297-305.
9. An $(\lambda_2^1, \lambda_2^1, \lambda_0^1)$-saturated ideal on $w_1$, *Logic Colloquium 1980* (Prague), North Holland (1982), 173-180.
10. Products of infinitely many perfect trees, *Journal of the London Mathematical Society* 29 (1984), 385-396.
11. Precipitousness in forcing extensions, *Israel Journal of Math* 48 (1984), 97-108.
12. Random reals and Souslin trees, *Proc. Amer. Math. Soc.* 100 (1987), 531-534.
13. Some downwards transfer properties for $\aleph_2$, *Advances in Math* 67 (1988), 230-238 (with M. Foreman).
14. Sack's reals and Martin's Axiom, *Fundamenta Mathematica* 133 (1989), 161-168 (with T. Carlson).
15. The left-distributive law and the freeness of an algebra of elementary embeddings, *Advances in Math* 91 (1992), 209-231.
16. A division algorithm for the free left-distributive algebra, *Logic Colloquium '90*, Springer (1993), 155-162.
17. The algebra of elementary embeddings of a rank into itself, *Advances in Math* 110 (1995), 334-346.
18. Brail group actions on left-distributive structures, and well orderings in the braid groups, *Jour. Pure and Applied Algebra* 108 (1996), 81-98.
19. Implication between strong large cardinal axioms, *Annals of Pure and Applied Logic* 90 (1997), 79-90.
20. Reflections of elementary embedding axioms on the $L[v_{\lambda+1}]$ hierarchy, *Annals of Pure and Applied Logic* 107 (2001), 227-238.

*Work in Progress*

21. Certain very large cardinals are not created in small forcing extensions, *Annals of Pure and Applied Logic*, to appear.
22. In reference 10 is a conjecture (the strong infinite Halpern-Lauchli problem) strengthening the results of that paper. Laver has proved a result in the direction of that strengthening, from a measurable cardinal.
23. New large cardinal axioms, in preparation (see research profile).

**Teaching:** Laver has had seven PhD students. Achievements of this group include Steve Grantham's analysis of a many-board version of Galvin's racing pawn game, Carl Danby's solution of a 20-year-old Erdös partition problem, Janet Barnett's solution of a problem of Herink on Martin's axiom, Manny Knill's career as one of the world's experts on quantum computations, David Larne's shorter proof of Dehornoy's irreflexivity theorem and his analysis of left-distributivities plus idempotence, Rene Schipperus' work on countable partitions which improves Darby's results and which won Rene the award for best PhD thesis of 1999 in logic (joint with one other), and Sheila Miller's extension of the division algorithm for a one-generated free left-distribution algebra to a many-generated one.

**Vision Statement:** It was Gödel's vision that every mathematical statement can be decided in ZFC plus a strong enough large cardinal axiom. For the continuum hypothesis, there is hope hat certain large cardinal axioms, stronger than the standard one, will decide.

# LEHMANN, Daniel

**Specialties:** Modal and temporal logic, nonmonotonic logic, belief revision and update, quantum logic.

**Born:** 29 April 1946 in Boulogne-Billancourt (France)

**Educated:** Hebrew University of Jerusalem, Ph.D. Computer Science 1978, M. Sc. Computer Science 1971; Universite de Paris, DEA Numerical Analysis 1969; Ecole Polytechnique (Paris) Diploma in Engineering 1967.

**Dissertation:** *Categories for fixpoint Semantics*; supervisor Eliahu Shamir.

**Regular Academic or Research Appointments:** PROFESSOR OF COMPUTER SCIENCE, EMERITUS, HEBREW UNIVERSITY, 2006–; Professor of Computer Science, Hebrew University, 1991–2006; Associate Professor of Computer Science, Hebrew University, 1984-1991; Senior Lecturer in Computer Science, Hebrew University, 1979–1984; Assistant Professor of Mathematics, University of Southern California, 1977–1979; Senior Research Fellow in Computer Science, University of Warwick (UK), 1975–1977; Visiting Assistant Professor in Applied Mathematics, Brown University, 1974–1975; Instructor in Computer Science, Hebrew University, 1971–1974.

**Visiting Academic or Research Appointments:** Visiting Professor of Computer Science, Paris 2 and Paris 11 Winter and Spring 2006; Visiting Researcher, Dept. of Computing King's College, London, Fall 2005; Fellow of the Institute for Advanced Studies, Jerusalem, 2004–2005; Visiting Professor of Computer Science, Stanford University, 1998–1999; Visiting Researcher, Electronics Lab, Ecole Superieure de Physique et Chimie Industrielles de la ville de Paris, Fall 1992; Visiting Professor at LITP, CNRS, Paris, 1991–1992; Associate Researcher at LITP, CNRS, Paris, Summer and Fall 1986; Visiting Associate Professor in Computer Science, Brandeis University, 1984–1985; Visiting Scientist at Mathematical Centre, Amsterdam, Summer 1981.

**Research Profile:** Daniel Lehmann has contributed to many sub-areas of Computer Science and only that part of his work which is connected to Logic will be described here. In an effort to formalize proofs of correctness for programs and algorithms, he has developed and studied modal logics that include modalities for more than one modal dimension: temporal and epistemic modalities, or temporal and probabilistic modalities. Such systems were shown capable of a formal description of temporal epistemic puzzles, such as the "muddy foreheads puzzle".

He is best known for his seminal work, with collaborators, on the axiomatization and proof theory of nonmonotonic logic, the sort of logic found necessary by Artificial Intelligence to describe commonsense reasoning. Such logics are intimately connected with Conditional Logic and with Theory Revision. Three sets of properties, defining respectively Cumulative, Preferential and Rational logics were identified as central in the classification of the many nonmonotonic logical systems. He initiated the study of the relation of those properties to patterns of human commonsense reasoning. A particularly original aspect of cumulative logic is that, even in the absence of connectives it is non-trivial. Engesser and Gabbay proposed, in 2002, to look at Quantum Logic as a nonmonotonic logic. Daniel Lehmann is currently developing this insight in an effort to provide a logical view meaningful for Physics. He is pursuing the study of the (nonmonotonic) logic defined by projections in a Hilbert space and of more general structures suitable for Quantum Physics. He has also contributed to Algorithmic Mechanism Design.

**Main Publications:**

1. "Reasoning with Time and Chance", with Saharon Shelah, Information and Control, 53(3) (June 1982) pp. 165-198.
2. "Knowledge, Belief and Time", with Sarit Kraus, Theoretical Computer Science, 58(1-3) (June 1988) pp. 155-174.
3. "Nonmonotonic Reasoning, Preferential Models and Cumulative Logics", with Sarit Kraus and Menachem Magidor, Artificial Intelligence, 44(1-2), (July 1990) pp. 167-207.
4. "Rationality, Transitivity and Contraposition", with Michael Freund and Paul Morris, Artificial Intelligence, 52(2) (Dec. 1991) pp. 191-203.
5. "What does a Conditional Knowledge Base Entail?", with Menachem Magidor, Artificial Intelligence, 55(1) (May 1992) pp. 1-60.
6. "Nonmonotonic Inference Operations", with Michael Freund, Bulletin of the IGPL, 1(1) (July 1993) pp. 23-68.
7. "Nonmonotonic Reasoning: from Finitary Relations to Infinitary Inference Operations", with Michael Freund, Studia Logica, 53(2) (1994) pp. 161-201.
8. "Deductive Nonmonotonic Inference Operations: Antitonic Representations", with Yuri Kaluzhny, Journal of Logic and Computation, 5(1) (1995) pp. 111-122.
9. "Another Perspective on Default Reasoning", Annals of Mathematics and Artificial Intelligence, 15(1) (1995) pp. 61-82.
10. "On Negation Rationality", with Michael Freund, Journal of Logic and Computation, 6(2) (1996) pp. 263-269.
11. "Stereotypical Reasoning: Logical Properties", Logic Journal of the IGPL, 6(1) (1998) pp. 49-58.
12. "Introducing the Mathematical Category of Artificial Perceptions", with Zippora Arzi-Gonczarowski, Annals of Mathematics and Artificial Intelligence, 23 (1998) pp. 267-298.
13. "From Environments to Representations – A Mathematical Theory of Artificial Perceptions", with Zippora Arzi-Gonczarowski, Artificial Intelligence, 102(2) (1998) pp. 187-247.
14. "Preferred History Semantics for Iterated Updates", with Shai Berger and Karl Schlechta, Journal of Logic and Computation, 9(6) (1999) pp. 817-833.
15. "Distance Semantics for Belief Revision", with Menachem Magidor and Karl Schlechta, Journal of Symbolic Logic, 66(1) (March 2001) pp. 295-317.
16. "Nonmonotonic Logics and Semantics", Journal of Logic and Computation, 11(2) (June 2001) pp. 229-256.
17. "Algebras of Measurements: the Logical Structure of Quantum Mechanics", with Kurt Engesser and Dov M. Gabbay, International Journal of Theoretical Physics, 45(4) (April 2006) pp. 698-723.
18. "A Presentation of Quantum Logic Based on an *and then* Connective", Journal of Logic and Computation 18(1) (Feb. 2008) pp.59-76. DOI: 10.1093/logcom/exm054.
19. "Connectives in Cumulative Logic" in Pillars of Computer Science, Essays dedicated to Boris (Boaz) Trakhtenbrot on the occasion of his $85^{th}$ birthday (Avron and al. eds) pp. 424-440. LNCS 4800, Springer-Verlag, 2008.
20. "Quantic Superpositions and the Geometry of

Complex Hilbert Spaces", International Journal of Theoretical Physics 47(5) (May 2008) pp. 1333-1353.

21. A new approach to Quantum Logic with K. Engesser and D. M. Gabbay, Studies in Logic, Mathematical Logic and Foundations Volume 8, College Publications 2007.

A full list of publications and links to the articles can be found on Lehmann's home page: http://www.cs.huji.ac.il/$\sim$lehmann

*Work in Progress*

22. "Similarity-Projection structures: the logical geometry of Quantum Physics", submitted to International Journal of Theoretical Physics.

**Service to the Profession:** Editor, with K. Engesser and D. M. Gabbay of the Handbook of Quantum Logic and Quantum Structures, 2007– (Elsevier); Editorial Board member, Journal of Applied Logic; Lecturer, Workshop on Logic, Language, Information and Computation, Fortaleza (Brasil), 1997; Lecturer, $5^{th}$ European Summer School in Logic, Language and Information, Lisbon (Portugal), 1993.

**Teaching:** Lehmann has had 9 Ph.D. students in Artificial Intelligence, Algorithms and Theoretical Computer Science, among them Sarit Kraus and Chaim (Craig) Gotsman.

**Vision Statement:** Computer Science has been the main driving force behind the development of new logical systems in the last 30 years. I believe logical methods still have a lot to offer to CS, in the areas of program specification and proofs of correctness and commonsense reasoning. But the time is ripe for CS to repay its debt to Logic in using the insights gained on many exotic logical systems to questions of interest to other scientific disciplines.

# LEMPP, Steffen

**Specialties:** Mathematical logic, esp. computability theory and its applications to model theory, algebra, proof theory and computer science.

**Born:** 1959 in Germany.

**Educated:** University of Chicago, PhD, 1986; University of Chicago, MS, 1983; University of Chicago, 1981-1986; University of Karlsruhe and University of Bonn, Germany, 1978-1981.

**Dissertation:** *Topics in recursively enumerable sets and degrees*; supervisor, Robert I. Soare.

**Regular Academic or Research Appointments:** PROFESSOR, UNIVERSITY OF WISCONSIN-MADISON, 1996-; Associate Professor, University of Wisconsin-Madison, 1992-1996; Assistant Professor, University of Wisconsin-Madison, 1988-1992; Gibbs Instructor, Yale University, 1986-1988.

**Visiting Academic or Research Appointments:** Member, Institute for Mathematical Sciences, National University of Singapore, 2005; Mercator Guest Professorship, University of Heidelberg, 2002-2003; CNR Research Visit, Sienna, Italy, 1998; Sabbatical, University of Leeds, England, 1996.

**Research Profile:** Steffen Lempp's primary research interest is computability in its various aspects, both in classical computability theory, in particular degree structures, and in applications of computability to model theory, algebra, proof theory, and computer science. Some particular problems I am currently working on include algebraic structures of, and decidability and fragments of theories of, degree structures (esp. the c.e. Turing degrees, the d.c.e. and n-c.e. Turing degrees, and the $\Sigma^0_2$ enumeration degrees), lattice embeddings into the computably enumerable Turing degrees, computable linear orders and Boolean algebras, characterization of computable algebraic structures by classical invariants (e.g., Ketonen invariants and Ulm invariants), degrees of models of $\aleph_1$-categorical theories, proof-theoretical strength of Ramsey's Theorem for pairs and related combinatorial principles.

**Main Publications:**

1. A high strongly noncappable degree, Journal of Symbolic Logic, Vol. 53, 1988, pp. 174-187.
2. A limit on relative genericity in the recursively enumerable degrees, with Theodore A. Slaman, Journal of Symbolic Logic, Vol. 54, 1989, pp. 376-395.
3. The d.r.e. degrees are not dense, with S. Barry Cooper, Leo Harrington, Alistair H. Lachlan, and Robert I. Soare, Annals of Pure and Applied Logic, Vol. 55, 1991, pp. 125-151.
4. The existential theory of the poset of r.e. degrees with a predicate for single jump reducibility, with Manuel Lerman, Journal of Symbolic Logic, Vol. 57, 1992, pp. 1120-1130.
5. Lattice embeddings into the r.e. degrees preserving 1, with Klaus Ambos-Spies and Manuel Lerman, in: "Logic and Philosophy of Science: Papers from the 9th International Congress of Logic, Methodology, and Philosophy of Science", D. Prawitz and D. Westerst°ahl eds., Kluwer Academic Publishers, Dordrecht, Boston, 1994, pp. 179-198.
6. There is no plus-capping degree, with Rodney G. Downey, Archive for Mathematical Logic, Vol. 33, 1994, pp. 109-119.
7. A general framework for priority arguments, with Manuel Lerman, Bulletin of Symbolic Logic, Vol. 1, 1995, pp. 189-201.

8. Interpolating d.r.e. and REA degrees between r.e. degrees, with Marat M. Arslanov and Richard A. Shore, Annals of Pure and Applied Logic, Vol. 78, 1996, pp. 29-56.

9. An extended Lachlan Splitting Theorem, with Sui Yuefei, Annals of Pure and Applied Logic, Vol. 79, 1996, pp. 53-59.

10. Decidability of the two-quantifier theory of the recursively enumerable weak truthtable degrees and other distributive upper semi-lattices, with Klaus Ambos-Spies, Peter A. Fejer, and Manuel Lerman, Journal of Symbolic Logic, Vol. 61, 1996, pp. 880-905.

11. Infinite versions of some problems from finite complexity theory, with Jeffry L. Hirst, Notre Dame Journal of Formal Logic, Vol. 37, 1996, pp. 545-553.

12. On isolating r.e. and isolated d.r.e. degrees, with Marat M. Arslanov and Richard A. Shore, Annals of Pure and Applied Logic, Vol. 78, 1996, pp. 29-56.

13. Iterated trees of strategies and priority arguments, with Manuel Lerman, Archive of Mathematical Logic, Vol. 36, 1997, pp. 297-312.

14. Contiguity and distributivity in the enumerable Turing degrees, with Rodney G. Downey, Journal of Symbolic Logic, Vol. 62, 1997, pp. 1215-1240; with Corrigendum, Journal of Symbolic Logic, Vol. 67, 2002, pp. 1579-1580.

15. A finite lattice without critical triple that cannot be embedded into the enumerable Turing degrees, with Manuel Lerman, Annals of Pure and Applied Logic, Vol. 87, 1997, pp. 167-185.

16. Decidability and undecidability in the enumerable Turing degrees, in: "Proceedings of the Sixth Asian Logic Conference, Beijing, China", C. T. Chong, Q. Feng, D. Ding, Q. Huang, M. Yasugi eds., World Scientific, Singapore University Press, Singapore, 1998, pp. 151-161.

17. Infima in the recursively enumerable wtt-degrees, with Richard Blaylock and Rodney G. Downey, Notre Dame Journal of Formal Logic, Notre Dame Journal of Mathematics, Vol. 38, 1997, pp. 406-419.

18. Friedberg numberings of families of n-computably enumerable sets, with Sergey S. Goncharov and D. Reed Solomon, Algebra and Logic, Vol. 41, 2002, pp. 81-86.

19. The Lindenbaum algebra of the theory of the class of all finite models, with Mikhail G. Peretyat'kin and D. Reed Solomon, Journal of Mathematical Logic, Vol. 2, 2002, pp. 145-225.

20. Computable categoricity of trees of finite height, with Charles F. D. McCoy, Russell G. Miller, and D. Reed Solomon, Journal of Symbolic Logic, Vol. 70, 2005, pp. 151-215.

A full list can be found at http://www.math.wisc.edu/$\sim$lempp/.

**Service to the Profession:** Programm Committee Member, conference on "Theory and Applications of Models of Computation 2006", Beijing, China, 2006; Co-organizer, special session in model theory and computability theory, AMS Sectional Meeting, Notre Dame, 2006; Program Committee member, ASL European Summer Meeting, Torino, Italy, 2004; Logic Editor, Transactions of the AMS and Memoirs of the AMS, 2003-; Co-organizer, Workshop on Computability and Logic, Heidelberg, Germany, 2003; Co-organizer, special session in computability theory, ASL European Summer Meeting, M¨unster, Germany, 2002; Co-organizer, Oberwolfach Meeting in Computability Theory, 2001; Editor, Lecture Notes in Logic, 1999-; Program Committee Member, International Conference on Mathematical Logic, 1999; Executive Committee Member, Association for Symbolic Logic, 1999-2001; Co-organizer, Special Session on Computability Theory, AMS, 1997; Co-organizer, Special Session on Computability Theory, European Logic Colloquium, 1997; Co-organizer, Workshop on Recursion and Complexity Theory, 1997; Organizing Committee and Co-Chair, Annual Meeting, Association for Symbolic Logic, 1996; Chair, ASL Committee on Translations and ASL Subcommitte on Translations from Russian and Other Slavic Languages, 1995-2001; Program Committee Chair, Association for Symbolic Logic, Winter, 1993; Editor, Journal of Symbolic Logic, 1993-1998, Coordinating Editor, 1996-1997; Founder and Organizer, Southern Wisconsin Logic Colloquium, 1990-;

**Teaching:** Doctoral theses supervised include Peter A. Cholak, Deborah S. Kaddah, Michael A. Jahn, Lisa R. Galminas and Steven D. Leonhardi.

**Honours and Awards:** Vilas Research Award (University of Wisconsin–Madison), 2000-2002; U.S.-New Zealand Binational NSF Grant with Rodney G. Downey, Richard A. Shore, and Michael Stob, 199-1995; Postdoctoral Fellowship, Mathematical Sciences Research Institute, Berkeley, California, 1989-1990; Individual NSF Grants, 1987-2008.

## LEVI, Isaac

**Specialties:** Probability logic, decision theory, belief logics, conditional logics.

**Born:** 30, June, 1930 in New York, New York, USA.

**Educated:** Columbia University, PhD, 1957, New York University, BA, 1951.

**Dissertation:** *The Epistemology of Moritz Schlick*; supervisor, Ernest Nagel.

**Regular Academic or Research Appointments:** JOHN DEWEY PROFESSOR OF PHILOSOPHY EMERITUS, COLUMBIA UNIVERSITY, 2003–; John Dewey Professor of Philosophy, Columbia

University, 1992-2003; Chairman, Philosophy Department, Columbia University, 1973-1976, 1982-1992; Professor, Philosophy, Columbia University, 1970-1992; Professor, Philosophy, Case Western Reserve University, 1967-1970; Chairman, Philosophy Department, Case Western Reserve University, 1968-1970; Associate Professor, Philosophy, Case Western Reserve University, 1964-1967; Assistant Professor, City College of New York, 1962-1964; Assistant Professor, Philosophy, Western Reserve University, 1958-1962; Instructor, Philosophy, Case Western Reserve University, 1957-1958.

**Visiting Academic or Research Appointments:** Visiting Professor of Philosophy, Carnegie Mellon University, 2000; Visiting Fellow, Wolfson College, Cambridge University, 1997; Visiting Fellow, Institute of Advanced Studies of the Hebrew University of Jerusalem, 1984; Visiting Scholar, Darwin College, Cambridge University, 1993; Visiting Fellow, All Souls College, Oxford University, 1988; Visiting Scholar, Darwin College, Cambridge University, 1980; Visiting Fellow, RSSS, Australian National University, 1987; Visiting Scholar, Corpus Christi College, Cambridge University, 1972-1973; Guggenheim Fellow and Fulbright Research Scholar, Department of Philosophy, The London School of Economics, 1966-1967.

**Research Profile:** The guiding preoccupation of most of Levi's work has been the elaboration of a normative model of problem solving inquiry where the object of justification is not the current point of view but change in point of view and justification is decision theoretic. The account elaborates on some of the fundamental insights of the Charles Peirce and John Dewey.

Levi proposed an account of rational choice that provides for rational choice that departs from wide spread requirement that preference for one option over another should be independent of the other options available. States of probability judgment are represented by convex sets of probability functions and evaluations of consequences of options are represented by convex sets of utility functions. Procedures are explored for matching probability and utility functions for the purpose of deriving expected utility functions for options. The evaluation of options is then representable by a set of expected utility functions. An option is admissible for choice if and only if it optimal according to some expected utility function from such a set. Additional criteria may then be invoked to reduce the set of E-admissible options to some subset.

The theory of rational choice thus elaborated is then deployed to provide an account of inductive expansion and of contraction of states of full belief for efforts to obtain new error-free and valuable information. Levi claims that the common features of scientific inquiry ought to be committed to the pursuit of such goals. The account of inductive expansion (a primitive version of which was first presented in *Gambling with Truth* in 1967) generates a family of inductive expansion rules that are nonmonotonic and parameterized by a boldness parameter. This family differs significantly from those generally found in the literature. Likewise the theory of contraction departs from the proposals commonly found in the literature. The relevance of these proposals to an account of conditionals is explored. A systematically epistemic account of both so called "indicative" and "subjunctive" conditionals is developed as an alternative to the widely received closest worlds accounts.

These and other technical ideas are deployed to elaborate a version of the belief-doubt model of inquiry pioneered by the classical American Pragmatists. Levi has, in addition, commented on the contributions of C.S. Peirce to theories of statistical inference and on the ideas of John Dewey concerning inquiry in ethics.

Levi has written on the similarities and differences between his account of individual decision making and Sen's views on social choice and has published critiques of other authors who have written on epistemology, decision making, and probability judgment.

**Main Publications:**

1. *Gambling with Truth* Knopf (1967) reissued in paper in 1973 by MIT Press.
2. *The Enterprise of Knowledge*, MIT (1980). Paperback (1983).
3. *Decisions and Revisions*, Cambridge University Press (1984).
4. *Hard Choices*, Cambridge University Press (1986). Paperback (1990).
5. *The Fixation of Belief and Its Undoing*, Cambridge University Press (1991).
6. *For the Sake of the Argument*, Cambridge University Press (1996).
7. *The Covenant of Reason*, Cambridge University Press (1997).
8. *Mild Contraction*: Oxford University Press (2004).

**Service to the Profession:** Editorial Board Member, *Journal of Philosophy*.

**Teaching:** During my career I have taught a wide range of courses including courses in elementary logic and ethics and philosophy of science. In later years, my teaching became restricted to decision theory, probability and induction and theory of knowledge at several levels of sophistication. My best students, most notably Teddy Seidenfeld and

Horacio Arló Costa, are far more accomplished masters of logic, probability theory and philosophical logic than I am and have been constant sources of stimulation for my own thinking.

**Vision Statement:** Although I have been engrossed with topics related to philosophical and applied logic, I have never thought of myself as a logician. I rarely even try to prove representation theorems. I do not write computer programs. I am not a statistician nor am I an economic modeler. But my interests demand that I have some familiarity with the achievements of others concerning these matters. I cannot, however, forecast future developments in Logic.

**Honours and Awards:** PhD *honoris causa*, University of Lund, 1988; Member Elect, American Academy of Arts and Sciences, 1986; Medal, University of Helsinki, 1980.

## LEVY, Azriel

**Specialties:** Set theory, mathematical logic

**Born:** 14 December 1934 in Haifa, Israel.

**Educated:** Hebrew University of Jerusalem, MSc Mathematics, 1956; PhD Mathematics, 1958.

**Dissertation:** *Contributions to the metamathematics of set theory*; supervisor, Abraham A. Fraenkel.

**Regular Academic or Research Appointments:** PROFESSOR OF MATHEMATICS, HEBREW UNIVERSITY, 1968-2002; Associate Professor of Mathematics, Hebrew University, 1964-1968; Senior Lecturer of Mathematics, Hebrew University, 1961-1964.

**Visiting Academic or Research Appointments:** Visiting Professor, University of California at Los Angeles, 1976-1977; Visiting Professor, Yale University, 1971-1972; Visiting Associate Professor, Stanford University, 1965-1966; Visiting Assistant Professor, University of California at Berkeley, 1959-1961; Foreign Postdoctoral Sloan Fellow, Massachusetts Institute of Technology, 1958-1959.

**Research Profile:** Azriel Levy did fundamental work in set theory when it was transmuting into a modern, sophisticated field of mathematics, a formative period of over a decade straddling Cohen's 1963 founding of forcing. His work is on the following subjects: the use of principles of reflection to characterize large cardinals, interdependence of versions of the axiom of choice, the quantifier hierarchy of the formulas of set theory, the definability and the logical complexity of various notions of set theory.

**Main Publications:**

1. On Ackermann's set theory, *The Journal of Symbolic Logic 24 (1959)*, 154-166.
2. Axiom schemata of strong infinity in axiomatic set theory, *Pacific Journal of Mathematics 10 (1960)*, 223-238.
3. A hierarchy of formulas in set theory, *Memoirs of the American Mathematical Society* 57 (1965), 76pp.
4. Definability in axiomatic set theory I, in: Yehoshua Bar-Hillel (editor), *Logic, Methodology and Philosophy of Science, Proceedings of the 1964 International Congress at Jerusalem*, North-Holland, Amsterdam, 1965, 127-151.
5. The effectivity of existential statements in axiomatic set theory, *Information Sciences 1 (1969)*, 119-130.
6. Basic Set Theory, Springer-Verlag, Berlin, 1979, 391 pp.; reprinted by Dover Publications, 2003.

**Service to the Profession:** Editor, *Archive of Mathematical Logic*, 1988-2003; Chair, Union of Professors and Senior Lecturers at the Hebrew University, 1978-1980, 1990-1993; Director, Landau Research Institute in Mathematical Analysis, 1987-1989; Chair, Institute of Mathematics and Computer Science, Hebrew University, 1984-1987; Associate Dean for Instruction, Faculty of Mathematics and Natural Sciences, Hebrew University, 1981-1984; Chair, Coordinating Council of the Academic Staff Unions at the Universities in Israel, 1978-1980; Dean, Faculty of Mathematics and Natural Sciences, Hebrew University, 1972-1975; Editor, *Journal of Symbolic Logic*, 1966-1972; Editor, *Israel Journal of Mathematics*, 1968-1970; Chair, Institute of Mathematics, Hebrew University, 1968-1970.

**Teaching:** Directed and co-directed ten PhD theses in the fields of Logic, Set theory and Mathematical Education. He has taught many mathematics courses, especially on logic, set theory, and linear algebra.

## LIFSCHITZ, Vladimir

**Specialties:** Commonsense and nonmonotonic reasoning, logic programming.

**Born:** 30 May 1947 in Leningrad, USSR (now St. Petersburg, Russia).

**Educated:** Leningrad State University (now St. Petersburg State University), Mathematics degree; Leningrad Branch (now St. Petersburg Branch) of Steklov Mathematical Institute, Kandidat degree Physical and Mathematical Sciences.

**Dissertation:** *Constructive Counterparts of Gödel's Completeness Theorem*; supervisor, Nikolay Shanin.

**Web Address:** http://www.cs.utexas.edu/users/vl

**Regular Academic or Research Appointments:** GOTTESMAN FAMILY CENTENNIAL PROFESSOR IN COMPUTER SCIENCES, UNIVERSITY OF TEXAS AT AUSTIN, 1990–. Senior Research Associate, Stanford University, 1985-1990; Associate Professor, University of Texas at El Paso, 1982-1985; Assistant Professor, University of Texas at El Paso, 1979-1982.

**Research Profile:** Vladimir Lifschitz is interested in logic-based artificial intelligence and in computational logic. He studied nonmonotonic reasoning, applied it to reasoning about actions, and designed several declarative languages for representing actions. His work on the semantics of negation in Prolog became a foundation for the declarative approach to combinatorial search problems called answer set programming. He investigated mathematical foundations of this programming method and applied it to computational problems in VLSI design and in historical linguistics. His current work is motivated by the problem of generality in knowledge representation.

**Main Publications:**

1. "Computing circumscription", in *Proceedings of International Joint Conference on Artificial Intelligence (IJCAI-85)*, 1985, 121–127.
2. "Pointwise circumscription", in *Readings in Nonmonotonic Reasoning*, Morgan Kaufmann, 1987, 179–193.
3. "The stable model semantics for logic programming" (with M. Gelfond), in *Logic Programming: Proceedings of the Fifth International Conference and Symposium*, 1988, 1070–1080.
4. "Classical negation in logic programs and disjunctive databases" (with M. Gelfond), *New Generation Computing*, 9 (1991), 365–385.
5. "Answer sets in general nonmonotonic reasoning (preliminary report)" (with T. Woo), in *Proceedings of International Conference on Principles of Knowledge Representation and Reasoning (KR-92)*, 1992, 603–614.
6. "Representing action and change by logic programs" (with M. Gelfond), *Journal of Logic Programming*, 17 (1993), 301–322.
7. "Circumscription", in *Handbook of Logic in Logic Programming and Artificial Intelligence*, Vol. 3, Oxford University Press, 1994, 298–352.
8. "Minimal belief and negation as failure", in *Artificial Intelligence*, 70 (1994), 53–72.
9. "Splitting a logic program", in *Proceedings of International Conference on Logic Programming (ICLP-94)*, 1994, 23–37.
10. "Nested abnormality theories", *Artificial Intelligence*, 74 (1995), 351–365.
11. "Foundations of logic programming", in *Principles of Knowledge Representation*, CSLI Publications, 1996, 69–128.
12. "Nested expressions in logic programs" (with L. R. Tang and H. Turner), *Annals of Mathematics and Artificial Intelligence*, 25 (1999), 369–389.
13. "Missionaries and cannibals in the Causal Calculator", in *Proceedings of International Conference on Principles of Knowledge Representation and Reasoning (KR-2000)*, 2000, 85–96.
14. "Wire routing and satisfiability planning" (with E. Erdem and M. Wong), in *Proceedings of International Conference on Computational Logic (CL-2000)*, 2000, 822–836.
15. "Getting to the airport: the oldest planning problem in AI" (with N. McCain, E. Remolina and A. Tacchella), in: *Logic-Based Artificial Intelligence*, Kluwer, 2000, 147–165.
16. "Strongly equivalent logic programs" (with D. Pearce and A. Valverde), *ACM Transactions on Computational Logic*, 2 (2001), 526–541.
17. "Answer set programming and plan generation", *Artificial Intelligence*, 138 (2002), 39–54.
18. "Reconstructing the evolutionary history of Indo-European languages using answer set programming" (with E. Erdem, L. Nakhleh, and D. Ringe), in: *Proceedings of International Symposium on Practical Aspects of Declarative Languages (PADL-03)*, 2003, 160–176.
19. "Weight constraints as nested expressions" (with P. Ferraris), *Theory and Practice of Logic Programming*, 5 (2005), 45–74.
20. "Why are there so many loop formulas?" (with A. Razborov), *ACM Transactions on Computational Logic*, to appear.

**Service to the Profession:** Member of the Executive Committee, Association for Logic Programming, 2000-2004; Area Editor, *ACM Transactions on Computational Logic*, 1999-; Editorial Advisor, *Theory and Practice of Logic Programming*, 1999-.

**Teaching:** Lifschitz practices the Socratic method of teaching, which aims at educating by means of questions. "Rather than knowing the correct rules of thought theoretically, one must have them assimilated into one's flesh and blood for instant and instinctive use. Therefore, for the schooling on one's powers of thought only the practice of thinking is really useful" (Pólya and Szegö, *Problems and Theorems in Analysys*). "Tell me and I forget. Show me and I remember. Involve me and I understand" (Chinese proverb).

**Vision Statement:** The main achievements of logic so far have been related to the analysis of mathematical proofs. In the future, logicians will be turning more and more to the investigation of commonsense reasoning and reasoning about computational systems.

**Honours and Awards:** A Most Influential Paper in 20 Year Award from the Association for Logic Programming (joint with M. Gelfond), 2004; Teaching Excellence Award from the College of Natural Sciences, University of Texas at Austin, 2000; Fellow, American Association for Artificial Intelligence, 1991; Publishers' Prize, Twelfth International Joint Conference on Artificial Intelligence, 1991; Publishers' Prize, Tenth International Joint Conference on Artificial Intelligence, 1987.

# LOVELAND, Donald

**Specialties:** Computational logic.

**Born:** 26 December 1934 in Rochester, New York, USA.

**Educated:** New York University, PhD Mathematics, 1964; Massachusetts Institute of Technology, SM Mathematics, 1958; Oberlin College, AB Physics, 1956.

**Dissertation:** *Recursively Random Sequences*; supervisor, Martin Davis.

**Regular Academic or Research Appointments:** PROFESSOR OF COMPUTER SCIENCE, EMERITUS, DUKE UNIVERSITY, 2001–; Professor of Computer Science, Duke University, 1973-2001; Chair, Computer Science Department, 1973-1978, 1991-1992, 1998-1999; Associate Professor of Mathematics and Computer Science, Carnegie-Mellon University, 1970-1973; Assistant Professor of Mathematics and Computer Science, Carnegie-Mellon University, 1967-1969; Assistant Professor of Mathematics, New York University, 1964-1969; Mathematician Programmer, IBM T.J. Watson Research Center, 1958-1959.

**Visiting Academic or Research Appointments:** IBM Distinguished Faculty Visitor, IBM T.J. Watson Research Center, 1979-1980.

**Research Profile:** Donald Loveland has made fundamental contributions to the areas of automated deduction and logic programming. He has also made contributions to the areas of recursive function theory, algorithmic complexity theory and expert systems/rule-based reasoning. In 1960, he and George Logemann devised and implemented a variant of the Davis-Putnam first-order logic refutation procedure that has become known as the Davis-Putnam-Logemann-Loveland (DPLL) refutation procedure. He developed the Model Elimination (ME) refutation procedure in the middle 1960s. In 1968 he devised a linear resolution refinement of general resolution. His book on automated theorem proving (entry 11 in the list below) provided a rigorous logic presentation of the resolution variants that had appeared in the literature in the decade after 1965, and included the relationship of ME to linear resolution (co-discovered by Kowalski and Kuehner). The logic programming language Prolog resulted from the restriction of an ME implementation by Roussel and Colmerauer to Horn clauses. Linear resolution provides the basic paradigm for describing Prolog's logic component. In the mid-1980s Loveland proposed near-Horn Prolog as a logic programming system for the disjunctive logic programming subarea. This extended the Prolog concept to full first-order logic (under a normal form mapping). Earlier, his PhD thesis studied the Kleene hierarchy complexity of "recursively random" sequences. He contributed the notion of uniform complexity to the algorithmic complexity field. In the 1980s he made contributions to the algorithms area with work on binary testing and the test-and-treatment problem, and also to the problem of knowledge evaluation within the expert systems area (with coauthor M. Valtorta).

**Main Publications:**

1. Empirical explorations of the geometry-theorem proving machine (with J.R. Hansen and H. Gelernter, principal author), Proc. of the Western Joint Computer Conf., 17 (1960), 143-147.
2. A machine program for theorem-proving (with M. Davis and G. Logemann), Communications of the ACM 5 (1962), 394-397.
3. A new interpretation of the von Mises' concept of random sequence, Zeit. Math. Logik and Grundlagen der Mathematik 12 (1966), 279-294.
4. The Kleene hierarchy classification of recursively random sequences, Transactions of the American Mathematical Society 125 (1966), 497-510.
5. Mechanical theorem-proving by Model Elimination, Jour. of the ACM 15 (1968), 236-251.
6. A simplified format for the Model Elimination theorem-proving procedure, Journal of the ACM 16 (1969), 349-363.
7. A variant of the Kolmogorov concept of complexity, Information and Control 15 (1969), 510-526.
8. On minimal-program complexity measures, Conf. Record of the ACM Symposium on Theory of Computing (1969), 61-65.
9. A unifying view of some linear herbrand procedures, Jour. of the ACM 19 (1972), 366-384.
10. A hole in goal trees: some guidance from resolution theory (with M. Stickel), Proc. of the Third Int'l Joint Conf. on Artificial Intelligence (1976), 153-161. Reprinted in IEEE transactions on Computing C-25

(1976), 335-341.

11. Automated Theorem Proving: A Logical Basis, North-Holland, Amsterdam (1978), xiii+405.

12. Presburger arithmetic with bounded quantifier alternation (with C.R. Reddy), Proc. of the ACM Symposium on the Theory of Computing (1978), 320-325.

13. Detecting ambiguity: an example of knowledge evaluation (with M. Valtorta), Proc. of the Eighth Int'l Joint Conf. on Artificial Intelligence (1983), 182-184. Reprinted in Validating and Verifying Knowledge-based Systems (U.G. Gupta, ed.), IEEE Computer Society Press, Los Alamitos (1991), 391-395.

14. Automated Theorem Proving: After 25 Years (with W.W. Bledsoe, eds.), Contemporary Mathematics, American Math. Society, Providence (1984), ix+360.

15. Performance bounds for binary testing with arbitrary weights, Acta Informatica 22 (1985), 111-114.

16. Near-Horn Prolog and beyond, Jour. of Automated Reasoning 7 (1991), 1-26.

17. SATCHMORE: SATCHMO with Relevancy (with D.W. Reed and D.S. Wilson), Jour. of Automated Reasoning 14 (1995), 325-351.

18. Near-Horn Prolog and the ancestry family of procedures (with D. Reed), Annals of Mathematics and Artificial Intelligence 14 (1995), 225-249.

19. Proof procedures for logic programming (with G. Nadathur), Handbook of Logic in Artificial Intelligence and Logic Programming (D. Gabbay, C. Hogger and J.A. Robinson, eds.), Vol. 5, Clarendon Press, Oxford (1997), 163-234.

20. Automated Deduction: looking ahead, AI Magazine 20 (Spring 1999), 77-98.

For a full publication list see http://www.cs.duke.edu/~dwl/CV/

**Service to the Profession:** Member, Editorial Board, Journal of Automated Reasoning, 1983-2001; Trustee, CADE (Conf. on Automated Deduction), Inc., 1994-1997; Co-organizer, Conf. on Disjunctive Logic Programming and Databases: Nonmonotonic Aspects, Saarbrucken, 1996; Organizer, Workshop on Future Directions of Automated Deduction, Chicago, 1996; Member, Advisory Board, Univ. of Pennsylvania AI Center, 1991-1995; Organizer, Japanese-American Workshop on Automated Theorem Proving, Duke Univ., 1994; Managing Editor, book series Symbolic Computation – Artificial Intelligence, Springer-Verlag, 1984-1992; Member, Editorial Board, Artificial Intelligence Jour., 1983-1992; Chairman, Committee for prospective Jour. of Computational Logic (became Jour. of Automated Reasoning), 1981-1982.

**Teaching:** PhD thesis advisor for Robert Daley (1971 CMU), Susan Gerhart (coadvisor) (1972 CMU), C. Ramu Reddy (1978 Duke), David Mutchler (1986 Duke), Marco Valtorta (1987 Duke), Owen Astrachan (1992 Duke), David Reed (1992 Duke), Timothy Gegg-Harrison (1993 Duke); strongly influenced the PhD work of Mark Stickel (CMU); advised seven AM and MS students.

**Vision Statement:** Automated and semi-automated deduction systems are becoming important in hardware and software specification/verification. This will intensify as more parallelism is sought in programs, as humans are poor parallelizers. Logic programming ideas can help here also. Computational logic will also play an increasing role in computer security protocols.

**Honours and Awards:** Herbrand Award for Distinguished Contributions to Automated Reasoning, 2001; Fellow, ACM, 1999-; Fellow, American Association of Artificial Intelligence, 1993-.

# M

## MCCARTHY, John

**Specialties:** Logic of common sense, logic of program correctness.

**Born:** 4 September 1927.

**Educated:** Princeton University, PhD Mathematics 1951; California Institute of Technology, BS Mathematics, 1948.

**Dissertation:** *Projection Operators and Partial Differential Equations* ; supervisor, Solomon Lefschetz.

**Regular Academic or Research Appointments:** PROFESSOR OF COMPUTER SCIENCE, STANFORD UNIVERSITY, 1962–, Emeritus 2001-; Assistant Professor, Communication Sciences, MIT, 1958, Associate Professor, 1961; Assistant Professor of Mathematics, Dartmouth College, 1955; Acting Assistant Professor, Mathematics, Stanford University, 1953; Higgins Instructor, Mathematics, Princeton University, 1951.

**Visiting Academic or Research Appointments:** Visiting Professor, University of Texas, 1987.

**Research Profile:** Analysis including differential equations, differential geometry, mathematical economics; formalization of common sense knowledge and reasoning in mathematical logic, necessary extensions to logic for formalizing common sense including nonmonotonic reasoning (circumscription), concepts as objects, contexts as objects, and approximately defined entities; symbolic computation (Lisp), mathematical theory of computation (proving programs correct), time sharing computer systems.

**Main Publications:**

1. "Fictionality and the Logic of Relations", *The Southern Journal of Philosophy*, 7 (1969), 51–63.
2. *The Logic of Fiction: A Philosophical Sounding of Deviant Logic*, The Hague and Paris: Mouton and Co., 1974.
3. "Programs with Common Sense", *Proceedings of Teddington Conference on the Mechanization of Thought Processes* H.M. Stationery Office 1959.
4. "Recursive Functions of Symbolic Expressions and their Computation by Machine," *Comm. ACM*, April 1960.
5. "Time-Sharing Computing Systems," in *Management and the Computer of the Future*, Martin Greenberger (ed.), MIT Press.
6. "A Basis for a Mathematical Theory of Computation", in P. Braffort and D. Hirschberg (eds.), *Computer Programming and Formal Systems*, North-Holland Publishing Co., Amsterdam, pp. 33-70.
7. "Towards a Mathematical Science of Computation", in Proc. IFIP Congress 62, North-Holland, Amsterdam.
8. "Some Philosophical Problems from the Standpoint of Artificial Intelligence", in D. Michie (ed), *Machine Intelligence 4*, American Elsevier, New York, NY. (with P. J. Hayes)
9. "Ascribing Mental Qualities to Machines" in *Philosophical Perspectives in Artificial Intelligence*, Ringle, Martin (ed.), Harvester Press, July 1979.
10. "First Order Theories of Individual Concepts and Propositions", in Michie, Donald (ed.) *Machine Intelligence 9*, (University of Edinburgh Press, Edinburgh).
11. "Circumscription - A Form of Non-Monotonic Reasoning", *Artificial Intelligence*, Volume 13, Numbers 1,2, April.
12. "Applications of Circumscription to Formalizing Common Sense Knowledge" *Artificial Intelligence*, April 1986
13. "Artificial Intelligence and Logic" in Thomason, Richmond (ed.) Philosophical Logic and Artificial Intelligence (Dordrecht ; Kluwer Academic, c1989).
14. *Formalizing Common Sense*, Ablex, Norwood, New Jersey
15. "Notes on Formalizing Context" IJCAI-93.
16. "Approximate objects and approximate theories", KR2002
17. "Actions and other events in situation calculus" in Principles of knowledge representation and reasoning: Proceedings of the eighth international conference (KR2002)
*Work in Progress*
18. "Simple Deterministic Free Will"
19. "The Philosophy of AI and the AI of Philosophy"

**Teaching:** I have taught courses in logical artificial intelligence and in proving that computer programs meet their specifications.

**Vision Statement:** Leibniz, Boole, and Frege all expected logic to incorporate common sense knowledge and reasoning. I'd like to see this happen. For this logic requires extension to include nonmonotonic reasoning, incompletely defined entities, and probably considerably more. My own work has included the situation calculus formalization of action, the circumscription method of nonmonotonic reasoning, formalization of concepts as objects, formalization of contexts as objects, and a study of elaboration tolerance of theories. The vision is to establish the logical component of human-level common sense.

**Honours and Awards:** Honorary Fellow, Royal Society of Edinburgh, 2005; Benjamin Franklin Medal, 2003; National Medal of Science, 1990; Kyoto Prize, 1987; Research Excellence Award, International Conference on Artificial Intelligence, 1985; Turing Award, Associaton for Computing Machinery, 1971.

# MCCUNE, William W.

**Specialties:** Automated Reasoning and Logic

**Born:**

**Educated:** B.A. Mathematics 1976 University of Vermont, M.S. Computer Science 1982 Ph.D. Computer Science 1984 Northwestern University Adviser Lawrence Henschen

**Regular Academic or Research Appointments:** ARGONNE NATIONAL LABORATORY 1984– ; SENIOR RESEARCH FELLOW, ARGONNE-UNIVERSITY OF CHICAGO COMPUTATION INSTITUTE, 1999–.

**Research Profile:** Developed powerful automated reasoning programs and then used them to attack challenging problems and open questions in various logics. Of especial significance is his monograph (with R. Padmanabhan), in which he uses automated deduction to answer numerous questions in equational logic, and his answering the Robbins question, whose answer had eluded mathematicians and logicians for six decades. Also found the first single axioms for the left and right group calculi, thereby solving two long-standing open problems posed by the late logician C. A. Meredith; answered open questions in combinatory logic concerning the presence or absence of fixed properties for various sets of combinators; used automated reasoning and model generation programs to answer many problems in so-called quantum logics; and developed clauses for Gödel's axioms for set theory in first-order logic.

Author of approximately 50 publications of which the following are representative.

**Main Publications:**

Books
1. W. McCune and R. Padmanabhan, Automated Deduction in Equational Logic and Cubic Curves, Lecture Notes in Artificial Intelligence, vol. 1095, Berlin, Springer-Verlag, 1996.
2. W. McCune, ed., Proceedings of the 14th International Conference on Automated Deduction, Lecture Notes in Artificial Intelligence, vol. 1249, Berlin, Springer-Verlag, 1997.

Articles

3. R. Boyer, E. Lusk, W. McCune, R. Overbeek, M. Stickel, and L. Wos. Set theory in first-order logic: Clauses for Gödel's axioms. J. Automated Reasoning, 2(3):287-327, 1986.
4. W. McCune and L. Wos. A case study in automated theorem proving: Searching for sages in combinatory logic. J. Automated Reasoning, 3(1):91-107, 1987
5. L. Wos and W. McCune. Challenge problems focusing on equality and combinatory logic: Evaluating automated theorem-proving programs. In E. Lusk and R. Overbeek, editors, Proceedings of the 9th International Conference on Automated Deduction, Lecture Notes in Computer Science, Vol. 310, Berlin, Springer-Verlag, pp. 714-729, 1988.
6. L. Wos, S. Winker, W. McCune, R. Overbeek, E. Lusk, R. Stevens, and R. Butler. Automated reasoning contributes to mathematics and logic. In M. Stickel, editor, Proceedings of the 10th International Conference on Automated Deduction, Lecture Notes in Artificial Intelligence, Vol. 449, Berlin, Springer-Verlag, pp. 485-499, 1990.
7. L. Wos and W. McCune. Automated theorem proving and logic programming: A natural symbiosis. J. Logic Programming, 11(1):1-53, July 1991.
8. L. Wos and W. McCune. The application of automated reasoning to questions in mathematics and logic. Annals of Mathematics and Artificial Intelligence, 5:321-370, 1992.
9. W. McCune and L. Wos. Application of automated deduction to the search for single axioms for exponent groups. In A. Voronkov, editor, Logic Programming and Automated Reasoning, LNAI Vol. 624, Berlin, Springer-Verlag, pp. 131-136, 1992.
10. W. McCune. Single axioms for the left group and right group calculi. Notre Dame J. Formal Logic, 34(1):132-139, 1993.
11. W. McCune. Single axioms for groups and Abelian groups with various operations. J. Automated Reasoning, 10(1):1-13, 1993.
12. W. McCune, Automatic proofs and counterexamples for some ortholattice identities. Information Processig Letters, 65:285-291, 1998.
13. W. McCune. Solution of the Robbins problem. J. Automated Reasoning, 19(3):263-276, 1997.
14. W. McCune and O. Shumsky. IVY: A preprocessor and proof checker for first-order logic. In M. Kaufmann, P. Manolios, and J Moore, editors, Computer-Aided Reasoning: ACL2 Case Studies, chapter 16. Kluwer Academic, 2000.
15. W. McCune, R. Veroff, B. Fitelson, K. Harris, A. Feist, and L. Wos, Short Single Axioms for Boolean Algebra, J. Automated Reasoning, to appear 2002.

**Service to the Profession:** Editor, Journal of Automated Reasoning; Board of Trustees, Conference on Automated Deduction, 1996-2000; Program Chair, 14th International Conference on Automated Deduction (CADE-14), 1997.

**Honours and Awards:** Honors include winner

of the Herbrand Award for Distinguished Contributions to Automated Reasoning, 2000; Winner CADE/CASC Automated Theorem Proving Competition: 1996, first place, equational, second place, mixed, and 1999: first place, first-order satisfiability; recipient of the Royal E. Cabell Research Fellowship, Northwestern University, 1983-1984.

## MCGEE, Vann

**Specialties:** Philosophical logic

**Born:** 10 March 1949 in Charlotte, North Carolina, USA.

**Educated:** University of California at Berkeley, PhD, Logic and the Methodology of Science, 1985; University of North Carolina at Chapel Hill, MA, Philosophy, 1978; Harvard University, AB, Philosophy, 1972.

**Dissertation:** *Truth and Necessity in Partially Interpreted Languages*, supervisor, Charles Chihara.

**Regular Academic or Research Appointments:** PROFESSOR MASSACHUSETTS INSTITUTE OF TECHNOLOGY, DEPARTMENT OF LINGUISTICS AND PHILOSOPHY, 1996; Associate Professor, Rutgers University, 1990-1996. Assistant Professor, University of Arizona, 1985-1990.

**Research Profile:** Tarski's *Wahrheitsbegriff* provides theories of truth for a wide variety of formal languages, but, because of its requirement that the semantic theory for a language be developed within an essentially richer metalanguage, its methods are inapplicable to natural languages. One badly wants to understand the notion of truth as it applies to everyday language, but Tarski's methods don't show us how. A natural-language theory of truth that is adequate in the sense of Convention T is out of the question, on account of the liar paradox, but one would still hope for some means for talking about truth in natural language coherently. A strategy for this problem that I have been pursuing is to incorporate the paradoxes of self-reference into a general understanding of meaningful but semantically deviant statements that takes account of such phenomena as the *sorites* paradox about vagueness, W. V. Quine's arguments for the inscrutability of reference, and the Peter Unger's problem of the many. The most promising approach is to suppose that speakers' usage picks out, not a single intended model of the language, but a family of intended models, and that a sentence counts as true, false, or unsettled according as it is true in all, none, or some but not all of them. Such an approach would appear to provide only the left-to-right direction of Tarski's (T)-schema, sentences of the form "If 'Snow is white' is true, then snow is white," but the left-to-right direction is, sadly, enough to generate antinomy in the form of Montague's paradox. I am confident that I'm very close to finding a way around this obstacle, but I've felt that way for twenty years.

The talk of "models" in the preceding paragraph comes with a grain of salt. Models are ordinarily understood as having set-sized domains, and semantics of natural language requires models that are larger than any set. George Boolos's work on plural quantification provides techniques for applying model-theoretic methods to domains of unlimited size. Boolos's ideas have important implications, it seems to me, for the philosophy of mathematics — they provide an attractive answer to some prominent arguments for mathematical skepticism — and also in providing a logical framework in which to pursue the Aristotelian program of a fully general science of being that takes all things, of whatever sort, into its domain. I have been trying to develop some of these implications.

I have also been working on conditionals, with particular emphasis on their connections with the theory of rational choices and with subjective probability. I've also been looking at objective probability as it's found in quantum theory, but the subject is so difficult that I've not made much progress.

**Main Publications:**

1. A Counterexample to *Modus Ponens*, *Journal of Philosophy* 82 (1985), pp. 462-471.
2. Applying Kripke's Theory of Truth, *Journal of Philosophy* 86 (1989), pp. 530-539.
3. *Truth, Vagueness, and Paradox: An Essay on the Logic of Truth* (Indianapolis: Hackett, 1991).
4. Maximal Consistent Sets of Instances of Tarski's Schema (T), *Journal of Philosophical Logic* 21 (1992), pp. 235-241.
5. On the Degrees of Unsolvability of Modal Predicate Logics of Provability, *Journal of Symbolic Logic* 59 (1994), pp. 253-261.
6. Learning the Impossible, in Brian Skyrms and Ellery Eells, eds., *Probability and Conditionals* (New York and Cambridge: Cambridge University Press, 1994), pp. 177-199.
7. Distinctions Without a Difference (with Brian P. McLaughlin), *Southern Journal of Philosophy* 33 supplement (1995) (Spindel Conference volume for 1994), pp. 203-252.
8. Logical Operations, *Journal of Philosophical Logic* 25 (1996), pp. 567-80.
9. How We Learn Mathematical Language, *The Philosophical Review* 106 (1997), pp. 35-68.
10. The Complexity of the Modal Predicate Logic of

'True in Every Transitive Model of ZFC,' *Journal of Symbolic Logic* 62 (1997), pp. 1371-1378.

11. An Airtight Dutch Book, *Analysis* 59 (1999), pp. 257-65. Reprinted in Patrick Grim, Kenneth Baynes, and Gary Mar, eds., *The Philosopher's Annual*, v. 22 (Stanford, California: CSLI Publications, 2000), pp. 155-164.

12. 'Everything,' in Gila Sher and Richard Tieszen, eds., *Between Logic and Intuition* (New York and Cambridge: Cambridge University Press, 2000), pp. 54-78.

13. The Lessons of the Many (with Brian P. McLauglin), *Philosophical Topics* 28 (2000), pp. 128-151.

14. Truth by Default, *Philosophia Mathematica* 9 (2001), pp. 5-20.

15. Tarski's Staggering Existential Assumptions, *Synthese* 142 (2004), pp. 371-387.

16. Ramsey's Dialetheism, in Graham Priest, J. C. Beall, and Bradley Armour-Garb, eds., *The*

17. *Law of Non-Contradiction* (Oxford: Oxford University Press, 2004), pp. 276-291.

18. Inscrutability and Its Discontents, *Noûs* 39 (2005), pp. 397-425.

19. In Praise of the Free Lunch, to appear in Vincent F. Hendricks, Stig Andur Pedersen, and Thomas Bollander, eds., *Self-Reference* (Stanford, California: CSLI).

20. There's a Rule for Everything, to appear in Agustín Rayo and Gabriel Uzquiano, eds., *Absolute Generality* (Oxford: Oxford University Press).

*Work in Progress*

21. There are Many Things, to appear in a *festschrift* volume for Robert Stalnaker, edited by Judith Jarvis Thomson.

**Service to the Profession:** Editorial board, Notre Dame Journal of Formal Logic, 1989-; Editorial board, Philosophical Studies, 1990-2005; Chair, ASL Oversight Committee, Journal of Philosophical Logic, 2000-2005; Coordinating editor, Journal of Philosophical Logic, 1995-2000; ASL Council, 1994-2000; Editorial board, Encyclopedia of Philosophy Supplement; National Academy of Science delegate, Tenth International Congress of Logic, Methodology, and Philosophy of Science, Florence, 1995.

**Teaching:** I teach logic, philosophy of language, philosophy of mathematics, and decision theory at the undergraduate and graduate levels. I was dissertation director for Marian David, Anna Vitola, Stephen Webb, Gabriel Uzquiano, Agustín Rayo, and Peter Koellner.

**Vision Statement:** In spite of the fact that no other logic is likely to provide nearly as rich a model theory or proof theory, I suspect that the reign of the first-order predicate calculus is coming to an end. My guess is that a big new direction in logic will be to develop ways of talking about collective properties of individuals that don't require reducing collective properties to properties of collections, taking as a starting point George Boolos's work on plural quantification.

**Honours and Awards:** Johnsonian prize for *Truth, Vagueness, and Paradox,* 1988; An Airtight Dutch Book selected for the *Philosopher's Annual*, 1999.

## MCLARTY, Colin Slator

**Specialties:** Categorical logic, topos theory.

**Born:** 12 July 1951 in Lancaster PA, USA.

**Educated:** Case Western Reserve University, Cleveland, USA, PhD Philosophy, 1980; Case Institute of Technology, Cleveland, USA, BS Mathematics, 1972.

**Dissertation:** *Things and Things in Themselves: The Logic of Reference in Leibniz, Lambert, and Kant*; supervisor, Raymond J. Nelson.

**Regular Academic or Research Appointments:** ASSOCIATE PROFESSOR AND CHAIR, DEPARTMENT OF PHILOSOPHY, CASE WESTERN RESERVE UNIVERSITY, 1993–; Associate Professor Mathematics, CWRU, 1993-; Assistant Professor, Philosophy, CWRU, 1986-1993.

**Visiting Academic or Research Appointments:** Visiting Associate Professor, University of Notre Dame, 2002; Visiting Scholar, Mathematics, Harvard University 1995-1997.

**Research Profile:** McLarty has worked in the proof theory and model theory of categorical logic and on axiomatic category theoretic foundations for mathematics. His philosophical papers relate all of the technical work to conceptual questions in the history of mathematics, in foundations of mathematics, and in structuralist philosophical conceptions of mathematics. He has given complete formalizations of some weak doctrines of logic; for example, left exact logic which uses only equations between partially defined operators where the domain of any partial operator must be given by equations in earlier operators. Other work axiomatizes certain relations between toposes, or relations of toposes to differential geometry or recursive functions. His axiomatics for the category of categories attempts to systematize questions of independence and definability in that theory by general methods for giving variant interpretations. Issues of constructivism, non-well-founded sets, and a universal set are related to categorical foundations for mathematics in other of his work.

His historical work focuses on the constantly interwoven progress of algebra, topology, number theory, logic, and foundations through the

20th century, notably by case studies of Poincaré, Brouwer (as topologist more than as logician), Noether, Eilenberg and Mac Lane, Serre and Grothendieck, and Lawvere. The further reaches of geometry and number theory have inspired work in axiomatics and model theory throughout this time although textbooks on logic often elide them in favor of set theory and elementary number theory more purely suited to standard formal methods.

**Main Publications:**

1. Elementary Categories, Elementary Toposes, Oxford, 1992.
2. "The last mathematician from Hilbert's Göttingen: Saunders Mac Lane as a philosopher of mathematics" forthcoming, *British Journal for the Philosophy of Science*.
3. "Two aspects of constructivism in category theory" forthcoming, *Philosophia Scientiae*.
4. "Emmy Noether's 'Set Theoretic' Topology: From Dedekind to the rise of functors", in J. Gray and J. Ferreirós eds *The Architecture of Modern Mathematics: Essays in history and philosophy*, Oxford, 2006.
5. "The Rising Sea: Grothendieck on simplicity and generality I" in J. Gray and K. Parshall eds. *Episodes in the History of Recent Algebra*, American Mathematical Society, 2006.
6. "Every Grothendieck topos has a one-way site", *Theory and Applications of Categories*, **16**, (2006) pp. 123–26.
7. "Mathematical platonism versus gathering the dead: What Socrates teaches Glaucon", *Philosophia Mathematica* 13 (2005) pp. 115–34.
8. "Learning from Questions on Categorical Foundations", *Philosophia Mathematica*, 13 (2005) pp. 44–60.
9. "Semantics for first and higher order realizability", in Anderson and Zeleny eds. \emph{Logic, Meaning, and Computation}, Kluwer Academic 2001, 353–64.
10. "Poincar'e: Mathematics & Logic & Intuition", {Philosophia Mathematica}{5} (1997) 97–115.
11. "Category theory in real time", {Philosophia Mathematica} {2} (1994) 36–44.
12. "Numbers can be just what they have to", {Noˆus} {27} (1993), 487–98.
13. "Anti-foundation and self-reference", {Journal of Philosophical Logic} {22} (1993) 19–28.
14. "Failure of cartesian closedness in NF", {Journal of Symbolic Logic} {57} (1992) 555–56. Reprinted in Follesdal ed. {The Philosophy of Quine}, Garland, 2000, vol.5, 109–11.
15. "Axiomatizing a category of categories", {Journal of Symbolic Logic} {56} (1991) 1243–60.
16. "The uses and abuses of the history of topos theory", {British Journal for the Philosophy of Science}, {41} (1990) 351–75.
17. "Defining sets as sets of points of spaces", {Journal of Philosophical Logic}, {17} (1988) 75–90.
18. "Elementary axioms for canonical points of toposes", {Journal of Symbolic Logic}, {52} (1987) 202–04.
19. "Left exact logic", Journal of Pure and Applied Algebra, \textbf{41} (1986) 63–66.
20. "Local, and some global, results in synthetic differential geometry", in A. Kock ed. {Category Theoretic Methods in Geometry}, (Aarhus Denmark: Aarhus Universitet, 1983), 226–56.

**Service to the Profession:** Editorial Board, Philosophia Mathematica, 2005-; Editorial Advisory Board, "Advanced Studies in Mathematics and Logic" for Polimetrica Publishers, 2005-;

**Teaching:** Undergraduate teaching.

**Vision Statement:** Logic should and will continue to formalize the working methods of mathematics. Mathematics should and will continue producing new methods. Examples today include "tameness" in topology, which model theorists have begun to capture, and the uses of extremely large-scale functorial organization in geometry and number theory, which category theorists continue to simplify. The challenge is to formalize at once the naive simplicity of the intuitive concepts and their practical power (which may be quite different from proof theoretic strength). I look forward to more of this and so to greater integration of explicit logic in working mathematics.

**Honours and Awards:** Director, National Endowment for the Humanities Summer Seminar: "Proofs and refutations in mathematics today", 2001; Science, Society and Technology Studies Scholars Award, National Science Foundation (USA) for "Alexander Grothendieck and the history of homology theory", 1995-1998.

## MCNAUGHTON, Robert

**Specialties:** Theoretical computer science, logic, foundations of mathematics.

**Born:** 13 March 1924 in Brooklyn, NY, USA.

**Educated:** Harvard University, PhD Philosophy, 1951; Columbia College, BA Philosophy, 1948;

**Dissertation:** *On establishing the consistency of systems*; supervisors, W.V. Quine and Hao Wang.

**Regular Academic or Research Appointments:** EMERITUS PROFESSOR OF COMPUTER SCIENCE, RENSSELAER POLYTECHNIC INST., 1989-; Professor of Mathematics and Computer Science, RPI, 1966-1989; Visiting Associate Professor, MIT, 1964-1966; Associate Professor of Computer Science, Moore School of Electrical

Engineering, University of Pennsylvania, 1963-1964; Assistant Professor there, 1957-1963; Assistant Professor of Philosophy, Stanford University, 1954-1957; Instructor, Philosophy, University of Michigan, 1953-1954, Research in Logic, U.S. Navy Contract, Illinois Institute of Technology, 1952-1953; Instructor, Philosophy, Ohio State University, 1951-1952.

**Research Profile:** Robert McNaughton was a philosopher logician from 1951 to 1957. In 1957 he became a computer scientist, doing what he called "Computer Logic" at first, and settling to work in theoretical computer science thereafter.

His work in pure logic began with the proof of a fundamental theorem about infinite-valued logic, supplying a necessary and sufficient condition that a given real-valued function corresponds to some logical formula. His doctoral dissertation was a study of consistency proofs for logic systems; this work was philosophical as well as proof-theoretic.

Some of his work in computer science would be of interest to logicians. The very start of this work was in switching theory, largely with the problem of minimizing the size of a logic formula equivalent to a given truth function (thereby minimizing the circuit realizing that truth function). Later he became interested in the problem of writing behavioral formulas describing the behavior of finite automata, going beyond truth-functional formulas to formulas with quantifiers. Still later, he worked in the areas of Turing machines, theory of computability and formal languages.

His research in computer science also included results in the star height of regular expressions, the infinite history of a finite automaton (following Richard Buchi); games of infinite duration; counter-free automata (i.e., star-free finite automata); the semigroup classification of finite automata; semi-Thue systems; formal languages; and the combinatorics on words.

**Main Publications:**

1. Les Systemes Axiomatiques de la Theorie des Ensembles, Gauthier-Villars, Paris, (55 pages) 1953 (with Hao Wang).
2. The Theory of Automata, A Survey, Advances in Computers (F. L. Alt, ed.,), Vol. II, 379-421, Academic Press, 1961.
3. Badly Timed Elements and Well Timed Nets, Moore School of Electrical Engineering, University of Pennsylvania, Philadelphia, (153 pages) 1964.
4. Counter-free Automata, M.I.T. Press monograph series, (160 pages) 1971 (with S. Papert).
5. Elementary Computability, Formal Languages and Automata, (textbook for the RPI course 66.405 with the same name) Prentice Hall, January 1982 (393 + xvii pages).
6. "A Theorem About Infinite-Valued Sentential Logic," Journal of Symbolic Logic, 16, 1-13, 1951.
7. "Some Formal Relative Consistency Proofs," J. Symbolic Logic, 18, 135-144, 1953.
8. "A Non-Standard Truth Definition," Proc. Am. Math. Soc., 5, 505-509, 1954.
9. "The Folded Tree," J. Franklin Institute, 189, 9-24, 115-126, 1955 (with A.W. Burks, C.H. Pollmer, D.W. Warren, and J.B. Wright).
10. "Axiomatic Systems, Conceptual Schemes, and the Consistency of Mathematical Theories," Philosophy of Science, 21, 44-53, 1954.
11. "Conceptual Schemes in Set Theory," Philosophical Review, 66, 66-80, 1957.
12. "Regular Expressions and State Graphs for Automata," Trans. IRE, POEC, Vol. EC-9, 39-47, 1960 (with H. Yamada). Reprinted in E.F. Moore,ed., Sequential Machines, Selected Papers, pp. 157-174, Addison-Wesley, 1964.
13. "Undefinability of Addition from one Unary Operator," Trans. Am. Math. Soc., 117 329-337, 1965.
14. "Testing and Generating Infinite Sequences by a Finite Automaton," Information and Control, 9, 521-530, 1966.
15. "Parenthesis Grammars," Journal of the Association for Computing Machinery, 14, 490-500, 1967.
16. "The Burnside Problem for Semigroups," J. Algebra, 34, 292-299, 1975 (with Y. Zalcstein).
17. "Special Monoids and Special Thue Systems," J. Algebra, 108, 248-255, 1987 (with P. Narendran).
18. "Church-Rosser Thue Systems and Formal Languages," J. Assoc. Computing Machinery, 35, 324-344, 1988 (with P. Narendran and F. Otto).
19. "Buchi's Sequential Calculus," in The Collected Works of J. Richard Buchi, Saunders MacLane and Dirk Siefkes, eds., Springer-Verlag, 1990.

**Teaching:** McNaughton was a mainstay of the computer science offering in theoretical computer science both at the Moore School (from 1957 to 1964) and at RPI (from 1966 to 1989). He has had eleven PhD students in Computer Science, including John Corcoran (who went on to become a well known philosopher logician), Hisao Yamada, Robert O. Winder, David Hannay, Tony dos Reis, Paliath Narendran and Robert McCloskey.

**Vision Statement:** Many of the theoretical problems in computer science have a basis in logic. One need only consider the annual meeting of "Logic in Computer Science", which has been going strong since the mid 1980's. Its content has shown the dependency of some important aspects of computer development on the logical foundations of computation.

**Honours and Awards:** In 1954 McNaughton received (with A.W. Burks, C.H. Pollmer, D.W. Warren and J.B. Wright) the Levy medal from the Franklin Institute for the research paper, "The folded tree," which had been published in the Journal of the Franklin Institute. Contact information for Robert McNaughton: Department of Computer Science, Rensselaer Polytechnic Institute, Troy, NY 12180-3590. Electronic

## MADDY, Penelope

**Specialties:** Philosophy of logic, set theory, philosophy of mathematics.

**Born:** 4 July 1950 in Tulsa, Oklahoma, USA.

**Educated:** Princeton University, PhD Philosophy, 1979; University of California, Berkeley, BA Mathematics, 1972.

**Dissertation:** *Set Theoretic Realism*; supervisor, John Burgess.

**Regular Academic or Research Appointments:** PROFESSOR OF LOGIC AND PHILOSOPHY OF SCIENCE, UNIVERSITY OF CALIFORNIA, IRVINE, 1998–; Professor of Philosophy, University of California, Irvine, 1989-1998; Associate Professor of Philosophy, University of California, Irvine, 1987-1989; Associate Professor of Philosophy, University of Illinois, Chicago, 1983-1987; Assistant Professor of Philosophy, University of Notre Dame, 1978-1983.

**Research Profile:** Maddy's work revolves around methodological questions in the foundations of set theory; in particular, what a solution to the Continuum Problem would look like. Her 'Believing the axioms' (1988) surveys arguments given for and against now-accepted axioms and new axiom candidates, from ZFC to large cardinals and determinacy, and poses the challenge of assessing their rationality. In *Realism in Mathematics* (1990), she defends a version of the view that the Continuum Problem is a legitimate question and that Cantor's hypothesis is either true or false in an objective world of sets, and frames the search for new axioms in this context. *Naturalism in Mathematics* (1997) uncovers some of the shortcomings of her earlier realism and argues that methodological issues in set theory – whether or not to allow impredicative definitions, whether or not to accept the Axiom of Choice, how to assess axiom candidates – turn not on metaphysics, realist or otherwise, but rather on intra-mathematical considerations peculiar to the norms and goals of set theoretic practice. (These ideas are illustrated in a case against Gödel's Axiom of Constructability.) She has since been occupied with spelling out the form of naturalism presupposed by this position – called 'second philosophy' to distinguish it from myriad other naturalistic views – and developing a second philosophy of logic. With these in place, she returns to the familiar philosophical questions of mathematical truth and existence that were set aside in *Naturalism* (as irrelevant to methodological decisions).

**Main Publications:**

1. 'Believing the axioms', *Journal of Symbolic Logic* 53 (1988), pp. ???
2. *Realism in Mathematics*, (Oxford: Oxford University Press, 1990).
3. *Naturalism in Mathematics*, (Oxford: Oxford University Press, 1997).

*Work in Progress*

4. *Second Philosophy*, to appear from Oxford University Press.

**Service to the Profession:** Editorial Board, *Bulletin of Symbolic Logic*, 2003-; Vice President of the Association for Symbolic Logic, 2001-2004, Executive Committee, 1993-1996, 2001-2004, Council Member, 2004-; *Journal of Symbolic Logic*, editor for book reviews in philosophy, 1995-2000; Executive Committee, Philosophy of Science Association, 1994-1996; Editorial Board, *Philosophia Mathematica*, 1993-; Executive Committee, American Philosophical Association, Pacific Division, 1993-1995; Editorial Board, *Journal of Philosophical Logic*, 1985-; Editorial Board, *Notre Dame Journal of Formal Logic*, 1978, Managing Editor, 1979-84.

**Teaching:** PhD advisor to Donna Summerfield (deceased), Christopher Menzel (Texas A&M), Don Fallis (University of Arizona), Patricia Marino (Waterloo), and Teri Merrick (Azusa Pacific).

**Vision Statement:** What I'd most like to see is a stable solution to the Continuum Problem!

**Honours and Awards:** Lakatos Prize for *Naturalism in Mathematics*, 2002; Patrick Romanell Lecture on Philosophical Naturalism, APA, 2001; Elected to American Academy of Arts and Sciences, 1998; NSF Research Grants 1994-1995, 1990-1991, 1988-1989, 1986; AAUW Fellowship, 1982-1983; Marshall Scholarship, 1972-1973; UC Berkeley, Mathematics Department Citation, 1972; Westinghouse Scholarship, 1968-1972.

# MAGIDOR, Menachem

**Specialties:** Mathematical logic: set theory, model theory, applications to computer science: non-monotonic reasoning, semantics of distributed processes, object-oriented programming.

**Born:** January 24 1946 in Petah Tikva, Israel.

**Educated:** Hebrew University of Jerusalem, PhD Mathematics, 1973; Hebrew University of Jerusalem, MSc Mathematics, 1967; Hebrew University of Jerusalem, BSc Mathematics & Physics, 1965.

**Dissertation:** *On Super-compact Cardinals*; *supervisor:* Azriel Levi.

**Regular Academic or Research Appointments:** PRESIDENT, HEBREW UNIVERSITY OF JERUSALEM, 1997–, PROFESSOR, MATHEMATICS, 1983–, Associate Professor, 1978-1983; Associate Professor, Mathematics, Ben Gurion University, Beer Sheva, 1977-1978; Senior Lecturer, Mathematics, Ben Gurion University, 1975-1977; Lecturer, Mathematics, University of California, Berkeley, 1973-1975; Assistant Professor, Mathematics, University of Colorado, Boulder, 1972-1973.

**Visiting Academic or Research Appointments:** Visiting Professor, Mathematics, California Institute of Technology, 1997; Visiting Professor, Mathematics, University of California, Irvine, 1996-1997; Visiting Professor, Mathematics, Ohio State University, Columbus, 1992; Visiting Professor, Mathematics, MIT, 1991; Visiting Fellow, Mathematical Science Research Institute, Berkeley, 1989-1990; Visiting Professor, Mathematics, California Institute of Technology, 1987; Visiting Professor, Mathematics, University of California, Los Angeles, 1986-1987; Visiting Associate Professor, Mathematics, California Institute of Technology, 1982; Visiting Associate Professor, Mathematics, University of California, LA, 1981-1982; Fellow, Institute for Advanced Studies, Jerusalem (group on Model Theory), 1980-1981.

**Research Profile:** In set theory, the work of Menachem Magidor has concentrated around the impact of Strong Axioms of infinity ("Large Cardinals") of the Universe of Sets. A major theme was the determination of the consistency strength of various statements in Set Theory. A central subject of his investigation was around the Singular Cardinals Problem, where Magidor was the first to develop a model in which $\aleph_\omega$ is the first cardinal violating G.C.H. using super-compact cardinals (later the consistency strengths was lowered by the results of Wooden, and the best result in Magidor-Gitik).

Another development was the formulation of a maximal version of the Martin Axiom (Martin Maximum = MM) together with Foreman and Shelah. The proof of the consistency of MM from large cardinals and its implications figured very significant results in Descriptive Set Theory leading to the exact classification of the consistency strength of the Axiom of Determining AD by Martin-Steel and Woodin). A related subject is the combinatories of singular cardinals which includes a systematic study of inter-relation between square-like principles, stationary sets, reflection, etc. (This is mainly joint work with Cummings and Foreman.) Also, Magidor contributed to the study of Abelian Groups and their connection with Set Theory (joint works with Shelah and with Fuchs).

In Model Theory, Magidor was involved in the study of generalized quantifiers and their expressive power. A noteworthy example is the introduction of what became known as the Magidor-Malitz Quantifier. Also the study of the classification of the uncountable admissible sets which satisfies a version of Barwise Completeness and Compactness Theorem falls under this category.

In application to Computer Science, Magidor was involved in studies of the Semantics of Distributed Systems, and the Object-Oriented Programming. But most significantly was the study of Non-Monotonic Designing, trying to create models for common sense reasoning. In particular, Magidor was involved, together with Kraus and Lehmann, in introducing the notion of cumulative logic and the study of their properties under the system which became known as KLM.

**Main Publications:**

1. "On the Role of Supercompact and Extendible Cardinals in Logic" *Israel Journal of Mathematics* 10 (1971) 147-157.
2. "How Large is the First Strongly Compact Cardinal?" *Annals of Mathematical Logic* 10 (1976) 33-57.
3. "Compact Extensions of L(Q)" *Annals of Mathematical Logic* 11 (1977) 217-261. (with J. Malitz)
4. "On the Singular Cardinals Problem I" *Israel Journal of Mathematics* 28.
5. "On the Singular Cardinals Problem II" *Annals of Mathematics* 106 (1977) 517-547, (1977) 1-31.
6. "The Evolution of Large Cardinals Axioms in Set Theory" in *"Higher Set Theory"* (G. Muller and D. Scott, Ed). Lecture Notes in Mathematics. *Springer Verlag*, 1978, pp. 93-275. (with A. Kanamori)
7. "Precipitous Ideals" *Journal of Symbolic Logic* 45 (1980) 1-8. (With T.J. Jech, W. Mitchell and K. Prikry)
8. "Reflecting Stationary Sets", *Journal of Symbolic Logic* 47 (1982) 755-771.
9. "Countably Decomposable Admissible Sets" *Annals of Pure and Applied Logic* 26 (1984) 287-361. (With S. Shelah and J. Stavi)

10. "Martin Maximum, Saturated Ideals and Non Regular Ultrafilters, part I" *Annals of Mathematics* 127 (1988) 1-47. (With M. Foreman and S. Shelah), part II, 127 (1988) 521-545.

11. "Representing Sets of Ordinals as Countable Unions of Sets in the Core Model" *Transactions of the American Mathematical Society* 317 (1990) 91-126.

12. "Non-monotonic Reasoning, Preferential Models and Cumulative Logic" *Artificial Intelligence* 44 (1990) 167-207. (With S. Kraus and D. Lehmann)

13. "Shelah Pcf Theory and its Applications" *Annals of Pure and Applies Logic* 50 (1990) 207-254. (With M.R. Burke and S. Shelah)

14. "What Does a Conditional Knowledge Base Entail?" *Artificial Intelligence* 55 (1992) 1-60. (With D. Lehmann)

15. "Universally Baire of Sets of Reals" in *"Set Theory of the Continuum"* M.S.R.I. research publications no 26, Berkeley 1992 (With Q. Feng and H. Woodin) pp. 203-242.

16. "The Singular Cardinals Problem Revisited" in *"Set Theory of the Continuum"* M.S.R.I. research publications no 26, Berkeley 1992 (H. Judah and H. Woodin, Ed.) pp. 243-279. (With M. Gitik)

17. "Buttler Groups of Arbitrary Cardinality" *Israel Journal of Mathematics* 84 (1993) 239-263. (With L. Fuchs)

18. "When Does Almost Free Imply Free?" *Journal of the American Mathematical Society* 7 (1994) pp. 769-830. (With S. Shelah)

19. "Mutually stationary sequences of sets and the non-saturation of the non-stationary ideal on $p/sb/varkappa (/lamda)$" *Acta Math* 186 (2001), no. 2, pp. 271-300. (With M. Foreman)

20. "Squares, scales and stationary reflection" *J. Math Log.* 1 (2001), pp. 35-98. (With J. Cummings and M. Foreman)

*Work in Progress*

21. Skolem Lowerkiem– Tarski Theorem for Generalized Quantifiers (with J. Vaananen)

22. Forcing Axiom and Square-like Principles

23. Cardinal Arithmetic (with U. Abraham)

**Service to the Profession:** Editorial Board Member, Gruyter Series in Logic and its Applications, 1997-; President, Association of Symbolic Logic, 1996-1998; Vice President, Association of Symbolic Logic, 1995-1996; Editorial Board Member, Perspectives in Mathematical Logic (Book Series of the Association of Symbolic Logic), 1987-1998; Executive Editor, Israel Journal of Mathematics, 1979-1981.

**Teaching:** Magidor supervised or co-supervised the theses of 13 PhD students in Set Theory, Computer Science and Mathematical Education.

**Vision Statement:** I believe that even given the great advances in Mathematical Logic over the last century, the challenge of the mathematical modeling of human reasoning, both in common sense, everyday reasoning and in formal mathematical reasoning, is still to be tackled.

In Set Theory I believe that many of the independent problems (like the continuous hypothesis) can and will be settled on the basis of naturally acceptable axioms.

## MAKINSON, David Clement

**Specialties:** Nonmonotonic logic, logic of belief change, input/output logics, logic of directives and norms.

**Born:** 27 August 1941 in Sydney, Australia.

**Educated:** Oxford University, DPhil, 1965; University of Sydney, Philosophy BA Hons, 1961.

**Dissertation:** *Rules of truth for modal logic*; supervisor, Michael Dummett

**Regular Academic or Research Appointments:** SENIOR RESEARCH FELLOW, DEPARTMENT OF COMPUTER SCIENCE, KING'S COLLEGE, LONDON, 2001-2006; Programme Specialist, Division of Social Science Research and Policy, UNESCO, Paris, 1980-2001; Assistant through Full Professor of Philosophy, American University of Beirut, Lebanon, 1965-83.

**Visiting Academic or Research Appointments:** Visiting Professor, Chinese Academy of Sciences, Guizhou, China, 2006; Visiting Professor, Department of Mathematics, University of Manchester, UK, 2006; Visiting Researcher, Department of Computer Science University of New South Wales, 2005; Visiting Professor, University of Bahia Blanca, Argentina, 1970 and 1999; Visiting Professor, University of Sao Paulo, Brazil, 1972.

**Research Profile:** Early work for the DPhil dissertation (1965) and following years was mainly in modal logics (including invention of the maximal consistent sets method for proving completeness theorems in them), with also forays elsewhere (such as the first formulation of the paradox of the preface). In the 1970s, many differing areas explored. In the 1980s, created and developed the so-called AGM account of the logic of belief change with Alchourrón and Gärdenfors. From the 1990s to the present, the principal focus was on the study of nonmonotonic reasoning, leading eventually to the book Bridges from Classical to Nonmonotonic Logic in 2005. Overlapping all this in the period 1970 to the present, worked on various aspects of the logic of directives (alias deontic logic), leading to the creation (with van der Torre) of input/output logics. Most recently in 2004-2005, work has led to the creation of a new concept of 'logical friendliness'. Current work is on the no-man's land between probabilistic and qualitative consequence (with James Hawthorne).

**Main Publications:**

1. The paradox of the preface, Analysis 25 (1965) 205-207.
2. On some completeness theorems in modal logic, Zeitschrift für Math. Logik und Grundl. der Math. 12 (1966) 379-384.
3. A normal modal calculus between T and S4 without the finite model property, The Journal of Symbolic Logic 34 (1969) 35-38.
4. With Carlos Alchourrón, Hierarchies of regulations and their logic, in Hilpinen ed., New Studies in Deontic Logic (Dordrecht: Reidel, 1981, 125-148).
5. With Carlos Alchourrón, On the logic of theory change: contraction functions and their associated revision functions, Theoria 48 (1982) 14-37.
6. With Carlos Alchourrón and Peter Gärdenfors, On the logic of theory change: partial meet contraction and revision functions, The Journal of Symbolic Logic 50 (1985) 510-530.
7. With Carlos Alchourrón, On the logic of theory change: safe contraction, Studia Logica 44 (1985) 405-422.
8. On the formal representation of rights relations: remarks on the work of Stig Kanger and Lars Lindahl, The Journal of Philosophical Logic 15 (1986) 403-425.
9. Rights of peoples: point of view of a logician, in J.Crawford ed., The Rights of Peoples (Oxford University Press, 1988, 69-92).
10. General theory of cumulative inference, in M.Reinfrank & others eds, Nonmonotonic Reasoning (Berlin: Springer-Verlag, Lecture Notes on Artificial Intelligence n° 346, 1989, 1-17).
11. With Peter Gärdenfors, Relations between the logic of theory change and nonmonotonic logic, in Fuhrmann & Morreau eds, The Logic of Theory Change (Berlin: Springer, 1991, 185-205).
12. With Karl Schlechta, Floating conclusions and zombie paths: two deep difficulties in the directly skeptical approach to defeasible inheritance nets, Artificial Intelligence 48 (1991) 199-209.
13. Five faces of minimality, Studia Logica 52 (1993) 339-379.
14. "General Patterns in Nonmonotonic Reasoning", in Handbook of Logic in Artificial Intelligence and Logic Programming, vol. 3, ed. Gabbay, Hogger and Robinson, Oxford University Press, 1994, pages 35-110.
15. Combinatorial versus decision-theoretic components of impossibility theorems, Theory and Decision 40 (1996) 181-190.
16. With Ramón Pino Pérez and Hassan Bezzazi Beyond rational monotony: on some strong non-Horn conditions for nonmonotonic inference operations Journal of Logic and Computation 7 (1997) 605-632.
17. With Leendert van der Torre, "Input/output logics". Journal of Philosophical Logic 29 (2000) 383-408.
18. With Leendert van der Torre, "Permission from an input/output perspective". Journal of Philosophical Logic 32, 2003, 391-416.
19. "Natural deduction and logically pure derivations". PhiNews, April 2004 http://www.phinews.ruc.dk
20. Bridges from Classical to Nonmonotonic Logic (London: King's College Publications. Series: Texts in Computing, vol 5, 2005) ISBN 1-904987-00-1.
21. "Friendliness for logicians" pp 259-292 in We Will Show Them: Essays in Honour of Dov Gabbay, vol 2, ed. Sergei Artemov, Howard Barringer, Artur Garcez, Luis Lamb, and John Woods. King's College Publications, October 2005.

*Work in Progress*

22. With James Hawthorne: "On the qualitative/quantitative watershed for rules of uncertain inference."

**Service to the Profession:** Board Member, Journal of Logic and Computation; Board Member, Journal of Logic; Board Member, Language and Information Board Member, Studia Logica; Board Member, Journal of Applied Non-Classical Logics; Board Member, Journal of Applied Logic; Reviewer, Math. Reviews, Reviewer, Zentralblatt Math.; Editorial Board, Trends in Logic and Studies in Logic and Practical Reasoning; Editor-in-Chief, UNESCO's International Social Science Journal, 1995-2001; Co-Editor, UNESCO's World Social Science Report, with Ali Kazancigil, 1999.

**Teaching:** Have taught logic to students in Lebanon, Argentina, Brazil, and the UK, and have always enjoyed the experience, whether on the introductory or advanced level. Logic should be fascinating and fun, for both the teacher and the student.

**Vision Statement:** Logic is not just about deduction. Indeed, it is not just about inference, because there are other kinds of knowledge management needing attention, such as belief change. This means that we need to go beyond the perspectives of classical logic. However, the task should not be thought of as the invention of new non-classical ones. Rather, it is a matter of developing more sophisticated and varied ways of deploying the good old classical basis.

**Honours and Awards:** Membre associé, Centre des Recherches en Epistemologie Appliquée (CREA) Paris.

# MAKSIMOVA, Larisa

**Specialties:** Mathematical logic, non-classical logics: relevant, intermediate, positive, modal, temporal logics; algebraic logic; universal algebra.

**Born:** 5 November 1943 in Novosibirsk, Russia.

**Educated:** Novosibirsk State University, Professor, 1993; Institute of Mathematics (Novosibirsk), Doctor of Physical and Mathematical Sciences (habilitation), 1986; Novosibirsk State University, Docent, 1972; Novosibirsk State University, Candidate of Physical and Mathematical Sciences (equivalent of PhD), 1968; Novosibirsk State University, 1965. **Dissertation Theses: Habilitation:** *Decidable properties of superintuitionistic and modal logic;*

**Dissertation:** *Logical calculi of rigorous implication*; supervisor, Anatolii Maltsev.

**Regular Academic or Research Appointments:** LEADING RESEARCHER, SOBOLEV INSTITUTE OF MATHEMATICS 1986– AND PROFESSOR OF MATHEMATICS, NOVOSIBIRSK STATE UNIVERSITY 1992–; Senior Researcher, Institute of Mathematics, Siberian Branch of Russian Academy of Sciences, 1979-1986; Researcher, Institute of Mathematics, Siberian Branch of Russian Academy of Sciences, 1967-1979; Associate Professor, Novosibirsk State University, 1981-1985; Associate Professor, Novosibirsk State University, 1969-1974; Research Probationer, Institute of Mathematics, Siberian Branch of Russian Academy of Sciences, 1965-1967.

**Visiting Academic or Research Appointments:** Visiting Professor, King's College London, 2003, 2004; Visiting Professor, Uppsala University, 1998-1999; Visiting Professor, Japan Advanced Institute of Science and Technology, 1995; Postdoctoral Fellow, Institute of Mathematics, Warsaw University 1972-1973.

**Research Profile:** In her earlier papers L. Maksimova found algebraic semantics for relevance logics, adequate relational semantics for the logic E of entailment, and proved a number of separation theorems for relevant logics and representation theorems for structures with implication.

At the same time she started a study of superintuitionistic logics and proved that there are just three pretabular superintuitionistic logics (1972). A logic is called tabular if it can be characterized by finitely many finite algebras; pretabular logic is maximal non-tabular logic. The number of pretabular normal extensions of the well-known modal S4 logic equals five (1975). As a consequence, decidability of tabularity and pre-tabularity problems for superintuitionistic logics and for extensions of S4 was obtained.

Many of her papers are devoted to the problem of interpolation in non-classical logics. Interpolation property is one of the most significant and interesting properties of logical theories. L. Maksimova made a significant contribution to the theory of interpolation and definability in nonclassical logics. In particular, she obtained a full description of extensions of the intuitionistic propositional logic Int with interpolation property and proved decidability of interpolation over Int and over S4. Analogous results are obtained for the projective Beth property PBP, a stronger analog of Beth's theorem on implicit definability, for superintuitionistic, positive and some modal logics.

A general algebraic approach is worked out in her investigations. Duality theory for modal and Heyting algebras is developed. Algebraic equivalents of many important properties of logical theories were found. The methods are proposed that combine algebraic and semantical tools and allow one to solve logical and algebraic problems at the same time. In particular, tabularity, pre-tabularity and amalgamation property are base-decidable in varieties of Heyting algebras and of closure algebras.

**Main Publications:**

1. Gabbay D.M., Maksimova L. Interpolation and Definability: Modal and Intuitionistic Logics. Oxford Logic Guides v. 46, Oxford University Press, Oxford, 2005, 509 pp.
2. I.Lavrov, L.Maksimova. Problems in Set Theory, Mathematical Logic and the theory of Algorithms. In Russian: 1974, 1985, 1991, 2001,2002, Fizmatlit, Moscow; in Hungarian: 1988; in English: The University Series in Mathematics, Kluwer/ Plenum Publishers, New York, 2003, 282 pp.; in Polish: 2004.
3. L.Maksimova. Complexity of some problems in positive and related calculi, Theoretical Computer Science, 303, no. 1 (2003), 171–185.
4. L.L.Maksimova. Implicit definability and positive logics. Algebra and Logic, 42, no. 1 (2003), 65–93.
5. L.Maksimova. Restricted interpolation in modal logics. Advances in Modal Logic, Volume 4, King's College Publications, London 2003, 297–311.
6. L.L.Maksimova. Restricted amalgamation and projective Beth property in equational logic. Algebra and Logic, 42, no. 6 (2003), 712–726.
7. L.Maksimova. Complexity of interpolation and related properties in positive calculi, J. Symbolic Logic, 67, no. 1 (2002), 397–408.
8. L.Maksimova. Intuitionistic Logic and Implicit Definability. Annals of Pure and Applied Logic, 105 (2000), 83–102.
9. L.Maksimova. Strongly Decidable Properties of Modal and Intuitionistic Calculi. Logic Journal of IGPL, 8, no. 6 (2000), 797–819
10. L.Maksimova. Interrelations of Algebraic, Semantical and Logical Properties for Superintuitionistic and Modal Logics. In: Logic, Algebra and Computer Science, Banach Center Publications, v. 46, Institute of

Mathematics, Polish Academy of Sciences, Warszawa 1999, 159–168.

11. L.Maksimova. Modal logics and varieties of modal algebras: the Beth properties, interpolation and amalgamation", Algebra and Logic, 31, no. 2 (1992), 145–166.

12. L.Maksimova. Temporal logics with the operator "the next" do not have interpolation or the Beth property, Siberian Mathematical Journal, 32, no. 6 (1991), 109–113.

13. L.Maksimova. Interpolation theorems in modal logics and amalgamable varieties of topological boolean algebras, Algebra and Logic, 18, no. 5 (1979), 556–586.

14. L.Maksimova. Craig's theorem in superintuitionistic logics and amalgamable varieties of pseudoboolean algebras, Algebra and Logic, 16, no. 6 (1977), 643–681.

15. L.Maksimova. Pretabular superintuitionistic logics. Algebra and Logic, 11 (1972), 448–570.

*Work in Progress*

Investigation of different variants of interpolation and of the Beth property in various families of logics including modal, substructured and paraconsistent logics. The purpose is to classify implicit definability properties including various versions of the interpolation property and of the Beth definability property.

At the present time L. Maksimova is working on INTAS project "Algebraic and deduction methods in non-classical logic and their applications to Computer Science" 2005-2007. She is Russian team leader of this project, coordinator of the task: Explicit and implicit definability in modal and related logics.

16. L.Maksimova. Projective Beth property in extensions of Grzegorczyk logic. Studia Logica, v. 82, no. 1 (2006), 27 pp.

17. L.Maksimova. Definability and interpolation in non-classical logics. Studia Logica, 82, no. 2 (2006), 271-291.

18. L.Maksimova. Decidable properties of logical calculi and of varieties of algebras. In: V.Stoltenberg-Hansen and J.Vaananen (Eds), Logic Colloquium '03, Ser. Lecture Notes in Logic 24, Association for Symbolic Logic, 2006, 21 pp.

19. D.M.Gabbay, L.L.Maksimova. Interpolation and Definability. In: D.Gabbay and Guenthner (eds.) Handbook of Philosophical Logic, v. 14, Kluwer, The Netherlands, to appear.

**Service to the Profession:** Member of the Council for Awarding Scientific Degrees, Sobolev Institute of Mathematics, Novosibirsk, 1989–; Collecting Editor, *Journal of Applied Non-Classical Logics*, 1991–2003; Member of Honorary Consulting Board of *Studia Logica*, 1993—1999; Member of Advisory Board, *Studia Logica*, 1976—1989; Program Committee Member, International Annual conference Maltsev Readings, Novosibirsk, 1997-2005; Co-chair of Workshop, Non-Standard Logics and Logical Aspects of Computer Science, Kanazawa, Japan, 1994.

**Teaching:** Since 1965 L.Maksimova has been teaching Mathematical Logic and Mathematics at Novosibirsk State University. Now she is a (half-time) professor of Department of Algebra and Logic. She gives lecture courses and seminars in mathematical logic and recursion theory for first and second year students, courses and seminars on Non-Classical Logics for Master and PhD students and she gave lecture courses on Logical Foundations of Programming for MsC students. She gave short courses at the International St. Banach Center, Warsaw, Poland in 1991 and 1996; and a tutorial on modal logic at Logic Colloquium '94, Clermont-Ferrand, France. Also she had a graduate course on modal and temporal logic in Uppsala University in 1999. She has had seven PhD students in mathematical logic and algebra, including Vladimir Rybakov and Sergei Mardaev. Now she is supervising several PhD and MsC students at Sobolev Institute of Mathematics and Novosibirsk State University.

**Vision Statement:** Logic is a foundation for Mathematics and Computer Science. For me, the most interesting is to provide general methods working not only for particular logics and theories but also for big families of logical systems.

**Honours and Awards:** State scientific grant of Russian Federation for distinguished scientists, 1994-1996, 1997-1999, 2000-2003.

## MANNA, Zohar

**Specialties:** Formal methods, theory of computation, logics of programs, automated deduction, temporal logic.

**Born:** 17 January 1939 in Haifa, Israel.

educated Carnegie Mellon University, Pittsburgh, PA, Computer Science PhD, 1968; Technion, Haifa, Israel, Mathematics MSc, 1965; Technion, Haifa, Israel, Mathematics BSc, 1962.

**Dissertation:** *Termination of Algorithms*; supervisors, Robert Floyd and Alan Perlis.

**Regular Academic or Research Appointments:** PROFESSOR, STANFORD UNIVERSITY, 1978–; Professor, Weizmann Institute, 1972-1995; Assistant Professor, Stanford University, 1968-1972.

**Research Profile:** Zohar Manna's research spans the theory of computation, automated deduction, temporal logic, and the specification and verification of concurrent and reactive systems. Known for his clear expositions, Manna has authored or co-authored a book and numerous papers in each of these areas.

In his early work, Manna focused on modeling programs and specifying their properties in the predicate calculus, building on the work of Hoare and Floyd in formalizing proofs of partial and total correctness of programs. This work culminated in the publishing of his highly acclaimed book, *Mathematical Theory of Computation*, which was subsequently translated into six languages in addition to English.

With Jean Vuillemin and Adi Shamir, Manna developed the fixpoint approach to the theory of computation. With Richard Waldinger, he pioneered the method of synthesizing programs from first-order proofs of properties. Together they expounded on automated deduction and synthesis in their book, *The Deductive Foundations of Computer Programming*.

In 1980, Manna began his long collaboration with Amir Pnueli in which they developed the deductive approach to verifying concurrent and reactive systems. The principles of the method include modeling systems in first-order logic, specifying system properties in linear temporal logic, and proving that systems satisfy their specifications by reducing a system and its property to a set of first-order verification conditions. Besides coauthoring with Pnueli the series of books, *Temporal Verification of Reactive Systems:* (1) *Specification*, (2) *Safety*, and the unpublished (3) *Progress*, Zohar Manna lead the development of the Stanford Temporal Prover (STeP), a proof assistant for proving properties of reactive and concurrent systems. With several students in the 1990s, Manna extended the deductive approach to modeling and verifying properties of continuous, timed, and hybrid systems.

Manna's recent work focuses on automating tasks in deductive verification. With his students, he has developed new foundations for developing procedures to synthesize program invariants and ranking functions; and he has pushed the boundaries on decidable fragments of first-order logic to automate deciding the validity of verification conditions.

**Main Publications:**

1. (with R. Waldinger) Toward Automatic Program Synthesis. *Commun. ACM* 14(3): 151-165 (1971)
2. (with J. Vuillemin) Fixpoint Approach to the Theory of Computation. *Commun. ACM* 15(7): 528-536 (1972)
3. *Mathematical Theory of Computation*, McGraw-Hill, New York, NY, 1974, 448pp. [translations: Japanese (1976), Italian (1978), Russian (1979), Czech (1981), Hungarian (1982), Bulgarian (1985)]
4. (with S. Katz) A Closer Look at Termination. *Acta Inf.* 5: 333-352 (1975)
5. (with A. Shamir) The Optimal Fixedpoint of Recursive Programs. STOC 1975: 194-206
6. (with Amir Pnueli) The Modal Logic of Programs. ICALP 1979: 385-409
7. (with R. Waldinger) A Deductive Approach to Program Synthesis. *ACM Trans. Program. Lang. Syst.* 2(1): 90-121 (1980)
8. (with P. Wolper) Synthesis of Communicating Processes from Temporal Logic Specifications. Logic of Programs 1981: 253-281
9. (with A. Pnueli) A Hierarchy of Temporal Properties. PODC 1990: 377-410
10. (with A. Pnueli) Completing the Temporal Picture. *Theoretical Computer Science* 83(1): 97-130 (1991)
11. (with T. Henzinger, A. Pnueli) Temporal Proof Methodologies for Real-time Systems. POPL 1991: 353-366
12. (with A. Pnueli) *Temporal Verification of Reactive Systems: Specification*, Springer-Verlag, New York, NY, 1991, 448pp.
13. (with A. Pnueli) Verifying Hybrid Systems. Hybrid Systems 1992: 4-35
14. (with R. Waldinger) Fundamentals of Deductive Program Synthesis. *IEEE Transactions on Software Engineering* 18(8): 674-704 (1992)
15. (with R. Waldinger) *The Deductive Foundations of Computer Programming*, Addison-Wesley Pub., Reading, MA, 1993, 717pp.
16. (with A. Browne, H. Sipma) Generalized Temporal Verification Diagrams. FSTTCS 1995: 484-498
17. (with A. Pnueli) *Temporal Verification of Reactive Systems: Safety*, Springer-Verlag, New York, NY, 1995, 514pp.
18. (with A. Browne, H. Sipma, T. Uribe) Visual Abstractions for Temporal Verification. AMAST 1998: 28-41
19. (with N. Bjorner, A. Browne, M. Colon, B. Finkbeiner, H. Sipma, T. Uribe) Verifying Temporal Properties of Reactive Systems: A STeP Tutorial. *Formal Methods in System Design* 16(3): 227-270 (2000)
*Work in Progress*
20. (with A. Bradley) *The Calculus of Computation: Decision Procedures with Applications to Verification*, Springer-Verlag, Berlin, Germany, 2007

**Service to the Profession:** Associate Editor, *Acta Informatica*; Associate Editor, *Theoretical Computer Science*; Board Member, International Institute for Software Technology, United Nations University.

**Teaching:** Among Manna's 30 students are professors at Ecole Normale Supérieure; The Technion; Weizmann Institute; Tel Aviv University; Carnegie Mellon University; Université de Liege; De Montfort University; University of California, Santa Cruz; University of Pennsylvania; University of California, Berkeley; Ecole Polytechnique Fédérale de Lausanne; Grand Valley State University; Universitat des Saarlandes; and University of Colorado, Boulder. "Verification: Theory and Practice", a symposium in honor of Manna's 64th birthday, was held in Taormina, Sicily, June 29 - July 4, 2003. The diversity of the topics of the accompanying symposium volume (Lecture Notes in Computer Science 2772) – which address temporal logics, model checking, games, security, and program analysis, to name a few areas – testifies to Manna's impact on the development of the field.

**Honours and Awards:** Fullbright Fellowship, 2002; *Doctor honoris causa*, Ecole Normale Supérieure de Cachan, France, 2002; Fellow, Association for Computing Machinery, 1993; F. L. Bauer Price, Technical University Munich, Germany, 1992; Guggenheim Fellowship, 1981; AMC Programming Systems and Languages Award, 1974.

# MARCUS, Ruth Barcan. Previously, BARCAN, Ruth C.

**Specialties:** Formal systems of quantified modal logic including interpretations and philosophical applications. Philosophy of logic, issues in semantics, philosophy of language, epistemology and metaphysics.

**Born:** August 1921 in New York City, NY, USA.

**Educated:** Yale University, PhD, 1946; Yale University, MA, 1942; New York University, Mathematics and Philosophy, BA, 1941.

**Dissertation:** *A Strict Functional Calculus*; supervisor, Frederic B. Fitch.

**Regular Academic or Research Appointments:** SENIKOR RESEARCH SCHOLAR AND HALLECK PROFESSOR OF PHILOSOPHY, EMERITUS, PHILOSOPHY, YALE UNIVERSITY, 1992–; Halleck Professor of Philosophy, Yale University, 1972-1991, Professor, Philosophy, Northwestern University, 1970-1972, Professor and Chair of the Department, University of Illinois at Chicago, 1964-1970, Assistant to Associate Professor (part time), Roosevelt University, 1959-1963; Research Associate, Institute for Human Relations, Yale University, 1945-1947.

**Visiting Academic or Research Appointments:** Visiting Distinguished Professor, University of California, Irvine, 1995-2000; Visiting Fellow, Clare Hall, Cambridge, Trinity Term, 1998; Visiting Fellow, Wolfson College, Trinity Terms, 1995, 1996; Mellon Senior Fellow, National Humanities Center, 1992-1993, Rockefeller Foundation Fellow at Bellagio, 1973, 1990; Fellow, Institute for Advanced Study, Edinburgh, 1983; Fellow, Center for Advanced Study in the Behavioral Sciences, Stanford, 1979; Fellow Center for Advanced Study, University of Illinois, 1968; Fellow, National Science Foundation, 1963-1964.

**Research Profile:** Marcus' publications, 1946 and 1947 present the first formal systems of quantified modal logic (QML). In QML, the necessity of identity is provable. That outcome, which denied contingent identities, was initially seen as paradoxical but was later accepted in the context of an appropriate interpretation which she presented in 1961 and 1962. This requires the distinction between genuine proper names and descriptions. Proper names refer directly without mediation. (She called genuine proper names "tags"). This initiated the theory of direct reference. At that time her position on the distinction was called by Quine a "red herring" and was slow to be accepted. The historical chain account of proper name transmission over time was later proposed by Peter Geach in "The Perils of Pauline", *Revue of Metaphysics,* 1969. These views were presented again by Kripke, 1972, 1980 in *Naming and Necessity*.

Among the theorems of her QML (often unnoticed) is a substitution theorem which *proscribes* in a principled way the substitution of non necessary equivalences such as material equivalence in the scope of a necessity operator. A controversial feature of her QML is "the Barcan Formula" i.e. "$\Diamond(Ex)Ax \to (Ex) \Diamond Ax$". It is shown (1962) that on a model theoretic interpretation where domains of interpretation are constant, the Barcan formula is provable. In 1975, 1985-1986 and 1997, she argues on linguistic and empirical grounds that the constant domains of interpretation should be restricted to actual objects. With the rejection of possibilia, the Barcan Formula holds. The Barcan formula and the status of so called *possibilia* are still being debated.

A substitutional semantics for the quantifiers is also considered (1961-1962), but is not urged. In 1972, Marcus shows Quine's claim that substitutional semantics leads to contradiction is fallacious. She considers substitutional semantics useful for certain applications as in the case of fictional discourse. In 1967 Marcus developed an account of Aristotelian essentialism and some variants within a formal framework of QML, 1967. It is argued that Aristotelian essentialism is a plausible notion in common use and not, as claimed "invidious". She conjectured that QML is not "committed" to essentialism in the sense that there must be essentialist truths, a claim that Terence Parsons proved in *Philosophical Review,* 1969. However, in 1971, Marcus argued that essential attributes such as kind properties are commonplace and central to an understanding of physical modalities and nomological theories. In 1980 and 1996 Marcus wrote papers on ethics which challenged some deontic logics. She shows on a standard view of consistency, that a moral code is consistent if there is some possible set of circumstances in which all the rules are obeyable. In the actual world, even a single rule can generate dilemmas. She argues that such conflicts do not wholly erase a failed obligation.

Between 1981 and 1995, Marcus developed an account of belief with consequences for epistemic logic. It treats belief as relating an agent and a possible state of affairs where states of affairs are structures of possible features or arrangements of actual objects. In believing, the agent behaves as if that possible state of affairs obtains. Contrary to the disquotation account of belief, assenting to a sentence is just one of the behaviors which may be a mark of believing. A consequence for epistemic logic is that an agent cannot believe an impossibility. Marcus has also written on the nature of extensionality (1963) on classes and collections in modal languages (1973), on puzzles about iterated modalities (1966). She has also written on historical topics such as Spinoza's ontological proof (1986), Russell's post-1920's views on identity and individuation (1986), and Quine's vacillation and displeasure with modal logic and semantics of modalities.

**Main Publications:**

1. *The Logical Enterprise,* eds. R.B. Marcus, A. Anderson and R. Martin, Yale Press, 1975.
2. *Modalities,* Oxford University press, 1993, 1995.
3. **Articles published under Ruth C. Barcan**
4. "A Functional Calculus of First Order Based on Strict Implication", *Journal of Symbolic Logic,* 11, 1946.
5. "The Deduction Theorem in a Functional Calculus of First Order Based on Strict Implication", *Journal of Symbolic Logic,* 11, 1946
6. "The Identity of Individuals in a Strict Functional Calculus of Second Order", *Journal of Symbolic Logic,* 12, 1947.

**Published under "Ruth Barcan Marcus" sometimes indexed under "Barcan-Marcus"**

7. "Strict Implication, Deducibility and the Deduction Theorem", *Journal of Symbolic Logic, 18, 1953.*
8. "Extensionality", *Mind,* LXIX ns. 273, 1963. Reprinted in *Reference and Modality,* ed. L. Linsky, OUP, 1971 Italian Translation of *Reference and Modality.*
9. "Modalities and Intensional Languages", *Synthese,* XIII, 4, 1961.Reprinted in*Boston Studies in Philosophy of Science,* ed. M. Wartofsky, Holland, 1963. *Contemporary Readings in Logical Theory,* eds. I. Copi, J. Gould Macmillan, 1967. Hungarian translation of the above, Gondolat, Budapest, 1985. *Readings in Semantics,* eds. F. Zabeh *et al.* 1974. *Readings in Philosophical logic,* eds. I. Copi and J. Gould St. Martins press, 1978.
10. 'Interpreting Quantification", *Inquiry,* 5, no 3, 1962.
11. "Essentialism in Modal Logic", *Nous* 1, 1967.
12. "Classes, Collections, and Individuals", *American Philosophical Quarterly, July, 1974.*
13. "Does the Principle of Substititutivity Rest on a Mistake" in *The Logical Enterprise,* eds. A. Anderson, R.B. Marcus, R. Martin, Yale, 1975.
14. "Nominalism and the substitutional Quantifier", *The Monist,* Vol. 61, 1978. Reprinted in *Philosophy of Logic,* ed. Dale Jacquette, Blackwell, 20.
15. "Moral Dilemmas and Consistency", *Journal of Philosophy,* LXXVII, 3, 1980.
16. "Rationality and Believing the Impossible", *Journal of Philosophy,* LXXV, no 6, 1983.
17. "Pissibilia and Possible Worlds", *Grazer Philosophische Studien,* ed. Rudolf Haller Vols. 25, 26 1985-1986.
18. "A Backward Look at Quine's Animadversions on Modalities", in *Perspectives on Quine,* eds. R. Gibson and R. Barrett, Oxford Press, 1990. Reprinted in *Philosophy of Logic,* ed. Dale Jacquette, Blackwell, 2002.
19. "Some Revisionary Proposals About Belief and Believing", *Philosophy and Phenomenological Research,* 1990. Reprinted in *Causality, Method and Modality,* ed. Gordon Brittain, Kluwer, 1991.
20. "Ontological Implications of the Barcan Formula", *Philosophia Handbook of Metaphysics and Ontology,* 1992.
21. Interview in *Formal Philosophy,* editors V. Hendricks and J. Symons *Auto Press,* 2005.

*Work in Progress*

22. Book length manuscript on Epistemology, Belief, and Reference.

**Service to the Profession:** APA Delegate, American Council of Learned Societies, 2006-2009; Institut International de Philosophie, elected member, President, 1989-1992, President Honoraire 1993-; Steering Committee, International Federation of Philosophy Societies, 1985-2000; Program Chair, 1983 Meeting in Salzburg, Austria and editor with P. Weingartner and G. Dorn of the Proceedings, North-Holland, 1986; Conference Board Member, Mathematical Sciences, 1983-1986; Chair, U.S. National Committee of the International Union of Logic, Methodology and Philosophy of Science, 1977-1979; Department Chair, Philosophy Department at the University of Illinois, Chicago, 1963-1969; Various Offices, American Philosophical Association (APA), 1961-1983 without interruption, including President of the

Western Division 1975-1976 and Chair of the National Board of Officers, 1977-1983; Various Offices, Association for Symbolic Logic, President, 1983-1986, Council member 1983-1990, 1961-;

Intermittent service on editorial boards. Partial list: Editorial Board Member, Journal of Symbolic Logic; Editorial Board Member, Philosophical Studies; Editorial Board Member, Philosophers Annual; Editorial Board Member, Mid-west Studies in Philosophy; Editorial Board Member, The Monist.

Intermittent service on foundation panels: Panel Member, National Science Foundation; Editorial Board Member, National Endowment for the Humanities; Editorial Board Member, Rockefeller Foundation Fulbright Committee, 1972-1974.

Outside advisor or examiner of academic programs or departments, partial list: Princeton University Chair of Visiting Committee 1977-1985, MIT 1978-1984, Caltech 1981-1985, University of California, diverse campuses 1979, 1985, 1986. Cornell Society of Fellows 1979-1982, Duke University 1987, Columbia University 1989, University of Virginia, 1992, Ohio State University 1996-1999, Carnegie Mellon 2001, University of Texas at Austin 2002, Brown University 2003.

**Teaching:** Former students both graduate and undergraduate are now to be found teacher, *inter alia,* at Dartmouth, Lehigh, NYU, Pittsburgh, Rutgers, SUNY-Purchase, Texas-Austin, Toronto, and UCLA. They range from undergraduates to graduates who wrote dissertations under supervision. Also served as a grader for the Yale Physics Department during WW2.

**Vision Statement:** "Many philosophers continue to remain aloof from research in the sciences at peril to the disciplines of logic and philosophy. There is also a failure of historical perspective among some logicians and analytical philosophers which should be corrected. This is not just in the interest of historical accuracy. Misrepresentation of history sometimes closes off fruitful areas of research"

**Honours and Awards:** Lauener Prize "for an outstanding oeuvre", to be presented in 2008 in Bern, Switzerland; Wilbur Cross Medal, Yale University, 2000; Honorary Doctor of Humane Letters, University of Illinois, 1995; Institut International de Philosophie, President Honoraire, 1993; Medal of the College de France, 1986-; American Academy of Arts and Sciences, Fellow, 1977-; Machette Award to Yale in honor of Professor Ruth Barcan Marcus' contributions to the profession, 1986; Guggenheim Fellow, 1953; Ph Beta Kappa, 1941; Conferences or symposia celebrating Ruth B. Marcus, UCLA 1998 and APA twice; Two festschrifts: *Modality, Morality and Belief,* eds. W. Sinnott-Armstrong, D. Raffman, N. Asher, Cambridge, 1995. and *Dialectica,* vol. 53, 1999, ed. H. Lauener. See also visiting research fellowships.

## MARES, Edwin David

**Specialties:** Non-classical logic, relevant logic, modal logic, substructural logics, paraconsistent logic, logic and probability theory, belief revision, philosophy of logic.

**Born:** 14 December 1961 in Toronto, Ontario, Canada.

**Educated:** Indiana University, PhD; McMaster University BA (Hons).

**Dissertation:** *The Logic of Fictional Discourse*; supervisor, J. Michael Dunn.

**Regular Academic or Research Appointments:** ASSOCIATE PROFESSOR AND READER, VICTORIA UNIVERSITY OF WELLINGTON (NZ), 2004–; Senior Lecturer, Victoria University of Wellington, 1997-2004; Lecturer, Victoria University of Wellington, 1993-1997.

**Visiting Academic or Research Appointments:** Visiting Assistant Professor, University of Victoria (Canada), 1992-1993; Visiting Assistant Professor, Dalhousie University, 1992; Visiting Fellow, Social Sciences and Humanities Council of Canada Postdoc, Australian National University, 1989-1991.

**Research Profile:** Most of Ed Mares' work has been in relevant logic and its interpretation. In his book, *Relevant Logic: A Philosophical Interpretation*, and in the various introductions to relevant logic that he has written, he has tried to provide an interpretation of relevant logic that most mainstream philosophers could understand and with which many would feel sympathetic. Alone and with others (such as Bob Meyer, Rob Goldblatt, Charles Morgan, and Paul McNamara), he has worked on the technical development of relevant logic and other logics, such as modal, deontic, and intuitionist logic. In the mid-1990s, together with Andre Fuhrmann, he developed a theory of the counterfactual conditional based on relevant logic. He has also developed a theory of belief revision based on paraconsistent logic.

**Main Publications:**

1. *Relevant Logic: A Philosophical Interpretation*, Cambridge: Cambridge University Press, 2004

2. (with Stuart Brock) *Realism and Antirealism*, Stocksfield, UK: Acumen, 2007

3. (with Robert Goldblatt) "A General Semantics for Quantified Modal Logic" in G. Govenatory, I. Hodkinson, and Y. Venema (eds.), *Advances in Modal Logic*, volume 6, London: College Publications, 2006, 227-246

4. (with Goldblatt) "An Alternative Semantics for Quantified Relevant Logic" *The Journal of Symbolic Logic* 71 (2006):163-187

5. "A Paraconsistent Theory of Belief Revision" *Erkenntnis* 56 (2002) pp 229-246

6. "Relevance Logic" in Dale Jacquette (ed.), *Companion to Philosophical Logic*, Oxford: Blackwell, 2002, 602-627

7. (with R.K. Meyer) "Relevant Logics" in Lou Goble (ed.), *Guide to Philosophical Logic*, Oxford: Blackwell, 2001, 280-308

8. "The Incompleteness of RGL" *Studia Logica* 65 (2000) pp 315-322

9. "Paraconsistent Probability Theory and Paraconsistent Bayseanism" *Logique et analyse* 160 (1997), 375-384

10. "Who's Afraid of Impossible Worlds?" *Notre Dame Journal of Formal Logic* 38 (1997), 516-526

11. "Relevant Logic and the Theory of Information" *Synthese* 109 (1997), 345-360

12. "A Star-Free Semantics for R" *The Journal of Symbolic Logic* 60 (1995): 579-90

13. (with Andre Fuhrmann) "A Relevant Theory of Conditionals" *Journal of Philosophical Logic* 24 (1995): 645-65

14. (with R.K. Meyer) "The Semantics of Entailment 0" in K. Dosen and P. Schroeder-Heister (eds.), *Substructural Logics*, Oxford: Oxford University Press, 2003, 239-258

15. (with R.K. Meyer) "The Admissibility of Gamma in R4" *Notre Dame Journal of Formal Logic*

*Work in Progress*

16. "Identity in Relevant Logic"

17. "Russell's Multiple Relations Theory of Judgment: Yet Another Reconstruction"

18. "Information and Inference"

**Service to the Profession:** Panel Member, Marsden Fund, 2007-; Editorial Board Member, Australasian Journal of Philosophy, 2004-; Editorial Board Member, Studia Logica, 2000-; Reviews Editor, Studia Logica, 2000-2002; Member, Australasian Committee of the Association of Symbolic Logic, 1999-2004.

**Teaching:** He teaches and supervises projects in a wide variety of areas in logic and philosophy, from ethics to formal logic. His PhD students include: David Hadorn, Marjan Kljakovic, Ismay Barwell and his MA students include: Beate Elsner, Lin Woolaston, Tony Smith, Mike Hurst.

**Vision Statement:** Although fascinating, the technical nature of logic has alienated many philosophers and driven logic out of many philosophy departments. I would like to see a reconciliation between logic and philosophy and see logic integrated into more mainstream work in philosophy and more philosophical questions raised about contemporary logic (in particular about substructural logics). In its technical development, logic has gone from strength to strength and I cannot see that abating any time soon.

**Honours and Awards:** (with R. Goldblatt) Fellowship, Marsden Fund, New Zealand Government, 2006-2008; President, Australasian Association of Philosophy (NZ), 2006; President, Australasian Association of Logic, 2001; Postdoctoral Fellow, Social Sciences and Humanities Research Council of Canada, 1989-1991; Doctoral Fellow, Social Sciences and Humanities Research Council of Canada, 1987-1988.

## MARKER, David

**Specialties:** Model Theory and its applications

**Born:** 29 January 1958 in Schenectady, New York, USA.

**Educated:** Yale University, PhD Mathematics, 1983; Union College, BS Mathematics, 1980.

**Dissertation:** *Degrees of Models of Arithmetic*; supervisor, Angus Macintyre.

**Regular Academic or Research Appointments:** PROFESSOR, UNIVERSITY OF ILLINOIS AT CHICAGO, 1994–; Associate Professor, University of Illinois at Chicago, 1989-1994; Assistant Professor, University of Illinois at Chicago, 1985-1989; National Science Foundation Postdoctoral Research Fellow, University of California at Berkeley, 1983-1985.

**Visiting Academic or Research Appointments:** Visiting Research Fellow, Merton College, Oxford, 2007; Member, Mathematical Sciences Research Institute, 1998; Visiting Professor, University of Notre Dame, 1994; Member, Mathematical Sciences Research Institute, 1989.

**Research Profile:** David Marker's research has focused on model theory and it applications to study definability in concrete mathematical structures. He has been particularly interested in: o-minimality and its applications to real algebraic and real analytic geometry, the real and complex fields with exponentiation, and differential algebra.

**Main Publications:**

1. A remark on Zilber's pseudoexponentiation. J. Symbolic Logic 71 (2006), no. 3, 791—798.

2. (with M. Messmer and A. Pillay) Model theory of fields. Second edition. Lecture Notes in Logic, 5. Association for Symbolic Logic, La Jolla, CA; A K Peters, Ltd., Wellesley, MA, 2006.
3. Model theory. An introduction. Graduate Texts in Mathematics, 217. Springer-Verlag, New York, 2002.
4. Manin kernels. Connections between model theory and algebraic and analytic geometry, 1–21, Quad. Mat., 6, Dept. Math., Seconda Univ. Napoli, Caserta, 2000.
5. (with D. Macpherson and C. Steinhorn) Weakly o-minimal structures and real closed fields. Trans. Amer. Math. Soc. 352 (2000), no. 12, 5435—5483.
6. (with L. van den Dries and A. Macintyre) Logarithmic-exponential series. Proceedings of the International Conference "Analyse & Logique" (Mons, 1997). Ann. Pure Appl. Logic 111 (2001), no. 1-2, 61–113.
7. (with L. van den Dries and A. Macintyre) Logarithmic-exponential power series. J. London Math. Soc. (2) 56 (1997), no. 3, 417–434.
8. (with L. van den Dries and A. Macintyre) The elementary theory of restricted analytic fields with exponentiation. Ann. of Math. (2) 140 (1994), no. 1, 183–205.
9. (with C. Steinhorn) Definable types in o-minimal theories. J. Symbolic Logic 59 (1994), no. 1, 185–198.
10. Semialgebraic expansions of C. Trans. Amer. Math. Soc. 320 (1990), no. 2, 581–592.
11. (with A. Pillay) Reducts of C which contain $+$. J. Symbolic Logic 55 (1990), no. 3, 1243–1251.
12. (with L. Harrington and S. Shelah) Borel orderings. Trans. Amer. Math. Soc. 310 (1988), no. 1, 293–302.
13. (with A. Macintyre) Degrees of recursively saturated models. Trans. Amer. Math. Soc. 282 (1984), no. 2, 539–554.
14. Degrees of models of true arithmetic. Proceedings of the Herbrand symposium (Marseilles, 1981), 233–242, Stud. Logic Found. Math., 107, North-Holland, Amsterdam, 1982.

**Service to the Profession:** Editorial Board Member, Archive for Mathematical Logic, 2004-; Advisory Editor, Annals of Pure and Applied Logic, 1998-, Editorial Board, 1996-; Editor, Journal for Symbolic Logic, 1994-2000; Managing Editor, Association for Symbolic Logic Lecture Notes in Logic, 2005-2006; Association for Symbolic Logic, Publisher, 2006-, Member, Executive Committee of the Association for Symbolic Logic, 1997-1999; Head, Department of Mathematics, Statistics, and Computer Science, University of Illinois at Chicago 2007-.

**Teaching:** Marker has supervised eight PhD students and taught numerous graduate level courses in mathematical logic. His model theory courses provided the basis for his textbook *Model Theory: An Introduction* and his chapters in *Model Theory of Fields*. In 2004 he was recognized with the University of Illinois at Chicago Award for Excellence in Teaching.

**Vision Statement:** I believe that Model Theory provides a powerful tool for studying definability and independence in important mathematical contexts. I anticipate the variety and depth of applications will grow dramatically in the coming years.

**Honours and Awards:** Shoenfield Prize, for Expository Writing, Association for Symbolic Logic, 2007; University of Illinois Scholar, 1995; Research Fellowship, American Mathematical Society Centennial, 1994-1996; Postdoctoral Research Fellowship, National Science Foundation, 1983-1985.

# MATHIAS, Adrian Richard David

**Specialties:** Set theory, its applications to symbolic dynamics, axiomatics, sociology of logic.

**Born:** 12 February 1944 in Llanidloes, Wales, Great Britain.

**Educated:** University of Cambridge, PhD, 1970; University of Cambridge, MA, 1969; University of Cambridge, BA, 1965.

**Dissertation:** *On a generalization of Ramsey's Theorem*; supervisor, Ronald Jensen.

**Regular Academic or Research Appointments:** PROFESSOR, UNIVERSITÉ DE LA REUNION, France Outre-Mer, Associé, 1999, Titulaire 2000, Première Classe, 2006- ; Fellow of Peterhouse, Cambridge, Research, 1969, Supernumerary, 1970, Official, 1972-1990; Lector, Trinity College, Cambridge, 1980-1989; Assistant Lecturer, Cambridge, 1970-1975.

**Visiting Academic or Research Appointments:** Sanford Professor, Universidad de los Andes, Bogota, 1997-1998; Member, Centre de Recerca Matematica, Barcelona, 1993-1996, 2003-2004; Research Fellow, Merton College, Oxford, 1995; Dauergast, Mathematisches Forschungsinstitut, Oberwolfach, 1992-1993; Professor, University of Warsaw, 1991-1992; Professor, University of California, Berkeley, 1991; Member, MSRI, Berkeley, 1989-1990; Senior Logician, Odyssey Research Associates, Ithaca, NY, 1985; Professor, Simon Fraser University, Canada, 1983; Lecturer, Technical University, Berlin, 1980; Wissenschaftlicher Angestellter, Freiburg im Breisgau, 1979-1980; Lecturer, University of Wisconsin, Madison, 1968-1969; Research Associate, Stanford, 1967-1968.

Short academic visits: Monash 1969, Banach Centre Warsaw 1973, Calgary 1974, IVIC, Caracas 1976; Harvard 1982; CalTech 1983; Cornell 1985; Sienna 1986; Bonn, Caen, 1993; MSRI

1993, 1996; Oberwolfach 1996/7; Humboldt Universitaet, Berlin, 1999; Universitat de Barcelona 2006, 2007.

**Research Profile:** A. R. D. Mathias' first papers studied set-theoretic properties contradicting the Axiom of Choice. The principal result of Mathias' thesis was that in a celebrated model of Solovay, all families of sets of integers are completely Ramsey; the principal new lemma was that any infinite subset of what is now called a Mathias real is also a Mathias real. The paper "Happy Families" gives generalizations of the main properties of Mathias forcing to the context of what are now called selective co-ideals, and numerous applications to the set theory of the real line, such as the result that every analytic filter is feeble, (meaning that it is projectible by a finite-to-one function to the Frechet filter).

These ideas also generalized to the study of large cardinals, yielding a characterization of Prikry generic sequences, and, in joint work with Henle, to establishing properties of strong partition cardinals. This work, in a further collaboration with Henle and Woodin, established the necessity of the hypothesis that V=L(R) in the Kechris-Woodin result concerning the derivation of the axiom of determinacy from the existence of numerous strong partition cardinals.

Mathias' encounter with the Barcelona symbolic dynamics group led to a series of papers solving certain iteration questions by set-theoretic methods.

Latterly Mathias has been collaborating with Woodin and Hauser in a book that aims to expound Woodin's equiconsistency results for the axiom of determinacy.

Mathias' recent work has concerned weak systems of set theory and their use in analyzing the foundational ideas of Mac Lane, Bourbaki, Gandy, Jensen and Devlin. At present he is exploring the properties of the class of rudimentarily recursive functions, a subclass of the primitive recursive set functions of Jensen and Karp, the study of which, he believes, will enlarge our understanding of Cohen's forcing relation, and of the proper balance between set-theoretic and category-theoretic approaches to the foundations of mathematics.

Besides the above purely mathematical work, A. R. D. Mathias has published several critical essays concerning philosophical, sociological and pedagogical aspects of logic.

**Main Publications:**

1. Solution of problems of Choquet and Puritz, in Logic Colloquium '70, ed. W. Hodges, Springer Lecture Notes in Mathematics 255, (1972) 204-210.
2. On sequences generic in the sense of Prikry, J.Austr. Math. Soc 15 (1973) 409-414.
3. The order extension principle, in Proc. Symp. Pure Mathematics, Volume XIII, Part II, ed T. Jech, (AMS 1974) 179-183.
4. Happy Families, Ann. Math. Logic 12 (1977) 59-111.
5. O sharp and the p-point problem, in Higher Set Theory, ed. G. Mueller and D. Scott, Springer Lecture Notes in Mathematics, 669, 375-384.
6. Surrealist Landscape with Figures, Periodica Hungarica 10 (1980) 109-175.
7. (with J.M.Henle) Supercontinuity, Math. Proc. Cam. Phil. Soc. 92 (1982) 1-15.
8. Unsound ordinals, Math. Proc. Cam. Phil. Soc. 96 (1984) 391-411.
9. (with J.M.Henle and W.H.Woodin) A barren extension, Proc. VIth Latin American Logic Coll.oquium, Caracas 1983, (edited C. Di Prisco) Springer Lecture Notes in Mathematics, 1130, (1985) 195-207.
10. Logic and Terror, Physis Riv.Internaz. Storia Sci. 28 (1991), 557-578.
11. The Ignorance of Bourbaki, Mathematical Intelligencer 14 (1992), 4-13. [also published, elsewhere, in Hungarian and Spanish translations.]
12. What is Mac Lane missing? in Set Theory of the Continuum, (edited H. Judah, W. Just and H. Woodin) MSRI Research Publications Volume 26, Springer-Verlag, 1992.
13. Strong statements of analysis, Bull. London Math. Soc., 32 (2000), 513-526.
14. Delays, Recurrence and Ordinals, Proc. London Math. Soc. (3) 82 (2001) 257-298.
15. Slim Models of Zermelo Set Theory, JSL 66 (2001), 487-496.
16. The Strength of Mac Lane Set Theory, APAL 110 (2001), 107-234.
17. A term of length 4,523,659,424,929, Synthese 133 (2002) 75-86.
18. Choosing an attacker by a local derivation, Acta Universitatis Carolinae – Math. et Phys., 45 (2004) 67-73.
19. Analytic Sets under Attack, Math Proc Cam Phil Soc 138 (2005), 465-485.
20. Weak systems of Gandy, Jensen and Devlin, in Set Theory: Centre de Recerca Matemàtica, Barcelona 2003-4, ed; J. Bagaria and S. Todorcevic, Trends in Mathematics, Birkhaeuser Verlag, Basel, 2006, 149-224.

A complete list may be found at http://www.dpmms.cam.ac.uk/~$ardm and at http://www.univ-reunion.fr/~ardm

*Work in Progress*

21. (with J. Bagaria and C. Casacuberta) "Epireflections and supercompact cardinals"; "Rudimentary recursion"; "The banning of formal logic from a French national competitive examination."
22. (with W. Hugh Woodin and Kai Hauser) The Axiom of Determinacy; Admissibility, Constructibility and

Forcing; Danish Lectures on the Foundations of Mathematics.

**Service to the Profession:** President, Commission des Specialistes, Section 25, Université de la Réunion and Scientific Director of the research team ERMIT, 2002;-Co-Organiser, CRM, Set Theory Meetings and Co-Editor of the Proceedings of CRM 1996, 1996, 1997; Editor, (Mathematical) Proceedings of the Cambridge Philosophical Society, 1972-1974; Director, two three-week Summer Schools, Cambridge, 1971, 1978; Campaigner, Logic Trust (a British registered charity) and Logfit (a UK government funding initiative); Composer, *La Marxa del Centre de Recerca Matemàtica*, a March written for the opening of the new buildings of the CRM at Bellaterra, Catalonia, and dedicated to its Director, Manuel Castellet, and his staff.

**Teaching:** Numerous graduate courses on set theory, model theory and recursion theory taught at Cambridge; wholly or partly supervised seven successful Cambridge PhD candidates: David Guaspari, Aki Kanamori, Thomas Forster, Robert Seeley, Neil Tennant, James Cummings, David Seetapun. Courses on fundamental techniques of set theory taught at Monash, Warsaw, Berlin, Sienna, Berkeley, Barcelona, Bogota, Oxford, and Reunion. Courses on Axiom of Determinacy taught at Warsaw, Caen, Bonn, Luminy, Barcelona, Bogota, and Reunion.

**Vision Statement:** "In my view the chief foundational problem to be faced in logic today is the damaging conflict between the set-theoretical and category-theoretical views of mathematics. I hope to clarify their relationship by the further study of very weak systems of set theory; that study might also contribute to the reworking and consolidation of the explosion of results in set theory that began in the 1960's."

**Honours and Awards:** Honorary Fellow, Department of Mathematics, University College of Wales, Aberystwyth; Honorary Fellow, Department of Mathematics, University College, London; Recognised Lecturer, Faculty of Mathematics, University of Cambridge, 1986-1990.

# MENDELSON, Elliott

**Specialties:** Axiomatic set theory, general logic.

**Born:** 24 May 1931 in New York, NY, USA.

**Educated:** Cornell University, PhD Mathematics, 1955; Cornell University, MA Mathematics, 1954; Columbia University, BA, 1952.

**Dissertation:** *The Independence of a Weak Axiom of Choice*; supervisor, Barkley Rosser.

**Regular Academic or Research Appointments:** PROFESSOR EMERITUS, MATHEMATICS, QUEENS COLLEGE AND THE GRADUATE CENTRE OF THE CITY UNIVERSITY OF NEW YORK, 2000–; Professor, Mathematics, Queens College, 1964-2000; (Also served as professor at the Graduate Center of the City University of New York from the late 1960's until 2000, first in the Ph.D, Program in Mathematics and later in the Ph.D. Program in Philosophy); Associate Professor, Mathematics, Queens College, 1961-1964; Ritt Instructor, Mathematics, Columbia University, 1958-1961; Junior Fellow, Society of Fellows, Harvard University, 1956-1958; Research Instructor, Mathematics, University of Chicago, 1955-1956.

**Visiting Academic or Research Appointments:** Visiting Scholar, Wolfson College, Oxford University, 1990; Visiting Professor, University of Siena, Italy, 1983; Visiting Professor, University of Pennsylvania, 1972.

**Research Profile:** His research has been in wide areas of logic and set theory. In the latter, he has done the most work on consistency and independence proofs. His earliest work, during his master's thesis, involved the use of Gödel's Second Theorem to prove independence results. His doctoral thesis proved the independence of a weak form of the axiom of choice. It did not include the full axiom of regularity, although it did include some weaker forms and did not use urelements. An independence proof that took care of regularity was obtained later by Paul Cohen using his new forcing techniques. Some of his research after that was incorporated into his mathematical logic textbook. He did some work on nonstandard models of number theory, but was disappointed by the failure to find significant applications. More recently, he has been interested in topics on the borderline between logic and philosophy of mathematics, for example, issues related to Church's Thesis. He also has wandered off recently into game theory, although not primarily into the parts related to logic. That work resulted in a book, *Introducing Game Theory*, published in 2004.

**Main Publications:**

1. Some proofs of independence in axiomatic set theory. JSL, 21, 1956, 291-303.
2. The independence of a weak axiom of choice. JSL, 21, 1956, 350-366.
3. The axiom of Fundierung and the axiom of choice. Archiv für mathematische Logik und Grundlagenforschung, 4, 1958, 65-70.

4. On a class of universal ordered sets. Proc. Amer. Math. Soc., 9, 1958, 712-713.
5. A semantic proof of the eliminability of descriptions. Zeitschrift für math. Logik und Grundlagenforschung, 6, 1960, 199-200.
6. On non-standard models for number theory. In Essays on the Foundations of Mathematics, Magnes, Jerusalem, 1961, 259-268.
7. On some recent criticism of Church's Thesis. Notre Dame J. of Formal Logic, IV, 1963, 201-205.
8. Introduction to Boolean Algebra and Switching Circuits. Schaum, McGraw-Hill, 1970.
9. Number Systems and the Foundations of Analysis. Academic Press, 1973.
10. Strong axioms of infinity. Atti degli Incontri di Logica Matematica, Vol. 2, Scuola di Specializzazione in Logica Matematica, Universitá di Siena, 1986, 75-166.
11. Infinity in set theory. In: L'infinito Nella Scienza, Istituto della Enciclopedia Italiana, 1987, 127-140.
12. Second thoughts about Church's Thesis and mathematical proofs. J. of Philosophy, 1990, 225-233.
13. Introduction to Mathematical Logic. Fourth Edition. Chapman & Hall / CRC Press, 1997. (The First, Second, and Third Editions were published, respectively, in 1964, 1979, and 1987.)
14. Introducing Game Theory and Its Applications. Chapman & Hall / CRC Press, 2004.
15. On the impossibility of proving the "hard-half" of Church's Thesis. In: Church's Thesis After 70 Years. Ontos Verlag, 2006, 304-309.

*Work in Progress*

16. I am now working on the manuscript for the Fifth Edition of Introduction to Mathematical Logic. I hope to improve the treatment of Gödel's Second Incompleteness Theorem, if that proves to be possible without enlarging that chapter enormously. Moreover, I plan to include an entirely new chapter, devoted to modal logic. There is also a need to add more interesting exercises and to enlarge the appendix of solutions. In addition, after having taught a course on Computability in the 2006 Summer Session, I am toying with the idea of writing an introductory text on the subject.

**Service to the Profession:** Comitato internazionale di lettura, Archimede, Rivista Trimestrale, Italy, 2003-; Reviewer, Mathematical Reviews and the Zentralblatt für Mathematik, "over many years"-; Consulting Editor, Notre Dame J. of Formal Logic, ?-2004.

**Teaching:** My teaching career has centered on teaching at the undergraduate and master's level. This has been very satisfying, but I missed the stimulation that teaching primarily at the doctoral level would have offered.

**Vision Statement:** We do not have at present any dominant inspiring theme or set of problems that compares to the Hilbert Programme or to the unresolved questions in set theory at the beginning of the 1930's. I am somewhat disturbed by a growing tendency among influential logicians to look for a solution in the opposite direction, by accepting a constructivism that simply dissolves these problems. Gödel attributed his most successful work to his Platonistic views, whereas, in his opinion, constructivistic blinders prevented other equally gifted mathematicians from seeing what he saw.

**Honours and Awards:** Listing, American Men of Science; Listing, Who's Who in America.

## MEYER, John-Jules Charles

**Specialties:** Applied logic, logic for artificial intelligence, deontic logic, epistemic logic, modal logic of intelligent agents.

**Born:** 17 November 1954 in The Hague, The Netherlands.

**Educated:** Vrije Universiteit Amsterdam, PhD Mathematics and Exact Sciences, 1985; Leyden University, Masters Mathematics with Computer Science and Digital Signal Processing, 1979.

**Dissertation:** *Programming Calculi Based on Fixed Point Transformations*; supervisor, Jaco W. de Bakker

**Regular Academic or Research Appointments:** FULL PROFESSOR OF ARTIFICIAL INTELLIGENCE, INSTITUTE OF INFORMATION & COMPUTING SCIENCES (ICS), UTRECHT UNIVERSITY, 1998–; Professor, Formal Methods for Programming, Utrecht University, 1993-1998; Professor Theoretical Computer Science, Nijmegen University, 1989-1993; Professor, "Logic for Distributed Systems and Artificial Intelligence," Vrije Universiteit Amsterdam, 1988-1993; Associate Professor Theoretical Computer Scientist, Vrije Universiteit Amsterdam, 1987-1988; Assistant Professor Theoretical Computer Scientist, Vrije Universiteit Amsterdam, 1985-1987; Lecturer Theoretical Computer Scientist, Vrije Universiteit Amsterdam, 1980-1985.

**Visiting Academic or Research Appointments:** Visiting Researcher Linköping University (IDA), Sweden, 1995.

**Research Profile:** Meyer's research has evolved over the years from the semantics of programming languages (his PhD work) to the use of logic in describing and specifying many kinds of knowledge and systems in philosophy and particularly AI, covering deontic logic, epistemic logic, dynamic logic, commonsense / non-monotonic reasoning, BDI logics for intelligent agents, logics for databases, analogous reasoning, game-theoretic reasoning, some linguistic issues, and, very recently, emotions in agent systems. In particular, in his early years of research, by combining his interest for deontic logic with his PhD work on programming semantics and logics, he realized that an Andersonian-like reduction of deontic logic to dynamic logic gave rise to a true logic of ought-to-do (rather than ought-to-be), which approach is later referred to in the literature as dynamic deontic logic. Later, he turned to the study of other 'applied' modal logics as well and epistemic logic in particular, together with his student Wiebe van der Hoek. This work gradually moved into the direction of modal logic descriptions of all kinds of mental attitudes of intelligent / rational agents. Since around 2000, he has worked with my students on the integration of logic and programming languages for intelligent agents, including issues of communication and co-ordination. Very recently, he has become very interested in other cognitive aspects of agents as well, such as emotions. He finds it particularly interesting how emotional behaviour of agents can be described using a (logical!) framework, and how emotions can be employed as a structuring device in an (improved) architecture of an agent-based system. Generally speaking, he has always been fascinated how to coin concepts into logical formalism. Besides that he loves to ponder (logical) paradoxes, such as in e.g. deontic and epistemic logic: the simpler they can be stated, the more intriguing and inspiring ...!

**Main Publications:**

1. J.-J.Ch. Meyer, A Different Approach to Deontic Logic: Deontic Logic Viewed as a Variant of Dynamic Logic, Notre Dame J. of Formal Logic 29(1), 1988, pp. 109-136.

2. R.J. Wieringa, J.-J.Ch. Meyer & H. Weigand, Specifying Dynamic and Deontic Integrity Constraints, Data & Knowledge Engineering 4(2), 1989, pp. 157-190

3. J.-J. Ch. Meyer & R.J. Wieringa (eds.), Deontic Logic in Computer Science: Normative System Specification, John Wiley & Sons Ltd., Chichester, 1993, xiv + 317 p.

4. W. van der Hoek, J.-J. Ch. Meyer & J. Treur, Temporalizing Epistemic Default Logic, Journal of Logic, Language and Information 7(3),1998, pp. 341-367.

5. B. van Linder, W. van der Hoek & J.-J. Ch. Meyer, Seeing is Believing (And So Are Hearing and Jumping), Journal of Logic, Language and Information 6, 1997, pp. 33-61.

6. J.-J. Ch. Meyer & W. van der Hoek, Epistemic Logic for AI and Computer Science, Cambridge Tracts in Theoretical Computer Science 41, Cambridge University Press, 1995.

7. F. Dignum, J.-J. Ch. Meyer , R.J. Wieringa & R. Kuiper, A Modal Approach to Intentions, Commitments and Obligations: Intention plus Commitment Yields Obligation, in: Deontic Logic, Agency and Normative Systems (Proc. DEON'96) (M.A. Brown & J. Carmo, eds.), Workshops in Computing, Springer, Berlin, 1996, pp. 80-97.

8. L.C. van der Gaag & J.-J. Ch. Meyer, Informational Independence: Models and Normal Forms, Int. J. of Intelligent Systems 13, 1998, pp. 83-109.

9. J.-J. Ch. Meyer, R.J. Wieringa & F.P.M. Dignum, The Role of Deontic Logic in the Specification of Information Systems, in: Logics for Databases and Information Systems (J. Chomicki & G. Saake, eds.), Kluwer, Boston/Dordrecht, 1998, pp. 71-115.

10. J.-J. Ch. Meyer, W. van der Hoek & B. van Linder, A Logical Approach to the Dynamics of Commitments, AI Journal 113, 1999, 1-40.

11. W. van der Hoek, B. van Linder & J.-J. Ch. Meyer, An Integrated Modal Approach to Rational Agents, in: M. Wooldridge & A. Rao (eds.), Foundations of Rational Agency, Applied Logic Series 14, Kluwer, Dordrecht, 1998, pp. 133-168.

12. J.-J. Ch. Meyer, Modal Epistemic and Doxastic Logic, in: Handbook of Philosophical Logic (2nd edition) (D. Gabbay & F. Guenthner, eds.) Vol. 10, Kluwer, Dordrecht, 2003, pp. 1-38.

13. R.M. van Eijk, F.S. de Boer, W. van der Hoek & J.-J. Ch. Meyer, Systems of Communicating Agents, in Proc. ECAI-98 (H. Prade, ed.), Brighton, Wiley, 1998, pp. 293-297.

14. J.-J. Ch. Meyer, Epistemic Logic, Chapter 9 of: The Blackwell Guide to Philosophical Logic (L. Goble, ed.), Blackwell Publishers, Oxford, UK, 2001, pp. 183-202.

15. K.V. Hindriks, F.S. de Boer, W. van der Hoek & J.-J. Ch. Meyer, Agent Programming in 3APL, in Int. J. of Autonomous Agents and Multi-Agent Systems 2(4), 1999, pp.357-401.

16. J.-J. Ch. Meyer, Dynamic Logic for Reasoning about Actions and Agents, in: Logic-Based Artificial Intelligence (J. Minker, ed.), Kluwer, Boston/Dordrecht, 2000, pp. 281-311.

17. J.-J. Ch. Meyer, Intelligent Agents: Issues and Logics, in: Logics for Emerging Applications of Databases (J. Chomicki, R. van der Meyden & G. Saake, eds.), Springer, Berlin, 2004, pp. 131-165.

18. J.-W. Roorda, W. van der Hoek & J.-J. Ch. Meyer, Iterated Belief Change in Multiple Agent Systems, in: Proc. of the $1^{st}$ Int. Joint Conf. on Autonomous Agents and Multiagent Systems (AAMAS2002) (C. Castelfranchi & W. L. Johnson, eds.), Bologna, Italy, ACM Press, 2002, pp. 889-896; full version under title ", Iterated Belief Change in Multi-Agent Systems", Logic Journal of the IGPL 11(2), 2003, pp. 223-246.

19. B.P. Harrenstein, W. van der Hoek, J.-J. Ch. Meyer & C. Witteveen, A Modal Characterization of Nash Equilibrium, Fundamenta Informaticae 57(2-4), 2003, pp. 281—321.

20. J.-J. Ch. Meyer, Reasoning about Emotional Agents, in Proc.$16^{th}$ European Conf. on Artif. Intell.

(ECAI 2004) (R. López de Mántaras & L. Saitta, eds.), IOS Press, 2004, pp. 129-133.

**Service to the Profession:** Editoral Board Member, Computing Letters, VSP/Brill, 2004-; Consulting Editor, Episteme, 2004; Board Member, Dutch Association for Theoretical Computer Science, NVTI, 1999-; Editoral Board Member, Journal of Intelligent Agents & Multi-Agent Systems, 1997-; Editorial Board Member, Data and Knowledge Engineering, 1996-; Scientific Director, Dutch National Graduate School for Information and Knowledge-Based Systems, SIKS, 1995-2005; Chairman, Board of The Dutch Association for Logic and Philosophy of the Science, VvL, 1995-2005; Executive Board Member, Dutch National Graduate School for Logic, OzsL, 1995-1997; Board member, Theoretical Aspects of Reasoning about Knowledge, TARK, 1994-2001.

Together with Roel Wieringa he initiated the DEON workshop series on "Deontic Logic in Computer Science" in 1991, designed to promote cooperation among scholars across disciplines who are interested in deontic logic and its use in computer science. The DEON workshops support research linking the formal-logical study of normative concepts and normative systems with computer science, artificial intelligence, philosophy, organisation theory and law. Nowadays DEON has turned into a regular biennial event.

**Teaching:** During his career he has taught several logic-related courses, varying from elementary propositional logic to logics for AI and programming verification, which is something he really enjoys. Until now he has had 24 students who got their PhDs with me as (co-)advisor, among whom Wiebe van der Hoek (currently professor at the University of Liverpool), Henry Prakken (currently professor at the University of Groningen) and Peter Flach (currently professor at the University of Bristol).

**Vision Statement:** I view logic as a wonderful tool to make your intuitions about concepts explicit and precise. Although over the years I've come to realize that when you want to devise working implemented (intelligent) systems one cannot merely employ logic and I've become more eclectic as to methods and techniques used to model and realize systems, I still think that logic is and will always be a very useful tool for many aspects of the representation of knowledge in philosophy and artificial intelligence.

**Honours and Awards:** Inclusion in Who's Who in Science and Engineering and Who's Who In The World, 1995-; ECCAI Selected Fellow by the European Coordinating Committee for Artificial Intelligence (ECCAI), 1995.

# MILLER, David William

**Specialties:** General metamathematics, abstract logic, axiomatics, application of logic to philosophical problems, philosophy of logic, teaching of logic.

**Born:** 19 August 1942 in Watford, Hertfordshire, UK.

**Educated:** Peterhouse (Cambridge), BA 1964, MA 1969; Stanford University 1967-1969; London School of Economics, MSc(Econ) 1965.

**Regular Academic or Research Appointments:** EMERITUS READER IN PHILOSOPHY, UNIVERSITY OF WARWICK, 2007–; Reader in Philosophy, University of Warwick, 1987–2007. Senior Lecturer in Philosophy, University of Warwick, 1976–1987; Lecturer in Philosophy, Univesity of Warwick, 1969–1976. University of Warwick, 1969-1976.

**Visiting Academic or Research Appointments:** Profesor invitado, Facultad de Ingenería, Universidad Nacional deColombia, 2005, 2007; Profesor invitado, Universidad Nacional de Córdoba, Argentina, 1999; Visiting Professor, Universidad Iberoamericana, Santa Fe, Mexico, 1997; Co-Director, Summer University Seminar on SCIENTIFIC METHODS FROM POSITIVISM TO POSTMODERNISM, Central European University, Budapest, 1996; Visiting Professor, Instituto de Estudos Avançados, Universidade de São Paulo, 1996; Visiting Professor, Departamento de Física, Universidad Autónoma Metropolitana—Iztapalapa, Mexico, 1994; Visiting Professor, Instituto de Estudos Avançados, Universidade de São Paulo, 1991, 1992; Visiting researcher, Laboratorio di Cibernetica [later Istituto di Cibernetica] del CNR, Arco Felice, in association with Istituto di Fisica Teorica, Università di Napoli, 1980, 1981, 1982; Visiting Professor of Philosophy, University of Arizona, 1979; Visiting Lecturer in Philosophy, University of Pennsylvania, 1974; Visiting Lecturer in Philosophy, University of Minnesota & Minnesota Center for the Philosophy of Science, 1971.

**Research Profile:** Much of Miller's work in logic has consisted of responses to, and continuations of, the logical contributions of his teacher Karl Popper to the epistemology and methodology of science. He has made numerous innovations, both critical and constructive, in the theory of verisimilitude or truthlikeness, and (more generally) in the theory of metric operations on lattices; (Publications ## 2–4, 6–7, and Chapters 10 and 11 of # 12, and Chapters 10 and 11 of # 16). Following, and in cooperation with, Popper he has developed further the approach to the axiomatization of the theory of probability that displays it as a generalization of the theory of Boolean algebra; much of this work is unpublished, but samples are available in #8, #9, and #13. One outcome of this work has been an extensive investigation of the calculus of theories (first sketched by Tarski), especially in abstract logics; see ##10, 11, 17. Other work has concentrated on exploiting the full force of Popper's suggestion that formal logic is the organon of criticism (rather than of proof); see especially #16, Chapters 3, 13, and 14. This has led to a sustained attack on the approach to logic and argument that is now widely promulgated under the misleading name of critical thinking # 15.

**Main Publications:**

1. *Introduction to Axiomatic Set Theory.* English translation of J.-L. Krivine, *Théorie Axiomatique des Ensembles.* Dordrecht: D. Reidel Publishing Company (1971).
2. Popper's Qualitative Theory of Verisimilitude. *The British Journal for the Philosophy of Science* 25 (1974), 166-177.
3. On the Comparison of False Theories by Their Bases. *The British Journal for the Philosophy of Science* 25 (1974), 178-188.
4. New Axioms for Boolean Geometry. *Bulletin of the Section of Logic* Institute of Philosophy & Sociology, Polish Academy of Sciences, Wrocław 6 (1977), 53-63.
5. The Uniqueness of Atomic Facts in Wittgenstein's *Tractatus. Theoria* 43 (1977), 174-185
6. On Distance from the Truth as a True Distance. In K.J.J. Hintikka, I. Niiniluoto, & E. Saarinen, editors (1978), pp. 415-435. *Essays on Mathematical and Philosophical Logic.* Dordrecht: D. Reidel Publishing Company.
7. A Geometry of Logic. In H.J. Skala, S. Termini, & E. Trillas, editors (1984), pp. 91-104. *Aspects of Vagueness.* Dordrecht: D. Reidel Publishing Company.
8. (with K.R. Popper) Deductive Dependence. *Actes IV Congrés Català de Lógica.* Barcelona: Universitat Politècnica de Catalunya & Universitat de Barcelona (1986), pp. 21-29.
9. (with K.R. Popper) Why Probabilistic Support Is Not Inductive. *Philosophical Transactions of the Royal Society of London*, Series A 321, 30/4/1987, 569-591.
10. An Open Problem in Tarski's Calculus of Deductive Systems. *Bulletin of the Section of Logic.* Institute of Philosophy & Sociology, Polish Academy of Sciences, Warsaw/Łódź 20 (1991), 36-43.
11. The Disposition of Complete Theories. *Coleção Documentos*, Série Lógica e Teoria da Ciência, 10 (1992). São Paulo: Instituto de Estudos Avançados, USP.
12. *Critical Rationalism. A Restatement & Defence.* Chicago & La Salle: Open Court Publishing Company (1994).
13. (with K.R. Popper) Contributions to the Formal Theory of Probability. In P.W. Humphreys, editor (1994). *Patrick Suppes: Scientific Philosopher.* Volume 1, pp. 3-21. Dordrecht: Kluwer Academic Publishers.
14. Word Games for Formal Logic. *Proceedings of the First International Congress on Tools for Teaching Logic* (2000), pp. 89-93. Salamanca: Universidad de Salamanca.
15. Do We Reason When We Think We Reason, or Do We Think? *Learning for Democracy* 1 (2005), 57-71.
16. *Out of Error. Further Essays on Critical Rationalism.* Aldershot: Ashgate Publishing Company (2006).
17. Some Restricted Lindenbaum Theorems Equivalent to the Axiom of Choice. *Logica Univesalis* 1 (2007), 183-199.

A full list of publications is available at Miller's home page http://www2.warwick.ac.uk/fac/soc/philosophy/staff/miller/.

*Work in Progress*

18. A book (*Truth Defined*) showing how truth (though not the term 'true sentence') can be given a consistent, universal, and materially adequate definition.
19. A book (*An Investigation of Consequence*) on Tarski's general metamathematics, especially the precise relationship of the calculus of theoreis to Brouwerian logic.
20. An extension to Brouwerian algebras of the ideas and results of 'A Geometry of Logic'.

**Service to the Profession:** Member, Scientific Committee, International Summer School on REASONING UNDER PARTIAL KNOWLEDGE, Foligno, Perugia, 2001–; Chair, Programme Committee, KARL POPPER 2002, Vienna; 1997–2002; Trustee, Karl Popper Charitable Trust, 1996–; British Logic Colloquium, Committee Member, 1984–1992, Secretary, 1993–2001; Committee Member, British National Committee for Logic, Methodology & Philosophy of Science, 1977–1986, 1990–1996, Chair, 1997–2001; Programme Committee Member, LOGIC COLLOQUIUM 1986, 1985–1986; Program Committee Member, Section 7, SECOND WORLD CONFERENCE ON MATHEMATICS AT THE SERVICE OF MAN, 1982.

**Teaching:** Miller was centrally involved since 1969 to 2007 in the teaching of logic in the Department of Philosophy at the University of Warwick, and at various times taught all the logic courses offered, from the most mathematical to the most philosophical. He supervised two PhD dissertations (Deryck Horton on measurement theory and Roman Tuziak on paraconsistent logic). His *Formal Logic Workbook*, a complement to Lemmon's *Beginning Logic*, was first issued for internal use in 1983 and has been reprinted 21 times.

**Vision Statement:** The remarkable mathematical developments of the last century have wonderfully clarified many philosophical problems. But the most important philosophical issue concerning logic, namely the crucial part that reasoning plays in the growth of knowledge, is still wrapped in inductivist and verificationist misconceptions, or is derided as antiquated logocentrism. A more modest and critical approach to reasoning and rationality, in the form of the critical rationalism of Karl Popper, offers the only consistent response to the perennial irrationalistic and relativistic despair of reason.

# MILNER, Arthur John Robin Gorell

**Specialties:** Semantics of programming languages, algebraic and logical models for concurrent communicating behaviour, computer-assisted formal reasoning.

**Born:** 13 January 1934 in Yealmpton, Devonshire, UK.

**Educated:** University of Cambridge, BA Mathematics and Moral Sciences.

**Regular Academic or Research Appointments:** EMERITUS PROFESSOR OF COMPUTER SCIENCE, UNIVERSITY OF CAMBRIDGE, 2001– . Research Professor, University of Cambridge, 1999–2001. Professor of Computer Science and Head of Department, University of Cambridge, 1996–1999. Professor of Computer Science, University of Cambridge, 1994–1999. Senior Research Fellow (Science and Engineering Research Council), University of Edinburgh, 1990–1994. Director of Laboratory for Foundations of Computer Science, University of Edinburgh, 1986–89. Professor of Computer Science, University of Edinburgh, 1984–1994. Lecturer – Senior Lecturer – Reader in Computer Science, University of Edinburgh, 1973–1984. Research Associate at Artificial Intelligence Project, Stanford University, USA, 1971–73. Lecturer in Mathematics and Computer Science, The City University, London, 1963–68.

**Visiting Academic or Research Appointments:** ^Ile de France Blaise Pascale Chaire de Recherche, Ecole Polytechnique, Paris, 2006–2007. Guest Professor in Computer Science, Aarhus University, Denmark, 1979–80.

**Research Profile:** Robin Milner began serious research at the University of Swansea in 1968, where he developed an interest in the semantics of programming languages, in computer assisted logical reasoning, and automatic logical proof. These interests developed at Stanford, where he began the computer system LCF (Logic for Computable Functions). At Stanford he also began to study semantic models for concurrent computation, initially founded in the domain theory of Dana Scott.

These interests continued at Edinburgh. The LCF system came to be controlled by a programming meta-language ML, in which a researcher could write tactics or strategies for proof, composing them from smaller ones using higher-order functions called *tacticals*; its meta-type theory ensured that the result was indeed a correct logical proof. The LCF system became a model for later reasoning systems developed elsewhere: HoL at Cambridge, NuPrl at Cornell, and Coq at Paris and Goteborg. ML became a fully-fledged programming language, used to this day in teaching, in research on computerised logic, and in some industrial applications. Its formal definition, both static (the type discipline) and dynamic (the evaluation discipline) uses the structural operational semantics originated by Plotkin; this makes it probably the most rigorously defined 'industrial strength' programming language.

Milner's main contribution to computation theory has been in calculi for communication processes. Over the last 25 years there were three stages of development in this research: CCS, allowing composition of processes with stable linkage; the $\pi$-calculus, allowing processes to change their own structure dynamically; and bigraphs, a generic calculus of which many previous calculi are instances, and which pays attention to manipulation of spatial structure as well as linkage. Behavioural equivalences and preorders among these calculi, such as bisimilarity, are typically characterised both algebraically and by modal logics. In the first decade of this century these calculi are proving useful in system modelling in biology, in business processes and in pervasive computing.

In all Milner's work, many colleagues and students have contributed. The work is not pure logic; it lies at the boundary between logic and computer science, and aims both to generate new ideas at this interface, and to smooth the passage of ideas from each to the other.

**Main Publications:**

1. **Edinburgh LCF; a Mechanized Logic of Computation**, (with M.J.C. Gordon and C.W. Wadsworth), *Lecture Notes in Computer Science* 78, Springer-Verlag, 1979 (159 pages).
2. **A Calculus for Communicating Systems**, *Lecture Notes in Computer Science* 92, Springer-Verlag, 1980 (171 pages).
3. **Communication and Concurrency**, Prentice Hall, 1989 (about 260 pages).

4. **Definition of Standard ML (Revised)**, (with M. Tofte, R. Harper, D. MacQueen), MIT Press, 1997 (about 100 pages).
5. **Communicating and Mobile Systems: the Pi Calculus**, Cambridge University Press, 1999 (160 pages).
6. Processes: a mathematical model of computing agents, *Studies in Logic and the Foundations of Mathematics* 80, North Holland, 1975, pp157–174.
7. Fully abstract models of typed $\lambda$-calculi, *Theoretical Computer Science* 4, 1, 1977, pp1–22.
8. A metalanguage for interactive proof in LCF, (with M. Gordon, L. Morris, M. Newey & C. Wadsworth), *Proc. 5th annual ACM SIGACT–SIGPLAN Symposium on Principles of Programming Languages*, 1978 (about 20 pages).
9. Synthesis of communicating behaviour, *Lecture Notes in Computer Science* 64, Springer Verlag, 1978, pp71–83.
10. Flowgraphs and Flow Algebras, *Journal of ACM* 26,4,1979, pp794–818.
11. Algebraic laws for nondeterminism and concurrency, (with M. Hennessy), *Journal of ACM* 32, 1, 1985, pp137–161.
12. The use of machines to assist in rigorous proof, *Phil.Trans. R.Soc.London A* 312, 1984, pp411–422.
13. A calculus of mobile processes, Parts I and II, (with J. Parrow and D. Walker), *Information and Computation*, 100, 1, 1992, pp1–77.
14. Elements of interaction, *Communications of ACM* 36, 1, 1993.
15. Bigraphical reactive systems, *Lecture Notes in Computer Science* 2154, 2001, pp16–35.
16. Axioms for bigraphical structure, *Mathematical Structures in Computer Science* 15, 2005, pp1005–1032.
17. Pure bigraphs: structure and dynamics, to appear in *Information and Computation*, 2006.
18. Link graphs, transitions and Petri nets, (with J.J. Leifer), to appear in *Mathematical Structures in Computer Science*, 2006.

*Work in Progress*

19. Papers on topics in bigraphs, e.g. on confluence, and on probabilistic behaviour. Also, probably, an expository text on pure bigraphs.
20. Papers on models for pervasive computing, and on informatics as science as well as technology.

**Service to the Profession:** Associate editor of several journals, 1975–. Founding Director, Laboratory for the Foundations of Computer Science (LFCS), Edinburgh University, 1986; LFCS is now famous as a source of computation theory, including related logics. UK scheme for Distinguished Dissertations in Computer Science, founding Chair of selection committee, 1990. UK Computing Research Committee, founding member, 2001.

**Teaching:** Twenty students attained PhD, 1975–2006. Several are now Professors and/or well-known researchers; one is a University President. A final-year course to Cambridge undergraduates, 1995–1999, led to the publication of my expository book on the $\pi$-calculus.

**Vision Statement:** Computer science arose from logic, It must become a hierarchy of models, in the senses of both logic and experimental science. Models are needed at different levels by theoreticians, software engineers and laymen. Logic is essential to cohere this hierarchy.

**Honours and Awards:**

- Foreign member, Acad'emie des Sciences (France), 2006.
- Distinguished Achievements Award, European Association for Theoretical Computer Science 2005.
- Royal Medal, Royal Society of Edinburgh, 2004
- Award for Achievement in Programming Languages, Association of Computing Machinery (USA), 2001.
- ITALGAS prize, Turin (Italy), 1994.
- Fellow, Association of Computing Machinery (USA), 1994.
- Fellow, Royal Society of Edinburgh, 1993.
- A.M. Turing Award, Association of Computing Machinery (USA), 1991.
- Distinguished Fellow, British Computer Society, 1988.
- Fellow, Royal Society of London, 1988.

## MINTS, Grigori

**Specialties:** Proof Theory and Non-classical Logics.

**Born:** 7 June 1939 in Leningrad, USSR

**Educated:** Leningrad University, Diploma, Leningrad University, Ph.D., Leningrad University, Sc.D.

**Dissertation:** *Predicate and Operator Theories in Constructive Mathematics*; supervisor, N. Shanin.

**Regular Academic or Research Appointments:** PROFESSOR OF PHILOSOPHY, STANFORD UNIVERSITY, 1991–. Courtesy Professor, Dept. of Mathematics, Stanford University 1997–; Courtesy Professor, Dept. of Computer Science, Stanford University 1992–; Mir Publishers, Translator, 1979–1985; Leading Researcher Institute of Cybernetics, Tallinn 1985–1991; Researcher, Steklov

Institute of Mathematics, St. Petersburg, 1961–1979.

**Visiting Academic or Research Appointments:** Visiting Professor, LM University, Munich, summer, 2005; Visiting Professor, Dept. of Mathematics, UC Berkeley, autumn, 2004.

**Research Profile:** Mints has done work in proof theory, foundations of mathematics, constructive mathematics, non-classical logic, automated deduction, and mathematical applications of logic.

**Main Publications:**

1. "Interpolation Theorems for Intuitionistic Logic". *Annals of Pure and Applied Logic*, 113(2002): 225–242.
2. "Completeness of indexed epsilon calculus", with D. Sarenac. *Archive for Mathematical Logic*, 42(2003): 617–625.
3. "Intuitionistic Frege systems are polynomially equivalent", with A. Kojevnikov. *Zap. Nauchn. Sem. S.-Peterburg. Otdel. Mat. Inst. Steklov. (POMI)* 316 (2004), Teor. Slozhn. Vychisl. 9: 129–146
4. More than 3300 published reviews.
*Work in Progress*
5. "Notes on Constructive Negation". In R. Kahle and P. Schroeder-Heister (eds.), *Proofs-Theoretic Semantics*, a special issue of *Synthese*. To appear 2005.

**Service to the Profession:** Member of the Council for the Association for Symbolic Logic, 1989–1993; Member, Committee on Translations, Association for Symbolic Logic, 1985–1993; Asessor, DLMPS, Union for History and Philosophy of Science, 1991–1995.

**Member of Editorial Boards:** *Journal of Philosophical Logic*, *Journal of Logic and Computation*, *Logic Journal of IGPL*.

**Teaching:** Professor Mints has taught classes in Modal Logic, Model Theory, First Order Logic, Computability and Logic, Proof Theory, Philosophy of Mathematics, Recursion Theory, and Freshman Seminars in Logic.

# MITCHELL, William John

**Specialties:** Set theory, forcing, inner models.

**Born:** 30 December 1943 in Minneapolis, Minnesota, USA.

**Educated:** : University of California, Berkeley, PhD, 1970; University of Wisconsin, Madison, BA, 1965; Carleton College, Northfield Minnesota.

**Dissertation:** *Aronszajn Trees and the Independence of the Transfer Property*; supervisor, Jack Silver.

**Regular Academic or Research Appointments:** PROFESSOR, UNIVERSITY OF FLORIDA, 1990–; Professor, Pennsylvania State University, 1979-1990; Assistant Professor, Rockefeller University, 1972-1977; Instructor, University of Chicago, 1970-1972.

**Visiting Academic or Research Appointments:** Visiting Professor, University of California, Los Angeles, 1988; Visiting Professor, California Institute of Technology, Los Angeles, 1988; Lady Davis Visiting Assoc. Professor, Hebrew University, Jerusalem, 1986; Visitor, Oxford University, 1985; Member, The Institute for Advanced Study, 1977–1978.

**Main Publications:**

1. Aronszajn Trees and the Independence of the Transfer Property, *Ann. Math. Logic* 5 (1972), 21-46.
2. Sets Constructible from sequences of Ultrafilters, *J. Symbolic Logic* 39) (1974), 57-66.
3. Ramsey Cardinals and Constructibility, *J. Symbolic Logic* 44 (1979), 260-266.
4. Hypermeasurable Cardinals, *Logic Colloquium '78*, (M. Boffa, D. Van Dalen, and K. McAloon, ed), North Holland, Amsterdam. 1979, 303-316.
5. (with T. Jech, M. Magidor, K. Prikry) Precipitous Ideals, *J. Symbolic Logic* 45 (1980), 1-8.
6. How Weak is the Closed Unbounded Ultrafilter, *Logic Colloquium '80* (D. Van Dalen, D. Lascar and J. Smiley, ed), North-Holland, Amsterdam, 1982, 209-230.
7. Indiscernibles, Skies and Ideals, *Proceedings of the 1983 Boulder Summer Conference in Set Theory* (J. Baumgartner, ed), *Contemporary Mathematics*, American Mathematical Society, 1984, 161-182
8. The Core Model for Sequences of Measures I, *Mathematical Proc. Cambridge Phil. Soc.* 95 (1984), 229-260.
9. Applications of the Covering Lemma for Sequences of Measures, *Trans. AMS* 299 (1987), 41-58.
10. Definable Singularity. *Trans. of the Amer. Math. Soc.* 327 (1991), 407-426.
11. On the Singular Cardinal Hypothesis, *Trans. of the Amer. Math. Soc.* 329 (1992), 507-530.
12. $\Sigma_3^1$ Absoluteness for Sequences of Measures, in *Proc. of the Workshop on Set Theory of the Continuum*, Math. Sci. Research Institute, Berkeley, CA, 1989, (H. Judah, W. Just, H. Woodin eds.), Springer-Verlag, 1992, 311-355.
13. An Infinitary Ramsey Property, *Ann. of Pure and Applied Logic* 57 (1992), 151-160.
14. (with John R. Steel) Fine Structure and Iteration Trees, *Lecture Notes in Logic* 3, Springer-Verlag, Berlin, 1994.
15. The core model up to a Woodin cardinal, in *Logic, methodology and philosophy of science, IX* (Uppsala, 1991), *Stud. Logic Found. Math* 134, North-Holland, Amsterdam, 1994, 157-175.
16. (with E. Schimmerling), Weak covering without

countable closure, *Math. Research Letters* 2 (1995), 595-609.

17. (with Moti Gitik). Indiscernible sequences for extenders, and the singular cardinal hypothesis, *Ann. Pure Appl. Logic* 82(1996), 273-316.

18. William J. Mitchell and Ralf Schindler. A universal extender model without large cardinals in V, *Journal of Symbolic Logic* 69 (2003), 371-386.

19. A weak variation of Shelah's $[I\omega_2]$, *Journal of Symbolic Logic* 69 (2004), 94-100.

**Work in Progress**

20.

21. $I[\omega_2]$ can be the nonstationary ideal on $Cof(\omega_1)$, Submitted to the *Transactions of the AMS*, 2006.

**Service to the Profession:** Journal of Symbolic Logic, Editor, 1997-2002, Coordinating Editor, Contributed Papers, 1998-2000.

**Teaching:** Students Pierre Matte, Jeffery Leaning, Omar De la Cruz, Diego Rojas-Reboledo.

# MOORE, J Strother

**Specialties:** Mechanical theorem proving.

**Born:** 11 September 1947 in Seminole, Oklahoma, USA.

**Educated:** University of Edinburgh, PhD; Massachusetts Institute of Technology, BS. *Dissertation: Computational Logic: Structure Sharing and Proof of Program Properties*; supervisor, R. M. Burstall.

**Regular Academic or Research Appointments:** DEPARTMENT CHAIR, DEPARTMENT OF COMPUTER SCIENCES, UNIVERSITY OF TEXAS AT AUSTIN, 2001; ADMIRAL B. R. INMAN PROFESSOR OF COMPUTING THEORY, DEPARTMENT OF COMPUTER SCIENCES, UNIVERSITY OF TEXAS AT AUSTIN, 1997–. Chief Scientist, Computational Logic, Inc., 1987-1996; Gottesman Family Centennial Professor, Department of Computer Sciences, University of Texas at Austin, 1985-1988; Associate Professor, Department of Computer Sciences, University of Texas at Austin, 1981-1984; Staff Scientist, Computer Science Laboratory, SRI International, 1981; Senior Research Mathematician, Computer Science Laboratory, SRI International, 1979-1981; Research Mathematician, Computer Science Laboratory, SRI International, 1976-1978; Research Mathematician, Computer Science Laboratory, Xerox Palo Alto Research Center, 1973-1976; Research Fellow, Department of Computational Logic, University of Edinburgh, 1973.

**Visiting Academic or Research Appointments:** Visiting Professor, School of Informatics, University of Edinburgh, 2005-; Lecturer, Marktoberdorf International Summer School, 1988, 2002, 2004; IBM Chaire Internationale D'Informatique, Universite de Liege, 1980.

**Research Profile:** Moore was among the pioneers in the fields of mechanical theorem proving and the use of formal mathematical logic in the mechanized verification of computer hardware and software. Along with Robert S. Boyer and later Matt Kaufmann, he wrote several influential mechanical theorem provers. Each supports a variant of applicative Lisp as a first-order theory with induction. Proof techniques first developed in the Boyer-Moore community include automatic induction, simplification in the presence of recursively defined functions, the integration of various decision procedures, the use of mechanically verified metafunctions, and efficient provisions for ground evaluation. The theorem provers are automatic but sensitive to previously proved theorems; the user guides the theorem prover most often by presenting it with a carefully designed sequence of lemmas to prove. Using these tools, Moore and his colleagues and students have produced the first mechanized proofs of Goedel's First Incompleteness Theorem, Gauss' Law of Quadratic Reciprocity, the Paris-Harrington Ramsey Theorem, and important properties of various computer hardware and software designs including a Motorola digital signal processor, floating point operations on processors by AMD and IBM, operating system kernels, and the Java bytecode verifier.

**Main Publications:**

1. Inductive Assertions and Operational Semantics, *CHARME 2003*, D. Geist, ed., pp. 298–303. Springer Verlag LNCS 2860, 2003.

2. Partial Functions in ACL2, with P. Manolios, *Journal of Automated Reasoning*, 31(2), 107–127, 2003.

3. Single-Threaded Objects in ACL2, with R. S. Boyer, S. Krishnamurthi, and C. R. Ramakrishnan, eds. *PADL 2002*, pp. 9–27. Springer Verlag LNCS 2257, 2002.

4. *Computer-Aided Reasoning: An Approach*, with M. Kaufmann and P. Manolios, Kluwer Academic Publishers, Boston, 2000.

5. Proving Theorems about Java-like Byte Code, in E.-R. Olderog and B. Steffen, eds. *Correct System Design — Recent Insights and Advances*, pp. 139–162. Springer Verlag LNCS 1710, 1999.

6. Structured Theory Development for a Mechanized Logic, with M. Kaufmann, *Journal of Automated Reasoning*, 26(2), 161–203, 2001.

7. A Mechanically Checked Proof of the Correctness of the Kernel of the AMD5K86 Floating-Point Division Algorithm, with T. Lynch and M. Kaufmann, *IEEE Trans. Comp.*, 47(9), 913–926, 1998.

8. *A Computational Logic Handbook*, with R. S. Boyer, Academic Press, London, 1997. Second Edition.

9. *Piton: A Mechanically Verified Assembly Level Language*, Kluwer Academic Publishers, Dordrecht, The Netherlands 1996.

10. A Formal Model of Asynchronous Communication and Its Use in Mechanically Verifying a Biphase Mark Protocol. *Formal Aspects of Computing*, 6(1), 60–91, 1994.

11. Functional Instantiation in First Order Logic, with R. S. Boyer, D. M. Goldschlag, and M. Kaufmann. In V. Lifschitz, ed. *Artificial Intelligence and Mathematical Theory of Computation: Papers in Honor of John McCarthy*, pp. 7–26, Academic Press, 1991.

12. Special Issue on System Verification, with W. R. Bevier, W. A. Hunt, and W. D. Young. *Journal of Automated Reasoning*, 5(4), 461–492, 1989.

13. The Addition of Bounded Quantification and Partial Functions to A Computational Logic and Its Theorem Prover, with R. S. Boyer. *Journal of Automated Reasoning*, 4(2), 117–172, 1988.

14. Integrating Decision Procedures into Heuristic Theorem Provers: A Case Study of Linear Arithmetic, with R. S. Boyer. *Machine Intelligence*, 11, Oxford University Press, 1988.

15. Metafunctions: Proving Them Correct and Using Them Efficiently as New Proof Procedures, with R. S. Boyer. In R. S. Boyer and J S. Moore, eds., *The Correctness Problem in Computer Science*, Academic Press, London, 1981.

16. A Computational Logic, with R. S. Boyer, Academic Press, New York, 1979.

17. A Fast String Searching Algorithm, with R. S. Boyer. *Communications of the Association for Computing Machinery*, 20(10), 762–772, 1979.

18. Proving Theorems about LISP Functions, with R. S. Boyer. Journal of the Association for Computing Machinery, 22(1), 129–144, 1975.

19. The Sharing of Structure in Theorem-proving Programs, with R. S. Boyer. In B. Meltzer and D. Michie, eds., *Machine Intelligence*, 7, pp. 101–116. Edinburgh University Press.

*Work in Progress*

20. ACL2: A Computational Logic for Applicative Common Lisp, with M. Kaufmann.

**Teaching:** Moore has supervised about a dozen PhD students; they constitute a critical segment of the emerging field of applied mechanized logic.

**Vision Statement:** I see mechanized theorem proving fundamentally altering mankind's relationship to computing machines as trained engineers and software designer – and eventually other scientists – turn to mechanized reasoning engines to help them solve abstract problems that are too complicated to be solved confidently by human thought alone.

**Honours and Awards:** Herbrand Award, Conference on Automated Deduction, 1999; Fellow, American Association for Artificial Intelligence, 1991; Current Prize in Automatic Theorem Proving, American Mathematical Society, 1991; John McCarthy Prize for Program Verification, 1983.

## MOORE, Robert C.

**Specialties:** Logic and artificial intelligence, logic and natural language semantics, logic of knowledge and action, autoepistemic logic.

**Born:** 28 July 1948 in Fort Worth, Texas, USA.

**Educated:** Massachusetts Institute of Technology, PhD Artificial Intelligence, 1979; Massachusetts Institute of Technology, MS Electrical Engineering and Computer Science, 1976; Massachusetts Institute of Technology, BS Electrical Engineering, 1971; Massachusetts Institute of Technology, BS Political Science, 1971.

**Dissertation:** *Reasoning about Knowledge and Action*; supervisor: Gerald Sussman.

**Regular Academic or Research Appointments:** PRINCIPAL RESEARCHER, MICROSOFT RESEARCH, 2006–; Senior Researcher, Microsoft Research, 1999-2006; Director, Research Institute for Advanced Computer Science, NASA Ames Research Center, 1998; Principal Scientist, SRI International, 1992-1998; Director, Natural Language Research Program, SRI International, 1989-1992; Staff Scientist and Principal Scientist, SRI International, 1987-1989; Director, Cambridge Computer Science Research Centre, SRI International, 1986-1987; Staff Scientist, SRI International, 1983-1986; Senior Computer Scientist, SRI International, 1980-1983; Computer Scientist, SRI International, 1977-1980.

**Visiting Academic or Research Appointments:** Visiting Lecturer, Stanford University, 2001-2002; Industrial Fellow Commoner, Churchill College, Cambridge, 1985-1987; Consulting Associate Professor, Computer Science, Stanford University, 1984-1985; Visiting Lecturer, Stanford University, 1983; Fellow, Center for Advanced Study in the Behavioral Sciences, 1979-1980.

**Research Profile:** Robert Moore's contributions to logic include being one of the first to suggest using modal logic to formalize programming language semantics, leading to the development of the dynamic logic of programs. Most of his work in logic, however, has been in applications of logic to problems in artificial intelligence and the semantics of natural language. In his dissertation research, Moore developed a bi-modal logic of knowledge and action, formalizing both the knowledge required in order to take certain actions, as well as the effects of taking action on one's state of knowledge. Moore is perhaps best known for his development of autoepistemic logic, which is a modal logic modeling reflection on one's own knowledge and belief. Autoepistemic logic has been one of the main formalisms used to study nonmonotonic reasoning, which refers to reasoning that allows withdrawing a conclusion when presented with additional information.

**Main Publications:**

1. *Logic and Representation*, CSLI Publications, Center for the Study of Language and Information, Stanford, California, 1995.
2. (with Jerry Hobbs, eds.) *Formal Theories of the Commonsense World*, Ablex Publishing Corp., Norwood, New Jersey, 1985.
3. *Reasoning from Incomplete Knowledge in a Procedural Deduction System*, Garland Publishing Inc., New York, New York, 1980.
4. "Events, Situations, and Adverbs," in *Challenges in Natural Language Processing*, Bates and Weischedel (eds.), Cambridge University Press, 1993.
5. "Autoepistemic Logic Revisited," *Artificial Intelligence*, Vol. 59, Nos. 1-2, February 1993.
6. "Unification-Based Semantic Interpretation," in *Proceedings, 27th Annual Meeting of the Association for Computational Linguistics*, June 1989.
7. "Propositional Attitudes and Russellian Propositions," in *Semantics and Contextual Expression*, Bartch, et al. (eds.), Foris Publications, 1989.
8. "Autoepistemic Logic," in *Non-Standard Logics for Automated Reasoning*, Smets, et al. (eds.), Academic Press, 1988.
9. "The Role of Logic in Artificial Intelligence," in *Intelligent Machinery: Theory and Practice*, Benson (ed.), Cambridge University Press, 1986.
10. "A Formal Theory of Knowledge and Action," in *Formal Theories of the Commonsense World*, Hobbs and Moore (eds.), Ablex Publishing Corp., 1985.
11. "Possible-World Semantics for Autoepistemic Logic," in *Proceedings, Workshop on Nonmonotonic Logic*, Mohonk Mountain House, New Paltz, New York, October 17-19, 1984.
12. "Semantical Considerations on Nonmonotonic Logic," *Artificial Intelligence*, Vol. 25, No. 1, January 1985.
13. "The Role of Logic in Knowledge Representation and Commonsense Reasoning," in *Proceedings, AAAI-82*, August 1982.
14. "Automatic Deduction for Commonsense Reasoning: An Overview," in *The Handbook of Artificial Intelligence*, Barr and Feigenbaum (eds.), William Kaufman, Inc., 1982.
15. (with Gary Hendrix) "Computational Models of Belief and the Semantics of Belief Sentences," in *Processes, Beliefs, and Questions*, Peters and Saarinen (eds.), D. Reidel Publishing Co., 1982.
16. "Problems in Logical Form," in *Proceedings, 19th Annual Meeting of the Association for Computational Linguistics*, June 1981.
17. "Reasoning about Knowledge and Action," in *Proceedings, IJCAI77*, August 1977.
18. "D-SCRIPT: A Computational Theory of Descriptions," *IEEE Transactions on Computers*, C-24, pp. 366-373, 1976.

**Service to the Profession:** Editorial Board Member, *Computational Intelligence*, 2003-; Editorial Board Member, *Artificial Intelligence*, 1984-2005; Editorial Board Member, *Computational Linguistics*, 1990-1992.

**Teaching:** All of Moore's regular appointments have been in nonacademic research positions, but he has taught three courses as a Visiting Lecturer at Stanford University, and he has supervised one doctoral student, Edwin Pednault, PhD 1987, Electrical Engineering, Stanford University.

**Vision Statement:** As logic became a branch of mathematics in the $20^{th}$ century, it came to be focused almost exclusively on the type of reasoning needed for mathematical proof: universally valid inference. It is my hope that in the future, logic will cast its net wider to encompass all forms of rational inference, which I believe we will need if the promise of intelligent information systems is ever to be fulfilled.

**Honours and Awards:** Fellow, Association for the Advancement of Artificial Intelligence, 1991-.

# MOSCHOVAKIS, Joan Rand

**Specialties:** Foundations of intuitionistic mathematics.

**Born:** 24 December, 1937 in Glendale, California, USA.

**Educated:** University of Wisconsin, PhD Mathematics, 1965, MS, 1961; University of California, BA Mathematics (Summa Cum Laude), 1959.

**Dissertation:** *Disjunction, existence and lambda-eliminability in formalized intuitionistic analysis*; supervisor, Stephen C. Kleene.

**Regular Academic or Research Appointments:** PROFESSOR EMERITA, MATHEMATICS, OCCIDENTAL COLLEGE, 1995; Professor, Mathematics, 1986-1995; Associate Professor, 1985-1986; Associate Professor (part time), 1976-1985; Assistant Professor (part time), 1970-1976; Assistant Professor, 1965-1967, 1969-1970.

**Visiting Academic or Research Appointments:** Associated Faculty, MPLA (Graduate Program in

Logic and Algorithms), Athens, Greece, 1998–; Visiting Lecturer, University of Bristol, England, 1969; Instructor, Mathematics, Oberlin College, 1963-1964.

**Research Profile:** Joan Rand Moschovakis became intrigued by the intuitionistic approach to logic and mathematics upon hearing a lecture by A. Heyting. Her graduate study coincided with Kleene and Vesley's completion of the book "Foundations of Intuitionistic Mathematics, Especially in Relation to Recursive Functions" and her PhD dissertation proved the disjunction and numerical existence properties for classically correct subtheories of Kleene's axiomatization of intuitionistic analysis. Inspired by R. Vesley's "A palatable substitute for to Kripke's Schema," she developed a modified function realizability interpretation to prove the consistency of intuitionistic analysis together with Vesley's Schema and a weak form of Church's Thesis asserting that there are no non-recursive functions. After adapting D. Scott's topological model to Kleene's formalism, she developed a classical model for an intuitionistic theory including Kreisel and Troelstra's axioms for lawless sequences. She has studied the classical and constructive arithmetical and analytical hierarchies in intuitionistic analysis extended by Markov's Principle and Krause's classical axiom of countable choice, a theory which cannot fail to contain every classically arithmetical (even every classically hyperarithmetical) function, while it preserves the constructive sense of the existential function quantifier by providing a recursive witness for every existential theorem.

**Main Publications:**

1. Disjunction and existence in formalized intuitionistic analysis, in J.N. Crossley (ed.), Sets, Models and Recursion Theory, Amsterdam (North-Holland), 1967.
2. Can there be no non-recursive functions?, Journal of Symbolic Logic 36 (1971), 309-315.
3. A topological interpretation of second-order intuitionistic arithmetic, Compositio Mathematica 26 (1973), 261-275.
4. Relative lawlessness in intuitionistic analysis, Journal of Symbolic Logic 52 (1987), 68-88.
5. More about relatively lawless sequences, Journal of Symbolic Logic 59 (1994), 813-829.
6. A classical view of the intuitionistic continuum, Annals of Pure and Applied Logic 81 (1996), 9-24.
7. Analyzing realizability by Troelsta's methods, Annals of Pure and Applied Logic 114 (2002), 203-225.
8. Classical and constructive hierarchies in extended intuitionistic analysis, Journal of Symbolic Logic 68 (2003), 1015-1043.
9. Intuitionistic Logic, entry in Stanford On-Line Encyclopedia of Philosophy (revised 2003).
10. The effect of Markov's Principle on the intuitionistic continuum, Proceedings of Oberwolfach Proof Theory Week (2005).
11. (with Garyfallia Vafeiadou, in Greek) Intuitionistic mathematics and logic, to appear.

*Work in Progress*

12. The logic of Brouwer and Heyting, entry in the Encyclopedia of the History of Philosophy, eds. D. Gabbay and J. Wood.
13. Notes on the Foundations of Constructive Mathematics.

**Service to the Profession:** Committee on Meetings in North America of the Association for Symbolic Logic, Committee Chair, 1997-1999, Committee Member, 1991-1999; Chair, Mathematics Department of Occidental College, 1985-1988; Reviewer, Mathematical Reviews, Mathematische Zeitschrift and the Journal of Symbolic Logic, 1966-1990. Referee for various journals, 1966–.

**Teaching:** At Occidental Professor Moschovakis enjoyed teaching calculus, analysis, topology, logic and set theory to able and motivated students. She was part of the faculty team which designed and taught the first Collegium, an interdisciplinary freshman curriculum. In addition to advising and teaching undergraduates, many of whom went on to graduate school in mathematics or in architecture, medicine, economics or law, she supervised some Master of Arts in Teaching candidates. Early retirement gave her the opportunity to teach graduate courses in constructive logic and mathematics in Athens, where she especially enjoyed supervising the Master's theses of Garyfallia Vafeiadou and Nikos Vaporis.

**Vision Statement:** "Applications of logic are increasingly important. The intuitionistic language and logic are capable of distinguishing constructive from nonconstructive content in any branch of mathematics. I would like to see this point of view developed as far as it will go."

**Honours and Awards:** University of Wisconsin Dissertation Fellowship, 1962–1963; National Science Foundation and University of Wisconsin Cooperative Fellowship, 1960-1961; Woodrow Wilson Fellowship, 1959-1960.

# MOSCHOVAKIS, Yiannis Nicholas

**Specialties:** Recursion theory, Descriptive set theory, Foundations of the theory of computation, Semantics of natural language.

**Born:** 18 January, 1938, in Athens, Greece.

**Educated:** Massachusetts Institute of Technology, B.S. and M.S., Mathematics, 1960; University of Wisconsin, Ph.D., Mathematics, 1963.

**Dissertation:** *Recursive Analysis*, supervisor, Stephen Cole Kleene

**Regular Academic or Research Appointments:** PROFESSOR OF MATHEMATICS, UCLA (UNIVERSITY OF CALIFORNIA, LOS ANGELES, 1971–; Professor of Mathematics, National and Kapodistrian University of Athens, 1996–2005 (and Emeritus Prof. since 2005); Assoc. Prof. of Mathematics, UCLA, 1968 –1971; Assist. Prof. of Mathematics, UCLA, 1964–1968; Benjamin Peirce Instructor, Harvard University, 1963–1964.

**Research Profile:** Moschovakis started out in constructive mathematics, and both his Master's and Doctoral Theses were in Recursive Analysis (Russian style). Next he worked in Kleene's recursion in higher types and the related subjects of abstract recursion and inductive definability from 1964 until about the late seventies and published the monograph 1 below; the methods he learned and developed in these areas are among the standard tools he has used in all his subsequent work. From 1967 until the late eighties he worked in Descriptive Set Theory, and especially in the so-called effective theory (which derives from the work of Kleene and folds recursion theory and the classical theory of pointsets into a single subject), and in the consequences of the Axiom of Definable (especially projective) Determinacy; his monograph 2 below on this topic was published in 1980. Since the mid-eighties, Moschovakis has been working on the foundations of the theory of computation; the applications of algorithms in the philosophy of language; and more recently, the derivation of lower complexity bounds (especially for problems in arithmetic and algebra) which apply to all algorithms — not just some, specific computation models.

**Main Publications:**

Books
1. *Elementary Induction on Abstract Structures*. North Holland, 1974. Reprinted by Dover Publications, 2008.
2. *Descriptive Set Theory*. North Holland, 1980.
3. *Notes on the Theory of Sets*, Second Edition, Springer, 2006.

Articles
4. Hyperanalytic predicates. *Transactions of the American Mathematical Society*, (129), 1967, 249-282.
5. (with J.W. Addison). Some consequences of the axiom of definable determinateness. *Proceedings of the National Academy of Sciences* (59), 1968, 708-712.
6. Abstract first order computability, I and II. *Transactions of the American Mathematical Society* (138), 1969, 427-464 and (138), 1968, 465-504.
7. Abstract computability and invariant definability. *Journal of Symbolic Logic*, (34), 1969, 605-633.
8. Determinacy and prewellorderings of the continuum. In *Mathematical Logic and Foundations of Set Theory* (Y. Bar-Hillel, ed.), Amsterdam-London, 1970, 24-62.
9. Uniformization in a playful universe. *Bulletin of the American Mathematical Society*, (77), 1971, 731-736.
10. Analytical definability in a playful universe. In *Logic, Methodology and Philosophy of Science IV* (P. Suppes, L. Henkin, A. Joja and G. Moisil eds.), North-Holland, 1974, 53-79.
11. (with Donald A. Martin and John R. Steel). The extent of definable scales. *Bulletin of the American Mathematical Society*, (6), 1982, 435-440.
12. Abstract recursion as a foundation for the theory of algorithms. In *Computation and Proof Theory*, Springer Lecture Notes in Mathematics vol 1104 (M.M. Richter et al. eds.), 1984, 289-362.
13. A model of concurrency with fair merge and full recursion. *Information and Computation*, (93), 1991, 114-171.
14. Sense and denotation as algorithm and value. In *Proceedings of the 1990 ASL Summer Meeting, Helsinki*. Lecture Notes in Logic, (2), 1993, 210-249.
15. Computable concurrent processes. *Theoretical Computer Science*, (139), 1995, 243-273.
16. On founding the theory of algorithms. In *Truth in Mathematics* (H. G. Dales and G. Oliveri, eds.), Clarendon Press, Oxford 1998, 71–104.
17. On primitive recursive algorithms and the greatest common divisor function, *Theoretical Computer Science*, (301), 2003, 1-30.
18. (with L. van den Dries). Is the Euclidean algorithm optimal among its peers? *Bulletin of Symbolic Logic*, (10), 2004, 390-418.
19. A logical calculus of meaning and synonymy. *Linguistics and Philosophy*, (29), 2006, 27-89.
20. (with L. van den Dries). Arithmetic complexity. To appear in *ACM Transactions on Computational Logic*.

**Service to the Profession:** In the UCLA Mathematics Department: Undergraduate Vice Chair, 1969–1970; Graduate Vice Chair, 1979–1980, 1982–1984; Chair, 1984–1987; Director of the Program in Computing, 1987–1989. In the Association for Symbolic Logic: Editor of the Journal of Symbolic Logic, 1977–1982; Vice President, 1983–1986, 1989–1992; Editor of the Perspectives in Mathematical Logic Series, 1989–1998; President, 1992–1995. In the American Mathematical Society: Member of the Steele Prize Committee, 1982–1985 (Chair for 1983–1985); Member of the Council, 1986–1989. In the National Science Foundation: Member of the Advisory Panel for Mathematics, 1980-1982; Member of the Selection Committee for Postdocs, 1981-1983. In Logic for Computer Science (LICS): Member of the Organizing Committee, 1990-1998; Member of the Program Committee, 1992. In the Graduate Program in Logic, Algorithms and Computation (Athens): Director, 1997–.

**Teaching:** Twenty three Ph.D. students (as of June 2008): twenty from Mathematics at UCLA, two from Computer Science at UCLA, one from the Graduate Program for Logic Algorithms and Computation at the University of Athens. A great deal of undergraduate teaching in UCLA since 1964; was awarded the Departmental Robert Sorgenfrey Teaching Award in 1996.

**Vision Statement:** Without claiming any originality, I think of logic as the study of truth, proof, computation and (especially) definability — which perhaps includes the first three notions: its most important problem is to identify those objects which can be defined (explicitly, implicitly, recursively, etc.) and to study their characteristic properties which stem from their definitions. This has been done most successfully in mathematics, but there is much promise in the currently evolving applications to computer science and to linguistics.

**Honours and Awards:** Tarski Lecturer at the University of Californis, Berkeley, 2008; Honorary doctorates in Mathematics from the University of Sofia (Bulgaria) in 2002 and the University of Athens in 1987; Corresponding Member of the Academy of Athens, 1980–; Fullbright Travelling Lecturer (Novosibirsk), March 1979; ICM Lecturer, 1974; Sloan Fellow, 1970–1972; Guggenheim Fellow, 1968–1969.

# MOSS, Lawrence Stuart

**Born:** 14 August 1959 in Los Angeles, California, USA.

**Educated:** BA (Mathematics, Summa Cum Laude) UCLA, 1979; MA (Mathematics) UCLA 1981; PhD (Mathematics) UCLA 1984.

**Dissertation:** *Power Set Recursion* ; supervisor Yiannis N. Moschovakis.

**Regular Academic or Research Appointments:** DIRECTOR, PROGRAM IN PURE AND APPLIED LOGIC; PROFESSOR OF MATHEMATICS, ADJUNCT PROFESSOR OF COMPUTER SCIENCE, INFORMATICS, LINGUISTICS, AND PHILOSOPHY, INDIANA UNIVERSITY, BLOOMINGTON, 1990–PRESENT. Postdoctoral Fellow, IBM Yorktown Heights, 1988–90. Assistant Professor of Mathematics, University of Michigan, Ann Arbor, 1985–88. Postdoctoral Fellow, Center for the Study of Language and Information, Stanford University, 1984–85.

**Visiting Academic or Research Appointments:** Johns Hopkins University Cognitive Science Department, 2000. University of Maryland Institute for Advanced Computer Studies, 2000-01. Computer Science Department, CUNY Graduate Center, New York, 1995–96. DIMACS, Rutgers University, 1995–96. Centre for Theoretical Studies, Charles University, Prague, Czechoslovakia, 1991. Mathematics Department, Tulane University, 1988. Max–Planck–Institut für Psycholinguistik, Nijmegen, Netherlands, 1984.

**Research Profile:** Moss has had a varied research career, so here are some of the topics with a short mention of some of the results. His dissertation was in an area of generalized recursion theory on the borderline of set theory called "power set recursion". This was perhaps the "largest" recursion theory developed: the computations are not absolute. His work emphasized equiconsistency results between inaccessible and Mahlo cardinals on the one hand, and some natural recursion-theoretic statements on the other. These statements come from a natural reducibility ordering on sets that comes from the subject itself.

A different line of work pertains to the model theory of graphs. He showed that there is a countable graph which isometrically embeds every countable graph, discussed variations of this and also connections to other model-theoretic issues.

Moss also worked with Edward L. Keenan on some aspects of generalized quantifiers; with Yuri Gurevich on the abstract state machine semantics of concurrency (based most directly on the language Occam); with Rohit Parikh on the connection of topology and epistemic logic, resulting in some logical systems that were subsequently taken up by others; with José Meseguer and Joseph Goguen on abstract data types (settling a conjecture in the area), with David E. Johnson on a series of papers that proposed a formalization of relational grammar, and a different set of papers proposing the use of evolving algebras for linguistics.

More recent areas include dynamic epistemic logic, and also areas of coalgebra. The work on dynamic epistemic logic was mainly done with Alexandru Baltag. It includes a proposal for a notion of epistemic action in Kripke models, and then it works out logical systems which are shown to be complete (with Slawomir Solecki). These systems have common knowlege operators and also modal operators for actions such as announcing sentences. In another paper, he shows (with Joseph Miller) that adding an iteration operator to logics of this type destroys the decidability.

Together with Jon Barwise, Moss wrote a book on the subject of non-wellfounded sets (hypersets). This book was intended to be both a popularization of the topic and also a spur to further research by others. Much of the subsequent work on areas close to the book has been carried out in the more algebraic setting of 'coalgebra'. This topic, like hypersets, is one that many people found their way to from the work of Peter Aczel. In coalgebra, Moss was the first to propose generalizations of modal logic to coalgebras of other functors on sets. This line of work has seen a lot of attention in recent years, resulting in variations and improvements on the original work. In a different direction, with his Ph.D. student Ignacio Viglizzo, he showed that functors on measurable spaces built from the probability measure functor, the constants, products, and coproducts all have final coalgebras. This work is connected to investigations in the economics literature on universal Harsanyi spaces, and to work in the computer science literature on probabilistic processes. He also investigated connections between the logic of recursion pioneered by Yiannis N. Moschovakis and coalgebra. This ongoing work is now generalizing the subject of recursive program schemes to the setting of coalgebra (with Stefan Milius).

**Main Publications:**

1. **Vicious Circles: On the Mathematics of Non-Wellfounded Phenomena**, co-authored with Jon Barwise. CSLI Lecture Notes Number 60, CSLI Publications, Stanford University, 1996. 390 p.
2. Completeness Theorems for Logics of Feature Structures. In Y. N. Moschovakis (ed.), *Logic From Computer Science*, MSRI Pubs. Vol. 21, Springer-Verlag, 1991, pp. 387–403.
3. Distanced Graphs. *Discrete Mathematics* 102, No. 3 (1992), pp. 287–305.
4. (with Andrew Dabrowski and Rohit Parikh) Topological Reasoning and the Logic of Knowledge. *Annals of Pure and Applied Logic* 78 (1996), no. 1-3, 73–110.
5. (with Jon Barwise) Modal Correspondence for Models, the *Journal of Philosophical Logic*, Vol. 27 (1998), 275–294.
6. (with A. J. C. Hurkens, Monica McArthur, Yiannis N. Moschovakis, and Glen Whitney) The Logic of Recursion Equations, the *Journal of Symbolic Logic*, Vol. 63, No. 2 (1998), 451–478.
7. Coalgebraic Logic, *Annals of Pure Applied Logic* 96 (1999), no. 1-3, 277–317.
8. Parametric Corecursion, *Theoretical Computer Science* 260 (1–2), 2001, 139–163.
9. Recursion and Corecursion Have the Same Equational Logic, *Theoretical Computer Science* 294 (2003), no. 1–2, 233–267.
10. (with Alexandru Baltag) Logics for Epistemic Programs, *Synthese* Vol. 139, issue 2 (2004), 165–224.
11. (with Ignacio Viglizzo) Final Coalgebras for Measurable Spaces. To appear in *Information and Computation*, special issue on papers from CMCS'04.
12. (with Stefan Milius) The Category Theoretic Solution of Recursive Program Schemes, to appear in the *Proceedings of CALCO 2005 (First Conference on Algebra and Coalgebra in Computer Science)*, Springer LNCS 2005.
13. (with Hans-Jörg Tiede), Modal Logic and Linguistics, to appear in P. Blackburn et al (eds.), *Handbook of Modal Logic*, Elsevier, 2006.
14. Applied Logic: A Manifesto, in D. Gabbay, S. Goncharov, and M. Zakharyaschev (eds.) *Mathematical Problems from Applied Logics I: New Logics for the XXIst Century*, Springer International Mathematical Series, to appear 2005, 22 pp.

*Work in Progress*

15. Natural Language, Natural Logic, Natural Deduction, ms. in preparation for a Festschrift for Uwe Mönnich, ms., 2005.
16. (with Alexandru Baltag and Sławomir Solecki) Logics for Epistemic Actions: Completeness, Decidability, Expressivity.

**Service to the Profession:** Editor, *Journal of Logic, Language, and Information*, 1996–present; *Logical Methods in Computer Science*, 2004–present; Editorial Board, the *Notre Dame Journal of Formal Logic*, 1992–present; *Grammars*, 1996–present; *Research on Language and Computation*, 1997–present; *Annals of Math., Computing and Teleinformatics*, 2003–present; *Logic and Logical Philosophy*, 2005–present; President, Association for the Mathematics of Language, 1999-2001. Director, NASSLLI'03. Beth Prize Committee, 1997-2005. Conference Chair: CMCS'02, FGMOL'01, ITALLC2, ASL Spring 1994. Some Conference Committees: ESSLLI'06 PC, Amsterdam Colloquium'06, LICS'05, AiML'02.

**Teaching:** Moss has taught classes in most areas of mathematical logic at the graduate level from 1985 onward. As his vision for logic changed, so did his teaching. For some years now, one of the main graduate logic classes at Indiana is in effect a course on logical systems. This class could eventually become the centerpiece of a logic curriculum that emphasizes applied topics. Moss also designs and teaches classes on Mathematics and Logic for Cognitive Science, Modal Logic (again, different from the standard presentations), and Natural Language Semantics. He has supervised seven Ph.D. students so far.

**Vision Statement:** I have a longer, more developed statement; please see my *Applied Logic: A Manifesto* mentioned in the list of publications above. That paper is an explanation of applied logic to a wider academic audience. My vision for logic as a whole is basically the same: it should/will be a subject that makes essential use of mathematical methods but is not principally directed towards the problems of mathematics. Instead, the logical enterprise will find its main motivation in problems from computer science (this already is the case), and also cognitive science, linguistics, philosophy, and other fields. Logic should/will stretch beyond its current boundaries to eventually interact with statistics and psychology, for example. Another point: many of the developing fields of inquiry across the sciences will be interdisciplinary pursuits. I look forward to seeing logic emerge as a field which is simultaneously one of the most "classical" and most contemporary, and also both a bridge between different subjects and a field of its own.

**Honours and Awards:** Rothrock Teaching Award, IU Mathematics Department, 2002. Teaching Excellence Recognition Award, Indiana University, 1998. UCLA Daus Prize for the outstanding graduate in mathematics, Member, Phi Beta Kappa Fraternity.

# MYCIELSKI, Jan

**Specialties:** Logic, set theory, model theory, universal algebra, topological algebra, games with perfect information, geometric topology, geometric measure theory.

**Born:** 7 February 1932 in Wisniowa and Wislokiem, county Rzeszow, Poland.

**Educated:** Institute of Mathematics of the Polish Acad. of Sciece, Docent, 1964; University of Wroclaw, PhD, 1957; MA, 1956.

**Dissertation:** PhD: *Applications of free groups to geometric constructions*; Docent thesis: *The Axiom of Determinacy*. Dissertation Supervisor: Stanislaw Hartman.

**Regular Academic or Research Appointments:** PROFESSOR EMERITUS, UNIVERSITY OF COLORADO, DEPARTMENT OF MATHEMATICS, 2001–, Professor, 1969-2001; Professor, Inst. of Math. Polish Ac.of Sci., 1968-1969, Docent, 1963-1968, Adjunct, 1958 - 1963.

**Visiting Academic or Research Appointments:** Visiting Professor, University of Warsaw, Poland, 1991; Long Term Visiting Scientist, Los Alamos National Laboratory, 1989-1990; Visiting Professor, University of Hawaii at Manoa, 1987; Guest Scientist, Institut des Hautes Etude Scietifique, 1977, 1978; Visiting Professor, University of Califoria-Berkeley, 1970; Visiting Professor, University of Colorado, 1967; Visiting Professor, Case-Western Reserve University, 1967; Visiting Assistant Professor, University of California, Bekeley, 1961-1962; Attache' de Recherche, Centre National de Recherche Scientifique, Paris, France, 1957-1958.

**Research Profile:** Jan Mycielski is interested in Logic and Foundations of Mathematics, geometric and topological problems, combinatorics, and brain science. He did some work that initiated or enhanced longer lines of research in the following areas. 1. Free groups and their applications to geometric construction such as paradoxical decompositions of spheres. 2. Theories of pursuit and evasion. 3. The Axiom of Determinacy and its consequences. 4. Constructions of independent sets in topological algebras. 5. The theory of Equationally Compact Algebras. 6. Lattices of interpretability types of theories. 7. Dynamic Approximation Theory; Linear Learning Theorems and mathematical models of the brain. 8. Finitistic interpretations of mathematics, locally finite theories with a criticism of Platonism and formalism. 9. Definition of the class of mathematical objects that have a potential for a physical interpretation. 10. Motivation and heuristic support of some systems of axioms of set theory.

**Main Publications:**

1. *About sets invariant with respect to denumerable changes*, Fund. Math. **45** (1958), 296 - 305.
2. (with S. Balcerzyk) *On faithful representations of free products of groups*, ibidem, **50** (1961), 63 - 71.
3. (with H. Steinhaus) *A mathematical axiom contradicting the axiom of choice*, Bull. Polish Acad. of Sci., Serie Math. Astr. et Phys. **10** (1962), 1 - 3.
4. *Some compactifications of general algebras*, Colloquium Mathematicum **13** (1964), 1 - 9.
5. *Almost every function is independent*, Fund. Math. **81** (1973), 43 - 48.
6. *Remarks on invariant measures in metric spaces*, Coll. Math. **32** (1974), 109 - 116.
7. *A lattice of interpretability types of theories*, The Journal of Symbolic Logic **42** (1977), 297 - 305.
8. *Locally finite theories*, ibidem, **51** (1986), 51 - 62.
9. *Theories of pursuit and evasion*, Journal of Optimization theory and its Applications, **56** (1988), 271 - 284.
10. *A learning theorem for linear operators*, Proc. Amer. Math. Soc. **103** (1988), 547 - 550.
11. (with P. Pudlak and A. Stern) *A lattice of chapters of mathematics*, Memoirs of the Amer. math Soc. **84**(426) (1990).

12. (with V. Faber) *Applications of learning theorems*, Fundamenta Informaticae **15** (1991), 145 - 167.

13. *Games with perfect information*, Chapter 3, Handbook of Game Theory, Vol. 1, Chapter 3, (Editors R. J. Aumann and S. Hart), Elsevier Sci. Publ. 1992.

14. (with R. Dougherty) *The prevalence of permutations with infinite cycles*, Fund. Math. **144** (1994), 89 - 94.

15. *Pure mathematics and physical reality*, Journal of Mathematical Sciences, 2005, and in Russian translation in Fundamentalnaya i Prikladnaya Matematika, **11,** no. 5 (2005).

16. *A system of axioms of set theory for the rationalists*, Notices of the Amer. Math, Soc. **53,** no. 2 (2006), 206 - 213.

*Work in Progress*

17. Generalization of a theorem of J. Hadamard about mappings of $R^n$ from the differentiable to the continuous category. 2. Applications of Mycielski's genus of closed irreducible 3-manifolds. 3. Philosophical applications of Hilbert's epsilon-symbols. 4. Conjectures on the process of learning which goes on in the neural network of the brain.

**Service to the Profession:** Editor, Games and Economic Behavior, 1991-2000; Editor, Fundamenta Mathematicae, 1984-; Editor, Contemporary Mathematics, AMC series of monographs, 1983-1989; Editor, Transaction of the American Mathematical Society, 1980-1985; Editor, Encyclopedia for Pure and Applied Mathematics (a series of Monographs), 1975-1995; Editor, Algebra Universalis, 1971-2000.

**Teaching:** Students of note: Prof. Bohdan Weglorz, Dr. James Fickett and Prof. James Lynch.

**Vision Statement:** Logic should enable us to build a computer to which one could communicate the human mathematical culture and which could use it to prove our conjectures as well as or better than a human beings. Studies of the brain should help us to achieve this goal.

**Honours and Awards:** University of Colorado, 78th Lecture on Research and Creative Work, 1990; W. Sierpinski Medal, Polish Mathematical Society and the University of Warsaw, 1990; Alfred Jurzykowski Foundation Award, 1978; Stefan Banach Prize, 1965.

# N

## NEEMAN, Itay

**Specialties:** Set theory.

**Born:** 9 December 1972 in Safed, Israel.

**Educated:** UCLA, PhD Mathematics, 1996; Oxford University, King's College London, BSc Mathematics, $1^{st}$ class honours, 1992; Tel Aviv University.

**Dissertation:** *Determinacy and Iteration Trees*; supervisor, John Steel.

**Regular Academic or Research Appointments:** PROFESSOR, MATHEMATICS, UCLA, 2006–; Associate Professor, Mathematics, UCLA, 2002-2006; Assistant Professor, Mathematics, UCLA, 2000-2002.

**Visiting Academic or Research Appointments:** Visiting Researcher, Microsoft Research, 2006; Fellow, Wissenschaftskolleg zu Berlin, 2005-2006; Junior Fellow, Harvard University, 1996-1999, 2000; Research Fellow, Humboldt Foundation, 1999.

**Research Profile:** Neeman's research concerns set theory, with emphasis on large cardinals and their use, especially in the study of determinacy. His most notable works address determinacy for games of variable countable length and games of length $\omega_1$; iterability for models with Woodin cardinals, and connections between iterable models and determinacy; inner model theory; and applications of inner models for large cardinals to the study of L(R) under determinacy.

**Main Publications:**

1. Itay Neeman, The determinacy of long games, vol. 7 of de Gruyter Series in Logic and its Application, Walter de Gruyter, Berlin, 2004.
2. Itay Neeman and John Steel, Counterexamples to the unique and cofinal branches hypotheses, J. of Symbolic Logic, vol. 71, pp. 977—988, 2006.
3. Itay Neeman, Unraveling $\Pi_1^1$ sets revisited, Israel J. of Mathematics, vol. 152, pp. 181—203, 2006.
4. Itay Neeman, An introduction to proofs of determinacy of long games, in Logic Colloquium '01, vol. 20 of Lecture Notes in Logic, pp. 43—86, Association of Symbolic Logic, Urbana, IL, 2005.
5. Donald A. Martin, Itay Neeman, and Marco Vervoort, The strength of Blackwell determinacy, J. of Symbolic Logic, vol. 68, pp. 615—636, 2003.
6. Itay Neeman, Optimal proofs of determinacy II, J. of Math. Logic, vol. 2, pp. 227—258, 2002.
7. Itay Neeman, Inner models in the region of a Woodin limit of Woodin cardinals, Ann. of Pure and Applied Logic, vol. 116, pp. 67—155, 2002.
8. Itay Neeman and Jindrich Zapletal, Proper forcing and $(L\mathbb{R})$, J. of Symbolic Logic, vol. 66, pp. 801—810, 2001.
9. Itay Neeman and John Steel, A weak Dodd-Jensen lemma, J. of Symbolic Logic, vol. 64, pp. 1285—1294, 1999.
10. Itay Neeman, Optimal proofs of determinacy, Bulletin of Symbolic Logic, vol. 1, pp. 327—339, 1995.

*Work in Progress*

11. Itay Neeman, Determinacy and large cardinals, to appear
12. Itay Neeman, Finite state automata and monadic definability of ordinals, to appear
13. Itay Neeman, Large cardinals and ultrafilters in $L(\mathbb{R})$, to appear
14. Itay Neeman, Games of length $\omega_1$, to appear
15. Alessandro Andretta, Greg Hjorth, and Itay Neeman, Effective cardinals of boldface pointclasses, to appear

**Service to the Profession:** Author, Survey Chapter on Proofs of Determinacy in $L(\mathbb{R})$, Handbook of Set Theory.

**Vision Statement:** "Logic is intrinsically important in mathematics and indeed in many other disciplines. It provides tools for the analysis of mathematical phenomena, and a foundational framework for the analysis of mathematics itself. I look forward to seeing both aspects develop further"

**Honours and Awards:** Invited Speaker, International Congress of Mathematicians, 2006; Fellow, Wissenschaftskolleg zu Berlin, 2005; NSF Career Award Recipient, National Science Foundation, 2000; Research Fellow, Alexander von Humboldt Foundation, 1999; Junior Fellow, Harvard University Society of Fellows, 1996; Sacks Prize Recipient, 1996.

## NÉMETI, István

**Specialties:** Mathematical logic, algebraic logic, logical foundations of spacetime theory and relativity, universal logic, philosophical logic, formal semantics, computer science and AI logics, foundation of mathematics.

**Born:** 17 August 1942 in Budapest, Hungary.

**Educated:** Hungarian Academy of Sciences, DSc Mathematics, 1987; Hungarian Academy of Sciences, Candidate's Degree Mathematics, 1977;

Eötvös Loránd University, PhD Mathematics, 1977; Technical University Budapest, MA Electrodynamics, 1966.

**Dissertation:** DSc: Free algebras and decidability in algebraic logic, PhD: Extending the universal algebraic notions of variety and related ones to partial algebras using abstract model theory and category theory, advisor: Gergely, T.

**Regular Academic or Research Appointments:** CHIEF SCIENTIFIC ADVISOR, RÉNY MATHEMATICAL INSTITUTE, BUDAPEST, 1977–; Scientific Advisor, Institute for Applied Computer Science, 1975-1976; Institute for Computer Science of the Ministry of Heavy Industries, 1970-1974; Institute for Designing the Electric Power System of the Ministry of Heavy Industries, 1966-1969; Instructor, Theoretical Electrodynamics, Technical University Budapest, 1962-1965.

**Visiting Academic or Research Appointments:** Visiting Professor, University of Amsterdam, 1998; Visiting Professor, University of California at Berkeley, 1991; Visiting Professor, Iowa State University, USA, 1987-1988; Visiting Professor, University of Waterloo, Canada, 1983-1984; Visiting Researcher, University of Edinburgh, Department of Computational Logic 1974-1975.

**Research Profile:** With Hajnal Andréka, there was a study about the above 3-way connection (logic, geometry, algebra) partly in the form of Tarskian approaches to algebraic logic (AL), and partly as the logical analysis of relativity theories (including general relativity GR) and spacetime geometry, including black hole physics. All parts of this 3-way connection go back to collaborations with Alfred Tarski's group [2,10,14].

He used the category of cylindric algebras (CA's) for representing the category of theories and interpretations acting between these theories as morphisms. In cooperation with Johan van Benthem, Bjarni Jónsson and Vaughan Pratt, among others, he refined the Tarskian tradition in algebraic logic in order to reinforce the so-called dynamic trend in logic. Dynamic logic or the logic of actions concentrates on the dynamic (even inductive and abductive) aspects of reasoning in several ways. Besides the dynamics of theory-formation, it also provides logics for reasoning about consequences of actions relevant for artificial intelligence (AI), program verification, logics of time, spacetime, and general relativity (GR) [4,5,6,20] and work in progress [1].

He made efforts to "refresh" Tarskian algebraic logic by making the interaction between logic and algebra stronger in the algebra-to-logic direction. He views algebraization of logic as a kind of abstraction. Generally, after the step of abstraction, one studies the so obtained abstract structures (e.g., AL). He, however, emphasizes that the results of this abstract study should be applied to the original "concrete" world, e.g., to logic. He initiated the study of relativized algebras, in particular relativized CA's, as an area worth studying in its own right. This study of new kinds of algebras of relations (namely relativized ones) led to new discoveries in pure logic. In particular, this catalyzed new kinds of applications in logic, and connected new trends and research interests in logic to methods of AL. He proved important properties of these relativized CA's, leading to the discovery of the very successful notions of guarded and packed fragments of first-order logic (FOL) [15]. He used AL to obtain new purely logical results. The latter include results on the finite-variable hierarchy of FOL, results on the schema version of FOL and its extensions (decidability and definability properties, interpolation, finitizability), and results on model theory of infinitary relational structures (algebraic model theory) [2,7-10,13,17]. The finite-variable hierarchy is well behaved in the new guarded fragment, while in the original FOL it is not, as proved in [15]. The same kind of approach was applied to relativity theory and its abstract counterpart spacetime geometry, by elaborating a duality theory between the observer-oriented (i.e., coordinate-dependent) and the geometrical (i.e., coordinate-independent) versions of relativity [20].

In category theoretic AL, he initiated "injectivity logic" and the injectivity approach to a very general conception of logic and applied it to the theory of partial algebras [3,6]. He contributed to the theories of cylindric, relation, polyadic, and other algebras of n-place relations establishing a new school (or at least invigorating the old one) [2,11,14,17].

He studies the applications of logic in various sciences, e.g., in the semantics of natural language, algebra, computer science and AI, theory of spacetime and relativity. In relativity, one of his aims is to make (even the most "esoteric" parts of) GR accessible for the nonspecialist readers (with a background in logic). He not only applies logic to GR, but also applies GR to the foundation of logic, pursuing a research direction suggested to them by the logician László Kalmár [19]. In his application of logic to relativity, he answers the "why type" questions (e.g., which axiom of (the logicized) GR is responsible for some interesting predictions/theorems of GR), using the same method as in his earlier work in reverse mathematics, where he studied the question of which axioms of set theory are responsible for which theorems of algebra and AL [12,16].

**Main Publications:**

1. The generalized completeness of Horn predicate logic as a programming language. Acta Cybernetica 4,1 (Szeged 1978), 3-10. (With Andréka, H.)
2. Cylindric Set Algebras. Lecture Notes in Mathematics vol. 883, Springer-Verlag, Berlin, 1981. vi +323 pp. (With Henkin, L., Monk, J. D., Tarski, A., Andréka, H.)
3. A general axiomatizability theorem formulated in terms of cone-injective subcategories. In: Universal Algebra (Proc. Coll. Esztergom 1977) Colloq. Math. Soc. J. Bolyai vol. 29, North-Holland, Amsterdam, 1981. pp. 13-35. (With Andréka, H.)
4. Nonstandard dynamic logic. (Invited paper.) In: Logics of Programs (Proc. Conf. New York, May 1981) Ed.: Kozen, D. Lecture Notes in Computer Science vol. 131, Springer-Verlag, Berlin, 1982. pp. 311-348.
5. A complete logic for reasoning about programs via nonstandard model theory. Theoretical Computer Science 117 (1982). Part I in no. 2, pp. 193-212, Part II in no. 3, pp. 259-278. (With Andréka, H., Sain, I.)
6. Generalization of the concept of variety and quasivariety to partial algebras through category theory. Dissertationes Mathematicae (Rozprawy Math.) no. 204. PWN - Polish Scientific Publishers, Warsaw, 1983. 51 pp. (With Andréka, H.)
7. Cylindric-relativized set algebras have strong amalgamation. Journal of Symbolic Logic 50 (1985), 689-700.
8. On varieties of cylindric algebras with applications to logic. Annals of Pure and Applied Logic 36 (1987), 235-277.
9. On cylindric algebraic model theory. In: Algebraic Logic and Universal Algebra in Computer Science (Proc. Conf. Ames 1988) Lecture Notes in Computer Science vol. 425, Springer-Verlag, Berlin, 1990. pp. 37-76.
10. Algebraic Logic. Colloq. Math. Soc. J. Bolyai vol. 54, North-Holland, Amsterdam, 1991. vi + 746 pp. (Edited with Andréka, H., Monk, J. D.)
11. A nonpermutational integral relation algebra. Michigan Math. J. 39 (1992), 371-384. (With Andréka, H., Düntsch, I.)
12. Connections between axioms of set theory and basic theorems of universal algebra. Journal of Symbolic Logic 59,3 (1994), 912-922. (With Andréka, H., Kurucz, Á.)
13. Fine-structure analysis of first order logic. In: Arrow Logic and Multi-Modal Logic, M. Marx, L. Pólos, and M. Masuch eds, CSLI Publications, Stanford, California, 1996. pp. 221-247.
14. Decision problems for equational theories of relation algebras. Memoirs of Amer. Math. Soc. vol. 126, no. 604, American Mathematical Society, Providence, Rhode Island, 1997. xiv+126pp. (With Andréka, H., Givant, S.)
15. Modal languages and bounded fragments of predicate logic. Journal of Philosophical Logic 27 (1998), 217-274. (With Andréka, H., van Benthem, J.)
16. Representability of pairing relation algebras depends on your ontology. Fundamenta Informaticae 44,4 (2000), 397-420. (With Kurucz, Á.)
17. Algebraic Logic. In: Handbook of Philosophical Logic, vol. 2, second edition, eds. D. M. Gabbay and F. Guenthner, Kluwer Academic Publishers, 2001. pp. 133-247. (With Andréka, H., Sain, I.)
18. Twin Paradox and the logical foundation of relativity theory. Foundation of Physics 36,5 (2006), 681-714. (With Madarász, J. X., Székely, G.)
19. Relativistic computers and the Turing barrier. Journal of Applied Mathematics and Computation 178 (2006), 118-142. (With Dávid, Gy.)
20. First-order logic foundation of relativity theories. In: New Logics for the XXIst Century II, Mathematical Problems from Applied Logics. International Mathematical Series Vol 5, Springer, to appear. (With Madarász, J. X., Székely, G.)

A full list of publications is available at Németi's homepage http://www.renyi.hu/~nemeti/

**Work in Progress**

21. A twist in the geometry of rotating black holes: seeking the cause of acausality; this is an in-depth mathematical analysis of some often mispresented features of some exotic GR spacetimes, with Andréka, H. and Wüthrich, C.
22. The weakest fragment of FOL still enjoying Gödel's incompleteness property.
23. A book on the logical analysis (in FOL) of general relativity theory presupposing no preliminaries except basics of logic, with Andréka, H. and Madarász, J. X.
24. Why do we stick with (many-sorted) FOL when studying foundational issues in mathematics and other scientific theories (e.g., relativity)?, with Andréka, H. This is a more carefully elaborated version of the Appendix of [Andréka, H., Madarász, J. X., Németi, I.: On the logical structure of relativity theories. http://www.math-inst.hu/pub/algebraic-logic/olsort.html]

**Service to the Profession:** Program Committee Member, Philosophical Interpretations of Relativity Theory, Budapest, 2007; Editorial Board Member, Logica Universalis, 2005-; Editorial Board Member, Advanced Studies in Mathematics and Logic book series, 2005-; Program Committee Member, Logic in Hungary, 2005; Editorial Board Member, Journal of Applied Logic, 2004-; Program Committee Member, János Bolyai Conference on Hyperbolic Geometry, Budapest, 2002; Program Committee Member, First Southern African School and Workshop on Logic, Universal Algebra, and Theoretical Computer Science, South Africa, 1999; Program Committee Member, International Conference on Temporal Logic, 1998; Program Committee Member, Frontiers of Combining Systems, 1998; Program Committee Member, International Conference on Temporal Logic, 1997; Program Committee Member, Mathematical Foundations of Computer Science, 1996; Program Committee Member, Fourth International Conference on Algebraic Methods and Software Technology (AMAST'95), 1995; Program Committee Member, First International Conference on Temporal Logic, 1994; Program Committee Member, Third International Conference on Algebraic Methods and Software Technology (AMAST), 1993; Editorial Board Member, Studia Logica, 1992-2003; Program Committee Member, European Summer Meeting of the Association for Symbolic Logic, Organizaed by J. Bolyai Mathematical Society and Symbolic Logic Department of Eötvös Loránd University, Veszprém, 1991; Program Committee Member, "Algebraic Methods in Logic and Their Computer Science Applications", $38^{th}$ Semester of Stefan Banach International Banach Mathematical Center, Warsaw, 1991; Program Committee Member, Mathematical Foundations of Computer Science'90, Banca Bistrica, 1990; Editorial Board Member, Journal of Applied Nonclassical Logic, 1989-; Program Committee Member, Conference on Algebraic Logic, Organizaed by J. Bolyai Mathematical Society, Budapest 1988; Editorial Board Member, Journal of Symbolic Computation, 1984-1990; Program Committee Member, Conference on Algebra, Combinatorics and Logic in Computer Science, Győr, Organizaed by J. Bolyai Mathematical Society, 1983; Program Committee Member, Conference on Universal Algebra, Organizaed by J. Bolyai Mathematical Society, Esztergom, 1977.

**Teaching:** Together with Hajnal Andréka, he has had 18 successful PhD students. In alphabetic order: Balázs Bíró (1986, co-author of S. Shelah), Buy Huy Hien (1981), Csaba Henk (present), Eva Hoogland (1996, MA, Amsterdam), Ágnes Kurucz (1997, London), Judit X. Madarász (2002), Zsuzsanna Márkusz (1989), Maarten Marx (1995, Amsterdam), Szabolcs Mikulás (1995, Amsterdam, London), Ana Pasztor (1979, University of Miami full professor), Ildikó Sain (1986), Gábor Sági (1999, co-author of S. Shelah), Tarek Sayed-Ahmed (2003, Cairo), György Serény (1986), András Simon (1998), Gergely Székely (present), Jenő Szigeti (1989), Csaba Tőke (2000, MA). They positively influenced the beginning of the scientific careers of Ben Hansen (Berkeley), Miklós Ferenczi, Viktor Gyuris, Richard J. Thompson (Berkeley), and others.

**Vision Statement:** Logic is the scienctific study of rational thinking or reasoning. As such, it includes a foundation of the scientific method, including the hypothetico-deductive account of science. So, logic is not only about the foundation of mathematics, but also of physics and other branches of science. A large portion of logic has been already formalized by using mathematics. In analogy with mathematical physics, mathematical logic should be conceived as research work in logic conducted by mathematical tools but aiming for all parts/aspects of logic (as its subject matter). Hence, mathematical logic should not be restricted to those parts of logic which are applied in mathematics. By viewing logic in this perspective much has been achieved and even more is to be done in the future.

**Honours and Awards:** Logic, Algebra, Relativity, Conference Dedicated to the Work of István Németi, http://www.renyi.hu/~n60/n60.body.html, 2002; Professorship, Széchenyi of Eötvös Loránd University, 2000-2003; László Kalmár Award, John von Neumann Computer Society, in recognition of work and results on unifying theoretical mathematics and computer science, 1979; "Gyula Farkas" Prize, Applying Mathematics, Mathematical Society János Bolyai, 1974; Scholarship, Receiving a Candidate's Degree, Hungarian Academy of Sciences, 1972-1975.

## NEPEJVODA, Nikolai N.

**Specialties:** Applied logic, constructive logics, theory of informatics, philosophy, linguistics.

**Born:** 19 June 1949 in Shelaevo, Kursk district, USSR (Russia).

**Educated:** Novosibirsk Computing Center of SAS USSR, DrSc Mathematics and Computer Science, 1988; Moscow State University, PhD Mathematics, 1974; Moscow State University, Mathematician, 1970.

**Dissertation:** *Impredicative Theories of Second Order with Unlimited Comprehenson Rule*; supervisors, Anderi A. Markov, Albert G. Dragalin; *Analysis and Methods of Demonstrative Programming in Constructive logics.*

**Regular Academic or Research Appointments:** PROFESSOR, MATHEMATICS AND COMPUTER SCIENCE, UDMURT STATE UNIVERSITY, 1992–; Chief of Applied Logic Scientific Venture, AN USSR, 1988-1992; Leading Research Fellow, Physical-Technical Institute of UrB AN USSR, 1983-1988; Senior Research Fellow, Mathematical Institute of Sibery AN USSR, 1980-1983; Associate Professor, Mathematics, Udmurt State University, 1973-1980.

**Visiting Academic or Research Appointments:** Main Research Fellow, Institute of Applied Mechanics, UrO RAN, 1992-; Visiting Professor, Computer Science, Tashkent State University of Informatics, 2006; Visiting Professor, Computer Science and Logic, Irkutsk State University, 2005; Visiting Professor, Computer Science, Novosibirsk State University, 2000-2002; Visiting Professor, Logic, Sanct-Petersbourg State University, 1999; Visiting Professor, Philosophy, Kaliningrad State University, 1999; Visiting Professor, Logic, Vladivostok State University, 1998. Earlier are not mentioned.

**Research Profile:** Nikolaj Nepejvoda has made fundamental contributions to constructive logics and their applications to computer science and to philosophy and has made significant contributions to mathematical linguistics, foundations of mathematics and proof theory. In constructive logics he is noted as inventor and developer of two new class of constructive logics: nilpotent and reversive ones. In applications of constructive logics to informatics he is one of inventors of homomorphism of proofs into programs, inventor of method of analysis programming concepts by means of constructive logic and above mentioned homomorphism, inventor and developer of constructive-logic based classifications of program styles. In proof theory he is noted for his pioneering work in predicative non-classical set theory which were used by S. Feferman in his concept of explicit mathematics and for his extensions of ***Goedel incompleteness theorem. In foundations of mathematics he is noted by his work on formal mereology, predicative mathematics and bridging NF and ZF. In (mathematical) philosophy he is one of founders (together with N. Belyakin) of the theory of informalizable notions and for his work in this topic. In methodology of science he is noted for his work of knowledge levels and logical approach to complex systems. In mathematical linguistics he is noted for his work on quasi-artificial languages. Almost all his work is in Russian and is very influential for the Russian speaking community of researchers and high level specialists in informatics.

**Main Publications:**

1. On embedding of Boolean algebras in Lindenbaum-Tarski algebras. Soviet Math. Doklady, Vol, 199:1, 1971.
2. A new notion of predicative truth and definability. Mathematical Notices, Vol 13:5, 1973.
3. On a generalisation of Kleene-Mostowski hierarchy. Soviet Math. Doklady, Vol, 212: 3, 1973.
4. Levelwise Beth models and realisability. In "Set theory and topology I", Izhevsk, 1977.
5. Correlation between natural deduction rules and operators in algorithmic high-level languages. Soviet Math. Doklady, 239:3, 1978.
6. On a method of constructing a correct program from correct subprograms. Journal of Programming, 11:1, 1979. (In Russian)
7. Application of proof theory to the problem of constructing correct programs. Cybernetics, 2, 1979. (In Russian)
8. A proof theoretical comparison of Program synthesis and Program Verification. In *6th Intern. Congress of Logic, Methodology and Philosophy of Science*, Hannover, 1979, v.1.
9. (with V. A. Smirnov and E. A. Paliutin). Logic at the international philosophical congress. Problems in Philosophy, 2, 1980. (In Russian)
10. The Logical Approach to Programming in *Logic, Methodology and Philosophy of Science VI*, Amsterdam, 1982.
11. Logical Approach to Programming. In *Lecture Notes in Computer Science*, 122, (1982).
12. Semantics of algorithmic languages (a survey). Surveys in Science and Technology, Theoretical Cybernetics, 1983. (In Russian)
13. The rule of unexpectedness and structural **goto**. Semiotics and Information Science, 23, 1984. (In Russian)
14. Logical approach as an alternative to system approach. In "Constructive Processes", Izhevsk, 1984.
15. Prefix semantic tableaus for modal logics. In "Many-valued Logics and Logical Consequence Theory", Institute of Philosphy, USSR Academy of Sciences, 1984. (In Russian)
16. Formalising informalisable notions: self-productive systems of theories. Semiotics and Information Science, 25, 1985. (In Russian)
17. Self productive systems of theories and systems of informalisable notions. in "Non classical logics", Institute of Philosphy, USSR Academy of Sciences, 1985. (In Russian)
18. Inferences in the form of graphs. Semiotics in Information Science, 26, 1985. (In Russian)
19. Constructive Logics, in *VIII Int. Congr. on Logic, Methodology and Philosophy of Science*, 1987.
20. Constructive logical means I: Generalised notion of logical calculus. Soviet Math. Izvestia, 5, (Theoretical Cybernetics), 1987. (In Russian)
21. Constructive logical means II: Intuitionistic logic and logics of program schemata. Soviet Math. Izvestia, 2, (Theoretical Cybernetics), 1988. (In Russian)
22. Formation of the notion of constructiveness in mathematics. In "Laws of development of contemporary mathematics", Moscow: Nauka 1988. (In Russian)
23. Some semantical constructions of constructive logics of program schemata. In "Computation Systems" Novisibirsk, v 129, 1989. (In Russian)
24. Propositional logics of program schemata, *Bulletin of the section of logic*, 1989, N 1.
25. Logical Approach as an alternative to system ap-

proach in mathematical description of systems. In "System Methods", Moscow, Nauka, 1991. (In Russian)

26. A bridge between constructive logic and computer programming, *Theoretical computer science* 90 (1991) 253-270.

27. First steps towards the theory of informalisable notions. In "Logical Investigations 1", Moscow, Nauka, 1993. (In Russian)

28. Incomplete Proof Structures and Their Applications, *Bulletin of the Section of Logic*, 1998, n 3.

29. On corellation between ZF and NF. In "International Conference on Mathematical Logic", Novsirbirsk, 1999. (In Russian)

30. Some analogues of partial and mixed computations in the Logical programming approach, *New Generation Computing, Japan*, 17 (1999), N 1.

31. Systemology 1. Proceedings of Scientific Seminar of Logical Centre, Russian Academy of Sciences, Moscow, 1999. (In Russian)

32. Levels of knowledge and skill. Proceedings of Scientific Seminar of Logical Centre, Russian Academy of Sciences, Moscow, 2001. (In Russian)

33. On a new class of constructive logics in logic and applications. Proceedigns of International Conference, Novosibirsk NGU, 2000. (In Russian)

34. Applied Logic, 2nd revised edition. Novosibirsk, NGU Press, 2000.

35. (with I. N. Skopin). Foundations of programming. Moskow-Izhevisk, 2003. (In Russian)

36. Styles and methods of programming. 8, 118-128, 2005.

37. Challenges of logic and mathematics of the 20th Century and mankind's "response". Problems in Philosophy, 9, 2005. (In Russian)

38. What matematics is necessary for computer scientists. Open Systems, 9, 2005. (In Russian)

See also my homepage http://www.uni.udm.ru/~nnn/

*Work in Progress*

39. A book on methodology of applying logic, Infosophia.

40. Series of popular papers on science of programming and applied logics for journal "Potential", Moscow Physical and Technical Institution (MPTI) (partially published)

41. A textbook on mathematical foundations of informatics.

42. Chaotic management: anti-process management theory based on logical approach.

**Service to the Profession:** Editor, Series of Books *Logical Investigations*, 1990-.

**Teaching:** Nepejvoda has been and remains a mainstay of the logic and informatic program at Udmurt University since his appointment to the faculty in 1973. He has had ten PhD students and two DrSc in the fields of Mathematics and Computer Science including A. Yashin. He was one of creators of concept and programs for new-born Informational Technologies Dept of Novosibirsk State University in 2000. For as long as 20 years Udmurt State University prepares specialists in informational analysis based on logic and non-numeric mathematic (original teaching program). The main idea of this specialty is despecialization and maximally deep fundamental education to provide specialists a system of knowledge which allows to adapt to any domain where innovative thinking is required and to use formal tools (including developing new programs and mental technologies) to investigate informalizable objects. Since 2006, this new specialty has become the backbone of the new-born Dept of Informational Technologies and Computing. Specialists trained are now leaders or leading experts in many innovative establishments in Izhevsk and throughout the world.

**Vision Statement:** Logic is a crucial subject that provides the chance to survive for our civilization. Only logic can bind together mathematics, humanism, informatics and religion (but not fanaticism!) Its methodological results are to be used to reconstruct all the systems of scientific knowledge and most of "common sense". Lack of logic or—what is worse—even a primitive understanding, whether based on Aristotelian syllogisms or on mathematical logic, is one of the deep roots of many problems (two of them are: admiring idols, including here democracy, totalitarism and freedom; mutual misunderstanding). Because constructive logics for different problems are contradictory there cannot be one true way of thinking or development. "Full unity of views can be only on cemetery" (Stalin, 1917). Thus I am antiglobalist.

**Honours and Awards:** Never bow to get them. I am an independent Russian bear.

## NERODE, Anil

**Specialties:** Mathematical Logic, Theoretical Computer Science, Applied Mathematics, Automata, Control Theory, Hybrid Systems, History of Mathematics.

**Born:** 04 June 1932 in Los Angeles California, USA.

**Educated:** The University of Chicago, BA (Liberal Arts, Hutchins Great Books Program), 1949; BS Mathematics, 1952; MS Mathematics, 1953; PhD Mathematics 1956. *Dissertation: Composita, Equations, and Freely Generated Algebra*; supervisor, Saunders MacLane.

**Regular Academic or Research Appointments:** GOLDWIN SMITH PROFESSOR OF MATHEMATICS, CORNELL UNIVERSITY, 1990–; Professor of Mathematics, Cornell University, 1965-1990; Associate Professor of Mathematics, Cornell University, 1962-1965; Assistant Professor of Mathematics, Cornell University, 1959-1962, Group Leader, Automata and Weapons Systems, Institute for Systems Research, University of Chicago, 1954-57.

**Visiting Academic or Research Appointments:** USEPA Distinguished Visiting Scientist, 1985-1987; Visiting Professor, University of California at San Diego, 1980, 1981, 1983, 1985; Visiting Professor, MIT, 1980; Visiting Professor, University of Chicago, 1976; Visiting Professor, Monash University, 1970, 1974-1975; Member, Institute for Advanced Study (with Kurt Gödel), 1962-1963; Member, Institute for Defense Analysis, Princeton, 1963; Visiting Assistant Professor of Mathematics (with Alfred Tarski), Berkeley, 1958-1959; National Science Foundation Postdoctoral Fellow, and Member, Institute for Advanced Study (with Kurt Gödel), 1957-1958.

**Research Profile:** Anil Nerode has made fundamental contributions to instantiation theory, recursive equivalence types and isols, computability, automata, computer science, computable mathematics, computable models of intuitionistic logics, non-monotonic logics, Finsler control theory, and hybrid systems. He has directed forty-five dissertations in these and other areas including models of set theory, lambda calculus, linear logic, history of logic, and applied mathematics. With Myhill he proved the fundamental Myhill-Nerode metatheorems which dominated the study of recursive equivalence types and isols. He showed, answering questions of Godel, that the isols embed all countable Diophantine correct models of arithmetic, and that the ring of arithmetic isolic integers modulo a non-principal prime ideal is a non-standard model of arithmetic. With Metakides he established the systematic use of priority methods in computable algebra, leading to a thirty-year effort by the Nerode and Ershov schools which culminated in the two volume "Handbook of Recursive Mathematics". He and R. Shore developed simple minimal pair coding methods for determining theories of degrees, which proved useful to a generation of workers. He and J. B. Remmel introduced the study of complexity-theoretic algebra. He and B. Khoussainov introduced the systematic study of automata theoretic algebra (automatic structures) which is undergoing a vast expansion at this time. In automata theory he early introduced the so-called "Nerode equivalence" which has played a fundamental and motivating role for about fifty years in research in systems theory and in automata theory. He and D. Wijesekera proved completeness of an intuitionistic version of dynamic logic. He and B. Khoussainov and H. Ishihara introduced the study of recursive models of intuitionistic logic, proving a completeness theorem for recursive theories. He and S. Ganguli established a complex analogue of the deduction theorem for predicate modal logic which led to the construction of recursive models of predicate modal logic and recursive completeness. He and J. B. Remmel and V. Marek developed a theory of non-monotonic rule systems encompassing many systems of non-monotonic logic which established many new theorems for these logics. Nerode has been a consultant for military projects, including control systems, for fifty years. Based on this background he and W. Kohn developed the theory of interacting reactive systems of discrete digital programs and continuous devices intended to control the behavior of complex systems. He named the area "hybrid systems" in 1992, held four conferences and published four volumes, and succeeded in forming a world wide community at the interface of logic, computer science, and control engineering. Nerode and Kohn developed their theory of hybrid systems based on the relaxed variational calculus on Finsler manifolds, for which applications have been patented and implemented and are in commercial use. This area is undergoing many interesting new developments ranging from real time financial prediction models to quantum control programs for nanoprocesses.

**Main Publications:**

1. S. Ganguli, A. Nerode, Effective Completeness Theorems for Modal Logic, 141-195, Ann. Pure and Applied Logic, 128 (1-3), 2004.
2. W. Kohn, V. Brayman, P. Cholewinski, A. Nerode, Control in Hybrid Systems, International Journal of Hybrid Systems 3, 2003.
3. W. Kohn, V. Brayman, A.l Nerode, Control Synthesis in Hybrid Systems with Finsler Dynamics, Houston Journal of Mathematics 28(2), 353-375, 2002.
4. Khoussainov, B.; Nerode, A., Automata Theory and Its Applications, Birkhauser, 430pp, 2001.
5. Kohn, W., Nerode, A., Remmel, J. B., Scalable Data and Sensor Fusion via Multiple Agent Hybrid Systems, LCNS 1567,Springer-Verlag,.pp.122-141, 1997.
6. Handbook of Recursive Mathematics. Vol. 1, 2., edited by Yu. L. Ershov, S. S. Goncharov, A. Nerode, J. B. Remmel and V. W. Marek. Studies in Logic and the Foundations of Mathematics, 138, 139. Elsevier, 1998
7. Nerode, A., Remmel, J. B., On the lattices of NP-subspaces of a polynomial time vector space over a finite field.. Ann. Pure Appl. Logic 81 no. 1-3, 125-170, 1996.
8. Marek, V., Nerode, A., Remmel, J. B.,Complexity of normal default logic and related modes of nonmonotonic reasoning, Proceedings of the 10th Annual Symposium of Logic in Computer Science, IEEE Computer

Society Press, pp.178-187, 1995.

9. Nerode, A., Shore, R., Logic for applications ($2^{nd}$ ed.) Springer-Verlag, New York, 1995.

10. Kohn, W.; Nerode, A., Remmel, J., Hybrid systems as Finsler manifolds: finite state control as approximation to connections, in Hybrid Systems, II , LNCS 999, 294-321,Springer-Verlag, 1994.

11. Kohn, W., Nerode, A., Models for hybrid systems: automata: topologies, controllability and observability, in Hybrid Systems II, LNCS 732, Springer-Verlag, 1993.

12. Marek, W.; Nerode, A.; Remmel, J. How complicated is the set of stable models of a recursive logic program? Ann. Pure Appl. Logic 56 (1992), no. 1-3, 119-135, 1992.

13. Nerode, A.; Remmel, J. B. Complexity-theoretic algebra: vector space bases, in Feasible Mathematics 293-319 (S. Buss and P. Scott, eds.) Birkhauser, 1990

14. Nerode, A. and Remmel, J. B., Complexity Theoretic Algebra I: vector spaces over finite fields, extended abstract with proofs, in Proc. Structures in Complexity (second annual conference), June 16-19, 1987, IEEE Computer Society, Silver Springs, Md., pp. 218-241, 1987.

15. Ash, C. J.; Nerode, A., Intrinsically recursive relations, in Aspects of Effective Algebra (J. N. Crossley, ed.) pp. 26-41, Upside Down A Book Co., Yarra Glen, Vic., Australia, 1981.

16. Nerode, Anil; Shore, Richard A. Second order logic and first order theories of reducibility orderings, The Kleene Symposium, pp. 181-200, Stud. Logic Found. Math.101, North-Holland, 1980.

17. Metakides, G.; Nerode, A. Effective content of field theory. Ann. Math. Logic 17, no. 3, 289-320, 1979.

18. Crossley, J. N.; Nerode, A.l, Combinatorial functors. Ergebnisse der Mathematik und ihrer Grenzgebiete, Band 81. Springer-Verlag, 146 pp, 1974

19. Nerode, A., Diophantine correct non-standard models in the isols, Ann. Math. (2) 84, 421-432, 1966.

20. Nerode, A. Extensions to isolic integers. Ann. of Math. (2) 75 1962 419-448. 1962: Nerode, A. Arithmetically isolated and nonstandard models. 1962 Proc. Sympos. Pure Math., Vol. V pp. 105-116 American Mathematical Society, Providence, R.I, 1962.

21. Nerode, Anil Extensions to isols. Ann. of Math. (2) 73 1961, 362-403, 1961.

22. Nerode, A. Linear automaton transformations. Proc. Amer. Math.Soc. 9 541-544, 1958.

*Work in Progress*

23. A. Nerode and W. Kohn, et al, Hybrid Systems, Finsler Geometry, and Optimal Control (Research Monograph)

24. A. Nerode, Logical Models for Agent Networks (Research Monograph)

**Service to the Profession:** Editor: Journal of Symbolic Logic, Ann. Pure and Applied Logic , Proc. AMS , Constraints, Inter. J. Hybrid Systems , Mathematics and Computer Modeling , Ann Math and AI, J. Pure and App. Algebra, and many others. Organizer: International Conferences in Logic, Computer Science, Engineering in the US, Russia, China, Southeast Asia, Australia, etc. He initiated the first several hybrid systems meetings starting 1991, and associated LNCS volumes. The area has become very popular in Engineering, Computer Science, and Mathematics. As director of the Army Research Office Center of Excellence at Cornell for 11 years, he organized over 100 meetings in diverse areas of mathematics and sponsored over 300 visitors in many areas of mathematics.

Advisory Board, NTT Japan, 2004-; Chair, Technical Advisory Board, Clearsight Systems, 2001-; Chebotarev Research Institute of Mathematics and Mechanics, Kazan State University, Universitetskaja 17, 420008, Kazan, Russia; International Advisory Board ,Centre for Discrete Mathematics and Theoretical Computer Science, University of Auckland, 1998-; International Advisory board, 1997-; Comm. on Science Policy, AMS, 1994-1997; Comm. on Mathematics, Nat. Academy of Sciences, 1994-1997; Vice President, AMS (elected), 1991-1994; Comm. on Committees, AMS, 1994-1997; Comm. on Science Policy, AMS, 1994-1997; Advisory Board, Army High Performance Computing Institute, U. of Minnesota, 1989-1992; Advisory Committee, Gauss Laboratory, University of Puerto Rico, 1989; Chair, Technical Advisory Panel for Global Change Program USEPA, 1988-1994; Advisory Board, Center for Intelligent Control, MIT-Brown-Harvard, 1987-92; International C. S. NRC group on computational mathematics, 1984; Comm. on Science Policy AMS, 1973-78; Advisory Panel for Math.l Sciences, US NSF, 1970-1973; Comm. on Applied Math. National Research Council, 1967-1970.

**Teaching:** On Arriving at Cornell in 1959, Nerode guided Gerald Sacks at Prof. Rosser's request. Among Nerode's own 44 students in mathematics, computer science, and applied mathematics are: R. I. Soare (CE sets, Chicago), M. Lerman (CE degress, Connecticut), Jeffrey Remmel (Computability theory, Combinatorics, UCSD), Terry Millar (Computable Model Theory, Wisconsin), Joe Miller (Computability Theory, Connecticut), Robert Jeroslow (Logic and OR, Georgia Tech, deceased) George Metakides (Recursive Mathematics, Director EEU Esprit, Univ. Patras), Louise Hay (Recursive Mathematics, Illinois at Chicago, deceased), Mark Scowcroft (Constructive Mathematics, Connecticut Weysleyan), Jim Lipton (Constructive mathematics, Computer Science), Connecticut Wesleyan, Charlotte Lynn (Boeing), Richard Tenney (Computer Science,

University of Mass. Boston (retired).

**Vision Statement:** He believes in the unity of science and therefore in the unity of mathematics. No matter what one learns in any mathematical field, it is likely it will prove useful elsewhere. He expects in the next 100 years a complete integration of logical methods and core mathematics, incorporating logics developed for computer science and engineering. He forwards to the melding of continuous and discrete mathematics in hybrid systems.

**Honours and Awards:** Distinguished Visiting Scientist, Expert Systems, USEPA 1980-82; National Science Foundation Postdoctoral fellow, 1957-1958.

# NIINILUOTO, Ilkka

**Specialties:** Philosophical logic, inductive logic, formal semantics.

**Born:** 12 March 1946 in Helsinki, Finland.

**Educated:** University of Helsinki, PhD Theoretical Philosophy, 1974; University of Helsinki, MA Mathematics, 1968.

**Dissertation:** *Conceptual Enrichment, Theories, and Inductive Systematization*, 1973; Supervisor, Jaakko Hintikka.

**Regular Academic or Research Appointments:** PROFESSOR, THEORETICAL PHILOSOPHY, UNIVERSITY OF HELSINKI, 1977–; Associate Professor, Foundational Mathematics, 1973-1977; Research Assistant, Academy of Finland, 1971-1973.

**Visiting Academic or Research Appointments:** Visiting Scholar, Stanford University, 1972.

**Research Profile:** In his studies in mathematics and philosophy, Niiniluoto's main interests were probability theory and mathematical logic. In his doctoral thesis, he applied Hintikka's system of inductive probabilities to study the role of theories and theoretical concepts in inductive inference. With Hintikka, Niiniluoto contributed to the axiomatic foundations of inductive logic. He has continued his work on induction and related forms of inference with studies in inductive-probabilistic explanation, analogical reasoning, and abduction.

In 1975 Niiniluoto started to study the concept of truthlikeness or verisimilitude. Popper's attempt to explicate the tricky notion of "closeness to the truth" had failed. Niiniluoto developed a new method of defining degrees of truthlikeness for various kinds qualitative and quantitative of statements by employing distance functions and metrics between states of affairs. For the general case of first-order theories, Hintikka's theory of distributive normal forms with constituents was applied. For the estimation of degrees of truthlikeness, when the target truth is unknown, expected values relative to inductive probabilities were proposed. This leads to the idea that scientific reasoning can be understood as an attempt to maximize expected verisimilitude. Niiniluoto's book *Truthlikeness* (1987) is still the most comprehensive and detailed exposition of the similarity approach to truthlikeness. Niiniluoto has always combined his logical work with philosophical applications. In *Is Science Progressive?* (1984), he analyzed theory-change and scientific progress in terms of increasing truthlikeness. In *Critical Scientific Realism* (1999), he further argued that truthlikeness is a key concept in developing a fallibilist epistemology and a realist philosophy of science.

Work on truthlikeness is related to classical debates on the concept of truth. In his historical and systematic studies on Tarski and Carnap, Niiniluoto has defended the semantic definition of truth as a form of the correspondence theory of truth.

Following the Finnish tradition of philosophical logic (G.H. von Wright, J. Hintikka), Niiniluoto has written essays on several branches of intensional logic: epistemic logic, logic of propositional attitudes (perception, memory, and imagination), modal logic, deontic logic, and the logic of norms.

**Main Publications:**

1. (with R. Tuomela) *Theoretical Concepts and Hypothetico-Inductive Inference*, D. Reidel, Dordrecht, 1973.
2. 'On the Truthlikeness of Generalizations', in R.E. Butts and J. Hintikka (eds.), *Basic Problems in Methodology and Linguistics, Part Three of the Proceedings of the Fifth International Congress of Logic, Methodology and Philosophy of Science, London, Ontario, 1975*, D. Reidel, Dordrecht, 1977, pp. 121-h147.
3. 'On a K-dimensional System of Inductive Logic', in F. Suppe and P. Asquith (eds.), *PSA 1976*, vol. 2, Philosophy of Science Association, East Lansing, 1977, pp. 425-h447.
4. 'Knowing that One Sees', in E. Saarinen, R. Hilpinen, I. Niiniluoto and M. Provence Hintikka (eds.), *Essays in Honor of Jaakko Hintikka*, D. Reidel, Dordrecht, 1979, pp. 249-282.
5. (ed. with R. Tuomela) *The Logic and Epistemology of Scientific Change*, Acta Philosophica Fennica, 30: 2h4, North-Holland, Amsterdam, 1979.
6. (with J. Hintikka) 'An Axiomatic Foundation for the Logic of Inductive Generalization', in R.C. Jeffrey (ed.), *Studies in Inductive Logic and Probability*, vol. 2, University of California Press, Berkeley, 1980, pp. 157h-181.
7. 'Statistical Explanation Reconsidered', *Synthese*

48 (1981), 437-472.
8. 'Remarks on the Logic of Perception', in I. Niiniluoto and E. Saarinen (eds.), *Intensional Logic: Theory and Applications*, Acta Philosophica Fennica 35, Helsinki, 1982, pp. 116–129.
9. 'Truth and Legal Norms', in N. MacCormick, S. Panov and L.L. Vallauri (eds.), *Conditions of Validity and Cognition in Modern Legal Thought*, Archiv für Rechts- und Sozialphilosophie (ARSP), Beiheft 25., Franz Steiner Verlag Wiesbaden GMBH, Stuttgart, 1985, pp. 168–190.
10. 'Hypothetical Imperatives and Conditional Obligations', Synthese 66:1 (1986), 111–134.
11. 'Theories, Approximations, and Idealizations', in R. Barcan Marcus, G.J.W. Dorn, and P. Weingartner (eds.), *Logic, Methodology and Philosophy of Science VII* (Proceedings of the Seventh International Congress, Salzburg, 1983), North-Holland, Amsterdam, 1986, pp. 255–289.
12. 'Imagination and Fiction', *Journal of Semantics* 4 (1986), 209–222.
13. *Truthlikeness*, D. Reidel, Dordrecht, 1987.
14. 'From Possibility to Probability: British Discussions on Modality in the Nineteenth Century', in S. Knuuttila (ed.), *Modern Modalities*, Kluwer Academic Publishers, Dordrecht, 1988, pp. 275–309.
15. 'Analogy and Similarity in Scientific Reasoning', in D.H. Helman (ed.), *Analogical Reasoning: Perspectives of Artificial Intelligence, Cognitive Science, and Philosophy*, Kluwer Academic Publishers, Dordrecht, 1988, pp. 271–298.
16. (with T. Aho) 'On the Logic of Memory', in L. Haaparanta, M. Kusch, and I. Niiniluoto (eds.), *Language, Knowledge, and Intentionality — Perspectives on the Philosophy of Jaakko Hintikka*, Acta Philosophica Fennica 49, Helsinki 1990, pp. 408-429.
17. 'Tarskian Truth as Correspondence — Replies to Some Objections', in J. Peregrin (ed.), *Truth and its Nature (if any)*, Kluwer, Dordrecht, 1999, pp. 91h-104.
18. 'Defending Abduction', *Philosophy of Science* 66 (Proceedings) (1999), S436-hS451.
19. 'Tarski's Definition and Truth-Makers', *Annals of Pure and Applied Logic* 126 (2004), 57-76.
20. 'Inductive Logic, Verisimilitude, and Machine Learning', in Petr Hájek, Luis Valdés-Villanueva, and Dag Westerståhl (eds.), *Logic, Methodology and Philosophy of Science: Proceedings of the Twelfth International Congress*, King's College Publications, London, 2005, pp. 295-314.

A more extensive list of publications is available at http://www.helsinki.fi/filosofia/filo/henk/niiniluoto.htm
*Work in Progress*
21. Abductive inference and truthlikeness.
22. Belief revision and truthlikeness.

**Service to the Profession:** Editor (with M. Sintonen and J. Wolenski), *Handbook of Epistemology*, 2004; Rector, University of Helsinki, 2003-2008; Chairman, Finnish Federation of Learned Societies, 2000-; Vice-Rector for Research, 1998-2003; Steering Committee Member, FISP, 1998-; Dean, Faculty of Arts, 1990-1991, 1993-1994; Department Chairman, Philosophy (Several Occasions); Editor, *Acta Philosophica Fennica*, 1980-; Editor, *Synthese*, 1977-1979; President, Philosophical Society of Finland, 1975-.

**Teaching:** In 1973-1977, Niiniluoto served as the first professor of mathematical logic at the Department of Mathematics, University of Helsinki. His teaching activities have been continued by Jouko Väänänen. Since 1977, Niiniluoto has educated several generations of masters and doctors at the Department of Philosophy. The main areas of his teaching have included logic, philosophy of science, epistemology, philosophy of technology, and philosophy of culture. He has also published several textbooks in these fields in Finnish.

**Vision Statement:** Since the $20^{th}$ century, logic has provided indispensable tools for analytic philosophy – fields like philosophy of mathematics, epistemology, philosophy of science, philosophy of language, and philosophy of law. New connections to computer science, cognitive science, and AI will be important in the future.

**Honours and Awards:** Academia Europaea 2004; Invited Member, Institute International de Philosophie, 1988-.

# NORMAN, Dag

**Specialties:** Computability theory, domain theory, semantics of algorithms.

**Born:** 15 May 1947 in Narvik, Norway.

**Educated:** University of Oslo, Dr Philosophy; University of Oslo, Cand Real (MSc). *Dissertation Characterization Problems in Higher Type Recursion Theory*; supervisor, Jens Erik Fenstad.

**Regular Academic or Research Appointments:** PROFESSOR, UNIVERSITY OF OSLO, 1992-; Associate Professor, University of Oslo, 1982-1991.

**Research Profile:** Dag Normann has been interested in computability in higher types through most of his career. He isolated the schemes for set recursion in relation with work on Kleene-computability relative to normal functionals of higher types. Then he got interested in the hereditarily continuous functionals, with several contributions in the late 70's and early 80's. Over the last years, he has been interested in applications of higher type computability theory to theoretical computer science, also considering typed computations over base structures like the reals and other topological spaces.

Over a short period, Normann gave contributions to Pi-1-2-logic and the theory of dilators and ptykes.

**Main Publications:**

1. With J. E. Fenstad On absolutely measurable sets Fundamentha Mathematica 81 pp. 91-98 (1974)
2. With J. Moldestad Models for recursion theory Journal of symb. logic (JSL) 41 pp. 719-729 (1976)
3. A continuous functional with noncollapsing hierarchy Journal of Symbolic Logic 43 pp. 487-491 (1978)
4. Set recursion In Fenstad & al.: Generalized Recursion Theory II North Holland, pp. 303-320 (1978)
5. Degrees of functionals Annals of Math. Logic 16 pp. 269-304 (1979)
6. With S. S. Wainer The one-section of a countable functional Journal of Symbolic Logic 45 pp. 549-562 (1980)
7. Recursion on the countable functionals Springer Lecture Notes in Mathematics 811 (1980)
8. The continuous functionals; computations, recursions and degrees Annals of Math. Log. 21 pp. 1-26 (1981)
9. Non-obtainable continuous functionals In Logic, Methodology and Philosophy of science VI. Proceedings of the sixth International Congress og Logic, Methodology and Philosophy of Science, Hannover 1979, pp. 241-249 (1982)
10. Formalizing the notion of total information In Petkov: Mathematical Logic Plenum Press, pp. 67-94 (1990)
11. With J.-Y. Girard Embeddability of ptykes Journal of Symbolic Logic 57 pp 659-676 (1992)
12. Closing the gap between the continuous functionals and recursion in $\frac{3}{E}$ Archives for Mathematical logic 36, pp. 269-287 (1997)
13. Computability over the partial continuous functionals Journal of Symbolic Logic (65) pp. 1133 - 1142, (2000)
14. The continuous functionals of finite types over the reals. In K. Keimel, G.Q.Zhang, Y. Liu and Y. Chen (eds.) Domains and Processes, pp. 103 – 124, Kluwer Academic Publishers (2001)
15. Hierarchies of total functionals over the reals. Theoretical Computer Science 316, pp 137 – 151 (2004)
16. Comparing hierarchies of total functionals Logical Methods in Computer Science, Volume 1, Issue 2, paper 4 (2005)

*Work in Progress*

17. On sequential functionals of type 3. Mathematical Structures in Computer Science
18. Computing with functionals- computability theory or computer science? Bulletin of Symbolic Logic
19. Definability and reducibility in higher types over the reals. Proccedings of Logic Colloquium '03

**Service to the Profession:** PC-member CiE, 2006; PC-member, Logic Colloquium, 2005; PC-member, CiE, 2005; Advisory Editor, Annals of Pure and Applied Logic 1989–2001.

**Teaching:** Normann has supervised four successful PhD students: Anna Salvesen, Lill Kristiansen, Reinert Rinvold and Geir Waagboe. Lill Kristiansen is still in academia, a full professor at NTNU in Trondheim.

**Vision Statement:** Logic can only survive as an active research area in interplay with mathematics, computer science, linguistics and other subjects. The challenge is to isolate problems of genuine interest to logic in this process.

# NUTE, Donald

**Specialties:** Conditional logic, deontic logic, nonmonotonic reasoning, logic and artificial intelligence.

**Born:** 12 August 1947 in Maysville, Kentucky, USA.

**Educated:** Indiana University, Bloomington, Indiana, PhD Philosophy, 1974;; University of Kentucky, BS Philosophy and Mathematics 1969.

**Dissertation:** *Identification and Demonstrative Reference*; supervisor, Hector Casteneda.

**Regular Academic or Research Appointments:** PROFESSOR, PHILOSOPHY, EMERITUS AND FRANKLIN COLLEGE PROFESSOR, EMERITUS, UNIVERSITY OF GEORGIA, 2004–; Franklin College Professor, University of Georgia, 2001-2004; Professor of Philosophy, University of Georgia, 1983-2004; Associate Professor, University of Georgia, 1978-1983; Assistant Professor, University of Georgia, 1973-1978; Associate Instructor, Indiana University, Bloomington, IN, 1970-1971.

**Visiting Academic or Research Appointments:** Research Associate, University of Antwerp, Belgium, 1989; Visiting Professor, University of Tubingen, West Germany, 1985; Research Associate, University of Stuttgart, West Germany, 1981.

**Research Profile:** Donald Nute has made contributions to conditional logic and nonmonotonic reasoning. His analysis of conterfactuals and tense and conditionals formed the foundation for further work in deontic logic, philosophy of language, and philosophy of science. He developed a new formalism for nonmonotonic reasoning called *defeasible logic* that differs from other approaches in providing a constructive proof theory. His work on defeasible logic has been applied in normative reasoning, epistemology, and artificial intelligence (automated reasoning systems.)

**Main Publications:**

1. Counterfactuals. *Notre Dame Journal of Formal Logic* **16** (1975):476-482.
2. Counterfactuals and the similarity of worlds. *Journal of Philosophy* **72** (1975):773-778.
3. An incompleteness theorem for conditional logic. *Notre Dame Journal of Formal Logic* **19** (1978):634-636.
4. Algebraic semantics for conditional logics. *Reports on Mathematical Logic* No. 10 (1978):79-101.
5. *Topics in Conditional Logic*. Philosophical Studies Monograph Series, D. Reidel Publishing Company, Dordrecht, Holland, 1980.
6. Logical relations. *Philosophical Studies* **46** (1984):41-56.
7. Conditional logic. *Handbook of Philosophical Logic, Vol. II*, ed. Franz Guenthner and Dov Gabbay, D. Reidel Publishing Company, Dordrecht, Holland, 1984:387-439.
8. Permission. *Journal of Philosophical Logic* **14** (1985):169-190.
9. LDR: a logic for defeasible reasoning. SNS-Bericht 86-11, University of Tübingen, 1986. Also appeared as ACMC Research Report 01-0013, The University of Georgia, 1986.
10. (with David Billington and Koen De Coster) A modular translation from defeasible nets to defeasible logics. *Journal of Experimental and Theoretical Artificial Intelligence* **2** (1990):151-177.
11. Defeasible logic and the frame problem. Henry Kyburg, Ronald Loui, and Greg Carlson (eds.), *Knowledge Representation and Defeasible Reasoning, Studies in Cognitive Systems*, Kluwer Academic Publishers, Boston, 1990:3-21.
12. Historical necessity and conditionals. *Noûs* **25** (1991):161-175.
13. Defeasible logic. In D. Gabbay and C. Hogger (eds.), *Handbook of Logic for Artificial Intelligence and Logic Programming, Vol. III*, Oxford University Press, 1994:353-395.
14. A decidable quantified defeasible logic. In D. Prawitz, B. Skyrms, and D. Westerstahl (eds.), *Logic, Methodology and Philosophy of Science IX*, Elsevier Science B. V., New York, 1994:263-284.
15. Apparent obligation. In Donald Nute (ed.), *Defeasible Deontic Logic: Essays in Nonmonotonic Normative Reasoning*. Kluwer Academic Publishers, Dordrecht, Holland, 1997, pp. 287–316.
16. (with Katrin Erk) Defeasible logic graphs I: theory. *Decision Support Systems* **22** (1998):277-293.
17. Defeasible logic: theory, implementation, and applications. In Oskar Bartenstein, Ulrich Geske, Markus Hannebauer, Osamu Yoshie (eds.), *Web Knowledge Management and Decision Support: Proceedings of INAP 2001, $14^{th}$ International Conference on Applications of Prolog (Revised Papers)*, Springer-Verlag, Berlin Heidelberg New York, 2003, pp. 151-169.
18. (with Frederick) Relating defeasible logic to the well-founded semantics for normal logic programs. In Juergen Dix and Anthony Hunter (Program Chairs,) *Proceedings of NMR'06: Eleventh International Workshop on Non-monotonic Reasoning, Windermere, U.K., May 30 - June 1, 2006*, http://cig.in.tu-clausthal.de/NMR06/

A full list of publications is available at Nute's home page http://donald.nute.ws.

**Work in Progress**

19. A series of papers or book on applications of defeasible logic in normative reasoning and epistemology.
20. Further investigation of the relationship between defeasible logic, other nonmonotonic formalisms, and logic programming.

**Service to the Profession:** Editorial Board, *Journal of Logic and Computation*, 199\*-; Editorial Board, *Minds and Machines*, 199\*-; President, Society for Exact Philosophy, 1994-1995; Founding Director, Artificial Intelligence Center, University of Georgia, 1994-2003; Vice President, Society for Exact Philosophy, 1993-1994; Head, Department of Philosophy, University of Georgia, 1989-1999.

**Teaching:** Besides teaching a wide range of courses in philosophy at the University of Georgia since his appointment to the faculty in 1973, Nute has also been a member of the faculties of artificial intelligence, cognitive science, and linguistics and taught courses in all of these areas. He has had four PhD students in philosophy and computer science, and nearly fifty MA and MS students in philosophy and artificial intelligence. In retirement, Nute continues to teach occasional courses in logic and artificial intelligence at the University of Georgia.

**Honours and Awards:** Franklin College Professor, University of Georgia, 2001-2004 (Emeritus 2004.)

# O

## OHLBACH, Hans Jürgen

**Specialties:** Automated Reasoning in classical and modal logics, description logics, computational modeling of geotemporal notions.

**Born:** 13 September 1952 in Bad Schwalbach, Germany.

**Educated:** Habilitation for Computer Science, University of Saarbrücken, Germany, 1994; PhD in Computer Science, University of Kaiserlautern, Germany, 1988; Dipoma in Nuclear Physics, University of Mainz, Germany, 1981.

**Dissertation:** *A Resolution Calculus for Modal Logics*; supervisor, Jörg Siekmann.

**Regular Academic or Research Appointments:** PROFESSOR FOR COMPUTER SCIENCE, LUDWIG-MAXIMILIANS UNIVERSITY, MUNICH, GERMANY, 2000–; Senior Lecturer, Department of Computer Science, King's College, London, 1998–2000; Senior Research Fellow, Department of Computer Science, Imperial College, London, 1996–1998; Deputy Director, Max-Planck Institute for Computer Science, Saarbrücken, Germany, 1991-1996; Research Assistant, University of Kaiserslautern, Germany, 1984-1991; Research Assistant, University of Karlsruhe, Germany, 1981–1984.

**Visiting Academic or Research Appointments:** Visiting Research Fellow, Australian National University (ANU), Canberra, 1988; Visiting Research Fellow at the University of Cape Town, 1994.

**Research Profile:** Hans Jürgen Ohlbach was one of the key developers of one of the earliest big resolution based automated theorem provers, the Markgraf Karl Refutation Procedure. His 'Terminator algorithm', which is an efficient graph based UR-resolution method, considerably improved the system. In his dissertation he investigated a new approach to Modal Logic theorem proving, by means of the 'functional translation'. Theorem proving by translation has then become a major field in non-classical logic theorem proving. In 1992 he developed, together with Dov Gabbay a resolution-based quantifier elimination algorithm for quantified predicates. The SCAN system was the first web-based theorem proving system and is still online at http://www.mpi-inf.mpg.de/departments/d2/software/SCAN. In Description Logics he introduced a new approach for dealing with number restrictions, the *atomic decomposition* method. An improved version has been integrated into the now commercialised RACER system. Currently he is working at modelling geotemporal and geospatial notions for semantic web applications.

**Main Publications:**

1. (with Grigorios Antoniou), Terminator, *Proc. of the 8th International Joint Conference on Artificial Intelligence*, (1983), 916—919.
2. A resolution calculus for modal logics, *Proc. of $9^{th}$ International Conference on Automated Deduction*, Springer LNCS 310, (1988), 500–516.
3. (with Jörg H. Siekmann), The Markgraf Karl refutation procedure, in *Computational Logic, Essays in Honor of Alan Robinson*, MIT Press (1991), 41–112.
4. (with Norbert Eisinger and Axel Präcklein), Reduction rules for resolution based systems, *Artificial Intelligence*, 50(2) (1991) 141–181.
5. Semantics based translation methods for modal logics, *Journal of Logic and Computation,* 1(5), (1991), 691-746.
6. (with Dov M. Gabbay), Quantifier elimination in second-order predicate logic, *South African Computer Journal*, 7 (1992) 35–43.
7. Translation methods for non-classical logics — an overview, *Bulletin of the Interest Group in Propositional and Predicate Logic*, 1(1) (1993), 69–90, 1993.
8. (with Chris Brink and Dov Gabbay), Towards Automating Duality, *Journal of Computers and Mathematics with Applications* 29(2), (1994), 73–90.
9. (with Franz Baader), A multi-dimensional terminological knowledge representation language, *Journal of Applied Non-Classical Logics*, 5(2), (1995) 153–198.
10. (with R. A. Schmidt and U.~Hustadt), Translating graded modalities into predicate logic, in *Proof Theory and Modal Logic,* Studies in Applied Logic Series, chapter 14, (1995) 245–285.
11. Elimination of self-resolving clauses, *Journal of Automated Reasoning* 20(3), (1998) 317–336.
12. (with Renate Schmidt), Functional translation and second-order frame properties of modal logics, *Journal of Logic and Computation* 7(5), (1997), 581–603.
13. (with Dov Gabbay), Calendar Logic, *Journal of Applied Non-Classical Logics*, 89:4 (1998), 291–324.
14. (with Jana Koehler), How to extend a formal system with a Boolean Algebra component, in *Automated Deduction. A Basis for Applications*, Volume III, (1998) 57–75.
15. (with Jana Koehler), Modal Logics, Description Logics and arithmetic reasoning. *Journal of Aritificial Intelligence* 109, (1999) 1–31.
16. (with Andreas Nonnengart and Andrzej Szałas),

Elimination of predicate quantifiers, in *Logic, Language and Reasoning — Essays in Honor of Dov Gabbay*, Kluwer (1999), 149–172.

17. Relations between fuzzy time intervals, in *Proceedings of 11th International Symposium on temporal representation and reasoning*, IEEE Computer Society (2004), 44–51.

18. Modelling periodic temporal notions by labelled partitionings – The PartLib Library, in *Essays in Honour of Dov Gabbay*, College Publications, King's College, London (2005), 453–498.

19. Computational treatment of temporal notions — the CTTN system, *Proceedings of PPSWR 2005*, LNCS 3208, (2005), 137–150.

20. (with François Bry, Bernhard Lorenz and Mike Rosner), A geospatial world model for the semantic web, in *Proceedings of Third Workshop on Principles and Practice of Semantic Web Reasoning*, LNCS 3703, (2005), 14-5-159.

*Work in Progress*

21. Development of the CTTN-system (Computational Treatment of Temporal Notions) and the CTSN-system (Computational Treatment of Spatial Notions). These systems model geotemporal and geospatial notions as close as possible to the reality. The goal is to bridge the gap between detailed calendrical and geographical computations and abstract symbolic temporal and spatial reasoning.

**Service to the Profession:** Deputy Coordinator of the EU-network of Excellence REWERSE, 2004–; Program Chairman, Principles and Practices of Semantic Web Reasoning (PPSWR), 2004; Program Chairman, UK Automated Reasoning Workshop, 2000; Vice President, Association of Automated Reasoning (AAR), 1998–; Steering Committee Member, Association of Automated Reasoning, Great Britain, 1998-2002; Local arrangements, International Conference on Formal and Applied Practical Reasoning, Bonn, 1996; Local arrangements, International Conference on Temporal Logic (ICTL) 1994; Program Chairman, German Workshop of Artificial Intelligence (GWAI), 1992; Bonn.

**Teaching:** Ohlbach has supervised about 30 diploma students and 10 PhD students, including Renate Schmidt, Ullrich Hustadt, Andreas Nonnengart. He was member of the School Teaching Committee, King's College, London from 1998–2000 and is currently head of the teaching committee of the dept. of Computer Science at LMU. His main task at the time being is the transition from the diploma system to the bachelor/master system.

**Vision Statement:** Logic has an enormous potential in the current Semantic Web activities. A first success story is the application of Description Logics in the Ontology Working Language (OWL) which has become industrial standard. But pure abstract logic is not sufficient in the long run. One must combine abstract logic with concrete theories about concrete phenomena in the world. Therefore I am working at geotemporal and geospatial information processing.

## ONO, Hiroakira

**Specialties:** Non-classical logic, algebra and logic.

**Born:** 12 October, 1942 in Nagoya, Japan.

**Educated:** University of Tokyo BSc 1966; University of Tokyo MSc 1968 Physics; Kyoto University PhD Mathematics 1973.

**Dissertation:** *A Study of Intermediate Predicate Logics*; supervisor, Satoru Takasu.

**Regular Academic or Research Appointments:** PROFESSOR, JAPAN ADVANCED INSTITUTE OF SCIENCE AND TECHNOLOGY, 1993–. Hiroshima University, Faculty of Engineering, 1990–1993; Hiroshima University, Faculty of Integrated Arts and Sciences, 1976–1990; Tsuda College, Department of Mathematics, 1973–1976; Kyoto University, Research Institute for Mathematical Science, 1968–1973.

**Visiting Academic or Research Appointments:** Visiting researcher, National ICT Australia, and RSISE Australian National University, 2004; Visiting professor, University of Florence, 2003; Visiting professor, University of Siena, 2001.

**Research Profile:** Ono has devoted almost twenty years to the syntactic and semantical study of substructural logics. Substructural logics include many of important nonclassical logics, like linear logic, relevant logics, Lukasiewicz's many-valued logics, fuzzy logics, Lambek's calculus and so on. The subject has now grown up to be one of major topics in non-classical logics. The main goal is to grasp various nonclassical logics within a uniform framework, and to study their properties as a whole.

Ono finds that a particularly interesting feature of this topic is that recent development of algebraic study of substructural logics has brought the subject into a closer relationship with studies of ordered algebraic structures, universal algebra, and abstract algebraic logic. Thus, we can expect a birth of a new interdisciplinary research linking logic to algebra. In 1999, Ono started a research project called "Algebra & Substructural Logics", organized two international workshops on this topics, and edited special issues in international journals.

His early concern was mainly in intermediate and modal logics. In particular, he had a good deal of interest in intermediate or modal predicate logics, which was the theme of his Ph.D. thesis, and published several papers on these topics. Though it is a quite important topic in non-classical logics, notorious incompleteness phenomena in predicate logics have hindered a rapid development of the study, and a lot of important problems still remain open. Ono has maintained a constant interest in the subject, and expects that a breakthrough will happen soon.

**Main Publications:**

1. "Kripke models and intermediate logics". *Publications of Research Institute for Mathematical Science.* Kyoto University, 6(1970): 461–476.
2. "A study of intermediate predicate logics". *Publications of Research Institute for Mathematical Science.* Kyoto University, 8(1973): 619–649.
3. "Intermediate propositional logics (A survey)", with T. Hosoi. *Journal of Tsuda College* 5(1973): 67–82.
4. "On some intuitionistic modal logics". *Publications of Research Institute for Mathematical Science.* Kyoto University, 13(1977): 687–722.
5. "On the size of refutation Kripke models for some linear modal and tense logics", with A. Nakamura. *Studia Logica*, 39(1980): 325–333.
6. "Logics without the contraction rule", with Y. Komori. *Journal of Symbolic Logic*, 50(1985): 169–201.
7. "Interpolation and the Robinson property for logics not closed under the Boolean operations". *Algebra Universalis*, 23(1986): 111–122.
8. "Relations between intuitionistic modal logics and intermediate predicate logics", with N.-Y. Suzuki. *Reports on Mathematical Logic*, 22(1988): 65–87.
9. "Structural rules and a logical hierarchy". In P.P. Petkov (ed.), *Mathematical Logic, Proceedings of the Summer School and Conference on Mathematical Logic, Heyting '88.* Plenum Press, New York, 1990, pp. 95–104.
10. "Intermediate predicate logics determined by ordinals", with P. Minari and M. Takano. *Journal of Symbolic Logic*, 55(1990): 1099–1124.
11. "Semantics for substructural logics". In K. Dosen and P. Schroeder-Heister (eds.), *Substructural Logics. Studies in Logic and Computation 2.* Oxford University Press, 1993, pp.259–291.
12. "The finite model property for BCK and BCIW", with R.K. Meyer. *Studia Logica*, 53(1994): 107–118.
13. "Decidability and the finite model property of substructural logics". In J. Ginzburg et al. (eds.), *Tbilisi Symposium on Logic, Language and Computation: Selected Papers (Studies in Logic, Language and Information).* CSLI, 1998, pp.263–274.
14. "Proof-theoretic methods for nonclassical logic — an introduction". In M. Takahashi, M. Okada and M. Dezani-Ciancaglini (eds.), *Theories of Types and Proofs (MSJ Memoirs 2).* Mathematical Society of Japan, 1998, pp.207–254.
15. "A syntactic approach to Maksimova's principle of variable separation for some substructural logics", with H. Naruse and Bayu Surarso. *Notre Dame Journal of Formal Logic*, 39(1998): 94–113.
16. Y. Tanaka and H. Ono, "Rasiowa-Sikorski lemma and Kripke-completeness of predicate and infinitary modal logics", with . In M. Zakharyaschev et al. (eds.), *Advances in Modal Logic 2.* CSLI Publications, 2001, pp.401–419.
17. "Substructural logics and residuated lattices — an introduction". In V.F. Hendricks and J. Malinowski (eds.), *50 Years of Studia Logica, Trends in Logic 21.* Kluwer Academic Publishers, 2003, pp. 193–228.
18. "Algebraic aspects of cut elimination", with F. Belardinelli and P. Jipsen. *Studia Logica*, 77(2004): 209–240.

*Work in Progress*

19. "Algebraization, parametrized local deduction theorem and interpolation for substructural logics over FL", with N. Galatos. *Studia Logica*, 2006. To appear.
20. "Glivenko theorems and other translations for substructural logics over FL", with N. Galatos. In preparation.

**Service to the Profession:** Member of the Committee on Membership of the Association for Symbolic Logic, 2003–2006; Council member of Mathematical Society of Japan, 2003–2004; Member of the Committee on Logic in Japan and East Asia of Association for Symbolic Logic, 1998–2008; Member of Advisory Board of *Advances in Modal Logic*, 1998–2002; Editor of *Reports on Mathematical Logic*, 1998–; Editor (advisary board) of *Annals of Pure and Applied Logic*, 1996–2000; Council member of Association for Symbolic Logic, 1993–1995; Editor of *Bulletin of the Section of Logic*, 1988–; Council member of Mathematical Society of Japan, 1983–1984; Editor of *Studia Logica*, 1983–.

**Teaching:** Ono has supervised more than ten Ph.D. students at JAIST. In each conference of *Advances in Modal Logic* until now, there have always been talks given by Ono's students. Also, Ono has published an introductory textbook on mathematical logic (in Japanese) in 1994, which is now the standard text in Japan.

**Vision Statement:** Logic will continue to be an important research subject related to foundational problems, as it has been in mathematics and computer science. Results and techniques of modern mathematics will be applied to mathematical logic far more than now.

## ORŁOWSKA, Ewa

**Specialties:** Mathematical logic, philosophical logic, logic in computer science, algebraic logic, non-classical logics.

**Born:** 15 December 1935 in Biaa Podlaska, Poland

**Educated:** Scientific title, Professor of Mathematics, 1993; University of Warsaw, Department of Mathematics, Mechanics and Computer Science, Habilitation 1978, *Thesis: Resolution systems and their applications*; University of Warsaw, Department of Mathematics, Mechanics and Computer Science, PhD with honors, 1971; *Dissertation: Theorem proving systems*; supervisor, Helena Rasiowa; University of Warsaw, Department of Mathematics and Physics, M.S. 1957.

**Regular Academic or Research Appointments:** PROFESSOR OF MATHEMATICS AND COMPUTER SCIENCE, NATIONAL INSTITUTE OF TELECOMMUNICATIONS, WARSAW, WARSAW 1996–; Professor of mathematics and computer science, Institute of Theoretical and Applied Computer Science of the Polish Academy of Sciences 1987-1996; Docent, Institute of Computer Science of the Polish Academy of Sciences 1980-1986; Adiunkt (assistant professor), Department of Mathematics, University of Warsaw 1971-1979; Research assistant, Department of Mathematics, University of Warsaw 1967-1971; Research assistant, Institute of Mathematical Machines and the Computation Centre of the Polish Academy of Sciences 1959-1966.

**Visiting Academic or Research Appointments:** Visiting professor, Ecole Normale Superieure de Cachan, Paris 2001; F. W. James Chair Professor, St. Francis Xavier University, Canada, 2000; Ulster University, Belfast, 1997; Université Claude Bernard-Laboratoire Logique, Mathematiques Discretes, Informatique, Lyon, 1994; Institut National Polytechnique de Grenoble, Laboratoire d'Informatique Fondamentale et d'Intelligence Artificielle, Grenoble, 1994; University of Cape Town, South Africa, 1993; Salzburg University, 1985, 1986; CNRS Laboratoire Langages et Systemes Informatiques, Toulouse, 1983.

**Research Profile:** Research work of Ewa Orłowska has been concerned with the following issues: Relationships between non-classical logics and algebras of relations; deduction systems in the style of Rasiowa and Sikorski for non-classical logics, in particular for modal, relevant and multiple-valued logics, and for various classes of algebras of relations; modal logics for reasoning with incomplete or uncertain information; Boolean algebras with various classes of additional operators, including Jónsson-Tarski operators, and their applications to reasoning with incomplete information; lattices with operators, in particular, with various kinds of modal-like operators and representation theory and correspondence theory for these classes of algebras. Ewa is also researching the relationship between duality and relational semantics.

**Main Publications:**

1. *Monograph*: Incomplete Information: Structure, Inference, Complexity (with S. Demri). EATCS Monographs in Theoretical Computer Science, Springer, 2002.
2. *Edited volume*: Logic at Work. Essays Dedicated to the Memory of Helena Rasiowa. Springer - Physica Verlag, Heidelberg, 1998.
3. Lattice-based modal algebras and modal logics (with Dimiter Vakarelov). In: P. Hajek, L. M. Valdés-Villanueva, and D. Westerstahl (eds) Logic, Methodology and Philosophy of Science. Proceedings of the $12^{th}$ International Congress. King's College London Publications, 2005, 147-170.
4. Boolean algebras arising from information systems (with Ivo Düntsch). Annals of Pure and Applied Logic, 127, No 1-3 Special issue, Provinces of Logic Determined, Essays in the memory of Alfred Tarski, edited by Z. Adamowicz, S. Artemov, D. Niwiński, E. Orłowska, A. Romanowska, and J. Woleński, 2004, 77-98.
5. Correspondence results for relational proof systems with applications to the Lambek calculus (with Wendy MacCaull). Studia Logica 71, 2002, 279-304.
6. Beyond modalities: sufficiency and mixed algebras (with Ivo Düntsch). In: E. Orlowska and A. Szalas (eds) Relational Methods for Computer Science Applications, Springer - Physica Verlag, Heidelberg, 2001, 263-285.
7. A proof system for contact relation algebras (with Ivo Düntsch). Journal of Philosophical Logic 29, 2000, 241-262.
8. Logics of complementarity in information systems (with Ivo Düntsch). Mathematical Logic Quarterly 46, 2000, 267-288.
9. A hierarchy of modal logics with relative accessibility relations (with Ph. Balbiani). Journal of Applied Non-Classical Logics 9, No 2-3, 1999, 303-328, special issue in the memory of George Gargov.
10. Equational reasoning in nonclassical logics (with M. Frias). Journal of Applied Non-Classical Logics 8, No 1-2, 1998, 27-66.
11. Relational formalisation of nonclassical logics. In: Brink, C., Kahl, W., and Schmidt, G. (eds) Relational Methods in Computer Science, Springer, Wien - New York, 1997, 90-105.
12. Logical analysis of demonic nondeterministic programs (with S. Demri). Theoretical Computer Science 166, 1996, 173-202.
13. A proof system for fork algebras and its applications to reasoning in logics based on intuitionism (with M. Frias). Logique et Analyse 150-151-152, 1995, 239-284.
14. Relational semantics for non-classical logics: Formulas are relations. In: Woleński, J. (ed) Philosophical

Logic in Poland. Kluwer, Dordrecht, 1994, 167-186.

15. Relational proof systems for relevant logics. Journal of Symbolic Logic 57, 1992, 1425-1440.

16. Kripke models with relative accessibility and their application to inferences from incomplete information. In: Mirkowska, G. and Rasiowa, H. (eds) Mathematical Problems in Computation Theory. Banach Centre Publications 21, 1988, 329-339.

17. Semantics of vague concepts. In: Dorn, G. and Weingartner, P. (eds) Foundations of Logic and Linguistics. Problems and Solutions. Selected contributions to the 7th International Congress of Logic, Methodology and Philosophy of Science. Plenum Press, London - New York, 1985, 465-482.

18. Theorem proving systems. Dissertationes Mathematicae CIII. Polish Scientific Publishers, Warsaw, 1973, 55 pp.

*Work in Progress*

19. A book 'Dual tableaux: Reasoning and Computing with Relations' (with Joanna Golińska-Pilarek) on deduction systems for non-classical logics in the style of Rasiowa and Sikorski; A book on relational semantics through duality (with Ingrid Rewitzky).

**Service to the Profession:** Editor, Trends in Logic – Studia Logica Library, 2004-; Associate Editor, Studia Logica, 1991-; Collecting Editor, Journal of Applied Non-Classical Logics, 1989-; Member, Editorial Board, Logic Journal of the Interest Group in Pure and Applied Logic, 1993-; Member, Editorial Board, Journal of Multiple-Valued Logic and Soft Computing 1997-; Member, Scientific Council of the Institute of Telecommunications, Warsaw, 1997-, and Vice Chair, 2002-; Chair of the Council of the Polish Association for Logic and Philosophy of Science, 2003-2005 and member of the Council 1993-1995 and 2000-2002; Member, Committee of the E. W. Beth Dissertation Prize, Foundation for Logic, Language, and Information, 2005; Assessor, Council of the Division of Logic, Methodology and Philosophy of Science of the International Union of History and Philosophy of Science, 1999-2003; President, Polish Association for Logic and Philosophy of Science, 1997-1999 and a member of the Executive Committee, 2000-2002; Member, Nominating Committee, Foundation of Logic, Language, and Information, 1994-1996; Member, Scientific Council of the Institute of Theoretical and Applied Computer Science of the Polish Academy of Sciences, 1986-1998; Chair of the Editorial Board, Studia Logica, 1989-1991.

**Vision Statement:** Logic provides a comprehensive frame for a general investigation of concepts. However, the influence and inspiration of computing sciences brings forward the need for greater flexibility in determining the formal-precise and the intuitive-uncertain. The relational approach to logic seems to be the correct answer that can provide the tools for modeling and reasoning about the multiple aspects of computation.

**Honours and Awards:** Invited Speaker, $12^{th}$ International Congress of Logic, Methodology and Philosophy of Science, 2003; Invited Speaker, $30^{th}$ IEEE International Symposium on Multiple Valued Logic, 2000; Committee Member, $8^{th}$ European Conference on Symbolic and Quantitative Approaches to Reasoning with Uncertainty, 2005; Committee Member, Tarski Centenary Conference, 2001; Program Co-Chair, $31^{st}$ IEEE International Symposium on Multiple Valued Logic, 2001; Co-Chair, Logics in Artificial Intelligence, European Workshop JELIA'96, 1996; Best Paper on Knowledge Representation Award (with L.Farinas del Cerro), European Conference on Artificial Intelligence, 1984; Polish Ministry of Science and Education award for the Ph.D dissertation, 1972.

# P

## PARIKH, Rohit Jivanlal

**Specialties:** Recursion theory, theory of proof, non-standard analysis, ultrafinitism, dynamic logic, logic of knowledge, philosophical logic.

**Born:** 20, November 1936 in Palanpur, Gujarat, India.

**Educated:** Harvard University, PhD Mathematics, 1962; Harvard College, AB with highest honors in Physics, 1957.

**Dissertation:** *Non-uniqueness in Transfinite Progressions*; supervisor, Hartley Rogers.

**Regular Academic or Research Appointments:** DISTINGUISHED PROFESSOR, CITY UNIVERSITY OF NEW YORK, 1982–; Professor, Mathematics, Boston University, 1972-1982; Associate Professor,Mathematics, Boston University, 1967-1972; Instructor, Bristol University, 1965-1967; Instructor, Panjab University, 1964-1965; Instructor, Stanford University 1961-1963.

**Visiting Academic or Research Appointments:** Visiting Professor, Mathematics, Courant Institute, 1981; Visiting Associate Professor, Mathematics, SUNY at Buffalo, 1971-1972; Visiting Appointment, Stanford; Visiting Appointment, TIFR-Bombay; Visiting Appointment, ETH-Zurich; Visiting Appointment, Caltech.

**Research Profile:** Rohit Parikh has worked in many areas in traditional logic like recursion theory or proof theory, as well as in areas which are associated with logic, but not necessarily part of it. This catholic attitude towards logic has resulted in work on topics like vagueness, untrafinitism, belief revision, logic of knowledge, game theory and social software. This last area seeks to combine techniques from logic, computer science (especially logic of programs) and game theory to understand the structure of social algorithms which we all participate in, often without realizing that they are indeed algorithms. Examples of such are elections, transport systems, lectures and conferences, and even monetary systems, all of which have properties of interest to those who are logically inclined. The success of this area is quite gratifying, with two conferences in Copenhagen (2004) and London (2005) devoted to it, as well as an upcoming conference (2006) in Utrecht.

**Main Publications:**

1. On Context Free Languages, *Journal of the Assoc. Comp. Mach.* **13** (1966) 570-81. Originally published in 1961 as a research report at RLE, MIT.
2. Existence and Feasibility in Arithmetic, *Jour. Symbolic Logic* **36** (1971) 494-508.
3. On the Length of Proofs, *Transactions of the Amer. Math. Soc.* **177** (1973) 29-36.
4. (With M. Parnes) Conditional Probability can be Defined for Arbitrary Pairs of Sets of Reals, *Advances in Math* **9** (1972) 520- 522.
5. (With D.H.J. de Jongh) Well Partial Orderings and Hierarchies, *Proc. Kon. Ned. Akad. Sci Series A* **80** (1977) 195- 207.
6. (With D. Kozen) An Elementary Completeness Proof for PDL *Theoretical Computer Science* **14** (1981) 113-118.
7. The Problem of Vague Predicates, in *Logic, Language and Method* Ed. Cohen and Wartofsky, Reidel (1982) 241-261.
8. The Logic of Games and its Applications, *Annals of Discrete Math.*, **24** (1985) 111-140.
9. (With R. Ramanujam) Distributed Processing and the Logic of Knowledge, in *Logics of Programs*, Springer Lecture Notes in Computer Science, **193** pp. 256-268.
10. Communication, Consensus and Knowledge, (with P. Krasucki), *Jour. Economic Theory* **52** (1990) pp. 178-189.
11. Knowledge and the Problem of Logical Omniscience *ISMIS- 87 (International Symp. on Methodology for Intelligent Systems)*, North Holland (1987) pp. 432-439.
12. Finite and Infinite Dialogues, in the *Proceedings of a Workshop on Logic from Computer Science*, Ed. Moschovakis, MSRI publications, Springer 1991 pp. 481-498.
13. Vagueness and Utility: the Semantics of Common Nouns in *Linguistics and Philosophy* **17** 1994, 521-35.
14. Topological Reasoning and The Logic of Knowledge (with Dabrowski and Moss) *Annals of Pure and Applied Logic* **78** (1996) 73-110.
15. Belief revision and language splitting, in *Proc. Logic, Language and Computation*, Ed. Moss, Ginzburg and de Rijke, CSLI 1999, pp. 266-278 (earlier version appeared in 1996 in the preliminary proceedings).
16. Social Software, *Synthese*, **132**, Sep 2002, 187-211.
17. (with Jouko Vaananen), Finite information logic, *Annals of Pure and Applied Logic*, **134** (2005) 83-93.
18. (With R. Ramanujam), A Knowledge based Semantics of Messages, *Jour. Logic, Language and Information*, **12** 2003, 453-467.
19. Levels of Knowledge, Games, and Group Action, *Research in Economics*, **57** 2003, 267-281.
20. (with Eric Pacuit and Eva Cogan) The logic of

knowledge based obligation, in *Knowledge, Rationality and Action*, 2006.

**Work in Progress**

21. Currently working on books in reasoning about knowledge as well as on social software. Also, working on the issue of logical omniscience on which substantial progress has been made.

**Service to the Profession:** Editor, *International Journal of the Foundations of Computer Science*, 1990-1995; Editor, *Journal of Philosophical Logic*, 2000-2003.

**Teaching:** The former doctoral students are David Ellerman, Tom Sibley, Shlomit Pinter, Paul Krasucki, Gilbert Ndjatou, Alessandra Carbone, Konstantinos Georgatos, Angela Weiss, Samir Chopra, Eric Pacuit and Samer Salame. People who were influenced by Parikh as graduate students include Rick Statman, Joseph Halpern, R. Ramanujam, Horacio Arlo Costa, Ruili Ye, Amy Greenwald, and Laxmi Parida.

**Vision Statement:** I think it is a scandal that Russell's paradox is still effectively unsolved after a hundred years. We do not need to worry about large cardinals, but need instead to worry about the fact that our notion of set is *conceptually deficient*. Apart from this I believe that both logic and philosophy are in a state of cowardly subservience to science, which is true as far as it goes, but whose language is severely limited – unable to analyze *propositional attitudes*, or the game theoretic notion of *agent*. This subservience leaves us in a state of smug satisfaction, but leaves fundamental problems unaddressed. I would suggest that people pay more attention to Zeno's paradoxes, to McTaggart's paper on Time, and perhaps also to the writings of the thirteenth-century Zen teacher Dogen Zenji in his Genjokoan.

**Honours and Awards:** William Lowell Putnam Mathematical Competition Prize Winner, 1955, 1956, 1957; Gibbs Prize, Bombay University, 1954.

# PARIS, Jeff

**Specialties:** Mathematical logic, philosophical logic, uncertain reasoning, models of arithmetic.

**Born:** 15 November 1944 in Sutton, Surrey, UK.

**Educated:** Manchester University, PhD Mathematics, 1969; Manchester University, BSc Mathematics, 1966. *Dissertation Boolean Valued Models and Large Cardinals*; supervisors, Robin Gandy and Mike Yates.

**Regular Academic or Research Appointments:** PROFESSOR, MANCHESTER UNIVERSITY, 1984-; Reader, Manchester University, 1974-1983; Lecturer, Manchester University, 1969–1974.

**Visiting Academic or Research Appointments:** Visiting Fellow, University of Amsterdam, 1996; Visiting Fellow, Center for Logic, Epistomology and the History of Science, Campinas, Brazil, 1989; Visiting Fellow, Catholic University of Santiago, 1985; Visiting Professor, Mathematical Institute of the Czech Academy of Science, 1982; Associate Professor, University of California, Berkeley, 1974; Associate Professor, Rockefeller University, NY, 1971.

**Research Profile:** Jeff Paris wrote his doctorial thesis on independence results for large cardinals in 1969 and continued to work largely in set theory, especially aspects of determinacy and large cardinals, until 1974. Following an earlier brief flirtation with models of arithmetic, he started in the mid 1970s to consider analogues of notions from set theory within such models. Over the next ten years this initial emphasis on models of full Peano Arithmetic changed to considering weaker fragments and their close connections with computational complexity. Around 1985 he accidentally became involved with a project to produce a medical expert system which led to an abiding interest in the rapidly developing area of uncertain reasoning. Despite his more practical introduction to the topic Paris's contribution has been mathematical, in particular in developing and investigating formal logics of rational or common sense reasoning.

**Main Publications:**

1. (with K. Kunen) Boolean extensions and measurable cardinals, *Annals Math. Logic*, 2 (1971), pp.359-378.

2. ZF proves $\Sigma_4^0$-determinateness, *Journal of Symbolic Logic*, 37, (1972), pp.661-667.

3. (with L. Kirby) Initial segments of models of Peano's axioms, *Proceedings of the Bierutowice Conference*, (1976), Springer Verlag Lecture Notes series, No.619, pp.211-226.

4. (with L. Harrington) A mathematical incompleteness in Peano Arithmetic, *Handbook of Mathematical Logic*, North Holland, (1977), pp.1133-1142.

5. (with L. Kirby) $\Sigma_n$-collection schemes in arithmetic, *Logic Colloquium '77 volume*, North Holland, (1978), pp.199-209.

6. Some independence results in Peano Arithmetic, *Journal of Symbolic Logic*, 43, (1978), pp.725-731.

7. A hierarchy of cuts in models of arithmetic, *Proceedings of the Logic Conference held in Karpacz, Poland*, (1979). North Holland, pp.312-337.

8. (with C. Dimitracopoulos) Truth definitions for $\Lambda_0$ formulae, *Monographie No.30 de L'Enseignement Math-*

*ematique, Universite de Geneve* (1982), pp.319-329.

9. (with L. Kirby) Accessible independence results for Peano Arithmetic, *Bulletin London Mathematical Society*, 14, (1982), pp.285-293.

10. (with A.J. Wilkie) Counting problems in bounded arithmetic, *Proceedings of the VI'th Latin American Logic Conference, Caracus* (1983), Springer-Verlag Lecture Notes Series, No.1130, pp.317-340.

11. (with A. Wilkie), Counting $\Lambda_0$ sets, *Fund. Math.* 127, (1986), pp 67-76.

12. (with A. Wilkie,) On the scheme of induction for bounded arithmetic formulas, *Annals of Pure and Applied Logic* 35, (1987), pp 261-302.

13. (with A.J. Wilkie & A.R. Woods) Provability of the pigeon hole principle and the existence of infinitely many primes, *Journal of Symbolic Logic* 53, (1988), pp.1235-1244.

14. (with A. Vencovska) A Note on the Inevitability of Maximum Entropy, *International Journal of Approximate Reasoning*, (1990) pp.183-224, Vol.4, No.3.

15. (with A. Vencovska), A Model of Belief', *Artificial Intelligence*, (1993), 64, pp.197-241.

16. *The Uncertain Reasoner's Companion, - A Mathematical Perspective*, Cambridge Tracts in Theoretical Computer Science 39, Cambridge University Press, (1994).

17. (with A.Vencovska) In defense of the Maximum Entropy Inference Process, *International Journal of Approximate Reasoning,* vol.17, no.1, 1997 pp 77-103.

18. Common sense and maximum entropy, *Synthese,* 117, (1999), pp 75-93.

19. (with H.Hosni) Rationality as conformity, *Synthese,* 144, (2005), pp 249-285.

20. (with C.J.Nix ), A Continuum of Inductive Methods arising from a Generalized Principle of Instantial Relevance", *Journal of Philosophical Logic*, Online First Issue, DOI:10,1007/s 10992-005-9003x, ISSN 0022-3611 (Paper) 1573-0433, 2005.

**Service to the Profession:** Member of the International Advisory Board of the Mathematical Institute of the Czech Academy of Sciences, 2004-; President of the British Logic Colloquium, 2001–2007.

**Teaching:** One of a handful of staff who have successfully run a research council funded MSc degree in Mathematical Logic and the Foundations of Computation at Manchester University for over 30 years. A career total of 15 PhD students amongst whom Laurie Kirby, Dionyssis Anapolitanos, Costas Dimitracopoulos, Alan Woods, Richard Kaye and Maged Wafy all went on to successful university careers.

**Vision Statement:** Many of the traditional central areas of logic such as set theory and recursion theory will lose their appeal. Instead ideas and thinking from logic will increasingly ripple out from the centre to have a crucial influence on foundational questions across the sciences and humanities.

**Honours and Awards:** Elected Fellow of the British Academy, 1999; SERC Senior Fellowship, 1989; London Mathematical Society Junior Whitehead Prize, 1983.

# PARSONS, Charles

**Specialties:** Philosophy of logic and mathematics, history of foundations of mathematics, mathematical logic.

**Born:** 13 April 1933 in Cambridge, Massachusetts, USA.

**Educated:** Harvard University, PhD, 1961; AM, 1956; AB, 1954; King's College, Cambridge, 1954-1955. Dissertation: *On Constructive Interpretation of Predicative Mathematics*; supervisor, Burton Dreben.

**Regular Academic or Research Appointments:** EDGAR PIERCE PROFESSOR OF PHILOSOPHY, EMERITUS, HARVARD UNIVERSITY, 2005–; Edgar Pierce Professor of Philosophy, Harvard University, 1991-2005; Professor of Philosophy, Harvard University, 1989-1991; Professor of Philosophy, Columbia University, 1969-1989 ; Columbia University, Chairman, Department of Philosophy, 1976-1979, 1985-1989; Associate Professor, Columbia University, 1965-69; Assistant Professor of Philosophy, Harvard University, 1962-65; Assistant Professor of Philosophy, Cornell University, 1961-62.

**Visiting Academic or Research Appointments:** Visiting Professor of Philosophy, UCLA, 2002, 2005; Fellow, Center for Advanced Study in the Behavioral Sciences, Stanford, 1994-1995; Guggenheim Fellow, 1986-87; Fellow, Netherlands Institute for Advanced Study, 1987; Visiting Fellow, All Souls College, Oxford, 1979-1980; NEH Fellow, 1979-1980; Visiting Professor, Rockefeller University, 1971-1972; Visiting Lecturer on Philosophy, Harvard University, 1968-69; Visiting Professor, University of Padua, 1997, 1983; Visiting Professor, University of Heidelberg, 1972; Junior Fellow, Society of Fellows, Harvard University, 1958-61.

**Research Profile:** Parsons' early work in his dissertation and in publications up to the early 1970s was in proof theory in the tradition of the Hilbert school, as it had been revived in the 1950s by Kreisel and Schütte. Papers and abstracts characterized the proof-theoretic strength of arithmetic with induction, or other schemata related to collection, restricted as to number of nested quantifiers allowed. Simultaneously he began work in philosophy of mathematics, concentrating first on historical figures, especially Kant and Frege. In the 1970s he turned to the concepts of set, class, and truth, and the semantic paradoxes. This work, including the historical papers, is collected in *Mathematics in Philosophy* (1983). Since then he has pursued a further philosophical project, in which the guiding ideas are a logical conception of object, a form of the structuralist view of mathematical objects, and a conception of intuition originating in attempts to understand Kant but more directly connected to Hilbert's ideas on finitism. Later, a concern with Reason and rational evidence has been added. (See item 19 below.) Parsons has continued to write on historical figures up to the living (e.g. Hilary Putnam), but the central one for many years has been Gödel, because of his work in the editing of Gödel's posthumous writings.

**Main Publications:**

1. The $\omega$-consistency of ramified analysis. **Archiv für mathematische Logik und Grundlagenforschung** 6 (1962), 30-34.
2. On a number-theoretic choice schema and its relation to induction. In John Myhill, Akiko Kino, and Richard E. Vesley (eds.), *Intuitionism and Proof Theory* (Proceedings of the summer conference, Buffalo 1968), pp. 549-563. Amsterdam: North-Holland, 1970.
3. On $n$-quantifier induction. **The Journal of Symbolic Logic** 37 (1972), 466-482.
4. Mathematical intuition. **Proceedings of the Aristotelian Society** 80 (1979-80), 145-168.
5. Intensional logic in extensional language. **The Journal of Symbolic Logic** 47 (1982), 289-328.
6. *Mathematics in Philosophy: Selected Essays.* Ithaca and London: Cornell University Press, 1983. Paperback edition (with minor corrections), 2005.
7. Objects and logic. **The Monist** 65 (1982), 491-516.
8. Intuition in constructive mathematics. In Jeremy Butterfield (ed.), *Language, Mind, and Logic*, pp. 211-229. Cambridge University Press, 1986.
9. The structuralist view of mathematical objects. **Synthese** 84 (1990), 303-346.
10. The impredicativity of induction. In Michael Detlefsen (ed.), *Proof, Logic, and Formalization*, pp. 139-161. London: Routledge, 1992. (Revised and expanded version of a paper published in 1983.)
11. Quine and Gödel on analyticity. In Paolo Leonardi and Marco Santambrogio (eds.), *On Quine: New Essays*, pp. 297-313. Cambridge University Press, 1995.
12. Intuition and number. In Alexander George (ed.), *Mathematics and Mind*, pp. 141-157. Oxford University Press, 1994.
13. Platonism and mathematical intuition in Kurt Gödel's thought. **The Bulletin of Symbolic Logic** 1 (1995), 44-74.
14. Reason and intuition. **Synthese** 125 (2000), 299-315.
15. Realism and the debate on impredicativity, 1917-1944. In Wilfried Sieg, Richard Sommer, and Carolyn Talcott (eds.), *Reflections on the Foundations of Mathematics: Essays in honor of Solomon Feferman*, pp. 372-389. Lecture Notes in Logic 15. Urbana, Ill.: Association for Symbolic Logic, and Natick: A. K. Peters, 2002.
16. Communication and the uniqueness of the natural numbers. In *Proceedings of the First Seminar on Philosophy of Mathematics in Iran*, pp. 1-30. Faculty of Mathematical Sciences, Shahid Beheshti University, Tehran, 2003.
17. (Edited, with Solomon Feferman, John W. Dawson, Jr., et al.) Kurt Gödel, *Collected Works*, volume III: *Unpublished Essays and Lectures*, and volumes IV-V: *Correspondence*. Oxford University Press, 1995, 2003.
18. The problem of absolute universality. In Agustín Rayo and Gabriel Uzquiano (eds.), *Absolute Generality*, pp. 203-219. Oxford University Press, 2006.
19. *Mathematical Thought and its Objects.* Cambridge University Press, 2008.

*Work in Progress*

20. Book of essays on historical figures, from Kant to Hao Wang and Hilary Putnam, with emphasis (not exclusive) on philosophy of mathematics.

**Service to the Profession:** Editorial Board, *Philosophia Mathematica*, 1992-; Editor, *Bulletin of Symbolic Logic*, 1994-1998; Editor, Kurt Gödel, *Collected Works*, 1987-2003; Editor, *Journal of Philosophy*, 1966-1990; Association for Symbolic Logic, President, 1989-1992, Vice President, 1986-1989, Secretary, 1971-1976, Executive Committee, 1969-1971.

**Teaching:** From 1965 to 1989 Parsons was the senior teacher of logic in philosophy at Columbia University and taught a variety of logic courses as well as Philosophy of Logic, seminars, and other philosophical courses. From 1989 he taught courses and seminars in philosophy of mathematics at Harvard, as well as Kant, modal logic, and other subjects. Among his PhD students were James Higginbotham, R. Gregory Taylor, Peter Ludlow, Richard Tieszen, and Gila Sher (Columbia), and Emily Carson, Michael Glanzberg, and Øystein Linnebo (Harvard).

**Vision Statement:** Solomon Feferman's statement for his own entry could hardly be improved upon. I would add that the position of logic in philosophy has become precarious. It seems able to survive and I hope it will do more than that since a philosophical perspective is important both for the foundations of mathematics and for other applications such as to problems about language.

**Honours and Awards:** Foreign Member, Norwegian Academy of Science and Letters, 2002-;

Membre titulaire, Institut International de Philosophie, 2001-; Fellow, American Academy of Arts and Sciences, 1982-.

## PERZANOWSKI, Jerzy Wacław

**Specialties:** Philosophical logic (theory of logical calculi, modal logic, paraconsistent logic), logical philosophy (ontologic, combination (or, in more general case, ontological) semantics, general theory of analysis and synthesis), and cognitive sciences (psychoontology).

**Born:** 23 April 1943 in Aix-les-Bains, France.

**Educated:** Jagiellonian University, Kraków, Poland, Habilitation philosophical logic and ontology, 1990; Jagiellonian University, PhD Logic, UJ, 1973; Jagiellonian University, MA Philosophy, 1965, MA Mathematics, 1968.

**Dissertation:** PhD. *The Deduction Theorems for the Modal Propositional Calculi Formalizad After the Manner of Lemmon*, supervisor: Kazimierz Pasenkiewicz.
*Habilitationschrift: Logiki modalne a filozofia (Modal logics and philosophy).*

**Regular Academic or Research Appointments:** PROFESSOR, LOGIC, PHILOSOPHY AND COGNITIVE SCIENCES (ORDINARIUS), JAGIELLONIAN UNIVERSITY, 2004–; Professor, Logic and Philosophy (ordinarius), Nicolas Copernicus University of Toruń, 1996-2004; Professor, Logic and Philosophy (extraordinarius), Jagiellonian University, 1992-2004; Professor, Logic and Philosophy (extraordinarius), Nicolas Copernicus University of Toruń, 1992-1995; Associate Professor, Logic and Philosophy (Dr. hab., Adjunct ), Jagiellonian University, 1990-1992; Assistant Professor, Logic and Philosophy (PhD, Adjunct ), Jagiellonian University, 1974-1990; Instructor, Logic (Assistant), Jagiellonian University, 1965-1973.

**Visiting Academic or Research Appointments:** Research Fellow, Internationale Akademie für Philosophie im Fürstentum Liechtenstein, 1991-1992; Visiting Professor, Logic, Universidade Federal da Paraiba, Brazil, 1991; Visiting Professor, Philosophy, Universität Salzburg, Austria, 1990; Research Fellow, Cambridge University, Great Britain, 1985.

**Research Profile:** Jerzy Perzanowski has made essential contributions to modal logic, paraconsistent logic, ontologic and logical philosophy. His work is chiefly devoted to the borderland between philosophy and logic. In modal logic he is noted for his work on general, transformational, deduction theorem for modal logics, in particular normal ones, as well as for his remarkable topography of modal logics with several applications, including logics of alethic modalities *True* and *False* and logics connected with the ontological argument. He introduced ontological (or combination) semantics for intensional logics, with a general theory of ontological modalities as its motivation and, next, key application. In paraconsistent logic he extended Stanisław Jaśkowski's modal approach to inconsistencies to a general framework, called parainconsistency; his work on M-counterparts of normal modal logical calculi was particularly influential. In logical philosophy he is noted for his fundamental work on the ontological argument, on the theory of analysis and synthesis, on the theory of being and qualities as well as on locative ontology and ontological melioration.

**Main Publications:**
1. The deduction theorems for the modal propositional calculi formalized after the manner of Lemmon, *Reports on Mathematical Logic*, 1 (1973), 1 – 12
2. On M-fragments and L-fragments of normal modal propositional logics, *Reports on Mathematical Logic*, 5 (1975), 63 – 72
3. Remarks on propositional embeddings and degrees, [in:] J. Perzanowski ed., *Essays on Philosophy and Logic*, Jagiellonian University Press, Kraków, 1987, 121 – 136
4. Byt (Being), *Studia Filozoficzne*, 6/7 (1988), 63 - 85
5. Logiki modalne a filozofia (Modal logics and philosophy), Jagiellonian University Press, Kraków, 1989, pp. 159
6. Towards Post-Tractatus Ontology, [in:] J. Brandl and R. Haller eds., *Wittgenstein - Towards a reevaluation. Proceedings of the 14th International Wittgenstein Symposium, 13th -20th August 1989, Kirchberg am Wechsel*, Verlag Hölder-Pichler-Tempsky, Wien, 1990, 185 - 199
7. Ontologies and Ontologics, [in:] E. ˉarnecka-Biay ed., *Logic Counts*, Kluwer Academic Publishers, Dordrecht-Boston-London, 1990, 23 – 42
8. Ontological Arguments II - Cartesian and Leibnizian, [in:] H. Burkhardt and B. Smith eds., *Handbook of Metaphysics and Ontology*, Philosophia Verlag, München, 1991, 625 – 633
9. What is non-Fregean in the Semantics of Wittgenstein's Tractatus and Why?, *Axiomathes*, 4.3, 1993, 327 – 342
10. Locative Ontology. Parts 1 - 3, *Logic and Logical Philosophy*, 1, 1993, 7 - 94
11. Teofilozofia Leibniza (Leibniz's Theophilosophy), [in:] *G. W. Leibniz: Pisma z teologii mistycznej*, Znak Publishers Co., Kraków, 1994, 243 – 351
12. Towards Psychoontology, [in:] R. Casati and G. White eds., *Philosophy and the Cognitive Sciences. Pro-

ceedings of the 16th International Wittgenstein Symposium, 15th - 21st August 1993, Kirchberg am Wechsel, Verlag Hölder-Pichler-Tempsky, Wien, 1994, 287 - 296
13. Reasons and Causes, [in:] J. Faye, U. Scheffler and M. Urchs, eds., *Logic and Causal Reasoning*, Akademie Verlag, Berlin, 1994, 169 – 189
14. O wskazanych przez Ch. Hartshorne'a modalnych krokach w dowodzie ontologicznym œw. Anzelma (On modal steps in the Ontological Argument of St. Anselm, indicated by Ch. Harshorne), [in:] J. Perzanowski and al., eds., *Filozofia / Logika: Filozofia logiczna, 1994*, N. Copernicus University Press (1995), 77 – 96
15. O modalnej logice parasymatrycznoœci KP i jej kuzynkach (On modal logic of parasymmetry KP and its cousins), [in:] J. Perzanowski and al. eds., *Filozofia / Logika: Filozofia logiczna, 1994*, N. Copernicus University Press (1995), 311 – 336
16. The Way of Truth, [in:] R. Poli and P. Simons eds., *Formal Ontology*, Kluwer Academic Publishers, Dordrecht-Boston-London (1996), 61 – 130
17. Combination Semantics for Intensional Logics. I Makings and Their Use in Making Combination Semantics, *Logique et Analyse*, 165/66 (1999), 181 - 203
18. Parainconsistency, or inconsistency tamed, investigated and exploited, *Logic and Logical Philosophy*, 9 (2001), 5 – 24
19. A Profile of Masonic Synthesis, *Logic and Logical Philosophy*, 11/12 (2003/2004), 167 – 189
20. Towards Combination Metaphysics, *Reports on Mathematical Logic*, 38 (2004), 93 - 116

Work in Progress
21. *Ontological Investigations*, a treatise on general ontology, including a theory of analysis and synthesis
22. *Locative Ontology*, an essay on pure relational theory of location
23. *Ontological Melioration*, an essay on ontologic of melioration
24. *Theologic*, a book on mathematical theory of God–like being
25. *Psychoontology*, a book on ontological framework for the realm of psychic phenomena
26. *An Outline of Classical Modal Logic, including a topography of modal logics with several royal applications*, An outline of modal logic putting emphasis on the lattice(s) of modal logics as well as on its intimate connection with philosophy

**Service to the Profession:** *Reports on Mathematical Logic*, Editor-in-Chief, 1985-, Executive Editor, 1973-1978; Founder and Editor-in-Chief, *Logic and Logical Philosophy*, 1993-2004; Chair, Department of Logic, Nicolas Copernicus University of Toruñ, 1992-2004; Member, Komitet Nauk Filozoficznych Polskiej Akademii Nauk, 1993-1996, 1999-2002.

**Teaching:** Perzanowski has been a mainstay in teaching logic and logical philosophy in changing logic program at Jagiellonian University and Nicolas Copernicus University since his appointment to the faculty respectively in 1965 and 1992. He has had five PhD students in the field of logic and philosophy: Tomasz Kowalski, Katarzyna Idziak, Marek Nasieniowski, Piotr Wasilewski, Janusz Czetwertyński.

**Vision Statement:** Leibnizian and Russellian program of formal development of philosophy by means of tools of modern logic was one of the main sources of the flowering of logic and analytical philosophy in the $20^{th}$ century. In recent years however the program itself and formal philosophy being its main product seems to suffer from scholastic fossilization. Remedy lies in return to the problems of *real* philosophy, including philosophical problems of science, done at once with *modification* and *enlarging* of suitable logical tools, which is the program and practice of logical philosophy.

**Honours and Awards:** Gold Cross of Merit, 1986; Knight's Cross of the Order, *Polonia Restituta*, 2004; Two Prizes of the Minister of Higher Education; Six Rector's Prizes.

# PIETARINEN, Ahti-Veikko Juhani

**Specialties:** Philosophy of logics, epistemic logic, game theory, game-theoretic semantics, independence-friendly logic, semantics & pragmatics, diagrammatic logics, Peirce's logic, logic & neuroscience, general history of logic.

**Born:** 4 November 1971, Espoo, Finland.

**Educated:** University of Helsinki, Ph.D. Theoretical Philosophy 2002; University of Turku, M.Sc. Computer Science 1997.

**Dissertation:** *Semantic Games in Logic and Language*, University of Helsinki; supervisor, Gabriel Sandu .

**Regular Academic or Research Appointments:** UNIVERSITY LECTURER, UNIVERSITY OF HELSINKI, JANUARY 2006–JULY 2008; Post-doctoral Fellow, Academy of Finland, August 2003–July 2005.

**Visiting Academic or Research Appointments:** Visiting Scholar, Chinese Academy of Social Sciences, 2006; ASLA-Fulbright Visiting Scholar, Institute for American Thought, Indiana University-Purdue University Indianapolis (IUPUI), 2005; Adjunct Professor, University of Helsinki, 2004; Adjunct Professor, University of Turku, 2004; Post-doctoral Fellow, Academy of Finland, 2003–2005; ;isiting Scholar, Hungarian Academy of Sciences, 2003-2004.

**Research Profile:** Pietarinen has contributed to the study of Peirce's logic, including diagrammatic and iconic logic, the development of modern logic, algebra of logic, and the theory of quantification. He has investigated independence-friendly logic, epistemic logic and general philosophy of logics, studied the variety of uses of game theory and game-theoretic semantics in logic and language, including evolutionary extensions and the semantics/pragmatics distinction. His publications include the philosophy of propositional attitude reports, ;he problem of intentional identity, possible-worlds semantics and imperfect information. Bringing logic and neuroscience closer to cognitive sciences and phenomenology, he has also written on the history of intellectual ideas of logic and language, Grice, game theory, significs movement and the early analytic philosophy.

**Main Publications:**

1. "Semantic games and generalised quantifiers" in A.-V. Pietarinen (ed.), *Game Theory and Linguistic Meaning* (Current Research in the Semantics/Pragmatics Interface 18), Oxford: Elsevier, 2007, 183-206.
2. *Signs of Logic: Peircean Themes on the Philosophy of Language, Games, and Communication* (Synthese Library 329), Dordrecht: Springer, 2006.
3. "On Peirce's late proof of pragmaticism" (with L. Snellman), in A.-V. Pietarinen and T. Aho (eds), *Truth and Games*, Acta Philosophica Fennica 79, Helsinki: Societas Philosophica Fennica, 2006, 275-288.
4. "Peirce's contributions to possible-worlds semantics", *Studia Logica* 82 (2006), 345-369.
5. "The evolution of semantics and language-games for meaning", *Interaction Studies: Social Behaviour and Communication in Biological and Artificial Systems* 7 (2006), 79-104.
6. "Peirce, Habermas and strategic dialogues: From pragmatism to the pragmatics of communication", *LODZ Papers in Pragmatics* 1 (2006), 197-222.
7. "Independence-friendly logic and incomplete information", in J. van Benthem et al. (eds), *The Age of Alternative Logics: Assessing Philosophy of Logic and Mathematics Today*, Dordrecht: Springer, 2006, 243-259.
8. "Cultivating habits of reason: Peirce and the *logica utens* vs. *logica docens* distinction", *The History of Philosophy Quarterly* 22 (2005), 357-372.
9. "Compositionality, relevance and Peirce's logic of existential graphs", *Axiomathes* 15 (2005), 513-540.
10. "Some games logic plays", in D. Vanderveken (ed.), *Logic, Thought and Action*, Dordrecht: Springer (2005), 409-431.
11. "Grice in the wake of Peirce", *Pragmatics & Cognition* 12 (2004), 395-415.
12. "Semantic games in logic and epistemology", in S. Rahman, J. Symons, D. Gabbay & J. P. Van Bendegem (eds), *Logic, Epistemology and the Unity of Science*, Dordrecht: Springer (2004), 57-103.
13. "Logic, language-games and ludics", *Acta Analytica* 18 (2003), 89-123.
14. "What do epistemic logic and cognitive science have to do with each other?", *Cognitive Systems Research* 4 (2003), 169-190.
15. "Peirce's game-theoretic ideas in logic", *Semiotica* 144 (2003), 33-47.
16. "Quantum logic and quantum theory in a game-theoretic perspective", *Open Systems & Information Dynamics* 9 (2002), 273-290.
17. "Propositional logic of imperfect information: Foundations and applications", *Notre Dame Journal of Formal Logic* 42 (2001), 193-210.
18. "Intentional identity revisited", *Nordic Journal of Philosophical Logic* 6 (2001), 144-188.
19. "Most even budged yet: some cases for game-theoretic semantics in natural language", *Theoretical Linguistics* 27 (2001), 20-54.
20. (with G. Sandu) "Partiality and games: Propositional logic", *Logic Journal of the Interest Group in Pure and Applied Logic* 9 (2001), 107-127.

A full list of publications is available at Pietarinen's home page http://www.helsinki.fi/~pietarin/

*Work in Progress*

21. "Significs and the origins of analytic philosophy", *Journal of the History of Ideas*.
22. "Peirce and the logic of image", *Semiotica*.
23. "Peirce's magic lantern: Moving pictures of thought", *Transactions of the Charles S. Peirce Society*.
24. "Moving pictures of thought II: Graphs, games, and pragmaticism's proof", *Semiotica*.
25. "On historical pragmatics and Peircean pragmatism", *Linguistics and the Human Sciences*.
26. "Information as epistemic states: Implicit vs. explicit information in epistemic logic and in neuroscience", *Logique & Analyse*.
27. "Which philosophy of mathematics is pragmaticism?", in M. Moore (ed.), *Peirce's Philosophy of Mathematics and Logic*, Open Court.
28. "Who plays games in philosophy?", B. Hale (ed.), *Chess and Philosophy*, Open Court.
29. "Abductive issues in Peirce's proof of pragmaticism", in O. Pombo (ed.), *Abduction and the Process of Scientific Discovery*.
30. "Towards the intellectual history of logic and games", in O. Majer, A.-V. Pietarinen and T. Tulenheimo (eds), *Logic and Games: Foundational Perspectives*.
31. *Studies in Epistemic Logic*, a book in progress.
32. *Significs: An Anthology*, a book in progress.
33. *Existential Graphs*, Software Package on Diagrammatic Logic.
34. *The Correspondence of Charles S. Peirce*, The Helsinki Peirce Research Centre.
35. *The Logic of the Future: Peirce's Unpublished Writings on Logic and Writings on Peirce's Logic*, a book in progress.
36. *The Basis of Pragmaticism* (in Finnish), a book in progress.

**Teaching:** Currently supervising three Ph.D. students.

**Vision Statement:** "The concept of logic broadens, and the logical investigation of the methodologies of special sciences becomes increasingly important. The vision of the unity of science is re-established with much improved logical tools and methods such as logical interaction theory. Logic of cognition is re-enacted in diagrammatic and iconic logics. The emphasis on 'symbolic' tends to recede and pragmatistic logic emerges as a rigorous discipline overcoming the strays of the $20^{th}$ century exploration."

**Honours and Awards:** Interviewed in *Signs and Meaning: Five Questions* (eds. P. Bundgaard & F. Stjernfelt); The University of Helsinki Research Funds, Excellence in Research Grant 2006-2008; The Jenny and Antti Wihuri Foundation, Personal Research Grant 2006; The Finnish Cultural Foundation, Research Grant 2005; The Fulbright Bilateral Commission Award 2005; The Academy of Finland Post-Doctoral Fellowship, University of Helsinki 2003-2005; First Prize, Peirce Essay Contest, Charles S. Peirce Society, 2003; The Ella and Georg Ehrnrooth Foundation, Personal Research Grant 2002; The Osk. Öflund Foundation, PRG 1998; The Helsingin Sanomat Centenary Foundation, PRG 1998; The Osk. Huttunen Foundation Fellowship 1997-2000.

## PNUELI, Amir

**Specialties:** Temporal logics: linear, branching, and partial order time structures; specification verification and synthesis of reactive, real-time and hybrid systems, deductive verification, abstraction and refinement.

**Born:** 22 April 1941 in Nahalal, Israel.

**Educated:** Weizmann Institute of Science, PhD Applied Math., 1967; Technion, BSc Mathematics, 1962.

**Dissertation:** *Tides in Simple Basins*; supervisor, Chaim L. Pekeris.

**Regular Academic or Research Appointments:** PROFESSOR, COMPUTER SCIENCE, NEW YORK UNIVERSITY, 1999–; Professor, Computer Science, Weizmann Institute of Science, 1981-2007; Associate Professor, Computer Science, Tel-Aviv University, 1973-1980; Senior Researcher, Computer Sciences, Weizmann Institute of Science, 1969-1972.

**Visiting Academic or Research Appointments:** Visiting Professor, Computer Science, Harvard University, 1983; Visiting Professor, Computer Science, University of Pennsylvania, 1976-1977; Visiting Researcher, Computer Science, Stanford University, 1968.

**Research Profile:** Amir Pnueli is mainly known for the introduction of temporal logic into Computer Science; his work on the application of temporal logic and similar formalisms for the specification and verification of reactive systems; and the identification of the class of "Reactive Systems" as systems whose formal specification, analysis, and verification require a distinctive approach. He also developed a detailed methodology, based on temporal logic, for the formal treatment of reactive system and extended this methodology into the realm of real-time systems. More recently, he introduced into formal analysis the models of hybrid systems with appropriate extension of the temporal-logic based methodology.

Beside his more theoretical work, concerning a complete axiom system and proof theory for program verification by temporal logic, he also contributed to algorithmic research in this area. He developed a deductive system for linear-time temporal logic and model-checking algorithms for the verification of temporal properties of finite-state systems. Together with David Harel, Pnueli worked on the semantics and implementation of Statecharts, a visual language for the specification, modeling, and prototyping of reactive systems. This language has been applied to avionics, transport, and electronic hardware systems. His current research interests involve synthesis of reactive modules, automatic verification of multi-process systems, and specification methods that combine transition systems with temporal logic.

Together with Zohar Manna, he is the author of a 3-volumes textbook on Temporal Logic and its application to Reactive Systems of which the first two volumes are:

1. Z. Manna & A. Pnueli: "The Temporal Logic of Reactive and Concurrent Systems: Specification", Springer-Verlag, 1991.

2. Z. Manna & A. Pnueli: "Temporal Verification of Reactive Systems: Safety", Springer-Verlag, 1995.

A draft of the third volume is available through the internet.

**Main Publications:**

1. Z. Manna and A. Pnueli. The Temporal Logic of Reactive and Concurrent Systems: Specification. Springer-Verlag, New York, 1991.

2. Z. Manna and A. Pnueli. Temporal Verification of Reactive Systems: Safety. Springer-Verlag, New York, 1995.
3. A. Pnueli, The temporal logic of programs, Proc. 18th IEEE Symp. Found. of Comp. Sci., 1977, pp. 46–57.
4. D. Gabbay, A. Pnueli, S. Shelah, and J. Stavi, On the temporal analysis of fairness, Proc. 7th ACM Symp. Princ. of Prog. Lang., 1980, pp. 163–173.
5. D. Gabbay, A. Pnueli, S. Shelah, and J. Stavi, On the temporal analysis of fairness, Proc. 7th ACM Symp. Princ. of Prog. Lang., 1980, pp. 163–173.
6. M. Sharir and A. Pnueli, Two approaches to interprocedural data-flow analysis, Program Flow Analysis: Theory and Applications (Jones and Muchnik, eds.), Prentice-Hall, 1981.
7. A. Pnueli, The temporal semantics of concurrent programs, Theoretical Computer Science 13, 1981, pp. 1–20.
8. D. Lehmann, A. Pnueli, and J. Stavi, Impartiality, justice and fairness: The ethics of concurrent termination, Proc. 8th Int. Colloq. Aut. Lang. Prog., Lec. Notes in Comp. Sci. 115, Springer-Verlag, 1981, pp. 264–277.
9. M. Ben-Ari, Z. Manna, and A. Pnueli, The temporal logic of branching time, Acta Informatica 20, 1983, pp. 207–226.
10. O. Lichtenstein and A. Pnueli, Checking that finite state concurrent programs satisfy their linear specification, Proc. 12th ACM Symp. Princ. of Prog. Lang., 1985, pp. 97–107.
11. O. Lichtenstein, A. Pnueli, and L. Zuck, The glory of the past, Proc. Conf. Logics of Programs, Lec. Notes in Comp. Sci. 193, Springer-Verlag, 1985, pp. 196–218.
12. A. Pnueli and R. Rosner, On the synthesis of a reactive module, Proc. 16th ACM Symp. Princ. of Prog. Lang., 1989, pp. 179–190.
13. D. Harel, H. Lachover, A. Naamad, A. Pnueli, M. Politi, R. Sherman, A. Shtull Trauring, and M. Trakhtenbrot, Statemate: a working environment for the development of complex reactive systems, IEEE Trans. Software Engin. 16, 1990, pp. 403–414.
14. A. Pnueli and R. Rosner, Distributed reactive systems are hard to synthesize, Proc. 31th IEEE Symp. Found. of Comp. Sci., 1990, pp. 746–757.
15. Z. Manna and A. Pnueli. Completing the temporal picture. Theor. Comp. Sci., 83(1):97–130, 1991.
16. Z. Manna and A. Pnueli. Models for reactivity. Acta Informatica, 30:609–678, 1993.
17. A. Pnueli and L.D. Zuck. Probabilistic verification. Information and Computation, 103(1):1–29, 1993.
18. E. Asarin, O. Maler, and A. Pnueli. Reachability analysis of dynamical systems having piecewise-constant derivatives. Theor. Comp. Sci., 138:35–66, 1995.
19. Y. Kesten and A. Pnueli. Verification by finitary abstraction. Information and Computation, a special issue on Compositionality, 163:203–243, 2000.
20. Yonit Kesten, Amir Pnueli: A compositional approach to CTL* verification. Theor. Comput. Sci. 331(2-3): 397-428 (2005).

**Service to the Profession:** Steering Committee Member, CAV (Computer Aided Verification) Conference, 1989-; Associate editor, "Science of Computer Programming", 1998-; Steering Committee Steering Committee Member, Workshop Series on Hybrid Systems, 1996-2003; Member, FTRTFT (Formal Techniques for Real-Time and Fault-Tolerant Systems), 1995-2000; Founder and First Chairman, Computer Science Department, Tel-Aviv University, 1973-1976; Associate Editor, "Journal of Logic and Computations"; Associate Editor, "Formal Methods in System Design"; Associate Editor, IGPL (Journal of the Interest Group in Pure and Applied Logics); Working Group Member, IFIP's WG2.2, Formal Description of Programming.

**Teaching:** He has supervised close to 30 PhD students, including:
Nissim Francez (1976, currently Prof. at the Technion),
Giora Slutzki (1977, currently Prof. at Iowa State),
Moredechai Ben-Ari (1981, Weizmann Institute),
Rivka Sherman (1984, Applied Materials),
Tmima Koren (Olshansky) (1986, Cysco),
Lenore Zuck (1986, University of Illinois, Chicago),
Shnuel Safra (1989, Tel Aviv University),
Oded Maler (1990, Verimag, Grenoble, France),
Orna Lichtenstien (1991, Holon Institute of Technology),
Roni Rosner (1991, Intel, Israel),
Asher Wilk (1993, Shankar College),
Yonit Kesten (1995, Ben Gurion University),
Monica Marcus (1997, U. of California San Diego),
Elad Shahar (2001, IBM Research Laboratory Haifa),
Sitvanit Ruah (2002, IBM Research Laboratory Haifa),
Ofer Shtrichman (2002, Technion),
Tamarah Arons (2003, Intel Haifa),
Raya Leviatan (2004),
Yoav Rodeh (2004, Tel Hai College),
Jessie Xu (2005, IBM Hawthorn),
Yi Fang (2005, Microsoft Research),
Ittai Balaban (2007, WorldEvolved LLC)

**Vision Statement:** "My main interest in Logic is as a tool for the specification, verification and analysis of Computer Programs. I strongly believe that Logic is a high-level programming language and will eventually provide powerful tools for the construction of correct complex programs."

**Honours and Awards:** Silver Professor of Computer Science, New York University, 2006; Member, Academia Europaea, Informatics Section, 2006; Member, European Academy of Sciences (EAS), 2004; Member, Israeli Academy of Science, 2001; Estrin Family Chair of Computer Science Holder, Weizmann Institute, 2000; Israel Prize, Category of Exact Sciences, Field of Computer Sciences, 2000; Nominated, Foreign Member, American National Academy of Engineering, 1999; Honorary Doctorate, Carl von Ossietzky Universitat Oldenburg in Germany, 2000; Honorary Doctorate, Universite Joseph Fourier, Grenoble, France, 1998; Honorary Doctorate, University of Uppsala, Sweden, 1997; ACM A.M. Turing Award, "For his seminal work introducing temporal logic into computing science and for outstanding contributions to program and system verification," 1996.

## POHLERS, Wolfram

**Specialties:** Proof Theory, Generalized Recursion Theory, Descriptive Set Theory

**Born:** 26, August 1943, in Leipzig, Germany.

**Educated:** Ludwig-Maximilians-University, Munich Habilitation (1978),
Ludwig-Maximilians-University, Munich, PhD (1973),
Ludwig-Maximilians-University, Munich, Diplom (1971)]

**Dissertation:** *Eine scharfe Grenze für die Herleitbarkeit der transfiniten Induktion in einem schwachen $\Pi_1^1$-Fragment der klassischen Analysis*, supervisor, Kurt Schütte

**Regular Academic or Research Appointments:** FULL PROFESSOR, DIRECTOR OF THE INSTITUTE FOR MATHEMATICAL LOGIC AND FOUNDATIONAL RESEARCH, WESTFÄLISCHE WILHELMS-UNIVERSITY, MÜNSTER (SINCE 1985), Associate Professor, Ludwig-Maximilians-University, Munich (1980 until 1985), Scientific Assistant (1973-1980), Ludwig-Maximilians-University, Munich, Scientific Collaborator (1971-1973), Ludwig-Maximilians-University, Munich.

**Visiting Academic or Research Appointments:** Visiting Professor, Ohio State University, Columbus OH, USA, 2005; Offer of the chair for Mathematical Logic from the University in Vienna, Austria, 1994, declined 1995; Visiting Scholar, Mathematical Sciences Research Institute, Berkeley, CA, USA,1989-1990; Lehrstuhlvertretung University of Freiburg, Germany, 1979.

**Research Profile:** Ordinal analysis of impredicative theories and its connection to generalized recursion theory and descriptive set theory

**Main Publications:**

1. "An upper bound for the provability of transfinite induction with N-times iterated inductive definitions". In "ISICL Proof Theory Symposium" edited by J. Diller and G.H. Müller, Lecture Notes in Mathematics 500, Springer ,1975, pp. 271-289
2. "Ordinals connected with formal theories for transfinitely iterated inductive definitions". The Journal of Symbolic Logic 43, 1978, pp. 161-182
3. "Provable wellorderings of formal theories for transfinitely iterated inductive definitions". Together with W. Buchholz. The Journal of Symbolic Logic 43, 1978, pp. 118-125
4. "Proof-theoretical analysis of $ID_\nu$ by the method of local predicativity". In Buchholz, Feferman, Pohlers, Sieg, "Iterated Inductive Definitions and Subsystems of Analysis: Recent Proof-Theoretical Studies", Lecture Notes in Mathematics 897, Springer 1981, pp. 261-357
5. "Ordinal notations based on a hierarchy of inaccessible cardinals". Annals of Pure and Applied Logic 33, 1987, pp 157-179
6. "Proof Theory: An introduction". Lecture Notes in Mathematics 1407, Springer, 1989, 213 pages
7. "Proof theory and ordinal analysis". Archive for Mathematical Logic 30, 1991, pp. 311-376
8. "Pure proof theory - aims, methods and results". The Bulletin of Symbolic Logic 2, 1996, pp. 159-188
9. "Applications of cut-free infinitary derivations to generalized recursion theory". Together with A. Beckmann. Annals of Pure and Applied Logic 94, 1998, pp. 7-19
10. "Subsystems of Set Theory and Second Order Number Theory". In "Handbook of Proof Theory", edited by S. Buss, Studies in Logic 137, Elsevier, 1998, pp. 209-335
*Work in Progress*
11. "Proof Theory. The first step into impredicativity". To appear in Universitext Springer. app. 380 pp.
12. "Proof Theory: The second step into impredicativity" in preparation.

**Service to the Profession:** Managing editor of the Archive for Mathematical Logic (Springer),
Member of the editorial board of Mathematical Logic Quarterly (Wiley-VCH) and the Journal of Applied Logic (Elsevier),
Member of the Editorial board of Series in Mathematical Logic (Ontos Verlag),
Review editor of the Bulletin for Symbolic Logic (until 2006),
Member of the ASL,
Chairman of the Committee for Logic in Europe (second term),
Member of the Vorstand of the DVMLG ( the German Association of Logicians),

Member of the DMV( German Association of Mathematicians).

**Teaching:** Students now in academic positions:
Gerhard Jäger, Berne;
Michael Rathjen, OSU and Leeds;
Andreas Weiermann, Utrecht;
Arnold Beckmann, Swansea.

**Vision Statement:** "I still believe that logic should be able to tell us something about the mathematical universe and the reach of mathematical methods in exploring it. Therefore I am exited about recent results in set theory. My vision is a symbiosis of methods of inner model theory and abstract (pure) proof theory to find the ordinal analysis for really strong systems even extending ZFC. Since ordinal analyses give raise to independence results for combinatorial principles it would be most interesting to see what theses principles will look like."

# POOLE, David

**Born:** 3 March 1958, Port Lincoln, South Australia.

**Specialties:** Artificial intelligence, computational logic, logic and probability, logical representations for decision making under uncertainty, probabilistic inference.

**Educated:** B.Sc., Flinders Univ. of South Australia (Maths), 1979
Ph.D., Australian National University (Computer Science) 1984
Postdoctoral Fellow, University of Waterloo, 1983-1985

**Dissertation:** The Theory of CES: A Complete Expert System, Robin Stanton (supervisor)

**Regular Academic or Research Appointments:** PROFESSOR, DEPARTMENT OF COMPUTER SCIENCE, THE UNIVERSITY OF BRITISH COLUMBIA, 1998–; Associate Professor, Department of Computer Science, The University of British Columbia, 1993–1998; Assistant Professor, Department of Computer Science, The University of British Columbia, 1988–1993; Assistant Professor, Department of Computer Science, The University of Waterloo, 1985–1988

**Research Profile:** Poole is known for his work on assumption-based reasoning, combining logic and probability, and automatic inference algorithms.

In the 1980s, Poole worked on assumption-based reasoning, in particular investigating a simple logical framework with facts written in first order logic and explicit assumables that can be used as long as they are consistent with the facts. This framework can be used for default reasoning when the assumables are normality assumptions used for predictions, and can be used for abduction when the assumptions are used to explain an observation and a design goal. This framework was implemented in the Theorist program. A common reasoning strategy was: when we make observations, we explain the observations then make predictions from these explanations. Different reasoning frameworks can be obtained by considering who chooses the assumptions: the agent, an adversary or nature.

In the early 1990s Poole developed probabilistic Horn abduction, a simple framework with independent probabilities on assumables and a logic program to give the consequences of the choices. Abduction from observations followed by prediction corresponded to reasoning in Bayesian networks. This showed that a Bayesian network can be interpreted as a deterministic system with (independent) stochastic inputs. It also showed how to extend Bayesian networks to a richer first-order language,

Probabilistic Horn abduction evolved into the independent choice logic which has a richer logic that includes negation as failure, multiple agents choosing assumptions, and various models of time, including the event calculus and the situation calculus.

When probabilistic Horn abduction was developed, the current inference algorithms were too complicated and needed to be extended to cover richer logic. With Nevin Zhang, he developed the variable elimination algorithm for probabilistic inference, and showed how it could be extended to implement causal independence and context-specific independence. he also developed various search algorithms for probabilistic inference.

In the early 2000s Poole developed a framework for lifted probabilistic inference. More recently he has worked on the probability of existence and identity and on combining probabilities with ontologies.

**Main Publications:**

1. David Poole, "Logical Generative Models for Probabilistic Reasoning about Existence, Roles and Identity", *Proc. Twenty Second AAAI Conference on AI (AAAI-07)*, July 2007.
2. David Poole and Alan Mackworth, "Dimensions of Complexity of Intelligent Agents", *International Symposium on Practical Cognitive Agents and Robots*, Perth, November 2006.
3. Rita Sharma and David Poole, "Probabilistic Reasoning with Hierarchically Structured Variables", *Proc. Nineteenth International Joint Conference on Artificial*

*Intelligence (IJCAI-05)*, Edinburgh, August 2005.

4. David Poole and Clinton Smyth, "Type Uncertainty in Ontologically-Grounded Qualitative Probabilistic Matching", *Eighth European Conference on Symbolic and Quantitative Approaches to Reasoning with Uncertainty (ECSQARU-2005)*, Barcelona, July 2005.

5. Craig Boutilier, Ronen I. Brafman, Carmel Domshlak, Holger Hoos, and David Poole "CP-nets: A Tool for Representing and Reasoning with Conditional Ceteris Paribus Preference Statements", Journal of AI Research, Volume 21, pages 135–191, February 2004.

6. David Poole, First-order probabilistic inference, *Proc. Eighteenth International Joint Conference on Artificial Intelligence (IJCAI-03)*, Acapulco, August 2003, 985–991.

7. David Poole and Nevin Lianwen Zhang, "Exploiting contextual independence in probabilistic inference", *Journal of Artificial Intelligence Research*, 18, 263-313, 2003.

8. D. Poole, "Abducing Through Negation as Failure: Stable models within the independent choice logic", *Journal of Logic Programming*, special issue on Abductive Logic Programming, 44, 5–35, 2000.

9. D. Poole, "Learning, Bayesian Probability, Graphical Models, and Abduction", in P.A. Flach and A.C. Kakas (Eds.), *Abduction and Induction: Essays on their Relation and Integration*, Kluwer, 2000, 153–168.

10. M. Horsch and D. Poole, "Estimating the Value of Computation", *Proc. Fifteenth Conference on Uncertainty in Artificial Intelligence (UAI-99)*, Stockholm, Sweden, pages 297–304, July 1999.

11. D. Poole, "Decision Theory, the Situation Calculus and Conditional Plans", Linköping Electronic Articles in Computer and Information Science, Vol 3 (1998):nr 8. http://www.ep.liu.se/ea/cis/1998/008/, June 15, 1998. *The Electronic Transactions on Artificial Intelligence*, Volume 2, 105–154, 1998.

12. D. Poole, A. Mackworth, and R. Goebel, *Computational Intelligence: A Logical Approach*, Oxford University Press, January 1998 (556 pages).

13. D. Poole, "Probabilistic Partial Evaluation: Exploiting rule structure in probabilistic inference", *Proc. Fifteenth International Joint Conference on Artificial Intelligence (IJCAI-97)*, Nagoya, Japan, pp. 1284–1291, 1997.

14. D. Poole, "The Independent Choice Logic for modelling multiple agents under uncertainty", *Artificial Intelligence*, special issue on Economic Principles of Multi-Agent Systems, 94, 7–56, 1997.

15. D. Poole, "Probabilistic conflicts in a search algorithm for estimating posterior probabilities in Bayesian networks", *Artificial Intelligence*, 88, 69–100, 1996.

16. N.L. Zhang and D. Poole, "Exploiting Causal Independence in Bayesian Network Inference", *Journal of Artificial Intelligence Research*, 5, 301–328, 1996.

17. D. Poole, "Probabilistic Horn abduction and Bayesian networks", *Artificial Intelligence*, 64(1), 81–129, 1993.

18. D. Poole, "The effect of knowledge on belief: conditioning, specificity and the lottery paradox in default reasoning", *Artificial Intelligence*, 49, 281-307, 1991. Republished in R. J. Brachman, H. J. Levesque and R. Reiter (Eds.), *Knowledge Representation*, MIT Press, 1991.

19. D. Poole, "A methodology for using a default and abductive reasoning system", *International Journal of Intelligent Systems*, 1990, 5(5), 521–548, 1990.

20. D. Poole, "Explanation and Prediction: An Architecture for Default and Abductive Reasoning", *Computational Intelligence* 5(2), 97-110, 1989.

21. D. L. Poole, "A Logical Framework for Default Reasoning", *Artificial Intelligence*, 36(1), 27–47, 1988.

*Work in Progress*

22. He is currently working on semantic science: the idea that data and theories can be published referring to formal ontologies. (Probabilistic) theories can then be tested on all available data and used on prediction on new cases. This involves developing theories that can be specified without knowing the individuals, and where we have to reason about existence and identity.

**Service to the Profession:** Poole was a program chair of the Uncertainty in AI (UAI) conference in 1994, and the general chair in 1995. He has served as the secretary of the Association for uncertainty in AI from 2004–2009.

Associate editor for the Journal of Artificial Intelligence Research (JAIR) 2000-2003, Artificial Intelligence Journal (AIJ) 2007-2011, and the International Journal of Approximate Reasoning 2005–. Member of the editorial boards of AAAI press (2006-2009), JAIR (1997-1999), New Generation Computing (1991-1997).

Elected Member of the American Association for Artificial Intelligence Executive Council (2000–2003)

He has been on program committees for IJCAI, UAI, AAAI, KR, ECSQARU, CAI, ECAI, and numerous workshops.

**Teaching:** Poole has graduated 12 Ph.D. students, 12 M.Sc students, and has supervised or co-supervised 9 postdoctoral fellows.

He was the coauthor of an AI textbook, *Computational Intelligence: A Logical Approach*, published by Oxford University Press, 1998. He is a co-developer of "CIspace: tools for learning computational intelligence", a set of interactive tools designed to learn the fundamental of AI. He wrote *cilog*, a logic programming language designed for teaching that allows for interactive exploration of proofs and search strategies, and includes ask the user mechanisms and probabilistic reasoning.

**Vision Statement:** I believe that practical reasoning is decision making, and that decision making is best modelled in terms of probabilities and utilities. Following the decision theory and game theory traditions, reasoning them becomes finding actions that maximize expected utility (taking into account other agents' reasoning when there are multiple agents). In the past most decision-theoretic reasoning was carried out with simple, essentially propositional, languages.

Logic gives us much richer languages in terms of individuals and relations. Individuals can be referred to by naming them in constant, indirectly using function symbols, and they can be quantified. I believe that using richer first-order languages with probabilities and utilities can capture the the language of thought or mentalese that logicians are searching for.

The main arguments for basing reasoning on probabilities and utilities are

- (1) probabilities are what you get from data (by observing and interacting with the world),

- (2) acting is gambling, and if an agent doesn't use probabilities and utilities in gambling it will lose to one that does, and

- (3) there is a well-defined principle for approximate reasoning: approximations can lead to a loss of utility, which lets us trade off thinking and acting.

Extending the language of probability to be richer is a major challenge (particularly when we need to be able to reason with the language and learn the representations).

**Honours and Awards:** Flinders University of South Australia University Medal 1978. Scholar Canadian Institute for Advanced Research 1992-1995. C. A. McDowell Medal for Excellence in Research U.B.C. 1994. Fellow of the Association for the Advancement of Artificial Intelligence (AAAI) 2000.

# PRADE, Henri

**Specialties:** Artificial intelligence, applied logic, uncertainty modeling.

**Born:** 20 August 1953 in Mulhouse, France.

**Educated:** Engineer, and then Doctor-Engineer degrees from Ecole Nationale SupÈrieure de l'AÈronautique et de l'Espace (Toulouse, France) in 1975 and in 1977; "Doctorat d'Etat" (1982) and "Habilitation ‡ Diriger des Recherches" (1986) both from Paul Sabatier University in Toulouse.

**Dissertation:** (Doctorat d'Etat): "ModËles MathÈmatiques de l'ImprÈcis et de l'Incertain en Vue d'Applications au Raisonnement Naturel".

**Regular Academic or Research Appointments:** "DIRECTEUR DE RECHERCHE" 1ST CLASS, AT C.N.R.S., IRIT ("INSTITUT DE RECHERCHE EN INFORMATIQUE DE TOULOUSE", TOULOUSE, FRANCE), 2002; "Directeur de Recherche", 2nd class, at C.N.R.S., IRIT, 1988-2002; "ChargÈ de Recherche" at C.N.R.S., Lab. LSI ("Langages et SystËmes Informatiques"), Toulouse, France, 1982-1988; "AttachÈ de Recherche" at C.N.R.S., Lab. LSI, 1979-1982.

**Visiting Academic or Research Appointments:** IRIA (Institut de Recherche en Informatique et Automatique) Post-doctoral Fellow, visiting scholar at Artificial Intelligence Lab., Stanford University, 1977-1978.

**Research Profile:** Henri Prade has been working on the handling of uncertainty in artificial intelligence for about twenty-five years, and more particularly on the modeling of different types of reasoning (reasoning under uncertainty, nonmonotonic reasoning and belief revision, similarity-based reasoning, reasoning under inconsistency). Jointly with Didier Dubois, he has contributed to the development of fuzzy sets and possibility theory (fuzzy interval analysis, typology of fuzzy if-then rules, study of possibilistic independence, relations between possibility and probability theories, ...), and their applications to approximate reasoning, since the late seventies. Possibility theory, initiated by Lotfi Zadeh in 1978, provides a framework for representing uncertainty in terms of a pair of dual measures of possibility and necessity (their duality expresses a graded version of the classical relationship between the modalities 'possibly' and 'necessarily'). Possibility measures are max-decomposable under disjunction, and are usually valued on the real unit interval, but any discrete linearly ordered scale may be used as valuation scale.

Regarding logic, Henri Prade (jointly with Didier Dubois) has mainly made three contributions: Developing possibilistic logic, relating logic of conditional objects with nonmonotonic reasoning, and introducing bipolarity in logic.

Possibilistic logic (whose first elements appear in Prade's French Doctorat d'Etat thesis in 1982, and whose name was coined in a 1987 paper) is a weighted logic that has been introduced for dealing with uncertain or prioritized information. Standard possibilistic logic expressions are classical logic formulas associated with weights, interpreted in the framework of possibility theory as lower bounds of necessity degrees. Possibilistic logic handles partial inconsistency by means of an inconsistency level that is associated with any possibilistic base (i.e. a set of possibilistic logic formulae). From a semantic point of view, a possibilistic base is understood as a possibility distribution representing the fuzzy set of models of the base. An interpretation is all the less possible as it falsifies formulae of higher degree. Soundness and completeness results in possibilistic logic generalize the ones of classical logic. A possibilistic logic base can be also equivalently represented by a possibilistic directed acyclic graph. Such a graph exhibits a conditional independence structure just like Bayesian nets do. Besides, a set of comparative constraints of the form "if $p$ is true then '$q$ true' is more possible than '$q$ false'" can also be encoded as a possibilistic logic base. This type of constraint can be used for representing default rules of the form "if $p$ then usually $q$". From such a set of constraints, two types of inference can be defined (depending if we consider all the possibility measures that are solutions of the set of constraints, or only the largest one which is unique), which have been shown to be closely related to the two consequence relations proposed by Lehmann and Magidor in their preferential approach to nonmonotonic reasoning.

Conditional objects, also called tri-events, are entities of form "$q$ given $p$", denoted "$q|p$", originally considered in the thirties by B. De Finetti when he discussed the foundations of conditioning in probability. A tri-event "$q|p$" may be true or false, when $p$ is true, depending if $q$ is true or false; if $p$ is false the tri-event receives a third truth-value whose meaning is "inapplicable". According to the chosen ordering of this third truth-value w. r. t. "true" and "false", different conjunctions (and disjunctions) can de defined between conditional objects. Semantic entailment between conditional objects is defined as "$q|p$" entails "$s|r$" if any interpretation that verifies "$q|p$" also verifies "$s|r$" and if any interpretation that falsifies "$s|r$" also falsifies "$q|p$". Then taking "inapplicable" as greater than "true" (itself greater than "false"), provides a ternary logic-based representation of Lehmann and Magidor' preferential entailment in nonmonotonic reasoning.

The idea of bipolarity in knowledge representation is closely related to the idea of stating separately a set of interpretations as impossible (negative information) and a set of interpretations whose possibility is "guaranteed" (positive information). Consistency requires that none impossible interpretation be also "guaranteed"possible. But in general, what is not impossible is not necessarily "guaranteed" possible. Classical logic by understanding "$p$ is true" as stating that the countermodels of $p$ are impossible, deal with the negative part of the information, while positive information obeys a reversed entailment (since only subsets of the models of p are also guaranteed to be possible if the models of $p$ are). Positive information then corresponds to the idea of accumulated observations. Several logical settings can be made bipolar, including possibilistic logic and modal logic.

**Main Publications:**

1. S. Benferhat, D. Dubois and H. Prade, Nonmonotonic reasoning, conditional objects and possibility theory. Artificial Intelligence, 92, 259-276, 1997.
2. S. Benferhat, D. Dubois and H. Prade, Possibilistic and standard probabilistic semantics of conditional knowledge bases, J. of Logic and Computation, 9, 873-895, 1999.
3. D. Dubois, F. Esteva, L. Godo and H. Prade, Fuzzy-set based logics – An history-oriented presentation of their main developments. In Handbook of the History of Logic, Volume 8: The Many Valued and Non-monotonic Turn in Logic, D. M. Gabbay and J. Woods, eds., 325-449, Elsevier, 2007.
4. D. Dubois, P. Hajek, and H. Prade, Knowledge-driven versus data-driven logics. Journal of Logic, Language, and Information, 9, 65-89, 2000.
5. D. Dubois, E. P. Klement and H. Prade, eds. Fuzzy Sets, Logics and Reasoning about Knowledge, Kluwer, Dordrecht, Vol. 15 in Applied Logic Series, 1999.
6. D. Dubois, J. Lang and H. Prade, Possibilistic logic. In: Handbook of Logic in Artificial Intelligence and Logic Programming, (D. M. Gabbay, C. J. Hogger, J. A. Robinson, eds.), Volume 3: Nonmonotonic Reasoning and Uncertain Reasoning (D. Nute, ed.), Oxford Univ. Press, 439-513, 1994.
7. D. Dubois and H. Prade, Conditional objects as nonmonotonic consequence relationships. IEEE Trans. on Systems, Man and Cybernetics, 24, 1724-1740, 1994.
8. D. Dubois, H. Prade. Possibility theory, probability theory and multiple-valued logics: A clarification. Annals of Mathematics and Artificial Intelligence, 32, 35-66, 2001
9. D. Dubois and H. Prade, Possibilistic logic: a retrospective and prospective view. Fuzzy Sets and Systems, 144, 3-23, 2004.

A full list of Prade's publications is available at http://www.irit.fr/-Publications-

*Work in Progress*

10. Double special issue of the Inter. J. of Intelligent Systems to appear on "Bipolar Representations of Information and Preference: 1. Cognition and Decision, 2. Reasoning and Learning", guest-edited by D. Dubois and H. Prade.
11. Multiple agent extensions of possibilistic logic; argumentative reasoning and decision; perception of causality.

**Service to the Profession:** Henri Prade has served as a member of the editorial board organization of several journals, including ACM Trans. on

Computational Logic (since 2000), Artificial Intelligence (since1997), Fundamenta Informaticae (since 1997), Fuzzy Sets and Systems (since1985), IEEE Trans. on Fuzzy Systems (since 1993), Inter. J. of Approximate Reasoning (since 1987), Inter. J. of Intelligent Systems (since1986), J. of Applied Non-Classical Logics (HermÈs) (since 1991), Journal of Applied Logic (Elsevier) (since 2003), Transactions on Rough Sets (since 2004). He has been a program chair of the1992 and 2002 IEEE Conferences on Fuzzy Systems, of the1998 European Conference on Artificial Intelligence, and of the1999 Uncertainty in Artificial Intelligence Conference, and has served on the program committees of many conferences.

**Teaching:** Being a full time CNRS researcher having no teaching duties, Henri Prade has only regularly taught a few advanced courses on reasoning under uncertainty in artificial intelligence for many years. He has advised twenty-five Ph. D. students in computer science, on different artificial intelligence topics, including JÈrÙme Lang, Salem Benferhat, HÈlËne Fargier, and Souhila Kaci.

**Vision Statement:** Research in logic has been mainly motivated in the 20th century by issues raised in relation to the foundations of mathematics. The artificial intelligence research program in the last quarter of this century has raised the need for formalizing various types of reasoning under incomplete, uncertain, fuzzy, or inconsistent information. This has completely renewed and put in a new perspective (in fact originated by George Boole in his book The Laws of Thought!) a significant part of the research in applied, (non-classical) logic. This trend will continue to develop in the forthcoming years and should have an impact not only on computerized information handling, but also, in the long range, in logic, in cognitive psychology, in linguistics, and in philosophy.

**Honours and Awards:** IBM-France Research Prize in Computer Science, 1988; Inter. Fuzzy Systems Assoc. (IFSA) Fellow, 1997; European Coordinating Committee for Artificial Intelligence (ECCAI) Fellow, 1999; ISI (Institute for Scientific Information) French Citation Laureate, 2001; Pioneer Award of the IEEE Neural Networks Society, 2002.

# PRATT, Vaughan Ronald

**Specialties:** Logics of programs, concurrency modelling, linear logic, computability theory.

**Born:** 12 April 1944 in Melbourne, Victoria, Australia.

**Educated:** University of Sydney BSc (Honours Mathematics and Physics) 1967, MSc 1970, Stanford University PhD 1972.

**Dissertation:** *Shellsort and Sorting Networks*; supervisor Donald Knuth.

**Regular Academic or Research Appointments:** PROFESSOR OF COMPUTER SCIENCE, STANFORD UNIVERSITY, 1981– (EMERITUS 2000–). Massachusetts Institute of Technology 1972-1982.

**Visiting Academic or Research Appointments:** Visiting professor, Stanford University, 1980-1981. Visiting scholar, University of Edinburgh, 1975. Visiting faculty, IBM Research Yorktown Heights, 1972.

**Research Profile:** Pratt's research interests in the early 1970s were in computational complexity: he showed that the primes were in NP [2], obtained a polynomial size complexity gap between monotone and general Boolean logic circuits [3] (subsequently increased to an exponential gap by A. Razborov), showed that parallel time is equivalent to sequential space via deterministic up-to-quadratic simulations of nondeterministic computation in each direction (with L. Stockmeyer) [4], and gave the first linear time pattern matcher (with Knuth and Morris) [6].

In the late 1970s Pratt's attention shifted to logics of sequential programs. He introduced dynamic logic as a serendipitous combination of relation algebra and modal logic [5], showed that both validity and satisfiability of propositional dynamic logic (PDL) were complete in deterministic exponential time [7] (the lower bound having previously been obtained by Fischer and Ladner), gave the first rigorous completeness proof for K. Segerberg's axiomatization of PDL [8] (in algebraic form, subsequently "dealgebraized" by D. Kozen and R. Parikh), and gave the first modal mu-calculus [9] (subsequently improved to the now-standard language $L_\mu$ by D. Kozen). He also diagnosed binding as the root cause of intractibility of logics of programs by showing that logics of programs (for both partial and total correctness, i.e. including halting) were decidable in exponential time in the absence of binding, i.e. absent quantifiers, assignment statements, and procedure definitions [10].

Much later Pratt defined a purely equational theory of conditional regular expressions, or action logic ACT [11]. Being equational, its conditionals are in the signature as nonlogical connectives, in contrast to D. Kozen's Kleene algebras, KA, a universal Horn theory where the conditionals are logical connectives. ACT is significant as a conservative extension of the equational theory REG of regular expressions having two properties neither of which hold for REG: ACT is finitely axiomatizable, and it is categorical in the sense of constraining star to be exactly reflexive transitive closure. (V. Redko showed non-finite-axiomatizability of REG in 1967, and J.H. Conway gave a four-element model of REG in 1971 for which star was not transitive closure.)

In the 1980s he turned to logics of concurrent behavior, where the main problem appeared to be a dearth of compelling models available as a semantic yardstick for soundness and completeness and an indicator of appropriate logical and nonlogical connectives for reasoning about concurrency. One of the models he developed, higher dimensional automata [12], has been relatively successful, having spawned a series of conferences around this topic initially organized by E. Goubault, with the seventh in San Francisco in 2005. However another model, based on Chu spaces [14], has not "taken hold" to date though Pratt remains hopeful — unlabelled event structures based on the four event states *before*, *during*, *after*, and *cancelled* [15] are strictly more expressive than unlabelled Petri nets, event structures, and higher dimensional automata.

At present he is working on a unification of Chu spaces and presheaves that he calls communes, having applications to ontology (in the sense of Carnap and Quine rather than as used in AI), Kripke structures, and relational databases. Pratt has also had a sporadic interest in constructive representations of the continuum, representing the field of reals as a ring of bounded integer polynomials modulo the polynomial $2x - 1$ (with D. Knuth), and the ordered set (and hence topological space) of reals as a final coalgebra (with D. Pavlović) [13]. Although the former has passed unnoticed, the latter has stimulated work by P. Freyd, M. Escardó, T. Leinster, and others.

**Main Publications:**

1. "Time Bounds for Selection" (with M. Blum, R. Floyd, R. Rivest, and R. Tarjan), *Journal of Computer and System Sciences*, 7:4, 448–461 (1973).
2. "Every Prime has a Succinct Certificate", *SIAM Journal on Computing*, 4:3, 214–220 (1975).
3. "The Power of Negative Thinking in Multiplying Boolean Matrices", *SIAM Journal on Computing*, 4:3, 326–330 (1975).
4. "A Characterization of the Power of Vector Machines" (with L. Stockmeyer), JCSS, 12, 2, 198–221 (1976).
5. "Semantical Considerations on Floyd-Hoare Logic", *Proc. 17th IEEE Symposium on Foundations of Computer Science*, 109–121 (1976).
6. "Fast Pattern Matching in Strings" (with D. Knuth and J. Morris), SIAM Journal on Computing, 6:2 323–350 (1977).
7. "A Near Optimal Method for Reasoning About Action", *Journal of Computer and System Sciences*, 20:2, 231–254, (1980).
8. "Dynamic Algebras and the Nature of Induction", *Proceedings 12th ACM Symposium on Theory of Computing*, 22–28 (1980)
9. "A Decidable Mu-Calculus", *Proc. 22nd IEEE Symposium on Foundations of Computer Science*, 421–427 (1981).
10. "Program Logic Without Binding is Decidable", *Proc. 8th Annual ACM Symposium on Principles of Programming Languages*, 159–163 (1981).
11. "Action Logic and Pure Induction", *Springer-Verlag Lecture Notes in Computer Science*, 478, 97–120 (1990).
12. "Modeling Concurrency with Geometry", *Proceedings 18th Annual ACM Symposium on Principles of Programming Languages*, 311–322 (1991).
13. "The continuum as a final coalgebra" (with D. Pavlović), *Theoretical Computer Science*, 280:1-2, 105–122 (2002).
14. "Chu spaces as a semantic bridge between linear logic and mathematics", *Theoretical Computer Science*, 294:3, 439–471 (2003).
15. "Transition and Cancellation in Concurrency and Branching Time", *Mathematical Structures in Computer Science*, 13:4, 485–529 (2003).

*Work in Progress*

16. *Multiplicative-additive linear logic (MALL) of Chu spaces.* Includes the Gustave function and hence is not a full model of sequential MALL. Main open problem: is it finitely axiomatizable?
17. *Communes.* A uniform and canonical unification of presheaves and Chu spaces organized into linearly distributive categories.
18. *Comonoids.* A cartesian closed sibling of topological spaces defined by joint continuity. Main open problem: is every $T_1$ comonoid discrete? (Seems to be an unusually difficult problem.)
19. *Chu spaces* (with D. Hughes). Monograph in preparation.

**Service to the Profession:** Editorial boards of *Mathematical Structures in Computer Science* (Cambridge University Press), *Applied Categorical Structures* (Kluwer Academic Press), and *Logic Journal of the Interest Group in Pure and Applied Logics* (Oxford University Press).

**Teaching:** Introduced and taught EECS 6.043 *Introduction to Algorithms*, as an undergraduate course at MIT 1973–1978; subsequently taught by C. Leiserson, R. Rivest and others, resulting in the well-known algorithms text by Cormen, Leiserson, and Rivest. Introduced and taught CS 353 *Algebraic Logic*, as an advanced graduate course at Stanford 1981–2002; course notes available on request.

*Students of Note: David Harel* (PhD 1978 MIT), Dean of Faculty of Mathematics and Computer Science 1998-2004, Head of Department of Applied Mathematics and Computer Science 1989-1995, author of a number of computer science texts. *Bob Streett* (PhD 1980 MIT), Inventor of Streett automata (sibling of Büchi automata and Rabin automata used in temporal logic). *Parham Aarabi* (PhD 2001 Stanford), Tenured at U. Toronto at age 28, Gordon R. Slemon Teaching of Design Award, ECE Departmental Teaching Award, IEEE Mac Van Valkenburg Early Career Teaching Award, Faculty of Engineering Early Career Teaching Award, ECE Professor of the Year Award, Canada Research Chair in Multi-Sensor Information Systems, Best Computer Engineering Professor Award, Ontario Distinguished Researcher Award, MIT Tech Review's Top 35 Innovators under 35. *Postdoctoral students and Research associates of Note: Dominic Hughes, Smuel Safra, Rob van Glabbeek*.

**Vision Statement:** A better rapprochement is urgently needed between logics informed by respectively set theory and category theory, which are evolving as though on different planets.

**Honours and Awards:** Fellow, Association for Computing Machinery, 1996.

# PRAWITZ, Dag (Hjalmar)

**Specialties:** Proof theory, foundations of logic and mathematics and philosophy of language.

**Born:** 16 May 1936 in Stockholm, Sweden.

**Educated:** Stockholm University, Filosofie Doktor, 1965; Stockholm University, Filosofie licentiat, 1960; Stockholm University, Filosofie Kandidat, 1957; Various Studies, Münster Universität, 1962; University of Wisconsin, Junior Student, 1955–1956. *Dissertation: Natural Deduction. A Proof-Theoretical Study*, supervisor, Wedberg.

**Regular Academic or Research Appointments:** PROFESSOR EMERITUS OF THEORETICAL PHILOSOPHY, STOCKHOLM UNIVERSITY, 2001–; Professor of Theoretical Philosophy, Stockholm University, 1976–2001; Professor of Philosophy, Oslo University, 1971–1977; Docent of Theoretical Philosophy, Stockholm University 1970–1971, 1965–1967; Docent of Theoretical Philosophy, Lund University 1967–1969.

**Visiting Academic or Research Appointments:** Researcg fekkiwm tge Ubstutyte if /advabced Studies, Università di Bologna, Italy, February–April, 2007; 'Professore al contratto' of Philosophy, Università degli Studi di Roma "La Sapienza", 1983; Fellow, Wolfson College, Oxford, 1975–1976; Visiting Associate Professor, Stanford University, 1969–1970; Visiting Professor, Michigan University, 1969; Visiting Assistant Professor, University of California, Los Angeles, 1964.

**Research Profile:** Dag Prawitz made his first research within Automated Deductions when that field was still in its infancy. Together with a small group of collaborators, he was one of the first to launch a complete proof procedure for first order predicate logic on a computer (see *Key Publications* 1). At the end of the fifties he improved this procedure in essential respects (see 2). He made some further contributions to the field in the sixties (see 7), but Proof Theory soon became his main field of research.

His interest there was not so much the developing of Hilbert's Program but rather the nature and properties of proof, what Prawitz called General Proof Theory. His main inspiration came from Gentzen's analysis of proofs that had resulted in Gentzen's systems of natural deduction and calculi of sequents with its Hauptsatz. Prawitz realized that also the systems of natural deduction enjoy a kind of Hauptsatz, more perspicuous than the one for the calculi of sequents, and that it depends on the fact that the rules of the systems satisfy a certain Inversion Principle. It allowed him to define specific reductions of proofs in natural deduction systems and to establish that each proof reduces to a significant normal form. The resulting Normalization Theorem for natural deduction systems implies the Hauptsatz for the sequent calculi, and brings out more clearly the significance of this kind of results and the features of the inference rules that it depends on (see 3).

Prawitz extended the Normalization Theorem to several other logical systems. The Hauptsatz for impredicative second and higher order logic, known as Takeuti's Conjecture, was established independently by Tait, Takahashi, and Prawitz by using model theoretic means (see 4, 5, and 8). Later, Normalization Theorems were obtained independently by Girard, Martin-Löf, and Prawitz by using Tait's notion of convertibility as adapted to natural deduction systems by Martin-Löf and to impredicative systems by Girard (see 9, appendix). Prawitz also obtained stronger forms of Normalization Theorems by further adapting Tait's notion of convertibility, called validity by Prawitz (see 9, appendix, 15, and 16).

Another feature of Gentzen's intuitionistic system of natural deduction that inspired Prawitz was the close correspondence of the introduction rules of the system and what may be called the constructive contents of the logical constants. Prawitz suggested making this more precise along two different lines. One line was to define a sentence as intuitionistically true when there is a construction of it that is definable in an extended typed lambda calculus, and then to note that the intuitionistic natural deductions can be mapped into the set of terms in that calculus (see 7). A second line was in terms of the above mentioned notion of validity of proofs, whose semantic nature was emphasized by Prawitz (and motivated the name "validity"). He extended the notion to deductive arguments in general to define what it is for such an argument to be correct in view of the constructive meaning of the logical constants as given by their introduction rules (see 10, 11, 19 and 20).

Such semantic matters became his main interest later in the seventies and has remained so. In particular, he investigated the completeness of the set of sentential operators when understood constructively (see 14), the conflict between classical and intuitionistic logic from the perspective of meaning theory (see 13), and the notions of truth and logical consequence from a constructive as well as a classical point of view (see 12, 17, and 18).

His main concern at present is a better understanding of deductive reasoning in general. He has argued in particular that we need to reconsider the nature of inference and what the validity of an inference consists in, if we are to account for the epistemic advance that may result from the drawing of a valid inference (see 21).

**Main Publications:**

1. A mechanical proof procedure and its realization in an electronic computer (together with H. Prawitz and N. Voghera), *Journal of the Association for Computing Machinery* 7, pp 102-128, 1960. (Reprinted in: *Automation of Reasoning 1, Classical Papers on Computational Logic*, pp 202-28, J. Siekmann and G. Wrightson (eds), Springer Verlag, 1983.)
2. An improved proof procedure, *Theoria* 26, pp 102-39, 1960. (Reprinted in: *Automation of Reasoning 1, Classical Papers on Computational Logic*, pp 162-201, J. Siekmann and G. Wrightson (eds), Springer Verlag, 1983.)
3. *Natural Deduction. A Proof-Theoretical Study*, Almqvist & Wiksell, Stockholm 1965. (Translated to Russian, Moscow 1997. To be reprinted by Dover Publications, 2006).
4. Completeness and Hauptsatz for second order logic, *Theoria* 33, pp 246-58, 1967.
5. Hauptsatz for higher order logic, *Journal of Symbolic Logic* 33, pp 452-57, 1969.
6. Advances and problems in mechanical proof procedures, in: *Machine Intelligence 4*, pp 59-71, B. Meltzer et al (eds), Edinburgh, 1969.
7. Constructive semantics, in: *Proceedings of the 1st Scandinavian Logic Symposium, Åbo 1968*, pp 96-114, Uppsala 1970.
8. Some results for intuitionistic logic with second order quantifiers, in: *Intuitionism and Proof Theory, Proc. of the Summer Conference at Buffalo*, pp 259-69, J. Myhill et al (eds), North-Holland,1970.
9. Ideas and results in proof theory, in: *Proceedings of the 2nd Scandinavian Logic Symposium*, pp 237-309, J. Fenstad (ed), North-Holland, 1971. (Translated into Italian: Idee e risultati nella teoria della dimostrazione in: *Teoria della dimostrazione*, pp127-204, D. Cagnoni (ed), Feltrinelli, 1981.)
10. Towards a foundation of general proof theory, in: *Logic, Methodology and Philosophy of Science IV*, pp 225-50, P. Suppes et al (eds), North Holland, 1973.
11. On the idea of a general proof theory, *Synthese* 27, pp 63-77, 1974. (Reprinted in: *A Philosophical Companion to First-Order Logic*, pp 212-24, R.I.G. Hughes (ed), Hackett, 1993. Translated into Italian: Sull'idea di una teoria generale della dimostrazione in: *Teoria della dimostrazione*, pp 205-20, D. Cagnoni (ed), Feltrinelli, 1981.)
12. Comments on Gentzen-type procedures and the classical notion of truth, in: *Proof Theory Symposium Kiel 1974*, pp 290-319, A. Dold et al (eds), Springer Verlag, 1975.
13. Meaning and proofs: On the conflict between classical and intuitionistic logic, *Theoria* 43, pp 2-40, 1977.
14. Proofs and the meaning and completeness of the logical constants, in J. Hintikka et al (eds), *Essays on Mathematical and Philosophical Logic*, pp 25-40, D. Reidel, Dordrecht, 1979. (Translated into German: Beweise und die Bedeutung und Vollständigkeit der logischen Konstanten, *Conceptus*, XVI, pp 3-44, 1982.)
15. Validity and normalizability of proofs in 1st and 2nd order classical and intuitionistic logic. in: *Atti del congresso nazionale di logica*, pp 11-36, Bibliopolis, Napoli, 1981.
16. Normalization of proofs in set theory, in: *Atti degli incontri di logica mathematica*, pp 357-71, Siena 1985.
17. Truth and objectivity from a verificationist point of view, in: *Truth in Mathematics*, pp 41-51, H. G. Dales et al (eds), Clarendon Press, Oxford, 1998.
18. Logical consequence from a constructive point of view, in: *The Oxford Handbook of Philosophy of Mathematics and Logic*, S. Shapiro (ed), 671-695, Oxford 2005.
19. Meaning approached via proofs, *Synthese* 148, no 3 (*Proof-Theoretic Semantics*, R. Kahle and P. Schroeder-Heister, eds), pp 507–524, 2006.
20. Pragmatism and Verificationism, in: *The Philosophy of Michael Dummett*, The Library of Living Philosophiers, Vol XXXI, R. E. Auxier and L. E. Hahn, eds., pp. 455–481, Open Court, Chicago.

*Forthcoming*

21. Validity of Inferences, to appear in the Proceed-

ings of the 2nd Launer Symposium on the Occasion of the Presentation of the Launer Prize 2006 to Dagfinn Føllesdal, held in Bern, 2006.

22. Proofs verifying programs and Programs Producing Proofs — a Conceptual analysis, to appear in *Deduction, Computation, Experiment, Exploring the Effectiveness of Proofs* (The Proceedings of a symposium with the same name helde in Bologna 2007).

23. An approach to general proof theory and a conjecture of completeness of intuitionistic logic revisited, to appear in a Proceedings of the conference *Natural Deduction*, held in Rio de Janeiro 2001.

**Service to the Profession:** $1^{st}$ Vice President, Division of Logic, Methodology and Philosophy of Science, International Union of the Philosophy and History of Science, 1991–1995; Chairman, Organizing Committee of the 9th International Congress of Logic, Methodology and Philosophy of Science, 1991; Vice President, Institut International de Philosophie, 1991-1994; Member, Editorial Board of the Journal of Logic and Computation, 1991-93; Chairman, Committee for the Rolf Schock Prize in Logic and Philosophy, 1991-1997; President, The Rolf Schock Foundation, 1988-97; Assessor, Division of Logic, Methodology and Philosophy of Science, International Union of the Philosophy and History of Science, 1987-1991; Editor, Thales Publishing House, 1985-2004; Member, Editorial Board of the Journal of Symbolic Computation, 1985-1990; Managing editor, Studies in Proof Theory, Bibliopolis, 1982-1995; Member, Editorial Board of the Journal of Philosophical Logic 1972-; Member, Council of the Association for Symbolic Logic, 1968-1969; Editor, Theoria 1967-1969.

**Teaching:** Dag Prawitz has been teaching logic at all levels at universities in Sweden, Norway, Italy, and USA for almost fifty years. He has published one textbook in logic, *ABC i symbolisk logik*, which has been used at many Swedish universities for the last thirty years. A number of his PhD students are now professors at universities in Sweden and other countries.

**Vision Statement:** Logic as a branch of Philosophy has still a mission. Its distinctive subject matter is deductive reasoning. It is this subject matter that makes it a philosophical rather than mathematical branch, although its methods are mathematical to a large extent.

**Honours and Awards:** Medal for Science 2007, awarded by the Institute of Advanced Studies, Università di Bologna, Italy; Foreign Member, Royal Norwegian Society of Sciences and Letters, 2000–; Member, Academia Europeae, 1989–; Foreign Member, Royal Norwegian Academy of Sciences, 1989–; Member, Royal Swedish Academy of Letters, History and Antiquities, 1987–, member of its governing board, 1995–1998; Member, Institut International de Philosophie, 1984–; Member, Royal Swedish Academy of Sciences, 1981–, member of its governing board, 1995–1998.

## PRIEST, Graham George

**Specialties:** Non-classical logic.

**Born:** 14 November 1948 in London, UK.

**Educated:** London School of Economics, 1974; University of Cambridge, MA, 1974; University of London, MSc, 1971; St. John's College, Cambridge University, BA, 1970.

**Dissertation:** *Type Theory in which Variables Range over Predicates*; supervisor, John Bell

**Regular Academic or Research Appointments:** BOYCE GIBSON PROFESSOR OF PHILOSOPHY, UNIVERSITY OF MELBOURNE, 2001–; Arché Professorial Fellow, Department of Logic and Metaphysics, University of St Andrews, 1999-; Professor of Philosophy, University of Queensland, 1988-2000; Lecturer, Senior Lecturer, Associate Professor, University of Western Australia, 1976-1988; Temporary Lecturer, Department of Logic and Metaphysics, University of St Andrews, 1974-1976.

**Visiting Academic or Research Appointments:** Visiting Professor, Central Institute for Higher Tibetan Studies, Sarnath, India, 2005; Visiting Professor, UC Berkeley, 2004; Visiting Professor, University of Kyoto, 2004; Visiting Lecturer, Sino-British-Australian Summerschool in Philosophy, Chengdu, China, 2002; Visiting Professor, Central Institute for Higher Tibetan Studies, Sarnath, India, 2001; Visiting Scholar, Department of Philosophy, New York University, 1997; Visiting Scholar, Graduate Center, City University of New York, 1997; Visiting Professor, Department of Philosophy, Indiana University, Bloomington, 1994; Visiting Professor, Institute of Advanced Studies, University of Sao Paulo, 1994; Visiting Fellow, Clare Hall, Cambridge University, 1990; Visiting Lecturer, Soviet Academy of Sciences, Moscow, 1989; Project Visitor, Automated Reasoning Project, Research School of Social Sciences, Australian National University, 1998; Research Fellow, Human Science Research Council, South Africa, 1987; Visiting Lecturer, Section of Logic, Polish Academy of Sciences, Lodz, Poland, 1986; Visiting Lecturer, Institute of Philosophy, Bulgarian Academy of Science, Sofia, Bulgaria,

1986; Visiting Professor, Department of Philosophy, University of Pittsburgh, 1982; Visiting Lecturer, Institute of Philosophy, Bulgarian Academy of Science, Sofia, Bulgaria, 1982; Visiting Fellow, Department of Philosophy, Research School of Social Sciences, Australian National University, 1982; Visiting Fellow, Department of Philosophy, Research School of Social Sciences, Australian National University, 1978.

**Research Profile:** Graham Priest has contributed to many areas of philosophy and logic. His major contribution has been to philosophical logic, and within that, to paraconsistent logic and dialetheism, plus all the areas that these affect.

**Main Publications:**

1. *In Contradiction*, first edition, Kluwer Academic Publishers, 1987; second edition, Cambridge University Press, 2006.
2. *Beyond the Limits of Thought*, first edition, Cambridge University Press, 1995; second edition, Oxford University Press, 2002.
3. *Introduction to Non-Classical Logic*, Cambridge University Press, 2001.
4. *Towards Non-Being*, Oxford University Press, 2006.
5. *Doubt Truth to be a Liar*, Oxford University Press, 2006.

Plus 10 edited collections and about 140 papers journals or books.

Work in ProgressA second volume of *Introduction to Non-Classical Logic*.

**Service to the Profession:** Chair of the Council, Australasian Association of Philosophy, 1996- ; 1998-2003 First Vice-President, International Union for Logic, Methodology and the Philosophy of Science, 1998-2003; Foundation Vice-President, Australasian Association for Cognitive Science, 1989-1990; President, Australasian Association of Philosophy, 1988; General Secretary, Australasian Association for Logic, 1988-95; President, Australasian Association for Logic, 1988.

**Teaching:** Dedicated and successful teacher at all levels throughout academic career.

**Vision Statement:** To see as much as possible.

**Honours and Awards:** LittD, University of Melbourne, 2002; Elected Fellow of the Australian Academy of Humanities, 1995; Elected life member of Clare Hall, Cambridge University, 1991.

# PUDLÁK, Pavel

**Specialties:** Proof theory, proof complexity.

**Born:** 30 May 1952, in Prague, Czechoslovakia

**Educated:** Charles University, Faculty of Mathematics and Physics, Prague, Czechoslovakia.

**Dissertation:** *Representation of Finite Lattices* (in Czech, 1977).

**Regular Academic or Research Appointments:** MATHEMATICAL INSTITUTE, CZECHOSLOVAK ACADEMY OF SCIENCES (NOW ACADEMY OF SCIENCES OF THE CZECH REPUBLIC), PRAGUE; SINCE 1976. Faculty of Mathematics and Physics, Charles University, Prague; since 2004.

**Visiting Academic or Research Appointments:** One semester positions in the USA: Vanderbilt University, Nashville, 1978; University of Colorado, Boulder, 1984; University of Illinois at Chicago, 1986; Emory University, Atlanta, 1991; Institute for Advanced Study, Princeton, 2000 Humboldt Fellowship at Dortmund University, Dortmund, Germany, 1992-93 and 1995; Short stays include: S. Banach Center, Warsaw, Poland; Mathematical Science Research Institute, Berkeley, California, USA; University of Bern, Switzerland; University of California, San Diego, USA; DIMACS - Rutgers University, USA; CNR, Pisa, Italy; IHES, Bures-sur-Ivette, France; Institute of Mathematics, Luminy, France; Max Planck Institute of Mathematics, Bonn, Germany.

**Research Profile:** Pavel Pudlák has worked mainly in Proof Complexity, which is an interdisciplinary field between logic and complexity theory. He also obtained results in algebra, combinatorics and computational complexity

**Main Publications:**

1. (with J. Tůma) Every finite lattice can be embedded in a finite partition lattice, Algebra Universalis, Vol.10, 1980, pp.74-95.
2. Cuts, consistency statements and interpretations, Journ. Symb. Logic Vol.50, 1985, pp.423-441.
3. On the length of proofs of finitistic consistency statements in first order theories. In: Logic Colloquium 84, North Holland P.C., 1986 pp.165-196.
4. (with J.Krajíček) The number of proof lines and the size of proofs in first order logic, Arch. Math. Log. Vol.27, 1988 pp.69-84.
5. (J. Krajíček) Propositional proof systems, the consistency of first order theories and the complexity of computations, JSL Vol.54, No.3, 1989, pp.1063-1079.
6. (with J. Krajíček and G. Takeuti) Bounded arithmetic and polynomial hierarchy, Annals of Pure and Applied Logic 52,(1991), pp.143-154.
7. (with A. Hajnal, W. Maass, M. Szegedy and G. Turan) Threshold circuits of bounded depth. Journ. of Comput. and System Science 46 (1993), pp.129-154.

8. (with P. Hájek) Metamathematics of first order arithmetic, Springer-Verlag/ASL Perspectives in Logic, 1993, 460 pp.

9. Logic and Complexity: Independence results and the complexity of propositional calculus, in Proc. Internat. Congress of Math., Zurich, 1994, pp.288-297.

10. (with J. Krajíček and A. Woods) Exponential lower bound to the size of bounded depth Frege proofs of the Pigeon Hole Principle, Random Structures and Algorithms, 7/1 (1995), pp.15-39.

11. (P. Beam, R. Impagliazzo, J. Krajíček and T. Pitassi) Lower bounds on Hilbert's Nullstellensatz and propositional proofs, Proc. London Math. Soc. (3) 73 (1996), pp.1-26.

12. Lower bounds for resolution and cutting planes proofs and monotone computations, J. of Symb. Logic 62(3), 1997, pp.981-998.

13. The lengths of proofs, Chapter VIII, Handbook of Proof Theory, S.R. Buss ed., Elsevier, 1998, pp.547-637.

14. (with N. Alon) Constructive lower bounds for off-diagonal Ramsey numbers, Israel Journ. of Math. 122 (2001), pp.243-251.

15. (with N. Alon) Equilateral sets in $l_p$, Geometric and Functional Analysis 13 (2003), pp. 467-482.

16. (with R. Paturi, M.E.Saks, and F. Zane) An improved exponential-time algorithm for k-SAT, Journal of the ACM Vol.52, No.3, 2005, pp.337-364

A full list is available at +www.math.cas.cz/~pudlak+

*Work in Progress*

17. a book on the logical foundations of mathematics and computational complexity.

**Service to the Profession:** Editor of the journals: Archive for Mathematical Logic, 1989-94; Information and Computation, 1989-96; Calcolo, since 1998; Computational Complexity, since 2003; Mathematical Logic Quarterly, since 2006. Editor of the ASL book series Perspectives in Logic, since 2002.

**Teaching:** PhD students: Jan Krajíček, Jiří Sgall, Marta Bílkova, Pavel Hrubeš.

**Vision Statement:** "Our inability to solve the deep problems in complexity theory, such as P vs NP, may simply be caused by their deep combinatorial complexity, but it may also be due to some fundamental obstacles of logical nature. I believe that the latter is quite likely and, therefore, we should study the connections of these problems to problems about provability and the lengths of proofs. This research may eventually help us to create better foundations of mathematics."

**Honours and Awards:**

The Prize of the Czechoslovak Academy of Sciences, 1979, (with J. Tůma).

The Bernard Bolzano Medal of the Academy of Sciences of the Czech Republic, 2001.

# PUTNAM, Hilary

**Specialties:** Recursion theory, Diophantine equations, arithmetic models for satisfiable formulas, recursive learning theory (trial and error predicates), hierarchy theory and the fine structure of L.

**Born:** 31 July 1926 in Chicago, Illinois, USA.

**Educated:** University of California at Los Angeles, PhD Philosophy, 1951; University of Pennsylvania, BA Philosophy, 1948.

**Dissertation:** *The Concept of Probability in Application to Finite Sequences*; supervisor, Hans Reichenbach.

**Regular Academic or Research Appointments:** COGAN UNIVERSITY PROFESSOR EMERITUS, HARVARD UNIVERSITY, 2000–; Cogan University Professor, Harvard University, 1995-2000; Walter Beverly Pearson Professor of Modern Mathematics and Mathematical Logic, Harvard University, 1976-1995; Professor of Philosophy, Harvard University, 1965-1976; Professor of the Philosophy of Science, Massachusetts Institute of Technology, 1961-1965; Associate Professor of Philosophy and Mathematics, Princeton University, 1960-1961; Assistant Professor of Philosophy, Princeton University, 1953-1960; Instructor in Philosophy, Northwestern University, 1952-1953.

**Visiting Academic or Research Appointments:** Visiting Professor of Philosophy, Tel Aviv University, 2006; Visiting Professor of Philosophy, Tel Aviv University, 2005; Visiting Professor of Philosophy, Tel Aviv University, 2004; Lecturer of the Year, Indian Council for Philosophical Research, 2002; Katz Visiting Professor of Philosophy University of Washington, 2002; Baruch Spinoza Professor of Philosophy, University of Amsterdam, 2001; Fellow of the Wissenschaftskolleg zu Berlin, 1994; Gifford Lecturer at the University of St. Andrews, 1990; National Endowment for the Humanities Fellow, 1982-1983; National Endowment the Humanities Fellow, 1976-76; National Science Foundation Fellow, 1968-1969, Harvard University; Guggenheim Fellow, University of Oxford and University of Paris, 1960-1961; Fellow of the Minnesota Center for the Philosophy of Science, University of Minnesota, 1958.

**Research Profile:** Hilary Putnam has made fundamental contributions to the study of Diophantine sets, and to the study of the degrees of unsolvability of sets in the hyperarithmetic hierarchy and in the hierarchy of sets of integers that are in

Gödel's model L (the "constructible sets of integers"), as well as to recursive learning theory. In the theory of Diophantine sets, he is known as being one of the group of four mathematicians (Martin Davis, Yuri Matiyasevich, Hilary Putnam, and Julia Robinson) whose work provided the "negative solution" to Hilbert's $10^{th}$ Problem, that is whose work showed that the problem was recursively unsolvable. Together with several of his students, he extended Spector's classical work on the hypererathmetic hierarchy far beyond that hierarchy, and, in a joint paper with George Boolos and subsequent papers published the first work on what came to be called "the fine structure of L". His work on "Trial and Error Predicates" was one of the first two papers on recursive learning theory (the other, which appeared in the same issue of the *J. Symbolic Logic*, was by Mark Gold.). His paper with Martin Davis on feasible computational methods in propositional calculus led to the Davis-Putnam ("DP") algorithm, which is still in wide use. Putnam has been a leading figure in the philosophy of logic and mathematics, as well as in general philosophy.

**Main Publications:**

1. Arithmetic models for consistent formulae of quantification theory, *J. Symbolic Logic* 22.1 (1957), 110-111.
2. Decidability and essential undecidability, *J. Symbolic Logic* 22.1 (1957), 39-54.
3. (with G. Kreisel), Eine Unableitbarkeitsbeweismethode für den Intuitionistischen Aussagenkalkül, *Archiv für Mathematische Logik und Grundlagenforschung* 3.1-2 (1957), 74-78.
4. An unsolvable problem in number theory, *J. Symbolic Logic* 25.3 (September 1960), 220-232.
5. (With Martin Davis) A computing procedure for quantification theory, *J. Association for Computing Machinery* 7.3 (1960), 201-215.
6. (With Martin Davis and Julia Robinson) The decision problem for exponential diophantine equations, *Annals of Mathematics* 74.3 (1961), 425-436.
7. Uniqueness ordinals in higher constructive number classes, in *Essays on the Foundations of Mathematics dedicated to A. A. Fraenkel on his Seventieth Anniversary* (Yoshua Bar-Hillel ed.), Magnes Press of The Hebrew University (1961), 190-206.
8. On hierarchies and systems of notations, *Proceedings of the American Mathematical Society* 15.1 (1964), 44-50.
9. (With David Luckham) On minimal and almost-minimal systems of notations, *Transactions of the American Mathematical Society* 119.1 (1965), 86-100.
10. (With Gustav Hensel) On the notational independence of various hierarchies of degrees of unsolvability, *J. Symbolic Logic* 30.1 (1965): 69-86.
11. Trial and error predicates and the solution to a problem of Mostowski, *J. Symbolic Logic* 30.1 (1965), 49-57.
12. (With George Boolos) Degrees of unsolvability of constructible sets of integers, *J. Symbolic Logic* 33.4 (1968), 497-513.
13. (With Gustav Hensel) Normal models and the field sigma-star, *Fundamenta Mathematicae* 64 (1969), 231-240.
14. (With Gustav Hensel and Richard Boyd) A recursion-theoretic characterization of the ramified analytical hierarchy, *Transactions of the American Mathematical Society* 141 (1969), 37-62.
15. (With Stephen Leeds) An intrinsic characterization of the hierarchy of constructible sets of integers, *Logic Colloquium '69* (R. O. Grandy and C. E. M. Yates, eds.), North-Holland (1971), 311-350.
16. Recursive functions and hierarchies. *American Mathematical Monthly, Supplement: Papers in the Foundations of Mathematics* 80.6, part 2 (1973), 68-86.
17. (With Joan Lukas) Systems of Notations and the Ramified Analytical Hierarchy, *J. Symbolic Logic* 39.2 (1974), 243-253.
18. Models and reality, *J. Symbolic Logic* 45.3 (1980), 464-482
19. Paradox revisited I: truth, II: A case of all or none? in *Between Logic and Intuition, Essays in Honor of Charles Parsons*, (Gila Sher and Richard Tieszen eds.), Cambridge University Press (2000), 3-26.
20. Nonstandard models and Kripke's proof of the Gödel theorem, *Notre Dame Journal of Formal Logic*, vol. 41.1 (2000), 53-58.

A full list of publications is available at http://www.pragmatism.org/putnam/

*Work in Progress*

21. Essays arguing that finitism, nominalism, and the various philosophies that would restrict mathematics to "predicative" means are not compatible with the sort of scientific realism that Putnam favors.

**Service to the Profession:** Past President of the Association for Symbolic Logic as well as of the Philosophy of Science Association and the American Philosophical Association (Eastern Division).

**Teaching:** Putnam has had fifteen PhD students who wrote their dissertations in mathematical logic (plus many more in pure philosophy) in the departments of mathematics and philosophy at Princeton University, the Massachusetts Institute of Technology and Harvard University, including George Boolos, Herbert Enderton, C. Ward Henson, Harold Hodes, David Isles, Raymond Smullyan and Lesley Tharp. He also influenced the Ph.D. work of Harvey Friedman.

**Vision Statement:** Although the dream that mathematical logic would either provide solutions to philosophical problems or show that they are "nonsensical", a dream that motivated, at least partly, Frege, Russell, Carnap and Quine, was certainly an unrealistic one, there is no doubt that the tools it provided enabled us both to state the issues far more clearly than was possible before these thinkers worked, and to raise new and fascinating problems. The work now being done at the frontiers of set theory will almost certainly transform our understanding of the subject in this century. Philosophically, there is much more to be done in understanding what Eugene Wigner famously called "the unreasonable effectiveness" of mathematics in physical science.

**Honours and Awards:** Doctorates of Philosophy from Cayetano Heredia University in Peru, 1980; University of Pennsylvania, 1985; Muhlenberg College, 1995; Kalamazoo College, 1995; University of Athens, 1998; Hebrew University in Jerusalem, 2004; University of Chicago, 2004; University of St. Andrew's, 2004. Corresponding Fellow, Institut de France, 2001; Fellow, American Philosophical Society, 1999; Fellow, American Academy of Arts and Letters, 1966; Corresponding Fellow, British Academy, 1978;.

# Q

## QI, Feng

**Specialties:** Theory of Mathematical Inequalities and Applications, Theory of Special Functions, Theory of Mean Values, Classical Analysis.

**Born:** 22 July 1965 in Xinxiang County, Xinxiang City, Henan Province, China.

**Educated:** Henan University, BS Mathematics, 1982; Xiamen University, MS Mathematics, 1989; University of Science and Technology of China, DS Mathematics, 1999.

**Dissertation:** *Studies on Problems in Topology and Geometry and on Weighted Abstracted Means* (Chinese); supervisor, Sen-Lin Xu.

**Regular Academic or Research Appointments:** SPECIALLY APPOINTED PROFESSOR AT COLLEGES AND UNIVERSITIES OF HENAN PROVINCE, EDUCATIONAL DEPARTMENT OF HENAN PROVINCE, CHINA, 2005–2009; Head, Department of Applied Mathematics and Informatics, Henan Polytechnic University, Henan Province, China, 2002-2005; Honorary Professor of Mathematics, Henan Normal University, 2006-2008; Head, Department of Fundamental Courses, Jiaozuo Institute of Technology, Henan Province, China, 1999-2002; Professor of Mathematics, Henan Polytechnic University, 1999-; Associate Professor of Mathematics, Jiaozuo Institute of Technology, 1995-1999; Assistant Professor of Mathematics, Jiaozuo Mining Institute, 1991-1995; Instructor of Mathematics, Jiaozuo Mining Institute, 1989–1991.

**Visiting Academic or Research Appointments:** Visiting Professor of Mathematics, Victoria University at Melbourne in Australia, 2006-2007; Visiting Professor of Mathematics, The University of Hong Kong, 2004; Visiting Professor of Mathematics, The University of Hong Kong, 2004; Visiting Professor of Mathematics, Victoria University at Melbourne in Australia, 2001.

**Research Profile:** From 1993, Professor Qi has published more than 150 papers in journals worldwide. These articles involve several research areas of mathematical sciences. From 1992, Dr. Feng Qi completed 9 research projects, including one for the National Natural Foundation of China. He wrote down explicitly the definition of the notion or terminology "logarithmically completely monotonic function" and proved that a logarithmically completely monotonic function is also completely monotonic on the positive half axis, but not conversely. It has also been showed that a Stieltjes transformation is also a logarithmically completely monotonic function. The logarithmically complete monotonic functions on the positive half axis can be characterized as the infinitely divisible completely monotonic functions. Some open problems posed by Dr. Qi stimulated the developments of Mathieu's series and inequalities. An integral inequality established by him has been generalized by more than 15 papers and applied to probability theory. The concept of the "generalized weighted mean values" established by Professor Qi has been applied to physics and information theory by some physicists. He generalized the well known Plya-Szegö's integral inequality posed in 1925 to cases of the higher derivatives, the mutiple integrals, and the weighted integrals. He proved the logarithmic convexity and Schur-convexity of the extended mean values.

**Main Publications:**

1. Feng Qi, *Logarithmic convexity of extended mean values*, Proceedings of the American Mathematical Society 130 (2002), no. 6, 1787–1796.

2. Feng Qi, *Generalized weighted mean values with two parameters*, Proceedings of the Royal Society of London Series A—Mathematical, Physical and Engineering Sciences 454 (1998), no.1978, 2723–2732.

3. Feng Qi, *Inequalities for a weighted multiple integral*, Journal of Mathematical Analysis and Applications 253 (2001), no. 2, 381–388.

4. Feng Qi, *Generalization of H. Alzer's inequality*, Journal of Mathematical Analysis and Applications 240 (1999), no. 1, 294–297.

5. Feng Qi, *A note on Schur-convexity of extended mean values*, Rocky Mountain Journal of Mathematics 35 (2005), no. 5, 1787–1793.

6. Feng Qi, *Inequalities for a multiple integral*, Acta Mathematica Hungarica 84 (1999), no. 1-2, 19–26.

7. Feng Qi, *Monotonicity results and inequalities for the gamma and incomplete gamma functions*, Mathematical Inequalities and Applications 5 (2002), no. 1, 61–67.

8. Feng Qi, Zong-Li Wei, and Qiao Yang, *Generalizations and refinements of Hermite-Hadamard's inequality*, Rocky Mountain Journal of Mathematics 35 (2005), no. 1, 235–251.

9. Feng Qi, Run-Qin Cui, Chao-Ping Chen, and Bai-Ni Guo, *Some completely monotonic functions involving polygamma functions and an application*, Journal of Mathematical Analysis and Applications 310 (2005), no. 1, 303–308.

10. Feng Qi, Pietro Cerone, and Sever S. Dragomir, *Some new Iyengar type inequalities*, Rocky Mountain

Journal of Mathematics 35 (2005), no. 3, 997–1015.

11. Feng Qi, József Sándor, Sever S. Dragomir, and Anthony Sofo, *Notes on the Schur-convexity of the extended mean values*, Taiwanese Journal of Mathematics 9 (2005), no. 3, 411–420.

12. Feng Qi and Chao-Ping Chen, *A complete monotonicity property of the gamma function*, Journal of Mathematical Analysis and Applications 296 (2004), no. 2, 603–607.

13. Feng Qi and Bai-Ni Guo, *An inequality between ratio of the extended logarithmic means and ratio of the exponential means*, Taiwanese Journal of Mathematics 7 (2003), no. 2, 229–237.

14. Feng Qi and Bai-Ni Guo, *On Steffensen pairs*, Journal of Mathematical Analysis and Applications 271 (2002), no. 2, 534–541.

15. Feng Qi, Li-Hong Cui, and Sen-Lin Xu, *Some inequalities constructed by Tchebysheff's integral inequality*, Mathematical Inequalities and Applications 2 (1999), no. 4, 517–528.

16. Feng Qi and Shi-Qin Zhang, *Note on monotonicity of generalized weighted mean values*, Proceedings of the Royal Society of London Series A—Mathematical, Physical and Engineering Sciences 455 (1999), no. 1989, 3259–3260.

17. Feng Qi and Sen-Lin Guo, *Inequalities for the incomplete gamma and related functions*, Mathematical Inequalities and Applications 2 (1999), no. 1, 47–53.

18. Feng Qi and Sen-Lin Xu, *The function $(b^x - a^x)/x$: Inequalities and properties*, Proceedings of the American Mathematical Society 126 (1998), no.11, 3355–3359.

19. Feng Qi and Qiu-Ming Luo, *A simple proof of monotonicity for extended mean values*, Journal of Mathematical Analysis and Applications 224 (1998), no.2, 356–359.

20. Feng Qi, *Several integral inequalities*, Journal of Inequalities in Pure and Applied Mathematics 1 (2000), no. 2, Article 19. Available online at http://jipam.vu.edu.au/article.php?sid=113

A full list of publications is available at Qi's home page http://rgmia.vu.edu.au/qi.html.
*Work in Progress*

21. A book on Plya-Szegö-Iyengar-Mahajani's integral inequality.

22. Construction and proofs of logarithmically completely monotonic functions and completely monotonic functions involving the gamma, psi, and polygamma functions.

23. Construction and proofs of inequalities involving some special functions including the gamma, psi, and polygamma functions and the Bernoulli, Euler numbers and polynomials.

**Service to the Profession:** *Service to the Profession:* Member, Editorial Board, *Journal of Mathematical Analysis and Approximation Theory*, 2005-; Member, Editorial Board, *Communications in Studies on Inequalities*, 2005-; Foundation Associate Editor, *Global Journal of Mathematics and Mathematical Sciences*, 2004-; Foundation Associate Editor, *International Journal of Pure and Applied Mathematical Sciences*, 2004-; Member, Editorial Board, *International Journal of Applied Mathematical Sciences*, 2004-; Foundation Editor, *Australian Journal of Mathematical Analysis and Applications*, 2004; Member, International Scientific Committee, First International Conference on Mathematical Inequalities and their Applications, held by the RGMIA, Victoria University, Melbourne, Australia, December 06-08, 2004; Member, Editorial Board, *Advanced Studies in Contemporary Mathematics*, 2003-; Reviewer, *Zentralblatt MATH*, European Mathematical Society, 2001-; Managing Editor, *Mathematics and Informatics Quarterly*, 2001-; Member, Scientific Program Committee, INEQUALITIES 2001 at University of the West, Timisoara, Romania, July 9-14, 2001; Member, Editorial Board, *International Journal of Mathematics and Mathematical Sciences*, 2000-; Foundation Editor, *Journal of Inequalities in Pure and Applied Mathematics*, 2000-; Member, Standing Council, Henan Mathematical Society, China, 2000-; Member, Editorial Board, *Journal of Henan Polytechnic University*, China, 1999-; Reviewer, *Mathematical Reviews*, American Mathematical Society, 1999-; Member, American Mathematical Society, 1999-; Member, Research Group in Mathematical Inequalities and Applications, 1998-.

**Teaching:** F. Qi has been a mainstay of the mathematical program at Henan Polytechnic University since his appointment to the university in 1989. As a subject leader of mathematics and as acting department head, Professor Qi masterminded and presided to acquire firstly the rights of educating postgraduates for Master Degree of Science in the Specialty of Pure Mathematics and the Specialty of Applied Mathematics at Henan Polytechnic University in 2003 and 2005 respectively. He has had two Master students, Jian Cao and Da-Wei Niu, in the fields of Mathematics. He has also influenced the work of mathematical education and research in the history of Henan Polytechnic University.

**Vision Statement:** No one can think, act, or speak without logic at any time. Logic exists in man's nature. Mathematics is a logical subject. Mathematics is the best systematic gymnastics of man's brain. Almost everyone should learn mathematics.

**Honours and Awards:** Award of Science and Technology for Youth of China, 2006; Award of Science and Technology for Youth of Henan Province, 2005; Prize for Production of Science and Technology of the Educational Department of Henan Province, 2005; Specially Appointed Professor at Colleges and Universities of Henan Province, Educational Department of Henan Province, 2005; Excellent Expert of Henan Province, the Committee of Henan Province of CCP and the Government of Henan Province, 2002; Award of Science and Technology for the Excellent Youth, Sun Yue-Qi's Fund of Science-Technology-Education, Developing Foundation of Science and Technology of China, 2000; Fostering Object of Academic and Technologic Leader Striding Century, 1999; Excellent Young Teacher of Henan Province, 1999; Excellent Young-Middle-Aged and Primary Teacher of Henan Province, 1994.

# R

## RABIN, Michael Oser

**Specialties:** Miller–Rabin primaltiy test, Rabin cryptosystem, oblivious transfer, Rabin–Karp string search algorithm, Nondeterminsitic finite automata.

**Born:** 1931 in Breslau, Germany (today in Poland).

**Educated:** MSc Hebrew University of Jerusalem, 1953, PhD Princeton University 1956.

**Regular Academic or Research Appointments:** THOMAS J. WATSON SR. PROFESSOR OF COMPUTER SCIENCE, HARVARD UNIVERSITY 1983–; Gordon McKay Professor of Computer Science, Harvard University, 1981–1983; Albert Einstein Chair, Hebrew University Israel, 1980–1999; Pro Rector, Hebrew University, Israel, 1976–1980; Rector, Hebrew University, Israel, 1972–75; Chairman, Computer Science, Hebrew University, Israel, 1970–71; Chairman, Institute of Mathematics, Hebrew University, 1964–66, Senior Lecturer, Associate Professor and Professor (1965), Hebrew University of Jerusalem, Israel, 1958–1964.

**Visiting Academic or Research Appointments:** Since 1961 Michael Rabin has held visiting positions at universities in the USA as well as France and England. His most recent appointments have been: Natch-Diplom Lectures in Mathematics, ETH, Zurich, 2000; Visiting Professor of Computer Science, Courant Institute, 2001, Visiting Professor of Computer Science, Columbia University, 2002; Visiting Professor of Computer Science, King's College London, 2004; Steward Fellow, Gonville and Caius College, Cambridge, 2004, Visiting Professor of Computer Science, Columbia University, 2007.

**Research Profile:** In 1959 jointly with Dana Scott, Rabin introduced finite automata and non-deterministic machines which have become a key concept in computational complexity theory, particularly with the description of complexity classes P and NP.

In 1969, Rabin proved that the second-order theory of in successors is decidable. A key component of the proof implicitly showed determinacy of parity games, which lie in the third level of the Borel hierarchy.

In 1975, Rabin also invented the Miller-Rabin primality test, a randomized algorithm that can determine very quickly (but with a tiny probability of error) whether a number is prime. Fast primality testing is key in the successful implementation of most public-key cryptography.

In 1979, Rabin invented the Rabin cryptosystem, the first asymmetric cryptosystem whose security was proved equivalent to the intractability of integer factorization.

In 1981, Rabin invented the technique of oblivious transfer, allowing a sender to transmit a message to a receiver where the receiver has some probability between 0 and 1 of learning the message, with the sender being unaware whether the receiver was able to do so.

In 1987, Rabin, together with Richard Karp, created one of the most well-known efficient string search algorithms, the Rabin-Karp string search algorithm, known for its rolling hash.

Rabin's more recent research has concentrated on computer security. He is currently the Thomas J. Watson Sr. Professor of Computer Science at Harvard University and Professor of Computer Science at Hebrew University. During the spring semester of 2007, he was a visiting professor at Columbia University teaching Introduction to Cryptography.

**Main Publications:**

1. Rabin, MO; Scott, D (April 1959). "Finite Automata and Their Decision Problems". *IBM Journal of Research and Development* 3 (2): 114–125.
2. Rabin, MO (1969). "Decidability of second order theories and automata on infinite trees". *Trans. AMS* 141: 1–35.
3. Rabin, MO (1976). "Probabilistic algorithms". *Algorithms and Complexity, Proc. Symp.*
4. Rabin, MO (1980). "Probabilistic algorithm for testing primality". *Journal of Number Theory* 12 (1): 128–138.
5. Rabin, MO (January 1979). "Digital signatures and public-key functions as intractable as factorization". MIT Laboratory of Computer Science Technical Report.
6. Rabin, Michael O. (1981), How to exchange secrets by oblivious transfer (Technical Report TR-81), Aiken Computation Laboratory: Harvard University,
7. Karp, RM; Rabin, MO (March 1987). "Efficient randomized pattern-matching algorithms". *IBM Journal of Research and Development* 31 (2): 249–260.

**Service to the Profession:** Member of editorial boards of *Journal of Computer and Systems Sciences*, *Journal of Combinatorial Theory*, *Journal of Algorithms*.

**Honours and Awards:** Academy memberships include:

- American Academy of Arts and Sciences (1975–);
- Israel Academy of Sciences and Humanities (1982–)
- Foreign Associate US National Academy of sciences (1984–)
- Foreign Member American Philosophical Society (1988–)
- Associé Étranger, French Academy of Sciences (1995–)
- Foreign Member Royal Society (2cd 007);
- Elected Member European Academy of Science (2007)

In 1976 the Turing Award was awarded jointly to Rabin and Dana scott. The award was granted

> For their joint paper "Finite Automata and Their Decision Problem," which introduced the idea of non-deterministic machines, which has proved to be an enormously valuable concept. Their (Scott and Rabin) classic paper has been a continuous source of inspiration for subsequent work in this field.

- The C. Weizmann Prize for Exact Sciences, 1960
- Rothschild Prize in Mathematics, 1974
- Harvey Prize in Science and Technology, 1980
- The Israel Prize in Exact Sciences/Computer Science, 1995
- IEEE Charles Babbage Award in Computer Science, 2000
- ASL Gödel Award Lecture, 2004
- ACM Kanellakis Theory and Practice Award, 2004
- The EMET Prize in Exact Sciences/Computer Science, 2004
- Best Teacher Award, Courant Institute of Mathematics, 1970

# RAHMAN, Shahid

**Specialties:** Dialogical logic, philosophy of logic, epistemology.

**Born:** 20 October 1956 in New Delhi, India, Argentinean and German citizenship.

**Educated:** Universidad Nacional del Sur, (Bahia Blanca, Argentina), Erlangen-Nürnberg Universität, Universität des Saarlandes (Germany). Studies in Mathematics, Psychology, Philosophy and Philology. PhD in Philosophy, Psychology and Philology, Universität des Saarlandes;

**Dissertation:** 1993 on Category theory and dialogical logic, supervisor: Kuno Lorenz. Habilitation 1997 on Hugh MacColl: Fictions and Connexive logic, Universität des Saarlandes.

**Regular Academic or Research Appointments:** FULL PROFESSOR OF LOGIC AND EPISTEMOLOGY, UNIVERSITY OF LILLE3 (HUMAN SCIENCES), FRANCE SINCE 2001

**Visiting Academic or Research Appointments:** Selection of recent visits as invited professor and researcher: Argentina (Bahia Blanca, Buenos Aires), Brasil (Campinas, Campos-Rio de Janeiro), Germany ( Erlangen-Nürnberg, Saarbruecken, MaxPlanckInstitut of Computersciences, Portugal (Lisboa), United States (Princeton)

**Research Profile:** Shahid Rahman extended dialogical logic to a general framework for developing and combining logics. Indeed; dialogical logic was in the view of their creators Paul Lorenzen and Kuno Lorenz thought as a "foundation" of intuitionistic logic. Rahman and his group extended the notion of dialogical logic to a general framework where they produced among others: free logic; standard modal logic; hybrid modal logic; non-normal logic, temporal logic; IF-logic, first order modal logic; paraconsistent logic, linear logic, relevant logic and connexive logic. One of the points of the approach is that dialogical logic could be used as bridge between a model theoretical and a proof theoretical approach and even as a method of generating tableaux-systems for logics with no clear model theoretical semantics such as linear logic and connexive logic.

**Main Publications:**

1. *Ueber Dialogue, Protologische Kategorient und anderen Seltenheiten*, P. Lang 1993
2. *New Perspectives in Dialogical Logic*. S S RAHMAN/ H. RÜCKERT, SYNTHESE 2001.
3. **Logic Epistemology and the Unity of Science**. S. RAHMAN ;DOV GABBAY ; JOHN SYMONS, J. P. VAN BENDEGEM. Kluwer 2004

4. *Truth, Unity and the Liar*. S. RAHMAN, T. TULENHEIMO, E. GENOT. In print: Springer 2008.

5. *Hugh MacColl an Overview of his Logical Work*, with Juan Redmond; College Pulbications. In print 2008..

6. *The Unity of Science in the Arabic Tradition : Science Logic Epistemology and their interactions* In print Springer, 2008

7. 2000: Hugh MacColl's criticism of Boole's formalization of traditional hypotheticals. In J. Gasser (ed.): *A Boole Anthology. Recent and Classical Studies in the Logic of George Boole*. Dordrecht / Boston / London: Kluwer, Synthese Library, 287-310

8. 2001: On Frege's Nightmare: Ways to combine paraconsistent and intuitionistic free logic. Dans H. Wansing (Hg.): *Essays on Non-Classical Logic*. London: World Scientific; 61-85.

9. 2002 The Dynamics of Adaptive Paraconsistency. Avec J. P. van Bendegem. InW. Carniell, M, Coniglio; I. M. Loffredo D'Ottaviano: *Paraconsistency*, New York: Marcel Dekker, pp. 295-321

10. 2004: Un desafío para la teorías cognitivas de la competencia lógica: Los fundamentos pragmáticos de la semántica de la lógica linear. In M. Beaumont Wrigley, *Festschrift for Marcelo Dascal*, Campinas: Manuscrito, 381-433..

11. 2005: On how to be a dialogician (with Laurent Keiff). In D. Vanderveken, *Logic, thought and Action* chez Kluwer-Springer, 2005, 359-408.

12. 2005: The Dialogic of just being diffferent. Hintikka's new approach to the notion of episteme and its impact on "second generation" dialogics. Dans D. Kolak and John Symons, *Quantifiers, Questions and Quantum Physics. Essays in Honour of Jaakko Hintikka*. Springer 2005, pp 57-187.

13. 2006: A non normal logic for a wonderful world and more. In J. van Benthem et alia *The Age of Alternative Logics*; chez Dordrecht: Kluwer-Springer, 311-334.

14. 2006 : *The Beetle in the Box: Exploring IF-Dialogues* (with Cedric Degremont). Dans : A. T. Pietarinen, A.Veikho, "Truth and Games. Essays in Honour of Gabriel Sandu", Helsinki: *Societas Philosophica Fennica*, 2006, ISBN 951-9264-57-4., 91-122.

15. 2006 : From Games to Dialogues and Back: Towards a General Frame for Validity (with Tero Tulenheimo). Dans Ondrej Majer, Ahti-Veikko Pietarinen and Tero Tulenheimo, *Games: Unifying Logic, Language and Philosophy: A Foundational Perspective* In print in Kluwer-Springer, 2008.

16. 2007 (with Juan Redmond): Hugh MacColl and the Birth of Logical Pluralism. In : J. Woods§D. Gabbay *Handbook of History of Logi*, vol. 4, Elsevier, in print 2008.

*Work in Progress*

17. A book on Dialogical and Game theoretical approach to first order Modal Logic. With Tero Tulenheimo (Helsinki)

18. A book on Poincaré's epistemology with Hassan Tahiri.

19. A book on Logic and Law with Patrice Canivez, Dov Gabbay and Alexandre Thiercelin

20. A book on the Realism anti-Realism debate with Laurent Keiff and Mathieu Marion.

**Service to the Profession:** Editor of the following book series

1. (With John Symons (Texas-El-Paso)

   *Logic, Epistemology and the Unity of Science, Springer.*

2. *(With Dov Gabbay (King's College)*
   Cahiers de Logique et d'Epistemologie, *College Publications.*

3. *(With Juan Redmond (Lille3)*
   Cuardernos de Logica, Epistemologia y Lenguaje, College Publications.

Member of Editorial Boards:
*Synthese* Kluwer-Springer
*Philosophia Scientiae*, Archives-Centre d'Etudes et de Recherche Henri-Poincaré.
*Nous*, publiée par Universidad Nacional de Buenos Aires

**Teaching:** Member and director of different research and teaching projects such as "Logical Modelling", IHPST (Paris 1), "La science et ses contextes" ( MSH-Nord Pas de Calais), "Logique et argumentation l'UMR 8163

Doctorate students:

Nicolas Clerbout, Une question de portée. La logique modale du subjonctif, la logique modale IF et le calcul Lambda comparés face au problème de la portée

Mathieu Fontaine: Rhétorique et engagement ontologique de l'objet de l'acte intensionnel Emmanuel Genot, Raisons, causes et explications. L'explication causale et rationalisante dans les contextes juridiques et historiques : un modèle logique

Marie-Hélène Gorisse, L'art du point de vue : le pluralisme épistémologique des Jainas . Laurent Keiff, Le pluralisme dialogique. Vers une dynamique. Eléments pour une étude des interactions entre la sémantique dialogique et certains contextes de l'activité rationnelle. Gildas Nzoukou, Sémantique et formalisme chez Russel, confrontés à la textualité des langues Bantu. Limites et ambiguïtés de la quantification dans la logique standard du 1er ordre. Juan Redmond, La dynamique de la fiction. Hassan Tahiri, La dynamique de la négation et la logique avec inconsistances : quelques conséquences scientifiques et épistémologiques Alexandre Thiercelin, Leibniz logicien et juriste. Réalité d'une logique juridique à partir de la doctrine juridique des conditions.

**Vision Statement:** Logic had a central role in the philosophy of science projects of the $20^{th}$ century. The historical and sociological criticisms of the 60ties banned logic from philosophy of science pointing out that science is more than a set of propositions linked by classical logic.. This does not mean that the links between logic and philosophy and history of science must be abandoned. Rather that the reflection on sciences must be approached with more and new sophisticated tools of logic able to deal with propositions in a structure the description of this structure and the dynamics of this structure.

The conjecture is that this will be possible by the confluence of; game theoretical approaches to logic, belief revision and game, decision and argumentation theories

**Honours and Awards:** Festschrift to appear 2008 withcontributions among others of J. van Benthem (Amsterdam), G. Bonannno (California), J. Dubucs (Paris 1); Dov Gabbay (K. Collegue), G. Priest (St Andrews), K. Lorenz (Saarbruecken), G. S. Read (St Andrews), G. Restall (Melbourne); A. Herzig (Toulouse), N. da Costa (S. Paulo) ; W. Carnielli (Campinas), M. Marion (Montreal), G. Sandu (Paris1); G. Sundholm (Leyden), T. Tulenheimo (Helsinki), H. Wansing (Dresden), A Pietarinnen (Helsinki), T. Tulenheimo (Helsinki) ; I. Wolenski (Cracau),

For online papers, teachings and research projects see http://stl.recherche.univ-lille3.fr/sitespersonnels/rahman/accueilrahman.html

# RAZBOROV, Alexander

**Specialties:** Applications of logic in computer science, proof complexity and bounded arithmetic, algorithmic problems in combinatorial group theory.

**Born:** 16 February 1963 in Belovo, USSR.

**Educated:** Steklov Mathematical Institute, PhD Mathematics, 1987; Moscow State University, BS Mathematics, 1985.

**Dissertation:** PhD: *On systems of equations in free groups*; Doctoral: *Lower Bounds in the Boolean Complexity*, 1991; supervisor, S. I. Adian

**Regular Academic or Research Appointments:** PRINCIPAL RESEARCHER, STEKLOV MATHEMATICAL INSTITUTE; Leading Researcher, Steklov Mathematical Institute, 1991-2000; Researcher, Steklov Mathematical Institute, 1987-1991.

**Visiting Academic or Research Appointments:** Visiting Professor, Institute for Advanced Study, Princeton, 2003-; Member, Institute for Advanced Study, Princeton, 2000-2003;

Visiting Researcher, Department of Computer Science, Princeton University, 1999-2000; Member, Institute for Advanced Study, Princeton, 1993-1994.

**Research Profile:** Alexander Razborov's work in mathematical logic and related areas includes contributions to the decidability of algebraic problems, computational complexity and proof complexity. The first topic is represented by the research on equations in free groups mostly conducted during his student years. Some of the ideas contained in that early work turned out to be influential for the recent solution of the famous *Tarski problem*. Razborov's research in computational and proof complexities has been greatly influenced by an even more famous open problem, *P vs. NP* question. He has made fundamental contributions to this question by solving its analogue for a number of restricted models of computation (such as monotone circuits or bounded-depth circuits over finite fields). In parallel with continuing these attempts, Razborov is employing tools and concepts from the classical mathematical logic for studying formal independence of this and other major open problems in *Complexity Theory* with respect to prescribed sets of techniques. Along these lines, jointly with S. Rudich, he developed the concept of *Natural Proofs* that by now has become the standard paradigm in that area. Closely related is Razborov's more recent work in Bounded Arithmetic and Propositional Proof Complexity.

He has observed that, like Natural Proofs, these frameworks also formalize all known lower bounds in circuit complexity and proposed a systematical study of the provability of the big P vs. NP question from these theories. Razborov has suggested metamathematical arguments that are specific to these weak theories, the most promising of them being based on the concept of a *pseudorandom generator* borrowed from cryptography and computational complexity. While pursuing these goals, Razborov has proved many results in proof complexity that are of considerable independent interest. Among others, they include the first non-trivial lower bounds for the *Polynomial Calculus* and resolution lower bounds for the *weak pigeon-hole principle*.

**Main Publications:**

1. A. A. Razborov. On systems of equations in a free group. Izvestiya AN SSSR, ser. matem., 48(4):779–

832, 1984. English Translation in Math. USSR Izvestiya, 25(1):115-162, 1985.
2. A. A. Razborov. Lower bounds for the monotone complexity of some boolean functions. Doklady Academii Nauk SSSR, 281(4):798–801, 1985. English Translation in Soviet Math. Dokl., 31:354-357, 1985.
3. A. A. Razborov. Lower bounds of monotone complexity of the logical permanent function. Matematicheskie Zametki, 37(6):887–900, 1985. English Translation in Mathem. Notes of the Academy of Sci. of the USSR, 37:485-493, 1985.
4. A. A. Razborov. Lower bounds for monotone complexity of boolean functions. In Proceedings of the International Congress of Mathematicians, volume 2, pages 1478–1487, Berkeley, California, USA, 1986. In Russian. For the English translation see Amer. Math. Soc. Transl., 147(2):75-84, 1990.
5. A. Razborov. Lower bounds on the size of bounded-depth networks over a complete basis with logical addition. Mathematical Notes of the Academy of Sciences of the USSR, 41(4):598–607, 1987. English translation in 41:4, pages 333-338.
6. A. Razborov. An equivalence between second order bounded domain bounded arithmetic and first order bounded arithmetic. In P. Clote and J. Krajcek, editors, Arithmetic, Proof Theory and Computational Complexity, pages 247–277. Oxford University Press, 1992.
7. A. Razborov. Bounded Arithmetic and lower bounds in Boolean complexity. In P. Clote and J. Remmel, editors, Feasible Mathematics II. Progress in Computer Science and Applied Logic, vol. 13, pages 344–386. Birkhauser, 1995.
8. A. Razborov and S. Rudich. Natural proofs. Journal of Computer and System Sciences, 55(1):24–35, 1997.
9. A. Razborov. Unprovability of lower bounds on the circuit size in certain fragments of bounded arithmetic. Izvestiya of the RAN, 59(1):201–224, 1995.
10. A. Razborov. Lower bounds for the polynomial calculus. Computational Complexity, 7:291–324, 1998.
11. M. Alekhnovich, E. Ben-Sasson, A. Razborov, and A. Wigderson. Pseudorandom generators in propositional proof complexity. SIAM Journal on Computing, 34(1):67–88, 2004.
12. A. Razborov. Resolution lower bounds for perfect matching principles. Journal of Computer and System Sciences, 69(1):3–27, 2004.
13. A. Razborov. Quantum communication complexity of symmetric predicates. Izvestiya: Mathematics, 67(1):145–159, 2003.
14. A. Razborov. Guessing more secrets via list decoding. Internet Mathematics, 2(1):21–30, 2005.
*Work in Progress*
15. Continuation of work on lower bounds and independence results in proof complexity and bounded arithmetic (specifically, further investigation of pseudorandom generators in this context); extremal combinatorics.

**Service to the Profession:** Board Member, Banff International Research Station, 2005-; Jury Member, EACSL Ackermann Annual Award for Outstanding Dissertation, 2005-; Member, ASL Committee on Prizes and Awards, 2002-; Assessor, Division of Logic, Methodology and Philosophy of Science; Board Member, Commission on PhD and Doctoral Theses of Russian Federation; Executive Member, European Association for Computer Science Logic, 1997-. Member, Commission on Applied Mathematics of European Mathematical Union; Member, Steering Committee of the Computational Complexity Conference, 2000-2003; Editorial Board, *Izvestiya of the Russian Academy of Sci., ser. mathem*; Editorial Board, *Combinatorica*; Editorial Board, *Computational Complexity*; Editorial Board, *Theoretical Computer Science*; Editorial Board, *Combinatorics*; Editorial Board, *Probability and Computing*; Editorial Board, *Electronic Colloquium on Computational Complexity*.

**Teaching:** Undergraduate students: M. Alekhnovich, V. Podolski. Graduate students: A. Nogin, O. Verbitsky.

**Vision Statement:** I expect (and am working toward) ever increasing amount of interaction between mathematical logic and adjacent disciplines like Theoretical Computer Science and Discrete Mathematics.

**Honours and Awards:** Russian Academy of Science (corresponding member), 2000; Academia Europea, 1993; Rolf Nevanlinna Prize of the International Mathematical Union, 1990.

# READ, Stephen Louis

**Specialties:** Medieval Logic, Philosophical Logic, Relevance Logic, Inferentialism.

**Born:** 18 August 1947 in Nailsworth, Gloucestershire, England.

**Educated:** Oxford D.Phil., Bristol M.Sc., Keele B.A.

**Dissertation:** *A Philosophical Grammar*; supervisor Pieter Seuren.

**Regular Academic or Research Appointments:** READER IN HISTORY AND PHILOSOPHY OF LOGIC, UNIVERSITY OF ST ANDREWS, 2000–; Senior Lecturer in Logic and Metaphysics, University of St Andrews, 1989-2000. Lecturer in Logic and Metaphysics, University of St Andrews, 1972-1989.

**Research Profile:** Stephen Read's approach to logic has always been historically informed, polemical and philosophically questioning. Early work in medieval logic on the theory of supposition and its relation to quantification theory has lately evolved into research into identifying a notion of truth which can handle the semantic paradoxes. Dissatisfaction with apologia for the implicational paradoxes arising from the standard account of logical consequence inspired a longstanding interest in relevance logic. Coupled to a belief that logical pluralism and dialetheism are a nonsense, this inspired attempts to show that the correct account of truth-preservation underpins a univocal paraconsistent consequence relation. This generic consequence relation lays the foundation for the specific consequences embodied in the logical constants by an inferentialist, proof-theoretic account of their meaning, rejecting any semantic account which relies on a plurality of possible or impossible worlds.

**Main Publications:**

1. "Merely Confused Supposition", (with G.Priest) *Franciscan Studies*, 40 (1983), 265-97.
2. *Relevant Logic*, Blackwells 1988, viii + 199.
3. "Thomas of Cleves and Collective Supposition", *Vivarium*, 29 (1991), 50-84.
4. "The Slingshot Argument", *Logique et Analyse*, 143-4 (1993), 195-218.
5. "Formal and Material Consequence", *Journal of Philosophical Logic*, 23 (1994), 247-65.
6. *Thinking about Logic: an introduction to the philosophy of logic*, Oxford U.P.: Oxford Paperbacks University Series (OPUS), 1995, viii + 262.
7. "Reach's Puzzle and Quotation", *Acta Analytica*, 19 (1997), 9-16.
8. "Hugh MacColl and the Algebra of Strict Implication", *Nordic Journal of Philosophical Logic*, 3 (1998), 59-83.
9. "How is Material Supposition Possible?", *Medieval Philosophy and Theology*, 7 (1999), 1-20.
10. "Harmony and Autonomy in Classical Logic", *Journal of Philosophical Logic*, 29 (2000), 123-154.
11. "Truthmakers and the Disjunction Thesis", *Mind*, 109 (2000), 67-79.
12. *Concepts: the treatises of Thomas of Cleves and Paul of Gelria: an edition of the texts with a systematic introduction*, (with E.P. Bos) Edition Peeters (Louvain), 2001, xii + 147.
13. "Self-Reference and Validity Revisited", an expanded version of 'Self-Reference and Validity' (*Synthese*, 42 (1979), pp. 265-74), in *Medieval Formal Logic*, ed. Mikko Yrjönsuuri, Kluwer 2001, 183-96.
14. "The Liar Paradox from John Buridan back to Thomas Bradwardine", *Vivarium*, 40 (2002), 189-218.
15. "Intentionality—Meinongianism and the Medievals", (with Graham Priest) *Australasian Journal of Philosophy*, 82 (2004), 416-35.
16. "Logical Consequence as Truth-preservation", *Logique et Analyse*, 183-4 (2003), 479-93.
17. "The Unity of the Fact", *Philosophy*, 80 (2005), 317-42.

*Work in Progress*

18. "The Philosophy of Alternative Logics", (with A. Aberdein) forthcoming in *The Development of Modern Logic*, ed. L. Haaparanta, Oxford University Press.
19. "Logical Monism", forthcoming in *A Logical Approach to Philosophy: Essays in Honour of Graham Solomon*, ed. David DeVidi and Tim Kenyon, Western Ontario Series in Philosophy of Science, Springer.
20. "The Truth-Schema and the Liar", forthcoming in *Essays on the Liar Paradox*, ed. S. Rahman, Springer.

**Service to the Profession:** *The Philosophical Quarterly*: Editorial Chairman, 1999-2005.

**Teaching:** Teaching has been an important and significant part of Stephen Read's career, and has often formed the basis for research. It is essential to convey the basic concepts and ideas to students, as much as the techniques. It is also essential to show that logic should not be seen as a dogmatic subject but zetetic, challenging assumptions, creative and open-minded.

**Vision Statement:** In history of logic, there are still texts which are unedited (e.g., Maulfelt, Strode), untranslated (e.g., Ockham), and insufficiently studied (e.g., consequences, obligations). In philosophical logic, we still lack fully satisfactory accounts of truth, conditionals, consequence, counterfactuals, tense and paradoxes.

# RESCHER, Nicholas

**Specialties:** Symbolic logic, philosophical logic, history of logic.

**Born:** 15 July 1928 in Hagen, Germany.

**Educated:** Princeton University, PhD Philosophy, 1951; Queens College NY, BS (with honors in mathematics), 1949;

**Dissertation:** *The Cosmology of Leibniz*; supervisor, Ledger Wood (in Philosophy but studied logic in Mathematics with Alonzo Church).

**Regular Academic or Research Appointments:** UNIVERSITY PROFESSOR OF PHILOSOPHY, UNIVERSITY OF PITTSBURGH, 1970–; Chair, Department of Philosophy, University of Pittsburgh, 1980-1981; Professor of Philosophy, University of Pittsburgh, 1961-1970; Co-Chair, Center for Philosophy of Science, University of Pittsburgh, 1961-; Associate Professor of Philosophy, Lehigh University, 1952-1961; Research Mathematician, RAND Corporation, 1954-1957; Instructor, Princeton University, 1951-1952.

**Visiting Academic or Research Appointments:** Distinguished Visiting Professor, Ohio University,

2002-2003; Distinguished Visiting Professor, University of Konstanz, 1984; Visiting Lecturer, Oxford University, Summer Terms of 1968-1983; University of Rochester, Spring 1979; Catholic University of America, 1971-1972; University of Western Ontario, 1970-1974; Temple University, 1968-1969.

**Research Profile:** Several logical devices are associated with his name including the Rescher Quantifier in symbolic logic; the Dienes-Rescher Inference Engine in nonstandard logic, and the Rescher-Manor mechanism in non-monotonic reasoning theory. He also initiated the idea of autodescriptivity for logical systems. More broadly, Rescher has contributed to virtually every area of nonstandard logic and philosophical logic, specifically including though not confined to modal logic, many-valued logic, temporal logic, epistemic logic, the logic of conditionals, and the theory of hypothetical reasoning.

**Main Publications:**

1. *Al-Farabi's Short Commentary on Aristotle's "Prior Analytics."* Translated from the Arabic, with Introduction and Notes, University of Pittsburgh Press, 1963.
2. *Studies in the History of Arabic Logic*, University of Pittsburgh Press, 1963.
3. *The Development of Arabic Logic,* University of Pittsburgh Press, 1964.
4. *Galen and the Syllogism: An Examination of the Claim that Galen Originated the Fourt Figure of the Syllogism*, University of Pittsburgh Press, 1966.
5. *Temporal Modalities in Arabic Logic*, D. Reidel, 1966; Supplementary Series of *Foundations of Language*.
6. *Hypothetical Reasoning*, North-Holland Publishing Co.; "Studies in Logic" series edited by L.E.J. Brouwer, E.W. Beth and A. Heyting, 1964.
7. *An Introduction to Logic*, St. Martin's Press, 1964.
8. *Topics in Philosophical Logic*, D. Reidel, 1968; Synthese Library.
9. *Many-Valued Logic*, McGraw-Hill, 1969; reprinted: Gregg Revivals, 1993.
10. *Temporal Logic*, Springer-Verlag, 1971, co-authored with Alastair Urquhart.
11. *A Theory of Possibility*, Basil Blackwell, 1975; co-published in the USA by the University of Pittsburgh Press.
12. *Plausible Reasoning*, Van Gorcum, 1976.
13. *Dialectics: A Controversy-Oriented Approach to the Theory of Knowledge*, State University of New York Press, 1977; translated into Japanese as *Taiwa No Roni*, Kinokuniya Press, 1981.
14. *The Logic of Inconsistency: A Study in Nonstandard Possible-World Semantics and Ontology*, Basil Blackwell), 1979; APQ Library of Philosophy; co-authored with Robert Brandom; published in the USA by Rowman & Littlefield, 1979.
15. *Induction*, Basil Blackwell, 1980; co-published in the USA by the University of Pittsburgh Press; translated into German as *Induktion*, Philosophia Verlag, 1986.
16. *Paradoxes*, Open Court Publishing Co., 2001.
17. *Epistemic Logic*, University of Pittsburgh Press, 2004.
18. *Conditionals*. Boston (MIT Press) 2005.

*Work in Progress*

19. A collaborative study with Patrick Grim of non-Cantorean, and seemingly inconsistent totalistic collectivities (such as "the set of all sets").
20. A study of the role of error in information management.

**Service to the Profession:** Editor, American Philosophical Quarterly; President, Charles Sanders Peirce Society, 1983-1986; President, G. W. Leibniz Society of America, 1983-1986; President, American Philosophical Association, Eastern Division, 1989-1990; President, American Catholic Philosophical Association, 2003-2004; President, American Metaphysical Society, 2004-2005.

**Teaching:** Rescher has had the privilege of teaching some excellent logicians at Pitt over the years including Brian Skyrms, Bas van Fraassen, Alasdair Urquhart, Tom Vinci, Arnold Vander Nat, and others.

**Vision Statement:** Plato wanted no-one to study philosophy who was not something of a geometer. I myself incline to seeing logic as replacing geometry in this regard as a sine qua non for effective philosophizing.

**Honours and Awards:** DPhil honoris causa, Fernuniversität Hagen, Germany, 2001; DSc hc, Queens College of the City University of New York, 1999; Member, European Academy of Arts and Sciences: Academia Europaea, 1997; Doctor honoris causa, University of Constance, Germany, 1996; LHD (Doctor of Humane Letters), Lehigh University, 1993; Doctor honoris causa, National Autonomous University of Córdoba, Argentina, 1992; President's Distinguished Research Award, University of Pittsburgh; Medal of Merit for Distinguished Scholarship, University of Helsinki, Finland, 1990; Member, Academie Internationale de Philosphie des Sciences, 1984; Alexander von Humboldt Humanities Prize, 1984; Member, Pennsylvania Academy of Science, 1976; Member, Institut International de Philosophie, 1971; Guggenheim Foundation Fellow, 1970; LHD, Loyola University of Chicago; Member, Royal Asiatic Society of Great Britain and Ireland, 1961; Ford Foundation Fellow, 1959.

# RESTALL, Greg

**Specialties:** Non-classical logics, philosophy of logic.

**Born:** 11 January 1969 in Brisbane, Australia.

**Educated:** University of Queensland, B. Sc. (Honours in Mathematics) 1989; Ph. D. (Philosophy) 1994.

**Dissertation:** *On Logics Without Contraction*; supervisor Graham Priest.

**Regular Academic or Research Appointments:** ASSOCIATE PROFESSOR, PHILOSOPHY DEPARTMENT, UNIVERSITY OF MELBOURNE, 2002–. Senior Lecturer, Philosophy Department, Macquarie University, 1996–2002. Australian Research Council Postdoctoral Research Fellow, Automated Reasoning Project, Australian National University, 1994–1996.

**Visiting Academic or Research Appointments:** Dan Taylor Fellow, University of Otago, 2005. Visiting Fellow, Wolfson College, Oxford, 2005.

**Research Profile:** By nature a pluralist, Restall works on both formal and philosophical aspects in logic. His research spans both the proof theory and semantics of substructural and other non-classical logics.

**Main Publications:**

1. *An Introduction to Substructural Logics*, Routledge, 2000.
2. "Relevance Logic," with J. Michael Dunn, pages 1–136 in the *Handbook of Philosophical Logic*, volume 6, second edition, Dov Gabbay and Franz Guenther (editors), Kluwer Academic Publishers, 2002.
3. *Logic*, Routledge, 2006.
4. *Logical Pluralism*, with JC Beall, Oxford University Press, 2006.

*Work in Progress*

5. *Proof and Counterexample*, a philosophical introduction and exploration of proof theory. See http://consequently.org/edit/page/Proof_and_Counterexample.

**Service to the Profession:** Member of the Editorial Board of the *Notre Dame Journal of Formal Logic* (2005– ); Editor, *Journal of Philosophical Logic*, (2005–2007); Editor, *Review of Symbolic Logic*, (2007–); Founding Editor, *Australasian Journal of Logic*, (2004– ).

**Teaching:** Restall teaches logic throughout the curriculum, from an introductory level to research students. My book *Logic* (Routledge, 2005) is a philosophically motivated introduction to propositional and predicate logic, arising out of his introductory logic lectures. He supervises a lively community of postgraduate logic students at the University of Melbourne.

**Vision Statement:** Logic is essentially interdisciplinary: insights are to be gained from its connections to mathematics, computer science, linguistics and, not least, philosophy. Logic has an important role to play in philosophy, and philosophical considerations give rise to new, important problems in logic.

**Honours and Awards:** Fellow, Australian Academy of the Humanities, 2004.

# S

## SACKS, Gerald E.

**Specialties:** Recursion theory, model theory.

**Born:** 22 March 1933 in New York, New York, USA.

**Educated:** Cornell University.

**Dissertation:** *Degrees of Unsolvability*; supervisor, Barkley Rosser.

**Regular Academic or Research Appointments:** PROFESSOR, MATH, HARVARD UNIVERSITY; Professor, Math, Massachusetts Institute of Technology; Asst Professor, Math, Cornell University.

**Research Profile:** Sacks has been involved in several areas of research. These include sets, Turing degrees, infinite injury arguments, splitting, density, jump inversion, alpha-recursion, perfect forcing, recursion in finite types, k-sections, scattered theories.

**Main Publications:**

1. Degrees of Unsolvability, Annals of Math. Study 55, Princeton. University Press, 174 pp. (1963, second edition 1966).
2. Saturated Model Theory, Benjamin (Reading) 1972, 335 pp. (Russian Translation, Moscow 1976, 190 pp.)
3. Higher Recursion Theory, Springer-Verlag (Heidelberg) 1990, 344 pp.
4. Selected Logic Papers, World Scientific (London, Singapore) 1999, xviii + 431 pp.
5. Mathematical Logic in the 20th Century, World Scientific (London, Singapore) 2003, 693 pp.
6. On the degrees less than $0\prime$, Annals of Math. 77 (1963) 211-231.
7. Recursive enumerability and the jump operator, Trans. Amer. Math. Soc. 108 (1963) 223-239.
8. The recursively enumerable degrees are dense, Annals of Math. 80 (1964) 300-312.
9. Measure-theoretic uniformity in set theory and recursion theory, Trans. Amer. Math. Soc. 142 (1969) 381-420.
10. Forcing with perfect closed sets, Proc. Symp. Pure Math., vol. 13, Part 1, Axiomatic Set Theory (1971) 331-355.
11. The $\alpha$-finite injury method (with S, G. Simpson), Ann. Math. Logic 4 (1972) 343-367.
12. The 1-section of a type n object, Generalized Recursion Theory, (edited by J. E. Fenstad and P. G. Hinman) (1974) 81-93.
13. Countable admissible ordinals and hyperdegrees, Adv. Math. 20 (1976) 213-262.
14. The k-section of a type n object, Amer. Jour. Of Math. 99 (1977) 901-917.
15. Post's problem in E-recursion, Proc. Symp. Pure Math., vol. 42 (1985), 177-193.
16. The limits of E-recursive enumerability, Ann. Pure Appl. Logic 31 (1986) 87-120.
17. Effective forcing versus proper forcing, Ann. Pure Appl. Logic 81 (1996) 177-185.
18. Bounds on weak scattering, to appear in Notre Dame Jour. Formal Log. pps. 1-29. (preprint available on author's web site).
19. Atomic models higher up (with Jessica Millar), submitted to Ann. Pure Appl. Logic, preprint available, pps. 1-35.

*Work in Progress*

20. Models of long sentences, in preparation.

**Teaching:** Sacks has had twenty-nine PhD students.

**Vision Statement:** More and more applications of model theory to number theory and geometry.

## SAGÜILLO, José Miguel

**Specialties:** Mathematical logic, philosophy of logic, history of logic, philosophical logic, argumentation and logic.

**Born:** 17 February 1958 in Buenos Aires, Argentina.

**Educated:** University of Santiago de Compostela, PhD Philosophy, 1990; State University of New York at Buffalo, MA Philosophy, 1984; University of Santiago de Compostela, BS Philosophy, 1980.

**Dissertation:** *Systems of modal Logic, physical necessity, and counterfactuals*; supervisor, Rafael Beneyto Torres.

**Regular Academic or Research Appointments:** PROFESSOR, UNIVERSITY OF SANTIAGO DE COMPOSTELA IN SPAIN, LOGIC AND PHILOSOPHY OF SCIENCE, 2002–; Associate Professor, University of Santiago de Compostela in Spain, 1989-2002; Adjunct Professor, University of Santiago de Compostela in Spain, 1983-1989; Assistant Professor, University of Santiago de Compostela in Spain, 1980-1981.

**Visiting Academic or Research Appointments:** Visiting Scholar, Philosophy, University of California at Berkeley, 2007; Visiting Scholar, Philosophy, SUNY at Buffalo, 1998; Visiting Scholar, Philosophy, SUNY at Buffalo, 1994; Visiting

Scholar, Philosophy, SUNY at Buffalo, 1993; Visiting Scholar, Philosophy/Linacre College, Oxford University, 1990; Visiting Scholar, Philosophy/Linacre College, Oxford University, 1989.

**Research Profile:** José M. Sagüillo has focused his research on foundational issues of logic with particular interest in the relation of logical consequence as the cornerstone of our deductive methods. His approach has been both systematic and historical with emphasis on methodological practice. One important question addressed has been identifying which of the concepts of logical consequence, whether the model-theoretic or the information-theoretic, support an adequate explanation of the processes that we actually use to establish whether a given conclusion follows from given premises in the tradition of classical mathematics. Related methodological issues investigated are the presuppositions and implications of choices of the domain of investigation of a science and the universe of discourse of the formalization of a science, emphasizing that the universe of discourse (of a given discourse) is important in determining which propositions are expressed by which sentences. He has also study "negative" phenomena, where reason gets lost and goes out of its normal modes, such as in the case of category mistakes, argumentative fallacies and paradoxical argumentations.

**Main Publications:**

1. "Art, Logic and Scientific Rhetoric". *Reason and Rhetoric* (R. Harré ed.). E. Mellen Press. Lewiston/Queenston/Lampeter, 41-67, 1993.
2. "Paradoxical Argumentations". *Proceedings of the III International Conference on Argumentation* 1994, Vol II *Analysis and Evaluation*. (Ed. by Frans H. Van Eemeren *et al.*), Amsterdam, 13-22.
3. "Validez y Semántica Representacional". *Theoria* 1995, 24, 103-120.
4. "El Contexto del Descubrimiento de la Lógica". *Agora* 1995a, 14, 1, 99-118.
5. "Logical consequence revisited". *Bulletin of Symbolic Logic* 1997, 2, 216-241.
6. "Fitch's problem and the knowability paradox: Logical and Philosophical Remarks". *Logica Trianguli* 1997a, 1, 73-91. Written with C. Martínez Vidal and J. Vilanova.
7. "Domains of sciences, universes of discourse and omega arguments". *History and Philosophy of Logic* 2000, 20, 267-290.
8. "Quine on logical truth and consequence". *Agora* 2001, 20, 1, 139-156.
9. "Conceptions of logical implication". *Logica Trianguli* 2002, 6, 41-67.
10. "Validez y consecuencia lógica. La concepción clásica". Forthcoming in *Filosofía de la Lógica* (María J. Frápoli ed.) Madrid: Tecnos 2007.
11. "On a new account of the Liar: Comments on Stephen Read". Forthcoming in *Logic, Epistemology and the Unity of Science*. Holland: Springer 2007.
12. "Methodological practice and complementary concepts of logical consequence: Tarski and Corcoran. Forthcoming in *History and Philosophy of Logic* 2007.

*Work in Progress*

13. A book on logical and mathematical thinking including heuristic and apodictic methods.
14. An article revisiting category mistakes proposing a new diagnosis of the issue.
15. An article on different notions of information.
16. A book on the mistakes in reasoning from the viewpoint of classical logic, addressing category mistakes, fallacies and paradoxes.

**Service to the Profession:** Scientific Committee Member, Center of Advance Studies, University of Santiago de Compostela, 2002-; Co-Organizer, Congress *Formal Theories and Empirical Theories*, University of Santiago de Compostela, 14-16 November 2002; Co-Organizer, *Coloquio Compostelano de Lógica y Filosofía Analítica*, Department of Logic and Moral Philosophy, University of Santiago de Compostela, 2001-2007; Co-Organizer, International Congress *Analytic Philosophy at the turn of the Millenium*, 1-4 December 1999; Chief-Editor, Newsletter of the Society of Logic, Methodology and Philosophy of Science, Spain, 1997-2000; Co-Editor, Newsletter of the Society of Logic, Methodology and Philosophy of Science, Spain, 1993-1996; University of Santiago de Compostela Press, 1992-2004; Vice-Dean, Faculty of Philosophy, University of Santiago de Compostela, 1992-1995; Chief Editor, Editorial Committee Member, Journal *Agora*.

**Teaching:** Sagüillo teaches undergraduate courses in Logic, Set-Theory, and Argumentation and Rhetoric, and graduate courses on Foundational issues, such as logical consequence, truth, information-theoretic logic, argumentation, and philosophy of mathematics.

**Vision Statement:** Mathematical logic is a construct of two branches of mathematics, string theory and model-theory. Sometimes logical phenomena that cannot be captured in this picture is covered up or kept out of sight. Logical practice can provide important insights into the actual capabilities of our logical theories pointing out existing gaps and suggesting ways to fulfill them.

# SAYWARD, Charles

**Specialties:** Philosophy of logic, philosophy of mathematics.

**Born:** 1 August 1937 in Lewiston, Maine, USA.

**Educated:** Cornell University, PhD, 1964; Bates College, BA, 1959.

**Dissertation:** *What is Pragmatics?*; supervisor, Max Black.

**Regular Academic or Research Appointments:** PROFESSOR, PHILOSOPHY, UNIVERSITY OF NEBRASKA-LINCOLN, 1974-; Associate Professor, Philosophy, University of Nebraska-Lincoln, 1969-1974; Assistant Professor, Philosophy, University of Nebraska-Lincoln, 1963-1969; Instructor, Vanderbilt University, 1962-1963.

**Research Profile:** Charles Sayward, in collaboration with Stephen Voss, contributed to the study of types and categories in 1980. At various times Sayward contributed to the study of the paradoxes, most recently arguing that the so-called liar paradox is not really a paradox; instead, we have a provable falsehood being an instance of Convention T, thus proving that Convention T is a false universal generalization. The situation is held to be analogous with a provable falsehood being an instance of another universal generalization, Frege's Basic Law V. In neither case do we have paradoxes. Rather, we have false universal generalizations. Since the early 1990s, in collaboration with Philip Hugly, Sayward brought together two ideas relating to the ontological status of arithmetic, ideas that hopefully connect together to form a useful perspective in the philosophy of mathematics. These ideas are (i) that pure arithmetic, taken in isolation from the use of arithmetical signs in empirical judgment, is an activity for which a formalist account (roughly, one on which arithmetic is strictly a system of signs) is about right, and (ii) that arithmetical signs nonetheless have meanings, but only in and through belonging to a system of signs with empirical application. The view that emerges from the combination of these ideas is called a philosophy of arithmetic.

**Main Publications:**

1. "The Structure of Type Theory," 1980, *Journal of Philosophy*, 77: 241-259. Co-authored with S. Voss.
2. "Completeness Theorems for Two Propositional Logics in which Propositional Identity and Coentailment Diverge," 1981, *Notre Dame Journal of Formal Logic*, 22: 269-282. Co-authored with P. Hugly.
3. "Indenumerability and Substitutional Quantification," 1982, *Notre Dame Journal of Formal Logic*, 23: 358-366. Co-authored with P. Hugly.
4. "Can a Language Have Indenumerably Many Expressions?" 1983, *History and Philosophy of Logic*, 4: 73-82. Co-authored with P. Hugly.
5. "What is a Second Order Theory Committed to?" 1983, *Erkenntnis*, 20: 79-91.
6. "Do We Need Quantification?" 1984, *Notre Dame Journal of Formal Logic*, 25: 289-302. Co-authored with P. Hugly.
7. "Prior's Theory of Truth," 1987, *Analysis*, 47: 83-87.
8. "Can There be a Proof that an Unprovable Sentence of Arithmetic is True?" 1989, *Dialectica*, 43: 289-292.
9. "Mathematical Relativism," 1989, *History and Philosophy of Logic*, 10: 53-65. Co-authored with Philip Hugly.
10. "Are All Tautologies True?" 1989, *Logique et Analyse*, 125-126: 3-14. Co-authored with Philip Hugly.
11. "Moral Relativism and Deontic Logic," 1990, *Synthese*, 85: 139-152. Co-authored with Philip Hugly.
12. "Four Views of Arithmetical Truth," 1990, *Philosophical Quarterly*, 40: 155-168.
13. "Definite Descriptions, Negation and Necessitation," 1993, *Russell* 13: 36-47.
14. "Quantifying Over the Reals," *Synthese*, 101: 53-64. Co-authored with Philip Hugly
15. *Intentionality and Truth: An Essay on the Philosophy of Arthur Prior*. 1996. Synthese Library 255. Dordrecht: Kluwer Academic Publications. Co-authored with Philip Hugly.
16. "Convention T and Basic Law V," 2002, *Analysis*, 62: 289-292.
17. "Remarks on Peano Arithmetic," 2000, *Russell: The Journal of Bertrand Russell Studies*, 20: 27-32.
18. "On Some Much Maligned Remarks of Wittgenstein on Gödel," 2001, *Philosophical Investigations*, 20: 262-270.
19. "A Conversation about Numbers and Knowledge," 2002, *American Philosophical Quarterly*.
20. *Arithmetic and Ontology: A Non-Realist Philosophy of Arithmetic*, 2006, volume 90 of the *Monographs-in-Debate* subseries of *Poznan Studies in Science and Humanities*. Co-authored by Philip Hugly. Edited by Pieranna Garavaso.

*Work in Progress*

21. *Logic for Philosophers*. Textbook.
22. "A Fregean Conception of Singular Existence". Article.
23. "Quine and his Critics on Truth-Functionality and Extensionality." Article.
24. "Malcolm and Schlick on Private Languages". Article.
25. "What Truth is There in Psychological Egoism?" Article.

**Service to the Profession:** Associate Editor, *Journal of Philosophical Research*, 1992-1994.

**Vision Statement:** W. V. Quine held that when we employ 'For some $x$, $Fx$' we are saying 'There exists at least one object such that it $F$s' and what it presupposes is a set of objects the $F$-ing of any one of which would satisfy the open formula. These are the values of the variable. Thus the slogan Quine made famous: "To be is to be a value of a variable". The prevailing attitude of current philosophers of mathematics, both realist and antirealist, is to agree with Quine. The agreement is almost always implicit. That is, while the philosopher does not come right out and say Quine's criterion is being used, his or her philosophy would be unintelligible unless this were so. Against this attitude Arthur Prior wrote several years ago "I doubt whether any dogma, even of empiricism, has ever been quite so muddling as the dogma that to be is to be a value of a bound variable" (Prior, *Objects of Thought*, p. 48). I think that Prior was right about quantification. I hope that some day this shall be generally recognized.

# SCEDROV, Andre

**Specialties:** Mathematical logic, logic in computer science, foundations of mathematics, foundations of computer science.

**Born:** 1 August 1955 in Zagreb, Croatia.

**Educated:** State University of New York at Buffalo, PhD Mathematics, 1981; University of Zagreb, BS Mathematics, 1977.

**Dissertation:** *Sheaves and Forcing and Their Metamathematical Applications,* supervisor, John Myhill.

**Regular Academic or Research Appointments:** PROFESSOR, MATHEMATICS AND COMPUTER AND INFORMATION SCIENCE, UNIVERSITY OF PENNSYLVANIA, 1992–; Associate Professor, Mathematics and Computer and Information Science, University of Pennsylvania, 1988-1992; Assistant Professor, Mathematics, University of Pennsylvania, 1982-1988; T.H. Hildebrandt Research Assistant Professor, Mathematics, University of Michigan, Ann Arbor, 1981-1982.

**Visiting Academic or Research Appointments:** Visiting scholar, Keio University, Tokyo, Japan, 2004, 2001, 2000, 1998; Visiting scholar, Instituto Superior Técnico, Lisbon, Portugal, 2002; Fellow, Mittag-Leffler Institute, Djursholm, Sweden, 2001; Visiting scholar, Stanford University, 1998; Visiting Professor, Keio University, Tokyo, Japan, 1997; Fellow, Isaac Newton Institute for Mathematical Sciences, Cambridge, England, 1995; Visiting scholar, Stanford University, 1995; Visiting Scholar, SRI International, Menlo Park, California, 1995; Visiting scholar, CNRS Laboratoire de Mathématiques Discrètes, Marseille, France, 1995; Visiting scholar, Rijksuniversiteit Utrecht, The Netherlands, 1993; Visiting scholar, Université Paris 7, France, 1992; Visiting Associate Professor of Computer Science, Stanford University, 1989-1990; Visiting scholar, Université Catholique de Louvain, Louvain-La-Neuve, Belgium, 1988; Visiting scientist, Mathematical Sciences Institute, Cornell University, 1987; Visiting scholar, University of Sydney, Australia, 1986; Visiting Assistant Professor of Mathematics, Ohio State University, 1986; Visiting scholar, McGill University, Montreal, Canada, 1985; Visiting scholar, Università degli Studi di Milano, 1982.

**Research Profile:** Andre Scedrov's contributions range from mathematical logic and the foundations of mathematics to applied logic and the foundations of computer science, and most recently foundations of computer security. His dissertation and early work was in categorical logic. His early contributions also include work on intuitionistic formal systems, including type theory and set theory. Scedrov's interest in type theory and in the propositions-as-types correspondence (the Curry-Howard isomorphism) gradually led him to theory and semantics of polymorphic types and typed programming languages in computer science, to proof-theoretic foundations of logic programming, and to linear logic. Scedrov contributed to the understanding of computational expressiveness of linear logic and its fragments. His recent work is on the formal analysis of network security protocols in computer security. The work of Scedrov and collaborators on formal analysis of the Kerberos authentication protocol discovered a structural, protocol-level flaw in the Kerberos public-key extension and caused a Microsoft security patch for Windows 2000 and Windows XP operating systems. The work contributed to the reformulation of the international standard for Kerberos public-key extension and provided security proofs of the corrected version of the protocol. The methodology developed in this work is applicable to a wide range of network security protocols.

**Main Publications:**

1. *Forcing and classifying topoi.* Memoirs of the American Mathematical Society 295, Providence, RI, 1984, $x + 93$ pp.

2. (with L.A. Harrington, M.D. Morley, and S.G. Simpson, eds.), *Harvey Friedman's the Foundations of Mathematics. Studies in Logic and the Foundations of mathematics,* North-Holland, Amsterdam, 1985, $xvi + 408$ pp.

3. (with H. Friedman), Arithmetic transfinite induction and recursive well-orderings.*Advances in Mathematics* 56 (1985) 283-294.

4. (with H. Friedman), The lack of definable witnesses and provably recursive functions in intuitionistic set theories. *Advances in Mathematics* 57 (1985) 1-13.

5. Diagonalization of continuous matrices as a representation of intuitionistic reals. *Annals of Pure and Applied Logic* 30 (1986) 201-206.
6. (with P.J. Freyd and H. Friedman), Lindenbaum algebras of intuitionistic theories and free categories. *Annals of Pure and Applied Logic* 35 (1987) 167-172.
7. (with A. Blass), *Freyd's Models for the Independence of the Axiom of Choice. Memoirs of the American Mathematical Society* 404, Providence, RI, 1989, viii + 134 pp.
8. (with E.S. Bainbridge, P.J. Freyd, P.J. Scott), Functorial polymorphism. *Theoretical Computer Science* 70 (1990) 35-64.
9. (with P.J. Freyd), *Categories, Allegories. North-Holland Mathematical Library,* North-Holland, Amsterdam, 1990, xviii + 296 pp.
10. (with D. Miller, G. Nadathur, and F. Pfenning), Uniform proofs as a foundation for logic programming. *Annals of Pure and Applied Logic* 51 (1991) 125-157.
11. (with J.-Y. Girard and P.J. Scott), Bounded linear logic: A modular approach to polynomial time computability. *Theoretical Computer Science* 97 (1992) 1-66.
12. (with P.D. Lincoln, J.C. Mitchell, and N. Shankar), Decision problems for propositional linear logic. *Annals of Pure and Applied Logic* 56 (1992) 239-311.
13. (with P.D. Lincoln), First-order linear logic without modalities is NEXPTIME-hard. *Theoretical Computer Science* 135 (1994) 139-154.
14. (with L. Cardelli, J.C. Mitchell, and S. Martini), An extension of system $F$ with subtyping. *Information and Computation* 109 (1994) 4-56.
15. (with P.D. Lincoln and J.C. Mitchell), Optimization complexity of linear logic proof games. *Theoretical Computer Science* 227 (1999) 299-331.
16. (with M. Kanovich and M. Okada), Phase semantics for light linear logic. *Theoretical Computer Science* 294 (2003) 525-549.
17. (with N. Durgin, P.D. Lincoln, and J.C. Mitchell), Multiset rewriting and the complexity of bounded security protocols. *Journal of Computer Security* 12 (2004) 247-311.
18. (with R. Chadha, J.C. Mitchell, and V. Shmatikov), Contract signing, optimism, and advantage. *Journal of Logic and Algebraic Programming* 64 (2005) 189-218.
19. (with J.C. Mitchell, A. Ramanathan, and V. Teague), A probabilistic polynomial-time process calculus for the analysis of cryptographic protocols. *Theoretical Computer Science* 353 (2006) 118-164.
20. (with F. Butler, I. Cervesato, A. Jaggard, and C. Walstad), Formal Analysis of Kerberos 5. *Theoretical Computer Science* 367 (2006) 57-87.

A more complete list of recent publications is available on Scedrov's home page http://www.cis.upenn.edu/~scedrov/

*Work in Progress*
21. Security analysis of public-key Kerberos (with I. Cervesato, A. Jaggard, J.-K. Tsay, and C. Walstad) and cryptographically sound security proofs for basic and public-key Kerberos (with M. Backes, I. Cervesato, A. Jaggard, J.-K. Tsay).
22. (with A. Datta, A. Derek, J.C. Mitchell, and A. Ramanathan), Games and the impossibility of realizable ideal functionality in the Universal Composability framework for computational cryptography.
23. Computational soundness of formal encryption in the presence of key cycles (with P. Adão, G. Bana, and J. Herzog in the passive adversary case and with M. Backes and B. Pfitzmann for the active adversary).
24. (with I. Cervesato), Relating state-based and process-based concurrency through linear logic.

**Service to the Profession:** Program Committee Member, IEEE Computer Security Foundations Symposium, 2007; Editor, *Journal of Computer Security*, 2005-; Program Committee Member, ACM Conference on Computer and Communications Security, 2004-2006; Program Committee Member, IEEE Computer Security Foundations Workshop, 2003-2006; Program Chair, Association for Symbolic Logic Annual Meeting,2001; Managing Editor, *Annals of Pure and Applied Logic,* 1999-2004; Program Co-Chair, Mathematical Foundations of Programming Semantics, 1999; Executive Committee Member, Association for Symbolic Logic, 1998-2001; Advisory Board Member, IEEE Symposium on Logic in Computer Science, 1997-; Editor, *Perspectives in Logic,* 1997-2007; Coordinating Editor, *Journal of Symbolic Logic,* 1994-1995; Chair Nominating Committee, Association for Symbolic Logic, 1994; Advisory Editor, *Annals of Pure and Applied Logic,* 1993-1999; Organizing Committee Member, IEEE Symposium on Logic in Computer Science, 1992-1997; Program Chair, IEEE Symposium on Logic in Computer Science, 1992; Council Member, Association for Symbolic Logic, 1990-1995; Editor, *Mathematical Structures in Computer Sciences*, 1989-2004; Editor, *Journal of Symbolic Logic,* 1988-1995.

**Teaching:** Through his activities in the Penn Logic and Computation Group, Scedrov has collaborated with colleagues from Computer and Information Science and from Philosophy to introduce, teach, and coordinate a number of graduate and undergraduate courses on logic and on mathematical methods in computer science. With colleagues from Philosophy and Linguistics, Scedrov has recently established an Undergraduate Major and Minor program in Logic, Information, and Computation. Scedrov's PhD students include Richard Blute, Moez Alimohamed, Peter Selinger, Rohit Chadha, and Gergely Bana. Scedrov has also influenced the PhD work of Gopalan Nadathur and Aaron Jaggard at Penn, Nancy Durgin, Ajith Ramanathan, and Vanessa Teague at Stanford, Masahiro Hamano, Kazushige Terui, and Koji Hasebe at Keio University, Tokyo, Japan, and Pedro Adão at Instituto Superior Técnico, Lisbon, Portugal.

**Vision Statement:** The methodology developed in the work in logic and the foundations of mathematics also leads to insights in the foundations of computer science, where foundational work has a direct bearing on the field and can in some cases also be of quite practical value. Formal logic and its methods continue to have a profound impact in computer science.

**Honours and Awards:** Senior Fellow, Japan Society for the Promotion of Science, 1997; Centennial Research Fellow, American Mathematical Society, 1993.

# SCHIMMERLING, Ernest

**Specialties:** Set theory.

**Born:** 7 April 1963 in Buenos Aires, Argentina.

**Educated:** University of California, Los Angeles, PhD Mathematics, 1992; Berkeley, MA Mathematics, 1987; Berkeley, AB Mathematics, 1984.

**Dissertation:** *"Combinatorial principles in the core model for one Woodin cardinal"*; Supervisor, John R. Steel

**Regular Academic or Research Appointments:** ASSOCIATE PROFESSOR, MATHEMATICAL SCIENCES, CMU, 2002–; Assistant Professor, Mathematical Sciences, CMU, 1999-2002; Assistant Professor (On Leave), Mathematics, University of Connecticut, 1998-1999.

**Visiting Academic or Research Appointments:** Fulbright Scholar, University of Paris 7, 2006; Professeur Invité, University of Paris 7, 2005; Visiteur, IHÉS, 2002; Visiting Assistant Professor, CMU, 1998-1999; Visiting Assistant Professor, University of California Irvine, 1996-1998; NSF Postdoctoral Fellow, MIT, 1993-1996; Research Fellow, Berkeley, 1993; Postdoctoral Scholar, University of California, Los Angeles, 1992.

**Research Profile:** Schimmerling works in set theory. More specifically, his research is on core models and their fine structure, which is a subfield of set theory that was pioneered by Jensen. Core models are generalizations of Gödel's constructible universe that arose in the study of large cardinal axioms. The field today builds on the fundamental discoveries of Dodd, Jensen, Martin, Mitchell, Steel and Woodin. Schimmerling is interested in pure core model theory as well as its applications to and interactions with other areas, such as forcing, infinitary combinatorics and descriptive set theory.

**Main Publications:**

1. Schimmerling, "Combinatorial principles in the core model for one Woodin cardinal", Annals of Pure and Applied Logic 74 (1995) 153-201.
2. Mitchell, Schimmerling and Steel, "The covering lemma up to a Woodin cardinal", Annals of Pure and Applied Logic 84 (1997) 219-255.
3. Mitchell and Schimmerling, "Weak covering without countable closure", Mathematical Research Letters 2 (1995) 595-609.
4. Schimmerling and Steel, "The maximality of the core model", Transactions of the American Mathematical Society 351 (1999) 3119-3141.
5. Schimmerling and Zeman, "Characterization of $\square_\kappa$ in core models", Journal of Mathematical Logic 4 (2004) 1-72.

**Service to the Profession:** Special session, AMS, Oxford OH, 2007; Conference Organizing Committee Member, Appalachian Set Theory Workshops, 2006-; Conference Organizing Committee Member, Mini-colloquium, DMV, Bonn, Germany, 2006; Editorial Book Series Board Member, Ontos Verlag Mathematical Logic, 2004-; Conference Organizing Committee Member, Petit groupe de travail, CIRM, Marseille, France, 2004; Conference Organizing Committee Member, Special session, ASL, Pittsburgh, 2004; Conference Organizing Committee Member, ARCC workshop, AIM, Palo Alto, 2004.

**Teaching:** Schimmerling teaches graduate courses for the Pure and Applied Logic Program at Carnegie Mellon University.

**Honours and Awards:** Fulbright Award 2006, NSF DMS Awards, 1998-1999, 2000-2003, 2004-2007; Featured Review, AMS Mathematical Reviews, 1995-1996; Reviews of Outstanding Recent Books and Papers," NSF Postdoctoral Research Fellowship, 1993-1996.

# SCHINDLER, Ralf

**Specialties:** Set theory.

**Born:** 19 February 1965 in Erlangen, Germany.

**Educated:** University of Bonn, PhD Mathematics 1996; University of Munich, MA Logic and Philosophy of Science 1992.

**Dissertation:** *The core model up to one strong cardinal*; Supervisor: Peter Koepke, Bonn.

**Regular Academic or Research Appointments:** FULL PROFESSOR AT THE UNIVERSITY OF MÜNSTER, 2003– PRESENT; Assistent Professor at the University of Vienna, 1999–2003.

**Visiting Academic or Research Appointments:** Research fellow at UC Berkeley, 1997–1999.

**Research Profile:** Ralf Schindler has made contributions to set theory, in particular inner model theory. He works on the construction of core models and their application in Descriptive Set Theory, Combinatorics, and Forcing Axioms.

**Main Publications:**

1. (joint with M. Foreman and M. Magidor) The consistency strength of successive cardinals with the tree property, Journal of Symbolic Logic 66 (2001), pp. 1837–1847.
2. (joint with K. Hauser) Projective uniformization revisited, Annals of Pure and Appl. Logic 103 (2000), pp. 109–153.
3. Strong cardinals and sets of reals in $L_{\omega_1}(R)$, Mathematical Logic Quarterly 45 (1999), pp. 361–369.
4. Coding into K by reasonable forcing, Transactions of the Amer. Math. Soc 353 (2000), pp. 479 - 489.
5. (joint with S. Friedman) Universally Baire sets and definable well-orderings of the reals, Journal of Symbolic Logic 68 (2003), pp. 1065 - 1081.
6. (joint with J. Steel and M. Zeman) Deconstructing inner model theory, Journal of Symbolic Logic 67 (2002), pp. 721 - 736.
7. Proper forcing and remarkable cardinals, Bulletin of Symbolic Logic 6 (2000), pp. 176 - 184.
8. Proper forcing and remarkable cardinals II, Journal of Symbolic Logic 66 (2001), pp. 1481- 1492.
9. The core model for almost linear iterations, Annals of Pure and Appl. Logic 116 (2002), pp. 207 - 274.
10. Iterates of the core model, Journal of Symbolic Logic 71 (2006), pp.241-251.
11. Mutual stationarity in the core model, Proceedings of LC2001, in: "Logic colloquium 01" (Baaz et al., eds., Lecture Notes in Logic 20, 2005), pp.386-401.
12. (joint with B. Mitchell) A universal extender model without large cardinals in V, J. Symb. Logic 69 (2004), pp. 371-386.
13. (joint with M. Gitik and S. Shelah), Pcf theory and Woodin cardinals, Proceedings of LC2002 (Chatzidakis et al., eds.), Lecture Notes in Logic 27, pp. 172-205.
14. (joint with P. Koepke) Homogeneously Souslin sets in small inner models, Archive for Math. Logic 45 (2006), pp.53-61.
15. (joint with M. Zeman) Fine structure theory, a chapter for the Handbook of Set Theory (Foreman, Kanamori, Magidor, eds.), to appear.
16. Semi-proper forcing, remarkable cardinals, and Bounded Martin's Maximum, Mathematical Logic Quarterly 50 (6) (2004), pp. 527 - 532.
17. Bounded Martin's Maximum and strong cardinals, in: Set theory. Centre de recerca Matematica, Barcelona 2003-4 (Bagaria, Todorcevic, eds.), Basel 2006, pp. 401-406.
18. Core models in the presence of Woodin cardinals, J. Symb. Logic 71 (2006), pp. 1145-1154.

*Work in Progress*
19. (with J. Steel) The self-iterability of L[E].
20. (with R. Jensen, E. Schimmerling, and J. Steel) The strength of PFA.

**Service to the Profession:** Editor of the Archive for Mathematical Logic; Editor of the ontos series in Mathematical Logic.

**Vision Statement:** Set theory is about investigating important truths about the mathematical universe. Set theoretic statements are true or false, but the truth value might be hard to decide in particular cases. Our understanding of the local and global structure of the universe has constantly been improving, so that many questions which are independent from ZFC have gotten firm answers by now. Many other questions will be answered in the future.

# SCHIRN, Matthias

**Specialties:** Philosophy of logic and mathematics, history of modern logic.

**Born:** 3 October 1944 in Weidenau, Germany.

**Educated:** University of Regensburg, Dr. phil. habil. 1985. *Habilitationsschrift*: *Studien zu Freges Philosophie der Mathematik*; supervisor, Franz von Kutschera; University of Freiburg i.Br., Dr. phil. 1974.

**Dissertation:** *Identität und Synonymie. Logisch-semantische Untersuchungen unter Berücksichtigung der sprachlichen Verständigungspraxis*; supervisor, Fernando Inciarte.

**Regular Academic or Research Appointments:** PROFESSOR OF PHILOSOPHY, UNIVERSITY OF MUNICH, 1987–; Akademischer Rat, University of Regensburg 1981–1983.

**Visiting Academic or Research Appointments:** Visiting Professor of Philosophy: Federal University of Paraíba, February-March 2005; Federal University of Ceará, September-October 2003; National University of Costa Rica, September-October 2000; Catholic University of São Paulo, September–October 1998; National Autonomous University of Mexico, October–November 1997; University of Santiago de Compostela, April–May 1995; National Autonomous University of Mexico, March–April 1994; National University of Buenos Aires, March–April 1992; State University of Campinas, March 1991; University of

Minnesota, Twin Cities, Fall Term 1989; University of Puerto Rico, Río Piedras, March–April 1989; Research Fellow of Deutsche Forschungsgemeinschaft 1986–1987, St. John's College, University of Oxford; Visiting Professor of Philosophy, University of Osnabrück, Winter Term 1985–1986; Research Fellow of Deutsche Forschungsgemeinschaft 1978–1980, University of California at Berkeley, St. John's College, University of Oxford, Harvard University, Wolfson College, University of Oxford; Visiting Lecturer of Philosophy, University of Cambridge, Michaelmas and Lent Term 1977–1978; Visiting Assistant Professor of Philosophy, Michigan State University, academic year 1976–1977; Visiting Lecturer of Philosophy, University of Oxford, Hilary and Trinity Term 1976.

**Research Profile:** Matthias Schirn has made contributions to Frege's logic and philosophy of mathematics and to Hilbert's proof theory. In a series of papers co-authored with K.-G. Niebergall on Hilbert's programme, Schirn and Niebergall were primarily concerned to present Hilbert's proof-theoretic approach in a new light. As to Hilbert's classical position in the 1920s, they argued that his finitist metamathematics need not be recursively enumerable and that, contrary to wide-spread opinion, it may be weaker than (QF-IA). The fact that in advocating finitist metamathematics Hilbert did not dispense with assumptions of infinity was shown to be the Achilles' heel of his approach in the 1920s. Moreover, Schirn and Niebergall provided evidence that it was only in the first volume of Hilbert's and Bernays's work *Grundlagen der Mathematik* of 1934 that metamathematics was considered to be at least as strong as PRA, while in the second volume the original standards of finitism such as intuitive evidence and unquestionable soundness were thrown overboard by making formalized metamathematics at least as strong as PA. Another important result of Schirn's and Niebergall's research was that PA proves its own consistency indirectly in one step and that recursively enumerable extensions of (QF-IA) prove their own consistency indirectly in one step. Schirn and Niebergall are also known for their attempted refutation of W.W. Tait's thesis that all finitist reasoning is essentially primitive recursive.

Regarding his research on Frege's logic and philosophy of mathematics, Schirn has more recently focused on Frege's introduction of logical objects and the problem of referential indeterminacy to which it gives rise. He has also sought to shed new light on the relation between Hume's Principle and Axiom V and on Frege's attitude towards the latter. One result attained by critically examining certain aspects of C. Wright's neologicism was that in *Die Grundlagen der Arithmetik* of 1884 Frege hardly could have recognized Hume's Principle as a primitive truth of logic and employed it as an axiom governing the cardinality operator as a primitive sign. Another point argued for was that Hume's Principle resists characterization as an analytic truth.

**Main Publications:**

1. (as editor) *Sprachhandlung — Existenz — Wahrheit. Hauptthemen der sprachanalytischen Philosophie*, Frommann-Holzboog, Stuttgart (1974).
2. *Identität und Synonymie*, Frommann-Holzboog, Stuttgart (1975).
3. (as editor) *Studien zu Frege — Studies on Frege*, vols. 1–3, Frommann-Holzboog, Stuttgart (1976).
4. Semantik der assertorischen Sätze, *Göttingsche Gelehrte Anzeigen* 231 (1979), 288–310.
5. Begriff und Begriffsumfang. Zu Freges Anzahldefinition in den *Grundlagen der Arithmetik*, *History and Philosophy of Logic* 4 (1983), 117–143.
6. Semantische Vollständigkeit, Wertverlaufsnamen und Freges Kontextprinzip, *Grazer Philosophische Studien* 23 (1985), 79–104.
7. Review essay on G. Frege, *Die Grundlagen der Arithmetik*, centenary edition, ed. C. Thiel, Hamburg 1986, *The Journal of Symbolic Logic* 53 (1988), 993–999.
8. Frege on the Purpose and Fruitfulness of Definitions, *Logique et Analyse* 125–26 (1989), 61–80.
9. Frege's objects of a quite special kind, *Erkenntnis* 32 (1990), 27–60.
10. Kant's Theorie der geometrischen Erkenntnis und die nichteuklidische Geometrie, *Kant-Studien* 82 (1991), 1–28.
11. Gottlob Frege (1848–1925), *Philosophy of Language. An International Handbook of Contemporary Research*, ed. M. Dascal et. al., Walter de Gruyter, Berlin, New York (1992), 467–494.
12. Frege y los nombres de cursos de valores, *Theoria* 21 (1994), 109–133.
13. (as editor) *Frege: Importance and Legacy*, Walter de Gruyter, Berlin, New York (1996).
14. Frege on the Foundations of Arithmetic and Geometry, in M. Schirn, *Frege: Importance and Legacy*, 1–42.
15. On Frege's introduction of cardinal numbers as logical objects, in M. Schirn, *Frege: Importance and Legacy*, 114–173.
16. (as editor) *The Philosophy of Mathematics Today*, Clarendon Press, Oxford (1998).
17. (with K.-G. Niebergall), Hilbert's finitism and the notion of infinity, in M. Schirn, *The Philosophy of Mathematics Today*, 271–306.
18. (with K.-G. Niebergall), Extensions of the finitist point of view, *History and Philosophy of Logic* 22 (2001), 135–161.

19. (with K.-G. Niebergall), Hilbert's programme and Gödel's theorems, *Dialectica* 56 (2002), 347–370.
20. Fregean abstraction, referential indeterminacy and the logical foundations of arithmetic, *Erkenntnis* 56 (2003), 1–31.
21. (with K.-G. Niebergall), Finitism = PRA? On a thesis of W.W. Tait, *Reports on Mathematical Logic* 39 (2005), 3–24.
22. Hume's Principle and Axiom V reconsidered: critical reflections on Frege and his interpreters, *Synthese* 148 (2006), 171–227.
23. Concepts, extensions, and Frege's logicist project, *Mind* 115 (2006), 983–1005.
24. (with G. Imaguire), Problemas Fundamentais da Filosofia Analitica da Linguagem, Edições Loyola, São Paulo (2008).

*Work in Progress*
25. *Logic and the Foundations of Mathematics. Essays on Frege.*
26. *Freges Philosophie der Mathematik.*

**Honours and Awards:** Elected candidate for a distinguished "Fiebiger-professorship" in Lower Saxony (University of Osnabrück) and Bavaria (University of Munich), 1986.

# SCHLECHTA, Karl

**Specialties:** Semantics of non-monotonic and related logics

**Born:** 7 January 1948 in Kronberg, Germany

**Educated:** Diplom 1980 Math, Heidelberg; PhD 1988 Math Berlin; Habilitation 1992 Comp.Sci., Hamburg.

**Dissertation:** *A Model of CH without Souslin Trees and without Kurepa Trees in ccc-Extensions*; supervisor, S. Koppelberg

**Regular Academic or Research Appointments:** PROFESSOR FOR COMPUTER SCIENCE, MARSEILLE, FRANCE, 1993–; delegation DR CNRS, Institut des Sciences Cognitives, Lyon, France, 2000–2002; research positions at IBM Stuttgart, Institute for Computer Science, University of Hamburg, DFKI Saarbruecken, GMD Bonn, 1988–1993.

**Research Profile:** Schlechta's main interests are in: Representation and impossibility results for preferential logic, theory revision and update, counterfactual conditionals — basic semantical concepts like distance and size — development of new proof methods for representation results.

**Main Publications:**
1. K. Schlechta: "Some results on classical preferential models", Journal of Logic and Computation, Oxford, Vol.2, No.6 (1992), p. 675-686
2. K. Schlechta, "Directly Sceptical Inheritance cannot Capture the Intersection of Extensions", Journal of Logic and Computation, Oxford, Vol.3, No.5, pp.455-467, 1995
3. K. Schlechta: "Defaults as generalized quantifiers", Journal of Logic and Computation, Oxford, Vol.5, No.4, p.473-494, 1995
4. K. Schlechta: "Some completeness results for stoppered and ranked classical preferential models", Journal of Logic and Computation, Oxford, Vol. 6, No. 4, pp. 599-622, 1996
5. K. Schlechta : "Nonmonotonic logics - Basic Concepts, Results, and Techniques" Springer Lecture Notes series, LNAI 1187, Jan. 1997
6. K. Schlechta: "New techniques and completeness results for preferential structures", Journal of Symbolic Logic, Vol.65, No.2, pp.719-746, 2000
7. K. Schlechta: "Unrestricted preferential structures", Journal of Logic and Computation, Vol.10, No.4, pp.573-581, 2000
8. D. Lehmann, M. Magidor, and K. Schlechta: "Distance Semantics for Belief Revision", Journal of Symbolic Logic, Vol.66, No. 1, March 2001, p. 295-317
9. K. Schlechta: "Coherent Systems", Elsevier, Amsterdam, 2004
10. D. Gabbay and K. Schlechta, "Cumulativity without closure of the domain under finite unions", to appear in Review of Symbolic Logic

**Teaching:** J.Ben-Naim (PhD student)

**Vision Statement:** Intuitively reasonable formal semantics seem one of the best ways to progress in nonclassical logics. Kripke semantics and modal logic are good examples.

# SCHMERL, James Henry

**Specialties:** Model theory, models of Peano arithmetic.

**Born:** 7 April 1940 in Oakland, California, USA.

**Educated:** University of California, Berkeley, PhD Mathematics, 1971; University of California, Berkeley, MA Mathematics, 1963; University of California, Berkeley, AB Mathematics 1962.

**Dissertation:** *On k-like models for inaccessible k*; supervisor, Robert Vaught.

**Regular Academic or Research Appointments:** PROFESSOR EMERITUS, UNIVERSITY OF CONNECTICUT, 2002–; Professor, University of Connecticut, Storrs, 1983–2002; Associate Professor, University of Connecticut, Storrs, 1978–1883; Assistant Professor, University of Connecticut, Storrs, 1972–1978; Gibbs Instructor, Yale University, 1970–1972.

**Visiting Academic or Research Appointments:** New Mexico State University, Las Cruces, 2000;

Arizona State University, Tempe, 1996; New Mexico State University, Las Cruces, 1990; University of California, San Diego, 1978-1979.

**Research Profile:** Schmerl's research interests in model theory have emphasized the interplay between model theory and combinatorics. Perhaps this is most noticed in his work on nonstandard models of Peano Arithmetic, culminating in his 2006 book on the subject. Topics in the study of models of PA to which he has significantly contributed include substructure lattices, automorphism groups, indiscernibles, higher order logics, and hyperreals. Another area with some combinatorial flavor is his work on partially ordered sets, especially having to do with countable categoricity, but there is also some work on jump numbers which seem to have no model theoretic content. Effectiveness of combinatorial theorems and constructions, such as graph coloring, have been among Schmerl's interests, as also has been the more modern incarnation in reverse mathematics.

**Main Publications:**

1. On power-like models for hyperinaccessible cardinals, *J. Symb. Logic* 37 (1972), 531-537. (with S. Shelah)
2. Countable homogeneous partially ordered sets, *Algebra Universalis* 9 (1979), 317-321.
3. On the role of Ramsey quantifiers in first order arithmetic, *J. Symb. Logic* 47 (1982), 423-435. (with S. Simpson)
4. Recursively saturated models generated by indiscernibles, *Notre Dame J. Formal Logic*, 26 (1985), 99-105.
5. Transfer theorems and their applications to logics. In *Model-Theoretic Logics*, Perspectives Math. Logic, Springer, New York, 1985, pp. 177-209.
6. The chromatic number of graphs which induce neither $K_{1,3}$ nor $K_{5-e}$, *Discrete Math.* 58 (1986), 253-262. (with H. A. Kierstead)
7. Peano arithmetic and hyper-Ramsey logic, *Trans. Amer. Math. Soc.* 296 (1986), 481-505.
8. Isomorphic incidence algebras, *Adv. Math.* 84 (1990), 226-236. (with M. Parmenter and E. Spiegel)
9. Coinductive $\aleph_0$-categorical theories, *J. Symb. Logic* 55 (1990), 1130-1137.
10. Making the hyperreal line both saturated and complete, *J. Symb. Logic* 56 (1991), 1016-1025. (with H. J. Keisler)
11. Finite substructure lattices of models of Peano Arithmetic, *Proc. Amer. Math. Soc.* 117 (1993), 833-838.
12. Critically indecomposable partially ordered sets, graphs, tournaments, and other binary relational structures, *Discrete Math.* 113 (1993), 191-205. (with W. T. Trotter)
13. Tiling space with notched cubes, *Discrete Math.* 133 (1994), 225-235.
14. The automorphism group of an arithmetically saturated model of Peano Arithmetic, *J. London Math. Soc.* 52, Series 2, (1995), 235-244. (with R. Kossak)
15. Countable partitions of Euclidean space, *Math. Proc. Cambridge Philos. Soc.* 120 (1996), 7-12.
16. What's the difference? *Annals of Pure Applied Logic* 93 (1998), 255-261.
17. Avoidable algebraic subsets of Euclidean space, *Trans. Amer. Math. Soc.* 352 (2000), 2479-2489.
18. Reverse mathematics and graph coloring. In *Reverse Mathematics 2001*, Lecture Notes in Logic, 21. Assoc. Symb. Logic, LaJolla, CA 2005, 331- 348.
19. *The Structure of Models of Peano Arithmetic*, Oxford Logic Guides, 50. Oxford Science Publications. The Clarendon Press, Oxford 2006. xiv + 311 pp. (with Roman Kossak)

*Work in Progress*

20. Tennenbaum's theorem and recursive models (a paper that discusses various ways that Tennenbaum's theorem on the nonrecursiveness of nonstandard models of Peano Arithmetic can and cannot be generalized, emphasizing the role of wqo).

**Teaching:** Ermek Nurkhaidorov received his PhD under Schmerl's direction in 2004.

**Vision Statement:** I would like to see more interest in nonstandard models of Peano Arithmetic. It would be very satisfying if the question of which finite lattices can appear as substructure lattices for nonstandard models of Peano Arithmetic could be answered. I believe an answer might have repercussions extending beyond the question itself.

# SCHROEDER-HEISTER, Peter

**Specialties:** Proof Theory, Computational Logic, Philosophy of Logic, History of Logic

**Born:** 2 March 1953 in Düren, Rhineland, Germany.

**Educated:** University of Bonn 'Staatsexamen' (Philosophy and Mathematics) 1977, University of Bonn Dr. phil. (Logic and Foundations of Mathematics) 1981, University of Konstanz Dr. phil. habil. (Philosophy) 1988.

**Dissertation:** *Untersuchungen zur regellogischen Deutung von Aussagenverknüpfungen ('Investigations into the rule-based interpretation of logical connectives')*; supervisors Gisbert Hasenjaeger, Dag Prawitz.

**Regular Academic or Research Appointments:** PROFESSOR OF LOGIC AND PHILOSOPHY OF LANGUAGE, UNIVERSITY OF TÜBINGEN, DEPARTMENT OF COMPUTER SCIENCE AND DEPARTMENT OF PHILOSOPHY, 1991–, Department of Computational Linguistics, 1989–1991; Lecturer and research associate, University of Konstanz, 1978–1989.

**Visiting Academic or Research Appointments:** Visiting Scholar, University of California, Irvine, 2005; Ohio State University, Columbus, 2003; Beijing University, 2002, Queen Mary and Westfield College, London 1999/2000, Imperial College, London 1997/1998; Chalmers University of Technology, Gothenburg, 1996; University of St Andrews, 1995 and 1985, University of Berne, 1994; University of Stockholm, 1987 and 1984, Institute for Advanced Studies in the Humanities, Edinburgh, 1983.

**Research Profile:** Schroeder-Heister's main area of interest is general proof theory and its application to questions of meaning, validity and consequence. He calls this subject *proof-theoretic semantics*, thus expressing that the proof-theoretic approach competes with the more common model-theoretic approaches to semantics. In contrast to Dag Prawitz, who developed the idea of a general proof theory, he favours an approach which is based on rules rather than whole proofs being the philosophically primary subjects of semantical considerations.

In the early 1980s, Schroeder-Heister developed a general schema for introduction and elimination rules for logical connectives and quantifiers, using rules of arbitrary finite levels. In joint work with Lars Hallnäs he used these ideas in a proof-theoretic approach to logic programming with iterated implications. Using an idea that they call *definitional reflection*, reading program clauses as definitions, they extended this approach to a conception of logic programming which allows the introduction of assumptions during the execution of a program.

Philosophically, these ideas led Schroeder-Heister to a general concept of definitional reasoning that permits a novel treatment of circular phenomena and paradoxical reasoning. This means in particular that sequent calculi are considered as reasoning systems in their own right, with the notion of an assumption being given equal emphasis to that of an assertion.

These approaches have been combined with aspects of substructural logic, in particular by considering contraction-free logics and Lambek calculi. The term *substructural logic*, which is due to Kosta Došen, was coined by a conference with that title in Tübingen in 1990, jointly organized by Došen and Schroeder-Heister.

Furthermore, Schroeder-Heister is interested in the history of modern logic, where he showed a special interest in the works of Gottlob Frege and Gerhard Gentzen, and their relationship with each other. He traced the origins of rule-based semantics back to Frege and Paul Hertz and found related ideas also in Karl Popper's logical writings.

Besides logic, Schroeder-Heister has also worked in general philosophy and in experimental psychology.

**Main Publications:**

1. "A natural extension of natural deduction". *Journal of Symbolic Logic* 49 (1984): 1284–1300.
2. "Generalized rules for quantifiers and the completeness of the intuitionistic operators $\wedge, \vee, \rightarrow, \bot, \forall, \exists$". In: M. M. Richter, E. Börger, W. Oberschelp, B. Schinzel, W. Thomas (eds.), *Computation and Proof Theory. Proceedings of the Logic Colloquium held in Aachen, July 18-23, 1983, Part II*. Springer, Berlin, 1984, pp. 399–426 (Lecture Notes in Mathematics, Vol. 1104).
3. "Popper's theory of deductive inference and the concept of a logical constant", *History and Philosophy of Logic* 5 (1984): 79–110.
4. "Proof-theoretic validity and the completeness of intuitionistic logic". In G. Dorn, P. Weingartner (eds.), *Foundations of Logic and Linguistics: Problems and Their Solutions*. Plenum Press, New York, 1985, pp. 43–87.
5. "A model-theoretic reconstruction of Frege's permutation argument". *Notre Dame Journal of Formal Logic* 28 (1987): 69–79.
6. "Uniqueness, definability and interpolation", with K. Došen. *Journal of Symbolic Logic* 53 (1988): 554–570.
7. "Reduction, representation and commensurability of theories", with F. Schaefer. *Philosophy of Science* 56 (1989): 130–157.
8. "A proof-theoretic approach to logic programming. I. Clauses as rules, II. Programs as definitions", with L. Hallnäs. *Journal of Logic and Computation* 1 (1990): 261—283, 635–660.
9. "Rules of definitional reflection". In: *Proceedings of the 8th Annual IEEE Symposium on Logic in Computer Science (Montreal 1993)*. IEEE Press, Los Alamitos, 1993, pp. 222–232.
10. *Substructural Logics*, edited with K. Došen, Clarendon Press, Oxford, 1993.
11. "Definitional reflection and the completion". In R. Dyckhoff (ed.), *Extensions of Logic Programming. Fourth International Workshop, St. Andrews, Scotland, April 1993, Proceedings*. Springer, Berlin, 1994, pp. 333–347 (Lecture Notes in Artificial Intelligence, Vol. 798).
12. "Classical Lambek logic", with J. Hudelmaier. In P. Baumgartner, R. Hähnle, J. Posegga (eds.), *Theorem Proving with Analytic Tableaux and Related Methods. 4th International Workshop, TABLEAUX '95 (St. Goar, May 7-10, 1995), Proceedings*. Springer, Berlin, 1995, pp. 247–262 (Lecture Notes in Artificial Intelligence, Vol. 918).
13. "Frege and the resolution calculus". *History and Philosophy of Logic* 18 (1997): 95–108.
14. "Resolution and the origins of structural reason-

ing: Early proof-theoretic ideas of Hertz and Gentzen". *Bulletin of Symbolic Logic* 8 (2002): 246–265.
15. "On the notion of assumption in logical systems". In R. Bluhm, C. Nimtz (eds.), *Selected Papers Contributed to the Sections of GAP5, Fifth International Congress of the Society for Analytical Philosophy*, Bielefeld, 22-26 September 2003, mentis, Paderborn, 2004, pp. 27–48.
16. "Frege's Permutation Argument Revisited", with K. F. Wehmeier. *Synthese* 147 (2005): 43–61

**Work in Progress**
17. "Validity Concepts in Proof-Theoretic Semantics". In R. Kahle, P. Schroeder-Heister (eds.), *Proof-Theoretic Semantics*, special issue of *Synthese*, 2006, to appear.
18. "Popper's structuralist theory of logic". In I. Jarvie, K. Milford, D. Miller (eds.), *Karl Popper: A Centenary Assessment. Volume III: Science and Social Science*, Ashgate, Aldershot, 2006, to appear.
19. "A survey of definitional reflection", with L. Hallnäs, in preparation.
20. *Proof-Theoretic Semantics*, monograph, in preparation.

**Service to the Profession:** Member Editorial Board: Notre Dame Journal of Formal Logic, 1992–2003; History and Philosophy of Logic, 1993–.

**Vision Statement:** Schroeder-Heister believes that the borderlines between mathematical logic, philosophical logic, logic in linguistics and logic in computer science will become increasingly irrelevant. Proof-theoretical and computational approaches will merge with model-theoretic ideas and alternative proposals (dynamical, probabilistic, game-theoretical ...). Logic has a bright future as a fundamental discipline relevant to many fields.

# SCHWICHTENBERG, Helmut

**Specialties:** Proof theory, lambda calculus, recursion theory, applications of
logic to computer science.

**Born:** 1942 in Sagan/Schlesien.

**Educated:** Universitaet Muenster, Habilitation fuer Mathematik, 1974; Mathematisches Institut, Universitaet Muenster, Dr.rer.nat, 1968; Berlin (Freie Universitaet), Muenster, 1961-1968.

**Dissertation:** *Eine Klassifikation der mehrfachrekursiven Funktionen*, Universitaet Muenster, 1968; supervisor, D. Roedding.

**Regular Academic or Research Appointments:** PROFESSOR, ORDINARIUS, MATHEMATISCHES INSTITUT, UNIVERSITAET MUENCHEN, 1978–; Wiss. Rat. und Professor, Mathematisches Institut, Universitaet Heidelberg, 1974-1978.

**Visiting Academic or Research Appointments:** Participant, Logic Year at Mittag-Leffler Institute, Sweden, 2001; Visiting Professor, Stanford University, 2000; Researcher, Carnegie-Mellon-University, Pittsburgh, 1986-1987; Member, Bayerische Akademie der Wissenschaften; Researcher, Stanford University, with Feferman and Kreisel, 1971-1972, 1981-1982.

**Research Profile:** Helmut Schwichtenberg has been involved in various research projects, on topics including

**Main Publications:**

1. (with Anne S. Troelstra): Basic Proof Theory, Cambridge University Press, Second edition 2000.
2. New Developments in Proofs and Computations. In: New Computational Paradigms (B. Cooper, B. Löwe, A. Sorbi, eds.), 2008.
3. Realizability interpretation of proofs in constructive analysis. To appear in ToCS, 2007.
4. Recursion on the partial continuous functionals. In: Logic Colloquium '05, (C. Dimitracopoulos and L. Newelski and D. Normann and J. Steel, eds.), 2006.
5. Minlog. In: The Seventeen Provers of the World (F. Wiedijk, ed.), 2006
6. An arithmetic for polynomial-time computation. In: Theoretical Computer Science, 2006.
7. Program extraction from normalization proofs (with U. Berger, S. Berghofer and P. Letouzey). In Studia Logica, 2006.
8. A direct proof of the equivalence between Brouwer's fan theorem and K"onig's lemma with a uniqueness hypothesis. In: Journal of Universal Computer Science, 2005.
9. An arithmetic for non-size-increasing polynomial time computation (with K. Aehlig, U. Berger and M. Hofmann). In: Theoretical Computer Science, 2004.
10. Refined Program Extraction from Classical Proofs (with Ulrich Berger and Wilfried Buchholz). Annals of Pure and Applied Logic 114 (2002), pp. 3-25.
11. The Warshall Algorithm and Dickson's Lemma: Two Examples of Realistic Program Extraction (with Ulrich Berger and Monika Seisenberger). Journal of Automated Reasoning Vol. 26, 2001, pp. 205-221.
12. A syntactical analysis of non-size-increasing polynomial time computation (with Klaus Aehlig). Proc. Fifteenth Annual IEEE Symposium on Logic in Computer Science (LICS'2000), pp. 84-91.
13. Term rewriting for normalization by evaluation (with U. Berger and M. Eberl). Information and Computation Vol. 183, 2003, pp. 19-42. Classifying recursive functions. In: Handbook of Computability Theory, Editor E. Griffor, Elsevier Science Amsterdam 1999, pp. 533-586.

14. Higher Type Recursion, Ramification and Polynomial Time (with S. Bellantoni and K.-H. Niggl). Annals of Pure and Applied Logic Vol. 104, 2000, pp. 17-30.

15. Dialectica interpretation of well-founded induction, Math. Logic. Quarterly, 2008.

http://www.mathematik.uni-muenchen.de/~schwicht

**Honours and Awards:** Invited Speaker, Computability in Europe, Logical approaches to computational barriers, Swansea, 2006; Invited Speaker, ASL European Logic Colloqium, Athens, 2005; Grant, Max-Kade-Foundation, 1971-1972.

# SCOTT, Dana Stewart

**Specialties:** Theoretical computer science, mathematical logic, philosophical logic.

**Educated:** University of California, Berkeley, BA in Mathematics, 1954. PhD University of Princeton, 1958.

**Dissertation:** *Convergent Sequences of Complete Theories*; supervisor Alonzo Church.

**Regular Academic or Research Appointments:** EMERITUS HILLMAN UNIVERSITY PROFESSOR OF COMPUTER SCIENCE, PHILOSOPHY, AND MATHEMATICAL LOGIC AT CARNEGIE MELLON UNIVERSITY. Carnegie Mellon University, 1981–2003; Oxford University, 1972–1981; Stanford, Amsterdam and Princeton, 1963–72; University of Califorinia, Berkeley, 1960–1963; University of Chicago, 1958–1960;

**Research Profile:** His research career has spanned computer science, mathematics, and philosophy, and has been characterized by a marriage of a concern for elucidating fundamental concepts in the manner of informal rigor, with a cultivation of mathematically hard problems that bear on these concepts. His work on automata theory earned him the ACM Turing Award in 1976, while his collaborative work with Christopher Strachey in the 1970s laid the foundations of modern approaches to the semantics of programming languages. He has worked also on modal logic, topology, and category theory. He is the editor-in-chief of the new journal Logical Methods in Computer Science.

In 1959, he published a joint paper with Michael O. Rabin, a colleague from Princeton, entitled Finite Automata and Their Decision Problem, which introduced the idea of nondeterministic machines to automata theory. This work led to the joint bestowal of the Turing Award on the two, for the introduction of this fundamental concept of computational complexity theory.

During his time at Berkely, Scott began working on modal logic in this period, beginning a collaboration with John Lemmon. Scott was especially interested in tense logic and the connection to the treatment of time in natural-language semantics, and began collaborating with Richard Montague. Later, Scott and Montague were independently to discover an important generalisation of Kripke semantics for modal and tense logic called Scott-Montague semantics. John Lemmon and Scott began work on a modal-logic textbook that was interrupted by Lemmon's death in 1966. Scott circulated the incomplete monograph amongst colleagues, introducing a number of important techniques in the semantics of model theory, most importantly presenting a refinement of canonical model that became standard, and introducing the technique of constructing models through filtrations, both of which are core concepts in modern Kripke semantics. Scott eventually published the work as An Introduction to Modal Logic.

Whilst at Oxford, Scott worked closely with Christopher Strachey, and the two managed, despite intense administrative pressures, to oversee a great deal of fundamental work on providing a mathematical foundation for the semantics of programming languages, the work for which Scott is best known. Together, their work constitutes the Scott-Strachey approach to denotational semantics; it constitutes one of the most influential pieces of work in theoretical computer science and can perhaps be regarded as founding one of the major schools of computer science. One of Scott's largest contributions is his formulation of domain theory, allowing programs involving recursive functions and looping-control constructs to be given a denotational semantics. Additionally, he provided a foundation for the understanding of infinitary and continuous information through domain theory and his theory of information systems.

**Main Publications:**

1. With Michael O. Rabin, 1959. *Finite Automata and Their Decision Problem.*
2. 1967. A proof of the independence of the continuum hypothesis. *Mathematical Systems Theory* 1:89-111.
3. 1970. 'Advice on modal logic'. In *Philosophical Problems in Logic*, ed. K. Lambert, pages 143-173.
4. With John Lemmon, 1977. *An Introduction to Modal Logic*. Oxford: Blackwell.
5. Semantics of $\lambda$-calculus. Outline of the mathematical theory of computation, 1970.

6. Models for the λ-calculus, Manuscript (unpublished, 1969.
7. 1976. Data types as lattices, *SIAM J. Comput.* 5, 522–587.
8. 1971 with Christopher Strachey. Toward a mathematical semantics for computer languages, *Proc. Symp. on Computers and Automata*, Polytechnic Institute of Brooklyn, 21, 19–46.
9. 1996. A new category? Domains, Spaces, and Equivalence Relations.

**Honours and Awards:** Dana Scott is a fellow of the following academy/associations.

- Academia Europaea
- American Association for the Advancement of Science
- American Academy of Arts and Sciences
- Association for Computing Machinery
- British Academy
- Finnish Academy of Sciences and Letters
- New York Academy of Sciences
- U.S. National Academy of Sciences

He has also held research fellowships with:

- Bell Telephone Fellow, Princeton University 1956-57
- Miller Institute Fellow, University of California, Berkeley 1960-61
- Alfred P. Sloan Research Fellow 1963-65
- Guggenheim Foundation Fellow 1978-79
- Visiting Scientist, Xerox Palo Alto Research Center 1978-79
- Professorial Fellow, Merton College, Oxford 1972-81
- Visiting Professor, Institut Mittag-Leffler, Sweden 2001
- Humbolt Stiftung Senior Visiting Scientist, Munich, Germany 2003

In 1976 Scott was awarded the ACM Turing Award. Leroy P. Steele Prize, 1972. Inducted as Fellow of the Association for Computing Machinery, 1994. The 1990 Harold Pender Award for his application of concepts from logic and algebra to the development of mathematical semantics of programming languages; In 1997 Rolf Schock Prize in logic and philosophy from the Royal Swedish Academy of Sciences for his conceptually oriented logical works, especially the creation of domain theory, which has made it possible to extend Tarski's semantical paradigm to programming languages as well as to construct models of Curry's combinatory logic and Church's calculus of lambda conversion; and in 2001 the Bolzano Prize for Merit in the Mathematical Sciences awarded by the Czech Academy of Sciences. In 2007, Scott received 2007 EATCS Award for his contribution to theoretical computer science.

## SEGERBERG, Krister

**Specialties:** Modal logic, including epistemic logic, deontic logic, the logic of action.

**Born:** April 1936 in Skövde, Sweden.

**Educated:** Stanford University, Philosophy PhD, 1971; Uppsala University, fil. kand., 1961, fil. lic. 1965, Theoretical Philosophy fil. dr., docent, 1968; Columbia College, AB Mathematics, 1959.

**Dissertation:** Uppsala dissertation: *Some Results in modal logic*. Stanford dissertation: *An essay in classical modal logic*; supervisor, Dana Scott.

**Regular Academic or Research Appointments:** PROFESSOR OF THEORETICAL PHILOSOPHY, EMERITUS, UPPSALA UNIVERSITY, 2001–; PROFESSOR OF PHILOSOPHY, EMERITUS, UNIVERSITY OF AUCKLAND, 1992–; Professor of Theoretical Philosophy, Uppsala University, 1991-2001; Professor of Philosophy, University of Auckland, 1980-1992; Professor of Philosophy, Åbo Academy, 1972-1979.

**Visiting Academic or Research Appointments:** Fellow, N.I.A.S. (The Netherlands Institute for Advanced Study), 2006-2007; E.W. Beth Visiting Professor of Logic, University of Amsterdam, 2006; Visiting Professor of Philosophy, Stanford University, 2004-2005; Visiting Professor of Philosophy, UCLA, 2004; Visiting Professor, University of Southern California, 2003; Fellow, S.C.A.S.S.S. (Swedish Collegium for Advanced Study in the Social Sciences), 2001-2002; Visiting Professor, U.C. Irvine, 1991, Visiting Professor, University of Oslo, 1990; Visiting Fellow in R.S.S.S., ANU Canberra, 1985; Visiting Professor, University of Parma, 1981; Visitor in the Academy of Sciences of the Georgian S.S.R., Tbilisi, 1978; Visiting Professor, University of Kansas, Lawrence, 1976-1977; Visiting Professor, University of Calgary, 1977.

**Research Profile:** Segerberg has spent his entire career in a relatively narrow field: modal logic, although modal logic in a wide sense. In earlier years, he was absorbed by its purely mathematical aspects but he has devoted himself in later years to applications of modal logic to philosophically important concepts such as rational belief, norms, and change.

**Main Publications:**

1. "Decidability of S4.1." *Theoria*, vol. 34 (1968), pp.7-20.
2. "Propositional logics related to Heyting's and Johansson's." *Theoria*, vol. 34, pp. 26-61.
3. "Modal logics with linear alternative relations." *Theoria*, vol. 36 (1970), pp. 301-322.
4. *An essay in classical modal logic*. Uppsala: The Philosophical Society, 1971.
5. "Two-dimensional modal logic." *Journal of philosophical logic*, vol. 2 (1973), pp. 77-96.
6. "A completeness theorem in the modal logic of programs." In *Universal algebra and applications*, edited by T. Tradczyk, pp. 31-46. Banach Center Publicaitons, vol. 9. Warsaw: PWN, 1982.
7. *Classical propositional operators: an exercise in the foundations of logic*. Oxford: Clarendon Press, 1982.
8. "Routines." *Synthese*, vol. 65 (1985), pp. 185-210.
9. "Bringing it about." *Journal of philosophical logic*, vol. 18 (1989), pp. 327-347.
10. "Validity and satisfaction in imperative logic." *Notre Dame journal of formal logic*, vol. 31 (1990), pp. 203-221.
11. "A model existence theorem in infinitary modal propositional logic." *Journal of philosophical logic*, vol. 23 (1994), pp. 337-367.
12. "A festival of facts." *Logic and logical philosophy*, vol. 2 (1994), pp. 9-22.
13. (With Brian F. Chellas) "Modal logics in the vicinity of S1." *Notre Dame journal of formal logic*, vol. 37 (1996), pp. 1-24.
14. "To do and not to do" In *Logic and reality: Essays on the legacy of Arthur Prior*, edited by B.J. Copeland, pp. 301-313. Oxford: Clarendon Press, 1996.
15. "Belief revision along the lines of Lindström and Rabinowicz." *Fundamenta informaticæ*, vol. 32 (1997), pp. 183-191.
16. "Irrevocable belief revision in dynamic doxastic logic." *Notre Dame journal of formal logic*, vol. 39 (1997), pp. 287-306.
17. "Two traditions in the logic of belief: bringing them together." In *Logic, language and reasoning: essays in honour of Dov Gabbay*, edited by Hans Jürgen Ohlbach & Uwe Reyle, pp. 135-147. Dordrecht, The Netherlands: Kluwer, 1999.
18. "Default logic as dynamic doxastic logic." *Erkenntnis*, vol. 50 (1999)l, pp. 333-352. Reprinted in "Default logic as dynamic doxastic logic." In *Dynamics and management of reasoning processes*, edited by J.-J. Ch. Meyer & J. Treur, pp. 159-176. *Handbook of defeasible reasoning and uncertainty management systems*, vol. 6. Dordrecht, The Netherlands: Kluwer: 2001.
19. "The basic dynamic doxastic logic of AGM." In *Fronteirs in belief revision*, edited by Mary-Anne Williams and Hans Rott, pp.57-84. Applied Logic Series, vol. 22. Dordrecht, The Netherlands: Kluwer, 2001.
20. "Outline of a logic of action." In *Advances in modal logic*, vol. 3, edited by F. Wolter, H. Wansing, M. de Rijke & M. Zakharyaschev, pp. 365-287. River Edge, New Jersey & London: World Scientific, 2002.
21. "Moore problems in full dynamic doxastic logic." In *Essays in logic and ontology: dedicated to Jerzy Perzanowski*, edited by J. Malinowski and A. Pietruszczak, pp 11-25. Proznan Studies in the Philosophy of the Sciences and the Humanities, vol. 50. Amsterdam/Atlanta, GA: Rodopi, 2003.
22. "Deconstruction of epistemic logic." In *Knowledge and belief / Wissen und Glauben*, pp. 103-119. Proceedings of the $26^{th}$ International Wittgenstein Symposium, Kirchberg am Wechsel, 2003, edited by Winfred Löffler & Paul Weingartner. Vienna: Österreichischer Bundesverlag & Hölder-Pichler-Tempsky, Vienna, 2004.

*Work in Progress*

23. *Work in Progress*: A book on dynamic doxastic logic and dynamic deontic logic.

**Service to the Profession:** Member of Standing Committee for Logic in Australasia, Association of Symbolic Logic, 1983-1989, Member of the Council, 1983-1986; President, Australasian Association of Logic, 1985-1986; Chair, National Committee for Logic, Methodology and Philosophy (of the Royal Swedish Academy of Sciences), 1985-1986; Committee for the Rolf Schock Prize for Philosophy and Logic, Royal Swedish Academy of Sciences, 1996-2004; Editor, *Theoria*, 1969-; Editor, *Journal of philosophical logic*, 1994-2001; Collecting Editor, *Studia logica*, 1976-1992.

**Teaching:** Segerberg has four PhD students.

**Vision Statement:** Logic was born in Philoosphy, came of age in Mathematics, and is now thriving in Computer Science. This development has affected-and is affecting-its nature. But it remains an indispensable tool for some very important areas of philosophical analysis.

**Honours and Awards:** Daniel Taylor Visiting Fellow, University of Otago, 1994; M.A.S.U.A. (Mid-American State Universities Association), Distinguished Foreign Visiting Fellow, 1988; Andrew Mellon Postdoctoral Fellow, University of Pittsburgh, 1971-1972.

# SELDIN, Jonathan P.

**Specialties:** Mathematical logic, combinatory logic and lambda-calculus, proof theory, philosophy of mathematics and logic, history of logic.

**Born:** 30 January 1942 in New York City, New York, USA.

**Educated:** University of Amsterdam, Dr. Mat. Nat. Sci., 1968; The Pennsylvania State University, MA Mathematics, 1966; Oberlin College, BA Cum Laude Mathematics, 1964.

**Dissertation:** *Studies in Illative Combinatory Logic;* supervisor, Haskell B. Curry; co-referent, Arend Heyting.

**Regular Academic or Research Appointments:** ASSOCIATE PROFESSOR, MATHEMATICS AND COMPUTER SCIENCE, UNIVERSITY OF LETHBRIDGE, 2002–; Assistant Professor, Mathematics and Computer Science, University of Lethbridge, 1999-2002; Adjunct Assistant Professor, Mathematics, Concordia University (Montreal), 1983-1999; Logician, Odyssey Research Associates, 1986-1989; Assistant Professor, Mathematics, Southern Illinois University, 1969-1976; Lecturer, Pure Mathematics, University College of Swansea, University of Wales, 1968-1969.

**Visiting Academic or Research Appointments:** Visiting Lecturer, Concordia University, Montreal, 1979-1983; Visiting Researcher, Mathematical Institute, Oxford University, 1976-1979; Senior Visiting Fellowship, British Science Research Council and Visiting Fellow at Wolfson College, Oxford University, 1976-1977.

**Research Profile:** Jonathan P. Seldin has made significant contributions to lambda-calculus and combinatory logic, including both systems with types. He began his work on Curry's program of finding systems of logic in the ordinary sense based on these two formalisms. Among these systems are counted systems of what Curry called *functionality*, or, as it is now called, typeassignment. This work ultimately led Seldin to work on the *Calculus of Constructions* of Coquand and Huet, a formalism which forms the basis for the proof assistant Coq. Seldin's work on this system, in addition to work on some of the basic proof theory, included proving the consistency of sets of unprovable assumptions that are useful for practical applications for a system like Coq. Seldin has also obtained results in proof theory, including a normalization result for the intermediate logic MH, some normalization results for systems of classical logic, and a result which shows that the strength of a system of formal logic depends on its rules for implication (and negation, which can be defined in terms of implication). Recently, Seldin has become interested in applications of the history of mathematics to the teaching of mathematics, especially the history of the notion of proof in mathematics, and also in some possible philosophical implications of Gödel's Incompleteness Theorems outside of the field of mathematical logic.

**Main Publications:**

1. (with Haskell B. Curry and J. Roger Hindley) *Combinatory Logic,* Vol. II (Amsterdam: North-Holland, 1972).
2. (with Bruce Lercher and J. Roger Hindley) *Introduction to Combinatory Logic* (Cambridge University Press, 1972).
3. A sequent calculus for type assignment, Journal of Symbolic Logic, 42 (1977) 11–28.
4. A sequent calculus formulation of type assignment with equality rules for the $\lambda\beta$-calculus, Journal of Symbolic Logic, 43 (1978) 643–649.
5. Progress report on generalized functionality, Annals of Mathematical Logic, 17 (1979) 29–59.
6. Curry's program, in *To H. B. Curry: Essays on Combinatory Logic, Lambda Calculus and Formalism,* edited by J. Roger Hindley and Jonathan P. Seldin (London, New York, et al: Academic Press, 1980), pp. 3–33.
7. (with J. Roger Hindley) *Introduction to Combinators and $\lambda$-Calculus* (Cambridge University Press, 1986).
8. On the proof theory of the intermediate logic MH, Journal of Symbolic Logic, 51 (1986) 626–647.
9. Normalization and excluded middle, I, Studia Logica, 48 (1989) 194–217.
10. (with Martin W. Bunder and J. Roger Hindley) On adding $(\xi)$ to weak equality in combinatory logic, Journal of Symbolic Logic, 54 (1989) 590–607.
11. Coquand's calculus of constructions: a mathematical foundation for a proof development system, Formal Aspects of Computing 4 (1992) 425–441.
12. On the proof theory of Coquand's calculus of constructions, Annals of Pure and Applied Logic, 83 (1997) 23–101.
13. On the role of implication in formal logic, Journal of Symbolic Logic, 65 (2000) 1076–1114.
14. A Gentzen-style sequent calculus of constructions with expansion rules, Theoretical Computer Science, 243 (2000) 199–215.
15. On lists and other abstract data types in the calculus of constructions, Mathematical Structures in Computer Science, 10 (2000) 261–276.
16. Extensional set equality in the calculus of constructions, Journal of Logic and Computation, 11 (2001) 483–493.
17. Interpreting HOL in the calculus of constructions, Journal of Applied Logic, 2 (2004) 173–189.
18. (with Martin W. Bunder) Variants of the basic calculus of constructions, Journal of Applied Logic, 2 (2004) 191–217.
19. Gödel, Kuhn, Popper, and Feyerabend, in *Karl Popper: A Centenary Assessment, Selected Papers from Karl Popper 2002 (Vienna, 3–7 July 2002),* vol. II, edited by Ian Jarvie, Karl Milford, and David Miller (Ashgate, 2007).

20. The Logic of Church and Curry, in *Handbook of the History of Logic: Logic from Russell to Church*, vol. 5, edited by Dov Gabbay and John Woods, (Amsterdam: Elsevier/North-Holland, to appear in 2007).

A full list of publications is available at Seldin's home page http://www.cs.uleth.ca/~seldin/.

*Work in Progress*

21. (with J. Roger Hindley) *Lambda-calculus and Combinators, an Introduction*, (Cambridge University Press, 2008), new version of item 7 above.
22. Beta-strong reduction in combinatory logic.
23. Thoughts on teaching elementary mathematics.
24. On the relation between Church-style typing and Curry-style typing.
25. Project to publish the writings of H. B. Curry, possibly including previously unpublished material. Project to begin with Curry's philosophical works.
26. Project to examine the consequences of Gödel's Incompleteness Theorem and related results to areas outside mathematics and mathematical logic. The results in question show that there are formal theories that cannot be completely and correctly characterized by means of strictly formal rules. If the theories and rules considered are not strictly formal, it is not, in general, possible to give proofs of similar results, but in most cases there is also no proof of completeness, and so it does not follow that it can be assumed that the theories can be completely and correctly characterized by means of rules. Examples occur in philosophy (characterization of the notions of right and wrong by means of rules, for example) and the law. In the case of the law, the relation between such results and Gödel's Incompleteness Theorems and the theory of computing raises the question of whether the application of ideas from the practice of computer programming can improve the process of making laws.

**Service to the Profession:** Committee Member, Mathematical Association of America Secondary School Lecturers, 1974-1977; Co-Editor, with Roger Hindley, *To H. B. Curry: Essays on Combinatory Logic, Lambda Calculus and Formalism* (London, New York, *et al*: Academic Press, 1980); Guest Editor, special issue of *Mathematical Structures in Computer Science* (volume 9 number 4) in honor of Roger Hindley, 1997-1999.

**Teaching:** Seldin has taught the complete range of elementary mathematics undergraduate courses and also some upper level undergraduate courses in mathematics and elementary courses in computer science courses at four different universities during his career. He has also taught some graduate level courses and supervised three master's students.

**Vision Statement:** Mathematical Logic as a subject began in the nineteenth century as a study closely related to the foundations of mathematics. In the twentieth century, its role expanded, as it came to include the theory of effective computability and then became increasingly important in theoretical computer science. Logicians such as John von Neumann were involved in the development of computers from the beginning with the ENIAC project during World War II, and since the development of FORTRAN, mathematical logic has influenced the development of programming languages. This influence has increased dramatically since computers became big enough and powerful enough to do serious symbolic computation. I think this connection between mathematical logic and theoretical computer science will continue to grow as time goes on. I also think some of the most important results in the field, Gödel's Incompleteness Theorem and related results, suggest important limits on our ability to use rules to completely and correctly characterize subjects. This is only one of the many contributions modern mathematical logic is likely to make to our culture as time goes on.

# SHAPIRO, Stewart

**Specialties:** Philosophical logic, philosophy of logic, logic of vagueness.

**Born:** 15 June, 1951 in Youngstown, Ohio, USA.

**Educated:** State University of New York at Buffalo, PhD, 1978; State University of New York at Buffalo, MA, 1975; Case Western University, Cleveland, Ohio, BA, 1973.

**Dissertation:** *The Philosophical Background of Computability*; supervisor, John Corcoran.

**Regular Academic or Research Appointments:** PHILOSOPHY, OHIO STATE UNIVERSITY, 1978- , Newark, Professor, 1991; Columbus Campus, 2001; O'Donnell Professor of Philosophy, 2002-; Professor of Philosophy, Department of Logic and Metaphysics, University of St. Andrews, St. Andrews, Scotland, 1996-1997.

**Visiting Academic or Research Appointments:** Arché Research Centre, Department of Logic and Metaphysics, University of St. Andrews, St. Andrews, Scotland, Professorial Fellow, Quarter-Time appointment, 1999-; Visiting Fellow, Center for Philosophy of Science, University of Pittsburgh, 1987-88; Visiting Lecturer, Department of Philosophy, The Hebrew University of Jerusalem, 1982.

**Research Profile:** I work in the philosophy of mathematics, logic, and the philosophy of logic. My publications include a defense of second-order logic, an articulation and defense of structuralism in the philosophy of mathematics, the logic and semantics of vagueness, computability, the abstractionist neo-logicist foundations of mathematics, intuitionism, paradox, and paraconisistent logic.

**Main Publications:**

1. *Vagueness in context*, Oxford, Oxford University Press, 2006.
2. *Thinking about mathematics: the philosophy of mathematics*, Oxford, Oxford University Press, 2000.
3. *Philosophy of mathematics: Structure and ontology*, Oxford, Oxford University Press, 1997; reissued in paperback, Autumn 2000.
4. *Foundations without foundationalism: A case for second-order logic*, Oxford Logic Guides 17, Oxford, Oxford University Press, 1991, reissued in paperback, Summer 2000.
5. "Categories, structures, and the Frege-Hilbert controversy: the status of meta-metamathematics", *Philosophia Mathematica (3) 13* (2005), 61-77.
6. "Logical consequence, proof theory, and model theory", *Oxford handbook for the philosophy of mathematics and logic*, edited by Stewart Shapiro, Oxford, Oxford University Press, 2005, 651-670.
7. "Simple truth, contradiction, and consistency", *The law of non-contradiction*, edited by Graham Priest and J. C. Beall, Oxford, Oxford University Press, 2004, 336-354.
8. "All sets great and small: and I do mean ALL", *Philosophical Perspectives 17* (2003), 467-490.
9. "The guru, the logician, and the deflationist", *Noûs 37* (2003), 113-132.
10. "Incompleteness and inconsistency", *Mind 111* (2002), 817-832.
11. "Frege meets Dedekind: a neo-logicist treatment of real analysis", *Notre Dame Journal of Formal Logic 41 (2000)*, 335-364.
12. "The status of logic", *New essays on the a priori*, edited by Paul Boghossian and Christopher Peacocke, Oxford University Press, 2000, 333-366; reprinted in part as "Quine on Logic", Logica Yearbook 1999, edited by Timothy Childers, Prague, Czech Academy Publishing House, 11-21.
13. "Incompleteness, mechanism, and optimism", *Bulletin of Symbolic Logic 4* (1998), 273-302.
14. "Proof and truth: Through thick and thin", *Journal of Philosophy 95* (1998), 493-521.

Work in Progress

15. "All things indefinitely extensible" (with Crispin Wright), *Unrestricted quantification*, edited by Agustin Rayo and Gabriel Uzquiano.
16. "Structure and identity" in *Modality and identity*, edited by Fraser MacBride, Oxford University Press, forthcoming.
17. "We hold these truths to be self-evident: But what do we mean by that? The rationalism of Frege and Zermelo."

**Service to the Profession:** Editorial Board, *Journal of Symbolic Logic*; Editorial Board, *Philosophia Mathematica*; Editorial Board, *Notre Dame Journal of Formal Logic*; Editorial Board, *Philosophical Quarterly*.

**Edited volumes:**

Editor, *Oxford handbook of the philosophy of logic and mathematics*, Oxford University Press, 2005; Executive committee, Association for Symbolic Logic, 2002-2005; Editor, Special issue of *History and Philosophy of Logic 20* (2000), a Festschrift for John Corcoran (edited with Michael Scanlan); Editor, Special issue of *Philosophia Mathematica (3) 8* (2000), devoted to abstraction and neo-logicism; Editor, Two special issues of *Philosophia Mathematica (3) 7* (1999), 9 (2001), devoted to the proceedings of a conference in memory of George Boolos, held at Notre Dame, 1998; Editor, *The limits of logic: Second-order logic and the Skolem paradox*, The international research library of philosophy, Dartmouth Publishing Company, 1996; Editor, *Special issue of Philosophia Mathematica (3) 4* (1996), devoted to structuralism. Contributors: P. Benacerraf, G. Hellman, B. Hale, C. Parsons, M. Resnik, S. Shapiro; Editor, *Intensional Mathematics, Studies in Logic and the Foundations of Mathematics 113*, Amsterdam, North Holland Publishing Company, 1985, Contributors: S. Shapiro, J. Myhill, N. D. Goodman, A. Scedrov, V. Lifschitz, R. Flagg, R. Smullyan.

**Teaching:** Taught philosophy and logic, at all levels, for 27 years. My PhD students include Beth Cohen, Rick Dewitt, Barbara Scholz (directed), Jill Dieterle (directed), Pierluigi Miraglia (directed), David Lightner, Michele Friend (directed), Roy Cook (directed), Jon Cogburn, Joseph Salerno, Patrice Philie, Lars Gunderson, Julian Cole (directed), and William Melanson (directed).

**Vision Statement:** To articulate and defend defensible versions of foundational programs in mathematics, philosophies that account for the logic, semantics, and epistemology of mathematics, and to apply the techniques of logic beyond mathematics, to the logic and semantics of vagueness, demonstratives, and the like.

**Honours and Awards:** University Distinguished Scholar Award, Ohio State University, 2003.

# SHEHTMAN, Valentin B.

**Specialties:** Modal logic, intuitionistic logic.

**Born:** 27 March 1953 in Moscow, Russia (formerly, USSR).

**Educated:** Moscow State Univerisity, Habilitation Mathematics (Russian degree of Doctor of Physical and Mathematical Sciences), 2000;

**Dissertation:** *Dissertation: Modal Logics of Topological Spaces.* Steklov Mathematical Institute, St. Petersburg (Leningrad) branch, PhD Mathematics (Russian degree of Candidate of Physical and Mathematical Sciences), 1985.

**Dissertation:** *Application of Kripke models to superintuitionistic and modal logics.* Moscow State University, Postgraduate studies in mathematics, 1977-1980, supervisor, Andrei A. Markov. Moscow State Pedagogical Institute, MSc in Mathematics, 1975.

**Regular Academic or Research Appointments:** PROFESSOR, MATHEMATICS, MOSCOW STATE UNIVERSITY, 2000–; LEADING RESEARCHER, INSTITUTE OF INFORMATION TRANSMISSION PROBLEMS, RUSSIAN ACADEMY OF SCIENCES, 2000–; RESEARCHER, PONCELET MATHEMATICAL LABORATORY OF CNRS AND INDEPENDENT UNIVERSITY OF MOSCOW, 2001–; Senior Research Fellow, Institute of Information Transmission Problems, Russian Academy of Sciences, 1993-2001; Senior Research Fellow, Moscow State University, 1995–2001; Leading Researcher, Institute of New Technologies, Moscow, 1990–1993; Senior Research Fellow, Research Institute of General Moscow Planning, 1984–1990; Senior Engineer, Research Institute of Patent Information, 1982–1984; High School Teacher, Mathematics, Moscow, 1975–1977, 1981–1982.

**Visiting Academic or Research Appointments:** Visiting Professor, King's College London, 2001–; Visiting Researcher, King's College London, 1998-2000; Visiting Researcher, Imperial College London, 1996-1998; Visiting Professor, University of Marseille, 2006; Visiting Researcher, University of Utrecht, 2006; Visiting Researcher, IRIT, Toulouse, 2002-2006; Visiting Researcher, LORIA, Nancy, 2004; Visiting Professor, University Paul Sabatier of Toulouse, 2001; Postdoctoral fellow, University of Amsterdam, 1991-1992; Visiting Researcher, Hiroshima University, 1991; Visiting Researcher, University of Amsterdam, 1990.

**Research Profile:** Valentin Shehtman worked in mathematics of modal and intuitionistic logics. His most essential contributions in this field are in model theory and decision problems. He constructed the first examples of intermediate logics incomplete in Kripke semantics and of finitely axiomatizable undecidable intermediate logics. He introduced and investigated products of modal logics (also in works with Dov Gabbay), a natural kind of combined modal logics. He showed that topological semantics of intermediate propositional logics is stronger than Kripke semantics and studied different kinds of compactness in topological semantics for modal and intermediate logics.

He found axiomatizations for derivational modal logics of Euclidean spaces (McKinsey – Tarski problems). He investigated modal logics of Minkowski time-space and related modal logics of regions and intervals (also in works with Ilya Shapirovsky). He developed a general method of finite model property proofs in modal logic – filtration via bisimulation. In joint works with Dmitrij Skvortsov he developed new kinds of semantics for first-order modal logics, such as Kripke bundle and simplicial (metaframe) semantics, and proved completeness and incompleteness results.

**Main Publications:**

1. Incomplete propositional logics (Russian), *Dokl. Akad. Nauk SSSR*, 235 (1977), 542-545.
2. An undecidable superintuitionistic propositional calculus (Russian), *Dokl. Akad. Nauk SSSR*, 240 (1978), 549-552.
3. Two-dimensional modal logics (Russian), *Mat. Zametki* 23 (1978), 759-772.
4. Undecidable propositional calculi (Russian), in *Voprosy kibernetiki. Neklassicheskie logiki i ix prilozheniya*, Moscow (1982), 74-116.
5. Modal logics of domains on the real plane. *Studia Logica*, 42 (1983), 63-80
6. (with D. Skvortsov) Logics of some Kripke frames connected with Medvedev notion of informational types. *Studia Logica*, 45(1986), 101-118
7. (with D. Skvortsov) Semantics of non-classical first order predicate logics, in *Mathematical Logic. Proceedings of Summer School and Conference on Mathematical Logic held Sept.13-23 in Chaika, Bulgaria.* Plenum Press (1990), 105-116.
8. Derived sets in Euclidean spaces and modal logic. Preprint. University of Amsterdam,1990, ITLI Prepublication Series, X-90-05.
9. A logic with progressive tenses, in *Diamonds and Defaults: Studies in Pure and Applied Intensional Logic* (M. De Rijke, ed.) Kluwer Academic Publishers (1993), 255-285.
10. (with D. P. Skvortsov) Maximal Kripke-type semantics for modal and superintuitionistic predicate logics. *Annals of Pure and Applied Logic*, 63(1993), 69-101.
11. On strong neighbourhood completeness of modal and intermediate propositional logics (Part I), in *Advances in Modal Logic, v.1*, (M.Kracht, M. de Rijke, H. Wansing and M. Zakharyaschev, eds), CSLI Publications (1998), 209-222.
12. (with D. Gabbay) Products of modal logics, I. *Logic Journal of the IGPL*, 6(1998), 73-146.
13. "Everywhere" and "here". *Journal of Applied Non-Classical Logics*, 9 (1999), 369-380.

14. (with I. Shapirovsky) Chronological future modality in Minkowski spacetime, in *Advances in Modal Logic*, Volume 4. King's College Publications (2003), 437-459.

15. (with I. Shapirovsky) Modal logics of regions and Minkowski spacetime. *Journal of Logic and Computation*, 15(2005), 559-574.

16. Filtration via bismulation, in *Advances in Modal Logic*, v. 5 (R. Schmidt et al., eds.), King's College Publications (2005), 289-308.

17. On neighbourhood semantics 30 years later, in *We Will Show Them! Essays in Honour of Dov Gabbay*, v. 2. College Publications (2005), 663-692.

18. (with A.V. Chagrov) Algorithmic aspects of propositional tense logics, in *Lecture Notes in Computer Science*, v. 933, Springer (1995), 442-455.

19. (with D. Gabbay) Products of modal logics. II. Relativised quantifiers in classical logic. *Logic Journal of the IGPL*, v. 8 (2000), No. 2, p. 165-210.

20. (with D. Gabbay) Products of modal logics. III. Products of modal and temporal logics. *Studia Logica*, v. 72 (2002), No. 2, p. 181-208.

21. *Quantification in Nonclassical Logics*, volume 1 (with D. Gabbay and D. Skvortsov), Elsevier, 2008, to appear, see http://lpcs.math.msu.su/~shehtman

*Work in Progress*

22. A paper on derivational modal logics.

23. (with P. Balbiani, S. Kikot, I. Shapirovsky) A paper on Sahlqvist-type modal logics

**Service to the Profession:** Editor, Journal of Applied Non-classical Logics, 2003-; Advising Editor, Studia Logica, 1995-1999.

**Teaching:** V. Shehtman taught different courses at Moscow University, especially in mathematical logic. He had several successful PhD students: Alexey Kravtsov, Ilya Shapirovsky, Stanislav Kikot, Andrey Kudinov, and Timofei Shatrov.

**Vision Statement:** Together with a general trend towards application in all fields of human knowledge, logic will essentially develop its mathematical tools and methods, especially those related to geometry and topology. This will move mathematical logic closer to contemporary mathematics and physics and help its work in foundation of science started in $20^{th}$ century.

# SHELAH, Saharon

**Specialties:** Model theory, Set theory

**Born:** 3 July 1945 in Jerusalem, Israel.

**Educated:** Educated: Tel Aviv University, BSc Applied Mathematics 1964; Hebrew University, MSc Mathematics 1968; Hebrew University, PhD (Summa Cum Laude) 1970.

**Dissertation:** *Categoricity of classes of models*; supervisor, M. O. Rabin.

**Regular Academic or Research Appointments:** DISTINGUISHED VISITING PROFESSOR (PERMANENT POSITION, 2 MONTHS A YEAR), MATHEMATICS DEPARTMENT, RUTGERS UNIVERSITY, 1986–; ABRAHAM ROBINSON PROFESSOR OF MATHEMATICAL LOGIC, INSTITUTE OF MATHEMATICS, HEBREW UNIVERSITY OF JERUSALEM, 1974–; Associate Professor, Institute of Mathematics, Hebrew University of Jerusalem, 1972-1974; Assistant Professor, Institute of Mathematics, Hebrew University of Jerusalem, 1971-1972; Assistant Professor (tenure track), University of California Los Angeles, 1970-1971; Lecturer, Mathematics Department, Princeton University, 1969-1970.

**Visiting Academic or Research Appointments:** Visiting Professor, Rutgers University, 1985-1986; Visiting Professor, Simon Fraser University, 1985; Visiting Professor, University of Michigan, EECS and Mathematics, 1984-1985; Visiting Professor, University of California Berkeley, 1982; Visiting Professor, University of California Berkeley, 1978; Visiting Professor, University of Wisconsin at Madison, 1977-1978.

**Main Publications:**

1. *Classification theory and the number of nonisomorphic models*, in the series "Studies in Logic and the Foundations of Mathematics", North-Holland Publishing Co., Amsterdam, 1990 (second edition), 1978 (first edition).

2. *Proper forcing*, in the series "Lecture Notes in Mathematics", Springer-Verlag, Berlin-New York, 1982.

3. *Proper and improper forcing*, in the series "Perspectives in Mathematical Logic", Springer, 1998.

4. *Cardinal Arithmetic*, (eds.: Dov M. Gabbai, Angus Macintyre, Dana Scott), in the series: "Oxford Logic Guides", Oxford University Press, 1994.

*Work in Progress*

5. *Non-structure theory*, Oxford University Press, accepted.

6. *Universal classes*, in preparation.

**Teaching:** The following students have completed their PhD under my supervision: Matti Rubin, Uri Abraham, Shai Ben David, Rami Grossberg, Menachem Kojman, Sami Lifsches, Ziv Shami, Alex Usvyatsov.

**Honours and Awards:**

- Erdös Prize (1977)

- Rothschild Prize in Mathematics (1982)

- C. Karp Prize (1983)

- SIAM 1991 George Polya Prize in Applications of Combinatorial Mathematics
- Israel Prize in Mathematics (1998)
- JAMS Prize (of the Japanese Association of Mathematical Sciences, 1999)
- Janos Bolyai Prize of the Hungarian Academy of Sciences (2000)
- Wolf Foundation Prize in Mathematics (2001)

## SHOHAM, Yoav

**Specialties:** Temporal logics, non-monotonic logics, modal logics of knowledge and other mental attitudes, belief revision.

**Born:** 22 January 1956 in Haifa, Israel.

**Educated:** Yale University, PhD Computer Science, 1987; Technion Israel Istitute of Technology, BA Computer Science, 1982.

**Dissertation:** *Time and Causation from the Standpoint of Artificial Intelligence*; supervisor, Drew McDermott.

**Regular Academic or Research Appointments:** PROFESSOR, STANFORD UNIVERSITY, 2001–; Associate Professor, Stanford University, 1993-2001; Assistant Professor, Stanford University, 1987-2001; Postdotoral Fellow, Weizmann Institute of Science, 1987.

**Research Profile:** Yoav Shoham's work is motivated by problems in artificial intelligence, in particular the formalization of commonsense reasoning. His logic-related work falls into four overlapping areas: temporal logic, nonmonotonic logic, modal logics of knowledge and belief, and belief revision. In the first area he developed several temporal logics of time which combined the point-based and interval-based perspectives. Also, together with Joseph Halpern he developed an interval-based modal temporal logic. In nonmonotonic he pioneered the preference-based, semantic approach to non-monotonic logics. Additionally, together with Fangzhen Lin he developed foundations which relate fixpoint-based nonmonotonic logics and preference-based ones. In the third area, he developed several approaches to relating knowledge and belief, approaches which shed light on each of them in isolation. Finally, in the areas of belief revision he made a number of contributions, including (with Alvaro del Val) the reduction of belief update to the belief revision.

**Main Publications:**

1. Temporal Logics in AI. Y. Shoham, Journal of Artificial Intelligence 33(1), pp. 89-104, 1987.
2. Nonmonotonic Logics: meaning and utility. Y. Shoham, Proceedings of 10th IJCAI, 388-393, Milan, 1987.
3. Chronological Ignorance: Experiments in Nonmonotonic Temporal Reasoning. Y. Shoham, Journal of Artificial Intelligence 36(3), pp. 279-331, 1988.
4. Epistemic Semantics for Fixpoint Nonmonotonic Logics. F. Lin and Y. Shoham, Proceedings of TARK III, Monterey, 1990.
5. A Propositional Modal Logic of Time Intervals. J.Y. Halpern and Y. Shoham, Journal of the ACM 38(4), pp. 935-962, 1991.
6. Provably Correct Theories of Action. F. Lin and Y. Shoham, Proceedings of AAAI, Anaheim, 1991.
7. Varieties of Context, Y. Shoham, in V. Lifschitz (ed.), Artificial Intelligence amd Mathematical Thery of Computation: Papers in honor of John McCarthy, Academic Press, 1991.
8. A Logic of Knowledge and Justified Assumptions. F. Lin and Y. Shoham, Journal of Artificial Intelligence 57(2-3), pp. 271-290, 1992.
9. Agent Oriented Programming. Y. Shoham, Journal of Artificial Intelligence 60 (1), pp. 51-92, 1993.
10. Deriving Properties of Belief Update from Theories of Action. A. Del Val and Y. Shoham, Journal of Logic, Language and Information, 1994.
11. Belief as Defeasible Knowledge. Y. Moses and Y. Shoham, Journal of Artificial Intelligence, 1994.
12. A Unified View of Belief Revision and Update. A. Del Val and Y. Shoham, Journal of Logic and Computation, 1994.
13. Nonmonotonic Temporal Reasoning. E.J. Sandwall and Y. Shoham, in the Handbook of Login in Artificial Intelligence and Logic Programming (D. Gabbai, ed.), Elsevier, 1995.
14. Logics of Knowledge and Robot Motion Planning. R. Brafman, J-C. Latombe, Y. Moses and Y. Shoham, Journal of the ACM, 1996.
15. From Belief Revision to Belief Fusion. P. Maynard-Reid II and Y. Shoham, Proceedings of LOFT-98, Torino, 1998.
16. On the Knowledge Requirements of Tasks. R.I. Brafman, J.Y. Halpern and Y. Shoham, Journal of Artificial Intelligence 98(1-2), pp.317-350, January 1998.

**Service to the Profession:** Associate Editor, Journal of Artificial Intelligence Research, 1999-; Program chair, Seventh Conference on Theoretical Aspects of Rationality and Knowledge (TARK), 1996; Editorial board, Journal of Artificial Intelligence, 1993-; Consulting editor, Journal of Logic, Language and Information, 1991-; Board of Directors, The Corporation for Theoretical Aspects of Reasoning about Knowledge, Cambridge, MA (a non-profit organization), 1991-.

**Teaching:** Students whose dissertation topic was logic-related include Ronen Brafman, Fangzhen

Lin, Pedrito Maynard-Zhang Miami, Sarah Rebecca Thomas, and Alvaro del Val.

# SHORE, Richard A.

**Specialties:** Mathematical logic, recursion (computability) theory, effective mathematics, recursive model theory, reverse mathematics.

**Born:** 18, August, 1946 in Boston, Massachusetts, USA.

**Educated:** MIT, PhD Mathematics, 1972; Harvard University, AB Mathematics, 1968.

**Dissertation:** *Priority Arguments in Alpha-Recursion Theory*; supervisor, Gerald Sacks.

**Regular Academic or Research Appointments:** PROFESSOR OF MATHEMATICS, CORNELL UNIVERSITY, 1983–; Associate Professor of Mathematics, Cornell University, 1978-1983; Assistant Professor of Mathematics, Cornell University, 1974-1978; Assistant Professor of Mathematics, University of Illinois, Chicago, 1977; Instructor of Mathematics, University of Chicago, 1972-1974.

**Visiting Academic or Research Appointments:** Member, IMS, National University of Singapore, 2005; Visiting Scholar, Harvard University, 2002; Distinguished Visiting Professor, National University of Singapore, 2000; Visiting Scholar, MIT, 1997; Visiting Scholar, Harvard University, 1997; Member, MSRI, 1989-1990; Visiting Professor, University of Sienna, 1987; Visiting Professor, University of Chicago, 1987; Visiting Professor, Hebrew University of Jerusalem, 1982-1983; Visiting Associate Professor, MIT, 1980; Visiting Associate Professor, University of Connecticut, Storrs, 1979.

**Research Profile:** Shore's major research interests have centered around analyzing the structures of relative complexity of computation of functions on the natural numbers. The primary measure of such complexity is given by Turing reducibility: $f$ is easier to compute than $g$, $f \leq_T g$, if there is a (Turing) machine which can compute $f$ if it is given access to the values of $g$. He has also worked with various other interesting measures of such complexity that are defined by restricting the resources available primarily in terms of access to $g$. The general thrust of the work has been to show that these structures are as complicated as possible both algebraically and logically (in terms of the complexity of the decision problems for their theories). These results also allow one to differentiate among different notions of relative complexity in terms of the orderings they define. Another major theme in my work has been the relationship between these notions of computational complexity and ones based on the difficulty of defining functions in arithmetic. Restricting the computational resources more directly in terms of time or space leads from recursion theory into complexity theory. Relaxing the restrictions by allowing various infinitary procedures leads instead into generalized recursion theory or set theory.

The methods developed in these investigations are also useful in determining the effective content of standard mathematical theorems (when can existence proofs be made effective) and the inherent difficulty of combinatorial theorems in proof theoretic terms. The first area concerns computable mathematics and model theory. Decidability, interpretability and computational complexity are the major themes. The second is the realm of reverse mathematics. Here Shore has been particularly interested in finding and analyzing mathematical theorems whose proof theoretic strength is different than any of the standard systems and developing computability theoretic notions that serve to (proof-theoretically) differentiate between such mathematical principles.]

**Main Publications:**

1. The recursively enumerable alpha-degrees are dense, *Annals of Mathematical Logic* **9** (1976), 123-155.
2. The homogeneity conjecture, *Proceedings of the National Academy of Sciences* **76** (1979), 4218-4219.
3. The theory of the degrees below $\mathbf{0}'$, *Journal of the London Mathematical Society* **24** (1981), 1-14.
4. An algebraic decomposition of the recursively enumerable degrees and the coincidence of several degree classes with the promptly simple degrees, *Transactions of the American Mathematical Society* **281** (1984), 109-128 (with K. Ambos-Spies, C. Jockusch and R. Soare).
5. Pseudo-jump operators II: transfinite iterations hierarchies and minimal covers, *Journal of Symbolic Logic* **49** (1984), 1205-1236 (with C. Jockusch).
6. The degrees of unsolvability: the ordering of functions by relative computability, in *Proceedings of the International Congress of Mathematicians (Warsaw) 1983*, PWN-Polish Scientific Publishers, Warsaw 1984, Vol. 1, 337-346.
7. Initial segments of the Turing degrees of size $\aleph_1$, *Israel Journal of Mathematics* **55** (1986), 1-51 (with U. Abraham).
8. A non-inversion theorem for the jump operator, *Annals of Pure and Applied Logic* **40** (1988), 277-303.
9. On the strength of König's duality theorem for infinite bipartite graphs, *Journal of Combinatorial Theory (B)* **54** (1992), 257-290 (with R. Aharoni and M. Magidor).
10. On the strength of Fraïssé's conjecture, in *Logical Methods*, J. N. C. Crossley, J. Remmel, R. A. Shore and M. Sweedler, eds., Birkhäuser, Boston, 1993, 782-813.

11. *Logic for Applications*, Texts and Monographs in Computer Science, Springer-Verlag, New York, 1993 (with A. Nerode); $2^{nd}$ ed., Graduate Texts in Computer Science, Springer-Verlag, New York, 1997 (with A. Nerode).

12. Interpretability and definability in the recursively enumerable degrees, *Proc. Lon. Math. Soc.* (3) **77** (1998), 241-291 (with A. Nies and T. Slaman).

13. Reasoning about common knowledge with infinitely many agents, *Information and Computation* **191** (2004), 1-40 (with J. Halpern).

14. Solutions of the Goncharov-Millar and degree spectra problems in the theory of computable models, *Dokl. Akad. Nauk SSSR* **371** (2000) 30–31 (Russian), English version: *Doklady Mathematics* **61** (2000), 178-179 (with B. Khoussainov).

15. Natural definability in degree structures, in *Computability Theory and Its Applications: Current Trends and Open Problems*, P. Cholak, S. Lempp, M. Lerman and R. A. Shore eds., *Contemporary Mathematics*, AMS, Providence RI, 2000, 255-272.

16. Defining the Turing jump, *Math. Research Letters* **6** (1999), 711-722 (with T. Slaman).

17. Computable Structures: Presentations Matter, in *In the scope of logic, methodology and the philosophy of science}*, Int. Congress of LMPS, Cracow, August 1999, P. Gardenfors, J. Wolenski and K. Kijania-Placek eds., Synthese Library **315**, Kluwer Academic Publishers, Dordrecht, 2002, vol. 1, 81-95.

18. The prospects for mathematical logic in the twenty-first century, *Bulletin of Symbolic Logic* **7** (2001), 169-196 (with S. Buss, A. Kechris and A. Pillay).

19. Undecidability of the AE-theory of $\mathcal{R}(\leq ,\vee ,\wedge )$, *Transactions of the American Mathematical Society*, **356** (2004), 3025-3067 (with A. Nies and R. Miller).

20. Combinatorial principles weaker than Ramsey's theorem for pairs, to appear (with D. Hirschfeldt).

A full list of publications with many papers in electronic versions is available at Shore's home page http://www.math.cornell.edu/~shore/publications.html

**Service to the Profession:** Council Member, Association for Symbolic Logic, 1984-; Editor, *Studies in Logic and the Foundations of Mathematics*, North-Holland, 1996-; Board Member, Project Euclid , 2002-; President, Association for Symbolic Logic, 2001-2004; Managing Editor, *Bulletin of Symbolic Logic*, 1993-2000; Coordinator of Editorial Board, *Journal of Symbolic Logic*, 1989-1991; Editor, *Journal of Symbolic Logic*, 1984-1993; Consulting Editor, *Journal of Symbolic Logic*, 1980-1983.

**Teaching:** Shore has had 11 PhD students and 10 postdoctoral fellows. His students include two winners of the Sacks prize for the best thesis in logic worldwide: Denis Hirschfeldt (1999) who now has tenure at the University of Chicago and Antonio Montalbán (2005) who is now a Dickson Instructor at the University of Chicago. Two other of his students won NSF fellowships: Christine Haught (1985), Loyola University of Chicago, and Reed Solomon (1998), University of Connecticut. Others include Yang Yue (National University of Singapore), Walker White (CS, Cornell) and Noam Greenberg (Notre Dame and Wellington, postdoc). In addition, he has actively served on many committees in CS and one in Philosophy. His postdocs include K. Ambos-Spies (Heidelberg), D. Yang (Academica Sinica), P. Fejer (CS, U. Mass., Boston), P. Cholak (U. Wisc., Madison), A. Nies (U. Auckland), R. Miller (CUNY, Queens) and B. Csima (Waterloo).

**Honours and Awards:** Continuous NSF research support since 1973. Various international and binational research grants for projects in Greece, Italy, Israel, New Zealand and Latin America.

## SIEG, Wilfried

**Specialties:** Proof theory and automated proofs, philosophy of mathematics, history of modern logic, foundations of computability.

**Born:** 1 July 1945 in Lünen, Westfalia, Germany.

**Educated:** Stanford University, PhD Philosophy, 1977; Wilhelms-University Münster, MS Mathematics and Logic, 1971; Free University in Berlin, BS Mathematics and Physics, 1969.

**Dissertation:** *Trees in Metamathematics (Theories of inductive definitions & subsystems of analysis)*; supervisor, Solomon Feferman.

**Regular Academic or Research Appointments:** PROFESSOR OF PHILOSOPHY AND MATHEMATICAL LOGIC, CARNEGIE MELLON UNIVERSITY, 1985–; Assistant and Associate Professor of Philosophy, Columbia University, 1977-1985; Research Assistant, Institute of Mathematical Studies in the Social Sciences, Stanford University, 1975-1977.

**Visiting Academic or Research Appointments:** Visiting Professor, Philosophy, University of Bologna, 2003-2006; Visiting Professor, Computer Science, University of Milan, 1992; Visiting Professor, Mathematics, University of Siena, 1988; Visiting Professor, Mathematical Logic, Ludwig-Maximilians-University, Munich, 1987-1988; Visiting Assistant Professor, Philosophy, Stanford University, 1981-1982.

**Research Profile:** Sieg has made significant contributions in each of his primary research areas.

*Proof Theory*: He has investigated (impredicative) theories for analysis and their reduction to intuitionistic theories of constructive number classes, fragments of arithmetic and the extraction of computational information from classical theories through "Herbrand analyses", and intercalation calculi and their use in the automated, direct search for normal natural deduction proofs in classical and non-classical logics. The latter work has led to a very effective, strategically guided method for finding such proofs, and that method has been implemented in the program AProS. The program is available at http://www.phil.cmu.edu/projects/apros/index.php. (Publications 1-7.)

*Foundations of Mathematics*: The proof theoretic work on theories for analysis is deeply connected to broad foundational issues. Sieg is noted for his related historical and philosophical contributions. He has analyzed the evolution from Hilbert's "existential axiomatics" towards the finitist program and their roots in the radical $19^{th}$ century transformation of mathematics, in novel ways. The latter comes to the fore especially in Dedekind's influential work. Finally, he articulated two aspects of mathematical experience and joined them into a particular philosophical position he calls "reductive structuralism". His historical and philosophical contributions have been complemented by extensive editorial work concerning Gödel, Hilbert, and Bernays. (Publications 8-14.)

*Foundations of Computability*: A deep understanding of computability is necessary for assessing the character of the most important metamathematical results that affect reflections on the methodology of mathematics, namely, the undecidability and incompleteness theorems. Sieg provided a detailed history of "effective calculability" isolating in a precise way the unique character of Turing's work. He generalized Turing's machines and arguments to K-graph machines, whose computations properly encompass Kolmogorov's algorithms. He also investigated Gandy machines as "discrete mechanical devices" that compute in parallel. Most importantly, he formulated axioms for human and machine computability, all of which hold for Turing machines, and proved a representation theorem: any model of these axioms allows only computations that are reducible to Turing machine computations. (Publications 15-21.)

**Main Publications:**

1. *Iterated Inductive Definitions and Subsystems of Analysis – Recent Proof- theoretical Studies* (with W. Buchholz, S. Feferman, and W. Pohlers); Springer Lecture Notes in Mathematics 897, 1981.

2. Fragments of arithmetic; Annals of Pure and Applied Logic 28, 1985, 33-71.

3. A note on polynomial time computable arithmetic (with W. Buchholz); Contemporary Mathematics 106, 1990, 51-56.

4. Herbrand analyses; Archive for Mathematical Logic 30, 1991, 409-441.

5. Normal natural deduction proofs (in classical logic) (with J. Byrnes); Studia Logica 60, 1998, 67-106.

6. Normal natural deduction proofs (in non-classical logics) (with S. Cittadini); Springer Lecture Notes in Computer Science 2605, 2005, 169-191.

7. Automated search for Gödel's proofs (with C. Field); Annals for Pure and Applied Logic 133, 2005, 319-338.

8. Foundations of analysis and proof theory; Synthese 60, 1984, 159-200.

9. Hilbert's program sixty years later; J. Symbolic Logic 53, 1988, 290-300.

10. Relative consistency and accessible domains; Synthese 84, 1990, 259-297.

11. Hilbert's Programs: 1917-1922; B. Symbolic Logic 5, 1999, 1-44.

12. Beyond Hilbert's reach? In: *Reading Natural Philosophy* (D. Malament, ed.), Open Court, 2002, 363-405.

13. Dedekind's analysis of number: systems and axioms (with D. Schlimm); Synthese 147, 2005, 121-170.

14. Hilbert and Bernays's "Grundlagen der Mathematik" (with M. Ravaglia); in: *Landmark Writings in Western Mathematics 1640-1940* (I. Grattan- Guinness, ed.), Elsevier, 2005, 981-999.

15. Mechanical procedures and mathematical experience; in: Mathematics & Mind (A. George, ed.), Oxford University Press, 1994, 71-117.

16. Paper machines (with D. Mundici); Philosophia Mathematica 3, 1995, 5-30.

17. Step by recursive step: Church's analysis of effective calculability; B. Symbolic Logic 3, 1997, 154-180.

18. Calculations by man and machine: conceptual analysis; in: *Reflections on the Foundations of Mathematics* (W. Sieg, R. Sommer, and C. Talcott, eds.), Notes in Logic 15, ASL, 2002, 390-409.

19. Calculations by man and machine: mathematical presentation; in: *Proc. Cracov Inter. Congress on Logic, Methodology, and Philosophy of Science*, Kluwer, 2002, 247-262.

20. Gödel on computability; Philosophia Mathematica, to appear in June 2006, 19 pp.

21. Church without dogma: axioms for computability; to appear in: *New Computational Paradigms* (B. Cooper, B. Loewe, and A. Sorbi, eds.), Springer.

*Work in Progress*

22. Essay on "Computability", providing a conceptual analysis of the classical notion and an historical account of its evolution (to be published in: *The Philosophy of Mathematics*, A. Irvine (ed.), Elsevier/Springer);

23. Hilbert Project: editing two volumes of Hilbert's unpublished lectures on logic and arithmetic from 1895 to 1917, respectively 1917/18 to 1933 (to be published

by Springer);

24. Bernays Project: a bilingual, two-volume edition of Bernays's essays on the philosophy of mathematics (to be published by Open Court);

25. AProS Project: extending automated proof search from pure logic to elementary parts of set theory and developing a web-based course "Proofs, functions, and computations";

26. Book on philosophy of mathematics: Reductive structuralism.

**Service to the Profession:** Editor, Journal of Philosophy, 2002-; Editor of "Paul Bernays' Essays on the Philosophy of Mathematics," with Tait, 2000-; Editor, Synthese, 1999-; Member of the ASL Meeting Committee, 1997-2002; Editor, Gödel's Collected Works vol. IV and V, with Dawson, Feferman, Goldfarb, Parsons, 1996-2003; Chairman of the Sectional Program Committee for the $8^{th}$ International Congress of Logic, Methodology, and Philosophy of Science, Florence, 1995; Editor of "David Hilbert's Lectures on the Foundations of Mathematics and Physics," with Ewald, Hallett, and Majer, 1992-; Editor, Quarterly of Mathematical Logic, 1992-; Editor, Studia Logica, 1992-2002-; ASL Committee Member on Logic Education, 1991-2002; Executive Committee Member and Council of the Association for Symbolic Logic, 1990-1993; Editor of the Oxford University Press Series "Logic and Computation in Philosophy," with Glymour and Seidenfeld, 1988-.

**Teaching:** Sieg has been a mainstay for teaching (mathematical) logic and philosophy of mathematics at Carnegie Mellon, from Freshman Seminars to Graduate courses. He was the founder of the Logic & Computation Program at Carnegie Mellon and directed it from 1985 to 1994. With a number of colleagues in mathematics and computer science, he also co-founded the university's interdepartmental Pure and Applied Logic Program. He has supervised four PhD students (Guglielmo Tamburrini, while still at Columbia University, John Byrnes, Barbara Kauffmann, and Mark Ravaglia); he has guided fifteen MS students to original research that frequently led to a joint publication.

Since the summer 2003 Sieg has been developing a fully web-based introduction to logic, called *Logic & Proofs*. The course is accessible at http://www.cmu.edu/oli/. It is a highly interactive presentation of sentential and quantificational logic with a strong emphasis on strategic thinking, i.e., on strategically guided construction of natural deduction proofs. Arguments are constructed in the Proof Lab, a sophisticated environment where students can argue forward and backward; students receive detailed feedback on mistakes and how to correct them. The course is being expanded in two different ways: first, the content is going to include elementary set theory and computability theory; second, the interaction of students with the Proof Lab is going to be dramatically enriched through dynamic automated tutoring. This tutoring uses the automated theorem prover AProS.

**Vision Statement:** The two most distinctive directions of logical work in the $20^{th}$ century – sustained meta-theoretic investigations and deep connections to fields outside of philosophy, in particular, to mathematics, linguistics, and computer science – will continue to be pursued. I also think that the trend of enriching foundational reflections by detailed historical perspectives will be further strengthened.

As to directly logical work, a more adequate structure theory of mathematical proofs will be developed. Such a theory has to be concerned with the sophisticated logical structure of individual proofs together with a local axiomatic context that makes for intelligible and explanatory proofs. I call this new field of study *dynamic axiomatics*, as it expands traditional axiomatics by heuristics for the discovery of proofs in a part of mathematics; the heuristics are based on "leading ideas" for that part of mathematics.

Corresponding interdisciplinary connections suggest themselves. After all, detailed developments in dynamic axiomatics can serve as bases for computer implementations to investigate how well the heuristics restrict search, and how fruitful they are for proving theorems that were not part of the original analysis, when isolating the leading ideas. In particular, connections to cognitive psychology should be established in order to inform, and be informed by, investigations of the core of (logical and mathematical) thinking. Such work will ensure that logic impacts education in a broad and significant way.

**Honours and Awards:** Sieg's collaborative projects have been supported by awards from the Hewlett Foundation (web-based logic course in the context of Carnegie Mellon's Open Learning Initiative), the National Science Foundation (automated proof search), the NEH (Bernays edition), the Sloan Foundation (Gödel's Collected Works), and the DFG, i.e., the German National Science Foundation (Hilbert's unpublished lectures).

# SIEKMANN, Jörg

**Specialties:** Artificial intelligence, deduction systems and automated reasoning, unification theory, e-learning mathematics.

**Born:** 5 August 1941

**Educated:** Mathematics and Physics at Göttingen University, M.Sc in Comp Sci at Essex University, Ph D. in AI from Essex University

**Dissertation:** *Unification and Matching in Equational Theories*

**Regular Academic or Research Appointments:** SENIOR PROFESSOR SAARLAND UNIVERSITY

**Research Profile:** Joörg Siemann worked as an assistant and associate professor in Karlsruhe and Kaiserslautern, and headed several research groups working on artificial intelligence at the Saarland University and at the German Research Center for Artificial Intelligence (DFKI) in Saarbrücken, of which he is one of the founding directors. The research groups comprise automated reasoning for mathematics, the Competence Centre for e-Learning, formal methods, multiagent systems, and the IT-Security Evaluation Center. Siekmann is the chairman of the collaborative research center "Resource-adaptive Cognitive Processes" (SFB 378) funded by the German Science Foundation (DFG). He is past chairman of the network of excellence on computational logic (CoLogNet), and Vice President of the International Federation on Computational Logic IFCoLog.

He is now a senior professor at Saarland University and director at the DFKI, where he heads the Competence Center for e-Learning (CCeL) at the DFKI. His main research interests are Artificial Intelligence, Automated Reasoning, and e-Learning for Mathematics. JS has been active in founding three companies, in which he holds some shares and patents.

**Main Publications:**

1. J. Siekmann, P. Szabo. *A noetherian and confluent rewriting system for idempotent semigroups*, Semigroup Forum 25, 1982, 83-110.
2. R. Book, J. Siekmann. *On Unification: Equational theories are not bounded* Journal of Symbolic Computation 2, 1986, 317-324.
3. A. Herold, J. Siekmann. *Unification in abelian semigroups* Journal of Automated Reasoning 3 (3), 1987, 247-283.
4. J. Siekmann, P. Szabo. *The undecidability of the DA-unification problem* Journal of Symbolic Logic, 54 (2), 1989, 402-414.
5. J. Siekmann. *Unification Theory* Journal of Symbolic Computation 7, 1989, 207-274.
6. F. Baader, H.-J. Bürckert, B. Hollunder, W. Nutt, and J. Siekmann. *Concept Logics*. In J. W. Lloyd (Ed.), Computational Logic, Symposium Proceedings, Brussels, November 1990, Springer-Verlag, 177-201.
7. C. Beierle, U. Hedtstück, U. Pletat, P. Schmitt, J. Siekmann. *An Order sorted Logic for Knowledge Representation* Journal of Artificial Intelligence, vol. 55, 1992.
8. J. Siekmann, M. Kohlhase, E. Melis. *OMEGA: Ein mathematisches Assistenzsystem* Kognitionswissenschaft, vol 7, No. 3, p 101-105, 1998. English in: JFAK, Essays dedicated to J. von Benthem, Amsterdam, Univ. Press, 1999.
9. E. Melis, J. Siekmann. *Knowledge-based proof planning* Journal of Artificial Intelligence, vol 115, p 65-105, 1999.
10. J. Siekmann, G. Wrightson. *Strong Completeness of A. Kowalski's Connection Graph Proof Procedure* Springer Lecture Notes on AI, vol 2408, pp 231, 2002.
11. E. Melis, A. Meier, J. Siekmann *Proofplanning with multiple Strategies* Journal of Artificial Intelligence, vol 172,pp656-658,2007.
12. J. Siekmann, Ch. Benzmüller, S. Autexier. *Computer supported mathematics with OMEGA* Journal of applied Logic, vol 4 (4), 2006.

**Teaching:** Mathematical Logic, Artificial Intelligence, Automated Reasoning, Cognitive Science and Educational Technologies. JS supervised more than 80 PhD students, many of whom are now professors in logic and AI.

**Vision Statement:** Principia Mathematica Mechanico: Mathematical Assistant Systems will provide common computersupport for the working mathematician.

## SIMPSON, Stephen G.

**Specialties:** Foundations of mathematics, recursion theory, axiomatic set theory, combinatorics, models of arithmetic, degrees of unsolvability, subsystems of second-order arithmetic, Reverse Mathematics, algorithmic randomness, mass problems.

**Born:** 8 September 1945 in Allentown, Pennsylvania, USA.

**Educated:** B.A. (summa cum laude), M.S., Mathematics, Lehigh University, 1966. Ph.D., Mathematics, Massachusetts Institute of Technology, 1971,

**Dissertation:** *Admissible Ordinals and Recursion Theory*, supervisor Gerald E. Sacks.

**Regular Academic or Research Appointments:** PROFESSOR OF MATHEMATICS, PENNSYLVANIA STATE UNIVERSITY, 1980–; Associate Professor, Pennsylvania State University, 1977–1980; Assistant Professor, Pennsylvania State University, 1975–1977.

**Visiting Academic or Research Appointments:** Postdoctoral appointments at Yale, Berkeley, and Oxford. Visiting faculty appointments at Chicago,

Connecticut, Paris, Munich, Stanford, Illinois, and Tennessee.

**Research Profile:** From a young age I have been passionate about the foundations of mathematics. A theme of my research has been to adapt techniques of modern mathematical logic to the study of issues and programs in foundations of mathematics.

At the beginning of my professional career I found it expedient to focus on technical topics within mathematical logic. My earliest research was in generalized recursion theory. Jointly with my thesis adviser Gerald E. Sacks I proved the existence of incomparable $\alpha$-recursively enumerable $\alpha$-degrees where $\alpha$ is an arbitrary admissible ordinal. I also proved some other results in admissible recursion theory.

In the 1970s I turned to classical recursion theory including Turing degrees and hyperdegrees. Some of my results revealed a close relationship between degrees of unsolvability and axiomatic set theory. For instance, I proved that there is a cone of minimal covers in the hyperdegrees if a measurable cardinal exists but not if $V = L$ or a forcing extension of $L$. Jointly with Carl G. Jockusch, Jr. I obtained a natural, Turing-degree-theoretical description of the ramified hierarchy. Later Harold Hodes extended this through the constructible hierarchy. In a rather influential paper I proved that the first-order theory of the partial ordering of Turing degrees is as complicated as possible. Namely, it is recursively isomorphic to true second-order arithmetic. In addition I worked with students and co-authors to develop a theory of so-called Kleene degrees in descriptive set theory. Here recursive and recursively enumerable sets are replaced by Borel and $\omega$-Souslin sets, and Turing reducibility is replaced by relative type-2 $\omega$-recursion with type-1 parameters.

Starting in 1969 under the influence of Georg Kreisel I became interested in subsystems of second-order arithmetic. For a long time I published nothing in this area, but during the 1980s and 1990s my interest blossomed into the program of Reverse Mathematics with many publications, students, and co-authors.

The purpose of Reverse Mathematics is to classify core mathematical theorems with respect to their logical strength, i.e., the axioms of second-order arithmetic which are needed in order to prove them. It turns out that precise classifications are often attainable. Moreover, case studies reveal that hundreds of theorems fall into a very small number of classes. For example, the Gödel completeness theorem, the Peano existence theorem for solutions of ordinary differential equations, and the prime ideal theorem for countable commutative rings, all fall into the same class. The five most important classes are known as the Big Five and correspond to Chapters II through VI of my book on this subject. Such results are of interest with respect to the study of restrictive foundational regimes such as computable analysis, finitistic reductionism in the style of Hilbert, predicativism in the style of Weyl and Feferman, and predicative reductionism. Reverse Mathematics helps us to determine which core mathematical theorems would be retained and which would be lost under such regimes. In addition, Reverse Mathematics is of value in the ongoing quest for a core mathematical problem which would require strong set-theoretical methods for its solution.

In 1997 I founded FOM, an electronic forum for lively discussion of issues and programs in the foundations of mathematics. As a result of an FOM debate with Robert I. Soare in 1999 I developed an interest in mass problems.

Mass problems are a rigorous elaboration of Kolmogorov's informal 1923 proposal to view intuitionistic logic as a calculus of problems. According to Medvedev 1955 and Muchnik 1963, a mass problem is an arbitrary set of real numbers. The elements of the set are to be regarded as solutions of the problem. A mass problem is said to be unsolvable if none of its solutions is Turing computable. A mass problem is said to be weakly reducible to another mass problem if each solution of the second problem can be used as a Turing oracle to compute a solution of the first problem. A weak degree is defined to be an equivalence class of mass problems under mutual weak reducibility.

Many unsolvable mathematical problems are best viewed as mass problems, and weak degrees are a measure of the amount of unsolvability inherent in such problems. I have discovered that the weak degrees of mass problems associated with nonempty effectively closed sets of real numbers form a rich structure. Moreover, they include many specific, natural degrees which are closely related to foundationally interesting topics. Among these topics are algorithmic randomness, algorithmic information theory, algorithmic dimension theory, almost everywhere domination, hyperarithmeticity, computational complexity, and Reverse Mathematics. Recently I have applied weak degrees to the study of symbolic dynamics. These results highlight a contrast with the recursively enumerable Turing degrees, where there are no known specific examples of natural degrees other than those of solvable problems and the halting problem.

**Main Publications:**

1. Gerald E. Sacks and Stephen G. Simpson, The $\alpha$-finite injury method, Annals of Mathematical Logic, 4, 1972, 343–367.
2. Stephen G. Simpson, Short course on admissible recursion theory, in: Generalized Recursion Theory II, edited by J.-E. Fenstad, R. O. Gandy and G. E. Sacks, North-Holland, Amsterdam, 1978, 355–390.
3. Stephen G. Simpson, Minimal covers and hyperdegrees, Transactions of the American Mathematical Society, 209, 1975, 45–64.
4. Carl G. Jockusch, Jr., and Stephen G. Simpson, A degree theoretic definition of the ramified analytical hierarchy, Annals of Mathematical Logic, 10, 1976, 1–32.
5. Stephen G. Simpson, First order theory of the degrees of recursive unsolvability, Annals of Mathematics, 105, 1977, 121–139.
6. Stephen G. Simpson and Galen Weitkamp, High and low Kleene degrees of coanalytic sets, Journal of Symbolic Logic, 47, 1982, 356–368.
7. James Schmerl and Stephen G. Simpson, On the role of Ramsey quantifiers in first order arithmetic, Journal of Symbolic Logic, 47, 1982, 15–27.
8. Timothy J. Carlson and Stephen G. Simpson, A dual form of Ramsey's Theorem, Advances in Mathematics, 53, 1984, 265–290.
9. Kurt Schütte and Stephen G. Simpson, Ein in der reinen Zahlentheorie unbeweisbarer Satz über endlichen Folgen von natürlichen Zahlen, Archiv für mathematische Logik und Grundlagen der Mathematik, 25, 1985, 75–89.
10. Stephen G. Simpson, Unprovable theorems and fast-growing functions, in: Logic and Combinatorics, edited by S. G. Simpson, Contemporary Mathematics, Volume 65, American Mathematical Society, 1987, 359–394.
11. Harvey Friedman, Kenneth McAloon, and Stephen G. Simpson, A finite combinatorial principle which is equivalent to the 1-consistency of predicative analysis, in: Logic Symposion I (Patras), edited by G. Metakides, North-Holland, Amsterdam, 1982, 197–220.
12. Harvey Friedman, Stephen G. Simpson, and Rick Smith, Countable algebra and set existence axioms, Annals of Pure and Applied Logic, 25, 1983, 141–181; Addendum, 28, 1985, 320–321.
13. Stephen G. Simpson, Which set existence axioms are needed to prove the Cauchy/Peano theorem of ordinary differential equations?, Journal of Symbolic Logic, 49, 1984, 783–802.
14. Stephen G. Simpson and Rick Smith, Factorization of polynomials and $\Sigma_1^0$ induction, Annals of Pure and Applied Logic, 31, 1986, 289–306.
15. Stephen G. Simpson, Partial realizations of Hilbert's Program, Journal of Symbolic Logic, 53, 1988, 349–363.
16. Stephen G. Simpson, Ordinal numbers and the Hilbert Basis Theorem, Journal of Symbolic Logic, 53, 1988, 961–974.
17. Stephen G. Simpson, On the strength of König's duality theorem for countable bipartite graphs, Journal of Symbolic Logic, 59, 1994, 113–123.
18. A. James Humphreys and Stephen G. Simpson, Separable Banach space theory needs strong set existence axioms, Transactions of the American Mathematical Society, 348, 1996, 4231–4255.
19. Stephen G. Simpson, Subsystems of Second Order Arithmetic, Perspectives in Mathematical Logic, Springer-Verlag, 1999, XIV + 445 pages.
20. Douglas K. Brown, Mariagnese Giusto, and Stephen G. Simpson, Vitali's theorem and WWKL, Archive for Mathematical Logic, 41, 2002, 191–206.
21. Carl Mummert and Stephen G. Simpson, Reverse mathematics and $\Pi_2^1$ comprehension, Bulletin of Symbolic Logic, 11, 2005, 526–533.
22. Stephen G. Simpson, Mass problems and randomness, Bulletin of Symbolic Logic, 11, 2005, 1–27.
23. Stephen G. Simpson, An extension of the recursively enumerable Turing degrees, Journal of the London Mathematical Society, 75, 2007, 287–297.
24. Natasha L. Dobrinen and Stephen G. Simpson, Almost everywhere domination, Journal of Symbolic Logic, 69, 2004, 914–922.
25. Stephen G. Simpson, Mass problems and almost everywhere domination, Mathematical Logic Quarterly, 53, 2007, 483–492.
26. Joshua A. Cole and Stephen G. Simpson, Mass problems and hyperarithmeticity, preprint, 20 pages, 28 November 2006, submitted for publication.
27. Stephen G. Simpson, Medvedev degrees of 2-dimensional subshifts of finite type, preprint, 8 pages, 1 May 2007, accepted 26 September 2007 for publication in Ergodic Theory and Dynamical Systems.
28. Stephen G. Simpson, Logic and mathematics, in: The Examined Life, Readings from Western Philosophy from Plato to Kant, edited by S. Rosen, Random House, 2000, 577–605.

Simpson's publication list and many of his papers are available at http://www.math.psu.edu/simpson/.

**Service to the Profession:** Referee for numerous journals. Editor or co-editor of several collections of papers. Organizer of several conferences. Associate editor of Mathematical Logic Quarterly. Member of several committees of the Association for Symbolic Logic. Founder and moderator of FOM, an electronic forum for lively discussion of issues and programs in the foundations of mathematics.

**Teaching:** Supervised sixteen Ph.D. advisees including John Steel (University of California at Berkeley), Peter Pappas (Vassar College), Jeffry L. Hirst (Appalachian State University), Fernando Ferreira (University of Lisbon), Alberto Marcone (University of Udine), and Carl Mummert (University of Michigan, postdoctoral appointment). Supervised four postdoctoral advisees including

Kazuyuki Tanaka (Tohoku University), Natasha Dobrinen (University of Denver), and Rebecca Weber (Dartmouth College). Served as a member of numerous Ph.D. committees in mathematics, computer science, philosophy, and economics.

**Honours and Awards:** At the Pennsylvania State University: Faculty Scholar Medal and Prize, Raymond F. Shibley Professorship, Cada and Susan Grove Award for Interdisciplinary Research. In the United States: Alfred P. Sloan Fellowship, numerous grants from the National Science Foundation. International: numerous invited addresses at international conferences; grants from the SRC (United Kingdom), CNRS (France), DFG (Germany), Templeton Foundation.

**Vision Statement:** Foundations of mathematics is an important subject which is of general intellectual and cultural interest. Mathematical logic needs to renew its connection to its historical roots in foundations of mathematics.

Currently Reverse Mathematics and other lines of foundationally motivated research suggest that strong set-theoretical axioms are destined to remain largely irrelevant to core mathematical practice. Hilbert's program of finitistic reductionism seems to have been largely vindicated. An attractive direction for future foundational research would be to study and clarify the interfaces between mathematics and the rest of human knowledge.

# SKYRMS, Brian

**Specialties:** Probability, induction, decision theory, game theory, conditionals.

**Born:** 11 March 1938 in Pittsburgh, Pennsylvania, USA.

**Educated:** University of Pittsburgh, PhD Philosophy, 1964; University of Pittsburgh, MA Philosophy, 1962; Lehigh University BA Economics 1960, BA Philosophy 1961.

**Dissertation:** *The Concept of Physical Necessity*; supervisor, Nicholas Rescher.

**Regular Academic or Research Appointments:** UCI DISTINGUISHED PROFESSOR OF LOGIC AND PHILOSOPHY OF SCIENCE, AND OF ECONOMICS, UNIVERSITY OF CALIFORNIA, IRVINE, 2007–; PROFESSOR, PHILOSOPHY, STANFORD UNIVERSITY, 2007–; Professor, University of California, Irvine, 1980; Professor, University of Illinois, Chicago, 1970; Associate Professor, University of Illinois at Chicago, 1968-1970; Assistant Professor, University of Illinois, Chicago, 1967-1968; Assistant Professor, University of Delaware, 1965-1966; Assistant Professor, San Fernando Valley State College, 1964-1965.

**Visiting Academic or Research Appointments:** Fellow, Center for Advanced Study in the Behavioral Sciences, Stanford, 1993-1994; Humanities Council Senior Fellow and Old Dominion Fellow and Guggenheim Fellow, Princeton University, 1987-1988; Visiting Professor of Philosophy, University of Michigan, 1966-1967.

**Research Profile:** Brian Skyrms has made contributions to decision theory, to game theory, and to the application of game theory to social philosophy. He has also worked on probability and induction and on the application of probability to causal reasoning and counterfactual conditionals.

**Main Publications:**

1. *Causal Necessity* Yale University Press: New Haven, 1980.
2. *Pragmatics and Empiricism* Yale University Press: New Haven, 1984.
3. *The Dynamics of Rational Deliberation* Harvard University Press: Cambridge, Mass., 1990.
4. *Evolution of the Social Contract* Cambridge University Press: Cambridge, 1996.
5. *The Stag Hunt and the Evolution of Social Structure* Cambridge University Press: Cambridge, Mass., 2004.
6. "Higher Order Degrees of Belief" in *Prospects for Pragmatism: Essays in Honor of F.P. Ramsey* ed. D.H. Mellor. Cambridge University Press: Cambridge, 1980.
7. "The Value of Knowledge" in *Scientific Theories* (Minnesota Studies in the Philosophy of Science vol. 14) ed. C.W. Savage. University of Minnesota Press: Minneapolis, 1990, 245-266.
8. "Updating, Supposing and MAXENT" in *Theory and Decision* 22 (1987) 225-246.
9. "Carnapian Inductive Logic for Markov Chains" *Erkenntnis* 35 (1991) 439-460.
10. "Analogy by Similarity in HyperCarnapian Inductive Logic" in *Philosophical Problems of the Internal and External Worlds: Essays On the Philosophy of Adolf Grunbaum* ed. Earman et al. University of Pittsburgh Press: Pittsburgh, 1993, 273-282.
11. "Logical Atomism and Combinatorial Possibility" *The Journal of Philosophy* 90 (1993) 219-232.
12. "Bayesian Projectibility" in *Grue: Essays on the New Riddle of Induction* ed. D. Stalker. Chicago: Open Court, 1994, 241-262.
13. "Carnapian Inductive Logic for a Value Continuum" in *The Philosophy of Science* (Midwest Studies in Philosophy Volume 18) ed. H. Wettstein. University of Notre Dame Press: South Bend, Indiana, 78-89.
14. "Strict Coherence, Sigma Coherence and the Metaphysics of Quantity" *Philosophical Studies* 77 (1995) 39-55.

15. "The Structure of Radical Probabilism" *Erkenntnis* 45 (1997) 285-297.
16. "Subjunctive Conditionals and Revealed Preference" *Philosophy of Science* 65 (1998) 545-574.
17. "Stability and Explanatory Significance of Some Simple Evolutionary Models" *Philosophy of Science* (March, 2000) 94-113.
18. "A Dynamic Model of Social Network Formation" (with Robin Pemantle) *Proceedings of the National Academy of Sciences of the U.S.A.* 97 (2000) 9340-9346.
19. "Learning to Take Turns" (with Peter Vanderschraaf) *Erkenntnis* 59 (2003) 311-348.
20. "Network Formation by Reinforcement Learning: the Long and the Medium Run" (with Robin Pemantle) *Mathematical Social Sciences* 48 (2004) 315-327.

**Service to the Profession:** Editorial Board, PNAS; President, Philosophy of Science Association, 2004-2006; President, American Philosophical Association (Pacific), 2000-2001; Governing Board, Western Center of the American Academy of Arts and Sciences, 2000-2004; Editor, Cambridge Studies in Probability, Induction and Decision Theory.

**Teaching:** Director of the Minor in History and Philosophy of Science, U. C. Irvine. Eleven PhD students.

**Honours and Awards:** Paul Silverman Award, 2006; Elected Fellow, American Association for the Advancement of Science, 2004; Lakatos Prize, *Evolution of the Social Contract*, 1999; Elected Fellow, National Academy of Sciences, 1999; Elected Fellow, American Academy of Arts and Sciences, 1994.

# SKVORTSOV, Dmitrij P.

**Specialties:** Intuitionistic and superintuitionistic logics, propositional and predicate.

**Born:** 9 November 1952, in Moscow, Russia (formerly, USSR).

**Educated:** PhD in Mathematics (Russian degree of Candidate of Physical and Mathematical Sciences), Moscow State University, 1980.

**Dissertation:** Interpretation of propositional and predicate formulas by means of finite and nonfinite problems. Postgraduate studies in mathematics, Moscow State University, 1975–1978; supervisor Vladimir A. Uspenskij.
MSc in Mathematics, Moscow State University, 1975.

**Regular Academic or Research Appointments:** SENIOR RESEARCHER, ALL-RUSSIAN INSTITUTE OF SCIENTIFIC AND TECHNICAL INFORMATION, RUSSIAN ACADEMY OF SCIENCES, 1986–; Junior researcher, All-Russian Institute of Scientific and Technical Information, Russian Academy of Sciences, 1978–1986.

**Visiting Academic or Research Appointments:** Visiting Researcher, King's College London, since 1998;
Visiting Researcher, Imperial College London, 1998.

**Research Profile:** Dmitrij Skvortsov worked in mathematics of superintuitionistic predicate logics.

His most essential contributions in this field are in model theory.

He constructed the first examples of non finitely axiomatizable intermediate predicate logics, characterized by Kripke frames based over finite partially ordered sets, and the first example of finitely axiomatizable intermediate predicate logics, whose intersection is not finitely axiomatizable. He proved that the intermediate predicate logic, characterized by Kripke frames based over all finite partially ordered sets, is not recursively axiomatizable, and intermediate predicate logics, characterized by Kripke frames based over some infinite partially ordered sets without infinite chains of some kinds (such as ordinals, dually well founded sets etc.) are non-arithmetical, and moreover, Pi-1-1 hard. In joint works with Valentin Shehtman he developed new kinds of semantics for first-order modal logics, such as Kripke bundle and simplicial (metaframe) semantics, and proved completeness and incompleteness results.

**Main Publications:**

1. Logic of infinite problems and Kripke models on atomic semilattices of sets, *Doklady Akad. Nauk SSSR*, v.245 (1979), No.4, pp. 798-801 (in Russian). Engish Translation: *Soviet Mathematics, Doklady*, v.20 (1979), No.2, pp. 360-363.
2. (with L.L.Maksimova, V.B.Shehtman) The impossibility of a finite axiomatization of Medvedev's logic of finitary problems, *Doklady Akad. Nauk SSSR*, v.245 (1979), No.5, pp. 1051-1054 (in Russian). Engish Translation: *Soviet Mathematics, Doklady*, v.20 (1979), No.2, pp. 394-398.
3. On interrelation between finite validity of some propositional formulas and their deducibility in Kreisel—Putnam's logic (in Russian), *Vestnik MGU [Moscow State University Herald], Ser.1: Mathematics, Mechanics*, (1980), No.3, pp. 29-32.
4. On axiomatizability of some intermediate predicate logics (summary), *Reports on mathematical logic*, v.22 (1988), pp. 115-116.
5. (with V. Shehtman) Semantics of non-classical first order predicate logics, in *Mathematical Logic. Proceedings of Summer School and Conference on Mathem.*

*Logic held Sept.13-23, 1988, in Chaika, Bulgaria.* Plenum Press (1990), 105-116.

6. (with V. Shehtman) Maximal Kripke-type semantics for modal and superintuitionistic predicate logics, *Annals of Pure and Applied Logic,* v. 63 (1993), No.1, 69-101.

7. On the predicate logic of finite Kripke frames, *Studia Logica,* v.54 (1995), No.1, pp. 79-88.

8. On finite intersections of intermediate predicate logics, in *Logic and Algebra* (A.Ursini and P.Agliano, eds.), (Proc. of Intern. Conf. on Logic and Algebra in memory of R.Magari, 1994), Marcel-Dekker, New York, 1996, pp. 667-688.

9. Non-axiomatizable second order intuitionistic propositional logic, *Annals of Pure and Applied Logic,* v.86 (1997), No.1, pp. 33-46.

10. Not every "tabular" predicate logic is finitely axiomatizable, *Studia Logica,* v.59 (1997), No.3, pp. 387-396.

11. On some Kripke complete and Kripke incomplete intermediate predicate logics, *Studia Logica,* v.61 (1998), No.2, pp. 281-292.

12. Remark on a finite axiomatization of finite intermediate propositional logics, *Journ. of Applied Nonclassical Logic,* v.9 (1999), No.2-3, pp. 381-386.

13. On the existence of continua of logics between some intermediate predicate logics, *Studia Logica,* v.64 (2000), No.2, pp. 257-270.

14. An incompleteness result for predicate extensions of intermediate propositional logics. *Advances in Modal Logic,* v.4 (P.Balbiani, N.-Y.Suzuki, F.Wolter, M.Zakharyaschev, ed.), (2003), pp. 461-474.

15. (with N.K.Vereshchagin, E.Z.Skvortsova, A.V.Chernov) Variants of realizability for propositional formulas and the logic of weak excluded middle, *Proceedings of the Steklov Institute of Mathematics,* v.242 (2003), pp.67-85.

16. On intermediate predicate logics of some finite Kripke frames, I: Levelwise uniform trees, *Studia Logica,* v.77 (2004), pp. 295-323.

17. The superintuitionistic predicate logic of finite Kripke frames is not recursively axiomatizable, *Journal of Symbolic Logic,* v.70 (2005), No.2, pp. 451-459.

18. On the superintuitionistic predicate logics of Kripke frames based on denumerable chains, *Algebraic and topological methods in non-classical logics, II, Barcelona, 15-18 June, Abstracts* (2005), pp. 73-74.

19. On non-axiomatizability of superintuitionistic predicate logics of some classes of well-founded and dually well-founded Kripke frames, *Journ. of Logic and Computation,* v.16 (2006), No.5, pp. 685 - 695.

20. On the predicate logic of linear Kripke frames and some of its extensions. *Studia Logica,* v. 81 (2005), pp. 261-282.

21. *Quantification in Nonclassical Logic,* Volume 1, (with D. Gabbay and V. Shehtman), Elsevier, 2008, to appear. See http://lpcs.math.msu.su/~shehtman

*Work in Progress*

22. A paper: On intermediate predicate logics of some finite Kripke frames, II (a continuation of reference 16), and two more papers on recursive and finite axiomatizability of intermediate predicate logics characterized (in Kripke semantics) by finite trees.

23. Papers: On Non-Axiomatizability of Some Superintuitionistic Predicate Logics,
I: Predicate Logics of Well-Ordered and Dually Well-Ordered Kripke Frames,
II: Sliced and dually well-founded Kripke frames,
III: Predicate Logics of Denumerable Chains
(continuation of references 4, 17, 19, 18).

24. Two papers on Kripke sheaf completeness and incompleteness of superintuitionistic predicate logic with weak versions of the axiom of constant domain (continuation of reference 20).

25. A paper on recursive axiomatizability of predicate logics of Kripke frames with nested domains based over the set of reals.

26. A paper: A remark on propositional Kripke frames sound for intuitionistic logic.

**Teaching:** D. Skvortsov taught (since 1980s) different courses at Russian State Humanitarian University, especially in mathematical logic and its applications.

## SMILEY, Timothy John

**Specialties:** Mathematical logic, philosophical logic.

**Born:** 13 November 1930 in London, England.

**Educated:** Gray's Inn, London, Called to the Bar, 1956; University of Cambridge, PhD, 1956; University of Cambridge, BA Mathematics, 1952; University of Fribourg, Switzerland, 1948 (studying with Prof. I. M. Bochenski).

**Dissertation:** *Natural Systems of Logic*; supervisor, S. W. P. Steen.

**Regular Academic or Research Appointments:** KNIGHTSBRIDGE PROFESSOR OF PHILOSOPHY, EMERITUS, UNIVERSITY OF CAMBRIDGE, 1998–; Knightbridge Professor of Philosophy, University of Cambridge, 1980-1998; Lecturer, Philosophy, University of Cambridge, 1962-1980; Assistant Lecturer, Philosophy, University of Cambridge, 1957-1962; Acting Master, Clare College, Cambridge, 2000; Acting Master, Clare College, 1984; Senior Tutor, Clare College, 1966-1969; Assistant Tutor, Clare College, 1959-1966; Research Fellow, Clare College, 1955-1959.

**Visiting Academic or Research Appointments:** Visiting Professor, Philosophy, Yale University, 1990; Visiting Professor, Philosophy, University of Notre Dame, 1986; Visiting Professor, Philosophy, Yale University, 1975; Visiting Professor,

Philosophy, University of Virginia, 1972; Radcliffe Fellow, University of Oxford, 1969-1971; Visiting Professor, Philosophy, Cornell University, 1964.

**Research Profile:** Smiley has had a lifelong engagement with Aristotle, arguing (contra Lukasiewicz) that syllogistic and modern logic are conceptually and methodologically continuous. He has used many-sorted quantification as a vehicle for comparing the two, and has shown that Aristotle not only had a notion of completeness but actually attempted a completeness proof. He has also developed a notion of 'relevant' implication to sustain Aristotle's key requirement that premises must form a chain of predications linking the terms of the conclusion.

He has always put the idea of consequence centre-stage, and much of his work explores the features of consequence relations, whether modal or formal or both or neither. With D. J. Shoesmith, he took up the anomaly that conventional arguments can have any number of premises but only one conclusion. In their multiple-conclusion logic an argument can have any number of conclusions, being valid if it is impossible for all the premises to be true and all the conclusions false, while multiple-conclusion rules of inference provide an elegant direct alternative to indirect natural-deduction rules. A more subversive move has been to ask whether deductions need to be logically valid. After all, we do not expect premises to be necessary truths, so why build necessity into the requirement that inferences from them should be truth-preserving? In the spirit of Bolzano, Smiley develops a theory of relative validity—validity relative to a presupposed argumentative basis, which might be anything from a legal code to Newtonian geometry.

Running through his work is a challenge to the uncritical orthodoxy that takes the current form of the predicate calculus as an exclusive paradigm for logic. It does not, for example, tolerate empty terms (the King of France, 1/0), and Smiley has used the treatment of functions as a test case to demonstrate that neither of the two most popular evasive manoeuvres, Russell's and Frege's, is sustainable. By contrast, once empty terms are admitted, descriptions and functions can readily be accommodated as they stand. In a similar spirit he and Alex Oliver have tackled the phenomenon of terms that denote more than one thing, e.g. plural definite descriptions (the men who surrounded the fort) or expressions for the values of many-valued functions ($\sqrt{4}$, log i). A common dodge is to replace them by singular terms standing for sets, but Oliver and Smiley have shown that this cannot succeed as a general strategy. Instead they generalize the syntax and semantics of predicate calculus to accommodate plural terms as they stand. Among other things they have argued that many-valued functions are legitimate and irreducible to single-valued ones. They have also shown that any satisfactory account of lists (as in 'Whitehead and Russell wrote *Principia*') requires another resource that orthodox predicate calculus excludes, namely multigrade predicates.

**Main Publications:**

1. Entailment and deducibility, *Proc. Aristotelian Soc.* 59 (1959), 233–54.
2. Sense without denotation, *Analysis* 20 (1960), 125–35.
3. Syllogism and quantification, *J. Symbolic Logic* 27 (1962), 58–72.
4. Relative necessity, *J. Symbolic Logic* 28 (1963), 113–34.
5. What is a syllogism?, *J. Philosophical Logic* 2 (1973), 136–54.
6. Does many-valued logic have any use?, in S. Korner, ed., *Philosophy of Logic* (Blackwell 1976), 74–88.
7. (with D. J. Shoesmith), *Multiple-conclusion logic* (CUP 1978), xiii + 396 pp.
8. The schematic fallacy, *Proc. Aristotelian Soc.* 83 (1982), 1–17.
9. Aristotle's completeness proof, *Ancient Philosophy* 14 (1994), 25–38.
10. A tale of two tortoises, *Mind* 104 (1995), 725–36.
11. Consequence, conceptions of, in E. J. Craig, ed., *Routledge Encyclopedia of Philosophy* (1998).
12. Rejection, *Analysis* 56 (1996), 1–9.
13. (with Alex Oliver) Strategies for a logic of plurals, *The Philosophical Quarterly* (2001), 289–306.
14. The theory of descriptions, in T. R. Baldwin & T. J. Smiley, eds, *Studies in the philosophy of logic and knowledge* (OUP 2003), 131–61.
15. (with Alex Oliver) Multigrade predicates, *Mind* 113 (2004), 609–81.
16. (with Alex Oliver) Plural descriptions and many-valued functions, *Mind* 114 (2005), 1039–68.
17. (with Alex Oliver) A modest logic of plurals, *J. Philosophical Logic* 35 (2006), 317–48.
18. (with Alex Oliver) What are sets and what are they for?, to appear in *Philosophical Perspectives* 20 (2006).

*Work in Progress*
19. A book on *The Logic of Plurals*, with Alex Oliver.
20. A book on *Aristotle and Modern Logic*.

**Service to the Profession:** Member and Chairman, Postgraduate Committee, Humanities Research Board, 1994-1996; Member of Council, Secretary for Postgraduate Studies, British Academy, 1992-1995; Member, Distinctions Committee, University of Oxford, 1995-1998; Chairman, Board of Scrutiny, University of Cambridge, 1995; Committee Member, Personal Professorships and Readerships, University of Cambridge, 1992-1995; President, Aristotelian Society, 1982-1983; Council Member, Association for Symbolic Logic 1979-1982; General Board of the Faculties Member, University of Cambridge, 1968-1972.

**Teaching:** Anyone who chooses to work at Cambridge must see teaching as their primary vocation. It was a congenial place to teach logic, since the legacy of a great Cambridge tradition in the subject had put it at the heart of the philosophy curriculum. Smiley's PhD students included Thomas Baldwin, Chris Brink, F. R. Drake, Susan Haack, Alex Oliver, D. J. Shoesmith and Neil Tennant. Others who passed through his hands were Ian Hacking, Imre Lakatos, Jonathan Lear and Crispin Wright.

**Vision Statement:** Under the umbrella of 'mathematical logic' I see a mass of mathematics-driven work without even a vestigial connection with the theory of argumentation. When the caravan has moved on, I hope logic will be left to return to its roots.

**Honours and Awards:** Fellow, British Academy, 1984.

# SPECKER, Ernst

**Specialties:** Topology, recursive analysis, combinatorial set theory, type theory, axiomatic set theory, Ramsey's theorem, arithmetic, logic of quantum mechanics.

**Born:** 11 February 1920.

**Educated:** ETHZ, Doctor scientiarum mathematicarum, 1949

**Regular Academic or Research Appointments:** ETHZ; UNIVERSITY OF GENEVA; University of Neufchatel; Cornell University; University of Cairo.

**Research Profile:** In 1953, he published "The axiom of choice in Quine's New Foundations for Mathematical Logic". Specker here proves the following important results which were long a subject for conjecture:

1. the axiom of infinity is probable in Quine's "New Foundations";

2. the axiom of choice is disprovable in "New Foundations";

3. the generalized continuum hypothesis is also disprovable in "New Foundations".

Specker's contributions to mathematics are reviewed in [5] where his 32 publications up to 1979 are divided into 10 categories: topology, recursive analysis, combinatorial set theory, type theory, axiomatic set theory, Ramsey's theorem, arithmetic, logic of quantum mechanics, algorithms, and miscellaneous.

In 1987 Specker gave his farewell lecture at the ETH Zurich Postmoderne Mathematik: Abschied vom Paradies?. This is published as [3] and presents problems of infinity in the form of a fairy tale. The volume [1] list 42 publications by Specker.

**Main Publications:**

1. The fundamental theorem of algebra in recursive analysis (1969).

2. Die Entwicklung der axiomatischen Mengenlehre (1978).

3. (with H Kull) Direct construction of mutually orthogonal Latin squares (1987).

4. Application of logic and combinatorics to enumeration problems (1988).

5. Publications: Selecta Ernst Specker, Birkhäuser 1990

# STALNAKER, Robert C.

**Specialties:** Modal and conditional logic, philosophical foundations and applications of logic, applications of logic to game theory, application of formal methods to linguistic semantics and pragmatics.

**Born:** 22 January 1940 in Princeton, New Jersey, USA.

**Educated:** Princeton University, PhD, 1965; Wesleyan University, BA, 1962.

**Dissertation:** "Historical Interpretation"; supervisor, Stuart Hampshire and C. G. Hempel.

**Regular Academic or Research Appointments:** PROFESSOR OF PHILOSOPHY, MASSACHUSETTS INSTITUTE OF TECHNOLOGY, 1988–; Professor of Philosophy, Cornell University, 1976-1988; Associate Professor of Philosophy, Cornell University, 1971-1976; Associate Professor of Philosophy, University of Illinois at Urbana-Champaign, 1968-1971; Assistant Professor of Philosophy, Yale University, 1967-68; Instructor in Philosophy, Yale University, 1965-1967.

**Research Profile:** Robert Stalnaker was one of the original developers of conditional logic and formal semantics, in the possible worlds framework, for counterfactual conditionals. He also contributed to the application of formal semantic methods to the semantics, and particularly the pragmatics of natural languages. He has made contributions to the epistemic foundations of game theory, belief revision theory, and the interpretation of nonmonotonic logic. Recent work concerns the application of epistemic logic to problems in epistemology, and the relation between technical and philosophical issues in quantified modal logic and semantics.

**Main Publications:**

1. "A Theory of Conditionals." *Studies in Logical Theory* (N. Rescher, ed.), Oxford, 1968, 98-112. (Reprinted in *Ifs: Conditionals, Belief, Decision, Chance, and Time* (edited by W. Harper, G. Pearce and R. Stalnaker). Dordrecht: D. Reidel, 1981, in E. Sosa(ed.), *Causation and Conditionals* (Oxford Readings in Philosophy), London: Oxford U. Press, 1975.) and in Frank Jackson (ed.), *Conditionals* (Oxford Readings in Philosophy), Oxford and New York: Oxford University Press, 1991.)
2. "Modality and Reference." (with R. H. Thomason) *Nous*, 2(1968), 359372.
3. "A Semantic Analysis of Conditional Logic." (with R. H. Thomason) *Theoria*,36 (1970).
4. "Possible Worlds." *Nous*, 10(1976), 6575.
5. "Complex Predicates." *The Monist*, 60(1977), 327-339.
6. "Assertion." *Syntax and Semantics*, 9(1978), 315-332.
7. "A Defense of Conditional Excluded Middle" *Ifs: Conditionals, Belief, Decision, Chance, and Time* (edited by W. Harper, G. Pearce and R. Stalnaker). Dordrecht: D. Reidel, 1981, 87104.
8. "The Problem of Logical Omniscience, I." *Synthese*, 89(1991), 425-440. (Reprinted in R. Stalnaker, *Context and Content: Essays on Intentionality in Speech and Thought*. Oxford, Oxford University Press, 1999.)
9. "Notes on Conditional Semantics," in Yoram Moses(ed.) *Proceedings of the Fourth Conference on Theoretical Aspects of Reasoning about Knowledge*, San Mateo, CA: Morgan Kaufmann Publishers, Inc., 1992, 316-327.
10. "A Note on Nonmonotonic Modal Logic, *Artificial Intelligence*, 64 (1993), 183-196.
11. "What is a Non-monotonic Consequence Relation?" *Fundamenta Informaticae*, 21 (1994).
12. "On the Evaluation of Solution Concepts," *Theory and Decision*, 37 (1994), 49-73. Revised version in *Epistemic Logic and the Theory of Games and Decisions*, ed. by M. O. L. Bacharach, L.-A. Gérard-Varet, P. Mongin and H. S. Shin. Kluwer Academic Publisher, 1997, 345-64.
13. "Conditionals as Random Variables" (with R. C. Jeffrey) in *Probability and Conditionals: belief revision and rational decision.* , ed. by B. Skyrms and E. Eells. Cambridge: Cambridge University Press, 1994, 31-46.
14. "The Interaction of Modality with Quantification and Identity," *Modality, Morality and Belief: essays in honor of Ruth Barcan Marcus*, ed. by W. Sinnott-Armstrong, D. Raffman, and N. Asher. Cambridge: Cambridge University Press, 1994, 12-28. (Reprinted in R. Stalnaker, *Ways a World Might Be: Metaphysical and Anti-metaphysical Essays*. Oxford, Oxford University Press, 2003.)
15. "Knowledge, Belief and Counterfactual Reasoning in Games," *Economics and Philosophy*, 12 (1996), 133-162.
16. "Extensive and Strategic Forms: Games and Models for Games." *Research in Economics*, 53 (1999), 293-319.
17. "The Problem of Logical Omniscience, II" in R. Stalnaker, *Context and Content: Essays on Intentionality in Speech and Thought*. Oxford, Oxford University Press, 1999.

*Work in Progress*

18. Logics of Knowledge and Belief; Counterfactual Propositions in Game Theory.

**Service to the Profession:** Executive Committee, American Philosophical Association, Eastern Div., 1992-1995; Associate Editor, *Linguistics & Philosophy*, 1989-1992; Member, Council for Philosophical Studies, 19861992; Coeditor, *Philosophical Review*, 197274, 197778, 198182, 1984-85.

**Teaching:** PhD students in philosophy at MIT include Paul Pietroski, Jason Stanley, Zoltan Szabo-Gendler, Josep Macia, Delia Graff.

**Vision Statement:** Developments in the past fifty years in modal, temporal, epistemic and deontic logic and semantics have had a significant impact in philosophy, and in the interaction of philosophy with related fields such as linguistics, cognitive science and economics. Philosophical problems can never be reduced to mathematical problems, or given purely technical solutions, and there will always remain a healthy tension between philosophical and technical problems and solutions, but properly used, the resources provided by logic and formal semantics can provide indispensable tools for the sharpening and clarifying of issues in all areas of philosophy.

**Honours and Awards:** Fellow, American Academy of Arts and Sciences, 1992-; NEH Fellowship, 19851986; NEH Fellowship at the Center for Advanced Study in the Behavioral Sciences, 197879; Guggenheim Fellowship, 197475.

# STATMAN, Richard

**Specialties:** Mathematical logic, theory of computation, theory of proofs.

**Born:** 2 April 1946 in New York, New York, USA.

**Educated:** Stanford University, PhD, 1974; Cambridge University, MA, 1974; Brooklyn College, BA, 1970;

**Dissertation:** *Structural Complexity of Proofs*; supervisor, Georg Kreisel.

**Regular Academic or Research Appointments:** PROFESSOR, CARNEGIE MELLON UNIVERSITY, 1988–; Associate Professor, Carnegie Mellon University, 1985-1988; Visiting Associate Professor, Carnegie Mellon University, 1983-1984; Associate Professor, Rutgers University, 1982-1985; Assistant Professor, Rutgers University, 1979-1982; Berry-Ramsey Senior Fellow, King's College, Cambridge, 1974-1976.

**Visiting Academic or Research Appointments:** Visiting Professor, University of Nijmegen, 1993; Visiting Scientist, INRIA (Paris), 1992-1993; Visiting Professor, Georgetown University, 1986; Visiting Associate Professor, Purdue University, 1982; Assistant Professor, University of Michigan, 1976-1979.

**Research Profile:** My area of research can be described hierarchically as :
    mathematical logic
    - theory of computation
    – theory of programming languages (computer science)
        — theory of functional programming
        —- lambda calculus / combinatory logic.
Lambda calculus is the study of certain computation rules or programs. From among those programs which can be applied to arguments and return values we single out those whose execution depends only on the fact that some of the data are themselves computation rules of the same sort. It is not obvious that there are any non-trivial examples of such rules. The rich deep structure of the lambda calculus had to be discovered by Church, Bernays, Curry, Kleene and those who followed them. Since then, many distinctions have been made, such as those between applicative and functional programming, and many quite different type systems have evolved.

There is currently a great deal of research into lambda calculus by the programming language, theorem proving and symbolic computing communities. In our work we are interested in the deep structure of pure lambda calculus with at most algebraic types. Roughly speaking my work falls into 6 general categories.

1. Simply typed lambda calculus and its elementary extensions
2. Evaluation, reduction and conversion strategies
3. Combinators and combinatory algebra
4. Computability of functions and invariants
5. Functional equations and unification
6. Connections to other branches of mathematics and computer science.

My work is principally co-Dana Scott ( co- as in complement).

**Main Publications:**

1. (with D. van Dalen) Equality in the presence of apartness, in Essays on Mathematical and Philosophical Logic, Hintikka et al., eds., Reidel (1978), pp. 95-116.
2. Proof search and speed-up in the predicate calculus, Annals of Mathematical Logic, 15, (1978), pp. 225-226
3. Lower bounds on Herbrand's theorem, Proceedings of the A.M.S., Vol. 75, No.1 (1979), pp. 104-107.
4. Intuitionistic propositional logic is polynomial space complete, Theoretical Computer Science, Vol. 9, No. 1 (1979), pp. 67-72.
5. The typed lambda-calculus is not elementary recursive, Theoretical Computer Science, Vol. 9, No. 1 (1979), pp. 73-81.
6. On the existence of closed terms in the typed lambda-calculus I, in: To H.B. Curry: Essays on Combinatory Logic, Lambda Calculus, and Formalism, J. P. Seldin and R. Hindley, eds., Academic Press (1980), pp. 511-534.
7. Completeness, invariance and lambda-definability, Journal of Symbolic Logic, Vol. 47, No. 1, March 1982, pp. 17-26.
8. Lambda-definable functionals and beta-eta conversion, Archiv fur Mathematische Logic und Grundlagenforschung, 22 (1982), pp., 1-6.
9. Logical relations in the typed lambda-calculus, Information and Control, Vol.65, No. 2/3, (June, 1985).
10. On translating lambda terms into combinators: the basis problem, Proceedings of 1986 Logic in Computer Science Conference, I.E.E.E., (June, 1986),pp. 378-382.
11. Equality between functionals revisited, in H. Friedman's Research on the Foundations of Mathematics, L. Harrington, M. Moreley, A. Scedrov, and S. Simpson, eds., North Holland Publishing Co., (1985).
12. The word problem for Smullyan's Lark Combinator is decidable, Journal for Symbolic Computation, 7 (1989), 103-112.
13. On sets of solutions to combinator equations, T.CS, 66, (1989).
14. Freyd's Hierarchy of Combinator Monoids Proceedings of 1991 Logic in Computer Science Conference I.E.E.E., (July, 1991).
15. (with P. Narendran , F. Pfenning) On the unification problem for Cartesian closed categories, Proceedings of 1993 Logic in Computer Science Conference, I.E.E.E., (July, 1993).

16. (with U. Deliquoro, A. Piperno) Retracts in the simply typed lambda beta-eta calculus, Proceedings of the Seventh Annual IEEE Symposium on Logic in Computer Science, Santa Cruz, California, IEEE Computer Society Press, 461-469 (1992).

17. Some examples of non-existent combinators, in A Collection of Contributions in Honour of Corrado Bohm on the Occasion of his 70th Birthday, edited by M. Dezani-Ciancaglini, S. Ronchi, Della Rocca, M. Venturini Zilli, Elsevier, 441-448 (1993). (Reprinted from Theoretical Computer Science, 121, (1993)).

18. Marginalia to a theorem of Jacopini, TLCA '99, LNCS 1581 (1999)

19. On the lambda Y calculus Proceedings of 2002 Logic in Computer Science Conference, I.E.E.E., (July,2002)

20. (with Benedetto Intrigila) The omega rule is pi-zero-two hard, Proceedings of 2004 Logic in Computer Science Conference, I.E.E.E., (July 2004)

**Service to the Profession:** Program Committee, Logic in Computer Science IEEE; Program Committee, Foundations of Computer Science IEEE; Program Committee, Symposium on Theory of Computing ACM; Program Committee, Typed Lambda Calculus and Applications.

**Teaching:** Statman is one of the founding members of the PAL (pure and applied logic) programme, and the ACO (algorithms, combinatorics, and optimization) programme at CMU. He also helped to develop the joint mathematics and computer science curricula based on the famous The Carnegie-Mellon Curriculum for Undergraduate Computer Science (Springer-Verlag, 1985).

**Vision Statement:** It is hard to imagine that logic will evolve as quickly in the next century as it did in the last. One should expect some very striking cross area applications in the $21^{st}$ century.

**Honours and Awards:** NSF Award, Software Engineering, "LambdaCalculus", 1996; NSF Award, Numeric, Symb, Geo Comp., "LambdaCalculus, type theory and automated theorem Proving," with Peter Andrews, 1992; NSF Award, Numeric, Symb, Geo Comp., "LambdaCalculus, type theory and automated theorem Proving," with Peter Andrews, 1990; NSF Award, Numeric, Symb, Geo Comp., "LambdaCalculus, type theory and automated theorem Proving," with Peter Andrews, 1987; NSF Award, Theoretical Comp. Sci., "Syntax and semantics of typed lambda calculus," 1983; NSF Award, Theoretical Comp. Sci., "Syntax and semantics of typed lambda calculus," 1979.

# STEPRĀNS, Juris

**Specialties:** Set theory, history of set theory, mathematical logic.

**Born:** 3 November 1953 in Montreal, Quebec, Canada.

**Educated:** University of Toronto, PhD Mathematics, 1982; University of Toronto, MSc Mathematics, 1978; University of Waterloo, BMath Pure Mathematics, 1977.

**Dissertation:** *Some Results in Set Theory*; supervisor, Franklin D. Tall.

**Regular Academic or Research Appointments:** DEPUTY DIRECTORY, FIELDS INSTITUTE, 2006–; PROFESSOR, MATHEMATICS, YORK UNIVERSITY, 1993–; Associate Professor, Mathematics, York University, 1988-1993; Assistant Professor, Mathematics, York University, 1982-1988.

**Visiting Academic or Research Appointments:** Visiting Professor, University of Wisconsin (Madison) 2004; Visiting Member, Fields Institute, 2002; Visiting Researcher, University of Latvia, 1994-1995; Visiting Researcher, Wroclaw University, 1989; Visiting Professor, University of Warsaw, 1988; Visiting researcher, Dartmouth College, 1982; Visiting Researcher, Rutgers University, frequently.

**Research Profile:** Juris Steprāns has focussed his research on the applications of set theory to other areas of mathematics. His theorem characterizing free abelian groups in terms of the existence of discrete norms is cited as a striking example of the essential use of set theoretic methods to solve a problem seeming to have no relation to set theory at all. Various of his papers dealing with set theoretic applications to group theory resulted in joint work with S. Shelah producing a ZFC example of a Banach space with few operators: few in the sense that all operators are the sum of a multiple of the identity and an operator with separable range even though the Banach space itself is not separable. This continues to be a starting point for currently very vigorous research into this area. His work with S. Shelah on the lack of non-trivial automorphisms under PFA is also widely cited. His interest in set theoretic questions related to the geometry of Euclidean space has produced progress on various aspects of cardinal invariants of the continuum as well as establishing links with developments in harmonic analysis.

**Main Publications:**

1. J. Steprāns, "The number of submodules", *Proc. London Math. Soc.*, 49, no. 3 (1984) 183–192.

2. J. Steprāns, "A characterization of free Abelian groups", *Proc. Amer. Math. Soc.*, 93 (1985) 347–349.
3. J. Steprāns, "Strong Q-sequences and versions of Martin's Axiom", *Can. J. Math.*, 37, no. 4 (1985) 740–746.
4. J. Steprāns and W.S. Watson, "Homeomorphisms of manifolds with prescribed behaviour on large dense sets", *Bull. London Math. Soc.*, 19 (1987) 305–310.
5. S. Shelah and J. Steprāns, "A Banach space with few operators", *Proc. Amer. Math. Soc.*, 104 (1988) 101–105.
6. S. Shelah and J. Steprāns, "PFA implies all automorphisms are trivial", *Proc. Amer. Math. Soc.*, 104 (1988) 1220–1225.
7. A. Dow and J. Steprāns, "Countable Frechét $\alpha_1$-spaces may be first countable", *Arch. für Math. Logik*, 32 (1992) 33–50.
8. A. Dow and J. Steprāns, "The $\sigma$-linkedness of the Measure Algebra", *Can. Math. Bull.* 37 (1993) 48–45.
9. J. Steprāns, "Martin's Axiom and the transitivity of P-points", *Israel Journal of Mathematics*, 83 (1993) 257–274.
10. J. Steprāns, "A very discontinuous Borel function", J.S.L, 58 (1993) 1268–1283.
11. S. Shelah and J. Steprāns, "Homogeneous Almost Disjoint Families", *Algebra Universalis*, 31 (1994) 196–203.
12. J. Steprāns, "Sum of Darboux and Continuous Functions", *Fund. Math.* 146 (1995), 107–120.
13. A.W. Miller and J. Steprāns, "Orthogonal Families of Real Sequences", *J.S.L.*, 63 (1998) 29–49.
14. J. Steprāns, Decomposing with Smooth Sets, *Trans. A.M.S.*, 351 (1999) 1461–1480.
15. J. Steprāns, "Unions of Rectifiable Curves and the Dimension of Banach Spaces", *J.S.L.*, 64 (1999) 701–726.
16. M. Hrusak, J. Steprāns, Y. Zhang, "Confinitary Groups, Almost Disjoint and Dominating Families", *J.S.L.*, 66 (2001) 1289–1276.
17. J. Steprāns, "The Autohomeomorphism Group of the Cech-Stone Compactification of the Integers", *Trans. A.M.S.*, 355 (2003) 4223–4240.
18. Jan Dijkstra, Jan van Mill and Juris Steprāns, "Complete Erdos Space is Unstable", *Math. Proc. Cambridge Philos. Soc.* 137 (2004), no. 2, 465–473.
19. S. Shelah and J. Steprāns, "Comparing the Uniformity Invariants of Null Sets of Different Measures", *Adv. Math.* 192 (2005), 403–426.
20. J. Steprāns, "Geometric Cardinal Invariants, Maximal Functions and a Measure Theoretic Pigeonhole Principle", *Bull. Symbolic Logic* 11 (2005), 517-525.

*Work in Progress*

21. Saharon Shelah and Juris Steprāns, "Possible Cardinalities of Maximal Abeliam Subgroups of Quotients of Permutation Groups of the Integers".
22. B. Kastermans, J.Steprāns, Y. Zhang, "Maximality of Analytic Families of Eventually Different Functions".
23. J. Steprāns, "Maximal Functions and the Additivity of Various Families of Null Sets".
24. J. Steprāns, a chapter in the Handbook of the History of Logic on the continuum in the $20^{th}$ century.
25. Initiation of a new direction of research into the set theoretic aspects of operator theory.

**Service to the Profession:** Editorial Board Member, Fields Institute, 2006-; Publications Chair, CMS, 2005-; Elected, CMS Board of Directors, 2003-2007; NSERC, Grant Selection Committee Member, 2002-2004, Grant Selection Committee Chair, 2004-2005; Latvian Council Member, Mathematics Professors, 2003-.

**Teaching:** Juris Steprāns has taught at York University and as a visiting scholar elsewehere since 1982. Among his accomplishments has been the development of a very successful undergraduate course introducing non-mathematics majors to the incompleteness phenomenon through readings on accessible popular literature on the subject. His three doctoral students to date are Mariusz Rabus, Michael Hrusak, and Vera Fischer.

**Vision Statement:** The development I would be most interested in seeing is a clearer elaboration of the connection between logic and the physical world.

**Honours and Awards:** Elected Fellow, Fields Institute, 2004.

# STOKHOF, Martin

**Specialties:** Formal semantics and pragmatics of natural language.

**Born:** 18 November 1950 in Amsterdam, The Netherlands.

**Educated:** Universiteit van Amsterdam, PhD Philosophy, 1984; Universiteit van Amsterdam, MA Philosophy, 1974.

**Dissertation:** (joint with Groenendijk)*Studies on the Semantics of Questions and the Pragmatics of Answers*; supervisors: R. Bartsch and J. van Benthem.

**Regular Academic or Research Appointments:** PROFESSOR, PHILOSOPHY OF LANGUAGE, UNIVERSITEIT VAN AMSTERDAM, 1998—; Professor, Philosophy of Language, Katholieke Universiteit Nijmegen, 1996–1998; Associate Professor, Philosophy of Language, Department of Philosophy, Universiteit van Amsterdam, 1988–1998; Assistant Professor, Logic and Semantics, Department of Computational Linguistics, Universiteit van Amsterdam, 1986–1992; Visiting Researcher, Department of Computational Linguistics, Katholieke Universiteit Brabant, 1985;

Assistant Professor, Philosophy of Language, Universiteit van Amsterdam, 1976–1988; Researcher, Dutch Organization for the Advancement of Pure Research (Z.W.O.), 1974–1978.

**Research Profile:** Martin Stokhof has worked on descriptive and theoretical issues in the formal semantics and pragmatics of natural language, and on the philosophy of Wittgenstein.

Together with J. Groenendijk he developed the so-called 'partition approach' to the semantics of questions, in the early 1980s. This approach makes a purely semantic analysis of questions possible, one that satisfies the same strict requirements as other branches of logical semantics. Such a semantic analysis can be embedded in a pragmatic, information-based analysis of various relations of answerhood.

Also with J. Groenendijk, and later with Frank Veltman and others, he initiated and explored the so-called 'dynamic approach' to meaning in natural language in the 1990s. This dynamic approach abandons the common reference and truth based analysis of natural language meaning that semantics inherited from classical logic, and treats meaning as 'information change potential' (making use of concepts from dynamic logic). This approach allows for a conceptual integration of semantics and pragmatics, and extends naturally to the analysis of larger discourses and of linguistic interactions.

At the end of the nineties his interests shifted to more theoretical issues, concerning the nature of semantic explanation and the central concepts of semantics, such as the role of interpretation, compositionality, and the distinction between grammatical and logical form.

His work on the philosophy of Ludwig Wittgenstein grew out of an interest in the place and role of ethics in Wittgenstein's early work. Defending the central role of ethics for the early Wittgenstein, he developed a reading of the latter's *Tractatus* that analyses the ontology as a derivative of the work's basic logical and semantic presuppositions together with the ineffability of its ethical doctrine.

**Main Publications:**

1. (With J. Groenendijk) A pragmatic analysis of specificity, in: F. Heny (ed), *Ambiguities in Intensional Contexts*, Dordrecht, Reidel, 1981, pp. 98-123
2. (With J. Groenendijk) Semantic analysis of wh-complements, *Linguistics and Philosophy*, 5(2), 1982, pp. 175-235
3. (With J. Groenendijk) Type-shifting rules and the semantics of interrogatives, G. Chierchia, B. Partee & R. Turner (eds), *Properties, Types and Meaning. Vol. II: Semantic Issues*, Dordrecht, Reidel, 1988, pp. 21-69 [Reprinted in P. Portner & B.H. Partee (eds), *Formal Semantics. The Essential Readings*, Oxford: Blackwell, 2002, pp. 421-456]
4. (With J. van Benthem, J. Groenendijk, D. de Jongh, H. Verkuyl) *Logic, Language and Meaning. Vol. I: Introduction to Logic*, Chicago, University of Chicago Press, 1990 [Spanish translation: *Introducción a la Lógica*, Buenos Aires: Eudeba, 2002]
5. (With J. van Benthem, J. Groenendijk, D. de Jongh, H. Verkuyl) *Logic, Language and Meaning. Vol. II: Intensional Logic and Logical Grammar*, Chicago, University of Chicago Press, 1990
6. (With J. Groenendijk) Dynamic Montague grammar, L. Kálmán & L. Pólos (eds), *Proceedings of the Second Symposion on Logic and Language*, Hajdúszoboszló, September 5-9, 1989, Budapest, Eötvös Loránd University, 1990, pp. 3-48
7. (With J. Groenendijk) Dynamic predicate logic, *Linguistics and Philosophy*, 14(1), 1991, pp 39-100 [Reprinted in: P. Grim, P. Ludlow, & G. Mar (eds), *The Philosopher's Annual*, Vol. XIV,1991, Atascadero, California: Ridgeview, 1993, pp. 67-128; and in: S. Davis & B. Gillon (eds), *Semantics: A Reader*, Oxford: Oxford University Press, 2004, pp. 263-305]
8. (With J. Groenendijk, F. Veltman) Coreference and modality, in: S. Lappin (ed), *Handbook of Contemporary Semantic Theory*, Oxford, Blackwell, 1996, pp. 179-216
9. (With J. Groenendijk), Questions, in: van Benthem & A. ter Meulen (eds), *Handbook of Logic and Language*, Amsterdam/Cambridge, Mass., Elsevier/MIT Press, 1997, pp. 1055-1124
10. (With J. Groenendijk) Meaning in motion, in: K. von Heusinger & U. Egli (eds), *Reference and Anaphora*, Kluwer, Dordrecht, 1999, pp. 47-76
11. Meaning, interpretation and semantics, in: D. Beaver & P. Scotto di Luzio (eds), *Words, Proofs, and Diagrams*, CSLI Publications, Stanford, 2002, pp. 217-240
12. *World and Life as One: Ethics and Ontology in Wittgenstein's Early Thought*, Stanford University Press, Stanford, 2002
13. (With J. Groenendijk), Why compositionality?, in: G. Carlson & Pelletier (eds), *Reference and Generality: The Partee Effect*, CSLI Publications, Stanford, 2005, pp. 83-106
14. (With J. van Eijck), The gamut of dynamic logics, in: D. Gabbay & Woods (eds), *Handbook of the History of Logic. Volume 7 - Logic and the Modalities in the Twentieth Century*, Elsevier, Amsterdam, 2006, pp. 499-600.
15. Hand or hammer? On formal and natural languages in semantics, *Journal of Indian Philosophy*, 35, 2007.

*Forthcoming work*

16. The architecture of meaning: Wittgenstein's Tractatus and formal semantics, to appear in D. Levy & E. Zamuner, eds., *Wittgenstein's Enduring Arguments*, London, Routledge.
17. (With H. Kamp), Information in natural language,

to appear in P. Adriaans an & J. van Benthem, eds., *Handbook of the Philosophy of Information*, Elsevier, Amsterdam, 2008.

A full list of publications is available at http://staff.science.uva.nl/~stokhof/publications.pdf

**Service to the Profession:** Editor, Logic and Language, *Stanford Encyclopedia of Philosophy*, 2004–; Member, Standing Committee for the Humanities ESF (European Science Foundation), 2004–; Chairman, HERA-Network of Humanities Councils in Europe, 2004–; Humanities Council NWO (Dutch Organization of Scientific Research), Member, 2003, Chairman, 2004–; Associate Editor, Semantics, *Linguistics and Philosophy*, 1990–; Scientific Director, Institute for Logic, Language and Computation ILLC, 1998–2004; Dutch Research School in Logic OZSL, Member, 1991, Deputy Director, 1991–1996, Chairman, 2001–2005; Organizer, Most of the bi-annual *Amsterdam Colloquium*, 1976–; Editorial Board Member, *Natural Language Semantics*, 1992–2006; Editorial Board Member, *Current Research in the Semantics/Pragmatics Interface* (Elsevier), 1997–; Editorial Board Member, *Journal for Research on Language and Computation*, 1997–; Editor, GRASS-series, Foris Publications/Mouton de Gruyter, 1984–1996.

**Teaching:** Stokhof has taught logic and formal semantics for many years, to philosophy students as well as to students in computational linguistics. Under his directorship, the ILLC started its international Master of Logic curriculum. Together with J. van Benthem, J. Groenendijk, D. de Jongh and H. Verkuyl he wrote the two Gamut textbooks on logic and logical grammar. He also wrote an introductory textbook in philosophy of language (in Dutch). He (co-)supervised 10 PhD students, among who are Fred Landman, Herman Hendriks and Paul Dekker.

**Vision Statement:** The application of logic in the semantics and pragmatics of natural language is one of the best examples of successful application of concepts and techniques from logic in another discipline. The underlying philosophical and methodological presuppositions of this development, however, have not yet been given the attention they deserve. In conjunction with the turn to cognitive (neuro)science an detailed investigation and assessment is very much needed.

Foundational programs were one of the main forces behind the flowering of logic in the $20^{th}$ century. The subsequent technical development of this subject both internally and through a variety of applications has been remarkable, but the motivating programs were battered under critical examination and largely left behind. This does not mean that foundational concerns must be abandoned. Rather, foundationally directed work retains its prime importance, only now to be approached with a greater clarity of aims and sophisticated use of a variety of tools of modern logic.

**Honours and Awards:** Elected member, Institut International de Philosophie, 2007; Elected Fellow, KNAW (Royal Dutch Academy of Sciences), 2006.

# SUPPES, Patrick

**Specialties:** Constructive foundations of mathematics, nonstandard analysis, axiomatizability questions, axiomatic analysis of developed scientific theories, logical inference in natural language, theory of brain computations

**Born:** 17, March, 1922 in Tulsa, Oklahoma, USA.

**Educated:** University of Chicago (B.S. 1943) and Columbia University (Ph.D. 1950).

**Dissertation:** *Action at A Distance* [in historical theories of physical force, including special relativity]; supervisor,

**Regular Academic or Research Appointments:** DIRECTOR AND FACULTY ADVISOR, EDUCATION PROGRAM FOR GIFTED YOUTH, STANFORD UNIVERSITY, 1992–; LUCIE STERN PROFESSOR OF PHILOSOPHY, EMERITUS, STANFORD UNIVERSITY 1992–; Lucie Stern Professor of Philosophy, Stanford University 1975–92; Professor by courtesy, Department of Psychology, Stanford University 1973–92; Professor by courtesy, School of Education, Stanford University 1967–92; Professor by courtesy, Department of Statistics, Stanford University 1960–92; Director, Institute for Mathematical Studies in the Social Sciences, Stanford University 1959–92; Professor, Department of Philosophy, Stanford University 1959-92; Assistant and Associate Professor, Department of Philosophy, Stanford University 1950–59; Lecturer (part-time), Columbia University 1948–50

**Visiting Academic or Research Appointments:** Visiting Professor, Collège de France, Paris, April 1988; Messenger Lecturer, Cornell University, Ithaca, New York, September 1981; Visiting Professor, Collège de France, Paris, November 1979; Howison Lecturer in Philosophy, University of California, Berkeley, 1979; Hägerström Lecturer, Uppsala University, Uppsala, Sweden, 1974; Fellow, John Simon Guggenheim Memorial Foundation, 1971-72.

**Research Profile:** Strong interest in all constructive aspects of mathematics from a foundational viewpoint. This includes free-variable axiomatizations of geometry and analysis, as well as negative results on axiomatizability.

**Main Publications:**

1. With J. C. C. McKinsey & A. C. Sugar. Axiomatic foundations of classical particle mechanics. *Journal of Rational Mechanics and Analysis*, 2 (1953), 253-272.
2. With H. Rubin. Transformations of systems of relativistic particle mechanics. *Pacific Journal of Mathematics*, 4 (1954), 563-601.
3. With D. Scott. Foundational aspects of theories of measurement. *Journal of Symbolic Logic*, 23 (1958), 113-128
4. Axioms for relativistic kinematics with or without parity. In L. Henkin, P. Suppes, & A. Tarski (Eds.), *The Axiomatic Method with Special Reference to Geometry and Physics*. Amsterdam: North-Holland (1959), 291-307.
5. Logics appropriate to empirical theories. In J. W. Addison, L. Henkin, & A. Tarski (Eds.), *Theory of Models*. Amsterdam: North-Holland (1965), 364-375.
6. The probabilistic argument for a non-classical logic of quantum mechanics. *Philosophy of Science*, 33 (1966), 14-21.
7. *Studies in the Methodology and Foundations of Science: Selected papers from 1951 to 1969*. Dordrecht: Reidel (1969), 473 pp.
8. Stimulus-response theory of finite automata. *Journal of Mathematical Psychology*, 6 (1969), 327-355.
9. Krantz, D. H., R. D. Luce, A. Tversky, & P. Suppes (1971). *Foundations of measurement, vol. I: additive and polynomial representations*; Suppes, P., D. H. Krantz, R. D. Luce, & A. Tversky (1989). *Foundations of measurement, vol. II: geometrical, threshold and probabilistic representations*; Luce, R. D., D. H. Krantz, P. Suppes, & A. Tversky (1990). *Foundations of measurement, vol. III: representation, axiomatization, and invariance*. New York: Academic Press. Reprinted by Dover Publications, Inc. (2007).
10. Semantics of context-free fragments of natural languages. In K. J. J. Hintikka, J. M. E. Moravcsik, & P. Suppes (Eds.), *Approaches to Natural Language*. Dordrecht: Reidel (1973), 370-394.
11. Congruence of meaning. *Proceedings and Addresses of the American Philosophical Association*, 46 (1973), 21-38.
12. With W. K. Estes. Foundations of stimulus sampling theory. In D. H. Krantz, R. C. Atkinson, R. D. Luce, & P. Suppes (Eds.), *Contemporary Developments in Mathematical Psychology, Vol. 1: Learning, memory, and thinking*. San Francisco: Freeman (1974), 163-183.
13. With M. Zanotti. Necessary and sufficient conditions for existence of a unique measure strictly agreeing with a qualitative probability ordering. *Journal of Philosophical Logic*, 5 (1976), 431-438.
14. *Logique du Probable*. Paris: Flammarion (1981), 136 pp. Italian translation by Alberto Artosi, *La logica del probabile, un approccio bayesiano alla razionalità*. Bologna, Italy: Cooperativa Libraria Universitaria Editrice Bologna, 1984, 145 pp.
15. With M. Zanotti. Necessary and sufficient qualitative axioms for conditional probability. *Zur Wahrscheinlichkeitstheorie verwandte Gebiete*, 60 (1982), 163-169
16. *Models and Methods in the Philosophy of Science: Selected essays*. Dordrecht: Kluwer Academic Publishers (1993), 525 pp.
17. With R. Chuaqui. Free-variable axiomatic foundations of infinitesimal analysis: A fragment with finitary consistency proof. *Journal of Symbolic Logic*, 60 (1995), 122-159.
18. With M. Zanotti. *Foundations of probability with applications. Selected papers, 1974–1995*. Cambridge (1996): Cambridge University Press.
19. With B. Han, J. Epelboim & Z.-L. Lu. Invariance between subjects of brain-wave representations of language. *Proceedings of the National Academy of Sciences USA*, 96 (1999), 12953-12958.
20. *Representation and Invariance of Scientific Structures*. Stanford, CA: CSLI Publications (2002), 636 pp.

Patrick Suppes' published articles are available at `http://suppes-corpus.stanford.edu`

*Work in Progress*

21. Free-variable system of constructive nonstandard analysis.
22. Computer theorem-checking program which will output an informed proof in a natural language.
23. Theory and analysis of brain computations (in the sense of system, not cell, neuroscience). Initial focus on the basic mechanism being phase-locking of weakly coupled oscillators.

**Service to the Profession:** Eleventh Annual Alfred Tarski Lectures, "Invariance and Meaning" and "A Physical Model of the Brain's Computation of Truth", University of California, Berkeley, 1999; President, Division of Logic, Methodology and Philosophy of Science, International Union of History and Philosophy of Science, 1975–79.

**Teaching:** Developed and taught Introduction to Logic course at Stanford University by computer, 1972–1992 (year of retirement), and also taught by computer, with online proof-checking, Axiomatic Set Theory 1974–1992.

**Vision Statement:** Much work on foundations of quantum computing and foundations of biological computation. The latter topic, more than quantum computing, will lead to an intimate mix of logic and physics in discovering how human brains compute.

**Honours and Awards:** Honorary Doctor of Science and Technology, Carnegie Mellon University, 2008; Corresponding Member, Brazilian Academy of Philosophy, 2006; Lauener Prize in Philosophy, Lauener Foundation, Basel, Switzerland, 2004; Lakatos Award Prize for book in philosophy of science (*Representation and Invariance of Scientific Structures*), London School of Economics and Political Science, 2003; Henry Chauncey Award for Distinguished Service to Assessment and Educational Science, Educational Testing Service, 2003; Barwise Prize, Committee on Philosophy and Computers, American Philosophy Association, 2002; Dottore (ad honorem) in Filosofia, University of Bologna, Italy, 1999; Doctor philosophiae honoris causa, Universität Regensburg, 1999; 1993 Louis Robinson Award, Educom, Washington, DC, 1993; National Medal of Science, 1990; Docteur Honoris Causa, Académie de Paris, Université René Descartes, 1982.; E. L. Thorndike Award for Distinguished Psychological Contribution to Education, American Psychological Association, 1979; Honorary Doctor's Degree in the Social Sciences, University of Nijmegen, The Netherlands, 1979; Columbia University Teachers College Medal for Distinguished Service, 1978; Distinguished Scientific Contribution Award, American Psychological Association, 1972.; American Educational Research Association, Phi Delta Kappa Meritorious Researcher Award, 1971; Palmer O. Johnson Memorial Award, American Educational Research Association, 1967.; Nicholas Murray Butler Medal in Silver, Columbia University, 1965; Social Science Research Council Research Award, 1959.

# T

## TAIT, William

**Specialties:** Mathematical logic, foundations of mathematics, philosophy of mathematics, history of philosophy of mathematics.

**Born:** 22 January 1929 in Freeport, New York, USA.

**Educated:** Yale, PhD Philosophy, 1958; Lehigh University, BA Philosophy, 1952.

**Dissertation:** *The Theory of Partial Recursive Operators;* supervisor: Fredrick Fitch.

**Regular Academic or Research Appointments:** PROFESSOR EMERITUS, PHILOSOPHY, UNIVERSITY OF CHICAGO, 1996–; Professor, Philosophy, University of Chicago, 1972-1996; Professor, Philosophy, University of Illinois, 1968-1971; Associate Professor, Philosophy, University of Illinois-Chicago, 1965-1968; Assistant Professor, Philosophy, Stanford University, 1959-1965; Instructor, Philosophy, Stanford University, 1958-1959.

**Visiting Academic or Research Appointments:** Visiting Professor, Philosophy, Notre Dame, 2004; Visiting Fellow, Wolfson College, Oxford University, 1981; Visiting Professor, Mathematics, University of Aarhus, Denmark, 1971-1972; Member, Center for Advanced Study, University of Illinois, 1966-1967; Member, School of Mathematics, Institute for Advance Study, Princeton, N.J.; Fulbright Scholar, Municipal University of Amsterdam, The Netherlands, 1954-1955.

**Research Profile:** William Tait worked primarily in the area of proof theory and constructive foundations of classical mathematics from the late 1950's through the early 1970's, at which time he decided (prematurely) that the latter area had hit a dead-end and turned to foundations/philosophy of mathematics and its history. He also studied and lectured on Shelah's stability theory in the 1970's, but accomplished nothing in this area aside from a quite simple treatment of the stability spectrum theorem (without invoking either Morely rank or forking). In the 1980's, influenced by papers of Zermelo, Gödel, and Reinhardt, he became interested in what large cardinal axioms could be obtained by genuine reflection principles and developed a theory of relatively consistent reflection principles for formulas containing parameters of order greater than 2. In recent years he has returned part-time to proof theory and in particular to the problem of cut-elimination for second-order number theory.

**Main Publications:**

1. Nested recursion. *Mathematische Annalen*, 143 (1961): 236-250.
2. Infinitely long terms of transfinite type, *Formal Systems and Recursive Functions: Proceedings of the Oxford Colloquium* (1963):176-185.
3. Functionals defined by transfinite recursion, *Journal of Symbolic Logic*, 30 (1965): 155-174.
4. The substitution method, *Journal of Symbolic Logic*, 30 (1965): 175-92.
5. A nonconstructive proof of Gentzen's Hauptsatz for second order predicate logic, *Bulletin of the AMS*, 72 (1966): 980-83.
6. Intensional interpretations of functionals of finite type I, *Journal of Symbolic Logic* 32 (1967): 198-212.
7. Normal derivability in classical logic, *The Syntax and Semantics of Languages*, J. Barwise (ed.), (Lecture Notes in Mathematics 72) Berlin: Springer-Verlag (1968): 204-36.
8. Applications of the cut elimination theorem to some subsystems of classical analysis, *Intuitionism and Proof Theory*, J. Myhill, A. Kino and R. Vesley (ed.), Amsterdam: North-Holland (1970): 475-88.
9. Normal form theorem for bar recursive functionals of finite type, *Proceedings of the Second Scandinavian Logic Symposium*, E. Fenstad (ed.), Amsterdam: North-Holland (1971): 353-68.
10. A realizability interpretation of the theory of species, *Logic Colloquium*, R. Parikh (ed.), (Lecture Notes in Mathematics 453), Berlin: Springer-Verlag (1974): 240-51.
11. Variable-free formalization of the Curry-Howard type theory, *Twenty-five Years of Constructive Type Theory* G. Sambin and J. Smith (ed.) Oxford: OUP 1998.
12. Gàdel's unpublished papers on foundations of mathematics, *Philosophia Mathematica* 9 (2001): 87-126.
13. The completeness of Heyting logic. *Journal of Symbolic Logic* 68 Number 3 (2003):751-763.
14. Constructing cardinals from below, *The Provenance of Pure Reason: Essays on the philosophy of mathematics and its history*, Chapter 6. Oxford University Press (2005).
15. Gàdel's reformulation of Gentzen's first consistency proof for arithmetic: The no-counterexample interpretation, *Bulletin of Symbolic Logic*, 11 no.2 (2005): 225-238.
16. Proof-theoretic semantics for classical mathematics, *Synthese* 148 (2006) pp. 603-622.
17. Gàdel's Correspondence on proof theory and constructive mathematics, *Philosophia Mathmatica* 14 (2006): 76-111.

18. Gödel's interpretation of intuitionist arithmetic, *Philosophia Mathemtica* 14 (2006): 208-28.

*Work in Progress*
19. Higher-order reflection principles in set theory.
20. Cut-elimination in subsystems of second-order number theory.
21. Game-theoretic interpretation of the sequence calculus and its generalizations.

**Service to the Profession:** Editorial Board Member, *Notre Dame Journal of Formal Logic*, 1989-; (Occasional) Referee, *Journal of Symbolic Logic*, 1961-; Program Committee Member, Multiple ASL Meetings in Chicago or Urbana, 1965-.

**Teaching:** Lectured in logic at Stanford, 1958-1965, at University of Illinois in Chicago, 1965-1970, at the University of Aarhus, 1970-1971, and at the University of Chicago, 1972-1996. Some students whom I was pleased to have in my courses and who later gained some prominence in the field of logic were Angus Macintyre (Stanford), Craig Smorynski (University of Illinois-Chicago) and Steve Awodey (University of Chicago), for none of whom, however, did I serve as thesis advisor.

**Vision Statement:** Aside from the area of choice of new axioms for set theory, big vision seems to me out of the picture for now. In proof theory, the main directions at present are in proof-mining, obtaining constructive content from classical proofs, a project initiated by G. Kreisel in the early 1950's and revived in recent years by U. Kohlenbach; the reverse mathematics of H. Friedman and S. Simpson; and cut-elimination (in the sense of the sequence calculus) for second-order number theory, where M. Rathjen has recently obtained results. I do not have a sense of the live areas or future directions of classical recursion theory or definability theory. In model theory, the subject of abstract elementary classes, introduced by S. Shelah, allowing for the extension of classical results of the model theory of first-order logic to L-omega1-omega (and maybe further), seems to me an interesting development. Aside from these meat-and-potatoes areas of logic, the application of logic or, at least, logic like structures in computer science and areas of cognitive science and epistemology seems to be a lively field—about which, however, I know very little.

**Honours and Awards:** Fellow, American Academy of Arts and Sciences, 2002; Guggenheim Fellow, 1968-1969; Fellowship, Center for Advanced Study, University of Illinois, 1966-1967; Grant, Institute for Advanced Study, 1961-2; Fulbright award, Municipal University of Amsterdam, 1954-1955.

# TALL, Franklin D.

**Specialties:** Set theoretic topology, set theory.

**Born:** 21 April 1944 in New York City, New York, USA.

**Educated:** Harvard College, AB Mathematics, 1964; University of Wisconsin at Madison, PhD Mathematics, 1969.

**Dissertation:** *Set-theoretic Consistency Results and Topological Theorems Concerning the Normal Moore Space Conjecture and Related Problems*; supervisor, Mary Ellen Rudin.

**Regular Academic or Research Appointments:** PROFESSOR, MATHEMATICS, UNIVERSITY OF TORONTO, 1980–; Associate Professor, Mathematics, University of Toronto, 1974–1980; Assistant Professor, Mathematics, University of Toronto, 1969–1974.

**Visiting Academic or Research Appointments:** Visiting Professor, University of São Paulo, 2001; Honorary Fellow, University of Wisconsin at Madison, 1997; Visiting Professor, University of São Paulo, 1997; Visiting Research Professor, Dartmouth College, 1996; Visiting Professor, University of São Paulo, 1990; Member, Mathematical Sciences Research Institute (Berkeley), 1989–1990; Visiting Research Professor, Dartmouth College, 1982–1983; Honorary Fellow, University of Wisconsin at Madison, 1976.

**Research Profile:** Tall was a co-founder of the modern field of set-theoretic topology, and has been at its forefront for almost 40 years, continuing to bring new developments in set theory to bear on topological problems. He was the first to use forcing directly on such problems, rather than translating them into set theory. Along with David Booth, he brought the study of cardinal invariants of the continuum into topology. His widely circulated, unpublished survey, *An alternative to the continuum hypothesis and its applications to topology*, introduced Martin's Axiom to the topological community. His seminal papers on the density topology and on irresolvable spaces re-invigorated research in those areas, establishing surprising connections with set theory. A major focus of his work has been the investigation of under what circumstances does normality imply collectionwise normality. This is the key to the Normal Moore Space Conjecture, to which he contributed important consistency results in his doctoral dissertation. One of the techniques he employed in his thesis led him to a counterexample in the theory of completeness properties designed for recognizing Baire spaces, which, surprisingly, has had applications in theoretical computer science. Later work on normality versus collectionwise normality introduced a useful framework for applying supercompact reflection to topology. In the 1990's, Tall turned his attention to the interaction of elementary submodels and topology, studying the elementary submodel topology with his student Lucia Junqueira. Among other results, this led to new insights about the real line. For the past few years, Tall has been investigating a class of models introduced by Todorcevic, in which one can simultaneously obtain the most useful topological consequences of the Proper Forcing Axiom as well as "normal implies collectionwise Hausdorff" consequences of V = L. This has led to the solution of a number of previously intractable topological problems, for example the consistency of locally compact perfectly normal spaces being paracompact.

**Main Publications:**

1. A counterexample in the theories of compactness and of metrization, *Koninkil. Akad. van Wetensch, Proc. Ser. A.*, Vol. 76 (1973) 471–474.
2. The Density Topology, *Pac. J. Math.* **62** (1976) 175–184.
3. Set-theoretic consistency results and topological theorems concerning the normal Moore space conjecture and related problems, *Dissert. Math.* **148** (1977) 1–53.
4. with K. Kunen, Between Martin's Axiom and Souslin's Hypothesis, *Fund. Math.* **102** (1979) 173–181.
5. Normality versus collectionwise normality, 685–732 in Handbook of Set-theoretic Topology, ed. K. Kunen and J. Vaughan, North-Holland, Amsterdam, 1984.
6. with K. Kunen and A. Szymański, Baire irresolvable spaces and ideal theory, *Ann. Math. Silesiana* **2** (14) (1986) 98–107.
7. with A. Dow and W. Weiss, New proofs of the consistency of the normal Moore space conjecture, I *Top. Appl.* **37** (1990) 33–51.
8. (editor), The work of Mary Ellen Rudin, *Ann. New York Acad. Sci.* **705** (1993).
9. Topological applications of generic huge embeddings, *Trans. Amer. Math. Soc.* **341** (1994) 45–68.
10. with L.R. Junqueira, The topology of elementary submodels, *Topology Appl.* **82** (1998) 239–266.
11. with R. Grunberg and L.R. Junqueira, Forcing and normality, *Topology Appl.* **84** (1998), 149–176.
12. with K. Kunen, The real line in elementary submodels of set theory, *J. Symb. Logic*, **65** (2000) 683–691.
13. If it looks and smells like the reals..., *Fund. Math.*, **163** (2000) 1–11.
14. with P. Koszmider, A Lindelöf space with no Lindelöf subspace of size $\aleph_1$, *Proc. Amer. Math. Soc.*, **130** (2002) 2777–2787.
15. with Y.Q. Qiao, Perfectly normal non-Archimedean non-metrizable spaces are generalized Souslin lines, *Proc. Amer. Math. Soc.*, **131** (2003) 3929–3936.
16. Compact spaces, elementary submodels, and the countable chain condition, II, *Topology Appl.* **153** (2006) 2703–2708.

*Work in Progress*

17. with P. Larson, Locally compact, perfectly normal spaces may all be paracompact, preprint.
18. Characterizing paracompactness in locally compact normal spaces, preprint.
19. PFA(S)[S] and small Dowker spaces, preprint.
20. PFA(S)[S]: more mutually consistent topological applications of PFA and V = L, in preparation.

**Service to the Profession:** Editor, Set-theoretic and General Topology, *Proceedings of the American Mathematical Society*, 1990-1996; Conference Organizer, Rothberger Conference, 1977; Conference Organizer, SETOP Conference, 1980; Co-organizer, Mary Ellen Rudin Conference, 1991.

**Teaching:** Tall founded the University of Toronto – York University Set Theory Seminar, which has been ongoing for more than 35 years, attracting numerous visitors and producing many Ph.D.'s. Tall himself has had twelve doctoral students, including Juris Steprâns, Stephen Watson, Maxim Burke, and Piotr Koszmider. Among his nine postdocs were Murray Bell, Claude Laflamme, and Paul Larson. Two NSERC University Research Fellows working with Tall were Gregory Moore and Alan Dow. In addition to those specifically mentioned above, ten others are tenured on faculties in Canada, the U.S., Israel, and Brazil.

**Vision Statement:** Logic will continue penetrating other branches of mathematics and computer science. Set-theoretic topologists will continue to be consumers — and occasional innovators — of set theory. It is to be hoped that new topological problems as compelling as the — largely solved — classic ones will bring fresh energy to the field, which has served as a model for the application of set theory to other branches of mathematics.

**Honours and Awards:** Member, Academy of Sciences of São Paulo, 1996–.

# TENNANT, Neil Wellesley

**Specialties:** Mathematical logic, philosophical logic, computational logic, belief revision.

**Born:** 1 March 1950 in Pietermaritzburg, Natal, South Africa.

**Educated:** University of Cambridge, PhD; University of Cambridge, BA Hons.

**Dissertation:** *Recursive Semantics for Knowledge and Belief*; supervisor, Timothy Smiley.

**Regular Academic or Research Appointments:**
PROFESSOR, PHILOSOPHY AND ADJUNCT PROFESSOR OF COGNITIVE SCIENCE, OHIO STATE UNIVERSITY, 1992-; Professor, Philosophy, Faculty of Arts, Australian National University, 1986-1991; Professor, Philosophy, University of Stirling, 1981-1986; Lecturer, Philosophy, University of Edinburgh, 1974-1981.

**Visiting Academic or Research Appointments:**
Visiting Fellow, Churchill College, Cambridge, 1993-1994; Visiting Professor, Philosophy, University of Michigan, 1986; Visiting Fellow, Research School of Social Sciences, Australian National University, 1985; Visiting Fellow, Center for Philosophy of Science, University of Pittsburgh, 1985; Alexander von Humboldt and British Academy Fellow, Institut für Wissenschaftstheorie und Statistik, University of Munich, 1979-1980; Visiting Professor, Philosophy, Dartmouth College, 1976.

**Research Profile:** Neil Tennant's research in logic has been driven by philosophical concerns about foundations, justification, and the a priori. His metaphysical and epistemological outlook is both naturalist and anti-realist. On the methodological side, he finds in natural deduction a fertile source of insights and methods for exact philosophizing; and he believes that computational implementability is an underappreciated constraint on foundational endeavors.

Much of Tennant's work in philosophical logic has been guided by an inferentialist approach to meaning. He has applied proof-theoretic techniques to throw light on such topics as paradox, relevance, knowability, the justification of deduction, cognitive significance, an anti-realist account of empirical discourse, and abstractionist foundations for arithmetic (his so-called 'constructive logicism').

Tennant has developed the proof theory of intuitionistic relevant logic, one of the earliest 'substructural' logics. He has argued that IR is adequate for all our logical needs in constructive mathematics and in hypothetico-deductive testing of scientific theories. He has also proved various normalization theorems for this system, which enable one to formulate powerful proof-search constraints on automated deduction.

Tennant's interests in automated deduction have led to his more recent research in the computational theory of belief-revision. Recently he has argued the shortcomings of AGM-theory and has offered an alternative, implementable theory of belief-revision in which rational systems of belief are represented as dependency-networks.

**Main Publications:**

1. The Taming of The True, Oxford University Press, 1997, xvii+465 pp. Paperback edition 2002.
2. Autologic, Edinburgh University Press, 1992, xiii+239 pp.
3. Anti-Realism and Logic: Truth as Eternal, Clarendon Library of Logic and Philosophy, Oxford University Press, 1987, xii+325 pp.
4. Philosophy, Evolution and Human Nature (with F. von Schilcher), Routledge and Kegan Paul, 1984, viii+283 pp.
5. Natural Logic, Edinburgh University Press, 1978, ix+196pp.; Japanese translation by T. Fujimura for Orion Press, 1981; second, revised edition, 1990.
6. 'On the Degeneracy of the Full AGM-Theory of Theory-Revision', Journal of Symbolic Logic, vol. 71, no. 2, 2006, pp. 661-676.
7. 'New Foundations for a Relational Theory of Theory-Revision', Journal of Philosophical Logic, vol. 35, 2006, pp. 489-528.
8. 'Theory-Contraction is NP-Complete', Logic Journal of the IGPL, vol. 11, no. 6, 2003, pp. 675-693.
9. 'Contracting Intuitionistic Theories', Studia Logica, 72, 2005, pp. 1-24.
10. 'Changing the Theory of Theory Change: Towards a Computational Approach', British Journal for Philosophy of Science , 45, 1994, pp. 865-897.
11. 'Relevance in Reasoning', in S. Shapiro, ed., Handbook of Philosophy of Logic and Mathematics, Oxford University Press, 2004, pp. 696–726.
12. 'Ultimate Normal Forms for Parallelized Natural Deductions, with Applications to Relevance and the Deep Isomorphism between Natural Deductions and Sequent Proofs', Logic Journal of the IGPL, vol. 10, no. 3, May 2002, pp. 1-39.
13. 'Frege's Content-Principle and Relevant Deducibility', Journal of Philosophical Logic, vol. 32, 2003, pp. 245-258.
14. 'A General Theory of Abstraction Operators', Philosophical Quarterly, vol. 54, no. 214, 2004, pp. 105-133.
15. 'Truth Table Logic, with a Survey of Embeddability Results', Notre Dame Journal of Formal Logic , 30, 1989, pp. 459-484.
16. 'Natural Deduction and Sequent Calculus for Intuitionistic Relevant Logic', Journal of Symbolic Logic 52, 1987, pp. 665-690.
17. 'Deflationism and the Gödel-Phenomena', Mind, vol. 111, 443, July 2002, pp. 551-582.
18. 'Anti-Realist Aporias', in Mind, Vol. 109, 436, October 2000, pp. 831-860.
19. (with D. C. McCarty) 'Skolem's Paradox and Constructivism', Journal of Philosophical Logic 16, 1987, pp. 165-202.
20. 'Deductive v. Expressive Power: A Pre-Gödelian Predicament', in Journal of Philosophy XCVII, no. 5, May 2000, pp. 257-277.
21. 'Negation, Absurdity and Contrariety', in D. Gabbay and H. Wansing (eds.), Negation, Kluwer, Dor-

drecht, 1999, pp. 199-222.

22. 'On the Necessary Existence of Numbers', Noûs, 31, 1997, pp. 307-336.

23. 'Transmission of Truth and Transitivity of Proof', in D. Gabbay (ed.), What is a Logical System?, Oxford University Press, 1994, pp. 161-177.

24. 'Minimal logic is adequate for Popperian science', in British Journal for Philosophy of Science 36, 1985, pp. 325-329.

25. 'Perfect validity, entailment and paraconsistency', Studia Logica XLIII, 1984, pp.179-198.

26. 'A proof-theoretic approach to entailment', Journal of Philosophical Logic 9, 1980, pp. 185-209.

**Service to the Profession:** Editor, American Philosophical Quarterly, 2004–2007.

**Teaching:** Distinguished Teaching Award, Ohio State University Office of Disability Services, for developing methods and materials to teach logic to blind students, 1999.

**Vision Statement:** "Logic will increasingly serve the needs of AI. Each branch of reasoning is sui generis, with its own primitive concepts and axioms. Computationally implementable proof-search methods will be appropriately tailored, to produce artificial geometers, algebraists, topologists, number-theorists, physicists, jurists, etc."

**Honours and Awards:** Distinguished University Scholar, Ohio State, 2004; Humanities Distinguished Professor in Philosophy, Ohio State, 2003; Fellow, Academy of Humanities of Australia, 1990; Australian Research Council Grant for research in computational logic, 1990.

# THOMASON, Richmond Hunt

**Specialties:** Modal and tense logic, logic of conditionals, applications of logic in linguistics, applications of logic in artificial intelligence.

**Born:** 5 October 1939 in Chicago, Illinois, USA.

**Educated:** Yale University, PhD Philosophy, 1965; Yale University, MA Philosophy, 1963; Wesleyan University, BA Philosophy and Mathematics, 1961.

**Dissertation:** *Studies in the logic of quantification*; supervisor, Frederic B. Fitch.

**Regular Academic or Research Appointments:** PROFESSOR, PHILOSOPHY, LINGUISTICS AND COMPUTER SCIENCE, UNIVERSITY OF MICHIGAN, 1999–; Acting Chair, Department of Linguistics, University of Pittsburgh, 1994; Co-Director, Intelligent Systems Program, University of Pittsburgh, 1987-1994; Professor, Linguistics and Intelligent Systems, University of Pittsburgh, 1990-1998; Professor, Philosophy and Linguistics, University of Pittsburgh, 1975-1989; Associate Professor, Philosophy and Linguistics, University of Pittsburgh, 1973-1975; Associate Professor, Philosophy, Yale University, 1969-1972; Assistant Professor, Philosophy, Yale University, 1965-1969.

**Visiting Academic or Research Appointments:** Instructor, Linguistic Society of America Summer Institute, Ohio State University, 1993; Instructor, Linguistic Society of America Summer Institute, University of Massachusetts, 1974.

**Research Profile:** Richmond Thomason's work in modal and tense logic began in 1963, and includes adaptation of the Henkin method of proving completeness to intuitionistic and modal logic, and subsequent work on the philosophical foundations of quantified modal logic, the logic of branching time, the logic of conditionals, and deontic logic.

In the early 1970s he became interested in the adaptation of logical techniques to natural language. He edited a posthumous collection of Richard Montague's papers, worked on extensions of Montague Grammar and various foundational issues, and especially on propositional attitudes.

In the mid 1980s, working with computer scientists at Carnegie Mellon University, he became interested in Artificial Intelligence, and especially in the use of logical techniques in knowledge representation and natural language processing. This interest led to projects in nonmonotonic logic, and especially the logic of inheritance networks, in the foundations of cooperative discourse, and in the formalization of commonsense knowledge.

These interests, as well as general interests in philosophical logic, inform Thomason's current projects. These projects include the formalization of semantic relations among words, issues in reasoning about time and change, the foundations of reasoning about context, and applications of contextualized abductive inference to the design of conversational agents. Information about all of these projects can be found in Thomason's web pages.

**Main Publications:**

1. "On the strong semantical completeness of the intuitionistic predicate calculus." *Journal of Symbolic Logic* 33 (1968), pp. 1-7.

2. "Some completeness results for modal predicate calculi." In *Recent Developments in Philosophical Problems in Logic*, K. Lambert, ed., D. Reidel, Dordrecht, 1970, pp. 20-40.

3. "Modal logic and metaphysics." In *The Logical Way of Doing Things*, K. Lambert, ed., Yale University Press, New Haven, 1969, pp. 119-146.

4. "A semantic analysis of conditional logic" (with R. Stalnaker). *Theoria* 36 (1970), pp. 23-42.

5. "Indeterminist time and truth-value gaps." *Theoria* 36 (1970), pp. 265-281.

6. "A semantic theory of adverbs" (with R. Stalnaker). *Linguistic Inquiry* 4 (1973), pp. 195-220.

7. "Introduction." In *Formal Philosophy: Selected Papers of Richard Montague*, R. Thomason, ed., Yale University Press, New Haven, 1974, pp. 1-69.

8. "Some extensions of Montague grammar." In *Montague Grammar*, B. Partee, ed., Academic Press, New York, 1976, pp. 77-117.

9. "A theory of conditionals in the context of branching time" (with A. Gupta). *Philosophical Review* 88 (1980), pp. 65-90. Reprinted in W. Harper, R. Stalnaker, and G. Pearce, eds., *Ifs*, D. Reidel, Dordrecht, 1981, pp. 299-322.

10. "A model theory for propositional attitudes. *Linguistics and Philosophy* 4 (1980), pp. 47-70.

11. "Deontic logic as founded on tense logic." In *New Studies in Deontic Logic*, R. Hilpinen, ed., D. Reidel, Dordrecht, 1981, pp. 165-176.

12. "Combinations of tense and modality." In *The Handbook of Philosophical Logic*, vol. 2, D. Gabbay and F. Guenthner, eds. D. Reidel Publishing Co., 1984, pp. 135-165.

13. "Accommodation, meaning, and implicature: interdisciplinary foundations for pragmatics." In *Intentions in communication*. P. Cohen, J. Morgan, and M. Pollack, eds., MIT Press, 1990, pp. 325-363.

14. "A skeptical theory of inheritance in nonmonotonic semantic nets" (with J. Horty and D. Touretzky)." *Artificial intelligence* 42 (1990), pp. 311-348.

15. "NETL and subsequent path-based inheritance theories." *Computers and mathematics with applications* 23 (1992), pp. 179-204. Reprinted in *Semantic networks in artificial intelligence*, F. Lehmann, ed., Pergamon Press, 1992, pp. 179-204.

16. "Nonmonotonicity in linguistics." In *Handbook of logic and language*, J. van Benthem and A. ter Meulen, eds., Elsevier, Amsterdam, 1997, pp. 777-831.

17. "Modeling the beliefs of other agents: achieving mutuality." In *Logic-Based Artificial Intelligence*, edited by Jack Minker, Kluwer Academic Publishers, 2000, pp. 375-403.

18. "Dynamic contextual intensional logic: logical foundations and an application." In *Modeling and Using Context: Fourth International and Interdisciplinary Conference*, P. Blackburn, C. Ghidini, R. Turner, and F. Giunchiglia, eds., Springer-Verlag, Berlin, 2003, pp. 328-341.

19. "Coordinating understanding and generation in an abductive approach to interpretation" (with Matthew Stone). *DIABRUCK 2003: Proceedings of the Seventh Workshop on the Semantics and Pragmatics of Dialogue*, Saarbrucken, 2003.

20. "Logic and artificial intelligence." In *The Stanford Encyclopedia of Philosophy*, Edward N. Zalta, ed., 2003. Available at http://plato.stanford.edu/archives/fall2003/entries/logic-ai/

*Work in Progress*

21. "Formalizing the semantics of derived words," 2006. http://www.eecs.umich.edu/~rthomaso/documents/nls/nmslite.pdf

22. "Lexical semantics for causal constructions," 2007. http://www.eecs.umich.edu/~rthomaso/documents/nls/cause.pdf

23. "Conditionals and action Logics," 2007. http://www.eecs.umich.edu/~rthomaso/documents/lai/ConditionalsandActionLogics.pdf

24. "Enlightened update: a computational architecture for presupposition and other pragmatic phenomena" (with Matthew Stone and David DeVault), 2006. http://www.pragmatics.osu.edu/links/events/enlightened-update.pdf

25. "Paradoxes of intensionality" (with Dustin Tucker), 2007. http://www.eecs.umich.edu/~rthomaso/documents/par-int/tucker-thomason.pdf

**Service to the Profession:** Program Co-Chair, Interdisciplinary Conference on Modeling and Using Contexts, Dundee, 2001; President, Linguistics and Philosophy Associates, 1993-2000; Program Committee Chairman, Joint Conference on Logic and Linguistics, Stanford University, 1987; Editor-in-Chief, Journal of Philosophical Logic, 1977-87

**Vision Statement:** The impact of logic on philosophy in the 20th century was profound. Logic has enabled new and refined techniques for deploying and critically examining philosophical arguments, making possible the development of philosophical theories that otherwise would have been impossible. One of the main purposes of the research tradition in philosophical logic was to develop logical theories applicable to nonmathematical domains of interest to philosophers. This tradition has been supplemented and to some extent superseded by research in knowledge representation. This work not only has practical value, but provides insights that are highly relevant to philosophical concerns. However, it has been largely neglected by the philosophical community. To sustain the central role of logic in philosophy it is important to make philosophers aware of relevant work by computer scientists. This work has the potential to transform philosophy in much the same way that philosophy was transformed in the early 20th century by developments in symbolic logic.

**Honours and Awards:** Nelson Professor of Philosophy, 2004-; Fellow, American Association for Artificial Intelligence, 1993-; Mellon Postdoctoral Fellow, University of Pittsburgh, 1972-1973; Mellon Postdoctoral Fellow, University of Pittsburgh, 1972-1973; Morse Research Fellow, Yale University, 1968-1969.

## TIURYN, Jerzy

**Specialties:** Mathematical logic, logic in computer science.

**Born:** 3 December 1950 in Warsaw, Poland.

**Educated:** Warsaw University, Scientific Ttitle of Professor, 1991; Warsaw University, Habilitation Mathematics, 1982; Warsaw university, PhD Mathematcis, 1975; Warsaw University, MS Mathematics, 1974.

**Dissertation:** *M-groupoids as a Tool to Investigate Mathematical Models of Computers and Programs*; supervisor, Helena Rasiowa.

**Regular Academic or Research Appointments:** FULL PROFESSOR, DEPARTMENT OF MATHEMATICS, INFORMATICS AND MECHANICS, WARSAW UNIVERSITY, 1992–; Associate Professor, Department of Mathematics, Informatics and Mechanics, Warsaw University, 1984–1992; Assistant Professor, Department of Mathematics, Informatics and Mechanics, Warsaw University, 1975–1984.

**Visiting Academic or Research Appointments:** Visiting Scientist, CNRS, Marseille, 2002; Senior Research Associate, Boston University, 1998; Senior Research Associate, Boston University, 1996; Visiting Professor of Computer Science, Northeastern University, Boston, 1995; Visiting Scientist, Ecole Normale Superieure, 1994; Senior Research Associate, Boston University, 1992, 1993, 1994; Visiting Professor, TH Darmstadt, 1992; Visiting Scientist, Northeastern University, 1991; Visiting Scientist, Boston University, 1990, 1989; Visiting Professor, Computer Science, Washington State University, Pullman, 1987–1989, 1984–1985; Visiting Scientist, University of Amsterdam, 1986, 1983; Visiting Associate Professor, Boston University, 1981; Visiting Scientist, Massachusetts Institute of Technology, 1981, 1980; Visiting Scientist, RWTH Aachen, 1978–1979.

**Research Profile:** Jerzy Tiuryn has made contributions to the logical aspects of theoretical computer science. The main areas of his research concentrated on the semantics of programming languages, verification of programs, expressive power of logics of programs, type theory and lambda calculus. In the area of semantics of programming languages his contribution consists in building an algebraic framework (termed *regular algebras*) for fixed point semantics of while programs. Tiuryn's contribution to the verification of programs consists in research on incompleteness of Hoare's logic, as well as in establishing a relationship between partial correctness and semantic equivalence of programs. In the study of logics of programs he is noted for establishing fundamental connections between computational power of programming languages over abstract structures, expressive power of first order logics of programs, and complexity theory. He has worked on the fundamental problem of an impact of nondeterminism and/or unbounded memory on the expressive power of logics of programs. In type theory and lambda calculus he is noted for the works on type reconstruction problems in polymorphic lambda calculus. In particular in establishing the complexity lower bounds of this problem for the full system, as well as for the functional programming language ML.

**Main Publications:**

1. Fixed points and algebras with infinitely long expressions. Part I: Regular algebras, *Fundamenta Informaticae* II, 1 (1978), 103–128.
2. Fixed points and algebras with infinitely long expressions. Part II: mu-clones of regular algebras, *Fundamenta Informaticae* III, 1 (1979), 317–335.
3. (with J. A. Bergstra and J. V. Tucker) Floyd's principle, correctness theories and program equivalence, *Theoretical Computer Science* 17 (1982), 113–149.
4. Unbounded program memory adds to the expressive power of first-order programming logics, *Information and Control* 60 (1984), 12–35.
5. Higher-order arrays and stacks in programming. An application of complexity theory to logics of programs, (J. Gruska and B. Rovan, eds.) *Proc. Math. Foundations of Computer Sci.*, Springer Verlag (1986), LNCS 233, 177–198.
6. (with A.J. Kfoury and P. Urzyczyn) On the computational power of universally polymorphic recursion, *Proc. 3rd IEEE Symp. Logic in Computer Sci.*, (1988) 72–81.
7. (with P. Urzyczyn) Some relationships between logics of programs and complexity theory, *Theoretical Computer Science* 60 (1988), 83–108.
8. A simplified proof of DDL<DL, *Information and Computation* 81 (1989), 1–12.
9. (with D. Kozen) Logics of Programs, (J. van Leeuven, ed.) *Handbook of Theoretical Computer Science*, Vol. B, *Formal Models and Semantics*, Elsevier Science Publ. and MIT Press, (1990), 789–840.
10. (with A.J. Kfoury and P. Urzyczyn) On the expressive power of finitely typed and universally polymorphic recursive procedures, *Theoretical Computer Science* 93 (1992) 1–41.
11. (with A.J. Kfoury and P. Urzyczyn) Type reconstruction in finite-rank fragments of the second-order lambda-calculus, *Information and Computation*

98, (1992), 228–257.

12. (with A.J. Kfoury and P. Urzyczyn) Type Reconstruction in the Presence of Polymorphic Recursion", *ACM Transactions on Programming Languages and Systems*, 15, No.2 (1993) 290–311.

13. (with A.J. Kfoury and P. Urzyczyn) The undecidability of the semi-unification problem, *Information and Computation* 102 (1993), 83–101.

14. (with A. Jung) A new characterization of lambda definability, (M. Bezem and J.F. Groote eds), *Proc. Intern. Conf. on Typed Lambda Calculi and Applications, TLCA '93*, Lecture Notes in Computer Science 664, Springer Verlag, (1993) 245–257.

15. (with V. Pratt) Satisfiability of Inequalities in a Poset, *Fundamenta Informaticae* 28 (1996) 165–182.

16. A Sequent Calculus for Subtyping Polymorphic Types (W. Penczek, A. Szalas, eds.), *Proc. 21st International Symposium on Math. Foundations of Computer Science*, Springer Verlag LNCS 1113 (1996) 135–155.

17. (with D. Harel and D. Kozen) *Dynamic Logic*, MIT Press (2000).

18. (with P. Urzyczyn) The subtyping problem for second-order types is undecidable, *Information and Computation* 79 (2002), 1–18.

19. (with V. Bono) Products and Polymorphic Types, *Fundamenta Informaticae* 51 (2002), 13–41.

20. (with D. Kozen) Substructural Logic and Partial Correctness, *ACM Transactions on Computational Logic* 4 (2003), 355–378.

*Work in Progress*

21. None in logic. Since 1999 Jerzy Tiuryn is involved in research on Computational Biology at Warsaw University.

**Service to the Profession:** Vice Dean, *Faculty of Mathematics, Informatics and Mechanics*, Warsaw University, 2005–; Program Committee Chair, Foundations of Software Science and Computation Structures, 2000; Conference Chair, $12^{th}$ IEEE Symp. Logic in Computer Science, 1997; Member, Polish National Committee of Mathematics, 1996–2000; Editor, *Theoretical Computer Science*, 1992–; Member, *Scientific Council of the European Association of Computer Science Logic*, 1992–1994; Editor, *Mathematical Structures in Computer Science*, 1990–; Editor, *Fundamenta Informaticae*, 1983–; Editor, *Information and Computation*, 1983–.

**Teaching:** Tiuryn has been involved in teaching logic and theoretical computer science at Warsaw University since his appointment to the faculty in 1974. Presently he is involved in teaching Computational Biology. He has had ten PhD students in the fields of Mathematics, Computer Science, and Computational Biology, including Pawel Urzyczyn, Damian Niwinski, Jerzy Tyszkiewicz, Igor Walukiewicz and Pawel Gorecki (in chronological order).

**Vision Statement:** Foundational challenges of Mathematics in the $20^{th}$ century were the driving force for most of the development of mathematical logic. Later in the $20^{th}$ century it was Computer Science which provided a fresh source of problems and motivations for new research directions in mathematical logic. Even though the progress was remarkable, many problems remain unsolved and this line of research should be continued. However, I feel that in addition to continuing research in foundations of Mathematics and Computer Science mathematical logic needs again a new source of motivations, perhaps from such a field of Science as Molecular Biology which is presently progressing at a remarkable speed and creates many new challenges to Mathematics and Computer Science. I hope that mathematical logic may find a niche to make its own valuable contributions to this area of Science.

**Honours and Awards:** Fellow, *Academia Europaea*, 1996–; Stanislaw Mazur Prize of Polish Mathematical Society, 1992.

# TRAKHTENBROT, Boris

(sometimes transcribed from Russian as Trahtenbrot, Trachtenbrot, Trajtenbrot)

**Specialties:** Mathematical logic, computability, theoretical computer science.

**Born:** 20 February 1921 in Brichevo, Moldova (then Romania).

**Educated:** Moldovian Pedagogical Institute 1945; Chernovtsy University (Ukraine) 1947; Mathematical Institute of the Ukranian Academy of Sciences, Kiev, PhD Mathematics 1950.

**Dissertation:** *Decidability problems for finite classes and definitions of finite sets*; supervisor, Piotr Sergeevich Novikov.
Mathematical Institute of the Siberian Branch of ht eUSSR Academy of Sciences, Novosibirsk, Soviet Doctor of Sciences, 1962. Dissertation: *Investigations on the Syntehsis of Finite Automata*.

**Regular Academic or Research Appointments:** PROFESSOR OF COMPUTER SCIENCE EMERITUS, TEL-AVIV UNIVERSITY, 1991–1996; Professor of Computer Science, Tel-Aviv University, 1981–1991; Senior researcher, then Head of Laboratory for Automata Theory and Mathematical Lingusitic, Novosibirsk, Mathematical Institute, Siberian branch of the USSR Academy of Sciences, 1960–1980; Professor of mathematics, Novosibirsk University, 1963–1978; Associate Professor, Polytechnical Institute, Penza, USSR, 1958–1960; Associate Professor, Head of

Mathematics Department, Pedagogical Institute, Penza, 1953–1958; Senior Lecturer, Pedagogical Institute, Penza, 1950–1953.

**Visiting Academic or Research Appointments:**
Visiting Professor, Uppsala University, Sept–Oct 1997; Visiting Professor, Massachusetts Institute of Technology, Nov 1990–Feb 1991; Visiting Professor, IBN Yorktown Heights, October 1990; Visiting Professor, München Technical University, September 1990; Visiting Professor, Imperial College London, July–August 1990; Visiting Professor, Gesellschaft für Mathematik und Datenverabeitung, Bonn, August 1987; Visiting Professor, Massachusetts Institute of Technology, Feb–June 1986; Visiting Professor, Carnegie Mellon University, september–December 1985; Visiting Professor, University Paris VI, Sept–Oct 1982; Visiting Professor, Massachusetts Institute of Technology, September–October 1981, Summer 1983 and 1984; Visiting Professor, Mathematical Institute of ht ePolish Academy of Sciences, Warsaw, October 1967; Senior researcher (part time) Research Institute of Control and Computing Machines (PNI-IUVM), Penza, USSR, 1958–1959.

**Research Profile:** Boris Trakhtenbrot has made fundamental contributions to logic and computability and has made significant contributions to the following areas of Theoretical Computer Science (TCS): Automata Theory, Complexity of comutations and algorithms, Logic and semantics of programs, concurrent and hybrid systmes. Actually, he was among the first logicians to make a steady transitition to logic in computer science.

**Logic and Computability.** The following results are widely recognised as basic for the theory of finite models:

1. the finite model version of Church's theorem about the undecidability of first-order logic;
2. the recursive inseparability of first-order tautologies from formulas refutable in finite models;
3. the connection between deductive incompleteness (in particular, of set theory) and recursive inseparability.

**Theoretical Computer Science.**

1. *Automata Theory*. The theroem (discovered independently also by R. Buechi and C. Elgot): finite-state automata and monadic second-order logic, interpreted over finite words, have the same expressive power; moreover, the trnasformation from formulas to automata and vice versa are effective. Hence, logical specification of automata beyond the propositional calculus and application of automata machinery to investigation of logical problems.

2. *Complexity Theory*. Initatory contribution to the elaboration of the conceptual framework and its analysis: signalising functin (aka complexity measures), gaps, brute-force search, frequencey computations, auto-reducibility.

3. *Logic and Semantics of Programs*. The study (with A. Meyer and J. Halpern) of Algol-like languages, whose essence is crystallised via a sutiably typed lambda-calculus. Connections iwth lgoical theories as a guide to design and correctness of programs.

4. *Foundational Models for Concurrent and Hybrid Systmes (HS)* with A. Rabinovich). Notably,

    a) Modular semantics for dataflow and for general nets of processes.

    b) Possibility and limits of adapting the classical heritage of Automata theory and Logic to systems that incorporate both discrete and continuous dynamics.

Finally, Trakhtenbrot has contributed to the history of modern logic and TCS with

- comparative analyses of Turing's and Church's works and
- history of Russian computer science.

**Main Publications:**

1. The impossibility of an algorithm for the decidability problem on finite classes. Doklady AN SSR 70, No. 4, 569–572, 1950.
2. On recursive spearability. Doklady AN SSR 88, No. 6, 953–956, 1953.
3. On the definition of finite set and the deductive incompleteness of set theory. Izvestia AN SSR 20, 569–582, 1956.
4. Algorithms and automatic computing machines, in the series "Topics in Mathematics", D. C. Heath and Company, Boston, 1963. (Russian, 1960)
5. The synthesis of logical nets whose operators are described in terms of monadic predicates. Doklady AN SSR 118, No. 4, 646–649, 1958.
6. Some constructions in the monadic predicate calculus. Doklady AN SSR 138, No. 2, 320–321, 1961.
7. Finite automata and the monadic predicate calculus. Doklady AN SS 140, No. 2, 326–329, 1961.

8. (with N. E. Kobrinski) Introduction to the theory of finite automata, in the series "Studies in Logic and the Foundations of Mathematics", North-Holland, Amsterdam, 1965. (Russian, 1962)

9. On frequency computation of recursive functions. Algebra i Logicka, Novosibirsk 1, No. 1, 25–32, 1963.

10. Turing computations with logarithmic delay. Algebra i Logika, Novosibirsk 3, No. 4, 33–48, 1964.

11. Complexity of computations an algorithms. Lectur eNotes edited by Novosibirsk University, 258pp, 1967. (In Russian)

12. On autoreducibility. Doklady AN SSR 192, No. 6, 1224–1227, 1970.

13. (with Ya. M. Barzdin') Finite automata (behaviour and syntehsis). In "Series Fundamental Studies in Computer Science", North Holland, Amsterdam, Elsevier New York, 1973. (Russian, 1970)

14. (with J. Halpern and A. Meyer) From denotational to operational and axiomatic semntics for Algol-like languages. In "Logics of Programs", proceedings 1983. Lecture notes in Computer Science No 164, pp. 474-500.

15. A survey on Russian approaches to PEREBOR (Brute Force Search). Algorithms Annals of the History of Computing, Vol. 6, no. 4, 384–400, 1984.

16. Comparing the Church and Turing appraoches: two prophetical messages. In "The Turing Universal Machin e— A Half-centrury Survey". Oxford Univesity Press, 1988. Also in Kammerer and Unverzagt, Berlin, 603–630, 1988.

17. (with A. Rabinovich) Behavior structures and nets of processes. In Fundamenta Infomaticae, Vol. xi, North Holland, Amsteram, 357–403, 1988.

18. (with A. Rabinovich) On nets, algebras and modularity. In "Theoreitcial Aspects of Computer Software". International Conference TACS'91, Sendai, Japan, 1991. LNCS 526.

19. (with A. Rabinovich) From finite automata toward hybrid systems. LNCS, 1997.

20. Understanding b asic automata theory in the continuous time setting. Fundamenta Informaticae, vol. 62, no1, 69–121, 2004.

*Work in Progress*

21. Revision of previous papers on compositional proofs for nets of processes and on conntinuous-time automata.

22. Recollections about TCS in USSR.

**Service to the Profession:** Membership of Editorial board of Cybernetics (GDR); Information and Control; Fundamenta Informaticae; Discrete Mathematics; Applied Discrete Mathematics.

**Teaching:** Sixteen PhD students including J. Barzdins, R. M. Freivalds, V. Sazonov, A. Rabinovich.

Syllabus of a course "Algorithms and computers" for Pedagogical institutes approved by the Russian Education Ministry (1959).

Conference "On teaching Introduction to Mathematical Logic and Algorthms" in Pedagogiccal Institutes, Penza, 1960.

Lecture notes for new courses:
Complexity of computations and algorithms (Novosibirsk University, 1967).
Topics in typed programming languages (Carnegie-Mellon Univesrity, Pittsburgh, 1985)
Automata and Hybrid Systems (Uppsala University, 1997)

**Honours and Awards:** Internationa Symposium on Theoretical Computer Science in Honour of Boris Trakhtenbrot, on the occasion of his retirement and 70th birthday, Tel Aviv, June 10–13, 1991.

Dr. Honoris Causa, Fr. Schiller University, Jena, Germany, 1997.

Award of the Council of the Computer Technology Institute (Greece), 2001.

# TROELSTRA, Anne Sjerp

**Specialties:** Mathematical logic, foundations of mathematics.

**Born:** 10 August 1939 in Maartensdijk, province Utrecht, Netherlands.

**Educated:** University of Amsterdam, PhD Mathematics, 1966.

**Dissertation:** *Intuitionistic General Topology*; supervisor, Arend Heijting.

**Regular Academic or Research Appointments:** PROFESSOR OF PURE MATHEMATICS ADN FOUNDATIONS OF MATHEMATICS, EMERITUS, UNIVERSITY OF AMSTERDAM, 2000–; Professor of Pure Mathematics and Foundations of Mathematics, University of Amsterdam, 1970-2000; Lector (Associate Professor) of Mathematics, University of Amsterdam, 1968-1970; Wetenschappelijk Medewerker (Assistant Professor), University of Amsterdam, 1964-1968.

**Visiting Academic or Research Appointments:** Visiting Professor, University of Berne, Switzerland, 1991; Visiting Professor, University of Siena, Italy, 1985-1985; Visiting Professor, University of Freiburg i.Br., Germany, 1974; Visiting Fellow, Wolfson College, Oxford, U.K., 1973-1974; Visiting Scholar, Stanford University, 1966-1967.

**Research Profile:** Anne Troelstra has contributed to mathematics, metamathematics and philosophy of constructivism (in particular intuitionism), and to proof theory.

In intuitionistic mathematics he has done work on set-theoretic topology and extensions of the real numbers. He has thoroughly analyzed the intuitionistic notion of choice sequence, a basic concept in intuitionistic analysis; together with G. Kreisel he has contributed to the axiomatic study of intuitionistic analysis. In particular, Kreisel and Troelstra were able to show that quantification over choice sequences can be eliminated by contextual definition in suitably rich formalisms.Troelstra studied the modelling within an intuitionistic metamathematical setting of many distinct notions of choice sequence by means of the especially simple case of lawless sequences. He made a systematic study and survey of metamathematical techniques used in the study of constructive formalisms, more especially of notions of realizability and functional interpretations.

He wrote, jointly with H. Schwichtenberg, a textbook on structural proof theory of first-order logic (structural proof theory is the part of proof theory which uses as its method and as its object of study the combinatorial properties of deduction systems).

In the philosophy of mathematics he has been particularly interested in the transition from an informal concept to a formalized mathematical theory. He also wrote some papers about the history of intuitionism.

**Main Publications:**

1. The theory of choice sequences, in: B. van Rootselaar, J.F.Staal (eds), *Logic, Methodology and Philosophy of Science III*, North-Holland Publ. Co., Amsterdam, 1968, 289-298.

2. One-point compactifications of intuitionistic locally compact spaces, *Fundamenta Mathematicae* 62 (1968), 75-93.

3. Notes on the intuitionistic theory of sequences (I), *Indagationes Mathematicae* 31 (1969), 430-440.

4. (with G. Kreisel) Formal systems for some branches of intuitionistic analysis, *Annals of Mathematical Logic* 1 (1970), 229-387.

5. (with D. van Dalen) Projections of lawless sequences, in: J. Myhill, A. Kino, R.E.Vesley (eds), *Intuitionism and Proof Theory*, North-Holland Publ. Co., Amsterdam 1970, 163-186.

6. Notes on the intuitionistic theory of sequences (II), (III) *Indagationes Mathematicae* 32 (1970), 99-109 and 245-252.

7. (with C.A. Smorynski, J.I. Zucker, W.A.Howard) *Metamathematical Investigation of Intuitionistic Arithmetic and Analysis*, Springer Verlag, Berlin, (1973). Chapters I-IV were written by A.S. Troelstra. (See also: Metamathematical Investigation of Intuitionistic Arithmetic and Analysis. Corrections to the first edition. ILLC Prepublication Series X-93-04, Universiteit van Amsterdam.)

8. Notes on intuitionistic second-order arithmetic, in: A. R. D. Mathias, H. Rogers (eds), *Cambridge Summer School in Mathematical Logic*, Springer Verlag, Berlin 1973, 171-205.

9. Note on the fan theorem, *The Journal of Symbolic Logic* 39 (1974), 584-596.

10. Some models for intuitionistic finite-type arithmetic with fan functional, *The Journal of Symbolic Logic* 42 (1977), 194-202.

11. A note on non-extensional operations in connection with continuity and recursiveness, *Indagationes Mathematicae* 39 (1977), 455-462.

12. (with G. F. van der Hoeven) Projections of lawless sequences II, in: M. Boffa, D. van Dalen and K. McAloon (eds), *Logic Colloquium '78*, North-Holland Publ. Co., Amsterdam 1979, 265-298.

13. Extended bar-induction of type 0, in: J. Barwise, H. J. Keisler and K. Kunen (eds), *The Kleene Symposium*, North-Holland Publ. Co., Amsterdam 1980, 277-316.

14. Intuitionistic extensions of the reals, *Nieuw Archief voor Wiskunde* (3) 28 (1980), 63-113.

15. On the origin and development of Brouwer's concept of choice sequence, in: A. S. Troelstra, D. van Dalen (eds), *The L. E. J.Brouwer Centenary Symposium*, North-Holland Publ.Co., Amsterdam 1982, 465-486.

16. Analyzing choice sequences, *Journal of Philosophical Logic* 12 (1983), 197-260.

17. Choice sequences and informal rigour, *Synthese* 62 (1985), 217-227.

18. (with D. van Dalen) *Constructivism in Mathematics. An Introduction*, North-Holland Publ. Co., Amsterdam, 1988. 2 volumes.

19. Realizability, in: S. Buss (editor), *Handbook of Proof Theory*, Elsevier, Amsterdam 1996.

20. (with H. Schwichtenberg) *Basic Proof Theory*, Cambridge University Press, Cambridge U.K. 1996. Second, revised and extended edition 2000.

**Service to the Profession:** Editor, *Studies in Logic and the Foundations of Mathematics*, published by Elsevier, 1970-2000; Editor, *Journal of Symbolic Logic* for Proof Theory and Constructivism, 1983-1991; Editor, Mathematical Logic and Foundations of *Indagationes Mathematicae*, 1970-2004; Organizer with others, Tagungen Mathematische Logik, Mathematisches Forschungsinstitut, Oberwolfach, Germany, 1988-2002.

**Teaching:** Troelstra has supervised 16 PhD students, some of them jointly with others. Among his PhD students are D. Leivant, I. Moerdijk, J. van Oosten, I. Bethke. In the period 1970-1980 he has been the principal person responsible for the creation of a program in mathematical logic at the University of Amsterdam.

**Vision Statement:** Programs in the foundations of mathematics have been the main driving force behind logic in the 20th century, but each of these has been discovered lacking as a basis for mathematics. The technical work originally engendered by these programs still continues, but now primarily carried forward by its own impetus. Nowadays the emphasis in logic has shifted to application in a diverse array of disciplines. However, each of these programs referred to represents certain aspects of mathematics, and as such they retain interest; but an integrated view is desirable. Also, in particular a better understanding of the transition from informal concepts to formal theories deserves further study.

**Honours and Awards:** F.L.Bauer Prize, Friends of the Technical University of Munich, for contributions to theoretical computer science, 1996; Elected Corresponding Member, Bavarian Academy of Sciences, 1996; Member, Royal Netherlands Academy of Sciences, 1976.

# U

## ULRICH, Dolph

**Specialties:** Philosophical logic, many-valued logics, modal logics, implicational logics.

**Born:** 21 August 1940 in Uhrichsville, Ohio, USA.

**Educated:** Wayne State University, PhD Philosophy, 1968; Oberlin College, BA Philosophy, 1963; Moravian College, 1959.

**Dissertation:** *Matrices for sentential calculi*; supervisor, J. Michael Dunn.

**Regular Academic or Research Appointments:** PROFESSOR OF PHILOSOPHY, PURDUE UNIVERSITY, 1984–; Associate Professor of Philosophy, Purdue University, 1971-1984; Assistant Professor of Philosophy, Purdue University, 1968-1971; Instructor, Purdue University, 1967-1968.

**Visiting Academic or Research Appointments:** Visiting Research Fellow, Automated Reasoning Project, Institute for Advanced Studies, Research School of Social Sciences, Australian National University, Spring, 1988.

**Research Profile:** Dolph Ulrich ("Ted") works primarily on classical and non-classical sentential calculi. He has made lasting contributions to the abstract study of matrix semantics having provided, for example, such things as a revealing pair of necessary conditions jointly sufficient for such a logic to be characterized by a single finite matrix and an interesting sufficient condition for a logic to have the finite model property—both of which have proven to be widely useful—and several results concerned with the conditions under which, for systems with the finite model property, there do or do not exist recursive bounds on the size of countermodels. Included in his work on modal logics are an elegant denumerable matrix characteristic for the strict-implicational fragment C5 of S5 and the development of relational semantics for calculi of pure strict implication which led to axiomatizations of the strict-implicational fragments of numerous extensions of S4 and to an infinite sequence of extensions of C5 incomplete with respect to every class of frames. He has also advanced our knowledge of a diversity of additional specific systems, for instance the classical equivalential calculus, and has answered a large variety of open questions posed by other logicians, many in print and others in personal correspondence with colleagues. The set of open problems he himself suggested for further investigation in his "Legacy" paper (entry 18 in the list of selected publications below) has already produced some interesting solutions from a number of researchers.

Ulrich is currently developing general methods for discovering short single axioms for various logics of pure implication and has, thus far in the process, quadrupled our stock of such axioms with, among others, ten of the twelve known seventeen-symbol single axioms for implicational intuitionism, the shortest currently known single axioms for the implicational fragments of Łukasiewicz's infinite-valued sentential calculus and of Dummett's LC, and the first known single axioms for the corresponding fragments of the modal logic S3 and of R-Mingle in the relevance logic family.

**Main Publications:**

1. *Solution to a problem posed by Kalicki*, Proceedings of the American Mathematical Society, vol. 22 (1969), pp. 728-729.

2. *Some results concerning finite models for sentential calculi*, Notre Dame Journal of Formal Logic, vol. 13 (1972), pp. 363-368.

3. Elementary Symbolic Logic (with William Gustason), Holt, Rinehart and Winston, New York, 1973. Second (and current) edition: Waveland Press, Prospect Heights, 1989.

4. *Finitely-many-valued logics with infinitely-many-valued extensions: two examples*, Proceedings of the Fifth International Symposium on Multiple-valued Logic, Bloomington, 1975, pp. 406-411.

5. *Semantics for S4.1.2*, Notre Dame Journal of Formal Logic, vol. 19 (1978), pp. 461-464.

6. *RMLC: Solution to a problem left open by Lemmon*, Notre Dame Journal of Formal Logic, vol. 22 (1981), pp. 187-189.

7. *Strict implication in a sequence of extensions of S4*, Zeitschrift für mathematische Logik und Grundlagen der Mathematik, vol. 27 (1981), pp. 201-214.

8. *Answer to a question raised by Harrop*, Bulletin of the Section of Logic (Polish Academy of Sciences), vol. 11 (1982), pp. 140-141.

9. *The finite model property and recursive bounds on the size of countermodels*, Journal of Philosophical Logic, vol. 12 (1983), pp. 477-480.

10. *Answer to a question suggested by Schumm*, Zeitschrift für mathematische Logik und Grundlagen der Mathematik, vol. 30 (1984), pp. 385-387.

11. *A descending chain of incomplete extensions of implicational S5*, Zeitschrift für mathematische Logik und Grundlagen der Mathematik, vol. 31 (1985), pp. 201-208.

12. *On the characterization of sentential calculi by finite matrices*, Reports on Mathematical Logic, vol. 20 (1986), pp. 63-86.

13. *A five-valued model of the E-p-q theses*, Notre Dame Journal of Formal Logic, vol. 29 (1988), pp. 137-138.

14. *The nonexistence of finite characteristic matrices for subsystems of $R_1$*, Directions in Relevant Logic, ed. Jean Norman and Richard Sylvan, Kluwer Academic Publishers, Dordrecht, 1989, pp. 177-178.

15. *An integer-valued matrix characteristic for implicational S5*, Bulletin of the Section of Logic (Polish Academy of Sciences), vol. 19 (1990), pp. 87-91.

16. *The shortest possible length of the longest implicational axiom*, Journal of Philosophical Logic, vol. 25 (1996), pp. 101-108.

17. *New single axioms for positive implication*, Bulletin of the Section of Logic (University of Lódz), vol. 28 (1999), pp. 39-42. 18. *A legacy recalled and a tradition continued*, Journal of Automated Reasoning, vol. 27 (2001), pp. 97-122. Reprinted with silent corrections on the CD-ROM accompanying Automated Reasoning and the Discovery of Missing and Elegant Proofs, Larry Wos and Gail W. Pieper, Rinton Press, Paramus, 2003.

18. *XCB, the last of the shortest single axioms for the equivalential calculus* (with Larry Wos and Branden Fitelson), Bulletin of the Section of Logic (University of LódŸ), vol. 29 (2003), pp. 131-136.

19. *D-complete axioms for the classical equivalential calculus*, Bulletin of the Section of Logic (University of LódŸ), vol. 31 (2005), pp. 135-142.

*Work in Progress*

20. Several pieces on new single axioms for various sentential calculi, and another introducing the new two-base {*Cpp, CCCpqrCCCqstCCtqr*} for classical implication to answer an open question first posed by Prior half a century ago.

**Vision Statement:** I expect to see more attention paid to the discovery of simple and elegant proofs, and to the unification and generalization of scattered results. The on-line searchable version of Mathematical Reviews is a wonderful resource, and I look forward to the creation of more specialized websites for individual sub-areas of our discipline (e.g., basic encyclopedias of modal logics already exist) and to the increased development and use of sophisticated automated reasoning tools.

## URQUHART, Alasdair

**Specialties:** Non-classical logics, lattice theory, philosophy of logic, history of logic, theory of computation, computational complexity theory, complexity of propositional proofs.

**Born:** 20 December 1945 in Auchtermuchty, Fife, Scotland.

**Educated:** University of Edinburgh MA (Honours Philosophy) 1967; University of Pittsburgh PhD 1973.

**Dissertation:** *The Semantics of Entailment*; supervisor, Nuel D. Belnap Jr.

**Regular Academic or Research Appointments:** PROFESSOR, DEPARTMENTS OF PHILOSOPHY AND COMPUTER SCIENCE, UNIVERSITY OF TORONTO, 1986–. Lecturer, Assistant Professor and Associate Professor, Department of Philosophy, University of Toronto 1970–1986.

**Visiting Academic or Research Appointments:** Visiting fellow, Australian National University Research School of Social Sciences 1982. Visiting scholar, Mathematical Institute, Oxford 1976–1977.

**Research Profile:** Urquhart's early work was in the area of semantics for non-classical logics, particularly relevance logics, where his first contribution was the semilattice semantics for entailment and relevant implication, later generalized by Routley and Meyer.

In the 1970s, his research was in the area of algebraic logic and lattice theory. His efforts towards solving the word problem for modular lattices, though not successful in their original goals, led to a topological representation theory for general lattices, later developed and extended by other researchers. A great deal of his research in this period was based on Hilary Priestley's beautiful representation theory for distributive lattices.

His best result in non-classical logic is the undecidability of the propositional logics of entailment and relevant implication; the technique here was adapted from the Hutchinson/Lipshitz proof that the word problem for modular lattices is unsolvable, and uses the Von Neumann coordinatization theorem for continuous geometries. Other results in the area are the proof of failure of interpolation in relevance logics, and the theorem showing that the inherent complexity of the implication-conjunction fragment of relevant implication is that of the Ackermann function.

The non-interpolation result also uses geometrical techniques and depends on the existence of a finite geometry in which Desargues's law fails.

Since the early 1980s, his research has been in the area of computational complexity, and particularly the area of the complexity of propositional proofs.

Significant results include the first truly exponential lower bound for propositional resolution refutations, and the first exponential lower bound for cutfree Gentzen derivations; both of these extend the breakthrough result on resolution by Armin Haken. Another theme in this area is that of feasibly constructive proofs; in a long paper co-authored with Stephen A. Cook, functional interpretations are given for feasibly constructive arithmetic in which the provably total functions are all polynomial-time computable.

Urquhart has also worked in the history of early 20th century logic, and is the editor of Volume 4 of the Collected Papers of Bertrand Russell, a volume including the unpublished manuscripts that preceded the famous article "On Denoting."

Since 1999, Urquhart has devoted a great deal of time to an attempt to understand the fruitful but mathematically unrigorous techniques employed by physicists in the field of disordered systems, particularly spin glasses, and later used to great effect in many other areas such as combinatorial optimization and algorithms for satisfiability. The replica method in particular remains largely a mathematical mystery, and a deeper understanding of this technique would very likely lead to considerable progress in computer science and logic.

**Main Publications:**

1. *Temporal Logic*, with Nicholas Rescher. Springer Verlag, New York and Vienna, 1971.
2. "Semantics for relevant logics". *Journal of Symbolic Logic*, 37(1972): 159–169.
3. "An interpretation of many-valued logic". *Zeitschrift für mathematische logik und grundlagen der Mathematik*, 19(1973): 111–114.
4. "Free distributive pseudocomplemented lattices". *Algebra Universalis*, 3(1973): 13–15.
5. "A topological representation theory for lattices". *Algebra Universalis*, 8(1978): 45–58.
6. "Distributive lattices with a dual homomorphic operation II". *Studia Logica*, 40(1981–1984): 391–404.
7. "The undecidability of entailment and relevant implication". *Journal of Symbolic Logic*, 49(1984): 1059–1073.
8. "Hard examples for resolution". *Journal of the Association for Computing Machinery*, 34(1987): 209–219.
9. "The Complexity of Gentzen Systems for Propositional Logic". *Theoretical Computer Science*, 66(1989): 87–97.
10. "Failure of interpolation in relevant logics". *J. of Philosophical Logic*, 22(1993): 449–479.
11. "Functional interpretations of feasibly constructive arithmetic", with S.A. Cook. *Annals of Pure and Applied Logic*, 63(1993): 103–200.
12. *The Collected Papers of Bertrand Russell, Volume 4: Foundations of Logic 1903–05*, Editor. Routledge, 1994.
13. "The Complexity of the Hajos calculus", with Toniann Pitassi. *SIAM J. of Discrete Mathematics*, 8(1995): 464–483.
14. "The complexity of propositional proofs". *Bulletin of Symbolic Logic*, 1(1995), 425–467.
15. "The symmetry rule in propositional logic". *Discrete Applied Mathematics*, 96–97(1999): 177–193.
16. "The Complexity of Decision Procedures in Relevance Logic". *Journal of Symbolic Logic*, 64(1999): 1774–1802.
17. "The Complexity of Analytic Tableaux", with Noriko Arai and Toniann Pitassi. *Journal of Symbolic Logic*, 71(2006), 777–790.
18. "An Exponential Separation between Regular and General Resolution", with Mikhail Alekhnovich, Jan Johannsen, and Toniann Pitassi. *Theory of Computing*, 3(2007), 81–102.
19. "Synonymous Logics", with F.J. Pelletier. *Journal of Philosophical Logic*, 32(2003): 259–285.

*Work in Progress*

20. *Equilibrium States in Mean Field Models*. Monograph in preparation.

**Service to the Profession:** Consulting editor *Journal of Symbolic Logic* 1983–1989; Member, Executive Committee, Association for Symbolic Logic 1987–1990; Member, Advisory Editorial board, *Studia Logica* 1991–; Board of editors, *Russell* 1999–; Editor, *Lecture Notes in Logic* (Springer Verlag), 1994–1997; Editor, *Trends in Logic*, 1994–2003; Consulting editor, *Studia Logica*; Member of editorial board, *International Studies in the Philosophy of Science*; Editor, *Reports on Mathematical Logic*, 1999–; Editor for non-classical logics, *Stanford On-Line Encyclopedia*, 1998–2004; Chair, Local arrangements committee, ASL Annual Meeting, Toronto, May 1998; Editor, FOM moderating mailing list, 2000–; Society for Exact Philosophy, Vice-President 2000–2003, President 2003–2005; Reviews managing editor, *Bulletin of Symbolic Logic*, 2002–.

**Teaching:** Notable former students include Philip Kremer and Toniann Pitassi (both now at the University of Toronto), Oliver Schulte (Simon Fraser), Achille Varzi (Columbia), Jamie Tappenden (University of Michigan), Michael Soltys (McMaster) and Peter Koellner (Harvard).

**Vision Statement:** Urquhart believes that logic is most fruitful as a part of applied rather than pure mathematics. There has been an enormous expansion of logical applications in computer science, linguistics and cognitive science; such applications are the wave of the future.

# USPENSKIY, (also USPENSKY, OUSPENSKI), Vladimir Andreyevich

**Specialties:** Computability, mathematical logic, foundations of mathematics, philosophy of mathematics.

**Born:** 27 November 1930 in Moscow, Russia.

**Educated:** USSR Highest Certifying Comission, DrSci Mathematics, 1964; Moscow University, PhD Mathematics, 1955; Moscow University, MS Mathematics, 1952.

**Dissertation:** *On Computable Operations*; supervisor, Andrei Kolmogorov. DrSci Dissertation: Lectures on Computable Functions; no supervisor for a DrSci dissertation in USSR.

**Regular Academic or Research Appointments:** HEAD OF DEPARTMENT, MATHEMATICAL LOGIC AND THEORY OF ALGORITHMS, MOSCOW UNIVERSITY, 1993–; PRINCIPAL SCIENTIST, ALL-RUSSIAN INSTITUTE FOR SCIENTIFIC AND TECHNICAL INFORMATION, 1987–; PROFESSOR, MATHEMATICAL LOGIC, MOSCOW UNIVERSITY, 1966–; Associate Professor, Mathematical Logic, Moscow University, 1959-1965; Assistant Professor, Mathematics, Moscow University, 1955-1959.

**Visiting Academic or Research Appointments:** Visiting Researcher, Universidad de Chile, Santiago, Chile, 1998; Visiting Researcher, Ecole Normale Supérieure de Lyon, Lyon, France, 1997; Visiting Researcher, Centre for Mathematics and Computer Science (CWI), Amsterdam, Netherlands, 1993.

**Research Profile:** Under the guidance by Kolmogorov, Uspenskiy participated in elaboration of a general definition of an algorithm (the resulting notion is called sometimes "Kolmogorov machine" and sometimes "Kolmogorov–Uspenskiy machine"). His other works related to algorithmic foundations of Godel Incompleteness Theorem and of Probability and Information Theories. Uspenskiy is the author of the first Russian book on computability theory (1960) as well as the fist Russian book on non-standard analysis.

**Main Publications:**

1. Lectures on Computable Functions}, Moscow, 1960 (in Russian); translated in French as: V. A. Ouspenski, Leçons sur les fonctions calculables}, Hermann, Paris, 1966.
2. Post's Machine, Moscow, 1983.
3. Godel's Incompleteness Theorem, Moscow, 1987.
4. What is Nonstandard Analysis About?, Moscow, 1987. (In Russian.)
5. Algorithms: Main Ideas and Applications, Kluwer Acad. Publ., 1993. (With A. L. Semenov.)
6. What is the Axiomatic Method About?, Izhevsk, 2001. (In Russian.)
7. An Introductory Course in Mathematical Logic, Moscow, 2002. (With N. K. Vereshchagin, V. A. Plisko; in Russian.)
8. Non-Mathematical Writings, in two volumes, Moscow, 2002. (In Russian.)
9. On the definition of an algorithm // American Mathematical Society Translations, ser. 2, vol. 29 (1963), pp. 217-245. (With A. N. Kolmogorov.)
10. Algoritmhs and randomness // Theory of Probability and Its Applications, vol. 2 (1987), pp. 389–412. (With A. N. Kolmogorov.)
11. "Algorithm" and "Algorithms, Theory of" // Encyclopaedia of Mathematics, Kluwer Acad. Publ., vol. 1 (1987), pp. 131–133, 150–152.
12. Seven reflections in the topic of philosophy of mathematics // Zakonomernosti razvitiya sovremennoyj matematiki, Moscow, 1987, pp. 106–155 (in Russian).
13. Diagnostic propositional formulas // Moscow University Mathematics Bulletin, vol. 46 (1991), no. 3, pp. 8–12. (With V. Ye. Plisko.)
14. Kolmogorov and mathematical logic // The Journal of Symbolic Logic, vol. 57 (1992), no. 2, pp. 385–412.
15. Goedel's Incompleteness Theorem // Theoretical Computer Science, vol. 130 (1994), no. 2, pp. 239–319.
16. Sofjya Aleksandrovna Yanovskaya // Modern Logic, 1996 (October), vol. 6, no. 4 (October 1996), pp. 357-372. (With I. G. Bashmakova, S. S. Demidov.)
17. Mathematical logic in the former Soviet Union: brief history and current trends // M. L. Dalla Chiara et al. (eds.). Logic and Scientific Methods. Kluwer Acad. Publ., 1997. Pp. 457–483.
18. Wittgenstein and the foundations of mathematics // Voprosih filosofii (Journal of Philosophy), No.5 (1998), pp.85–97. (In Russian.)
19. The law of excluded middle and the law of double negation // A. N. Kolmogorov. Selected Works, vol. 1, Moscow, 2005. Pp. 445–454. (With V. Ye. Plisko; in Russian.)
20. Four algorithmic faces of randomness // Matematicheskoe Prosveschenie, ser. 3. No. 10. Moscow, 2006. Pp. 71–108. (In Russian.)

**Service to the Profession:** Organizing Committee Member, Session in Mathematical Logic Leader, International Conference "Kolmogorov and Modern Mathematics", 2003; "Semiotics and Informatics", Editor-in-Chief, 1993-2002, Deputy Editor-in-Chief, 1980-1991; Chair, Department of Mathematical Logic and the Theory of Algorithms, Moscow University, 1993-; Editor, "Problems of Information Transmission", 1965-.

Invited speaker to congresses, conferences and university seminars in Bulgaria, Chile, China, Czechoslovakia, Denmark, France, Germany, Italy, Netherlands, Sweden, USA, USSR.

**Teaching:** In the USSR, Uspenskiy was the first to teach computability theory and one of the key lecturers in mathematical logic. He has had twenty six PhD students in Mathematical Logic and two PhD students in Theoretical Linguistics. Four of his students in Mathematical Logic (Kanovei, Lyubetsky, Vereschagin, Vyugin) and one student in Linguistics (Razlogova) became Doctors of Science (the highest scientific degree in Russia).

**Vision Statement:** There is a profound relationship between computability and foundations. Some fundamental facts of mathematical logic turned out to have the algorithmic formulation (such as Tennenbaum Theorem on models of arithmetic) or an algorithmic basis (such as Godel's Incompleteness Theorem). In the future, some counterinfluence between logic and linguistics is expected.

# V

## VÄÄNÄNEN, Jouko

**Specialties:** Mathematical logic, set theory, model theory, foundations of mathematics, computer science logic.

**Born:** 3 September 1950 in Rovaniemi, Finland.

**Educated:** University of Manchester, PhD Mathematics, 1977; University of Helsinki, MSc Mathematics, 1973.

**Dissertation:** *Applications of set theory to generalized quantifiers*; supervisor, Peter Aczel.

**Regular Academic or Research Appointments:** PROFESSOR, MATHEMATICAL LOGIC AND FOUNDATIONS OF MATHEMATICS, UNIVERSITY OF AMSTERDAM, 2006–; Professor, Mathematics, University of Helsinki, 1998-; Associate Professor, Mathematics, University of Helsinki, 1983-1998; Assistant, University of Helsinki, 1978-1983.

**Visiting Academic or Research Appointments:** Scientific director, Mittag-Leffler Institute, Djursholm Sweden, 2000-2001; Visiting Scholar, University of Paris VII, 1997; Visiting Scholar, Florida State University, 1996-1997; Visiting Scholar, University of Freiburg, 1995; Visiting Scholar, Simon Fraser University, Vancouver, 1992-1993, 1990; Visiting Scholar, Stanford University, 1991-1992; Visiting Scholar, University of California at Santa Cruz, 1991-1992; Visiting Scholar, University of California at Los Angeles, 1988-1989.

**Research Profile:** The research of Väänänen is on the borderline between set theory and model theory. A typical case is his extensive study of the Härtig-quantifier which has taken him deep into set theory, although the starting point has been model theoretic.

The main contribution of Jouko Väänänen in abstract model theory is in the area of generalized quantifiers, where he has developed the topic both on finite structures, with a heavy use of combinatorics and with applications in computer science and linguistics, and on arbitrary, often uncountable, structures with a strong presence of set theoretical methods.

In set theoretic model theory he has developed a new game theoretical approach to studying uncountable structures, based on transfinite Ehrenfeucht-Fraïssé games. This led him to the set theoretic study of the structure of trees as generalizations of ordinals.

In infinitary model theory Väänänen has used games to bring a coherent approach to both generalized quantutifiers and infinitary logic. Especially in infinitary logic this led out of the cul de sac which the area had found itself in after the seminal early work by Karp, Lopez-Escobar, Scott and Barwise.

In set theory Väänänen introduced already in the late 70s in cooperation with J. Stavi maximality principles for CCC forcing which have subsequently become known as generic absoluteness principles and gained importance in connection with proper forcing.

Väänänen has a continued interest in mathematical properties of second order logic, and more generally, in questions of the foundations of mathematics. Recently he has developed the mathematics of a logical theory of dependence.

**Main Publications:**

1. *Two axioms of set theory with applications to logic.* **Ann. Acad. Sci. Fenn. Ser. A I. Math. Diss.** 20:1-19, 1978.
2. *Boolean valued models and generalized quantifiers*, **Annals of Mathematical Logic**, 79(1980), pages 193-225.
3. *The Hanf number of $L_{\omega_1\omega_1}$.* **Proceedings of American Mathematical Society** 79:294-297, 1980.
4. *Set theoretic definability of logics.* In J.Barwise and S.Feferman, editors, **Model Theoretic Logics**, pages 599-643, Springer, 1985.
5. *On Scott and Karp trees of uncountable model*, with Tapani Hyttinen. **Journal of Symbolic Logic** 55(3):897-908, 1990.
6. *On the number of automorphisms of uncountable models*, with Saharon Shelah and Heikki Tuuri, **Journal of Symbolic Logic** 58(1993), 1402-1418.
7. *The Ehrenfeucht-Fraïssé game of length $\omega_1$*, with Alan Mekler and Saharon Shelah, **Transactions of the American Mathematical Society** 339(1993), 567-580.
8. *Trees and $\Pi^1_1$-subsets of $^\omega 1\omega_1$*, with Alan Mekler, **Journal of Symbolic Logic** 58(1993), 1052-1070.
9. *Generalized quantifiers and pebble games on finite structures*, with Phokion Kolaitis, **Annals of Pure and Applied Logic** 74(1995), 23-75,
10. *Definability of polyadic lifts of generalized quantifiers*, with Lauri Hella and Dag Westerståhl, **Journal of Logic, Language and Information.** 6:305-335, 1997.
11. *Trees and Ehrenfeucht-Fraïssé games*, with Stevo Todorcevic, **Annals of Pure and Applied Logic**, 100, 69-97, 1999.
12. *Stationary sets and infinitary logic*, with Saharon

Shelah, **Journal of Symbolic Logic**, 65:1311-1320, 2000.
13. *Second order logic and foundations of mathematics*, **Bulletin of Symbolic Logic** 7 (2001), no. 4, 504-520.
14. *On the semantics of informational independence*, **Logic Journal of the Interest Group in Pure and Applied Logics** 10:3, 337-350, 2002.
15. *Reflection principles for the continuum*, with J. Stavi, **Logic and Algebra**, ed. Yi Zhang, pp. 59-84, Contemporary Mathematics, Vol 302, AMS, 2002.
16. *More on the Ehrenfeucht-Fraïssé game of length $\omega_1$*, with T. Hyttinen and S. Shelah, **Fund. Math.** 175 (2002), no. 1, 79-96.
17. *On the expressive power of monotone natural language quantifiers over finite models*, with Dag Westerståhl, **Journal of Philosophical Logic**, vol. 31 (2002), no. 4, 327-358.
18. *Games played on partial isomorphisms*, with B. Velickovic, **Archive for Mathematical Logic**, 43:1, 19-30, 2004.
19. *Finite information logic*, with R. Parikh, **Annals of Pure and Applied Logic**, 134, 83-93, 2005.
20. *Dependence Logic*, **Cambridge University Press**, 2007.

A full list of publications is available at Väänänen's home page http://math.helsinki.fi/logic/vaananen

*Work in Progress*

21. Games and models, a monograph on the Ehrenfeuch-Fraisse game, to appear in 2008.
22. Work with Saharon Shelah on the cofinality quantifier, on a new logic with a Lindström theorem, and on other topics in abstract model theory.

**Service to the Profession:** Treasurer and Executive Committee Member, European Mathematical Society 2007-; Committee Chairman, Logic in Europe, Association for Symbolic Logic, 2007-; Executive Committee Member, Association for Symbolic Logic, 2007-; Editorial Board Member, Notre Dame Journal of Formal Logic, 2006-; Editorial Board Member, Logica Universalis, 2006-; Senate Member, University of Helsinki, 2004-2006; Chair, Department of Mathematics and Statistics, University of Helsinki, 2004-2006.

**Teaching:** Väänänen has created a logic group in Helsinki with a regular MSc and PhD program in logic. He has had thirteen PhD students: Maaret Karttunen, Lauri Hella, Heikki Tuuri, Taneli Huuskonen, Kerkko Luosto, Jyrki Akkanen, Juha Nurmonen, Aapo Halko, Risto Kaila, Alex Hellsten, Juha Kontinen, Marta Garcia-Matos and Matti Pauna. He has also influenced the PhD work of Tapani Hyttinen.

**Vision Statement:** Different branches of mathematical logic, such as model theory, recursion theory, set theory and proof theory have had a tendency to become so specialized and in some cases so close to other areas of mathematics or computer science that they have almost ceased to be logic at all. I am interested in logic as a discipline. Fragmentation is a pity, despite the successes.

**Honours and Awards:** Elected, Finnish Academy of Science and Letters, 2002.

## VAKARELOV, Dimiter Ivanov

**Specialties:** Non-classical logic, applied modal logic.

**Born:** 18 April, 1938 in Plovdiv, Bulgaria.

**Educated:** Sofia University, MSc Mathematics 1965; Warsaw University, PhD in Logic 1967; Sofia University, Doctor of Mathematical Sciences 1997.

**Dissertation:** *Theory of negation in certain logical systems. Algebraic and semantical approach*; supervisor Helena Rasiowa.
Doctor of Mat. Sc.: *Applied Modal Logic: Modal logics in information science*.

**Regular Academic or Research Appointments:** PROFESSOR OF LOGIC, DEPARTMENT OF MATHEMATICAL LOGIC WITH LABORATORY FOR APPLIED LOGIC, FACULTY OF MATHEMATICS AND INFORMATICS, SOFIA UNIVERSITY, 1965-2008.

**Visiting Academic or Research Appointments:** Scientific visits in Warsaw University, University of Novosibirsk, Tokio University of Science and Technology, University Paris 13, University of Toulouse, Brock University, University of Antigonish, University of Witwatersrand, Johannesburg SA.

**Research Profile:** Dimiter Vakarelov made contributions to the theory of negation, many-valued logic, intuitionistic modal logic, Dynamic Logic, modal logics for information systems, modal logics of space, Arrow Logic, Sahlqvist theory, Region-based theory of space.

**Main Publications:**

1. "Semantics for $\omega+$-valued predicate Calculi", with L. Maksimova. *Bull. Acad. Polon. Sci. Ser. Math. Phis.*, 22(1974): 765-771.
2. "Notes on Constructive logic with strong negation". *Studia Logica*, 36(1977): 110-125.
3. "Intuitionistic modal logics incompatible with the law of excluded middle", *Studia Logica*, XL(2) (1981): 103-111.
4. "Consistency, Completeness and Negation". In: Gr. Priest, R. Routley and J. Norman (eds.) *Paraconsistent*

*Logic. Essays on the Inconsistent.* Analitica, Philosophia Verlag, Munhen, 1989, pp. 328-363.

5. "Modal logics for knowledge representation systems", *Theoretical Computer Science*, 90(1991): 433-456.

6. "Many-dimensional arrow structures. Arrow logics II". In: M. Marx, L. Polos and M. Masuch (eds.), *Arrow Logic and Multi-Modal Logic.* Studies in Logic Language and Information, Stanford CA, 1996, pp. 141-187.

7. "Hyper Arrow Structures. Arrow Logics III". In: M. Kracht, M. de Rijke, H. Wansing and M. Zakharyaschev Eds. *Advances in Modal Logic'96.* Studies in Logic, Language and Information, Stanford CA, 1997, pp. 253-273.

8. "Modal Logics for Incidence Geometries", with Balbiani, F., L. Farinas del Cerro and T. Tinchev. *Journal of Logic and Computation*, 7(1)(1997): 59-78.

9. "Information Systems, Similarity Relations and Modal Logics" In: E. Orlowska (ed.) *Incomplete Information: Rough Set Analysis.* Vol. 13 of *Studies in Fuzziness and Soft Computing.* Phisica-Verlag, Heidelberg and New York, 1998, pp. 492-550.

10. "Modal definability in languages with a finite number of propositional variables and a new extension of the Sahlqvist's class". In: Ph. Balbiani, Nobu-Yuki Suzuki, F. Wolter and M. Zakharyaschev (eds.) *Advances in Modal Logic*, 4, King's College Publications, 2003, pp. 499-518.

11. "PDL with Intersection of Programs: a Complete Axiomatization", with Ph. Balbiani. *Journal of Applied Non-Classical Logics*, 13(3-4) (2003): 231-276. 2003.

12. "Modal definability, solving equations in modal algebras and generalizations of the Ackermann Lemma". In: C. Dimitracopoulos (ed.) *Proceedings of 5th Panhellenic Logic Symposium*, July 25-28, 2005, Athens, pp. 182-189.

13. "Nelson's Negation on the Base of Weaker versions of Intuitionistic Negation".*Sudia Logica*, 80(2-3)(2005): 393-430.

14. "Elementary canonical formulae: extending Sahlkvist's theorem", with V. Goranko. *Annals of Pure and Applied Logic*, 141(2006): 180-217.

15. "Algorithmic Correspondence and Completeness in Modal Logic. I. The Core Algorithm SQEMA", with W. Conradie and V. Goranko. *Logical Methods in Computer Science* (2)(1:5)(2006): 1-26.

16. "Contact Algebras and Region-based Theory of Space: a Proximity Approach. I and II", with G. Dimov. *Fundamenta Informaticae*, 74 (2-3) (2006): 209-282.

17. "Modal Logics for Region-based Theories of Space", with Ph. Balbiani and T. Tinchev. *Fundamenta Informaticae*, 81(1-3)(2007): 29-82.

18. "Region-Based Theory of Space: Algebras of Regions, Representation theory, and Logics". In: D. Gabbay, S. Goncharov and M. Zakharyaschev (eds.) *Mathematical Problems from Applied Logic II. Logics for the XXIst Century*, Springer, 2007, pp. 267-348.
*Work in Progress*

19. "Modal Logics for mereotopological relations", with Y. Nenov.

20. "Mereotopological Representation of Tarski Consequence Relation.

**Service to the Profession:** Editor *Journal of Applied Non-Classical Logics*, in the previous Editorial Board of *Studia Logica*, member of the Program Committee of various conferences.

**Teaching:** Vakarelov's students include Solomon Passy, Tinko Tinchev, Valentin Goranko, Dimiter Guelev.

**Vision Statement:** I consider Non-classical Logic as an important part of Mathematical logic with significant applications in computer science, AI, linguistics and philosophy. Generalizing the classical two-valued logic in many different and reasonable ways it makes possible to understand deeper the nature of Logic.

# VAN BENDEGEM, Jean Paul

**Specialties:** Philosophical logic, foundations of mathematics, philosophy of mathematics

**Born:** 28 March 1953 in Ghent, Belgium.

**Educated:** University of Ghent, PhD Philosophy, 1983; University of Ghent, Graduate in Philosophy, 1979; University of Ghent, Graduate in Mathematics, 1976.

**Dissertation:** *Design of a Finite, Empirical Mathematics*; supervisors, Leo Apostel and Diderik Batens.

**Regular Academic or Research Appointments:** PROFESSOR LOGIC AND PHILOSOPHY OF SCIENCE, VRIJE UNIVERSITEIT BRUSSEL (FREE UNIVERSITY OF BRUSSELS), 1990– AND PART-TIME PROFESSOR UNIVERSITEIT GENT (UNIVERSITY OF GHENT), 1990–; Part-time professor, Rijksuniversitair Centrum Antwerpen (State University Center Antwerp), 1984-1990; Part-time professor, Vrije Universiteit Brussel (Free University of Brussels), 1985-1990; Postdoctoral researcher, Universiteit Gent (University of Ghent), 1979-1990.

**Visiting Academic or Research Appointments:** Department of Philosophy, University of Queensland, Brisbane, Queensland, Australia, 1993; Center for Philosophy of Science, University of Pittsburgh, Pennsylvania, USA, 1985.

**Research Profile:** In a first phase of his research Van Bendegem was interested in showing that strict finitism is a genuine possibility within the spectrum of foundational approaches of mathematics. This resulted in the book [4], a derivation from his PhD. However many open questions remained and, through the work of Graham Priest, a paraconsistent approach was now tried out, leading to a technically and philosophically more pleasing presentation (see, e.g., [7],[10]). Applications to physical problems, with a special interest for supertasks, have also been looked at (e.g., [9],[18]) as well as to historical-mathematical problems ([19], where an attempt is made to interpret infinitesimals using a paraconsistent approach.

In a second phase a study of actual mathematical practice was started. Although the first paper on the topic ([3]) treated the history of Fermat's Last Theorem from an epistemological-evolutionary perspective, this line of thinking was not maintained in later work. Rather the emphasis was on describing all the various factors and elements that enter into "real" mathematical practice, ranging from the quality of a proof to aesthetical considerations and the importance of heuristics. This has lead so far to a series of papers (see, e.g., [6],[8],[11],[17]), that will result in a book publication in the near future.

Part of the inspiration for the second phase came from dialogue logic. There are some early papers on this topic ([2]), but for some time it 'disappeared' into the background. However, after a meeting with Shahid Rahman, the topic was taken up again, this time from a paraconsistent view ([16], co-authored with Shahid Rahman).

In a third phase, corresponding to the last two or three years, a new idea has come up to try to integrate the work of both previous phases: the exploration of the possibilities of 'truly' alternative mathematics (in the sense of not corresponding to one of the alternatives known at present). First results are to be found in [13] and [14].

Since 1995 Van Bendegem is director of the *Center for Logic and Philosophy of Science* at the Vrije Universiteit Brussel (http://www.vub.ac.be/CLWF). Up to now about ten PhDs have been defended. Of special importance for the logic community is the work of Sonja Smets on quantum logic, the work of Bart van Kerkhove on mathematical practice and naturalism in mathematics, the forthcoming thesis of Patrick Allo on philosophy of information, and the forthcoming thesis of Helen de Cruz on the biological origins of mathematical capacities. A full list of publications of Van Bendegem and his researchers can be found at the above mentioned website.

**Main Publications:**

1. "Relevant derivability and classical derivability in Fitch-style and axiomatic formulations of relevant logics" (co-author: Diderik Batens). *Logique et Analyse*, 109, 1985, 21-31.
2. "Dialogue Logic and Problem-Solving". *Philosophica* 35, 1985, 113-134.
3. "Fermat's Last Theorem seen as an Exercise in Evolutionary Epistemology". In: Werner Callebaut & Rik Pinxten (eds.), *Evolutionary Epistemology*, Kluwer, Dordrecht, 1987, 337-363.
4. *Finite, Empirical Mathematics: Outline of a Model*. Works edited by the Faculty of Arts and Letters, State University Ghent, volume 174, Ghent, 1987.
5. (editor), Recent Issues in the Philosophy of Mathematics, volume I and volume II. Special issues of *Philosophica*, 42 and 43, 1989.
6. "Characteristics of Real Mathematical Proofs" In: A. Diaz, J. Echeverria & A. Ibarra (eds.), *Structures in Mathematical Theories*, Servicio Editorial Universidad del Pais Vasco, San Sebastian, 1990, 333-337.
7. "Strict, Yet Rich Finitism". In: Z.W. Wolkowski (ed.): *First International Symposium on Gödel's Theorems*, World Scientific, Singapore, 1993, 61-79.
8. (editor, co-editors: Sal Restivo & Ronald Fischer), *Math Worlds: New Directions in the Social Studies and Philosophy of Mathematics*. State University New York Press, New York, 1993.
9. "Ross' Paradox is an Impossible Super-Task". *The British Journal for the Philosophy of Science*, Vol. 45, 1994, 743-748.
10. "Strict Finitism as a Viable Alternative in the Foundations of Mathematics". *Logique et Analyse*, vol. 37, 145, 1994 (date of publication : 1996), 23-40.
11. "Mathematical Experiments and Mathematical Pictures". In: Igor Douven & Leon Horsten (eds.), *Realism in the Sciences. Proceedings of the Ernan McMullin Symposium Leuven 1995*. Louvain Philosophical Studies 10. Leuven University Press, Louvain, 1996, 203-216.
12. "'Knowledge of Philosophy - Nil' or Sherlock Holmes as Inspiration for Philosophers". *The Sherlock Holmes Journal*, Vol. 22, no. 4 (Eighty-Seventh Issue), 1996, 121-124.
13. "Analogy and Metaphor as Essentials Tools for the Working Mathematician". In: Fernand Hallyn (ed.), *Metaphor and Analogy in the Sciences*, (Origins: Studies in the Sources of Scientific Creativity), Kluwer Academic, Dordrecht, 2000, 105-123.
14. "Alternative Mathematics: The Vague Way". In: Décio Krause, Steven French & Francisco A. Doria (eds.), Festschrift in Honour of Newton C.A. da Costa on the Occasion of his Seventieth Birthday, *Synthese*, vol. 125, nos. 1-2, 2000, 19-31.
15. (editor, co-editors: Diderik Batens, Graham Priest & Chris Mortensen), *Frontiers of Paraconsistent Logic. Studies in Logic and Computation, volume 8*. King's College London Publications, Baldock, 2000.
16. "Paraconsistency and Dialogical Logic. Critical Examination and Further Explorations". Synthese, vol. 127, nos. 1-2, 2001, 35-55.
17. "The Creative Growth of Mathematics". *Philo-

*sophica*, vol. 63, 1, 1999 (date of publication: 2001), 119-152.
18. "Finitism in Geometry". *The Stanford Encyclopedia of Philosophy* (Spring 2002 Edition), Edward N. Zalta (ed.), http://plato.stanford.edu/entries/geometry-finitism/, The Metaphysics Research Lab at the Center for the Study of Language and Information, Stanford University, Stanford, CA, 2002.
19. "Inconsistencies in the history of mathematics: The case of infinitesimals". In: Joke Meheus (ed.): *Inconsistency in Science*. Dordrecht: Kluwer Academic Publishers, 2002, 43-57 (Origins: Studies in the Sources of Scientific Creativity, volume 2).
20. "The Unreasonable Richness of Mathematics" (co-author: Bart Van Kerkhove). *Journal of Cognition and Culture*, vol. 4, no. 3-4, 2004, pp. 525-549.
21. "The Collatz Conjecture: A Case Study in Mathematical Problem Solving". *Logic and Logical Philosophy*, vol. 14, 2005, pp. 7-23. Special issue "Patterns of Scientific Reasoning", guest-editor: Erik Weber.
22. "Mathematical Practice and Naturalist Epistemology: Structures with Potential for Interaction" (co-author: Bart Van Kerkhove). In: Gerhard Heinzmann & Manuel Rebuschi (eds.): "Aperçus philosophiques en logique et en mathématiques", *Philosophia Scientiae*, volume 9, cahier 2, 2005, pp. 61-78.
*Work in Progress*
23. (editor, co-editor: Bart Van Kerkhove), *Perspectives on Mathematical Practices. Bringing together Philosophy of Mathematics, Sociology of Mathematics, and Mathematics Education*. Dordrecht: Springer (formerly: Kluwer Academic), 2006.

**Service to the Profession:** President of the *National Center for Research in Logic*, 2001-; Editor, *Logique et Analyse*, 1988-; Committee Member in Philosophy, *Fund of Scientific Research in Flanders*, 1997-2006; Member of the Committee for the Evaluation of Philosophy Teaching in the Netherlands, 1995-1996; Member of the Committee of Philosophical Research in the Netherlands, 1999-2000; Managing editor, *Philosophica*, 1987-1995.

**Teaching:** As far as the teaching of Van Bendegem is concerned, one of the core aims is to 'preserve' and continue the work of Leo Apostel. This includes the Belgian logico-argumentational approach (Chaïm Perelman, Jean Ladrière and other founding members of the *National Center for Research in Logic*, founded in 1955) and the Dutch language approach (so-called 'Significs', Gerrit Mannoury and L.E.J. Brouwer as most prominent members). As far as students are concerned, there are at present two students who obtained their PhD quite recently (mentioned above in the *research profile*) and who promise to be excellent researchers: Sonja Smets and Bart Van Kerkhove.

**Vision Statement:** Logic is the *microscope* of human reasoning. It shows details one does not see in 'real life', but, at the same time, it should remain relevant for that very same 'real life', by improving the quality of discussions and argumentations in society at large.

**Honours and Awards:** Fulbright-Hays grant awarded by the Commission for Educational Exchange between the United States of America, Belgium and Luxembourg, 1985.

# VAN BENTHEM, Johannes Franciscus Abraham Karel

**Specialties:** Modal logic, epistemic logic, dynamic logic, logics of time and space, semantics of natural language: generalized quantifiers and categorial grammars, substructural proof theory, philosophical logic, logics of computation, information update and interaction, logic and games.

**Born:** 12 June 1949 in Rijswijk, Zuid-Holland, Netherlands.

**Educated:** University of Amsterdam, PhD Mathematics, 1977; MSc Mathematics, 1973; MA Philosophy, 1972; BSc Physics 1969.

**Dissertation:** *Modal Correspondence Theory*; supervisor Martin Löb.

**Web Address:** http://staff.science.uva.nl/~johan

**Regular Academic or Research Appointments:** UNIVERSITY PROFESSOR, PURE AND APPLIED LOGIC, UNIVERSITY OF AMSTERDAM, 2003–; PROFESSOR, PHILOSOPHY, STANFORD UNIVERSITY, 2005–; University of Amsterdam, Mathematics and Computer Science, 1986–2003; University of Groningen, Philosophy and Mathematics, 1977–1986; University of Amsterdam, Philosophy, 1973–1977.

**Visiting Academic or Research Appointments:** Visiting University Professor, Sun Yat-Sen University, Guangzhou, 2005–; Bonsall Chair, Stanford University, 1991–2005; Senior Researcher, Center for the Study of Language and Information, Stanford, 1988–; Simon Fraser University, Burnaby, Mathematics, 1984.

**Research Profile:** In the 1970s, van Benthem worked in the 'correspondence theory' of modal axioms and their definability in first- and higher-order logics. A spin-off was the theorem characterizing the modal language as consisting of just those first-order formulas that are invariant for bisimulation. Cf. *Modal Correspondence Theory*, 1977, *Modal Logic and Classical Logic*, 1983. Recent publications in this line are about modal model theory for infinitary languages (J. Barwise & J. van Benthem, 1999, 'Interpolation, Preservation, and Pebble Games', *Journal of Symbolic Logic* 64:2, 881–903), modal fixed-point languages ('Minimal Predicates, Fixed-Points, and Definability', *Journal of Symbolic Logic* 70:3, 696–712, 2005; 'Modal Frame Correspondences and Fixed-Points', *Studia Logica* 83:1, 133–155, 2006), and modal abstract model theory ('Lindström Theorems for Fragments of First-Order Logic', *Proceedings LICS 2007*, 280–292, with Balder ten Cate and Jouko Väänänen).

Around 1980 van Benthem turned to the philosophy of science ('The Logical Study of Science', *Synthese* 51, 1982, 431–472), and temporal point- and interval-ontology: *The Logic of Time*, 1983. A related theme are logics of topology, and geometry (M. Aiello & J. van Benthem, 'A Modal Walk Through Space', *Journal of Applied Non-Classical Logic* 12, 2004, 319–363; J. van Benthem & G. Bezhanishvili, 2007, 'Modal Logics of Space', *Handbook of Spatial Logics*, 217–298).

In the later 1980s, van Benthem switched to natural language. His *Essays in Logical Semantics* (1986) is about semantic universals and expressive power for generalized quantifiers, with classification theorems in terms of the 'Number Tree', monotonicity properties, and automata complexity. There is also a first calculus of 'natural logic' for reasoning directly on linguistic surface form, using type-theoretic derivation with monotonicity marking. This work led to a study of categorial grammars as resource-bounded substructural proof systems. Results include a semantics for meaning composition via 'linear lambda terms', the study of expressive power and semantic constraints in finite type hierarchies, a general account of logicality as permutation invariance, and systematic connections between proof-theoretic and grammatical issues. Most of this work is collected in the book *Language in Action*, 1991.

In the 1990s, van Benthem returned to modal logic, now with a view to computation. Results include a characterization of the regular operations on imperative programs as those that are 'safe for bisimulation', and a first-order model theory of process equivalences (J. van Benthem & J. Bergstra, 1995, 'Logic of Transition Systems', *Journal of Logic, Language and Information* 3, 247–283). Modal techniques also brought to light decidable 'core versions' of undecidable computational formalisms, such as 'arrow logic': a decidable version of relational algebra. Another side of this coin was the discovery of the 'Guarded Fragment', a new large decidable part of first-order logic (H. Andréka, J. van Benthem & I. Németi, 1998, 'Modal Logics and Bounded Fragments of Predicate Logic', *Journal of Philosophical Logic* 27, 217–274). This work is collected in the monograph *Exploring Logical Dynamics*, 1996.

Van Benthem's key interest since his Spinoza project *'Logic in Action'* (1996–2001) is 'logical dynamics', making actions of inference, observation, information update, belief revision, or preference change first-class citizens. These come together in the study of rational agency using *dynamic-epistemic logics* ('One is a Lonely Number', Z. Chatzidakis et al., eds., 2006, *Logic Colloquium '02*, A.K. Peters, Wellesley, 96–129; 'Logics of Communication and Change', *Information and Computation* 204, 2006, 1620–1662, with Jan van Eijck and Barteld Kooi; 'Dynamic Logic of Belief Change', *Journal of Applied Non-Classical Logics* 17, 2007, 129–155). In this view, logic is essentially about multi-agent communication, and 'intelligent interaction'. This leads to new interfaces between logic and game theory (*Logic in Games*, ILLC, Amsterdam; 'Games in Dynamic Epistemic Logic', *Bulletin of Economic Research* 53, 2001, 219–248; 'Extensive Games as Process Models', *Journal of Logic, Language and Information* 11, 2002, 289–313; 'Logic Games are Complete for Game Logics', *Studia Logica* 75, 2003, 183–203; 'Rational Dynamics', *International Game Theory Review* 9, 2007, 377–409). This new take on what logic is about also seems highly relevant to its past and future interfaces with philosophy ('Logic in Philosophy', in D. Jacquette, ed., 2007, *Handbook of the Philosophy of Logic*).

In Amsterdam, van Benthem holds the chair of Evert Willem Beth, created in 1950, whose broad view of logic in between philosophy, mathematics, computer science, linguistics, cognitive psychology, and even rational public debate, now lives on in the Institute for Logic, Language and Computation. His position at Stanford is at a similar interdisciplinary interface, against the backdrop of the Center for the Study of Language and Information, pioneered by Jon Barwise and others in the early 1980s.

**Main Publications:**

**Major book publications**
1. *The Logic of Time* (1983, 1991),
2. *Modal Logic and Classical Logic* (1985),
3. *Essays in Logical Semantics* (1986),
4. *Language in Action* (1991, 1995),
5. *Exploring Logical Dynamics* (1996),
6. *Logic in Games* (2001).

Some 350 papers in journals and books, plus some 50 items for a general public.

Textbooks: *Logic, Language and Meaning* (1982, 1991; with collective GAMUT; Dutch, English, and Spanish versions), *A Manual of Intensional Logic* (1988), *Logica voor Informatica* ('Logic for Computer Science', 1991, 2003; with four co-authors), and *Hoe Wiskunde Werkt* ('How Mathematics Works', 2004; with two co-authors).

A Chinese translation series of selected papers will appear in Beijing, starting 2008, under the title (*'A Door to Logic'*).

**Teaching:** Supervised 54 master's theses, and 60 Ph.D. dissertations in logic, broadly conceived. 43 Ph.D. students have tenured positions, some 17 as full professors, 10 are postdocs, 6 work in ICT and banking, and one is an independent artist. Courses taught range from pure to applied logic in many departments (philosophy, mathematics, computer science, linguistics, economics). Current major interest: producing new generation textbooks within disciplines based on modern logic, and spreading logic as a general culture item to broader audiences, outside of Academia.

**Service to the Profession:** Program Chair, First Chinese Conference on Logic, Rationality and Interaction, Beijing, 2007; Fellow, Games, Action, and Social Software, Netherlands Institute for Advanced Studies NIAS, 2006; Program Committee, First Indian Conference on Logic and its Relation with Other Disciplines, IIT Bombay, 2005; Area Chair, Philosophical Logic, $12^{th}$ International Congress of Logic, Methodology & Philosophy of Science, Oviedo, 2002; Program Director, First American Summer School in Logic, Language and Computation, Stanford, 2002; General Chair, TIME, Udine, 2001; Chair, Dutch National Program for Cognitive Science, 2001–2004; Program Chair *TARK*, Siena, 2001; Treasurer, Beth Foundation, Amsterdam, 2001–; Chair, Vienna Circle Archive, Amsterdam, 1999–; Vice-President, International Federation for Computational Logic, 1999–; General Program Chair, $11^{th}$ International Congress of Logic, Methodology & Philosophy of Science, Florence, 1997; Program Chair, CSLI Workshops in Logic, Language & Computation, Stanford, 1992–2002; First President, European Foundation for Logic, Language and Information FoLLI, 1989–1995; Founding Director, Institute for Logic, Language and Computation ILLC, Amsterdam, 1986–1998; President, Dutch Association of Logic, 1979–1984; Chair, Departments of Philosophy, Mathematics, and Computer Science, Amsterdam and Groningen, 1973–1993.

Editor, Handbook of the Philosophy of Information, 2008; Editor, Handbook of Spatial Logics, 2007; Editor, Handbook of Modal Logic, 2006; Editor-in-Chief, Texts in Logic and Games, 2006–; Managing Editor Synthese, 2005–; Managing Editor, Transactions on Computational Logic, 2005–; Editor, Handbook of Logic and Language, 1997; Managing Editor, Journal of Logic, Language and Information, 1991–1996; Managing Editor, Logic and Computation; 1990–1995; Managing Editor, Linguistics and Philosophy, 1989–1992; Coordinating Editor, Journal of Symbolic Logic, 1989–1993; Nominating Editor, The Philosopher's Annual, 1988–; Member Editorial Board of 15 further journals. Editor of 13 anthologies and proceedings.

**Vision Statement:** Logic is becoming a general study of inference, information flow, computation, and intelligent interaction, extending far beyond traditional concerns with 'valid consequence', and drawing upon many sources, from mathematics to cognitive science. We need this agenda expansion to, not just prove our consistency, but justify our existence today. I hope that this will lead to new fundamental insights on a par with the intellectual peaks of the 1930s which still awe us. Viewed in this way, logic is a core subject of benefit across Academia. My ambition is to make it even a part of basic intellectual education in high schools, and for a general public.

**Honours and Awards:** Honorary Member, European Foundation of Logic, Language and Information, 2005; University Professor, University of Amsterdam, 2003; Hollandsche Maatschappij van Wetenschappen, 2002; Institut International de Philosophie, 2001; Doctor h.c., Université de Liège, 1998; Dutch National Spinoza Prize, 1996–2001; Who's Who International (1995, 2004, 2007); Royal Dutch Academy of Arts and Sciences KNAW 1992; Academia Europaea, 1991.

# VAN DALEN, Dirk

**Specialties:** Intuitionistic logic, history and philosophy of mathematics and logic.

**Born:** 20 December 1932, Amsterdam, The Netherlands.

**Educated:** University of Amsterdam, Ph.D., masters (doctoral exam).

**Dissertation:** *Extension problems in intuitionistic plane projective geometry*; supervisor, Arend Heyting.

**Web Address:** www.phil.uu.nl/~dvdalen/

**Regular Academic or Research Appointments:**
PROFESSOR, HISTORICAL ASPECTS OF LOGIC AND THE PHILOSOPHY OF MATHEMATICS, UTRECHT UNIVERSITY PHIL. DEPT. 1997–2004. Professor, Logic and philosophy of mathematics, including the foundations of mathematics, Utrecht University Phil. Dept., 1979–1997; Lecturer, Utrecht University, Phil.Dept., 1967–1979; Ass. Prof. (hoofdmedewerker), 1960–1967, Utrecht University, Math. Dept.; Assistant, 1954–1958, University of Amsterdam Math.Dept.

**Visiting Academic or Research Appointments:**
Erskine Fellow, Canterbury University, New Zealand, September–October 2000; Visiting Prof., Oxford University, Math. Dept., 1974–1975; Instructor. MIT, 1964–1966; University of Colorado, Boulder, Math. Dept. summer 1965.

**Research Profile:** Intuitionistic mathematics and logic; intuitionistic treatment of traditional mathematical theories and structures, applications of metamathematics and model theory; investigation of fundamental notions such as choice sequences; modeltheory for intuitionistic theories, both metamathematical and 'natural'. History of the foundations of mathematics and logic; the history of the development of intuitionism; biographical research re L.E.J. Brouwer.

**Main Publications:**

1. *Mystic, Geometer, and Intuitionist: The Life of L.E.J. Brouwer Volume 2: Hope and Disillusion*. Oxford University Press, Oxford. 2005.
2. "Kolmogorov and Brouwer on constructive implication and the *Ex Falso* rule". *Russian Math. Surveys*, 59 (2004): 247–257.
3. "Arguments for the continuity principle", with M. van Atten. *BSL*, 8 (2002): 329–347.
4. "Brouwer and Weyl: The phenomenology and mathematics of the intuitive continuum", with M. van Atten, and R. Tieszen. *Philosophia Mathematica*, 10 (2002): 203–226.
5. "The Development of Brouwer's Intuitionism". In V.F. Hendricks, S.A. Pedersen, K.F. Jörgensen (eds), *Proof Theory. History and Philosophical Significance*. Synthese, vol. 292. Kluwer, Dordrecht, 2000, pp. 117–152.
6. "Brouwer and Fraenkel on Intuitionism". *BSL*, 6 (2000): 284–310.
7. *Mystic, Geometer, and Intuitionist: The Life of L.E.J. Brouwer. The Dawning Revolution*. Oxford University Press, Oxford, 1999.
8. "From Brouwerian Counter Examples to the Creating Subject". *Studia Logica*, 62 (1999): 305–314.
9. "Hermann Weyl's Intuitionistic Mathematics". *BSL*, 1 (1995): 145–169.
10. "The continuum and first-order intuitionistic logic". *JSL* 57 (1992): 1417–1424.
11. "The War of the Frogs and the Mice, or the Crisis of the Mathematische Annalen". *Mathematical Intelligencer*, 12 (1990): 17–31.
12. "Intuitionistic Free Abelian Groups", with F.J. de Vries. *Zeitschr. f. math. Logik und Grundlagen d. Math.*, 34 (1988): 3–12.
13. *Constructivism in Mathematics, vol. 1 & 2*, with A.S. Troelstra. *Studies in Logic*, no. 121. North-Holland, Amsterdam, 1988.
14. "Singleton reals". In D. van Dalen, D. Lascar, J. Smiley (eds.), *Logic Colloquium 1980*. North-Holland, 1982, pp. 83–94.
15. *Logic and Structure*. Springer Verlag, Berlin, 1980. viii+172 pp. Fourth expanded ed., 2004.
16. "Equality in the presence of apartness", with R. Statman. In Hintikka et al., (eds.), *Essays on Mathematical and Philosophical Logic*. Reidel, Dordrecht, 1978, pp. 95–116.
17. "An interpretation of intuitionistic analysis". *Annals Math. Logic*, 13 (1978): 1–43.
18. "The use of Kripke's schema as a reduction principle". *JSL*, 42 (1977): 238–240.
19. "Projections of Lawless Sequences", with A.S. Troelstra. *Intuitionism and Proof Theory (Proc. Summer Conference Buffalo 1968)*. North-Holland, Amsterdam, 1970, pp. 163–186.
20. "Extension problems in intuitionistic plane projective geometry". *Indagationes Mathematicae*, 25 (1963): 349–383.

*Work in Progress*

21. The editing of Brouwer's correspondence and unpublished foundational manuscripts; the work is part of the L.E.J. Brouwer project (Utrecht University).

**Service to the Profession:** Editor of: Synthese Library (active), History and Philosophy of Logic, 1993–2003; Annals of Pure and Applied Logic, 1983–1996.

Chairman of: the European Committee, ??–??; Nederlandse Vereniging voor Logica, 1967–1973.

Founding member of the European Ass. For Computer Science Logic, Vice chairman, 1992–1997.

**Teaching:** Standard courses, seminars and master classes in Utrecht, concentrated on Proof Theory, Categorical Logic and Intuitionistic logic. Furthermore Courses in Amsterdam, Antwerp, Leuven. Advanced course on Intuitionistic Logic in Oxford. Prominent Ph.D. students: M. van Atten (Intuitionism and Phenomenology), H.P. Barendregt (Lambda Calculus), J.A. Bergstra (Higher type recursion theory), L. van den Dries (Model Theory), J.W. Klop (Reduction in Combinatory systems), J. van Leeuwen (Automata theory) A. Visser (Provability).

**Vision Statement:** The trend towards applications in Computer Science will continue, complexity will become a standard theme. In model theory, more and more parts of traditional mathematics will be incorporated. Constructive and intuitionistic logic will contribute to the above through e.g., type theoretic systems. Here also the study of subsystems must be encouraged. Artemov's new Gödel interpretations promise improved analysis. Intuitionistic logic proper will develop its own 'reverse mathematics'. Furthermore the specific problems of the basic theories of arithmetic and the continuum (including e.g., topology) offer a rich variety of challenges.

**Honours and Awards:** Academy Medal of the Royal Dutch Ac. of Sciences, 2003, Knight of the Lion of the Netherlands, 1998.

# VAN DER HOEK, Wiebe

**Specialties:** Epistemic logic, modal logic, logics for multi-agent systems, logic and games.

**Born:** 15 March 1959 in Luxwoude, Friesland, Netherlands.

**Educated:** Free University of Amsterdam, PhD Mathematics and Computer Science, 1992; Rijks Uiversiteit Groningen, Netherlands, MSc Mathematics 1987; Ubbo Emmius, Groningen, Netherlands, BSc Mathematics, 1985; Ubbo Emmius, BSc in Dutch Language, Groningen, the Netherlands, 1985.

**Dissertation:** *Modalities for Reasoning about Knowledge and Quantities*; supervisor, John-Jules Meyer.

**Regular Academic or Research Appointments:** PROFESSOR, COMPUTER SCIENCE, UNIVERSITY OF LIVERPOOL, UK, 2002–; Senior Lecturer, Computer Science, Utrecht University, Netherlands, 2001-2002; Lecturer, Computer Science, Utrecht University, Netherlands, 1993-2001.

**Visiting Academic or Research Appointments:** Visiting Fellow, Otago University, New Zealand, 2002; Visiting Professor, University of Liverpool, 2001.

**Research Profile:** Wiebe van der Hoek made contributions to the area of epistemic and doxastic logic, rooted in modal logic. In particular, he worked on theories of individual and group knowledge, on the combination of knowledge and belief, and on theories of only knowing. He is now also working on dynamic epistemic logic, which studies the change of knowledge due to specific epistemic actions. He has also worked on using modal logic as a specification language for agent theories, and he published on agent programming languages. Recently he is active in the field of logics for cooperation, in particular to develop logics for agent systems that focus on the abilities of coalitions to achieve certain outcomes, given the information and abilities at hand.

**Main Publications:**

1. (with M. Pauly), "Modal Logic for Games and Information." To appear in Johan van Benthem, Patrick Blackburn, and Frank Wolter (eds), *Handbook of Modal Logic*, pp. 1181 - 1152, Elsevier, 2006
2. (with T. Ågotnes and M. Wooldridge), "On the Logic of Coalitional Games" In *Proceedings of the Fifth International Joint Conference on Autonomous Agents and Multi-agent Systems*, ACM Press, pp. 153–160, 2006.
3. (with A. Lomuscio, and M. Wooldridge), "On the Complexity of Practical ATL Model Checking." In *Proceedings of the Fifth International Joint Conference on Autonomous Agents and Multi-agent Systems*, ACM Press, pp. 201-208, 2006.
4. (with M. Wooldridge), "On obligations and normative ability: Towards a logical analysis of the social contract." *Journal of Applied Logic*, 4:3-4, pp. 396–420, 2005.
5. (with H. van Ditmarsch and B.P. Kooi), "Dynamic Epistemic Logic with Assignment" in *Proceedings of the Fourth International Joint Conference on Autonomous Agents and Multi-Agent Systems (AAMAS 05)*, ACM Inc, New York, vol. 1 pp 141–148, 2005.
6. (with M. Wooldridge), "On the Logic of Cooperation and Propositional Control." *Artificial Intelligence*, 64:1-2, pp. 81–119.
7. (with H. Aldewereld and J.-J.Ch. Meyer), "Rational Teams: Logical Aspects of Multi-Agent Systems." *Fundamenta Informaticae*, 63:2-3, pp. 159 – 183, 2004
8. "Knowledge, Rationality and Action," in *Proceedings of the Third International Joint Conference on Autonomous Agents and Multi Agent Systems*, (invited contribution) pp. 16–25, 2004.
9. (with W. Jamroga), "Agents that Know how to Play" *Fundamenta Informaticae*, 63:2-3, pp. 185 – 219, 2004.
10. (with P. Harrenstein, J-J. Ch. Meyer C. Witteveen), "A Modal Characterization of Nash Equilibrium" *Fundamenta Informaticae*, 57:2–4, pp. 281–321, 2003.
11. (with M. Wooldridge), "Cooperation, Knowledge, and Time: Alternating-time Temporal Epistemic Logic and its Applications" *Studia Logica*,75:1, pp. 125 – 157, 2003.
12. (with R. van Eijk, F. de Boer, & J.-J.Ch. Meyer), "A Fully Abstract Model for the Exchange of Information in Multi-Agent Systems," *Theoretical Computer Science*, 290:3 pp. 1753 – 1773, 2003.
13. (with R. van Eijk, F. de Boer, and J.-J.Ch. Meyer), "A Verification Framework for Agent Communication." *Autonomous Agents and Multi-Agent Systems*, 6, pp. 185–219, 2003.
14. (with E. Thijsse, "A General Approach to Multi-

Agent Minimal Knowledge: With Tools and Samples," *Studia Logica*, **72**,1, pp. 61–84, 2002.

15. (with P. Harrenstein, J.-J.Ch. Meyer and C. Witteveen), "*Boolean Games*" In Theoretical Aspects of Rationality and Knowledge (Proceedings of the eigth TARK conference), Morgan Kaufmann Publishers, pp. 287–298 (2001).

16. (with C. Witteveen), "Recovery of (non)monotonic theories." *Artificial Intelligence* (106)1 (1998) pp. 139–159.

17. (with J. Jaspars & E. Thijsse), "Honesty in Partial Modal Logic." *Studia Logica.*, **56**, 3 (1996) pp. 323-360.

18. (with J.-J.Ch. ), "*Epistemic Logic for Computer Science and Artificial Intelligence,*" Cambridge Tracts in Theoretical Computer Science 41, Cambridge University Press (1995) ISBN 0 521 46014.

19. (with M. de Rijke), "Counting Objects." *Journal of Logic and Computation*, **5**, 3 (1995) pp. 325-345.

20. "Systems for Knowledge and Beliefs." *Journal of Logic and Computation*, **3**, 2 (1993) pp. 173-195.

*Work in Progress*

21. A book on Dynamic Epistemic Logic (with Hans van Ditmarsch and Barteld Kooi).

22. A book on Reasoning about Cooperation (with Thomas Ågotnes and Mike Wooldridge.

**Service to the Profession:** General Chair, European Conference on Logics in AI (JELIA), 2006; Program Chair, European Workshop on Multi-Agent Systems (EUMAS), 2004; Editor, *Autonomous Agents and Multi-Agent Systems*, 2003- ; Founder and Editor-in-Chief, *Knowledge, Rationality and Action*, a subseries of *Synthese*, 2003-; General Chair, European Agent Systems Summer School (EASSS), 2001-2005; Organiser (with G. Bonanno), Conference on Logics and the Foundations of Game and Decision Theory (LOFT), 2000-; Editorial Board Member, *Studia Logica*, 1998-.

**Teaching:** Van der Hoek has taught many courses in logic, for students as diverse as from Computer Science, Cognitive AI, Mathematics, and Philosophy. He was a co-(supervisor) of 10 students, all in Computer Science.

**Vision Statement:** Logic is becoming more and more exciting, because logicians get engaged in more and more real life phenomena, including its dynamics and openness.

**Honours and Awards:** Fellow, *British Computer Society*, 2004-.

# VAN DER MEYDEN, Ron

**Specialties:** Logic in computer science, particularly applications of modal logic to distributed and multi-agent systems and computer security, database theory.

**Born:** 27 March 1960 in Amsterdam, Netherlands.

**Educated:** Rutgers University, PhD Computer Science, 1992, Sydney University, MA Mathematics, 1986, Sydney University, BA Mathematics, 1982.

**Dissertation:** *The complexity of querying indefinite information: defined relations, recursion and linear order*; supervisor, L. Thorne McCarty.

**Regular Academic or Research Appointments:** ASSOCIATE PROFESSOR, UNIVERSITY OF NEW SOUTH WALES, 1999–; PROGRAM LEADER, FORMAL METHODS PROGRAM, NATIONAL ICT AUSTRALIA, 2002–; Program Leader, Smart Personal Assistant Program, Smart Internet CRC, 2001; Lecturer, University of Technology, Sydney, 1995-1999.

**Visiting Academic or Research Appointments:** Visiting Scholar, New York University, 2006; Research Fellow, Weizmann Institute of Science, 1995; Researcher, NTT Basic Research Labs, Tokyo, 1993-1995; Visiting Fellow, University of New South Wales, 1992.

**Research Profile:** The unifying theme of Ron van der Meyden's research is the study of logical approaches to the problem of dealing with incomplete information as it arises in computational settings. His early work dealt with incomplete information databases, classifying the complexity of query answering in databases with second order forms of incompleteness representation such as facts stated in terms of recursively defined predicates.

More recent work has concentrated on applications of epistemic logic to distributed and multi-agent systems, with a particular focus on temporal aspects. Contributions in this area have included complete axiomatizations for combinations of epistemic logic and temporal logic, the development of algorithms and an implemented system MCK for model checking the logic of knowledge and time, and the application of this model checking theory to the study of problems in distributed computing and computer security. A further line of work, touching upon deontic logic, has dealt with logical representations for authorization in computer security settings.

**Main Publications:**

1. Synthesis of Distributed Systems from Knowledge-based Specifications, R. van der Meyden, T. Wilke, CONCUR 2005 - Concurrency Theory, 16th International Conference, San Francisco, Aug 2005, pp. 562-.

2. Symbolic Model Checking the Knowledge of the Dining Cryptographers, R. van der Meyden and K. Su, 17th IEEE Computer Security Foundations Workshop, Asilomar, June 2004, pp. 280-291.

3. Complete Axiomatizations for Reasoning about Knowledge and Branching Time, R. van der Meyden and K. Wong, *Studia Logica* Vol 75, pp. 93-123.

4. A logical reconstruction of SPKI, J.Y. Halpern and R. van der Meyden, Journal of Computer Security, Volume 11, Issue 4, 2003.

5. Modal Logics with a Hierarchy of Propositional Quantifiers, Kai Engelhardt, R. van der Meyden and Kaile Su, Advances in Modal Logic, Vol 4, World Scientific, 2003, pp. 9-30.

6. A Logic for Probability in Quantum Systems, R. van der Meyden and M. Patra, Proc. Computer Science Logic and 8th Kurt Gödel Colloquium, Vienna, Austria, 25th -30th August 2003, pp. 427-440.

7. Knowledge in Quantum Systems, R. van der Meyden and M. Patra, Theoretical Aspects of Knowledge and Rationality, Bloomington, Indiana, June 2003, pp 104 – 117.

8. Complete Axiomatizations for Reasoning about Knowledge and Time, J. Y. Halpern, R. van der Meyden and M. Y. Vardi, SIAM Journal on Computing, Vol 33, No. 3, 2004, pp. 674-703.

9. Knowledge in Multi-agent Systems: Initial Configurations and Broadcast, A.R. Lomuscio, R. van der Meyden, M. D. Ryan, ACM Transactions on Computational Logic Vol 1, No 2, October 2000.

10. A logic for SDSI's Linked Local Name Spaces, J. Y. Halpern and R. van der Meyden, *Journal of Computer Security*, vol. 9, number 1, 2, pp. 75 - 104, 2001.

11. Model Checking Knowledge and time in Systems with Perfect recall, R. van der Meyden and N. Shilov, Proc. Conf. on Foundations of Software Technology and Theoretical Computer Science, Madras, Dec 1999. Springer LNCS No. 1738, pp. 432-445.

12. Synthesis from Knowledge-based Specifications, R. van der Meyden and M. Y. Vardi, *Proc. CONCUR'98, 9th International Conf. on Concurrency Theory*, Springer LNCS No. 1466, Nice, Sept 98, pp. 34-49.

13. Logical Approaches to Incomplete Information: A Survey, R. van der Meyden, in Logics for Databases and Information Systems, J. Chomicki and G. Saake eds, Kluwer, 1998, pp. 309-358.

14. Finite State Implementations of Knowledge-based programs, R. van der Meyden, in *Proc. Conf. on Foundations of Software Technology and Theoretical Computer Science,* Hyderabad India, Dec 1996, Springer LNCS No. 1180, pp. 262-273.

15. Knowledge-based programs: on the Complexity of Perfect Recall in Finite Environments, R. van der Meyden, *Proceedings of the Conference on Theoretical Aspects of Reasoning about Knowledge*, Renesse, Netherlands, March 1996, pp. 31-50.

16. Common Knowledge and Update in Finite Environments, R. van der Meyden, *Information and Computation*, Vol 140, No. 2, Feb 1998, pp. 115-157.

17. The Complexity of Querying Indefinite Information about Linearly Ordered Domains, R. van der Meyden, *Journal of Computer and Systems Science* Vol 54, No. 1, Feb 1997, pp. 113-135.

18. The Dynamic Logic of Permission, R. van der Meyden, *Journal of Logic and Computation*, Vol 6, No. 3 pp. 465-479, 1996. A version of this paper appeared at the *IEEE Symposium on Logic in Computer Science,* Philadelphia, 1990.

19. Complexity Tailored Design: A new Design Methodology for Databases with Incomplete Information, T. Imielinksi, R. van der Meyden and K. Vadaparty, *Journal of Computer and Systems Science*, Vol 51, No. 3, Dec 1995, pp. 405-432.

20. Recursively Indefinite Databases, R. van der Meyden, *Theoretical Computer Science* 116 (1,2) (1993) pp. 151-194.

**Service to the Profession:** Program Chair, Conference on Theoretical Aspects of Reasoning about Knowledge, 2005; Research Committee Chair, School of Computer Science and Engineering, University of New South Wales; Contributor to Establishment, Smart Internet Cooperative Research Centre; Contributor to Establishment, National ICT Australia; Program Committee Member, Association for Computing Machinery; Program Committee Member, IEEE Computer Society.

**Vision Statement:** Logical Methods of reasoning about uncertainty in computational settings have only very recently started to come together with the rich mathematical traditions of reasoning about uncertainty such as probability theory. This is likely to be a very fertile domain, and we can expect to see significant advances in the coming years in areas such as probabilistic model checking and applications of logics of knowledge and probability in computer security. More speculatively, I am also enthusiastic about the potential for the development of formal methods for the verification of quantum computation and communication technologies

**Honours and Awards:** Best paper award, Advances in Modal Logic Conference, 2002; Barker Fellowship, 1995-1998.

# VAN FRAASSEN, Bastiaan Cornelis

**Specialties:** Free logic, modal logic, formal semantics, probabilistic semantics, vagueness, presuppositions, description theory, semantics of physical theories, quantum logic, logical aspects of probability.

**Born:** 5 April 1941 in Goes, Zeeland, the Netherlands.

**Educated:** University of Pittsburgh, PhD 1966. University of Pittsburgh, MA 1964. University of Alberta, BA (hon) 1963.

**Dissertation:** *Foundations of the Causal Theory of Time*, Supervisor: Adolf Grunbaum

**Web Address:** http://www.princeton.edu/~fraassen/

**Regular Academic or Research Appointments:** MCCOSH PROFESSOR OF PHILOSOPHY, PRINCETON UNIVERSITY, 1998–. Professor, Princeton University, Spring term, 1982–98; Professor, University of Southern California, 1976–81; Professor, University of Toronto, 1973–January 1982; Associate Professor, University of Toronto, 1969–73; Associate Professor, Yale University, 1968–69; Assistant Professor, Yale University, 1966–68.

**Visiting Academic or Research Appointments:** CREA/Ecole Polytechnique, Paris, Spring 2005, Directeur de Recherche, University of Paris 1 (Sorbonne), Spring 2005; Visiting Professor; Center for Advanced Study in the Behavioral Sciences (Stanford), 2003–2004; University of Hamburg, Carl Friedrich von Weizsaecker Lecturer 2003; University of Oxford, Trinity Term, 2001, Locke Lecturer; University of California at Santa Cruz, Visiting Fellow Spring 2000; Yale University, Fall 1999, Dwight H. Terry Lecturer; University of Louvain-la-Neuve, Fall 1999, Cardinal Mercier Chair; University of California at Davis, Oct–Dec 1990, Visiting Lecturer; Indiana University, Visiting Associate Professor, 1968–1969; West Virginia University, Visiting Assistant Professor, Spring Term 1966.

**Research Profile:** Van Fraassen has mainly published in philosophy of science but has contributed to logic under the following headings: free logic and free description theory (completeness and compactness proofs, for a semantics which does not specify referents outside the quantifier domain) modal and deontic logic (alternatives to possible world semantics) logic of probability judgments (including irreducible conditional probabilities) semantics for quantum logic and relevance logics (FDE and R-mingle) probabilistic semantics, subjective semantics, presuppositions, supervaluations, Liar-type paradoxes, theory of truth for non-bivalent languages.

**Main Publications:**

1. *Formal Semantics and Logic*. New York: Macmillan, 1971.
2. "Singular Terms, Truth-Value Gaps and Free Logic", *Journal of Philosophy*, **63** (1966) 481–495.
3. "The Completeness of Free Logic", *Zeitschrift fur Mathem. Logik und Grundlagen der Mathematik*, **12** (1966).
4. "Meaning Relations among Predicates", *Nous*, 1 (1967) 160–179.
5. "On Free Description Theory", (with Karel Lambert) *Zeitschrift fur Mathem. Logik und Grundlagen der Mathematik*, **13** (1967) 225–240.
6. "Presupposition, Implication, and Self-Reference", *Journal of Philosophy*, **65** (1968) 136–152.
7. "A Topological Proof of the Lowenheim-Skolem, Compactness, and Strong Completeness Theorems for Free Logic", *Zeitschrift fur Math. Logik und Grundlagen der Mathematik*, **14** (1968) 245–254.
8. "Compactness and Lowenheim–Skolem Proofs in Modal Logic", *Logique et Analyse*, **12** (1969) 167–178.
9. "Facts and Tautological Entailments", *Journal of Philosophy* **66** (1969) 477–487.
10. "Meaning Relations and Modalities", *Nous* **3** (1969) 155–167.
11. "On the Extension of Beth's Semantics of Physical Theories", *Philosophy of Science* **37** (1970) 325–334.
12. "Truth and Paradoxical Consequences" in R. Martin (ed.), *The Paradox of the Liar*. New Haven: Yale University Press, 1970; 13–23.
13. "The Logic of Conditional Obligation", *Journal Philosophical Logic* **1** (1972) 417–438.
14. "Values and the Heart's Command", *Journal of Philosophy* **70** (1973) 5–19.
15. "Extension, Intension, and Comprehension", in M. Munitz (ed.), *Logic and Ontology*, New York: University Press, 1973.
16. "Semantic Analysis of Quantum Logic", in C.A. Hooker (ed.), *Contemporary Research in the Foundations and Philosophy of Quantum Theory*. Dordrecht: Reidel, 1973; 80–113.
17. "Hidden Variables in Conditional Logic", *Theoria* **40** (1974) 176–190.
18. "The Only Necessity is Verbal Necessity", *Journal of Philosophy* **74** (1977) 71–85.
19. "Probabilistic Semantics Objectified, I", *Journal of Philosophical Logic*, **10** (1981) 371–394; Part II, **10** (1981) 495–510.
20. "Quantification as an Act of Mind", *Journal of Philosophical Logic* **11** (1982) 343–369.
21. "Identity in Intensional Logic: Subjective Semantics" *Versus* **44/45** (1986) 201–219.
22. "Fine-grained opinion, probability, and the logic of belief", *Journal of Philosophical Logic* **24** (1995) 349–377.
23. "Elgin on Lewis' Putnam's Paradox", *Journal of Philosophy* **94** (1997) 85–93.
24. "Modal interpretation of repeated measurement: Reply to Leeds and Healey", *Philosophy of Science* **64** (1997) 669–676.
25. "Conditionalization, A New Argument For", *Topoi* **18** (1999) 93–96.

*Work in Progress*

26. "The Day of the Dolphins: Puzzling Over Epis-

temic Partnership", in A. Irvine and K. Peacock (eds.), *Mistakes of Reason: Essays in Honour of John Woods*. University of Toronto Press, forthcoming fall 2005.

27. "Vague Expectation Loss", to be published in *Philosophical Studies*.

**Service to the Profession:** Conference Organizing Committees (recent), Princeton–Oxford Philosophy of Physics Conference, Princeton May 2002; Princeton–Rutgers Joint Conference on Gravity And The Quantum. May 1998; Princeton–Rutgers Joint Conference on the Relations between Epistemology and Philosophy of Science, May 1994; Philosophy of Quantum Mechanics Conference, Princeton University, May 1993; Symposium on the Interpretation of Quantum Theory, New York, April 1992; Symposium on Modern Physics, Joensuu, Finland, June 1990.

Editorial Positions (recent), Editorial Board, Contemporary Pragmatism, 2003–; Editorial Board, Logic, Epistemology And The Unity Of Science, 2003–; Editorial Board, Studies in the History and Philosophy of Science, 1998–; Editorial Advisory Board, Theoria (Spain) 1996–; Editorial Advisory Board, Studies in History and Philosophy of Modern Physics, 1995–; Editorial Board, Foundations of Science, 1993–; Editorial Board, Foundations of Physics Letters, 1988–; Board of Advisors, The Cambridge Dictionary of Philosophy, 1988–.

Professional Associations (recent), Advisory Committee to the Program Committee, Eastern Division, American Philosophical Association, 1996–9; Governing Board, Philosophy of Science Association, 1992–1994; President, Philosophy of Science Association, 1990–1992; Governing Board, Evert Willem Beth Stichting (The Netherlands) 1989–.

**Vision Statement:** The relevance of logic for van Fraassen is as handmaiden to philosophy (while recognizing its value in other quarters of course). Since standard truth and reference semantics has been naively incorporated into analytic metaphysics and epistemology, it seems to van Fraassen imperative to explore alternative semantics and pragmatics for familiar logics. The help derived from those familiar logics for the study of reasoning and rational opinion-management seems to have pretty well reached its limit and van Fraassen would advocate incorporating the formal study of probability into the logic curriculum for philosophers. Advanced courses should include a substantial introduction to algebraic methods.

**Honours and Awards:** University of Notre Dame, Doctor of Law, *honoris causa*, 2001; University of Lethbridge, Doctor of Letters, *honoris causa* 1999; Fellow, American Academy of Arts and Sciences, 1997–; Foreign Member, Royal Netherlands Academy of Arts and Sciences; 1995–. Howard T. Behrman Award for Distinguished Achievement in the Humanities, 1995 President, Philosophy of Science Association, 1992–94; Co-winner, Imre Lakatos Award for Books in Philosophy of Science, 1986; Distinguished Alumnus Award for Excellence, University of Pittsburgh, 1984; Co-winner, Franklin J. Matchette Prize for Philosophical Books, 1982; John Simon Guggenheim Memorial Fellowship, 1970–71.

## VELLEMAN, Daniel Jon

**Specialties:** Set theory, philosophy of mathematics.

**Born:** 10 August 1954 in Manhasset, New York, USA.

**Educated:** University of Wisconsin–Madison, PhD Mathematics, 1980; Dartmouth College, BA Mathematics, 1976.

**Dissertation:** *Morasses, Diamond, and Forcing*; supervisor, Kenneth Kunen.

**Regular Academic or Research Appointments:** PROFESSOR OF MATHEMATICS, AMHERST COLLEGE, 1992–; Associate Professor of Mathematics, Amherst College, 1987–1992; Assistant Professor of Mathematics, Amherst College, 1983–1987; Instructor, University of Texas at Austin, 1980–1983.

**Visiting Academic or Research Appointments:** Visiting Assistant Professor, Erindale College, University of Toronto, 1982.

**Research Profile:** Daniel Velleman has made contributions to the study of combinatorial principles related to the constructible universe, such as morasses, diamond sequences, and square sequences, and their relationship to forcing. He has also published papers on the philosophy of mathematics. He is the author of a textbook on how to write proofs, and also coauthor (with Alexander George) of a textbook on the philosophy of mathematics.

**Main Publications:**

1. Morasses, diamond, and forcing *Annals of Mathematical Logic* **23** (1982), pp. 199-281.
2. On a generalization of Jensen's $\Box_\kappa$, and strategic closure of partial orders, *Journal of Symbolic Logic* **48** (1983), pp. 1046-1052.
3. Simplified morasses, *Journal of Symbolic Logic* **49** (1984), pp. 257-271.
4. Simplified morasses with linear limits, *Journal of Symbolic Logic* **49** (1984), 1001-1021.

5. $\omega$−Morasses, and a weak form of Martin's axiom provable in ZFC, *Transactions of the American Mathematical Society* **285** (1984), pp. 617-627.
6. Souslin trees constructed from morasses, in *Proceedings of Boulder set theory conference, Contemporary Mathematics* **31** (Providence, American Mathematical Society, 1984), pp. 219-241.
7. On a combinatorial principle of Hajnal and Komjáth, *Journal of Symbolic Logic* **51** (1986), pp. 1056-1060.
8. Simplified gap-2 morasses, *Annals of Pure and Applied Logic* **34** (1987), pp. 171-208.
9. Gap-2 morasses of height $\omega$, *Journal of Symbolic Logic* **52** (1987), pp. 928-938.
10. Partitioning pairs of countable sets of ordinals, *Journal of Symbolic Logic* **55** (1990), pp. 1019-1021.
11. $\omega_3\omega_1 \rightarrow (\omega_3\omega_1, 3)^2$ Requires an inaccessible (with Lee Stanley and Charles Morgan), *Proceedings of the American Mathematical Society* **111** (1991), pp. 1105-1118.
12. On a topological construction of Juhasz and Shelah, *Journal of Symbolic Logic* **57** (1992), pp. 166-171.
13. Constructivism liberalized, *Philosophical Review* **102** (1993), pp. 59-84.
14. *How To Prove It: A Structured Approach*, Cambridge University Press, 1994.
15. Fermat's last theorem and Hilbert's program, *Mathematical Intelligencer* vol. 19 no. 1 (Winter 1997), pp. 64-67.
16. Two conceptions of natural number (with Alexander George), in *Truth in Mathematics*, Garth Dales and Gianluigi Oliveri (eds.), Oxford University Press, 1998, pp. 311–327.
17. Review of *The Principles of Mathematics Revisited*, by Jaakko Hintikka, *Mind* **108** (1999), pp. 170–179.
18. The mean value theorem in second order arithmetic (with Christopher Hardin), *Journal of Symbolic Logic* **66** (2001), pp. 1353–1358.
19. *Philosophies of Mathematics* (with Alexander George), Blackwell, 2002.
20. Variable declarations in natural deduction. *Annals of Pure and Applied Logic*, 144 (2006) pp. 133-146.

**Service to the Profession:** American Mathematical Monthly, Editor-elect 2006, Editor 2007–2011. Committee on Logic Education, Association for Symbolic Logic, 2002–; Editor, Dolciani Mathematical Expositions, 1999–2004; Member of Editorial Board, Dolciani Mathematical Expositions, 1998–1999; Member of Editorial Board, American Mathematical Monthly, 1997–2005; Chair, Department of Mathematics and Computer Science, Amherst College, 1987–1988, 1993–1996.

**Honours and Awards:** Carl B. Allendoerfer Award, 1996; Lester R. Ford Award, 1994; Honorable mention, Putnam Exam, 1975.

# VEROFF, Robert

**Specialties:** Automated Reasoning and Logic

**Born:**

**Educated:** A.B. Mathematics 1975 Cornell University, M.A. Applied Mathematics 1977 University of California at San Diego, Ph.D. Computer Science 1980 Northwestern University Adviser Lawrence Henschen

**Regular Academic or Research Appointments:** PROFESSOR UNIVERSITY OF NEW MEXICO 1996–; ASSOCIATE CHAIR, DEPARTMENT OF COMPUTER SCIENCE, UNIVERSITY OF NEW MEXICO 2000-PRESENT.

**Research Profile:** Developed (with his colleague L. Wos) the concept of linked inference, extended it to include equality, and used it to find shortest proofs for formulas in such areas as intuitionist logic; designed and implemented a weighting package for the environmental theorem prover AURA and developed a method for automated case analysis; introduced the idea for automatic unit deletion (a special case of automatic transformations in the inference process). Of particular note is his development of the concept of proof sketches and the hints strategy. Proof sketches have been used to answer challenging problems and open questions in various logics; the hints strategy has been used to find a shorter proof for a problem in sentential calculus and to assist in the proof that the axiom XCB is a shortest single axiom in equivalential calculus.

**Main Publications:**

Books
1. P. Helman and R. Veroff, Intermediate Problem Solving and Data Structures: Walls and Mirrors, Menlo Park, California, Benjamin Cummings Publishing Co., 1986; Modula-2 Edition 1988
2. F. Carrano, P. Helman, and R. Veroff, Data Abstraction and Problem Solving with C++: Walls and Mirrors (2nd ed.), Reading, Massachusetts, Addison-Wesley, 1998
3. R. Veroff, ed., Automated Reasoning and Its Applications: Essays in Honor of Larry Wos, Cambridge, Mass., MIT Press, 1997
Articles
4. L. Wos, S. Winker, R. Veroff, B. Smith, and L. Henschen, Questions concerning possible shortest single axioms in equivalential calculus: An application of automated theorem proving to infinite domains, Notre Dame J. Formal Logic, 24(2) 205-223, 1983
5. L. Wos, S. Winker, R. Veroff, B. Smith, and L. Henschen, A new use of an automated reasoning assistant: Open questions in equivalential calculus and the study of infinite domains, J. Artificial Intelligence, 22: 303-356, 1984

6. L. Wos, R. Veroff, B. Smith, and W. McCune, The linked inference principle, II: The users viewpoint, Proceedings of the Seventh International Conference on Automated Deduction, vol. 170, Lecture Notes in Computer Science, edited by R. E. Shostak, Springer-Verlag, New York, 1984, pp. 316-332

7. L. Wos and R. Veroff, Resolution, binary: Its nature, history, and impact on the use of computers, Encyclopedia of Artificial Intelligence, 2nd ed., John Wiley and Sons, pp. 1341-1353, 1992

8. R. Veroff and L. Wos, The linked inference principle, I: The formal treatment, J. Automated Reasoning 8, no. 2 (1992) 213-274

9. L. Wos and R. Veroff, Logical basis for the automation of reasoning: Case studies, in Handbook of Logic in Artificial Intelligence and Logic Programming, vol. 2, edited by D. M. Gabbay, C. J. Hogger, and J. A. Robinson, Oxford University Press, Oxford, 1994, pp. 1-40

10. R. Veroff, Using hints to increase the effectiveness of an automated reasoning program: Case studies, J. Automated Reasoning 16, no. 3 (1996) 223-239

11. R. Veroff, Finding shortest proofs: An application of linked inference rules, J. Automated Reasoning, 27 (2001) 123-139

12. R. Veroff, Solving open questions and other challenge problems using proof sketches, J. Automated Reasoning 27, no. 2 (2001) 157-176

13. P. Helman and R. Veroff, The application of automated reasoning to formal models of combinatorial optimization, Applied Mathematics and Computation, 120, no. 1-3 (2001) 175-194

14. W. McCune, R. Veroff, B. Fitelson, K. Harris, A. Feist, and L. Wos, Short single axioms for Boolean algebra, J. Automated Reasoning, to appear.

15. W. McCune, R. Padmadabhan, and R. Veroff, Yet another single law for lattices, Algebra Universalis, to appear.

**Service to the Profession:** Editor, Journal of Automated Reasoning

**Honours and Awards:** Honors and awards include Students' Faculty Recognition Award, University of New Mexico, 1998; Presidential Lectureship, University of New Mexico 1986-1988

# VELTMAN, Frank

**Specialties:** Logical analysis of natural language.

**Born:** 5 April 1949 in Eindhoven, The Netherlands

**Educated:** Universiteit van Amsterdam BSc Physics, 1971; Universiteit van Utrecht, MA Philosophy and MSc Mathematics, 1974; Universiteit van Amsterdam, PhD Philosophy, 1985. *Dissertation: Logics for Conditionals*; supervisors, Johan van Benthem, Hans Kamp.

**Regular Academic or Research Appointments:** PROFESSOR, LOGIC AND COGNITION, UNIVERSITEIT VAN AMSTERDAM, 2001-; Associate Professor, Philosophical Logic, Universiteit van Amsterdam, 1986–2001; Assistant Professor, Philosophical Logic, Universiteit van Amsterdam, 1978–1986; Assistant Professor, Philosophy, Erasmus Universiteit Rotterdam, 1974–1978.

**Visiting Academic or Research Appointments:** Stanford University, Department of Philosophy, 1998; Edinburgh University, Institute for Cognitive Science, 1997; Universität Tübingen, Forschungsstelle für natürlich-sprachliche Systeme, 1986, 1987.

**Research Profile:** Veltman's major research interest is in the logical analysis of natural language, with a particular interest in matters of mood and modality. He has published papers on conditionals, epistemic modalities, default reasoning, and vagueness.

Every now and then he makes an excursion outside his main field, like when he developed a semantics for the notion of relative interpretability.

**Main Publications:**

1. Data semantics and the pragmatics of indicative conditionals. In: E. Traugott et al. (eds.), *On conditionals*, Cambridge: Cambridge University Press, 1986, 147-167.

2. Provability logics for relative interpretability. In: P. Petkov (ed.), *Mathematical logic. Proceedings of the Heyting '88 Summer School*, New York: Plenum Press, 1990, 31-42, (with Dick de Jongh).

3. Coreference and Modality. In: S. Lappin (ed.), *The Handbook of Contemporary Semantic Theory*, Blackwell, Oxford, 1996, 179-213, (with Jeroen Groenendijk and Martin Stokhof).

4. Defaults in update semantics. *Journal of Philosophical Logic*, Vol. 25, 1996, 221-261.

5. Making counterfactual assumptions. *Journal of Semantics*, Vol. 22, 2005, 159-180.

More information (handouts etc.) on these topics can be found at `http://staff.science.uva.nl/~veltman/`

**Service to the Profession:** Member of Editorial Board *Journal of Philosophical Logic, Argumentation, Semantics & Pragmatics*; Scientific Director, *Institute for Logic, Language and Computation* (ILLC), Universiteit van Amsterdam, 2003-.

**Teaching:** So far Veltman has supervised forty two Master students and ten PhD students.

**Vision Statement:** "Logic florishes best when it interferes with other people's business – philosophy (Frege), mathematics (Gödel), computer science (Turing), linguistics (Montague). What will be next?"

**Honours and Awards:** The paper 'Defaults in Update Semantics' was selected as one of the ten best papers published in philosophy in 1996, and reprinted in *The Philosopher's Annual*, Vol. 19.

# VISSER, Albert

**Specialties:** Logics for provability & interpretability, interpretations, intuitionistic logic and arithmetic, dynamic semantics, philosophy of language.

**Born:** 19 December 1950 in Zwijndrecht, Zuid Holland, The Netherlands.

**Educated:** Twente University, Bachelor in Applied Mathematics, 1974; Utrecht University, MSc, Mathematics, 1976; Utrecht University, PhD, Logic, 1981.

**Dissertation:** *Aspects of Diagonalization & Provability*; supervisor Dirk van Dalen.

**Regular Academic or Research Appointments:** PROFESSOR OF LOGIC, PHILOSOPHY OF MATHEMATICS & EPISTEMOLOGY, UTRECHT UNIVERSITY, 1998–; Assistant professor in Logic, Utrecht University, 1981–1998.

**Visiting Academic or Research Appointments:** Assistant Professor, Stanford, 1982–1983.

**Research Profile:** Albert Visser considers himself to be both a logician and a philosopher. Like his teacher Craig Smoryński, he would have liked to be a gallant scientific amateur, but to his surprise, after some time, he found that he had become a specialist in several areas. We can roughly divide his work into two projects.

The first project is logics for arithmetical theories. He works on provability logic, where he was the first to notice that Solovay's Theorem could be viewed as a theorem on embedding algebras. The spectacular descendant of this idea is Shavrukov's Theorem on the embeddings of Magari Algebras in the Magari Algebra of Peano Arithmetic. Albert Visser worked on the question concerning the scope of Solovay's Theorem, but the great open questions concerning the provability logics of Buss' $S_2^1$ and of $I\Delta_0 + \Omega_1$ remain, alas, open to the present day.

Around 1988, he took up some ideas of Viteslav Švejdar and Franco Montagna, and started a project on Interpretability Logic. In this project, he closely collaborated with Dick de Jongh and Frank Veltman. Frank Veltman provided a semantics for interpretability logic. Albert Visser, subsequently, characterized the logic of all finitely axiomatized sequential extensions of $I\Delta_0 + \mathsf{supexp}$. Volodya Shavrukov and Alessandro Berarducci independently characterized the logic of all essentially reflexive theories. The question of the interpretability logic of 'all reasonable theories' remains open. During the last few years, Joost Joosten and Evan Goris discovered new principles of this logic —amazingly using modal semantics to find arithmetical principles.

Another strand of the logics of arithmetical theories project, is the study of logics for constructive theories. Albert Visser worked on the great open problem of the provability logic of Heyting's Arithmetic, HA. Some spin-off's of this project are the characterization of the closed fragment of HA and the characterization of the propositional admissible rules of HA. Dick de Jongh and Albert Visser adapted Shavrukov's Theorem on *embedding Magari Algebras* to a theorem concerning *embedding Heyting Algebras into the Lindenbaum Heyting algebra of* HA. Of the possible modal semantics of the provability logic of HA we know a lot, thanks to Rosalie Iemhoff's work which resulted in her PhD thesis.

Albert Visser is working on closed fragments of the provability logics of extensions of HA and, in particular, of Markov's Arithmetic MA. A great open problem in this area is propositional logics for realizability. Albert Visser hopes to contribute to this problem.

A quite different project is the study of dynamic semantics. A major idea here is that the distinction between semantics and syntax gets blurred. E.g., there is an analogy between an ordinary opening bracket and the existential quantifier dynamically construed. Visser worked with Kees Vermeulen on the project of constructing languages with a flat incremental syntax and semantics. The development of these ideas benefited a lot of the comments and insights of Marcus Kracht.

Another strand of the dynamic project is algebraization. Albert Visser contributed together with Marco Hollenberg to the study of 'dynamic algebras'. Regrettably, this project never caught on, but who knows what the future will bring.

Finally there is the idea to treat dynamic semantics in a category theoretical setting. This is the idea Albert Visser is currently working on.

Connected with the dynamical project, Albert Visser is working on the philosophy of syntax and semantics.

**Main Publications:**

1. "Fictionality and the Logic of Relations", *The Southern Journal of Philosophy*, 7 (1969), 51–63.
2. *The Logic of Fiction: A Philosophical Sounding of*

*Deviant Logic*, The Hague and Paris: Mouton and Co., 1974.

3. Numerations, λ–calculus & arithmetic, in: J.P. Seldin and J.R. Hindly (eds.), *To H.B. Curry: Essays on Combinatory Logic, Lambda Calculus and Formalism*, Academic Press, London, 1980, 259–284.

4. A propositional logic with explicit fixed points, *Studia Logica*, vol. 40, 1981, 155–175.

5. On the Completeness Principle, *Annals of Mathematical Logic*, vol. 22, 1982, 263–295.

6. Four valued semantics and the Liar, *Journal of Philosophical Logic*, vol. 13, 1984, 181–212.

7. The provability logics of recursively enumerable theories extending Peano Arithmetic at arbitrary theories extending Peano Arithmetic, *Journal of Philosophical Logic*, vol. 13, 1984, 97–113.

8. Peano's Smart Children: A Provability Logical Study of Systems with Built–in Consistency, *Notre Dame Journal of Formal Logic*, vol 30, 1989, 161–196.

9. Semantics and the Liar Paradox, in: D. Gabbay and F. Guenthner (eds.), *Handbook of Philosophical Logic, Volume IV, Topics in the Philosophy of Language*. Dordrecht, Reidel, 1989, 617–706. (A slightly improved version can be found in: D. Gabbay and F. Guenthner (eds.): *Handbook of Philosophical Logic 11*, 2nd edition. Springer, Heidelberg, 2004, 149-240.)

10. Interpretability Logic, in: P.P. Petkov (ed.), *Mathematical Logic*, Plenum Press, New York, 1990, 175–208.

11. An Inside View of *EXP*, *Journal of Symbolic Logic*, vol. 57, 1992, 131–165.

12. The unprovability of small inconsistency, *Archive for Mathematical Logic*, vol. 32, 1993, 275–298.

13. With Johan van Benthem, Dick de Jongh and Gerard Renardel: *NNIL*, a Study in Intuitionistic Propositional Logic, in: A. Ponse, M. De Rijke, Y. Venema (eds.), *Modal Logic and Process Algebra, a Bisimulation Perspective*, 1995, CSLI Publications, Lecture Notes, no. 53, 289–326.

14. Uniform Interpolation and layered Bisimulation, in: Petr Hájek (ed.), *Gödel '96, Logical Foundations of Mathematics, Computer Science and Physics —Kurt Gödel's Legacy*, 1996, Lecture Notes in Logic, Springer, Berlin, 139–164.

15. With Kees Vermeulen: Dynamic bracketing and discourse representation, *Notre Dame Journal of Formal Logic*, vol. 37, 1996, 321–365.

16. The Donkey and the Monoid. Dynamic semantics with control elements, *Journal of Logic Language and Information*, volume 11, number 1, 2002, 107–131.

17. With Giovanna d'Agostino, Finality regained: a coalgebraic study of Scott-sets and multisets, *Archive for Mathematical Logic*, volume 41, 2002, 267–298.

18. Substitutions of $\Sigma_1^0$-sentences: explorations between intuitionistic propositional logic and intuitionistic arithmetic, *Annals of Pure and Applied Logic*, special issue for the commemorative symposium, dedicated to Anne Troelstra, guest-editors Jaap van Oosten & Harold Schellinx, volume 114, 2002, 227–271.

19. Faith & falsity, in: *Annals of Pure and Applied Logic 131*, 2005, 103-131.

20. With Maartje de Jonge, No Escape from Vardanyan's Theorem, *Archive for Mathematical Logic*, vol. 45, 5, 2006, 539–554.

**Service to the Profession:** Member of the Board of the Werkgemeenschap Logica en Taalfilosofie, [1986–1988]; Member of the board of the Nederlandse Vereniging voor Logica, [1986–1989]; Chair of the board of the Nederlandse Vereniging voor Logica, [1989–1995]; Member of the board of the AIO–netwerk TLI (a committee responsible for the organization of PhD studies in Logic in the Netherlands), [1989–1991];

Chair of the board of the AIO–netwerk TLI, [1991–1994]; Member of the ad hoc Committee for Logic and Education of the Association for Symbolic Logic, [1991–1992]; Member of the editorial board of Editor of the *Notre Dame Journal of Formal Logic*, [1991–]; Editor of *the Journal of Symbolic Logic*, [1994–1998]; Member of the ESSLLI Standing Committee, [1994–1996]; Chair of the ESSLI Standing Comittee, [1996–1998]; Member of the ASL Committee on Logic in Europe, [1999–2003]; Scientific director of the OZSL (Dutch Research School in Logic), [2002–]; Editor of the *Journal of Philosophical Logic*, [2002–2007]; Member of the ASL Executive Committee, [2004–2006]; Member of the ASL Committee on Prizes and Awards, [2005–]; Member of the NWO Programme Committee for Cognition, [2005–]; Editor of the *Review of Symbolic Logic*, [2008–].

**Teaching:** Albert Visser was director of the AI Educational Program of Utrecht University from 1992 to 2006. He has taught on a wide spectrum of subjects: a.o., introduction to logic, arithmetic, recursion theory, dynamic semantics, philosophy of language. He was co-advisor/advisor of a.o. Rineke Verbrugge, Volodya Shavrukov, Domenico Zambella, Kees Vermeulen, Marco Hollenberg, Rosalie Iemhoff and Joost Joosten.

**Vision Statement:** Albert Visser believes in a healthy balance between pure logic —i.e., model theory, proof theory, recursion theory and set theory— and applied logic. He thinks that we are only at the beginning of the application of logic in linguistics. He also believes that the results of the applied projects, e.g. concerning computer science and linguistics will supply insights for the more traditional fields even concerning the most elementary parts of the tradition.

**Honours and Awards:** Member of the Royal Dutch Academy, KNAW, [2005–] .

# VOPĚNKA, Petr

**Specialties:** Mathematical logic, axiomatic set theory, foundations of mathematics, philosophy of mathematics, history of mathematics.

**Born:** 16 May 1935 in Prague, Czechoslovakia.

**Educated:** Charles University, Faculty of Mathematics and Physics, CSc (equivalent of PhD), 1962; Charles University, Faculty of Mathematics and Physics RNDr, 1958.

**Dissertation:** *Non-standard models of set theory.*

**Regular Academic or Research Appointments:** PROFESSOR, MATHEMATICS, EMERITUS, CHARLES UNIVERSITY, 2000–; PROFESSOR, MATHEMATICS, UNIVERSITY OF J. E. PURKYNĚ, 1995–; PROFESSOR, MATHEMATICS, WEST BOHEMIAN UNIVERSITY, 2004–; Professor, Mathematics, Charles University, 1990-2000; Docent (equivalent of Associate Professor), Mathematics, Charles University, 1965-1989; Assistant Professor, Mathematics, Charles University, 1958-1964.

**Visiting Academic or Research Appointments:** Political situation in former Czechoslovakia made visiting appointments impossible. The only abroad travels allowed were the short term visits. Lecturer, University Bordeaux, 2004; Lecturer, Université de Paris I, 2002; Lecturer, Logic Colloquium Wroclaw 1977; Invited Lecturer, International Congress of Mathematicians, Nice, 1970; Invited Lecturer, Third International Congress for Logic, Methodology and Philosophy of Science, Amsterdam, 1967; Lecturer, Moscow State University, 1966; Lecturer, University of Berkeley, 1963; Periodic Lecturer, Warsaw and Wroclaw, 1963-1970.

**Research Profile:** In 1960, Petr Vopěnka became acquainted with Gödel's paper "The consistency of the axiom of choice and the generalized continuum hypothesis." He realized that the main problem of axiomatic set theory in those days was to break the cage, where it was barred by the axiom V=L. Two ways of solution were available. 1) To add other sets to a standard countable model so that no new ordinals are created. This was the way successfully used by P. J. Cohen. 2) To add other sets in a similar way to a suitable submodel of a nonstandard model of set theory. This was Vopěnka's approach. In 1961, he constructed a nonstandard model of Gödel-Bernays theory by the method of ultraproduct [1], and in [2,3,4,5] he investigated its submodels. He tried then to fill in these models in a desired way. He was not lucky in the choice of these submodels. It was Cohen's remarkable result which inspired him for a correct choice of submodels of an ultraproduct. In 1964, he built up a general theory of so called ∇- models and presented it at A. Mostowski's seminar in Warsaw in December. The papers [6–9] are devoted to this theory. Vopěnka's models differ from the later so-called Boolean valued models in one respect only: The values of formulas are open sets in topological spaces. In fact, the models are the same [12].

In the period 1967-1972, Vopěnka's investigations concentrated to the theory, which he called the theory of semisets. If the axiom of constructibility does not hold, then these subsets of the class L, which are not members of L, are semisets. He founded axiomatic theory of semisets and then he looked for necessary and sufficient conditions under which these axioms describe just all subsets of the class L in some set theory which does not satisfy constructibility [15]. Byproducts of this research were various statements on large cardinals [10,13], formulation of a principle known today as Vopěnka's principle, new proof of Gödel's theorem [11], etc.

The study of Bolzano's works together with a deep knowledge of the contemporary set theory enabled Vopěnka to give a solution of a problem of truth in the Cantor set theory. It appeared in [14] and as an appendix in a Russian translation of [16]. However, Vopěnka finished his activities in the classical set theory in this period. Since 1973, inspired by Husserl, he is interested in the phenomenology of infinity. From here, the origins of an alternative set theory came. Alternative set theory studies and elaborates his phenomenological interpretation of infinity, and the corresponding mathematics [16,17]. Closely tied is his philosophical essay [18].

Since 1980, Vopěnka works also in history of mathematics with emphasis on the origins and development of mathematical ideas, notions and concepts. His findings are subject of two massive volumes [19] (geometry) and [20] (set theory).

**Main Publications:**

1. *A method of construction a non-standard model in Bernays-Gödel axiomatic set theory (in Russian)*, Dokl. Akad. Nauk SSSR (1962), 11–12
2. *Models of set theory (in Russian)*, Z. Math. Logik Grundlagen Math. 8 (1962), 281–292
3. *Construction of models for set theory by the method of ultraproduct (in Russian)*, Z. Math. Logik Grundlagen Math. 8 (1962), 293–304
4. *A construction of models of set theory by the method of a spectrum (in Russian)*, Z. Math. Logik Grundlagen Math. 9 (1963), 149–160
5. *Submodels of models of set theory (in Russian)*, Z.

Math. Logik Grundlagen Math. 10 (1964), 163–172

6. *The independence of Continuum Hypothesis*, Amer. Math. Soc. Transl. Ser. 2, 57(2)(1966), 85–112

7. *The limits of sheaves and applications on constructions of models*, Bull. Acad. Polon. Sci. Sér. Sci. Math. Astronom. Phys. 13 (1965), 189 - 192

8. *On $\nabla$-model of set theory*, Bull. Acad. Polon. Sci. Sér. Sci. Math. Astronom. Phys. 13 (1965), 267–272

9. *Properties of $\nabla$-model*, Bull. Acad. Polon. Sci. Sér. Sci. Math. Astronom. Phys. 13 (1965), 441–444

10. *The first measurable cardinal and the generalized continuum hypothesis*, Comment. Math. Univ. Carolinae 6 (1965), 367–370

11. *A new proof of the Gödel's result on nonprovability of consistency*, Bull. Acad. Polon. Sci. Sér. Sci. Math. Astronom. Phys. 14 (1966), 111–116

12. *General theory of $\nabla$-models*, Comment. Math. Univ. Carolinae 8 (1967), 145–170

13. *On strongly measurable cardinals* (jointly with K. Hrbáček), Bull. Acad. Polon. Sci. Sér. Sci. Math. Astronom. Phys. 14 (1966), 587–591

14. *Infinity, sets and possibility in the sense of B. Bolzano* (in Polish), Wiadomosci Matematyczne XXVI (1985), 171–204

15. The Theory of Semisets (jointly with P. Hájek), Academia Praha, North-Holland Publ. Co., Amsterdam, London, 1972, 332pp.

16. Mathematics in the Alternative Set Theory, Teubner Texte, Leipzig, 1979, 120pp. (Russian translation with appendices: Izd. Mir, Moskva, 1983, 148pp.)

17. An Introduction to Mathematics in the Alternative Set Theory (in Slovak), Alfa, Bratislava, 1989, 443pp. (Russian translation: Izd. Inst. Math., Novosibirsk 2004, 611pp.)

18. Meditations on the Foundations of Science (in Czech), Práh, Praha, 2001, 261pp.

19. The Cornerstone of European Education and Power (in Czech), Práh, Praha, 2000, 918pp.

20. Narration on the Beauty of the Neobaroque Mathematics (in Czech), Práh, Praha, 2004, 820pp.

**Service to the Profession:** Editorial Board Member, Comment. Math. Univ. Carolinae 1990-; Editorial Board Member, Zeitschrift J. Math. Logik und Grundlagen der Mathematik, 1962-1990; Honorary Chairman, Logic Colloquium, 1998; Minister of Education, Government of Czech Republic, 1990-1992; Vice-chancellor, Charles University, 1990.

**Teaching:** Petr Vopěnka was the leader of Czechoslovak logicians since the death of L. S. Rieger (1963). He founded a seminar on foundations of mathematics, which was his most influential teaching activity. Later it was called "Prague seminar on set theory."

The first group (1962–1972) consisted from B. Balcar, L. Bukovský, P. Hájek, K. Hrbáček, T. Jech, K. Příkrý, A. Sochor and P. Štěpánek. The second seminar, Seminar on alternative set theory, worked since 1973 till 1989 and its members were K. Čuda, J. Chudáček, J. Mlček, A. Vencovská and P. Zlatoš.

Moreover, the difficult political situation forced Vopěnka to open Seminar on Philosophy, since he felt necessary to offer a meeting place at the University for young philosophers who were fired from their jobs. This seminar worked in the period 1981–1989.

**Vision Statement:** "The contemporary state of mathematics and mathematical logic, based on the Cantor set theory, requires a coming back to the essential phenomenon, upon which it is built, the phenomenon of infinity."

**Honours and Awards:** "Medal of Merit" State Decoration, Awarded by Czech President Václav Havel, 28 October 1998; Dagmar and Václav Havel Foundation VIZE 97 Prize, 5 October 2004.

# W

## WAGON, Stan

Set theory and its connection with measure theory.

**Born:** 9 July 1951 in Montreal, Quebec, Canada.

**Educated:** Dartmouth College, Hanover, New Hampshire, PhD; McGill University, BSc.

**Dissertation:** *Decompositions of Saturated Ideals*; supervisor, James Baumgartner.

**Regular Academic or Research Appointments:** MACALESTER COLLEGE, 1990-; Smith College, 1975-1990.

**Main Publications:**

1. The Banach-Tarski Paradox, Cambridge University Press, New York, 1985.
2. Large free groups of isometries and their geometrical uses, with J. Mycielski, l'Enseignement Mathématique 30 (1984) 247-267.
3. A hyperbolic interpretation of the Banach–Tarski Paradox, The Mathematica Journal 3:4 (1993) 58-61.

**Honours and Awards:** Chauvenet Award, MAA 2002; Trever Evans Award for paper in Math Horizons, MAA, 1999; Ford Award for paper in American Math. Monthly, MAA, 1987.

## WANSING, Heinrich Theodor

**Specialties:** Philosophical logic, modal logic, non-classical logic.

**Born:** 20 March, 1963, in Ahaus, Westphalia, Germany.

**Educated:** FU Berlin MA (Philosophy) 1988, FU Berlin DPhil (Philosophy) 1992, University of Leipzig DPhil habil (Logic and Analytical Philosophy) 1997.

**Dissertation:** *The Logic of Information Structures*; supervisor David Pearce.

**Regular Academic or Research Appointments:** PROFESSOR OF PHILOSOPHY OF SCIENCE AND LOGIC, TU DRESDEN, INSTITUTE OF PHILOSOPHY, 1998–. University of Leipzig, Institute of Logic and Philosophy of Science, 1993–1998; University of Hamburg, Graduate School in Cognitive Science, 1993; University of Amsterdam, Institute of Logic, Language and Computation, 1992; FU Berlin, Institute of Philosophy, 1991.

**Visiting Academic or Research Appointments:** Visiting Scholar, Institute of Logic, Language and Computation, Amsterdam, 1988–89; Visiting Scholar, CSLI Stanford February/March 1994.

**Research Profile:** Heinrich Wansing is well-known for his work on non-classical logics, their philosophical motivation, their proof theory and their relational semantics. He has made contributions to epistemic logic, categorial grammar, substructural logic, sequent-style proof systems for modal logics, the idea of a proof-theoretic semantics, and logics with strong, constructive negation. In particular, he was the first to investigate substructural subsystems of constructive logics with strong negation and to develop a formulas-as-types notion of construction for such logics. Moreover, he extended the proof-theoretic semantics for intuitionistic logic and constructive logic with strong negation to substructural subsystems of these logics and, more recently, to modal logics with strong negation. In the 1990ies he took up Nuel Belnap's display calculus and made several contributions to further developing the modal display calculus. More recently, he suggested introducing the strong negation of programs into concurrent propositional dynamic logic and noted that connexive logic can be viewed as a branch of logics with constructive negation. Wansing has also written papers on applying the modal logic of agency to the problem of doxastic voluntarism in epistemology in order to obtain a logic of belief formation. In addition he is involved in a research project on generalized truth values and multilattices resulting from ordering relations on sets of generalized truth values.

**Main Publications:**

1. *The Logic of Information Structures*. In ???? (eds.), Springer Lecture Notes in AI 681. Springer-Verlag, Berlin, 1993.
2. *Displaying Modal Logic*. Kluwer Academic Publishers, Dordrecht, 1998.
3. "A General Possible Worlds Framework for Reasoning about Knowledge and Belief". *Studia Logica* 49(1990): 523–539, 50(1991): 359.
4. "Informational Interpretation of Substructural Propositional Logics". *Journal of Logic, Language and Information* 2(1993): 285–308.
5. "Functional Completeness for Subsystems of Intuitionistic Propositional Logic". *Journal of Philosophical Logic*, 22(1993): 303–321.
6. "Sequent Calculi for Normal Modal Propositional Logics". *Journal of Logic and Computation*, 4(1994): 125–142.

7. "Tarskian Structured Consequence Relations and Functional Completeness". *Mathematical Logic Quarterly*, 41(1995): 73–92.
8. "Translation of Hypersequents into Display Sequents". *Logic Journal of the IGPL*, 6(1998): 719–733.
9. "Predicate Logics on Display". *Studia Logica*, 62(1999): 49–75.
10. "Negation". In L. Goble (ed.), *The Blackwell Guide to Philosophical Logic*. Basil Blackwell Publishers, Cambridge MA, 2001, pp. 415–436.
11. "Diamonds are a Philosopher's Best Friends. The Knowability Paradox and Modal Epistemic Relevance Logic". *Journal of Philosophical Logic*, 31(2002): 591–612.
12. "Sequent Systems for Modal Logics". In D. Gabbay and F. Guenthner (eds.), *Handbook of Philosophical Logic, vol. 8*. Kluwer Academic Publishers, Dordrecht, 2002, pp. 61–145.
13. "Consequence, Counterparts and Substitution", with Sebastian Bauer. *The Monist*, 85(2002): 483–497.
14. "Constructive Predicate Logic and Constructive Modal Logic. Formal Duality versus Semantical Duality", with S.P. Odintsov. In V. Hendricks et al. (eds.), *First-Order Logic Revisited*. Logos Verlag, Berlin, 2004, pp. 269–286.
15. "Action-theoretic aspects of theory choice". In S. Rahman et al. (eds.), *Logic, Epistemology and the Unity of Science*. Kluwer Academic Publishers, Dordrecht, 2004, pp. 419–435.
16. "Some useful 16-valued logics. How a computer network should think", with Yaroslav Shramko. *Journal of Philosophical Logic*, 34(2005): 121–153.
17. "On the Negation of Action Types: Constructive Concurrent PDL". In P. Hájek, L. Valdes, and D. Westerstahl (eds.), *Proceedings of Logic Methodology and Philosophy of Science 12*. ???, ???, 2005. To appear.
18. "Logical connectives for constructive modal logic". *Synthese*, 2005. To appear.
19. "Connexive Modal Logic". In R. Schmidt et al. (eds.), *Advances in Modal Logic. vol. 5*. King's College Publications, London, 2005. To appear.
20. "Doxastic decisions, epistemic justification, and the logic of agency". *Philosophical Studies*, 2005. To appear.

**Service to the Profession:** Associate Editor *Studia Logica*, 1996–2000, Managing Editor *Studia Logica*, 2000–; Editorial Board *Reports on Mathematical Logic*, 1998–; Collecting Editor *Bulletin of the Section of Logic*, 2000–2004; Editorial Board *Journal of Applied Logic*, 2003–; Managing Editor book series *Trends in Logic*, Springer Verlag, 1999–; Editorial Board book series *Logic, Epistemology, and the Unity of Science*, Springer Verlag, 2004–; Editorial Board book series *Studies in Logic and Practical Reasoning*, Elsevier, 2004–; Editorial Board *Stanford Encyclopedia of Philosophy* area Philosophical Logic, 2004–; Founding member and member of the Steering Committee of the International Initiative *Advances in Modal Logic*, 1995–

**Teaching:** Heinrich Wansing teaches philosophical logic at undergraduate and graduate level. His teaching also comprises topics from philosophy of science, epistemology, and philosophy of language. He considers teaching as one of the best ways for obtaining an understanding of a subject matter.

**Vision Statement:** Wansing believes that logic is a *conditio sine qua non* of the scientific enterprise. The aim of logic is often described as separating the correct inferences from the incorrect ones. Taking into account that there are uncountably many logical systems, in Wansing's opinion the aim of logic as a discipline as well consists in separating the suitable logics form the unsuitable ones, relative to certain purposes and domains. This will involve increasing the expressive resources of formal languages and the development of criteria for comparing logical systems. Moreover, Wansing is convinced that a major role in this development will be played by investigating the philosophical foundations of non-classical logics and that this requires taking seriously modality and cognitive attitudes. In particular, he assumes that modal logics (also with non-classical bases) are indispensable for developing formal systems used to represent reasoning about actions and all kinds of propositional attitudes.

**Honours and Awards:** Runner-up, IGPL/FoLLI Prize for the Best Idea of the Year 1995.

# WEINGARTNER, Paul A.

**Specialties:** Philosophical logic, epistemic logic, modal logic, intensional logic, relevance logic.

**Born:** 8 June 1931 in Innsbruck, Tirol, Austria.

**Educated:** Habilitation, 1965; University of Innsbruck, PhD Philosophy (Major) and Physics (minor), 1961; Teacher's Training College, Innsbruck, Certificate of Primary Teacher 1952.

**Dissertation:** *Modes and Motives of Love;* supervisor, R. Strohal. Thesis for the certificate of habilitation: *Basic Questions on Truth.*

**Regular Academic or Research Appointments:** PROFESSOR OF PHILOSOPHY, UNIVERSITY OF SALZBURG, 1971–, EMERITUS, 1999–; CHAIRMAN OF THE *Institut fuer Wissenschaftstheorie* (INSTITUTE FOR THE PHILOSOPHY OF SCIENCE) AT THE INTERNATIONAL RESEARCH CENTER SALZBURG, 1972–; Associate Professor of Philosophy, University of Salzburg, 1970-1971; Chairman of the Section Foundations of

Logic and (Natural) Science at the *Institut fuer Wissenschaftstheorie*, 1967-1972; Assistant Professor of Philosophy, 1965-1971; Research Assistant at the *Institut fuer Wissenschaftstheorie*, 1962-1967.

**Visiting Academic or Research Appointments:**
Visiting Professor, University of California Irvine, 1978, 1990, 1991, 2000, 2001; Visiting Professor, University of Santos, of Florianopolis, of Sao Paulo, 1998; Visiting Professor, University of Brasilia, 1993, 1998; Visiting Professor, University of Manaus, Brazil, 1993; Visiting Professor, Center of Philosophy of Science at the University of Pittsburgh, 1984; Research Appointment, McGill University Montreal, Philosophy of Science Unit, 1980; Visiting Professor, National University of Canberra, 1978.
*Invited Guest Lectures:* At Austrian universities and at more than 100 foreign universities in Europe, USA, Canada, Russia, Australia, Chile, Brazil, South Korea and Japan.

**Research Profile:** Paul Weingartner has made contributions to matrix systems representing systems of Modal Logic (1968), systems of Intensional Logic (1973, 1981b), systems of Epistemic Logic (1982) and finite models of Set Theory (1973, 1975, 1981a). He further worked on problems of Philosophical Logic and Epistemology such as the Theory of Truth (2000a), the Paradoxes of the Liar (2000a, ch. 7; 2006), Tarski's Truth Condition (1999), Versimilitude (Approximation to Truth, 1987b), Analogy (1979) and the Theory of Definition (1989, 1991).

In his book *Basic Questions on Truth* (2000a) he investigated a wide range of contemporary problems involved in the definition of truth, in verisimilitude (approximation to truth) and in attributing the truth predicate to sentences or propositions, to rules and definitions and to scientific theories; he also connected these problems to important traditional problems.

In Relevance Logic he is noted for the construction of systems of relevance logic which are based on Classical Logic and distinguish between inferences (implications) which are valid and which are both, valid and relevant. The basic idea of this kind of relevance (which he developed together with Gerhard Schurz) is that no propositional variable or predicate in the conclusion of an inference must be replaceable (on one or more than one occurrences) by an arbitrary propositional variable or predicate salva validitate of the inference (1985, 1987b, 1993, 2000b, 2001, 2004).

Recently, Paul Weingartner contributed also to the domain of the *Logic of Religion* by proposing an axiomatic system of theodicy (2003).

**Main Publications:**

1. *Basic Questions on Truth*, Series Episteme 24, Kluwer, Dordrecht (2000a).
2. *Evil. Different Kinds of Evil in the Light of a Modern Theodicy*, Peter Lang, Frankfurt/M. (2003).
3. (with L. Schmetterer) *Gödel Remembered. Gödel-Symposium, Salzburg 1983,* Bibliopolis, Napoli (1987a).
4. *Alternative Logics. Do Sciences Need Them?*, Springer, Heidelberg (2004).
5. Modal Logics with Two Kinds of Necessity and Possibility, *Notre Dame Journal of Formal Logic* 9 (1968), 97-159.
6. A Predicate Calculus for Intensional Logic, *Journal of Philosophical Logic* 2 (1973), 220-303.
7. A Finite Approximation to Models of Set Theory, *Studia Logica* 34 (1975), 45-58.
8. Analogy Among Systems, *Dialictica* 33 (1979), 355-378.
9. Similarities and Differences Between the ∈ of Set-Theory and Part-Whole-Relations, Weinke, K. (ed.), *Logik, Ethik und Sprache. Festschrift für Rudolf Freundlich*. Oldenbourg, Vienna/Munich (1981a), 266-287.
10. A New Theory of Intension, Agassi, J./Cohen, R.S. (eds.), *Scientific Philosophy Today. Essays in Honor of Mario Bunge*, Reidel, Dordrecht (1981b), 439-464.
11. Conditions of Rationality for the Concepts Belief, Knowledge and Assumption, *Dialectica* 36 (1982), 243-263.
12. A Simple Relevance-Criterion for Natural Language and its Semantics, Dorn, G./Weingartner, P. (eds.), *Foundations of Logic and Linguistics: Problems and their Solutions.* [Selected Contributions to the $7^{th}$ International Congress of Logic, Methodology and Philosophy of Science (Salzburg 1983).], Plenum Press, New York (1985), 563-575.
13. (with G. Schurz) Verisimilitude Defined by Relevant Consequence-Elements. A new Reconstruction of Popper's Original Idea, Kuipers. Th. (ed.), *What is Closer-to-the-Truth?*, Rodopi, Amsterdam (1987b), 47-77.
14. Definitions in Russell, in the Vienna-Circle and in the Lvov-Warsaw School, Szaniawski, K. (ed.), *The Vienna Circle and the Lvov-Warsaw School*, Kluwer, Dordrecht (1989), 225-247.
15. A Note on Aristotle's Theory of Definition and Scientific Explanation, Spohn, W. (ed.), *Existence and Explanation*, Kluwer, Dordrecht (1991), 207-217.
16. A Logic for *QM* Based on Classical Logic, Maria de la Luiz Garcia Alonso/Moutsopoulos, E./Seel, G. (eds.), *L'art, la science et la métaphysique, Études offertes à André Mercier à l'occasion de son quatre-vingtéme anniversaire et recueillies au nom de l'académie internationale de philosophie de l'art*, Peter Lang, Bern (1993), 439-458.
17. Tarski's Truth Condition Revisited, *Vienna Circle Institute Yearbook* 6/1998, Kluwer, Dordrecht (1999), 193-201.

18. Reasons for Filtering Classical Logic, Batens, D. et al. (eds.), *Frontiers in Paraconsistent Logics. Proceedings of the First World Congress on Paraconsistency, Gent 1997*, Research Studies Press, London (2000b), 315-327.

19. Applications of Logic outside Logic and Mathematics: Do such Applications Force us to Deviate from Classical Logic?, Stelzner, W. (ed.), *Zwischen tradtioneller und moderner Logik*, Mentis Verlag, Paderborn (2001), 53-64.

20. A Solution for Different Types of Liar Paradoxes, Thiel, Ch. (ed.), *Operations and Constructions in Science. Proceedings of the Annual Meeting of the International Academy of the Philosophy of Science. Erlangen 2004*, Erlanger Forschungen A, Vol. 111 (2004), 95-105.

A full list of publications is available at www.sbg.ac.at/phs/people/weingartner.html

*Work in Progress*

21. A system of finite matrices is constructed which approximates the relevance logic described in publication [18.] The system is decidable, possesses a semantics and has the *Finite Model Property*. It solves paradoxes in different domains and also the most important ones coming up when Classical Logic is applied to Quantum Theory. It separates classically valid formulas from classically valid and relevant formulas and it contains a Modal Logic.

22. A book on the topics of "Omniscience, from a logical point of view" is almost finished. It contains an axiomatic system as a theory of omniscience.

**Service to the Profession:** Editorships, 22 volumes of Proceedings of International Conferences edited between 1966 and 2006, most recent: *Alternative Logics. Do Sciences Need them?* (Springer 2004); Chairman, Department of Philosophy, University of Salzburg. 1994-1999, 1988-1990, 1971-1979; Chairman, *Institut fuer* Wissenschaftstheorie, International Research Center Salzburg, 1972- ; Vice President, International Union for History and Philosophy of Science, Division of Logic Methodology and Philosophy of Science, 1983-1987.

**Teaching:** Paul Weingartner has contributed substantially to the development of Logic, Philosophy of Science and to the advancement of scientifically oriented philosophy in Austria since his appointment at the University of Salzburg in 1971. Together with Karel Lambert he established an exchange program between the departments of philosophy at the University of Salzburg and at the University of California, Irvine, in 1973. It started its exchange of scholars in 1975 and is running ever since for the benefit of the two departments.

Paul Weingartner has supervised 11 students for their diploma thesis (MA or MS), 12 students for their PhD thesis and 7 post doctoral students for their habilitation theses (including Edgar Morscher, Johannes Czermak, Georg Dorn, Gerhard Schurz) in the fields of Logic and Philosophy of Science. He had 14 research assistants at the *Institut fuer Wissenschaftstheorie*. Two books *Essays in Scientific Philosophy* (eds. E. Morscher, O. Neumaier, G. Zecha, 1981) and *Advances in Scientific Philosophy* (eds. G. Schurz, G. Dorn, 1991) were dedicated to Paul Weingartner acknowledging his work in logic, philosophy of science and scientifically oriented philosophy.

**Vision Statement:** The further development of Logic can benefit from its application to different fields of science; especially by investigating domains where methodological and conceptual difficulties are hindering progress. [THIS PARALLELS?] Like the development of mathematics has benefited from its application to physics and to many other sciences.

**Honours and Awards:** Membership of the New York Academy of Sciences, 1997; Honorary Doctorate (Dr.h.c.), Marie Curie Sklodowska University Lublin, Poland, 1995; Kardinal Innitzer Price for Philosophy of the Year, 1966; Postdoctoral Research Fellowship of the Alexander von Humboldt Foundation, 1964; British Council Postdoctoral Research Fellowship, 1962.

# WEINSTEIN, Scott

**Specialties:** Mathematical logic.

**Born:** 25 February 1948 in Brooklyn, New York, USA.

**Educated:** Rockefeller University, PhD, 1975; Princeton University, AB, 1969.

**Dissertation:** *Some applications of Kripke models to formal systems of intuitionistic analysis*; supervisor, Saul A. Kripke.

**Regular Academic or Research Appointments:** PROFESSOR, PHILOSOPHY AND COMPUTER AND INFORMATION SCIENCE, UNIVERSITY OF PENNSYLVANIA, 1987–; MATHEMATICS, 1994–.

**Visiting Academic or Research Appointments:** Visiting Fellow, Isaac Newton Institute for Mathematical Sciences, Cambridge, 2006; External Collaborator, Graduate Program in Logic and Algorithms, University of Athens, 1999-2000; Visiting Scientist, Center for Cognitive Science, Massachusetts Institute of Technology, 1984-1985.

**Research Profile:** Scott Weinstein has made contributions to the study of intuitionistic logic, computational learning theory, and finite model theory. In his dissertation, he used techniques from classical model theory to construct Kripke models of theories of free choice sequences, and applied these models to establish proof-theoretic closure properties and independence results for formal systems of intuitionistic analysis. In collaboration with Daniel Osherson and Michael Stob, he worked extensively on recursion-theoretic models of inductive inference, and applications of such models to the study of language acquisition, concept formation, and belief fixation. In the area of finite model theory, he collaborated with Anuj Dawar and Steven Lindell to further understanding of the expressive power of the finite variable fragments of infinitary logic over the class of finite structures, and with Eric Rosen on the study of preservation theorems for fragments of infinitary logic over finite structures. In collaboration with Peter Buneman and Wenfei Fan, he explored applications of finite model theory to database theory.

**Main Publications:**

1. Some Applications of Kripke Models to Formal Systems of Intuitionistic Analysis, Annals of Mathematical Logic, v. 16 (1979), pp. 1-32.
2. (with D. Osherson) "Criteria of Language Learning," Information and Control, v. 52 (1982), pp. 123-138.
3. (with D. Osherson and M. Stob) "Learning Strategies," Information and Control, v. 53 (1982), pp. 32-51.
4. "The Intended Interpretation of Intuitionistic Logic," Journal of Philosophical Logic, v. 12 (1983), pp. 261-270.
5. (with D. Osherson) "Identification in the Limit of First - Order Structures," The Journal of Philosophical Logic, v. 15(1986), pp. 55-81.
6. (with D. Osherson and M. Stob) "Aggregating Inductive Expertise," Information and Control, v. 70(1986), pp. 69-95.
7. (with D. Osherson and M. Stob) "Social Learning and Collective Choice," Synthese, v. 30(1987), pp. 319-347.
8. (with D. Osherson and M. Stob) "Mechanical Learners Pay a Price for Bayesianism," Journal of Symbolic Logic, v. 53(1988), pp. 1245-1251.
9. (with D. Osherson and M. Stob) "Synthesizing Inductive Expertise," Information and Computation, v. 77(1988), pp. 138-161.
10. (with D. Osherson) "Paradigms of Truth Detection," Journal of Philosophical Logic, v. 18(1989), pp. 1-42.
11. (with M. A. Papalaskari) "Minimal Consequence in Sentential Logic," Journal of Logic Programming, v. 7(1990), pp. 1-13.
12. (with H. Gaifman and D. Osherson) "A Reason for Theoretical Terms," Erkenntnis, v. 32, 1990, pp. 149-159.
13. (with D. Osherson and M. Stob) "A Universal Inductive Inference Machine," Journal of Symbolic Logic, v. 56(1991), pp. 661-672.
14. (with A. Dawar and S. Lindell) "Infinitary Logic and Inductive Definability over Finite Structures," Information and Computation, v. 119(1995), pp. 160-175.
15. (with E. Rosen) "Preservation Theorems in Finite Model Theory," in Daniel Leivant (ed.), Logic and Computational Complexity, Springer, 1995, pp. 480-502.
16. (with A. Dawar and S. Lindell) "First Order Logic, Fixed Point Logic, and Linear Order," in KleineBuening, H. (ed.), Computer Science Logic '95, Springer, 1996, pp. 161-177.
17. (with E. Rosen and S. Shelah) "k-Universal Finite Graphs," in Boppana, R. and Lynch, J. (eds.), Logic and Random Structures, American Mathematical Society, 1997, pp. 65-77.
18. (with A. Dawar, K. Doets, and S. Lindell) "Elementary Properties of the Finite Ranks," Mathematical Logic Quarterly, v. 44 (1998), pp. 349-353.
19. (with P. Buneman and W. Fan) "Path Constraints in Semistructured Databases," Journal of Computer and System Sciences, v. 61 (2000), pp. 146-193.
20. (with P. Buneman and W. Fan) "Interaction between Path and Type Constraints," ACM Transactions on Computational Logic, v. 4 (2003), pp. 530-577.

**Service to the Profession:** Program Committee Member, American Philosophical Association, Eastern Division, 2007-2009; Committee Member, Meetings in North America, Association for Symbolic Logic, 2003-2008; Executive Committee Member, Association for Symbolic Logic, 1989-1992; Executive Board Member, Society for Philosophy and Psychology, 1983-1986.

**Teaching:** Scott Weinstein has taught courses on logic at the University of Pennsylvania since 1975. He has been Director of the Logic, Information, and Computation Program in the School of Arts and Sciences at the University since its inception in 2003. He has had six PhD students in the fields of Computer and Information Science and Philosophy: Naoki Abe, James Cain, Anuj Dawar, Wenfei Fan, Zoran Markovic, and Eric Rosen.

## WEIR, Alan John

**Specialties:** Non-classical logics, paraconsistency, dialetheism, foundations and epistemology of logic.

**Born:** 19 October, 1955 in Johnstone, Scotland.

**Educated:** Oxford, MA, B.Phil, D.Phil; Edinburgh: MA;

**Dissertation:** *The Philosophy of Language of W. V. Quine*; supervisors, M. Woods, L. J. Cohen.

**Regular Academic or Research Appointments:** PROFESSOR OF PHILOSOPHY, GLASGOW UNIVERSITY; Professor, Senior Lecturer, Lecturer,

Queen's University Belfast, 1985-2006; Teaching Fellow, Balliol College Oxford, 1983-1985.

**Research Profile:** Recently Weir has worked on naïve set theory (in the sense of the naïve comprehension axiom or Frege's Basic Law V) and investigated logics under which it is not only not trivial (as in dialetheism) but also not inconsistent. He has extended this work to include the naïve theory of truth and deflationism. Other work includes investigating the logical framework of the 'neo-logicist' revival of logicism, in particular whether it is innocent of ontological commitment, either at first or second-order level. More generally he was written on the ontological commitments of the comprehension principle in second-order logic. He has also written on the philosophical foundations of intuitionistic logic, in particular proof-theoretic considerations which have been adduced in its favour in the Dummett/Prawitz tradition and on epistemological and foundational issues in logic– is it possible to justify or criticise logics, in what sense if logic rationally compelling?– and so forth.

**Main Publications:**

1. 2005a: 'Naïve Truth and Sophisticated Logic', in B.Armour-Garb, J.C. Beall, (eds.) *Deflationism and Paradox*, (Oxford: Oxford University Press 2005). pp. 218-49.
2. 2005b: 'On Kit Fine's *The Limits of Abstraction*', *Philosophical Studies* **122** (2005) pp. 333-348.
3. 2004a: 'There are no true contradictions', in G. Priest, J.C.Beall, (eds.) *The Law of Non-Contradiction; New Philosophical Essays*, (Oxford: Oxford University Press, 2004) Chapter 22, pp. 385-417.
4. 2004b: 'Naturalism Reconsidered' in Stewart Shapiro (ed.) *The Oxford Handbook of Philosophy of Mathematics and Logic*, (Oxford: Oxford University Press, 2004) Chapter 14, pp. 460-82.
5. 2003:'Neo-Fregeanism: An Embarrassment of Riches' *The Notre Dame Journal of Formal Logic*, **44** 2003, pp. 13-48. To be reprinted in *The Arché Papers on the Mathematics of Abstraction* ed. Roy Cook, Kluwer Academic Publishers.
6. 2000a'The Force of Reason: why is logic compelling?' *Logica Yearbook 1999* (Academy of Sciences of the Czech Republic, 2000), pp. 37-52.
7. 2000b: 'Token Relativism and the Liar'. *Analysis* 60 (2000) pp. 156-170.
8. 2000c ' "Neo-logicist" Logic is not Epistemically Innocent', (with Stewart Shapiro), *Philosophia Mathematica*. (3) 8 (2000) pp. 160-189. (joint author).
9. Translated as "La logica 'neologicista' non è epistemicamente innocente" in *Frege E Il Neologicismo* edited and translated by Andrea Pedeferri (Milan: FrancoAngeli, 2005).
10. 1999a 'New V, ZF and Abstraction' (with Stewart Shapiro), *Philosophia Mathematica*. 7 (1999) pp. 293-321.
11. 1999b 'Naïve Set Theory, Paraconsistency and Indeterminacy II, *Logique et Analyse,* 167-8 (1999) pp. 283-340. .
12. 1998a 'Naïve Set Theory, Paraconsistency and Indeterminacy I'. *Logique et Analyse* 161-163 (1998) pp. 219-266.
13. 1998b: 'Dummett on Impredicativity' *Grazer Philosophische Studien* 55 (1998). pp. 65-101.
14. 1998c 'Naïve Set Theory is Innocent!' *Mind* 107 1998. pp. 763-798. .
15. 1997: 'Two concepts of rationality?' *Cahiers de Psychologie Cognitive* 16 No. 1-2 (1997) pp. 238-246.
16. 1996a 'Ultra-Maximalist Minimalism!' *Analysis* 56 pp. 10-22.
17. 1996b 'On an Argument for Irrationalism' *Philosophical Papers* XXVpp. 95-114.
18. 1993 'Putnam, Gödel and Mathematical Realism', *International Journal of Philosophical Studies*, Vol. 1 No. 2 1993. pp. 255-285.
19. 1986a 'Dummett on Meaning and Classical Logic', *Mind*, Vol. 95, 1986, pp. 465-477.
20. 1986b 'Classical Harmony', *The Notre Dame Journal of Formal Logic*, Vol. 27, 1986, pp. 459-482.
21. 1983: 'Truth Conditions and Truth Values', *Analysis*, Vol. 43, 1983, pp. 176-180.

*Work in Progress*

22. *Truth through Proof: A Formalist Foundation for Mathematics* just completed;
23. *The Naïve Universe: Paradise Regained?* mostly on naïve set theory, arguing that the theory is in fact consistent, the 'blame' for the paradoxes lying with the overly strong logic, and that one can still extract standard mathematics from the theory in an appropriately weakened logic.

**Service to the Profession:** Panelist, UK Research Assessment Exercise Panel (Philosophy) for RAE 2008 exercise; Member, National Committee for History and Philosophy of Science Royal Irish Academy, 1993-1998; Executive Committee, Aristotelian Society, 1995-1998; Honorary Secretary, Irish Philosophical, 1998-2006.

**Teaching:** Weir's main logic teaching has been in elementary and, occasionally, intermediate level logic. None of his PhD students have written dissertations squarely on logic, though many, as is the norm in analytical philosophy, make use of formalisation.

**Vision Statement:** I am a philosopher not a logician, so my main interest is in foundational and epistemological issues. I think there is still a lot to say here. I am interested in particular in such questions as – can one justify or validate a logic, or criticise another? Is there only 'one true logic', at least relative to a given language and conceptual system? In what sense, if any, is logic rationally compelling or absolutely binding? I hope interest and discussion of these issues continues to be important, indeed increases in importance, among those working on the boundaries of philosophy and logic.

# WELCH, Philip David

**Specialties:** Set theory, problems in determinacy, mathematical logic, theories of truth, applications of logical methods in epistemology, semantics and the philosophy of mathematics.

**Born:** 6 January 1954 in Newport, Hants, UK.

**Educated:** University of Oxford, DPhil, 1979; University of Oxford, MSc Mathematical Logic, 1976; University College London, BSc, 1975.

**Dissertation:** *Combinatorial Principles in the Core Model*; supervisor, Robin Gandy.

**Regular Academic or Research Appointments:** PROFESSOR, MATHEMATICAL LOGIC, UNIVERSITY OF BRISTOL, 2003–; Professor, Mathematical Logic, Graduate School of Sciences, Kobe University, 1997-2000; Reader, Mathematical Logic, University of Bristol, 1996-2003; Lecturer, Pure Mathematics, University of Bristol, 1986-1995.

**Visiting Academic or Research Appointments:** Visiting Research Fellow, New College Oxford, 2006; Mercator Gastprofessur, Dept. of Mathematics, University of Bonn, 2002-2003; Gastprofessur Institut fuer Formal Logik, Vienna, 2000-2002;Visiting Professor, UCLA, 1993-1994; Visiting Research Fellow, University of Kobe, 1992;Visiting Professor, University of Nagoya, 1992;Visiting Assistant Professor, UCLA, 1984-1985; Royal Society European Research Fellowship, FU Berlin and University of Bonn 1983-1984; SERC Research Fellowship, Wolfson College, Oxford 1981-1983.

**Research Profile:** Welch has made contributions to the theory of inner models of set theory, in particular to problems of determinacy and combinatorial principles. He solved the last remaining open problem in determining the exact strength in the Hausdorff Difference Hierarchy on co-analytic sets, in terms of core model theory. He currently works on problems relating Jonsson cardinals to inner models, and on Global Square properties. He has made founding contributions in the theory of models of transfinite computation, with cross-over contributions in the revision theory of truth. He has worked on the strength of low levels of determinacy in analysis as it relates both to theories of truth and to transfinite machines. He currently heads projects with the UK's EPSRC in Set Theory and with the British Academy in logical applications to epistemology, and philosophy of mathematics. He is interested in the philosophy of intensionality, in particular, with its logical representation.

**Main Publications:**

1. '*Coding the Universe*' co-author with R.B.Jensen, (All Souls' College, Oxford) and A.Beller, (Dept.of Decision Sciences, Univ. of Penn.);London Mathematical Society Lecture Notes in Mathematics Series No.47, Cambridge University Press 1982, pp360.
2. '*Determinacy in the Difference Hierarchy of Co-analytic sets*'}, in the Annals of Pure and Applied Logic, 80, 1996, pp 69-108.
3. '*Doing without Determinacy - Aspects of Inner Models*' in the Proceedings of the Hull Logic Colloquium, Ed. F.Drake & J.Truss, North-Holland Studies in Logic Series, Amsterdam, 1988, 333-342.
4. '$\Sigma_3^1$-*absoluteness and the Second Uniform Indiscernible*', with J.R.Steel (UCLA), in Israeli Journal of Mathematics, 104, 1998, pp 157-190.
5. '*Friedman's trick: Minimality Arguments in the Infinite Time Turing degrees*', in '*Sets and Proofs*'} Ed. S.B.Cooper & J.K.Truss, London Math. Soc. Lecture Notes in Mathematics Series, 259, Cambridge University Press, 1999, pp. 425-436.
6. '*Some Remarks on the Maximality of Inner Models*' in Proceedings of the 1998 European Meeting of the Association for Symbolic Logic, Praha, Ed. P. Pudlak & S. Buss, Lecture Notes in Logic, 13, Kluwer, 1999, pp 516-540.
7. '*On Successors of Jónsson Cardinals*', with J. Vickers, in the Archive for Mathematical Logic, 39, No.6, 2000, pp 465-473.
8. '*The Length of Infinite Time Turing Machine Computations*', in the Bulletin of the London Math.Soc., 32, No. 2, Mar. 2000, pp 129-136.
9. '*Eventually Infinite Time Turing Machine Degrees*', in the J. of Symbolic Logic, 65, No.3, 2000, pp 1193-1203.
10. '*Elementary Embeddings from a submodel into the Universe*', with J.Vickers in the J. of Symbolic Logic, 66, No.3, Sep. 2001, pp 1090-1116.
11. '*On Possible Non-homeomorphic substructures of the real line,*' in Proceedings of the American Mathematical Society, 130, (2002), no. 9, 2771-2775.
12. '*Bounded Martin's Maximum, weak Erdös cardinals, and $\psi_-\{AC\}$*', with D. Asperó, in J. of Symbolic Logic, 67, No. 3, Sep. 2002, 1141-1152.
13. '*On Gupta-Belnap Revision Theories of truth, Kripkean Fixed points, and the Next Stable Set,*' a Communication in the Bulletin of Symbolic Logic, 7}, No.3, Sep.2001, 345-360.
14. '*Possible Worlds Semantics for Modal Notions conceived as Predicates*', with V. Halbach & H.Leitgeb, in the J. of Philosophical Logic, 32, No.3, April 2003, 179—223.

15. 'On Revision Operators', in the *J. of Symbolic Logic*, 68, No.3 June 2003, 689—711.

16. 'On Unfoldable cardinals, omega-closed cardinals, and the beginning of the Inner Model Hierarchy' in *Archive for Mathematical Logic*, 43, No. 4, 2004, 443-458.

17. 'On the strength of mutual stationarity', with P. Koepke, in *CRM Set Theory Year, Barcelona*, Ed. J. Bagaria, S. Todorcevic, Basel, Birkhäuser, Trends in Mathematics Series, 2006, pp307—318.

**Work in Progress**

18. 'Turing Unbound: the extent of computation in Malament-Hogarth spacetimes'. To appear in *British Journal for the Philosophy of Science*.

19. '$\Sigma^*$-Fine Structure', for the *Handbook of Set Theory*, Eds., M. Foreman, A. Kanamori, \& M. Magidor. pp78.

20. 'The Strength of the Inner Model Hypothesis, with S-D. Friedman, W.H. Woodin. To appear in *Journal of Symbolic Logic*, 73, No. 2, 391–400.

21. 'Necessities and Necessary Truths: A Prolegomena to the Use of Modal logic in the Analysis of Intensional Notions' with Volker Halbach. To appear in *Mind*.

**Service to the Profession:** Editorial Board Member, *Scientiae Mathematicae, Mathematica Japonica;* Subject Co-Editor, Philosophy of Mathematics, *Stanford Encyclopedia of Philosophy*.

**Teaching:** Welch has been responsible for the teaching of mathematical logic at the University of Bristol where he has been, intermittently, for over 20 years. He has, and currently has, 4 PhD students. During a three and half year period at the Graduate School of Science and Technology at the University of Kobe, Japan he was responsible for setting up a new Master and PhD program in Set Theory and Mathematical Logic. (The successor to his Chair there is T. Arai).

**Vision Statement:** "Methods of mathematical analysis have made their impact felt in a wide range of areas now called "Logic of X". I suspect that other areas are only just beginning to receive their full attention from mathematical logicians. As an example, "mathematical philosophy" - is the as-yet vague term one could attach to a variety of examples of mathematical analysis in philosophical areas; whether it be the mathematical (partial) construction (or reconstruction) of philosophers' system building attempts, or as attempts to bring clarity or rigour to notions or relations they use."

**Honours and Awards:** Inamori Foundation Young Researchers Award, 1998.

# WESTERSTåHL, Dag

**Specialties:** Model theory, generalized quantifiers, formal semantics.

**Born:** 5 January 1946, Stockholm, Sweden.

**Educated:** Gothenburg University, Docent Theoretical Philosophy, 1983; Gothenburg University, PhD Theoretical Philosophy, 1977; Gothenburg University, BA, 1967.

**Dissertation:** *Some Philosophical Aspects of Abstract Model Theory*, supervisor, Per Lindström.

**Regular Academic or Research Appointments:** PROFESSOR, THEORETICAL PHILOSOPHY, GOTHENBURG UNIVERSITY, 1998–; Researcher, Philosophy of Language, Swedish Council for Research in the Humanities and Social Sciences (HSFR), 1991-1996; Associate Professor, Theoretical Philosophy, Stockholm University, 1988-1998; Associate Professor, Philosophy, Gothenburg University, 1984-1988; Assistant Professor (forskarassistent), Practical and Theoretical Philosophy, Gothenburg University, 1978-1984.

**Visiting Academic or Research Appointments:** Fellow, Mittag-Leffler Institute, Djursholm Sweden, 2000, 2001; Visiting Scholar, Helsinki University, 1995, 1997; Visiting Scholar, University of Bologna, 1994; Visiting Researcher, ICOT, Tokyo, 1993; Visiting Scholar, University of Amsterdam, 1992; Visiting Research Associate, Stanford University, 1985-1986.

**Research Profile:** Continuing work by Per Lindström, Jon Barwise, and Johan van Benthem, Westerståhl's main focus so far has been the study of generalized quantifiers, in particular issues of expressive power and the application of quantifier theory to natural language semantics. Some of the mathematical part of this work has been done jointly with members of the Helsinki Logic Group (Jouko Väänänen, Lauri Hella), and some of the linguistic part with linguists (Ed Keenan, Stanley Peters). He is a leading expert in this field, with numerous research papers, handbook chapters, etc. over the years, culminating in the recently published book (with Peters) *Quantifiers in Language and Logic*. He is still active in this area. Westerståhl also works in other areas of logic and philosophy of language, most of them characterized by the application of mathematical-logical tools to linguistic issues. He worked for an extended period in situation semantics and situation theory. A number of more recent papers center around the notion of compositionality, in its abstract mathematical aspects (where he has built on and developed work by Wilfrid Hodges) as well as its methodological and linguistic aspects. A research interest that has been with him since the dissertation is the concept of a logical constant, where one of his recent ideas is that the notion of logicality or topic-neutrality, which can be spelled out in terms of various invariance conditions, is distinct from a more proof-theoretic notion of constancy, yet to be explored. Still another recent research topic concerns methodological aspects of the conflict between classical and intuitionistic mathematics.

**Main Publications:**

1. Some results on quantifiers, *Notre Dame Journal of Formal Logic* 25, 1984, 152–170.
2. Determiners and context sets, in J. van Benthem and A. ter Meulen (eds), *Generalized quantifiers in Natural Language*, Foris, Dordrecht, 1985, 45–71. Reprinted in J. Gutierrez Rexach (ed.) *Semantics: Critical Concepts in Linguistics, Volume II*, London/New York, Routledge, 2003, 127–151.
3. Logical constants in quantifier languages, *Linguistics and Philosophy* 8, 1985, 387–413.
4. Branching generalized quantifiers and natural language, in P. Gärdenfors (ed.), *Generalized Quantifiers. Linguistic and Logical Approaches*, D. Reidel, Dordrecht, 1987, 269—298.
5. Quantifiers in formal and natural languages, in D. Gabbay and F. Guenthner (eds), *Handbook of Philosophical Logic, vol. IV*, D. Reidel, Dordrecht, 1989, 1—131.
6. Parametric types and propositions in first-order situation theory, in R. Cooper et al. (eds.), *Situation Theory and its Applications, vol. 1*, CSLI Publications, Stanford, 1990, 193–230.
7. Relativization of quantifiers in finite models, in J. van der Does and J. van Eijck (eds.), *Generalized Quantifier Theory and Applications*, ILLC, University of Amsterdam, 1991, 187–205. Also in J. van der Does and J. van Eijck (eds.), *Quantifiers, Logic and Language*, CSLI Lecture Notes, Stanford, 1996, 375–383.
8. A situation-theoretic representation of text meaning: anaphora, quantification and negation (with B. Haglund and T. Lager), in P. Aczel et al. (eds.), *Situation Theory and its Applications, vol. 3*, CSLI Publications, Stanford, 1993, 375–408.
9. Predicate logic with flexibly binding operators and natural language semantics (with P. Pagin), *Journal of Logic, Language and Information* 2, 1993, 89–128.
10. Iterated quantifiers, in M. Kanazawa and C. Piñon (eds.), *Dynamics, Polarity and Quantification*, CSLI Lecture Notes, Stanford, 1994, 173–209.
11. Directions in generalized quantifier theory (with J. van Benthem), *Studia Logica* 55, 1995, 389–419.
12. Self-commuting quantifiers, *Journal of Symbolic Logic* 61, 1996, 212–224.
13. Generalized quantifiers in linguistics and logic (with E. Keenan), in J. van Benthem and A. ter Meulen (eds), *Handbook of Logic and Language*, Elsevier, Amsterdam 1997, 837–893.
14. Definability of polyadic lifts of generalized quantifiers (with L. Hella and J. Väänänen), *Journal of Logic, Language and Information* 6, 1997, 305–335.
15. On mathematical proofs of the vacuity of compositionality, *Linguistics and Philosophy* 21, 1998, 635–643.
16. On predicate logic as modal logic, in A. Cantini et al. (eds), *Logic and Foundations of Mathematics*, Kluwer, 1999, 195–207.
17. Does English really have resumptive quantification? (And do 'donkey' sentences really express it?) (with S. Peters), in D. Beaver et al. (eds), *The Construction of Meaning*, CSLI Publications, Stanford, 2002, 181–195.
18. On the compositionality of idioms; an abstract approach, in D. Barker-Plummer, D. Beaver, J. van Benthem, and P. Scotto di Luzio (eds.), *Words, Proofs, and Diagrams*, CSLI Publications, Stanford, 2002, 241–271.
19. On the expressive power of monotone natural language quantifiers over finite models (with J. Väänänen), *Journal of Philosophical Logic* 31, 2002, 327–358.
20. On the compositional extension problem, 2004, *Journal of Philosophical Logic* 33, 2004, 549–582.
21. *Quantifiers in Language and Logic* (with. S. Peters), Oxford University Press, Oxford and New York, 2006, xix + 528 pp.

*Work in Progress*

22. Reflections on the conflict between classical and intuitionistic mathematics (forthcoming).
23. A study of how compositionality can be combined with context-dependence and ambiguity (forthcoming).
24. A book on compositionality (with Peter Pagin).
25. Further development of several themes from the quantifier book with Peters, among them the study of possessive and exceptive quantifiers, and 'Post-complement vs. contrariety: a defense of the modern square of opposition' (forthcoming). Likewise:
26. A continued study of the notion of logical constant, in particular the constancy aspect.

**Service to the Profession:** Secretary-General, International Union of History and Philosophy of Science, Division of Logic, Methodology and Philosophy of Science (IUHPS/DLMPS); President, Swedish National Committee for Logic, Methodology and Philosophy of Science, 2003- ; Secretary-General, IUHPS, 2002-2005; ASL Council Member, 2001-2003; Secretary, Swedish National Committee for Logic, Methodology and Philosophy of Science, 1998-2002; Managing Editor, *Journal of Logic, Language and Information*, 1996-2001.

**Teaching:** Westerståhl was instrumental for establishing logic as an independent discipline at Gothenburg University, with Per Lindström as an initial central figure, and carrying on that tradition after Lindström's retirement. He has taught all kinds of philosophy and logic courses over the years, and supervised three PhD students. He is most known for his teaching on generalized quantifiers, about which he has lectured at universities all over the world and at least four summer schools.

**Vision Statement:** "Logic currently develops in

two directions: one increasingly technical, driven by standard mathematical concerns, one more philosophical, attempting to account for reasoning, cognition, computation with a variety of tools. For the latter, an increased concern with foundational issues seems desirable: What is logic? What is it supposed to do?"

**Honours and Awards:** Fellow, Royal Swedish Academy of Sciences, 2007-; Fellow, Royal Society of Arts and Sciences in Gothenburg (KVVS), 2001-.

# WIGDERSON, Avi

**Specialties:** Combinatorics and graph theory, quantum computation, theoretical computer logic, proof complexity.

**Born:** 9 September 1956 in Haifa, Israel.

**Educated:** Princeton University, PhD Computer Science, Department of Electrical Engineering and Computer Science, 1983; Technion Israel Institute of Technology, BSc Summa cum laude Computer Science, 1980.

**Dissertation:** *Studies in Combinatorial Complexity;* supervisor, R.J. Lipton.

**Regular Academic or Research Appointments:** PROFESSOR, MATHEMATICS, INSTITUTE FOR ADVANCE STUDY, 1999–; Computer Science Institute, Hebrew University, 1986–2003.

**Visiting Academic or Research Appointments:** Visiting Professor, Institute for Advanced Study, Princeton University, 1995-1996; Visiting Associate Professor, Department of Computer Science, Princeton University, 1990-1992; Fellow, Mathematical Sciences Research Institute, 1985-1986; Visiting Scientist, IBM Research, 1984-1985; Visiting Assistant Professor, Department of Computer Science, U.C. Berkeley, 1983-1984.

**Research Profile:** Avi Wigderson has made fundamental contributions to most areas of theoretical computer science, including efficient sequential and parallel algorithms, cryptography and distributed protocols, randomness and pseudorandomness, circuit and proof complexity.

**Main Publications:**

1. A. Wigderson, P, NP and Mathematics - a computational complexity perspective, *Proceedings of the ICM 06, to appear.*
2. S. Hoory, N. Linial, A. Wigderson, Expander Graphs and Their Applications, *Bulletin of the AMS, Vol 43, No 4, 439—561, 2006.*
3. B. Barak, A. Rao, R. Shaltiel, A. Wigderson, 2-Source Dispersers for Sub-Polynomial Entropy and Ramsey Graphs Beating the Frankl-Wilson Construction, *Proceedings of STOC 06, pp. 671-680, 2006.*
4. M. Alekhnovitch, E. Ben-Sasson, A. Razborov, A. Wigderson, Pseudorandom Generators in Propositional Proof Complexity, *SIAM Journal on Computing, Vol 34, Number 1, pp. 67-88, 2004.*
5. B. Barak, R. Impagliazzo, A. Wigderson, Extracting randomness using few independent sources, *Proc of the 45th FOCS 2004, pp 384-393, 2004. SICOMP, Vol 36, No 3, 2006.*
6. C-J Lu, O. Reingold, S. Vadhan, A. Wigderson, Extractors: Optimal up to Constant Factors, *35th Annual ACM Symposium, STOC 2003, pgs 602-611, 2003.*
7. M. Capalbo, O. Reingold, S. Vadhan, A. Wigderson, Randomness Conductors and Constant-Degree Expansion Beyond the Degree /2 Barrier, *Proc. of the 34th STOC, pp. 659-668, 2002.*
8. Reingold, S. Vadhan, A. Wigderson, Entropy Waves, the Zig-Zag Graph Product, and New Constant-Degree Expanders, *Annals of Mathematics, Vol. 155, No.1, pp. 157-187, 2002.*
9. Reingold, S. Vadhan, A. Wigderson, Entropy Waves, The Zig-Zag Graph Product, and New Constant-Degree Expanders and Extractors, *Proc. of the 41st FOCS, pp. 3-13, 2000.*
10. E. Ben-Sasson, A. Wigderson, Short Proofs are Narrow - Resolution made Simple, *Proc. of the 31th STOC, pp. 517-526, 1999.*
11. R. Impagliazzo, A. Wigderson, Randomness vs. Time: De-randomization under a uniform assumption, *Proc. of the 39th FOCS, pp.734-743, 1998.*
12. R. Impagliazzo, A. Wigderson, P=BPP unless E has Subexponential Circuits: Derandomizing the XOR Lemma, *Proc. of the 29th STOC, pp. 220-229, 1997.*
13. N. Nisan, A. Wigderson, Lower Bounds on Arithmetic Circuits via Partial Derivatives, *Preliminary version in Proc. of the 36th FOCS, pp. 16-25, 1995. Computational Complexity, Vol. 6, pp. 217-234, 1996.*
14. N. Nisan, A. Wigderson, Hardness vs. Randomness, *Journal of Computer Systems and Sciences, Vol. 49, No. 2, pp. 149-167, 1994.*
15. Y. Rabinovich, A. Sinclair, A. Wigderson, Quadratic Dynamical Systems, *Proc. of the 33rd FOCS conference, pp. 304-313, 1992.*
16. R. Raz, A. Wigderson, Monotone Circuits for Matching require Linear Depth, *Journal of the ACM, Vol. 39, pp. 736-744, 1992. Preliminary version in Proc. of the 22nd STOC, pp. 287-292, May 1990.*
17. Goldreich, S. Micali, A. Wigderson, Proofs that Yield Nothing but their Validity, or All Languages in NP have Zero-Knowledge Proof Systems, *Journal of the ACM, Vol. 38, No. 1, pp. 691-729, 1991*
18. M. Karchmer, A. Wigderson, Monotone Circuits for Connectivity require Super-Logarithmic Depth, *SIAM Journal on Discrete Mathematics, Vol. 3, No. 2, pp. 255-265, 1990.*
19. S. Goldwasser, M. Ben-Or, A. Wigderson,

Completeness Theorems for Non-cryptographic Fault-tolerant Distributed Computing, *Proc. of the 20th STOC*, pp. 1-10 May 1988.

20. Goldreich, S. Micali, A. Wigderson, How to Play any Mental Game, *Proc. of the 19th STOC, pp. 218-229, May 1987.*

21. R. Karp, E. Upfal, A. Wigderson, Constructing a Perfect Matching is in Random NC, *Combinatorica, Vol. 6, No. 1, pp. 35-48, 1986.*

22. E. Upfal, A. Wigderson, How to Share Memory in a Distributed System, *Journal of the ACM, Vol. 34, No. 1, pp.116-127, 1986.*

23. R. Karp, A. Wigderson, A Fast Parallel Algorithm for the Maximal Independent Set Problem, *Journal of the ACM, Vol. 32, No. 4, pp.762-773, October 1985.* Preliminary version in *Proc. of the 16th STOC, pp. 266-272, May 1984.*

A full list of publications is available at Wigderson's home page http://www.math.ias.edu/~avi

**Service to the Profession:** *Editorship*: Editorial Board Member, SIAM Journal on Discrete Mathematics; Editorial Board Member, Information and Computation; Editorial Board Member, Complexity Theory; Grant Proposal Referee, Israel Academy of Sciences; Grant Proposal Referee, U.S. National Science Foundation; Grant Proposal Referee, National Sciences and Engineering Council of Canada; Grant Proposal Referee, American-Israeli Binational Science Foundation; Referee, Journal of the ACM; Referee, SIAM Journal on Computing; Referee, Theoretical Computer Science; Referee, Journal of Algorithms; Referee, IEEE Transactions on Information Theory; Referee, Journal of Computer Systems and Sciences; Referee, Information Processing Letters; Referee, Information and Control; Referee, Science of Computer Programming; Referee, Acta Informatica; Referee, Algorithmica; Referee, Advances in Computing Research; Referee, Journal of Complexity; Referee, Combinatorica; Referee, Journal of Economic Theory; Book Reviewer, Addison Wesley.

**Teaching:** Combinatorics and Graph Theory, Lower Bound Techniques, Data Structures, Algorithms, Probabilistic Algorithms, Circuit Complexity, Introduction to Complexity Theory, Randomness in Computation, The Probabilistic Method, Proof Techniques in Complexity Theory, Expander Graphs and their Applications, The Efficient Universe.

Avi Wigderson has advised 14 PhD students and 6 MSc students. A full list can be found in http://www.math.ias.edu/~avi/students.html

**Vision Statement:** "Computation, broadly viewed as the evolution of an environment under simple, local rules, is at the heart of almost all natural and artificial processes.

The study of computation, primarily of how efficient it is (or could be made to be) under various natural resources, is thus key not only to Computer Science, but to all sciences and mathematics. The knowledge we already have today is but the tip of the iceberg of what the study of efficient computation will reveal."

**Honours and Awards:** Rolf Nevanlinna Prize, 1994; Thrice invited speaker, International Congress of Mathematicians, 1990, 1994, 2006.

# WÓJCICKI, Ryszard

**Specialties:** Formal logic, philosophical foundations of logic

**Born:** 30 October 1931 in Krzywicze Wielkie, Vilnius District, Poland.

**Educated:** Wrocław University, PhD Logic, 1962; Wrocław University, MD Mathematical Physics, 1959; Warsaw University, MD Philosophy, 1956; Wrocław University, BD Mathematical Physics, 1954.

**Dissertation:** *Analityczne komponenty definicji arbitralnych (Analytical components of definitions, in Polish)*; supervisor, Maria Kokoszyńska.

**Regular Academic or Research Appointments:** PROFESSOR, LOGIC AND PHILOSOPHY OF KNOWLEDGE, EMERITUS, 2003–; Professor, Logic and Philosophy of Knowledge, Institute of Philosophy and Sociology of the Polish Academy of Sciences, 1969-2002; Associate Professor, Logic, Wrocław University, Department of History and Social Sciences, 1962-1968; Assistant Professor, Logic, Wrocław University, Department of History and Social Sciences, 1959-1961.

**Visiting Academic or Research Appointments:** Guest Professor, University of Leipzig, 2002-2003; Visiting Fellow, University of Pittsburgh, Centre for Philosophy of Science, 1990; Guest Professor, Campinas University, Department of Mathematics and Statistics (Brazil), 1980; Visiting Scholar, Australian National University, Canberra, Australia, 1979; Visiting Scholar, Queen's Univ., Kingston, Canada, 1975-1976; Visiting Scholar, University of California, Berkeley, USA, 1964-1965.

**Research Profile:** Alfred Tarski's investigations into formal properties of logical consequence operations started in the thirties of the past century were undertaken by several Polish logicians (notably Jerzy Słupecki, Jerzy Łoś Roman Suszko, Andrzej Grzegorczyk) after the Second World

War. Most of Wójcicki's contributions to logic belong to that area.

The common view is that logical calculi (propositional logics, in particular) may be adequately characterized in terms of logical truth; a logical calculus may be identified with the set of its tautologies (logically true formulas). In a number of papers Wójcicki demonstrates that the difference between the above mentioned 'formulaic' notion of a logical calculus and an 'inferential' one, i.e. one under which a logical calculi is defined to be the set of logically valid inferences (logical consequence operation) is rather dramatic. In particular, he proved, that Gödel's theorem to the effect that classical connectives are definable in terms intuitionistic ones is not valid if the two logics are understood in the inferential way. Also, he proved that, if logic is understood in the inferential way, there are two non-trivial logics stronger than the three-valued Łukasiewicz logic. They have the same tautologies (in the formulaic sense they are identical with the classical two valued logic) but they do not have not the same rules of inference.

In the seventies, Wójcicki launched and then coordinated a project examining the algebraic properties of logical consequence operations. The chief idea was to extend the well known technique of algebraic treatment of logical problems, thus far applied to study logical calculi understood in the formulaic manner, to logical consequence operation (*a nie " operations"?*). The project resulted in a series of papers published (chiefly in *Studia Logica*) by J. Czelakowski, W. Dziobiak, G. Malinowski, M. Maduch, T. Prucnal, W. A. Pogorzelski, W. Rautenberg, R. Suszko, M. Tokarz, P. Wojtylak, A. Wroński, J. Zygmunt, R. Wójcicki and others. A unified and fairly complete presentation of both results thus obtained and others relevant to the research project (notably various H. Rasiowa's results concerning implicative logics) is Wójcicki's monograph *Theory of Logical Calculi; Basic Theory of Consequence Operation* (Kluwer, 1988). For the most recent investigations (W. J. Blok, J. Czelakowski, W. Dziobiak, J. M. Font, D. Pigozzi and others) in this area see Czelakowski's monograph *Protoalgebraic logics* (Kluwer, 2001).

Wójcicki's results concerning 'strongly finite' propositional calculi, i.e. logical calculi that might be adequately defined by means of finite logical matrices are applied in computer science analyzes. Also, his theorem stating that 'self-extensional logics' are exactly those logics which preserve truth at every 'reference point' in the properly defined set (see his 'Referential Semantics' for the relevant definitions) seems to be of considerable significance for investigations into possible applications of logic to computer science.

**Main Publications:**

1. 'Logical matrices strongly adequate for structural sentential calculi', *Bulletin de l'Académie Polonaise des Sciences, Série des sciences math. astr. et phys.*, **17**(1969), 333—335.

2. 'Some remarks on the consequence operation in sentential logics', *Fundamenta Mathematicae*, **68**(1970), 269—279.

3. 'On reconstructability of classical propositional logic in intuitionistic logic', Bulletin de l'Académie Polonaise des Sciences, Serie des sciences math. astr. et phys., **18**(1970), 421—422.

4. 'The degree of completeness of the finitely-valued propositional calculi', *Zeszyty Naukowe Uniwersytetu Jagiellońskiego, Prace z logiki*, **7**(1972), 77—85.

5. 'On matrix representations of consequence operations of Łukasiewicz's sentential calculi', *Zeitschrift Für Mathematische Logik und Grundlagen der Mathematik*, **19**(1973), 239—247.

6. 'Note on deducibility and many-valuedness', *J. Symbolic Logic* **39**(1974), 563—566.

7. 'The logics stronger than Łukasiewicz's three-valued sentential calculus; the notion of degree of maximality versus the notion of degree of completeness', *Studia Logica*, **34**(1974), 201—214.

8. 'A theorem on the finiteness of the degree of maximality of the $n$-valued Łukasiewicz logics', in R. Epstein (ed.), *Proceedings of the 1975 International Symposium on Multiple-Valued Logic, Bloomington, May 13 - 16, 1975*, 240—251.

9. 'Entailment semantics for T.', in A.I. Arruda, N.C.A. da Costa and A.M. Sette (eds.), *Proceedings of the Third Brazilian Conference on Mathematical Logic, Sociedade Brasilera de Logica*, 1980, 309—336.

10. 'Referential Matrix Semantics for Propositional Calculi' in *Logic, Methodology and Philosophy of Science VI, Proceedings of Sixth International Congress of Logic, Methodology and Philosophy of Science}*, Hanover, August, 1979, North Holland Publishing Company and Polish Scientific Publishers, 1982, 325—334.

11. *Lectures on Propositional Calculi*, Ossolineum, Wrocław 1984 (e-version of this book is available at the home page www.studialogica.org/wojcicki)

12. 'Situation semantics for non-Fregean logic', *Journal for Non-Classical Logic* **3**(1986), 33—69.

13. *Theory of Logical Calculi; Basic Theory of Consequence Operation*, vol. 199 of *Synthese Library*, Kluwer Academic Publishers, Dordrecht, 1988.

14. 'Referential semantics', in V. F. Hendricks, K. F. Jorgensen, S. A. Pedersen (eds.), *Knowledge Contributors*, Synthese Library Series, Kluwer Academic Publishers, Dordrecht.

15. (in collaboration with Jan Zygmunt), *Polish Logic in Postwar Period*.

A full list of publications is available at Wojcicki's home page www.studialogica.org/wojcicki
*Work in Progress*

16. An electronic version of an elementary course of logic (jointly with Adam Grobler, Elżbieta Kałuszyńska, Jerzy Pogonowski. Ireneusz Sierocki).

17. Philosophical foundations of logic. Tentatively planned as a series of essays that might eventually combine into a book. Logic has always been conceived as a science whose chief goal is to state the rules one must observe if one wishes to reason correctly. I plan to critically examine both this idea of logic and some its interpretations.

**Service to the Profession:** Editor-in-Chief, *Trends in Logic, Studia Logica Library*, 1993-; Steering Committee Member, Federation Internationale des Societes de Philosophie, 1992-; President, Executive Committee, Polish Association for Logic and Philosophy of Science, 1992–1995; $2^{nd}$ Vice-President, Executive Committee of the International Union for History and Philosophy of Science, Division for Logic, Methodology and Philosophy of Sciences, 1991-1995; Chief Coordinator, Research Program of the Polish Academy of Sciences *Formal Structure of Informal* Reasoning, 1986-1990, 1992-1994; Council Member, Association for Symbolic Logic, 1979-1983; *Studia Logica; An International Journal for Logic*, Editor-in-Chief, 1975–1980, Editorial Board Chairman, 1991–2004; Editorial Board Member, *Erkenntnis, An International Journal of Analytic Philosophy*, 1975-; Head, Section of Logic and Methodology of Science of the Institute of Philosophy and Sociology of the Polish Academy of Sciences, 1969-2000.

**Teaching:** He has had nine PhD students in logic and philosophy of knowledge (Wiesław Dziobiak, Grzegorz Malinowski, Marian Maduch, Marek Tokarz and others). He organized and was the director of international *Summer Schools for the Theory of Knowledge*, 1997, 1998 and 1999.

**Vision Statement:** The discovery of the first order logic ended investigations whose chief goal was to find out and explicitly state the principles of mathematical reasoning. What other forms of reasoning can be examined with the help of essentially the same tools and techniques that proved their usefulness in dealing with mathematical reasoning? It is not probabilistic reasoning, for it is based on laws of statistics not logic. Is it non-monotonic reasoning? Is it fuzzy reasoning? Rather to defend its autonomy, logic should merge into a larger discipline, formal theory of reasoning, becoming its central part.

**Honours and Awards:** Honorary Editor-in-Chief, *Studia Logica; An International Journal for Logic*, 2005-; Correspondent Member, Warsaw Scientific Association, 1996; Correspondent Member, Polish Academy of Sciences and Letters, 1992–;Visiting Fellow, Center for Philosophy of Science, University of Pittsburgh, 1990-.

## WOODIN, W. Hugh

**Specialties:** Mathematical logic, foundations of mathematics.

**Educated:** University of California, Berkeley, PhD Mathematics; California Institute of Technology, BS Mathematics, 1977.

**Dissertation:** Supervisor: Robert M. Solovay.

**Regular Academic or Research Appointments:** MATHEMATICS, UNIVERSITY OF CALIFORNIA, BERKELEY, DISTINGUISHED PROFESSOR, 2003–; Chair, Mathematics, 2002-2003, Professor, 1989-2003; Professor, Mathematics, California Institute of Technology, 1983-1989; Assistant Professor, Mathematics, California Institute of Technology, 1980-1983.

**Visiting Academic or Research Appointments:** Distinguished Visiting Professor, National University of Singapore, 2006, 2005, 2004, 2002; Tutorial Lecturer, IMS, Singapore, 2005; Miller Research Professor, 1997-1998.

**Main Publications:**

1. W. Hugh Woodin. The cardinals below $|[!1]<!1\,|$. Ann. Pure Appl. Logic, 140(1-3):161–232, 2006.

2. W. Hugh Woodin. The continuum hypothesis. In Logic Colloquium 2000, volume 19 of Lect. Notes Log., pages 143–197. Assoc. Symbol. Logic, Urbana, IL, 2005.

3. Q. Feng and W. H. Woodin. P-points in Qmax models. Ann. Pure Appl. Logic, 119(1-3):121–190, 2003.

4. E. Schimmerling and W. H. Woodin. The Jensen covering property. J. Symbolic Logic, 66(4):1505–1523, 2001.

5. Kai Hauser and W. HughWoodin. _13sets and _13singletons. J. Symbolic Logic, 64(2):590–616, 1999. [_13 = Pi]

6. Boban Velickovic and W. Hugh Woodin. Complexity of reals in inner models of set theory. Ann. Pure Appl. Logic, 92(3):283–295, 1998.

7. Theodore A. Slaman and W. Hugh Woodin. Extending partial orders to dense linear orders. Ann. Pure Appl. Logic, 94(1-3):253–261, 1998. Conference on Computability Theory (Oberwolfach, 1996).

8. W. Hugh Woodin. The universe constructed from a sequence of ordinals. Arch. Math. Logic, 35(5-6):371–383, 1996.

9. A. S. Kechris and W. H. Woodin. A strong boundedness theorem for dilators. Ann. Pure Appl. Logic, 52(1-2):93–97, 1991. International Symposium on Mathematical Logic and its Applications (Nagoya, 1988).

10. Haim Judah, Saharon Shelah, and W. H. Woodin. The Borel conjecture. Ann. Pure Appl. Logic, 50(3):255–269, 1990.

11. Hugh Woodin. Aspects of determinacy. In Logic, methodology and philosophy of science, VII (Salzburg, 1983), volume 114 of Stud. Logic Found. Math., pages 171–181. North-Holland, Amsterdam, 1986.

12. M. Gitik, M. Magidor, and H. Woodin. Two weak consequences of 0]. J. Symbolic Logic, 50(3):597–603, 1985. (that symbol is O with a musical "sharp")

13. Saharon Shelah and Hugh Woodin. Forcing the failure of CH by adding a real. J. Symbolic Logic, 49(4):1185–1189, 1984.

14. W. Hugh Woodin. The axiom of determinacy, forcing axioms, and the nonstationary ideal, volume 1 of de Gruyter Series in Logic and its Applications. Walter de Gruyter & Co., Berlin, 1999. 934 pages.

**Honours and Awards:** Elected Fellow, American Academy of Arts and Sciences, 2000; Senior Humboldt Research Award, 1996; Honorary Fellow, Department of Mathematics, University of Leeds, 1990; Carol Karp Prize, Association for Symbolic Logic, 1989; Sloan Research Fellow, 1985-1987; Presidential Young Investigator Award, 5 Year Tenure, 1985; Fellow Commoner, Peterhouse, Cambridge, Egland, 1984; NSF Principle Investigator, 1980-.

## WOODS, John Hayden

**Specialties:** Philosophical logic, logics of fiction, the logic of fallacies, conflict resolution logics, abductive logics, logics of practical reasoning, and the history of logic.

**Born:** 16 March 1937 in Barrie, Ontario, Canada.

**Educated:** University of Michigan, PhD Philosophy, 1965; University of Toronto, MA Philosophy, 1959; University of Toronto, BA Honours Philosophy, 1958.

**Dissertation:** *Entailment and the Paradoxes of Strict Implication*; supervisor Arthur W. Burks.

**Regular Academic or Research Appointments:** DIRECTOR, ABDUCTIVE SYSTEMS GROUP, UNIVERSITY OF BRITISH COLUMBIA, 2002-; ADJUNCT PROFESSOR, PHILOSOPHY, UNIVERSITY OF LETHBRIDGE, 2002–; Professor, Philosophy, University of Lethbridge, 1979-2002; Professor, Philosophy, University of Calgary, 1976-1979; Professor, Philosophy, University of Victoria, 1971-1976; Lecturer, Assistant Professor, Associate Professor, Philosophy, University of Toronto, 1962-1971; Instructor, Philosophy, University of Michigan, 1961-1962; Instructor, Philosophy, University of Toronto, 1958-1959.

**Visiting Academic or Research Appointments:** Charles S. Peirce Regular Visiting Professor of Logic, Group in Logic and Computation, King's College London, 2001-; Research Fellow, Engineering and Physical Sciences Research Council of the United Kingdom, 2003-2004, 2004-2005; Vonhoff Professor of Humanities, University of Groningen, 2001; Fellow, Netherlands Institute for Advanced Study, 1990; Visiting Professor, Department of Discourse Analysis, University of Amsterdam, 1987-2001; Visiting Scholar, Department of Philosophy, Stanford University, 1994; Visiting Professor of Philosophy, University of Groningen, 1987, 1988; Visiting Professor, Department of Philosophy, Stanford University, 1971; Visiting Scholar, Department of Philosophy, Stanford University, 1968-1969; Visiting Associate Professor, Department of Philosophy, Laurntian University, 1969; Visiting Associate Professor, Department of Philosophy, University of Michigan, 1967; Associate Professor, Summer Institute of the Linguistics Society of America, 1967; Associate Professor, Summer Institute of Philosophy, University of Calgary, 1965.

**Research Profile:** Woods is widely recognized for his pioneering contributions in a number of areas of logical theory and the history of logic. His "Fictionality and the logic of relations" (1969) was instrumental in launching the study of fictional discourse as an independent research programme in the formal semantics of natural languages. *The Logic of Fiction* (1974) consolidated this independence, arguing that the logic of fiction could not plausibly be accommodated within existing frameworks such as description theory, and free, existence-neutral, many-valued and Meinongean logics. Instead he advocated a modal treatment of the sentence operator "fictionally" harnassed to a substitutional interpretation of the quantifiers. During this same period Woods published a number of papers aimed at disarming Quine's objections to quantified modal logic.

Another substantial stimulus to research arose from Woods' collaboration with his former student, Douglas Walton, on a series of papers investigating the logic of fallacious reasoning. The so-called Woods-Walton Approach came to dominate fallacy theory in the period 1972 -1985. Most of the papers of the Woods-Walton Approach are collected in *Fallacies: Selected Papers 1972-1982* (1989). This work was in direct response to a rebuke launched by C.L. Hamblin's *Fallacies* (1970) at his fellow logicians for having abandoned the fallacies programme. Woods and Walton undertook to analyze each of the standard types of fallacy in its own non-classical framework, thus in some quarters drawing the twofold objection that their approach was too "logocentric" and disunified. Characteristic of their findings is that there are contexts in which most of the standard candidates aren't fallacies at all, a result that impelled them to the view that the natural theoretical home for the study of fallacies is a formal pragmatics.

By the mid-eighties, it was clear to Woods and Walton that their collaboration had gone as far as it profitably could, and each continued with solo work on the subject. Since then, Walton has moved progressively towards a more discourse analytic approach, whereas Woods has continued to regard fallacy theory as proper by a branch of logic.

In the early 1990s Woods formed a collaboration with Dov Gabbay for the purpose of probing the logical structure of practical reasoning. One of the innovations of their approach is to regard reasoning as taking its character from the type of agent who performs it. It is characteristic of practical agents to transact their cognitive agendas with comparatively scant resources, such as information, time, and storage and computational capacity. Practical agents are typified by human individuals, as contrasted with institutional agents such as NATO or the International Monetary Fund. Practical agents are faced with the necessity of tailoring their targets and the means for their attainment to these resource-limitations, and in consequence are motivated to use their resources economically. Two of the more prominent economies effected by practical agents is the evasion of irrelevant information and the mitigation of ignorance by way of abductive presumption. These are discussed in detail in, respectively, *Agenda Relevance: A Study in Formal Pragmatics* (2003) and *The Reach of Abduction: Insight and Trial* (2005).

Also in the 1990s, Woods began an enquiry into the phenomenon of deep disagreement, especially in the absence of empirical checkpoints. These efforts culminated in his *Paradox and Paraconsistency* (2003), in which resolution strategies re proposed for disagreements between classical and relevant logicians, modal and anti-modal logicians, standard and non-standard approaches to the paradoxes, dialetheic and consistentist logicians, among others. Prominent in this treatment are Woods' rejections of the method of analytic intuitions in such disciplines, and the theorist's presumedly privileged access to the requisite normative standards.

In 2001 Woods published *Aristotle's Earlier Logic*, which advances the thesis that Aristotle's writings in the early books of the *Organon* have a significantly greater theoretical importance than logicians usually – and steadfastly - suppose.

**Main Publications:**

1. "Fictionality and the logic of relations", *Southern Journal of Philosophy* 7 (1969), 51-63.
2. "Essentialism, self-identity and quantifying in, in *Identity and Individuation* (M.K. Munitz, ed.), New York University Press, New York (1971), 165-188.
3. "Descriptions, essences and quantified modal logic", *Journal of Philosophical Logic* 2 (1973), 29-34.
4. *The Logic of Fiction: A Philosophical Founding of Deviant Logic*, The Hague, Mouton (1974).
5. "Identity and modality". In a special number of *Philosophia*, V (1975), 69-120.
6. (with Douglas Walton), "Arresting circles in formal dialectic", *Journal of Philosophical Logic* 7 (1978) 73-90.
7. (with Douglas Walton) "Question begging and cumulativeness in dialectical games", *Noûs* 16 (1982).
8. "The relevance of relevant logic", in *Directions in Relevant Logic*, Jean Norman and Richard Sylvan (eds.), Dordrecht and Boston: Kluwer (1989), 77-86.
9. (with Douglas Walton), *Fallacies: Selected Papers 1972-1982*, Dordrecht: Foris (1989). Second edition, with a Foreword by Dale Jacquette, London: College Publications (2007).
10. "Semantic intuitions", in *Logic and Argumentation*, Johan van Benthem *et al.* (eds.), Amsterdam: North-Holland (1996), 177-208.
11. "A captious nicety of argument: The philosophy of W.V. Quine", in *The Philosophy of W.V. Quine* (Library of Living Philosophers), $2^{nd}$. ed., L.E. Hahn (ed.), La Salle, IL: Open Court, (1998), 687-725.
12. (with Dov Gabbay), "Non-cooperation in dialogue logic", *Synthese* 127 (2001), 161-186.
13. (with Dov Gabbay), "The new logic", *Logic Journal of the IGPL* 9 (2001), 157-190.
14. *Aristotle's Earlier Logic*, Oxford: Hermes Science (2001).
15. (with Dov Gabbay), "Normative models of rational agency: The disutility of some approaches", *Logic Journal of the IGPL* 11 (2003), 597-613.
16. *Paradox and Paraconsistency: Conflict Resolution in the Abstract Sciences*, Cambridge University Press, Cambridge (2003).
17. (with Dov Gabbay), *Agenda Relevance: A study in Formal Pragmatics*, volume I of Gabbay and Woods, of *A Practical Logic of Cognitive Systems*, Amsterdam: North-Holland, 2003).
18. *The Death of Argument: Fallacies in Agent-Based Reasoning*, Dordrecht and Boston: Kluwer (2004).
19. (with Dov Gabbay), *The Reach of Abduction: Insight and Trial*, volume 2 of Gabbay and Woods, *A Practical Logic of Cognitive Systems*, Amsterdam: North-Holland, 2005).
20. (with Dov Gabbay) "Advice on abductive logic", *Logic Journal of the IGPL* 14 (2006), 189-219.

A full list of publications is available at Woods' homepage www.johnwoods.ca

*Work in Progress*

21. *Seductions and Shortcuts: Error in the Cognitive*

*Economy,* to appear.

22. *Epistemic Bubbles and the Determinism of Belief,* to appear.

23. *Sherlock's Member: New Directions in the Semantics of Fiction,* to appear.

24. A paper with Peter Bruza and Dominic Widdows, "A quantum logic of down below", which investigates sublinguistic and subconscious reasoning. To appear in *Handbook of Quantum Logic,* Dov Gabbay and Kurt Engesser (eds.), Amsterdam: Elsevier.

25. A series of papers, some with Dov Gabbay and others of sole authorship, on logic and the law.

**Service to the Profession:** Editorial Board Member, *Logic Journal of the IGPL,* 2006-; Area Editor, *Journal of Applied Logic,* 2003-; Editor, *Studies in Logic and Practical Reasoning* ("The Red Series"), a North-Holland monograph series, 2002-; Co-Editor, *Argumentation Library,* Kluwer (now Springer) Monograph Series, 1998; President, Academy of Humanities and Social Sciences, Royal Society of Canada, 1996-1998; Co-Editor-in-Chief, *Argumentation,* 1994-; Editorial Board Member, *Argumentation,* 1986-1994; Editorial Board Member, *Informal Logic,* 1984-; President, Canadian Federation for the Humanities, 1981-1982; President and Vice-Chancellor, University of Lethbridge, 1979-1986; Dean, Faculty of Humanities, University of Calgary, 1976-1979; Associate Dean, Faculty of Arts and Science, University of Victoria, 1975;English-Language Editor, *Dialogue: The Canadian Philosophical Revue,* 1974-1981; Co-Editor, *Handbook of the History of Logic,* 11 volumes, Elsevier; Co-General Editor, *Handbook of the Philosophy of Science,* 16 volumes, Elsevier.

**Teaching:** In the first forty years of his career (1962-2002), Woods spent a total of 33 of them involved in university administration, from department head, associate dean, dean and president. For 28 of those 33 he was appointed to universities which, as a matter of policy, had no graduate programmes. In each case, Woods' involvement was by choice. His administrative appointments, all held at new universities, reflected his desire to see intellectual excellence institutionalized in new places. His attachment to undergraduate teaching betokened his conviction that graduate programmes often represent a disproportionate draw of resources away from undergraduates. Among Woods' graduate students at Toronto, Douglas Walton has achieved major prominence as an argumentation theorist. Among his undergraduate students, Tom Vinci is a Descartes specialist at Dalhousie University; Cheryl Misak is a Peirce scholar at the University of Toronto and Provost there; and Timothy Schroeder is a philosopher of mind at Ohio State. During his time in Amsterdam, Woods also taught and served in various advisory roles many of the next generation of "pragma-dialectical" argumentation theorists, including, Eveline Feteris, Francisca Snoek-Henkemans, Peter Houtlosser and Bart Garssen (all at the University of Amsterdam) as well as Henrike Jensen at the University of Leiden. His Groningen students over the years include philosophers Jeanne Peijnenburg, and Jan Albert van Laar, both appointed to their alma mater, as well as Jan-Willem Romeyn, a University of Amsterdam probability theorist.

**Vision Statement:** "Two things have happened to logic of late, each an indication of its great success these past 120 years, and each a harbinger of trouble. One is its sprawling multiplicity. The other is its emerging re-admittance of agents as load-bearing objects of theory. The first generates a pluralism in logic which puts pressure on its realist presumptions. The second threatens with psychologism. Before long, it will have to be determined how much trouble these really are."

**Honours and Awards:** Interviewed in *Masses of Formal Philosophy: Aim, Scope, Directions,* Vincent F. Hendricks and John Symons, eds., Automatic Press VIP, 2006; A Festschrift, *Mistakes of Reason: Essays in Honour of John Woods,* Kent D. Peacock and Andrew A. Irvine, eds., Toronto: University of Toronto Press, 2005; Alberta Centennial Gold Medal, 2005; Doctor of Arts, *honoris causa,* University of Lethbridge, 2003; The Queen's Golden Jubilee Medal, 2002; Ingrid Speaker Medal for Distinguished Research, University of Lethbridge, 1997; Doctor of Laws, *honoris causa,* Mount Allison University, 1997; Distinguished Teaching Award, University of Lethbridge, 1996; ISSA Prize from the International Society for the Study of Argumentation, for Distinguished Research, 1991; Fellow, the Royal Society of Canada, 1990-; Honorary Fellow, Bretton Hall College, Leeds University, 1984-; Phi Beta Kappa, 1962;

# WOOLDRIDGE, Michael John

**Specialties:** Logics of rational agency, logics for multi-agent systems.

**Born:** 26 August 1966 in Wakefield, Yorkshire, England.

**Educated:** University of Manchester, PhD Computation, 1992; CNAA, BSc with Honors Computer Science, 1989.

**Dissertation:** *On the Logical Modelling of Computational Multi-Agent Systems*; supervisor, Gregory O'Hare.

**Regular Academic or Research Appointments:** PROFESSOR, COMPUTER SCIENCE, UNIVERSITY OF LIVERPOOL, UK, 2000–; Reader, Electronic Engineering, Queen Mary & Westfield College, University of London, 1997-2000; Lecturer, Computing, Manchester Metropolitan University, UK, 1992-1996.

**Research Profile:** Michael Wooldridge is a computer scientist whose research area is multi-agent systems: the study of how computers can be designed to work (semi)autonomously with one another, in order to carry out tasks delegated to them by humans. Wooldridge uses logic as a language for the specification and verification of such systems; his early work focused on "logics of rational agency", that is, the development of logics which attempted to characterize the mental state of a rational agent with beliefs, desires, and intentions; later work focused on "cooperation logics", such as the use of Alternating-time Temporal Logic for the specification and verification of multi-agent protocols and mechanisms. He has also made contributions to the semantics of agent communication, and to the study of computational coalitional games and their complexity.

**Main Publications:**

1. *Intelligent Agents: Theory and Practice.* Knowledge Engineering Review 10(2), 1995. (With N R Jennings)
2. *The Cooperative Problem Solving Process.* Journal of Logic and Computation. 9(4) pages 563-592, 1999. (with N R Jennings)
3. *Reasoning about Rational Agents.* MIT Press, 2000.
4. *Semantic Issues in the Verification of Agent Communication Languages.* In *Journal of Autonomous Agents and Multi-Agent Systems.* 3(1):9-31, 2000.
5. *Logic for Mechanism Design - A Manifesto.* In *Proceedings of the 2003 Workshop on Game Theory and Decision Theory in Agent-based Systems (GTDT-2003),* Melbourne, Australia, July 2003. (with M Pauly)
6. *Cooperation, Knowledge, and Time: Alternating-time Temporal Epistemic Logic and its Applications.* In *Studia Logica,* 75(1):125-157, October 2003. (with W van der Hoek)
7. *On the Computational Complexity of Qualitative Coalitional Games.* In *Artificial Intelligence,* 158(1):27-73, September 2004. (with P E Dunne)
8. *The Complexity of Contract Negotiation.* In *Artificial Intelligence,* 164(1-2):23-46, May 2005. (with P E Dunne and M R Laurence)
9. *On the Logic of Cooperation and Propositional Control.* In *Artificial Intelligence,* 164(1-2):81–119, May 2005. (with W van der Hoek)

*Work in Progress*

10. *Delegation, cooperation, and control.* With W van der Hoek and D. Walther.
11. *Logic for automated mechanism design and analysis.* With M Pauly.

**Service to the Profession:** Corners Editor, Journal of Logic and Computation, 2005-; Editor-in-Chief, Journal of Autonomous agents & multiagent systems, 2003; Editor, Journal of Applied Logic, 2003-; Editor, Journal of Applied AI, 1995-.

**Teaching:** Wooldridge has taught on multi-agent systems at Liverpool and many international institutes, and wrote the first undergraduate text in this area: *An Introduction to Multiagent Systems* (John Wiley, 2002).

**Vision Statement:** My main interest is in how logic can be used to help us understand rational behaviour in humans in artificial, software agents. Often, the formal models that are developed by those in the computer science and artificial intelligence communities are logically naïve, or have strange, undesirable properties. The logic community has an enormous opportunity here to apply the toolkit of logical analysis to this problem. In the short term, I think there is a significant opportunity in the development of logics based on game-theoretic ideas, and in terms of technologies, I think model checking has much to offer.

**Honours and Awards:** Research Award, ACM SIGART Autonomous Agents, 2006.

## WOS, Larry

**Specialties:** Automated reasoning, mathematics and first-order logic, logic calculi.

**Born:** 13 July 1930 in Chicago, Illinois, USA.

**Educated:** University of Illinois at Urbana-Champaign, PhD Mathematics, 1957; University of Chicago, MS Mathematics, 1954; University of Chicago, BA 1950.

**Dissertation:** *On Commutative Prime Power Subgroups of the Norm*; supervisor, R. Baer.

**Regular Academic or Research Appointments:** SENIOR MATHEMATICIAN, MATHEMATICS AND COMPUTER SCIENCE DIVISION, ARGONNE NATIONAL LABORATORY, 1984-2006; Mathematician, Mathematics and Computer Science Division (formerly Applied Mathematics Division), Argonne National Laboratory, 1960-1984; Assistant Mathematician, Applied Mathematics Division, Argonne National Laboratory, 1957-1960.

**Visiting Academic or Research Appointments:** Visiting Professor, University of Illinois at Chicago, 1985; Adjunct Professor, University of Illinois at Urbana-Champaign, 1984-1989; Distinguished Visitor of the IEEE Computer Society 1982-1983, 1983-1984.

**Research Profile:** Larry Wos introduced the concept of strategy in the field of automated reasoning. He developed numerous strategies, including the set of support strategy (considered by many to be the most powerful strategy available), the kernel strategy (the first systematic strategy for searching for fixed point combinators), the resonance strategy (enabling the user to impart knowledge and intuition to a program), and the hot list strategy (allowing the program to revisit user-chosen items in the input). Wos formulated the linked inference principle (significantly generalizing various inference rules) and devised the procedure demodulation (essential for attacking problems involving equality). With his colleagues, he initiated the first systematic attack on open questions with an automated reasoning program. Among his successes are answers to open questions in combinatory logic and in various logic calculi, including equivalential calculus, two-valued sentential (classical propositional) calculus, the right-group calculus, and modal logic.

**Main Publications:**

1. L. Wos, R. Overbeek, E. Lusk, and J. Boyle, Automated Reasoning: Introduction and Applications, Prentice-Hall, 1984; 2nd ed. 1988.
2. L. Wos, Automated Reasoning: 33 Basic Research Problems, Prentice-Hall, 1988.
3. L. Wos, The Automation of Reasoning: An Experimenter's Notebook with OTTER Tutorial, Academic Press, 1996.
4. L. Wos and G. W. Pieper, A Fascinating Country in the World of Computing: Your Guide to Automated Reasoning, World Scientific, 1999.
5. L. Wos with G. W. Pieper, Collected Works of Larry Wos, 2 vols., World Scientific, 1999.
6. L. Wos, B. Fitelson, and G. W. Pieper, Automated Reasoning and the Discovery of Missing and Elegant Proofs, Rinton Press, 2003.
7. L. Wos, S. Winker, R. Veroff, B. Smith, and L. Henschen, Questions concerning possible shortest single axioms in equivalential calculus: An application of automated reasoning to infinite domains, Notre Dame J. Formal Logic, 24(2) 205-223, 1983.
8. R. Boyer, E. Lusk, W. McCune, R. Overbeek, M. Stickel, and L. Wos. Set theory in first-order logic: Clauses for Gödel's axioms. J. Automated Reasoning, 2(3) 287-327, 1986.
9. W. McCune and L. Wos. A case study in automated theorem proving: Searching for sages in combinatory logic. J. Automated Reasoning, 3(1) 91-107, 1987.
10. L. Wos and W. McCune. Challenge problems focusing on equality and combinatory logic: Evaluating automated theorem-proving programs. In E. Lusk and R. Overbeek, editors, Proceedings of the 9th International Conference on Automated Deduction, Lecture Notes in Computer Science, Vol. 310, Berlin, Springer-Verlag, pp. 714-729, 1988.
11. L. Wos, S. Winker, W. McCune, R. Overbeek, E. Lusk, R. Stevens, and R. Butler. Automated reasoning contributes to mathematics and logic. In M. Stickel, editor, Proceedings of the 10th International Conference on Automated Deduction, Lecture Notes in Artificial Intelligence, Vol. 449, Berlin, Springer-Verlag, pp. 485-499, 1990.
12. L. Wos, Meeting the challenge of fifty years of logic, J. Automated Reasoning, 6, 213-232, 1990.
13. L. Wos and W. McCune. Automated theorem proving and logic programming: A natural symbiosis. J. Logic Programming, 11(1) 1-53, July 1991.
14. R. Veroff and L. Wos, The linked inference principle, I: The formal treatment, J. Automated Reasoning 8(2) 213-274, 1992.
15. L. Wos and W. McCune, The application of automated reasoning to questions in mathematics and logic. Annals of Mathematics and Artificial Intelligence, 5, 321-370, 1992.
16. W. McCune and L. Wos, Application of automated deduction to the search for single axioms for exponent groups. In A. Voronkov, editor, Logic Programming and Automated Reasoning, LNAI Vol. 624, Berlin, Springer-Verlag, pp. 131-136, 1992.
17. L. Wos, The kernel strategy and its use for the study of combinatory logic, J. Automated Reasoning, 10(3) 287-343, 1993.
18. L. Wos, The resonance strategy, Computers and Mathematics with Applications, 29(2) 133-178, 1995
19. L. Wos with G. W. Pieper, Experiments with the hot list strategy, J. Automated Reasoning, 22(1) 1-44, 1999.
20. B. Fitelson and L. Wos, Axiomatic proofs through automated reasoning, Bulletin of the Section of Logic, 29(1/2) 329-356, 2000.
21. L. Wos, Conquering the Meredith Single Axiom, J. Automated Reasoning, 27(2) 175-199, 2001.
22. Z. Ernst, B. Fitelson, K. Harris, and L. Wos, A concise axiomatization of RM->, Bulletin of the Section of Logic, 30(4) 191-194, 2001.
23. L. Wos, D. Ulrich, and B. Fitelson, XCB, the last of the shortest single axioms for the classical equivalential calculus, Bulletin of the Section of Logic, 32(3) 129-134, 2002.
24. M. Beeson, R. Veroff, and L. Wos, Double-negation elimination in some propositional logics, Studia Logica, special issue on Negation in Constructive Logic, 80(2-3) 195-234, 2005.

A full list of publications is available at Wos's home page
http://www.mcs.anl.gov/~wos
*Work in Progress*

25. Application of automated reasoning to median al-

gebra.

26. Simplification of proofs in various logics, for example, the BCSK logic. This work includes discovery of axiom dependencies.
27. Proofs of axiom systems in infinite-valued logic.
28. New approach to proof sketches based on adding lemmas, getting a pseudo-proof, and then shortening the proof before removing the unwanted lemmas.

**Service to the Profession:** Editorial Board Member, Who's Who in Logic, 2002-; Editor, Journal of Automated Reasoning, 1992-; Editor-in-Chief and Founder, Journal of Automated Reasoning, 1983-1992; President, Association for Automated Reasoning, 1982-.

**Teaching:** Wos was an Adjunct Professor at the Department of Computer Science, University of Illinois, Urbana, for five years in the mid- to late 1980s. He was also a Visiting Professor in Mathematics, Statistics, and Computer Science at the University of Illinois at Chicago in autumn 1985, when he taught a course entitled "Automated Reasoning Applied to Answering Open Questions." Wos's first book, Automated Reasoning: Introduction and Applications, was—like many of his subsequent writings—designed to teach new researchers how to put a reasoning program to general use.

**Vision Statement:** The objective of automated reasoning is to write computer programs that assist in solving problems and in answering questions requiring reasoning. Automated reasoning embraces far more than automated theorem proving, from which it traces its origins. It extends from finding proofs in mathematics and in logic, to designing electronic circuits, to solving logical reasoning problems in many other areas of science and engineering. Key to the success of automated reasoning is the use of strategy and experimentation. *Strategy* is essential for directing and controlling the reasoning; *experimentation* on real problems and open questions is essential for validating ideas.

**Honours and Awards:** First Herbrand Award for Distinguished Contributions to Automated Reasoning, 1992; University of Chicago Distinguished Performance Award 1988; American Mathematical Society Automated Theorem Proving Prize (with S. Winker) for Current Achievement in Automated Theorem Proving 1983.

# WOLEŃSKI, Jan

**Specialties:** History of logic, philosophy of logic, applications of logic to philosophy.

**Born:** 21 September 1940 in Radom, Poland.

**Educated:** Jagiellonian University, Magister Philosophy 1964; Jagiellonian University, Magister Law 1963.

**Dissertation:** *Analytical Jurisprudence and Linguistic Philosophy (in Polish)* 1968; supervisor, Kazimierz Opałek, *Dissertation (habilitation): Logical Problems of Legal Interpretation* (in Polish) 1972.

**Regular Academic or Research Appointments:** PROFESSOR OF PHILOSOPHY, JAGIELLONIAN UNIVERSITY, 1990–; Docent, Jagiellonian University, 1988-1990; Docent , Technical University of Wrocław, 1979-1988; Docent, Jagiellonian University 1974-1978; Assistant Professor, Jagiellonian University, 1963-1974.

**Visiting Academic or Research Appointments:** Visiting Professor, Sun Yat-sen University, Guanghzao, China, 2008; Fellow, Netherlands Institute of Advanced Studies in the Humanities and Social Sciences, Wassenaar, 2003–2004; Fellow, Pittsburgh Center in History and Philosohy of Science, 1993; Visiting Professor, University of Salzburg, 1992; Fellow (supported by the Kaściuszko Foundation), University of Boston, 1992; Fellow (supported by the Kościuszko Foundation), University of Berkeley, 1989.

**Research Profile:** Jan Woleński has made several contributions to history of logic. He wrote a book on the development of logic in the Lvov-Warsaw School as well as numerous papers on particular logicians, like Frege, Russell, MacCol, Reichenbach, Carnap, Chwistek, Hintikka and others. Especially, he investigated the reception of Polish logicians abroad and foreign logicians in Poland. In the philosophy of logic he tried to show that the first-order thesis is correct if logic is conceived as the source of inferential tools, but not as a device for analysis of concepts. He considers philosophical logic as a preparation of logic for applications in philosophy. Hence, formal systems of logic, in particular, modal (in a broad sense) logic are taken by him as instruments for analysis of basic philosophical concepts, like being, reality, determinism, fatalism, knowledge, permission and so on. The problem of truth is his favorite topic. He has contributed to the question of a philosophical interpretation of Tarski's semantic theory of truth.

**Main Publications:**

1. *Logic and Philosophy in the Lvov-Warsaw School*, Kluwer Academic Publishers, Dordrecht 1989, XII+364 pp.
2. *Essays in the History of Logic and Logical Philosophy* [Wybów artykułów], Jagiellonian University Press, Kraków 1999, 279 pp.

3. *Lógica y filosofía*, Publicaciones de la Facultad de Teologia "San Dámaso", Madrid 2005 (together with P. Domínguez), 267 pp.

4. Deontic Logic and Consequence Operations, in *Action, Logic and Social Theory*, ed. by G. Holmström and A. Jones, Societas Philosophica Fennica, Helsinki 1985, pp. 314-326.

5. Logic and Falsity, in *Logica '94. Proceedings of the $8^{th}$ International Symposium*, ed. by T. Childers and O. Majer, Filosofia, Praha 1995, pp. 95-105. Logicism and the Concept of Logic, in *Mathematik und Logik. Frege Kolloquium 1993*, hrp. von I. Max und W. Stelzner, Walter de Gruyter, Berlin 1995, pp. 111-119.

6. Logic and Mathematics, in The *Foundational Debate. Constructivity in Mathematics and Physics*, ed. by W. De Pauli-Schimanovich, E. Köhler and F. Stadler, Kluwer Academic Publishers, Dordrecht 1995, pp. 197-210.

7. Logical Squares: Generalizations and Interpretations, in *Logica '95. Proceedings of the $9^{th}$ International Symposium*, ed. by T. Childers, P. Kolár and O. Majer, Filosofia, Praha 1996, pp. 67-75.

8. Semantic Conception of Truth as a Philosophical Theory, in *The Nature of Truth (If Any)*, ed. by J. Peregrin, Kluwer Academic Publishers, Dordrecht 1999, pp. 51-66. 3-178.

9. Metatheory of Logics and the Characterization Problem, in *A Companion to Philosophical Logic*, ed. by D. Jacquette, Blackwell Publishing, Oxford 2002, pp. 319-331.

10. Metalogical Properties, Being Logical and Being Formal, *Logic and Logical Philosophy* 10(2002), pp. 211-221.

11. First-Order Logic: (Philosophical) Pro and Contra, in *First-Order Logic Revisited*, ed. V. Hendricks, F. Neuhaus, S. A. Pedersen, U. Scheffler and H. Wansing, λογος, Berlin 2004, pp. 369-399.

12. Gödel, Tarski and Truth, *Revue Internationale de Philosophie* 59(2004), pp. 459-490.

13. The Status of Church's Thesis, w: *Church's Thesis After 70 Years*, ed. by A. Olszewski, J. Woleński and R. Janusz, Ontos Verlag, Frankfurt 2006, ss. 310-330; together with R. Murawski.

14. *Philosophical Logic in Poland*, Kluwer Academic Publishers, Dordrecht 1994.

15. *The Lvov-Warsaw School and Contemporary Philosophy*, Kluwer Academic Publishers, Dordrecht 1998; together with K. Kijania-Placek.

16. *Alfred Tarski and the Vienna Circle*, Kluwer Academic Publishers, Dordrecht 1999; together with E. Köhler.

17. *Handbook of Epistemology*, Kluwer Academic Publishers, Dodrecht 2004; together with I. Niiniluoto and M. Sintonen.

18. *Provinces of Logic Determined. Essays in the Memory of Alfred Tarski*. Parts I, II and III (An*nals of Pure and Applied Logic* v. 126 (2004)); Parts IV, V and VI (An*nals of Pure and Applied Logic* v. 127 (2004)); together with Z. Adamowicz, S. Artemov, D. Niwiński, E. Orłowska and A. Romanowska.

19. *Church's Thesis After 70 Years*, Ontos Verlag, Frankfurt 2006, ss. 310-330; together with ed. by A. Olszewski and R. Janusz.

**Work in Progress**

20. Semantics and Truth (a monograph)

**Service to the Profession:** Member, *Association of Symbolic Logic*, 2005–;President, *European Society of Analytic Philosophy*, 2005; Member, *American Mathematical Society*, 2001–; Member, *Polish Academy of Sciences and Humanities*, 2001; President, *Polish Association of Logic and Philosophy of Science*, 1999-2002; Main Organizer, *Organizations of Scientific Events:* $11^{th}$ International Congress of Logic and Philosophy of Science, Kraków, 1999; Editor, *Book Series*: Synthese Library; Editorial Board, *Axiomathes*; Editorial Board, *Dialectica*; Editorial Board, *History and Philosophy of Logic*; Editorial Board, *The Monist*; Editorial Board, *The Polish Journal of Philosophy*; Editorial Board, *Synthese*.

**Teaching:** Jan Woleński has taught elementary logic for students of various fields. He introduced the course of logical propedeutic for philosophers and conducted seminars in philosophical logic. He has trained, among others: Katarzyna Kijania-Placek, Tomasz Placek and Artur Rojszczak.

**Vision Statement:** There are three statements which beautifully summarize the importance of logic:

1. Dialectica est art artium et scientia scientiarum ad omnium aliarum scientiarum methodorum principia viam habent (Pertus Hispanus);

2. Logic is the moral of speech and thought (Jan Łukasiewicz);

3. Logic is a formal exposition of intuition (Stanisław Leśniewski).

These three opinions, taken together, show the role of logic everywhere: from ordinary life to the most abstract parts of mathematics and philosophy.

**Honours and Awards:** Consulting Editor, *Theoria*; Scientific Prize of the Prime Ministry of Poland 2002; Scientific Prize of City Kraków, 1996.

# Z

## ZEMAN, Martin

**Specialties:** Set theory

**Born:** 27 December 1964 in Bratislava, Slovakia.

**Educated:** Dr. Rer. Nat., Humboldt Universität zu Berlin, Berlin, Germany; Diplom, Charles University, Prague, Czech Republic.

**Dissertation:** The core model for nonoverlapping extenders and its applications; supervisor, Ronald B. Jensen.

**Regular Academic or Research Appointments:** ASSOCIATE PROFESSOR, UNIVERSITY OF CALIFORNIA AT IRVINE, 2005–; Assistant Professor, University of California at Irvine, 2001-2005; Vertragsassistent, Universität Wien, Wien, Austria, 1999-2001; Wissenschaftlicher Mitarbeiter, Humboldt Universität zu Berlin, Berlin, Germany, 1998-1999; Wissenschaftlicher Mitarbeiter, Rheinische Friedrich Wilhelms Universität, Bonn, Germany, 1997-1998.

**Visiting Academic or Research Appointments:** Wissenschaftskolleg zu Berlin (Visiting Fellow), Awarded and declined, Berlin, Germany, 2005-2006; Visiting Graduate Student, Worcester College, Oxford, Great Britain, 1992-1993.

**Research Profile:** The research of Martin Zeman focuses mainly on inner model theory, fine structure theory, large cardinals and infinitary combinatorics. The second area of interest is forcing with emphasis on applications in infinitary combinatorics.

The most significant series of results comprise a kind of "catalogue" of methods developed for fine structural analysis of extender models and for constructions of combinatorial principles in these models. Many of these methods were developed jointly with Ernest Schimmerling. This project can be viewed as a continuation of Jensen's groundbreaking analysis of Gödel's constructible universe. The most significant applications of these methods include constructions of square sequences and morasses, and a complete characterization of Jensen's principle square in extender models. This led directly to locating the large cardinal strength of the failure of the principle square at singular cardinals, and provided the motivation for construction of forcing extensions for the failure of square at singular cardinals, using a relatively weak large cardinal hypothesis that is believed to be very close to the optimal.

**Main Publications:**

1. Square in core models, Bulletin of Symbolic Logic Vol.7 No.3 (2001), 305-314 (with E. Schimmerling)

2. Characterization of $\kappa$ in core models, Journal of Mathematical Logic Vol.4 No.1 (2004), 1-12 (with E. Schimmerling)

3. Dodd Parameters and $\lambda$-indexing of extenders, Journal of Mathematical Logic Vol.4 No.1 (2004), 73-108

4. Cardinal transfer properties in extender models I, to appear in Annals of Pure and Applied Logic (with E. Schimmerling)

5. Inner Models and Large Cardinals, de Gruyter Series in Logic and its Applications, de Gruyter, Berlin 2002

*Work in Progress*

6. Cardinal transfer properties in extender models II (with E. Schimmerling); p-1 morasses with linear limits in extender models; Gap-2 morasses in extender models; Forcing the failure of square; Models that preserve mice (with A. Caicedo).

**Service to the Profession:** Program Committee Member, ASL Summer Meeting, Irvine, 2008; Co-Organizer, Special Session on Infinitary Combinatorics and Inner Model Theory, Joint AMS-ASL Meetings, Phoenix, January 7-10, 2007.

**Teaching:** Current PhD students: Sean Cox, Kyriakos Kypriotakis. These have expected completion of PhD studies in summer 2008. Undergraduate research project with Nam Trang (2006-2007), Postgraduate tutor of Yasuo Yoshinobu (Nagoya, Japan). Former interactions with the following students: Gunter Fuchs (PhD under R. Jensen), Jakob Kellner (PhD under M. Goldstern), Tomas Futas and David Schrittesser (both PhD under S. Friedman).

**Vision Statement:** "In coming decades, I would like to see much tighter connections between Logic and the rest of Mathematics."

**Honours and Awards:** Marie-Curie IntraEuropean Fellowship, Awarded and declined, European Commission, 2005-2006; Fellowship, Awarded and declined, Wissenschaftskolleg zu Berlin, Berlin, Germany, 2005-2006; Dervorguilla Scholarship, Balliol College, Oxford, Great Britain, 2003-2004; Soros/FCO Scholarship, Worcester College, Oxford, Great Britain, 2002-2003.

## ZILBER (ZIL'BER), Boris

**Specialties:** Model Theory.

**Born:** 14 March 1949 in Tashkent, Uzbekistan, USSR.

**Educated:** Leningrad University, Doctor of Science (Physics and Maths), 1986; Novosibirsk University, Candidate of Science (Physics and Maths), 1975; Novosibirsk University, Master Math, 1971.

**Dissertation:** *Uncountably categorical theories* (Doctor of Science), Weakly categorical groups and rings (Candidate of Science); supervisor, Taitslin Mikhail.

**Regular Academic or Research Appointments:** PROFESSOR, MATHEMATICAL LOGIC, UNIVERSITY OF OXFORD, UK, 1999-; Kemerovo University, Kemerovo, Russian Federation, Chairman, Professor, Lecturer, Geometry and Algebra, 1974-1999.

**Research Profile:** The main interest of Zilber has always been in Model Theory, studying mainly countable theories categorical in uncountable cardinals. The natural development of the subject led Zilber outside the context of pure model theory into a number of applications in broader mathematics. He was one of the first to realise the deep connections of the subject to Algebraic Geometry. In 1971-1975, Zilber in a series of papers laid the foundations to the theory of groups and algebras of finite Morley rank and established the analogy with the theory of algebraic groups, which is still the driving motive of studies in this field. In particular, in these papers it was first demonstrated and systemetically used that Morley rank is the dimension notion analogous to the dimension in algebraic geometry. In 1979-1986 Zilber developed the Geometric Stability Theory in finite Morley rank context, which is based on the notion of a pregeometry of a strongly minimal subsets and introduced types of pregeometry (trivial, locally modular or field-like) and studied their effect on the properties of the structure as a whole, including liason groups and 'analysability'. One of the results of the theory was the fine structure theorem for models of theories categorical in all infinite cardinals and the proof of the Finite Model Property (and so impossibility of finite axiomatization) for the theories. One of the striking applications of the method was the classification of finite homogeneous (combinatorial) geometries of dimension 7 and higher (without assuming the classification of finite simple groups). The main principle of Geometric Stability Theory was formulated by Zilber as the Trichotomy conjecture. This stated that any pregeometry of a strongly minimal set has to be of one of the three types cited above. This conjecture has been refuted in general (Hrushovski, 1989) but proved to be true in a variety of key classes. In particular the Trichotomy holds in the class of Zariski geometries (Hrushovski and Zilber, 1993). A lot of more recent work of Zilber is around Zariski geometries. His important observation that compact complex manifolds (and indeed spaces) are Zariski geometries opened the way to applications of Model Theory to the classification of compact manifolds. A cycle of papers in 1997-2004 aims at explaining the nature of Hrushovski counterexamples to the Trichotomy Conjecture in terms of classical analytic structures, such as the field of complex numbers with exponentiation. This study found a deep link between the Schanuel conjecture in transcendental number theory and Geometric Stability Theory, including nonelementary stability (Shelah's excellence). Another application of this work is in Diophantine Geometry. More recent works establish connections between Zariski geometries and noncommutative algebraic geometry.

**Main Publications:**

1. An example of two elementarily equivalent but non-isomorphic metabelian groups, Algebra i Logika, 10(1971), 309-315
2. Rings, whose theories are categorical, Algebra i Logika, 13(1973), 168-187
3. On the transcendense rank of formulas of $aleph_1$-categorical theories, Mat.Zametki, 15(1974),321-329
4. Groups and rings whose theories are categorical. Fundamenta Math., 95, (1977) 173-188. Transl. In: AMS Transl.(2)149 (1991)1-16.
5. Totally categorical theories: structural properties and the nonfinite axiomatizability. In: Model Theory of Algebra and Arithmetic (Proc. Conf.Karpacz, 1979), Lecture Notes in Math. v.834, Springer-Verlag, 1980, 381-410
6. Totally categorical structures and combinatorial geometries. Soviet Math.Dokl. 259 (1981) 1039-1041
7. Strongly minimal countably categorical theories, Siberian Math.J 21 (1980), 98-112
8. Strongly minimal countably categorical theories II Siberian Math.J. 25 (1984) no.3, 71-88
9. Strongly minimal countably categorical theories III.Siberian Math.J 25 (1984) no.4, 63-77
10. The structure of models of uncountably categorical theories. In: Proc.Int.Congr.of Mathematicians, 1983, Warszawa, PWN - North-Holland P.Co., Amsterdam - New York - Oxford,1984, v.1, 359-368.
11. Hereditarily transitive groups and quasi-Urbanik structures. In:Model Theory and its Applications, Proc. of Math.Inst.Sib.Branch Ac.Sci.USSR, Novosibirsk,1988, ed.Yu.Ershov, 58-77 (English translation: Amer.Math.Soc.Transl.(2) v.195, (1999) pp.165-18)
12. Towards the structural stability theory. In: Logic, Methodology And Philosophy of Science VIII. Proc 8th Int.Congress LMPS Moscow 1987, Elsevier Pub.Co,1989,North-Holland, Amsterdam, eds.Fenstad, Frolov, Hilpinen, 163-178

13. Uncountably Categorical Theories. Transl.of Math.Monographs 117, Providence, RI: AMS. 1993

14. Finite Homogeneous Geometries.In: Proc.6th Easter Conference in Model Theory, ed.B.I.Dahn et al.,186-208. Berlin: Humboldt-Univ.1988

15. (with E.Hrushovski) Zariski Geometries. Journal of AMS, 9(1996),1-56

16. Exponential sums equations and the Schanuel conjecture, J.London Math.Soc.(2) 65 (2002), 27-44

17. Raising to powers in algebraically closed fields. JML, v.3, no.2, 2003, 217-238

18. Analytic and pseudoanalytic structures In Logic Coll. 2000 Paris, Lect.Notes in Logic v. 19, 2005, eds. R.Cori, A.Razborov, S.Todorcevic and C.Wood, AK Peters, Wellesley, Mass. pp.392-408

19. Pseudo-exponentiation on algebraically closed fields of characteristic zero, Annals of Pure and Applied Logic,Vol 132 (2004) 1, pp 67-95

*Work in Progress*

20. Lecture Notes on Zariski Geometries.

**Service to the Profession:** Editorial Board Member, Selecta Mathematica, 2007-; Council Member, Association of Symbolic Logic, 1998-2002.

**Teaching:** PhD: E. Rabinovich, N. Peatfield, M. Gavrilovich, J. Kirby.

**Vision Statement:** "Model Theory has worked out the hierarchy of "logically perfect" structures on the top of which are structures of countable languages with uncountably categorical theories. The analysis of this hierarchy in concrete fields of mathematics yielded new insights and often new results. *The true world is based on logically perfect structures.* This must be applicable to mathematical physics as well."

**Honours and Awards:** Thomas and Yvonne Williams Lecturer, University of Pennsylvania, 2008; Senior Berwick Prize, London Math Society, 2004; Godel Lecturer, Association for Symbolic Logic, 2003; Tarski Lecturer, University of California at Berkeley, 2002.

Printed in the United States
216629BV00003B/2/P